VOLUME ONE HUNDRED AND FORTY SEVEN

Current Topics in DEVELOPMENTAL BIOLOGY

Emerging Model Systems in Developmental Biology

CURRENT TOPICS IN DEVELOPMENTAL BIOLOGY

"A meeting-ground for critical review and discussion of developmental processes"
A.A. Moscona and Alberto Monroy (Volume 1, 1966)

VOLUME ONE HUNDRED AND FORTY SEVEN

CURRENT TOPICS IN DEVELOPMENTAL BIOLOGY

Emerging Model Systems in Developmental Biology

Edited by

BOB GOLDSTEIN

Department of Biology and Lineberger Comprehensive Cancer Center, University of North Carolina at Chapel Hill, Chapel Hill, North Carolina, United States

MANSI SRIVASTAVA

Department of Organismic and Evolutionary Biology, Museum of Comparative Zoology, Harvard University, Cambridge, MA, United States

ELSEVIER

ACADEMIC PRESS

An imprint of Elsevier

Academic Press is an imprint of Elsevier
50 Hampshire Street, 5th Floor, Cambridge, MA 02139, United States
525 B Street, Suite 1650, San Diego, CA 92101, United States
The Boulevard, Langford Lane, Kidlington, Oxford OX5 1GB, United Kingdom
125 London Wall, London, EC2Y 5AS, United Kingdom

First edition 2022

ISBN: 978-0-12-820154-1
ISSN: 0070-2153

For information on all Academic Press publications
visit our website at https://www.elsevier.com/books-and-journals

Publisher: Zoe Kruze
Developmental Editor: Naiza Mendoza
Production Project Manager: Vijayaraj Purushothaman
Cover Designer: Greg Harris

Typeset by STRAIVE, India

Contents

Contributors

Ehab Abouheif
Department of Biology, McGill University, Montreal, QC, Canada

Michalis Averof
Institut de Génomique Fonctionnelle de Lyon, École Normale Supérieure de Lyon, Lyon; Centre National de la Recherche Scientifique (CNRS), France

Alexandra E. Bely
Department of Biology, University of Maryland, College Park, MD, United States

Scott Bennett
Marine Biological Laboratory, Woods Hole, MA, United States

David S. Booth
Chan Zuckerberg Biohub and Department of Biochemistry and Biophysics, University of California, San Francisco, CA, United States

John L. Bowman
School of Biological Sciences, Monash University, Melbourne, VIC, Australia

Dan Calzarette
Marine Biological Laboratory, Woods Hole, MA, United States

James J. Collins III
Department of Pharmacology, UT Southwestern Medical Center, Dallas, TX, United States

Richard R. Copley
Sorbonne Université, CNRS, Laboratoire de Biologie du Développement de Villefranche-sur-mer (LBDV), France

Katharine E. Criswell
Department of Zoology, University of Cambridge, Cambridge, United Kingdom

Doreen D. Cunningham
Department of Biology, Georgetown University, Washington, DC, United States

Jeremy Dasen
Department of Neuroscience and Physiology, NYU School of Medicine, Neuroscience Institute, NY, United States

Karen Echeverri
Bell Center for Regenerative Biology and Tissue Engineering, Marine Biological Laboratory, Woods Hole, MA, United States

Cassandra G. Extavour
Department of Organismic and Evolutionary Biology; Department of Molecular and Cellular Biology, Harvard University, Cambridge, MA; Howard Hughes Medical Institute, Chevy Chase, MD, United States

Jifeng Fei
Guangdong Provincial People's Hospital, Guangdong Academy of Medical Sciences, Guangzhou, China

Jens H. Fritzenwanker
Department of Biology, Georgetown University, Washington, DC, United States

J. Andrew Gillis
Department of Zoology, University of Cambridge, Cambridge, United Kingdom; Marine Biological Laboratory, Woods Hole, MA, United States

Bob Goldstein
Department of Biology and Lineberger Comprehensive Cancer Center, University of North Carolina at Chapel Hill, Chapel Hill, North Carolina, United States

Jessica Gray
Department of Systems Biology, Harvard Medical School, Boston, MA, United States

Jonathan Q. Henry
Marine Biological Laboratory, Marine Resources Center, Woods Hole, MA, United States

Veronica Hinman
Department of Biological Sciences, Carnegie Mellon University, Pittsburgh, PA, United States

Christine Hirschberger
Department of Zoology, University of Cambridge, Cambridge, United Kingdom

Linda Z. Holland
Marine Biology Research Division, Scripps Institution of Oceanography, University of California San Diego, La Jolla, CA, United States

Nicholas D. Holland
Marine Biology Research Division, Scripps Institution of Oceanography, University of California San Diego, La Jolla, CA, United States

Evelyn Houliston
Sorbonne Université, CNRS, Laboratoire de Biologie du Développement de Villefranche-sur-mer (LBDV), France

Yoshiyasu Ishimaru
Division of Bioscience and Bioindustry, Graduate School of Technology, Industrial and Social Sciences, Tokushima University, Tokushima City, Tokushima, Japan

Sarah Kerr
Wesleyan University, Middletown, CT, United States

Nicole King
Howard Hughes Medical Institute and Department of Molecular and Cell Biology, University of California, Berkeley, CA, United States

Lucas Leclère
Sorbonne Université, CNRS, Laboratoire de Biologie du Développement de Villefranche-sur-mer (LBDV), France

Christopher J. Lowe
Hopkins Marine Station, Department of Biology, Stanford University, Pacific Grove, CA, United States

Deirdre C. Lyons
U.C. San Diego, Scripps Institution of Oceanography, La Jolla, CA, United States

Mark Q. Martindale
Whitney Lab for Marine Bioscience, University of Florida, St. Augustine, FL, United States

Anne Meyer
Department of Biological Sciences, Carnegie Mellon University, Pittsburgh, PA, United States

Taro Mito
Division of Bioscience and Bioindustry, Graduate School of Technology, Industrial and Social Sciences, Tokushima University, Tokushima City, Tokushima, Japan

Tsuyoshi Momose
Sorbonne Université, CNRS, Laboratoire de Biologie du Développement de Villefranche-sur-mer (LBDV), France

Catriona Munro
Sorbonne Université, CNRS, Laboratoire de Biologie du Développement de Villefranche-sur-mer (LBDV); Center for Interdisciplinary Research in Biology, Collège de France, PSL Research University, Paris, France

Taro Nakamura
Division of Evolutionary Developmental Biology, National Institute for Basic Biology, Okazaki, Aichi, Japan

Phillip A. Newmark
Howard Hughes Medical Institute, Morgridge Institute for Research, Department of Integrative Biology, University of Wisconsin, Madison, WI, United States

Sumihare Noji
Division of Bioscience and Bioindustry, Graduate School of Technology, Industrial and Social Sciences, Tokushima University, Tokushima City, Tokushima, Japan

Mathilde Paris
Institut de Génomique Fonctionnelle de Lyon, École Normale Supérieure de Lyon, Lyon; Centre National de la Recherche Scientifique (CNRS), France

Nipam H. Patel
Marine Biological Laboratory, Woods Hole, MA; Department of Organismal Biology and Anatomy, University of Chicago, Chicago, IL, United States

Michael T. Raissig
Centre for Organismal Studies Heidelberg, Heidelberg University, Heidelberg, Germany; Institute of Plant Sciences, University of Bern, Bern, Switzerland

Jenaid Rees
Department of Zoology, University of Cambridge, Cambridge, United Kingdom

Tania Rozario
Center for Tropical and Emerging Global Diseases and Department of Genetics, University of Georgia, Athens, GA, United States

Alejandro Sánchez Alvarado
Howard Hughes Medical Institute, Stowers Institute for Medical Research, Kansas City, MO, United States

Elaine C. Seaver
Whitney Laboratory for Marine Bioscience, University of Florida, St. Augustine, FL, United States

Ashley W. Seifert
Department of Biology, University of Kentucky, Lexington, KY, United States; Department of Veterinary Anatomy and Physiology, University of Nairobi, Nairobi, Kenya

Victoria A. Sleight
School of Biological Sciences, University of Aberdeen, Aberdeen, United Kingdom

Mansi Srivastava
Department of Organismic and Evolutionary Biology, Museum of Comparative Zoology, Harvard University, Cambridge, MA, United States

Elly M. Tanaka
Institute of Molecular Pathology (IMP), Vienna Biocenter (VBC), Vienna, Austria

Peter Temple-Smith
Department of Obstetrics & Gynecology, Monash University, Clayton, VIC, Australia

Takahito Watanabe
Division of Bioscience and Bioindustry, Graduate School of Technology, Industrial and Social Sciences, Tokushima University, Tokushima City, Tokushima, Japan

David A. Weisblat
Department of Molecular & Cell Biology, University of California, Berkeley, CA, United States

Carsten Wolff
Marine Biological Laboratory, Woods Hole, MA, United States

Daniel P. Woods
Department of Plant Sciences, University of California, Davis, CA; Howard Hughes Medical Institute, Chevy Chase, MD, United States

Guillem Ylla
Department of Organismic and Evolutionary Biology, Harvard University, Cambridge, MA, United States; Laboratory of Bioinformatics and Genome Biology, Faculty of Biochemistry, Biophysics and Biotechnology, Jagiellonian University, Krakow, Poland

Christina Zakas
North Carolina State University, Raleigh, NC, United States

Preface

A full understanding of the biological principles upon which life operates requires the study of diverse systems. Prior to the establishment of "model" organisms in the 20th century, biologists worked on diverse species—whatever they could get their hands on, be it in their backyards or from the sea at marine stations.

The focus on model organisms such as fruit flies (*Drosophila melanogaster*), nematodes (*Caenorhabditis elegans*), zebrafish (*Danio rerio*), and mice (*Mus musculus*) has revealed many of the fundamental principles upon which genomes, genes, cells, and embryos operate. This work has also generated rigorous experimental and analytical frameworks as well as new tools and resources, which have enabled scientists once again to venture out to explore diverse research organisms. This apparent hourglass shape of species used in laboratory research over time leaves us at an exciting moment in biological research.

New or long-neglected research organisms offer some special opportunities. They allow us to study the mechanisms for biological processes that cannot be studied in well-established model systems. For example, model organisms like flies, nematodes, or mice lack extensive regeneration or pluripotent stem cells in adults, but these features are widespread across the tree of animal life. Additionally, even for processes that can be studied in established model systems, sampling of species from varied branches across the tree of life is necessary to build a thorough understanding of how life works. For example, although all animals make germ cells, studying germline formation only in flies and vertebrates is not sufficient to make inferences about the shared molecular/cellular principles or the evolution of germline formation. Comparative studies involving diverse species, or species in key phylogenetic positions, are foundational for evolutionary developmental biology (evo-devo).

Our objective in putting together this volume was to highlight a growing panel of research organisms that can illustrate different values that studying a new system brings to a field of study. We invited as authors some of the people who played central roles in the emergence of these model systems, and we encouraged them to tell their origin stories, i.e., the scientists' personal journeys to the model systems. We gave the authors great flexibility in the length and format of their contributions, and, therefore, the chapters themselves are diverse in style. Some authors have focused on explaining the system and the questions that the species they study can address; other authors have provided a narrative of their own journeys in seeking and establishing or reestablishing a new system; and some authors have contributed chapters that represent a combination of these two approaches.

This volume is by no means a comprehensive catalog of all new and exciting systems. We appreciate that there are many more emerging model organisms than we could fit in a volume and that other scientists will have other origin stories for their own work on those organisms and indeed on some of the organisms represented here, in cases where multiple scientists made major contributions to developing an emerging model.

We expect that the chapters in this volume will serve many functions. The chapters can be resources and potential inspiration for young trainees looking to undertake studies in nontraditional model systems. They can also serve as tools for educating funding agencies and reviewers about the value of these systems. The species covered by these chapters were chosen to show the phylogenetic breadth across which biologists are pursuing new systems in developmental biology and in some neighboring fields. These chapters should provide a definitive and formal introduction (or reintroduction in some cases) to some of the model systems that are poised to answer major questions in developmental biology and beyond.

Bob Goldstein
Department of Biology and Lineberger Comprehensive Cancer Center, The University of North Carolina at Chapel Hill, Chapel Hill, NC, United States

Mansi Srivastava
Department of Organismic and Evolutionary Biology, Museum of Comparative Zoology, Harvard University, Cambridge, MA, United States

CHAPTER ONE

The liverwort *Marchantia polymorpha*, a model for all ages

John L. Bowman*
School of Biological Sciences, Monash University, Melbourne, VIC, Australia
*Corresponding author: e-mail address: john.bowman@monash.edu

Contents

Abstract

The liverwort *Marchantia polymorpha* has been known to man for millennia due to its inclusion Greek herbals. Perhaps due to its familiarity and association with growth in, often, man-made disturbed habitats, it was readily used to address fundamental biological questions of the day, including elucidation of land plant life cycles in the late 18th century, the formulation of cell theory early in the 19th century and the discovery of the alternation of generations in land plants in the mid-19th century. Subsequently, *Marchantia* was used as model in botany classes. With the arrival of the molecular era, its organellar genomes, the chloroplast and mitochondrial, were some of the first to be sequenced from any plant. In the past two decades, molecular genetic tools have been applied such that genes may be manipulated seemingly at will. Here, are past, present, and some views to the future of *Marchantia* as a model.

Land plants dominate nearly all terrestrial ecosystems and their origin dramatically changed Earth's geochemical cycles, as well as the evolutionary trajectories of many other lineages of life. Land plants evolved from an ancestral freshwater charophycean alga, with the extant lineage of land plants arising near the base of the Dapingian in the mid-Ordovocian, ca 470 MYA (e.g., Delwiche & Cooper, 2015; Edwards, Morris, Richardson, & Kenrick, 2014; Strother & Taylor, 2018). Perhaps unsurprisingly, due to

Current Topics in Developmental Biology, Volume 147
ISSN 0070-2153
https://doi.org/10.1016/bs.ctdb.2021.12.009

the open ecological niches available, land plants rapidly diversified, with all six extant land plant lineages (liverworts, mosses, hornworts, lycophytes, ferns, seed plants) established by the Middle Devonian (Morris et al., 2018). By the end of the Devonian, complex ecosystems with metazoan herbivory and fungal and bacterial detritivory similar to those of the present had evolved. From a botanical perspective, other than the lack of flowers and recognizable species, one might not have felt "out of time" walking through the Late Devonian forests with layered canopies and large trees. To reconstruct the nature of the ancestral land plant and its subsequent evolution, the interrelationships of extant land plant lineages and sister lineages of charophycean algae must be resolved and combined with morphological and anatomical attributes gleaned from extant and fossil plants. Given the pivotal nature of the origin of a land flora, it has been a subject of investigation and speculation for more than a century, but only recently have hypotheses converged on a scenario that is generally agreed upon. This is in large part due to the development of relevant genetic and/or genomic land plant and charophycean algal model systems in which evolutionary questions can be addressed.

The transition to a terrestrial habitat necessitated physiological adaptations to cope with abiotic stressors (e.g., desiccation and UV radiation) as well as developmental innovations to disperse desiccation tolerant spores. For example, at a fundamental level, at least two major changes in life cycle histories must have accompanied the origin and diversification of land plants. First, the sister charophycean algae have a haplontic life cycle with a multicellular haploid phase and a single-celled diploid zygote. In contrast, land plants have complex multicellular bodies in both their haploid and diploid phases, and the evolution of a multicellular diploid embryo gives land plants their taxonomic name, Embryophyta. This innovation allowed the production of a large number of spores from a single aquatic fertilization event, an often-limiting factor in terrestrial habitats. Second, extant land plants may have life cycles in which the haploid phase is dominant, with the diploid phase largely dependent and ephemeral, as in all extant bryophytes (liverworts, mosses, hornworts), or a life cycle in which the diploid phase is dominant and the haploid phase transitory, as in all extant vascular plants (lycophytes, ferns, seed plants). Due to their being the source of much (or all) of our food, fiber, and timber almost all land plant model systems are seed plants, with nearly all models being angiosperms. However, in recent decades it has become apparent that having model systems in each of the six lineages would be useful to address fundamental questions concerning land plant evolution. Several recent reviews have focused on the liverwort *Marchantia polymorpha*, including ones covering anatomical details of the life cycle (Shimamura, 2016), the history of its use in experimental biology (Bowman, 2016), the development of genetic and genomic tools (Ishizaki et al., 2015; Ishizaki, Nishihama, Yamato, & Kohchi, 2016; Nishihama, Ishida, Urawa, Kamei, & Kohchi, 2016; Sauret-Güeto et al., 2020), nomenclatural issues (Bowman, Araki, et al., 2016), recent advances in understanding its physiology and development (Hisanaga et al., 2019; Kohchi, Yamato, Ishizaki, Yamaoka, & Nishihama, 2021; Romani & Moreno, 2021), and the coevolution of land plants with associated microbial symbionts and pathogens (Delaux & Schornack, 2021). In this review, I trace the history of using *Marchantia* as a model system and attempt to highlight both recent discoveries that have expanded our knowledge and areas in which additional investigation might be fruitful.

1. Why a liverwort and why *Marchantia*?

While *M. polymorpha*, hereafter *Marchantia*, has been utilized as a model system to investigate biological questions since the Enlightenment (discussed in more detail below), it's modern incarnation as model system dates to the past few decades. When cladistics began to be applied to the question of the origin and evolution of land plants, DNA sequence data was not widely available, and morphology and anatomy provided the bulk of the phylogenetically informative characters. On these criteria, liverworts were resolved as the first lineage of land plants to diverge from the others, largely on the basis of their lack of stomata, which are present is all other land plant lineages (e.g., Kenrick & Crane, 1997). Subsequently, when nucleic acid sequences became more plentiful, in many studies liverworts continued to be placed as the diverging from the most basal node within land plants (e.g., Qiu et al., 2006). It was with perspective in mind that two research groups, Takayuki Kohchi's (University of Kyoto) and my own (Monash University), proposed a *Marchantia* genome project to the US DOE Joint Genome Institute in 2007. During the time that this genome project took to reach fruition (published in 2017), our views of the relationships between the six land plant groups also evolved. Although nearly every possible topology has been proposed at some time, phylogenomic analyses have converged upon a consensus topology wherein the three bryophyte lineages are a clade and the three vascular plant lineages are a clade (e.g., de Sousa, Foster, Donoghue, Schneider, & Cox, 2019; Nishiyama et al., 2004; Puttick et al., 2018; Wickett et al., 2014). While this topology is the consensus, it is only supported by about half of individual gene trees, perhaps reflecting the rapid divergence of the six lineages from one another, especially those of the three bryophytes (Fig. 1). In this respect, each of the six extant lineages has

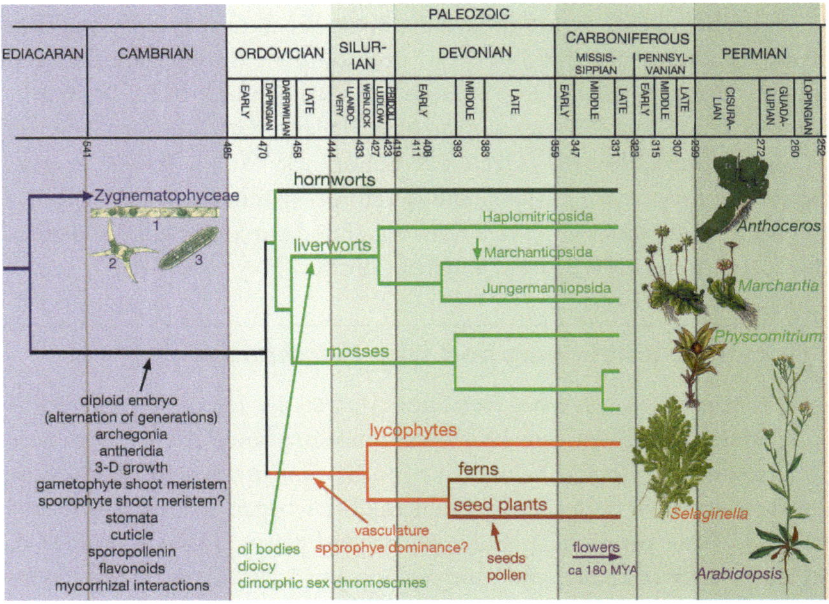

Fig. 1 The Phylogenetic context of *Marchantia*. The sister taxon to land plants is the Zygnematophyceae, a group that includes both multicellular filamentous forms, such as *Spirogyra* (1), and unicellular species, such as *Closterium* (2) and *Penium* (3). The six major extant lineages of land plants diverged by the middle Devonian, with the three lineages of bryophytes diverging from one another as early as the Ordovician. The three major extant lineages of both liverworts and mosses also diverged during this time frame. Representative models with genome sequence available of five land plant lineages are at the right—each of these is amenable to both forward and reverse genetic approaches, with the exception of *Selaginella* for which transgenic plants have not yet been reported. Key innovations during the evolution of land plants (*black*), liverworts (*green*), vascular plants (*red*), and seed plants (*dark red*) are listed, with allusions to many of these in the text. The green arrow on the Marchantiopsida lineage indicates the origin of derived characters, including complex thallus with air chambers, gametangiophores, and pegged rhizoids, found in *Marchantia*. Flowering plants, one of two seed plant lineages evolved later, with the blooming of the first flowers (*purple*) estimated to be around 180 MYA. The approximate divergence times of nodes in the trees were adapted from previous estimates of early land plant divergences (Morris et al., 2018), with the timing of the initial divergence between bryophytes and tracheophytes adjusted to 470 MYA corresponding to the origin of land plant cryptospore morphology (Strother & Taylor, 2018).

been on its own isolated evolutionary trajectory for the past 400+ million years (Bowman, 2013). If the currently accepted consensus topology is indeed a genuine reflection of evolutionary history, the liverworts do not occupy a unique position, but rather all bryophyte lineages are equally related to those of vascular plants, and as discussed elsewhere, resolution of monophyletic clades of bryophyte and vascular plants renders many ancestral attributes, including the nature of the life cycle, ambiguous (Bowman, Briginshaw, & Florent, 2019). Thus, in the reconstruction of ancestral character states, analyses should focus on comparisons amongst members of all six extant lineages.

The usefulness of a model system depends upon both its suitability for use in experimental (e.g., laboratory) studies and its unique combination of synapomorphies and derived characters that facilitate the investigation of particular biological questions (Fig. 1). While *Marchantia*'s modern incarnation as a model system is founded on its facile genetics and genomic simplicity, previous manifestations (Table 1) were likely primarily due to its familiarity as a colonizer of man-made disturbed habitats (Schuster, 1992). As such, *Marchantia* has been known to man for millennia, being one of the several hundred medicinal plants first described by Dioscorides and Pliny in the first century CE, and then continuously in herbals through to the Renaissance (Bowman, 2016). Following the Enlightenment, *Marchantia* was employed in studies to elucidate the bryophyte life cycle (Hedwig, 1783), and then with the advent of microscopy where spherical and chromatic aberrations were corrected, *Marchantia* was utilized for studies on cell theory (Mirbel, 1835), motility of bryophyte sperm (Unger, 1837), and Hofmeister's discovery of the alternation of generations, which unified land plants as a natural group (Hofmeister, 1862). During this time, related liverwort species were used as models to investigate the structure and function of shoot meristems, with the discovery that many shoot meristems possess a single apical cell from which the remainder of the plant body is derived (Nägeli, 1845).

The ubiquity of *Marchantia* led to its use in the classroom as a model for both botany and general microscopy from the mid 19th century onwards, with some enthusiastically endorsing liverworts to entice students: *But if you wish to elicit from your audience exclamations of delight, put on the slide a freshly ruptured spore-case of one of the larger Hepaticae—*Marchantia, Asterella, *or* Conocephalum *will answer. The lively way in which the spiral elaters toss the spores about is sure to "bring down the house"* (Blanchard, 1891). However, not all were enamored with the use *Marchantia* due to its derived features, and it was written in a review of Scott's botany textbook in Bohr, 1897:

Table 1 Landmarks in the history of *Marchantia*.

1st–17th centuries	Description of *Marchantia* as a treatment to prevent inflammation of wounds in Greek herbals	Bowman (2016)
1713	Jean Marchant provides a name	Marchant (1713)
1783	Hedwig's use to discover the bryophyte life cycle	Hedwig (1783)
1835	Mirbel's use in investigating cell theory	Mirbel (1835)
1837	Unger's discovery of motile sperm in bryophytes	Unger (1837)
1845	Nägeli's investigation of apical cells (in a related species)	Nägeli (1845)
1851	Hofmeister's elucidation of the alternation of generations	Hofmeister (1862)
1917	Allen's discovery of sex chromosomes in *Sphaerocarpos*	Allen (1917)
1928	Heitz's description of heterochromatin in *Pellia*	Heitz (1928)
1943	Burgeff's genetic studies of *Marchantia*	Burgeff (1943)
1986	*Marchantia paleacea* chloroplast genome sequence	Ohyama et al. (1986)
1992	*M. paleacea* mitochondrial genome sequence	Oda et al. (1992)
2007	JGI *Marchantia* genome project initiated	https://genome.jgi.doe.gov/portal/Marpha_FD/Marpha_FD.info.html
2008	Transformation via *Agrobacterium*	Ishizaki, Chiyoda, Yamato, and Kohchi (2008)
2013	Homologous recombination	Ishizaki, Johzuka-Hisatomi, Ishida, Iida, and Kohchi (2013)
2014	CRISPR/Cas9 genome editing	Sugano et al. (2014) and Sugano et al. (2018)
2016	Conditional knockouts	Nishihama et al. (2016)
2017	Nuclear genome sequence	Bowman et al. (2017)
2020	Chromosome based nuclear sequence	Montgomery et al. (2020)

"which its complex and highly specialised structure render it eminently unfit to occupy" (Bohr, 1897). Due to the influx and availability of plants sourced from all over the globe, this era encompassed the flourishing of gardening. *Marchantia* was noted in this context as well, but not in a good light, as it was described as a "troublesome weed" (Smee, 1872) and "the greatest enemy" (Robinson, 1910). However, even the gardeners occasionally marveled at its beauty, "*Although so troublesome a weed, it is welcome in small quantities, and is one of our finest natural plants*" (Smee, 1872).

With the advent of the chromosome theory of inheritance, the first plant sex chromosomes were discovered in a related liverwort, *Sphaerocarpos texanus* (Allen, 1917), and the first description of eukaryotic heterochromatin was made in the liverwort *Pellia epiphylla* (Heitz, 1928). Following Muller's demonstration that X-rays are mutagenic, *Marchantia* and related liverwort species were subjected to intensive genetic studies, many aimed at elucidating the biology of sex chromosomes (Bowman, 2016), with classical genetic experiments culminating in Burgeff's monograph on the genetics of *Marchantia* (Burgeff, 1943). In the molecular era, sequence of the complete chloroplast genome of *Marchantia paleacea* was the second determined for any plant species (Ohyama et al., 1986) and that of the *M. paleacea* mitochondrial genome was the first land plant sequence available (Oda et al., 1992). Note that these two genomes were originally published as those of *M. polymorpha*, but were later shown to be derived from *M. paleacea* cultures (Bowman et al., 2017). Thus, the use of *Marchantia* as a model system has paralleled advances in biological thought and technological advances over the past few centuries.

While most research over the past few decades on land plant biology has been concentrated on flowering plant species, there has been a recent resurgence of interest in phylogenetically diverse land plants to address issues of the origin and diversification of a land flora. For this, *Marchantia* is well suited as a model system due to its genetic and genomic attributes. Perhaps due to its "weedy" nature, *M. polymorpha* is easy to grow in the laboratory, with its life cycle completed in as little as 3 months. *M. polymorpha* is a dioicious species, i.e., with separate male and female plants, and as such crosses between lines are easily accomplished. As has been reviewed in more detail elsewhere (Ishizaki et al., 2016), the creation of transgenic plants via multiple methods (Ishizaki et al., 2008; Kubota, Ishizaki, Hosaka, & Kohchi, 2013; Tsuboyama, Nonaka, Exura, & Kodama, 2018), including insertion of multiple different transgenes simultaneously (Ishizaki et al., 2015), insertion of DNA via homologous recombination (Ishizaki, Johzuka-Hisatomi, et al., 2013), CRISPR-Cas9 mediated genome editing (Sugano et al., 2014; Sugano et al., 2018), and conditional alleles (Flores-Sandoval, Dierschke, Fisher, & Bowman, 2016; Nishihama et al., 2016), is routine, and in general, not limiting. Since the dominant generation is haploid, the time from the designing of guide RNAs for CRISPR-Cas9 genome editing to observing a mutant phenotype can be as little as two months, although subsequent genetic work would be required to confirm the genetic lesions. Thus, the creation of loss-of-function and gain-of-function alleles as well as reporter lines, either using histochemical or fluorescent proteins encoded by reporter genes, can be accomplished rapidly. Such ease of genetic manipulation facilitates construction of more sophisticated genetic combinations, such as conditional alleles to circumvent lethality of null alleles (Nishihama et al., 2016).

One advantageous feature that was foreshadowed by early work on specific gene families is the lack of whole genome duplications (WGDs) in *Marchantia*. This feature is characteristic of liverworts, which appear to lack ancient WGDs as a group. Thus, gene families, specifically those encoding regulatory proteins, have minimal numbers of paralogs, simplifying mutant studies. Further, as a consequence of the lack of WGDs, the regulatory genome of Marchantia may closely resemble that of the ancestral land plant (Bowman et al., 2017). It was postulated that the lack of ancient WGDs in Marchantia, and liverworts in general, was due to the early evolution of highly dimorphic sex chromosomes (Bowman et al., 2017; Iwasaki et al., 2021). Thus, the evolution of sex chromosomes may have facilitated preservation of ancestral genomic characteristics in this lineage. More than a half century ago Ohno proposed that a lack of WGDs might constrain subsequent evolution, which would then be canalized without a mechanism for large scale gene duplication (Ohno, 1970). Ohno suggested that the evolution of novelty would be more likely to arise from extant lineages that could undergo WGDs, as they would have the potential to possess a large number of paralogs that could subsequent neofunctionalize, leading to new evolutionary trajectories (Ohno, 1970). Given the lack of ancient WGDs in liverworts, it can be speculated that the gene regulatory networks of liverworts have been evolutionarily constrained, and thus may more closely resemble ancestral gene regulatory networks. Whether morphological evolution within the liverwort lineage has been similarly canalized is an open question.

The remainder of this review is an outline of the *Marchantia* life cycle, in which is embedded a discussion of synapomorphies and derived features (i.e., features and limitations), that make *Marchantia* a model system for the investigation of various biological processes. Only references to the original discoveries and studies in *Marchantia* in the post-genomic era are highlighted, and for reasons of space limitations, this review is focused on developmental and evolutionary questions. In addition, a non-exhaustive set of specific questions of interest, from the authors perspective, are in italics throughout the text.

2. The sexual life cycle

A detailed description of the modern view of the *Marchantia* life cycle is presented in Shimamura (2016), however, many morphological and anatomical details were worked out prior to the turn of the 20th century (Fig. 2). Some older references are highlighted here along with relevant ones subsequent to Shimamura (2016).

2.1 The haploid gametophyte generation

Dispersed haploid spores that initiate the gametophyte generation (**1**) will germinate in the presence of light and moisture, the first sign being a swelling of the spore (**2**) with rupture of the exospore preceding accumulation of chlorophyll (Leitgeb, 1877a; O'Hanlon, 1926). Addition of exogenous auxin, a phytohormone, can induce germination in the dark, in essence substituting for the light stimulus (Fitting, 1939). Auxin is arguably the most important phytohormone, and transcriptional changes downstream of auxin signaling affect nearly all of plant development, prominently including tropic responses to gravity and light. *Marchantia* has been instrumental as a model system to reveal that auxin transcriptional signaling evolved in the ancestral land plant and, due to the lack of genetic redundancy of auxin signaling components in *Marchantia,* it was clarified that auxin acts as a facilitator, rather than a determiner, of developmental decisions (Bowman et al., 2017; Flores-Sandoval et al., 2018; Flores-Sandoval, Eklund, & Bowman, 2015; Kato et al., 2015; Kato et al., 2017; Kato et al., 2020). This topic has recently been reviewed (Bowman, Flores-Sandoval, & Kato, 2021; Suzuki, Kohchi, & Nishihama, 2021).

The first division of the germinating spore is usually asymmetric (**3**) producing a basal cell that will differentiate into a rhizoid and an apical cell that will give rise to the sporeling shoot (Mirbel, 1835; O'Hanlon, 1926). *The genetic and physiological bases of this initial asymmetric division are not known*, although auxin (see above) is likely to play a role. The apical region of the sporeling continues to proliferate producing a mass of cells (**4**), and

Fig. 2 The sexual life cycle of *M. polymorpha*. In the green shaded region is the haploid gametophyte and in the blue region is the diploid sporophyte, with spore dispersal in aqua. Individual numbered stages described in the text. Drawings from Groenland (1854); Hedwig, 1783), Henfrey (1853), Hofmeister (1862), Leitgeb (1881), Marchant (1713), Mirbel (1835), Mottier (1891), Schwann and Schleiden (1847), Thuret (1851), and Unger (1837).

subsequently a single apical cell is specified in a region corresponding to the highest incident light (Hansel, 1876; Leitgeb, 1877a). The specification of an apical cell defines the prothallus (**5**), with the apical cell having two division planes such that a heart-shaped unistratose structure develops (Bischoff, 1835; Leitgeb, 1881; Menge, 1930). *The molecular mechanism by which a single* *cell is chosen to be the apical cell is unknown*, but again, light and likely auxin are both critical, as loss of auxin biosynthesis results either in a failure to specify or maintain an apical cell (Eklund et al., 2015).

Under optimal growth conditions the prothallus rapidly transitions into a complex thallus (characteristic of, and specific to, the class Marchantiopsida of liverworts), with dorsal air chambers and ventral scales and rhizoids, which will be described later. However, under suboptimal low light conditions, it may revert back to a filamentous protonemal growth until it reaches a more suitable light environment (Leitgeb, 1877a; Schostakowitsch, 1894). A similar extension of filamentous sporeling growth is observed when grown under red light (Téodoresco, 1929). During the transition from prothallus to complex thallus, the apical cell also transitions from possessing two cutting faces to four cleavage planes: dorsal, ventral and two lateral faces (Leitgeb, 1881; Mottier, 1891). *How the transition from 2-dimensional to 3-dimensional growth is regulated is unknown*, but it too is facilitated by auxin signaling since disruption of auxin signaling can results in plants failing to make the transition (Flores-Sandoval et al., 2015). While the precise pattern has not been observed in *Marchantia*, in *Blasia* a regular spiral pattern (lateral to dorsal to lateral to ventral) occurs (Leitgeb, 1874), and this likely to be the case for *Marchantia* (Apostolakos, Galatis, & Mitrakos, 1982) In longitudinal sections (**6**) the apical cell is conspicuous, but in transverse section, the apical cell is difficult to distinguish from immediate lateral derivatives (**7**), sometimes referred to as subapical cells (Leitgeb, 1879; Leitgeb, 1881; Mottier, 1891). However, both on anatomical and theoretical grounds, each apex harbors only a single apical cell (Douin, 1921; Mottier, 1891). While branching has not been directly observed due to lack of appropriate markers, it is likely to be pseudo-dichotomous, as has been observed in Jungermanniopsida simple thalloid and leafy liverworts (Kny, 1865; Leitgeb, 1877b), wherein the original apical cell is retained and a new apical cell is specified from a recent lateral derivative a cell or two distant from the original apical cell. *The molecular mechanisms of (pseudo)dichotomous branching are unknown*, however reduction in auxin levels can reduce branching (Eklund et al., 2015). Further, the land plant specific CLE peptide signaling systems modulate meristem function. Two distinct CLE peptide signaling systems, CLV3 and TDIF, existed in the ancestral land plant. In Marchantia, MpCLE1, a TDIF-type, restricts the proliferative activity of the gametophyte shoot meristem, in contrast to the stem cell promoting activity of TDIF in angiosperm vascular meristems (Hirakawa et al., 2019). Conversely, MpCLE2, a CLV3-type, promotes the accumulation of stem cells, likely apical cell derivatives, within the *Marchantia* gametophyte shoot meristem (Hirakawa et al., 2020). For instance, application of exogenous CLE2 peptide can cause an excessive accumulation of stem cells with the potential to become multiple apical cells, and conversely, loss-of-function alleles have a reduced number of meristematic cells (Hirakawa et al., 2020). *The regular periodic dichotomous branching of the vegetative thallus suggests a broader intercellular signaling network, perhaps involving an interaction between the CLE and auxin signaling pathways and as yet unknown other factors.* That the orthologous CLE signaling systems operate in angiosperm sporophyte meristems in the opposite manner, suggests that while gametophyte and sporophyte shoot meristems share some genetic components, their genetic wiring may be substantially different. *The relationship between the shoot meristems of the gametophyte and sporophyte generations raises the question as to whether the two generations are homologous,* i.e., *utilizing a shared ancestral genetic program, or antithetic, whereby they would have two distinct evolutionary origins and trajectories.*

Similar to the predicted ancestral liverwort, *Marchantia* is dioecious with sex determined chromosomally (see Bowman, 2016, for review of sex chromosomes). Sex chromosomes in plants were first discovered in a related liverwort, *S. texanus* (Allen, 1917), and it has been suggested that the ancestral liverwort possessed sex chromosomes (Berrie, 1963). Two of the four meiotic products inherit a U (X in the older literature) chromosome and develop as female (**8**), while the other two meiotic products inherit a V (Y) chromosome and develop as males (**9**). In phylogenetically divergent liverworts (*Marchantia*, *Sphaerocarpos*, *Pellia*) genetic experiments have implicated a U chromosome feminizer locus that is sufficient to specify female sex (Haupt, 1932; Knapp, 1935; Lorbeer, 1936; Heitz, 1949–1951). *Marchantia* has proven an excellent model for the investigation of sex determination and the evolution of sex chromosomes (Bowman et al., 2017; Montgomery et al., 2020; Yamato et al., 2007). In *M. polymorpha* the feminizer locus encodes a member of the plant-specific BASIC PENTACYSTEINE transcription factor family, Mp*BPCU* (Iwasaki et al., 2021). Surprisingly, Mp*BPCU* has a homolog (gametolog) on the V chromosome, Mp*BPCV*, and both genes act in the reproductive transition but only BPCU possesses a function in sex determination (Iwasaki et al., 2021). MpBPCU negatively regulates SUPRESSOR OF FEMINIZATION, a cis acting antisense non-coding RNA, that in turn negatively regulates MpFGMYB, an autosomal gene encoding a land plant specific MYB transcription factor that promotes feminization (Hisanaga et al., 2019; Iwasaki et al., 2021). Arabidopsis FGMYB orthologs were also shown to promote female gametophyte development (Hisanaga, Okahashi, et al., 2019), providing an example of discoveries made in *Marchantia* informing investigations in other more intensively studies taxa. *Whether the molecular identity of the feminizer is conserved throughout liverworts and beyond or whether there has been "turnover" of the sex determining gene are open questions.*

Gametangia are borne upon gametangiophores, which are specialized extensions of the dorsal-ventral thallus that grow orthotropically to facilitate gamete and spore dispersal (Bischoff, 1835; Hofmeister, 1862; Mirbel, 1835; Sachs, 1882). Gametangiophores evolved early in the Marchantiopsida lineage following the divergence of the Blasiales, and were subsequently lost in several taxa (Goebel, 1930). *Marchantia* is a long day plant (Benson-Evans, 1961; Wann, 1925), making the transition to reproduction in the spring in response to changes in light quality, i.e., red to far-red ratios (Courtoy, 1964), such that mature sporophytes are usually conspicuous in early summer in temperate climates (e.g., Bock, 1552). Experiments in *Marchantia* have demonstrated that the molecular machinery regulating the day-length dependent reproductive transition is at least partially shared across land plants (Kubota et al., 2014). For example, the single phytochrome protein, MpPHY, encoded in the genome acts as a red and far-red light receptor whose action is mediated via the downstream transcription factor MpPIF (Inoue et al., 2016). MpPHY is required for the reproduction transition, with loss-of-function alleles failing to initiate anatomical or morphological changes associated with the reproductive transition (Inoue, Nishihama, Araki, & Kohchi, 2019). Similarly, circadian clock genes conserved as well (Linde et al., 2017), with one output of the clock being rhythmic growth, or nyctinasty (Lagerkrantz, Billhardt, Rousku, Ljung, & Eklund, 2020).

During the reproductive transition, gametangiophores develop from one of two apices (**10**) produced from the most recent thallus bifurcation (Leitgeb, 1881). In *Marchantia* the gametangiophore capitula are the product of three (usually) rapid bifurcations forming eight meristems (Leitgeb, 1880b), each of which produce a series of gametangia, either archegonia or antheridia, from dorsal epidermal derivatives (**11**, **12**) (Hofmeister, 1862). These rapid bifurcations are then elevated via elongation of modified thallus growth, forming the gametangiophore which serves to elevate both the gametangia pre-fertilization and the sporophyte post-fertilization. The activity of a regulatory module involving a land plant-specific transcription factor, MpSPL2 and microRNA156/529 is required for proper gametangiophore development (Tsuzuki et al., 2019). Furthermore, the transition to reproductive growth and the specification of both male and female generative cells in *Marchantia* is dependent upon a land plant-specific bHLH transcription factor, MpBONOBO, which is expressed in archegonial and antheridial initial cells (Yamaoka et al., 2004; Yamaoka et al., 2018). BONOBO orthologs also have a role in generative cell specification in Arabidopsis male gametophytes (Yamaoka et al., 2018), highlighting the conservation of molecular mechanisms of reproductive development across land plants. *The molecular mechanisms by which environmental signals are transduced to activate BONOBO, and furthermore in only one apex of a recently bifurcated shoot, remain unknown.* Archegonia and antheridia are innovations of the ancestral land plant (Fig. 1) and the two structures are thought to be homologous (Goebel, 1902). *While no archegonia or antheridia specifying gene has yet been identified, the two structures might be specified by the activity of BNB and FGMYB, with BNB alone conferring antheridial identity and the combination of both conferring archegonial identity.*

Archegonia harboring a single egg cell ensheathed in unistratose archegonium wall (Hofmeister, 1862; Mirbel, 1835; Strasburger, 1869) (**13**) are produced in a row with the oldest towards the outside of the archegoniophores (**11**) (Bischoff, 1835; Mirbel, 1835) due to epinastic growth of the apex displacing the apical cell ventrally (Hofmeister, 1862; Leitgeb, 1881). Egg cells are, at least in part, specified by the activity of an RKD transcription factor, MpRKD, and in its absence, the cell that would normally differentiate into an egg cell does not arrest and instead continues to divide (Koi et al., 2016; Rövekamp, Bowman, & Grossniklaus, 2016). Likewise, antheridia (**14**) form in irregular rows, but with the oldest towards the center of the capitulum (**12**) (Hedwig, 1783; Schmidel, 1762). Sperm mother cells are formed by synchronized cells divisions within sectors of the antheridia (**14**). Motile sperm with two anterior flagella (**15**) (Unger, 1837) are extruded through antheridial pores due to swelling of the mucilage containing antheridial jacket cells when in contact with water (Bergdolt, 1926; Goebel, 1898; Strasburger, 1869). Sperm differentiation is regulated in part by the MYB transcription factor MpDUO (Higo et al., 2018), whose Arabidopsis ortholog also regulates sperm differentiation. The role of DUO in regulating sperm morphogenesis likely predates land plants, with its origin in an ancestral charophyte alga at the time that anisogamy evolved in the Streptophyte lineage (Higo et al., 2018). *The mechanism by which male gametes are specified remains unknown, as does the mechanisms by which MpRKD is activated in the egg cell.*

Fertilization is aquatic, but it is likely sperm swim only a short distance, with other mechanisms facilitating long distance events. Males produce gametangiophores earlier than females, and those of males elongate as antheridia are maturing, while those of females harbor mature egg cells before substantial elongation has occurred. In some Marchantiopsida sperm are forcibly ejected (Cavers, 1903; Peirce, 1902; Thuret, 1851), but not in *Marchantia*. However, the elevated antheridiophore forms a splash platform from which a limited amount of sperm may be dispersed short distances (Duckett & Pressel, 2009; Strasburger, 1869). However, it is not free spermatozoid cells that are ejected, but rather sperm that are still contained in their mother cell membrane (Strasburger, 1869), which allows the sperm to be transported via spread of a lipophilic layer on water surfaces (Muggoch & Walton, 1942). Thus, the 200;000+ sperm produced per antheridia (Bolleter, 1905; Johnson, 1904) could spread tens of meters from the source (Pressel & Duckett, 2019). The scales and rhizoids on the ventral thallus surface, as well as in the archegoniophore rhizoid furrow, allow rapid transport via capillary action (Duckett, Ligrone, Renzaglia, & Silvia, 2014; McConaha, 1941) to the top of the archegoniophores, such that the sperm may only have to swim >1 mm (Shimamura, 2016). Thus, in a natural setting, fertilization was observed to occur between male and female plants located almost 20 m apart (Pressel & Duckett, 2019). While the mucilage extruded by the archegonia is thought to provide chemotactic cues for sperm (Strasburger, 1869), *the precise nature of the chemicals involved is unknown* despite some chemicals acting as chemo-attractants in vitro (Åkerman, 1910; Lidforss, 1904-5).

2.2 The diploid sporophyte generation

Following fertilization, the diploid zygote representing the sporophyte generation expands to fill the venter (**16**) (Hofmeister, 1862). The genetic mechanism by which the diploid sporophyte gene expression program is activated likely involves paralogous homeodomain genes, as such a system is active diverse eukaryotes and likely represents a character of the ancestral eukaryote (Bowman, Sakakibara, Furumizu and Dierschke, 2016). In extant unicellular Viridiplantae, two paralogous TALE-homeodomain genes have mutually exclusive gamete specific expression patterns (Lee, Lin, Joo, & Goodenough, 2008). However, such systems may be more complex in multicellular lineages as there exists evidence that one, e.g., in the moss *Physcomitrium* (Horst et al., 2016; Sakakibara et al., 2013; Sakakibara, Nishiyama, Deguchi, & Hasebe, 2008), or both, e.g., in the brown alga *Ectocarpus* (Arun et al., 2019), TALE-homeodomain paralogs are expressed in both types of gametes. Indeed, in *M. polymorpha* one class of

TALE-homeodomain paralog, Mp*KNOX1*, is expressed in the egg cell and is absolutely required for activation of the diploid sporophyte program, while genes of the other class of TALE-homeodomain gene, Mp*BELL3* and Mp*BELL4*, are expressed in both the the egg cell and developing sperm (Dierschke et al., 2021; Hisanaga et al., 2021). Loss of either male or female BELL greatly reduces efficient of activation of the diploid program and loss in both parents resembles loss of Mp*KNOX1* (Dierschke et al., 2021; Hisanaga et al., 2021). *How the diploid sporophyte program is specifically activated post-fertilization in systems in which one or both TALE-homeodomain paralogs are expressed in both parents is unknown.*

The first division of the zygote is usually transverse, with the epibasal cell (the one closer to the archegonial neck) eventually giving rise to the capsule and the hypobasal cell (the one closer to the archegonial stalk) giving rise to the foot and short seta (Leclerc du Sablon, 1885). The next divisions are typically vertical resulting in a "quadrant" embryo (**17**). There is no conspicuous apical meristem, but rather, cell divisions are dispersed through the embryo (Kienitz-Gerloff, 1874), and after a couple weeks differences are evident between the cells of the foot and seta compared to the cells that will give rise to the capsule (Durand, 1908; Kienitz-Gerloff, 1874) (**18**). *Little is known about the genetic basis of patterning events during early sporophyte development,* i.e., *how the foot, seta and capsule are specified.* During the early growth of the embryo, both periclinal and anticlinal divisions in the gametophytic calyptra occur (**17**) such that its growth keeps pace with that of the ensheathed sporophyte (**19**) (Hofmeister, 1862). The developing sporophyte is also surrounded by a gametophytic pseudoperinath that develops post-fertilization and groups of archegonia derived from the same apical cell are surrounded by an involucre, such that three gametophytic tissues surround developing sporophytes (Mirbel, 1835). *The biochemical nature of communication pathways between the two generations remains to be explored.*

Subsequently, internal cells of the capsule begin to elongate in the vertical plane (**20**) and a subset of these cells, elaterocytes, are set aside to differentiate into diploid elaters (**21**) while the other cells are sporogenous and divide transversely to produce sporocytes (**22**) (Henfrey, 1853). In *Marchantia*, the sporogenous cells typically divide five times producing 32 sporocytes, each of which undergoes meiosis generating four haploid spores (Henfrey, 1853) (**23**), for a total of 128 spores per elater (O'Hanlon, 1926). Each sporophyte can generate approximately 300,000 spores (O'Hanlon, 1926). *Little is understood about the specification of elaterocytes* versus *sporogenous cells, or of the regulation of the number of divisions of the latter before their* *specification as meiocytes*. The foot is the conduit between the maternal gametophyte generation and the diploid sporophyte generation, providing both nutrients and water. While there are no plasmodesmatal connections between generations (Kelley, 1969), extensive ingrowths occur in presumed transfer cells of both generations (Ligrone, Duckett, & Renzaglia, 1993). *The signaling networks that must be involved in establishment of communication and nourishment channels between the generations are as yet unexplored. Further, whether there is any genomic imprinting associated with the dependency of the embryo on the female gametophyte is unknown*. Prior to capsule rupture the seta elongates, as in other liverworts, via cell elongation (Thomas, 1980), with the relatively short seta of *Marchantia* being a derived condition since the gametangiophores serve to elevate the sporophytes to facilitate spore dispersal.

The unistratose capsule wall ruptures (**20**, **24**) aided by annular cell wall thickenings, similar to other liverworts (Leclerc du Sablon, 1885), and hygroscopic elaters, which are dead at maturity, facilitate spore dispersal (**25**, **26**) (Marchant, 1713). *The identity of the presumed carbohydrate polymers responsible for the hygroscopic movements of the elaters are not yet identified*. Under natural conditions, spore longevity in *M. polymorpha* is limited, with viability dropping substantially between 12 and 18 months post dispersal (Inoue, 1960; O'Hanlon, 1926). The smooth spores coated in sporopollenin, the most durable biological polymer known, can be dispersed widely enabling the colonization of distant disturbed, often recently burnt, habitats (e.g., Skutch, 1929; Torrey, 1932). *The mechanism of sporopollenin deposition is enigmatic as a tapetum is lacking in liverworts*. Spores have been detected in rainwater (Pettersson, 1940) and transoceanic dispersals have occurred within the genus as evidenced by colonization of oceanic islands (Bischler-Causse, 1989; Young & Kläy, 1971).

2.3 The vegetative thallus of the gametophyte

M. polymorpha is a complex thalloid liverwort with the vegetative gametophyte having a distinct dorsal-ventral structure (Fig. 3). The dorsal side (**1**) is characterized by a latticework of air chambers, while the ventral side (**2**) is covered with a patterned network of scales and rhizoids. Air chambers develop close to the apical mersitem from the dorsal epidermal and epidermal cell layers (Barnes & Land, 1907) (**3**) with each chamber having a centrally located multistratose air pore (Nägeli, 1842) (**4**, **5**) and basal photosynthetic assimilatory filaments (**5**). The extent of air chamber development is influenced upon light quality and intensity (Stahl, 1882), moisture

Fig. 3 *M. polymorpha* vegetative thallus anatomy. Dorsal (1) and ventral (2) overviews of vegetative thalli. Development of air chambers and pores (3, 4) and a fully developed air chamber (5). Cross section through mature vegetative thallus (6); unistratose scales arise in 2–4 rows flanking the ventral midrib, enveloping bundles of pegged rhizoids (6, 7). Smooth rhizoids arise from near the ventral midrib, but are not shown in this view. Fours lines at left highlight oil body cells (6). Individual numbered images described in the text. Drawings from Leitgeb (1880a), Mirbel (1835), and Nägeli (1842).

(Ruge, 1893), and CO2 concentration (Téodoresco, 1899). The architecture of air chambers does not correspond to merophytes, implying a requirement of cell-cell communication to coordinate their development (Suzuki, Harrison, Shimamura, Kohchi, & Nishihama, 2020). The E3 ubiquitin ligase NOPPERABO1 is required for the initial stage of formation of the intercellular spaces, with mutants lacking both air chambers and pores (Ishizaki et al., 2013). *The intercellular communication pathways responsible for the stereotypic pattern of air chambers and pores is not known.* While mostly inconspicuous elsewhere, the deposition of a cuticular layer is prominent on the air pore and surrounding cells (Schönherr & Ziegler, 1975). As in flowering plants, a land plant specific MYB transcription factor, MpMIXTA, regulates cuticle biosynthesis genes in *Marchantia* (Xu et al., 2021).

Ventral parenchyma acts as a storage tissue, and is interspersed with oil body cells, a plesiomorphic character of liverworts (**6**). While many roles have been postulated in the past, in *Marchantia* they do not play a role in abiotic stress responses, but rather provide protection against herbivores (Kanazawa et al., 2020; Romani et al., 2020), as postulated more than a century previously (Stahl, 1888). As predicted from earlier studies that placed terpene synthases at the oil body membrane (Suire et al., 2000), the oil bodies of *Marchantia* are filled with sequiterpenes and bis-bibenzyl compounds (Romani et al., 2020; Takizawa et al., 2021; Tanaka et al., 2016). Two unrelated transcription factors, MpERF13 and MpC1HDZ are required for oil body formation (Kanazawa et al., 2020; Romani et al., 2020), and the involvement of a SNARE protein, MpSYP12B, suggests that oil body formation within a cell involves a cell biological process similar to a typical plant cell division, i.e., that the oil body membrane is analogous to the plasma membrane lipid bilayer (Kanazawa et al., 2020).

From the ventral epidermis are formed 3–4 pairs scales in rows flanking the midrib (Taylor, 1837) (**6**). The proper development of the scales requires the transcription factor MpLOS1, which is predominantly expressed in scales and differentiating ventral tissues but also affects shoot meristem activity non-autonomously (Naramoto et al., 2019). The ventral scales, as well as the ventral epidermis are often pigmented red to purple (Nagai, 1915), which is thought to provide protection from UVB radiation. The pigment, a cell wall-bound auronidin, was recently shown to be formed via a phenylpropanoid biosynthetic pathway distinct from that leading to anthocyanins, the primary red pigments in flowering plants (Berland et al., 2019). However, the auronidin biosynthetic pathway is regulated via MpMYB14, an ortholog of land plant specific MYB genes that regulate anthocyanain biosynthesis in angiosperms (Albert et al., 2018; Kubo et al., 2018). Mp*MYB14* gain-of-function mutants that have increased flavonoid production have increased UVB tolerance, with UVR8-mediated induction in response to increased UVB conserved across land plants (Clayton et al., 2018).

Two types of rhizoids are produced, smooth and pegged (Gasparrini, 1856), and both consist of single cells that can be millimeters in length. The large diameter smooth rhizoids are negatively phototropic, tip growing, and emanate from near the midrib and act in anchorage and nutrient absorption (Haberlandt, 1889; Molisch, 1884; Pfeffer, 1871). The identification of a common conserved module involving bHLH transcription factors (MpRSL1 in *Marchantia*) regulating genes encoding proteins involved in tip growth revealed molecular conservation between *Marchantia* rhizoid and Arabidopsis root hair development (Honkanen et al., 2016; Proust et al., 2016). Although *M. polymorpha* does not exhibit interactions with mycorrhizal fungi, in related species, such as *M. paleacea*, the smooth rhizoids provide an avenue of entry for mycorrhizal fungi (Radhakrishnan et al., 2020). The ability to interact with mycorrhizal fungi is ancestral, and is thought to have been crucial during the early terrestrial evolution of land plants (Rimington, Duckett, Field, Bidartondo, & Pressel, 2020; Strullu-Derrien, Selosse, Kenrick, & Martin, 2018). While the signaling pathway genes in *M. polymorpha* are reduced to pseudogenes, *M. paleacea* provides an excellent model to study mycorrhizal interactions by harnessing the genomic tools and technologies developed for *M. polymorpha* (Radhakrishnan et al., 2020).

In contrast to the smooth rhizoids, the pegged rhizoids (**7**) are dead at maturity (Cavers, 1904; Kamerling, 1897), and act in conjunction with the scales in capillary water transfer across the ventral surface of the thallus (Bowen, 1935; Duckett et al., 2014; McConaha, 1941). *Little is known about whether these cells undergo a typical programmed cell death program, or whether their cell wall thickenings are regulated in a similar manner as those of vascular cells that also undergo programmed cell death in vascular plants. As of yet, no mutant has been reported that differentially affects smooth or pegged rhizoids.*

2.4 Asexual reproduction

M. polymorpha can propagate asexually via the production of gemmae, which are produced in gemmae cups, which in turn are produced from the dorsal surface of the vegetative thallus (Fig. 4). Gemmae cup formation requires the function of the MYB transcription factor GCAM1 (Yasui et al., 2019) and the cytokinin signaling pathway (Aki et al., 2019; Flores-Sandoval et al., 2016). Gemmae develop from dorsal epidermal cells (Barnes & Land, 1908) just distal the apical meristem (**1**), initially via tip growth from single cells in the cup base (Hofmeister, 1862) (**2**). That each gemma is derived from a single epidermal cell allows the use of gemmae in "purifying" mutations generated by CRISPR-Cas9 genome editing occurring during sporeling development. Gemma initiation requires Mp*RSL1*, perhaps due to both rhizoids and gemma having their origin in outgrowths from single epidermal cells (Proust et al., 2016). Likewise, mucilage cells have a similar origin and also require Mp*RSL1* (Proust et al., 2016). In contrast, gemma initiation requires the ROP-GEF, KARRAPO, while neither rhizoids nor mucilage cells requires its function (Hiwatashi et al., 2019).

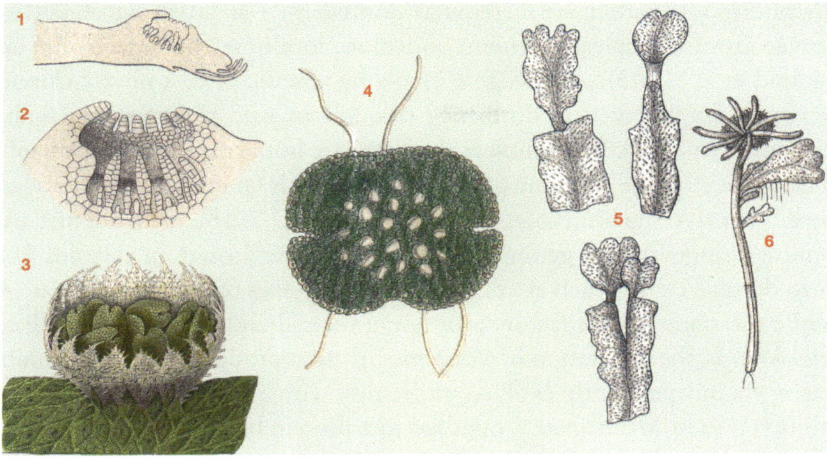

Fig. 4 Asexual reproduction in *M. polymorpha*. Developing gemmae cups and gemmae (1, 2). Mature gemmae cup with gemmae (3). Mature gemma (4); the central hyaline cells are rhizoid precursor cells, which may differentiate into rhizioids depending upon light and gravity cures. Regeneration can occur from almost all cells, except rhizoids; regeneration from vegetative thallus from which the shoot apex was removed (5), and from an excised archegoniophore (6). Individual numbered images described in the text. Drawings from Hofmeister (1862), and Mirbel (1835), and Vöchting (1885).

Hofmeister suggested that gemmae are not oriented at random, but have their surfaces perpendicular to the longitudinal axis of thallus (Hofmeister, 1862), suggestive of a polar signal emanating from the thallus, with the obvious candidate being auxin, which has been postulated to be transported from the apex into older tissues (Eklund et al., 2015; LaRue & Narayanaswami, 1957; Maravolo, 1976). Auxin transcriptional signaling is also important during pattern formation of developing gemmae, with its disruption leading to altered patterns of cell division planes and supernumerary shoot meristems (Kato et al., 2017).

Mature gemmae cups (**3**) act as splash cups with raindrops displacing the uppermost gemmae away from the parental plant (Brodie, 1951). For nearly the past two centuries the irreversible dorsiventrality induced in the initially "two-sided" lenticular gemmae (**4**) by light (primarily) and gravity has been investigated [see (Bowman, 2016) for review]. While gemmae reside in the cup they remain dormant, only resuming growth once displaced, with early experiments suggesting signals from outside the cup as well as within are required to impose and maintain dormancy (Molisch, 1922; Oppenheimer, 1922; Tarén, 1958). Genetic approaches have revealed that at least three phytohormones participate in gemmae dormancy — an auxin signal, perhaps derived from the apical meristem and more locally at the base of the cup (Eklund et al., 2015), and a signal involving abscisic acid, a phytohormone known to be involved in dormancy throughout land plants, acting within the gemmae in the cup (Eklund et al., 2018) are both required for dormancy. Finally, the ethylene signaling pathway may provide an intermediary between these other two signals (Ruge, 1893; Eklund et al., 2018). One the first anatomical changes during gemmae germination is the growth of rhizoids from rhizoid initial cells, which are dispersed on both sides of mature gemmae (**4**). While the transcription factors promoting rhizoid and root hair growth are orthologous, the generation of a dispersed pattern produced by lateral inhibition is via independently evolved molecules, with a microRNA performing this function in *Marchantia*, as opposed to a protein based system in root hair patterning (Honkanen, Thamm, Arteaga-Vazquez, & Dolan, 2018; Thamm, Saunders, & Dolan, 2020).

As in many other bryophytes, *M. polymorpha* exhibits a high capacity for regeneration from gametophytic tissues from which the shoot meristem has been removed (**5**, **6**) (Dickson, 1932; Kreh, 1909; Nagai, 1919; Vöchting, 1885). In *Marchantia*, light, primarily via perception by phytochrome of red light, promotes regeneration, with the process greatly reduced under far-red light or dark conditions (Nishihama et al., 2015). Consistent with auxin having a role in regeneration, an ARF transcription factor, MpARF3, that may negatively regulate auxin signaling in some capacity, is required for efficient regeneration (Flores-Sandoval et al., 2018). Indeed, a reduction is auxin signaling is required for induction of a land plant-specific AP2 transcription factor, MpLAXR, that subsequently promotes regeneration (Ishida et al., 2022). *The processes of shoot regeneration and apical cell specification in the shoot meristem share genetic programs and studies on each should inform the other.*

3. The future

In order to ascertain the ancestral or derived nature of characters it is imperative to develop model systems amenable to both forward and reverse genetics in as many of the six major lineages as feasible. Such models could help overcome a major bias in current analyses by establishing a broader baseline for comparisons, rather than assuming the character states, often derived, in angiosperms as the standard. While past incarnations of *Marchantia* as a model system were primarily due to its familiarity, the present manifestation is based on a combination of phylogenetic relationships, facile genetics and a regulatory genome that exhibits a paucity of genetic redundancy. This latter feature makes *Marchantia* well suited to large-scale experiments aimed at the identification of putatively ancestral land plant gene regulatory networks, as a baseline to determine how they have been modified and elaborated over evolutionary time. With its amenability to high throughput genetic manipulation and genomic editing *Marchantia* will be one of the languages of the Rosetta stone of land plants for some time to come.

Acknowledgments

This review was improved by the constructive comments of Facundo Romani and members of my lab. Research in J.L.B.s laboratory is supported by the Australian Research Council (DP200100225; DP210101423).

References

Åkerman, Å. (1910). Über die chemotaxis der *Marchantia*-spermatozoiden. *Zeitschrift für Botanik*, *2*, 94–103.

Aki, S. S., Mikami, T., Naramoto, S., Nishihama, R., Ishizaki, K., et al. (2019). Cytokinin signaling is essential for organ formation in *Marchantia polymorpha*. *Plant & Cell Physiology*, *60*, 1842–1854.

Albert, N. W., Thrimawithana, A. H., McGhie, T. K., Clayton, W. A., Deroles, S. C., et al. (2018). Genetic analysis of the liverwort *Marchantia polymorpha* reveals that R2R3MYB activation of flavonoid production in response to abiotic stress is an ancient character in land plants. *New Phytologist*, *218*, 554–566.

Allen, C. E. (1917). A chromosome difference correlated with sex differences in *Sphaerocarpos*. *Science*, *46*, 466–467.

Apostolakos, P., Galatis, B., & Mitrakos, K. (1982). Studies on the development of the air pores and air chambers of *Marchantia* paleacea I. Light-microscopy. *Annals of Botany*, *49*, 377–396.

Arun, A., Coelho, S. M., Peters, A. F., Bourdareau, S., Péres, L., et al. (2019). Convergent recruitment of TALE homeodomain life cycle regulators to direct sporophyte development in land plants and brown algae. *eLife*, *8*, e43101.

Barnes, C. R., & Land, W. J. G. (1907). Bryological papers I. the origin of air chambers—contributions from the Hull botanical laboratory, 100. *Botanical Gazette*, *44*, 197–213.

Barnes, C. R., & Land, W. J. G. (1908). Bryological papers II. The origin of the cupule of *Marchantia*—contributions from the hull botanical laboratory 120. *Botanical Gazette*, *46*, 401–409.

Benson-Evans, K. (1961). Environmental factors and bryophytes. *Nature*, *191*, 255–260.

Bergdolt, E. (1926). *Untersuchungen über Marchantiaceen*. Jena: Gustav Fischer.

Berland, H. N. W. A., Stavland, A., Jordheim, M., McGhie, T. K., et al. (2019). Auronidins are a previously unreported class of flavonoid pigments that challenges when anthocyanin biosynthesis evolved in plants. *Proceedings of the National Academy of Sciences the United States of America*, *116*, 20232–20239.

Berrie, G. K. (1963). Cytology and phylogeny of liverworts. *Evolution*, *17*, 347–357.

Bischler-Causse, H. (1989). *Marchantia L. The Asiatic and Oceanic taxa*. Berlin - Stuttgart: J. Cramer.

Bischoff, G. W. (1835). Bemerkungen über die Lebermoose vozüglich aus den gruppen der Marchantieen und Riccieen. *Nova Acta Physico Medica A C L, 17*(2), 909–1088.
Blanchard, F. (1891). Common objects available for exhibiting the power of the microscope. *The American Monthly Microscopical Journal, 12*, 138–139.
Bock, H. (1552). *De stirpium maxime*. Argenorati: Excudebat V. Rihelius.
Bohr, F. J. (1897). An introduction to structural botany physiologische pflanzenanatomie. *Nature, 55*, 457–458. https://doi.org/10.1038/055457a0.
Bolleter, E. (1905). *Fegatella conica* (L) Corda. Eine morphologisch-physiologische Monographie. *Beih. Bot. Centralbl., 18*, 327–408.
Bowen, E. J. (1935). A note on the conduction of water in *Fibriaria bleumeana*. *Annals of Botany, 49*, 844–848.
Bowman, J. L. (2013). Walkabout on the long branches of plant evolution. *Current Opinion in Plant Biology, 16*, 70–77.
Bowman, J. L. (2016). A brief history of Marchantia from Greece to genomics. *Plant and Cell Physiology, 57*, 210–229.
Bowman, J. L., Araki, T., Arteaga-Vazquez, M. A., Berger, F., Dolan, L., et al. (2016). The naming of names: Guidelines for gene nomenclature in *Marchantia*. *Plant and Cell Physiology, 57*, 257–261.
Bowman, J. L., Briginshaw, L. N., & Florent, S. N. (2019). Evolution and co-option of developmental regulatory networks in early land plants. *Current Topics in Developmental Biology, 131*, 35–53.
Bowman, J. L., Flores-Sandoval, E., & Kato, H. (2021). On the evolutionary origins of Land plant auxin biology. *Cold Spring Harbor Perspectives in Biology, a040048*.
Bowman, J. L., Kohchi, T., Yamato, K. T., Jenkins, J., Shu, S., et al. (2017). Insights into Land plant evolution garnered from the *Marchantia polymorpha* genome. *Cell, 171*, 287–304.
Bowman, J. L., Sakakibara, K., Furumizu, C., & Dierschke, T. (2016). Evolution in the cycles of life. *Annual Review of Genetics, 50*, 133–154.
Brodie, H. J. (1951). The splash-cup dispersal mechanism in plants. *Canadian Journal of Botany-Revue Canadienne De Botanique, 29*, 224–234.
Burgeff, H. (1943). *Genetische Studien an Marchantia*. Jena: Gustav Fischer.
Cavers, F. (1903). Explosive discharge of antherozoids in Hepaticae. *Torreya, 3*, 179–183.
Cavers, F. (1904). On the structure and biology of *Fegatella conica*. *Annals of Botany, 18*, 87–120.
Clayton, W. A., Albert, N. W., Thrimawithana, A. H., McGhie, T. K., Deroles, S. C., et al. (2018). UVR8-mediated induction of flavonoid biosynthesis for UVB tolerance is conserved between the liverwort *Marchantia polymorpha* and flowering plants. *Plant Journal, 96*, 503–517.
Courtoy, R. (1964). Recherches expérimentales sur les facteurs de la sexualisation des Thalles de *Marchantia polymorpha* L. *Bulletin de la Classes des Sciences, 50*, 1341–1359.
de Sousa, F., Foster, P. G., Donoghue, P. C. J., Schneider, H., & Cox, C. J. (2019). Nuclear protein phylogenies support the monophyly of the three bryophyte groups (Bryophyta Schimp.). *New Phytologist, 222*, 565–575.
Delaux, P.-M., & Schornack, S. (2021). Plant evolution driven by interactions with symbiotic and pathogenic microbes. *Science, 371*, eaba6605.
Delwiche, C. F., & Cooper, E. D. (2015). The evolutionary origin of a terrestrial Flora. *Current Biology, 25*, R899–R910.
Dickson, H. (1932). Polarity and the production of adventitious growing points in *Marchantia polymorpha*. *Annals of Botany, 46*, 683–701.
Dierschke, T., Flores-Sandoval, E., Rast-Somssich, M. I., Altoff, F., Zachgo, S., & Bowman, J. L. (2021). Gamete expression of TALE class HD genes activates the diploid sporophyte program in *Marchantia polymorpha*. *eLife, 10*, e57088.

Douin, R. (1921). Reserches sur les Marchantiées. *Revue générale de botanique, 33*, 34–62. 99–145; 190–209.
Duckett, J. G., Ligrone, R., Renzaglia, K., & Silvia, P. (2014). Pegged and smooth rhizoids in complex thalloid liverworts (Marchantiopsida): Structure, function and evolution. *Botanical Journal of the Linnean Society, 174*, 68–92.
Duckett, J. G., & Pressel, S. (2009). Extraordinary features of the reproductive biology of *Marchantia* at Thursley common. *Field Bryology, 97*, 2–11.
Durand, E. J. (1908). The development of the sexual organs and Sporogonium of *Marchantia polymorpha*. *Bulletin of the Torrey Botanical Club, 35*, 321–335.
Edwards, D., Morris, J. L., Richardson, J. B., & Kenrick, P. (2014). Cryptospores and cryptophytes reveal hidden diversity in early land floras. *New Phytologist, 202*, 50–78.
Eklund, D. M., Ishizaki, K., Flores-Sandoval, E., Kikuchi, S., Takebayashi, Y., et al. (2015). Auxin produced by the indole-3-pyruvate pathway regulates development and gemmae dormancy in the liverwort Marchantia polymorpha. *The Plant Cell, 27*, 1650–1669.
Eklund, D. M., Kanei, M., Flores-Sandoval, E., Ishizaki, K., Nishihama, R., et al. (2018). An evolutionarily conserved abscisic acid signaling pathway regulates dormancy in the liverwort *Marchantia polymorpha*. *Current Biology, 28*, 3691–3699.
Fitting, H. (1939). Untersuchungen über den Einfluss von Licht und Dunkelheit auf die Entwicklung von Moosen. I. Die Brutkörper der Marchantieen. *Jahrbucher fur Wissenschaftliche Botanik, 88*, 633–722.
Flores-Sandoval, E., Dierschke, T., Fisher, T. J., & Bowman, J. L. (2016). Efficient and inducible use of artificial MicroRNAs in *Marchantia polymorpha*. *Plant and Cell Physiology, 57*, 281–290.
Flores-Sandoval, E., Eklund, D. M., & Bowman, J. L. (2015). A simple auxin transcriptional response system regulates multiple morphogenetic processes in the liverwort *Marchantia polymorpha*. *PLoS Genetics, 11*, 1005207.
Flores-Sandoval, E., Eklund, D. M., Hong, S.-F., Alvarez, J. P., Fisher, T. J., et al. (2018). Class C ARFs evolved before the origin of land plants and antagonize differentiation and developmental transitions in *Marchantia polymorpha*. *New Phytologist, 218*, 1612–1630.
Gasparrini, G. (1856). *Ricerche sulla natura dei succiatori*. Napoli: Giuseppe Dura Libraio-Editore.
Goebel, K. (1898). Uber den Offnungsmechanismus der Moos-antheridien. *Annales du Jardin botanique de Buitenzorg Supplement, II*, 65–72.
Goebel, K. (1902). Morphologische und biologische Bemerkungen. 11. Ueber Homologien in der Entwickelung männlicher und weiblicher Geschlechtsorgane. *Flora, 90*, 279–305.
Goebel, K. (1930). *Organographie der Pflanzen*. Gustav Fischer, Jena: Insbesondere der Archegoniaten und Samenpflanzen.
Groenland, J. (1854). Mémoire sur la Germination de Quelques Hépatiques. *Annales de Sciences Naturelles Botanique, 4*(1), 5–29.
Haberlandt, G. (1889). Ueber das Längenwachsthum und den Geotropismus der Rhizoiden von Marchantia und Lunularia. *Össter Bot Zeit, 39*, 93–98.
Hansel, V. (1876). Über die Keimung der *Preissia commutata*. Sitzungsberichte der Kaiserlichen Akademie der Wissenschaften. Mathematisch-Naturwissenschaftliche Classe. Abt. 1. *Mineralogie, Botanik, Zoologie, Geologie und Paläontologie, 73*, 89–97.
Haupt, G. (1932). Beiträge zur Zytologie der Gattung Marchantia I. *Zeitschrift für Induktive Abstammungs- und Vererbungslehre, 62*, 367–428.
Hedwig, J. (1783). *Theoria Generationis et Fructificationis Plantarum Cryptogamicarum Linnaei, mere propriis Observationibus et Experimentis Superstructa*. Scientiarum, Petropoli: Academiae Imper.
Heitz, E. (1928). Das Heterochromatin der Moose I. Jahrbücher für wissenschaftliche. *Botanik, 69*, 762–818.

Heitz, E. (1949–1951). Weitere beobachtungen an den geschelchtsmutanten, insbesondere den intersexen von *Pellia neesiana*. *Portugaliae Acta Biologica, Serie A. Morfologia, Fisiologia, Genetica e Biologia Geral*, 353–366.

Henfrey, A. (1853). XII. On the development of the spores and elaters of Marchantia polymorpha. *Transactions of the Linnean Society*, *21*, 103–110. tab. XI.

Higo, A., Kawashima, T., Borg, M., Zhao, M., López-Vidriero, I., et al. (2018). Transcription factor DUO1 generated by neofunctionalization is associated with evolution of sperm differentiation in plants. *Nature Communications*, *9*, 5283.

Hirakawa, Y., Fujimoto, T., Ishida, S., Uchida, N., Sawa, S., et al. (2020). Induction of Multichotomous branching by CLAVATA peptide in *Marchantia polymorpha*. *Current Biology*, *30*, 3833–3840.

Hirakawa, Y., Uchida, N., Yamaguchi, Y. L., Tabata, R., Ishida, S., et al. (2019). Control of haploid meristem activity by CLE peptide signaling in *Marchantia polymorpha*. *PLoS Genetics*, *15*, e1007997.

Hisanaga, T., Fujimoto, S., Cui, Y., Sato, K., Sano, R., Yamaoka, S., et al. (2021). Deep evolutionary origin of gamete-directed zygote activation by KNOX/BELL transcription factors in green plants. *eLife*, *10*, e57090.

Hisanaga, T., Okahashi, K., Yamaoka, S., Kajiwara, T., Nishihama, R., et al. (2019). A cis-acting bidirectional transcription switch controls sexual dimorphism in the liverwort. *The EMBO Journal*, *38*, e100240.

Hisanaga, T., Yamaoka, S., Kawashima, T., Higo, A., Nakajima, K., et al. (2019). Building new insights in plant gametogenesis from an evolutionary perspective. *Nature Plants*, *5*, 663–669.

Hiwatashi, T., Goh, H., Yasui, Y., Koh, L. Q., Takami, H., et al. (2019). The RopGEF KARAPPO is essential for the initiation of vegetative reproduction in *Marchantia polymorpha*. *Current Biology*, *29*, 3525–3531.

Hofmeister, W. F. B. (1862). *On the germination, development, and fructification of the higher Cryptogamia, and on the fructification of the Coniferae*. London: Ray Society.

Honkanen, S., Jones, V. A. S., Morieri, G., Champion, C., Hetherington, A. J., et al. (2016). The mechanism forming the cell surface of tip-growing rooting cells is conserved among Land plants. *Current Biology*, *26*, 3238–3244.

Honkanen, S., Thamm, A., Arteaga-Vazquez, M. A., & Dolan, L. (2018). Negative regulation of conserved RSL class I bHLH transcription factors evolved independently among land plants. *eLife*, 7, e38529.

Horst, N. A., Katz, A., Pereman, I., Decker, E. L., Ohad, N., et al. (2016). A single homeobox gene triggers phase transition, embryogenesis and asexual reproduction. *Nature Plants*, *2*, 15209.

Inoue, H. (1960). Studies in spore germination and the earlier stages of gametophyte development in the Marchantiales. *The Journal of the Hattori Botanical Laboratory*, *23*, 148–191.

Inoue, K., Nishihama, R., Araki, T., & Kohchi, T. (2019). Reproductive induction is a far-red high irradiance response that is mediated by phytochrome and PHYTOCHROME INTERACTING FACTOR in *Marchantia polymorpha*. *Plant and Cell Physiology*, *60*, 1136–1145.

Inoue, K., Nishihama, R., Kataoka, H., Hosaka, M., Manabe, R., et al. (2016). Phytochrome signaling is mediated by PHYTOCHROME INTERACTING FACTOR in the liverwort *Marchantia polymorpha*. *Plant Cell*, *28*, 1406–1421.

Ishida, S., Suzuki, H., Iwaki, A., Kawamura, S., Yamaoka, S., Kojima, M., et al. (2022). Diminished auxin signaling triggers cellular reprogramming by inducing a regeneration factor in the liverwort Marchantia polymorpha. *Plant & Cell Physiology*, *63*. https://doi.org/10.1093/pcp/pcac004.

Ishizaki, K., Chiyoda, S., Yamato, K. T., & Kohchi, T. (2008). Agrobacterium-mediated transformation of the haploid liverwort *Marchantia polymorpha* L., an emerging model for plant biology. *Plant and Cell Physiology*, *49*, 1084–1091.

Ishizaki, K., Johzuka-Hisatomi, Y., Ishida, S., Iida, S., & Kohchi, T. (2013). Homologous recombination-mediated gene targeting in the liverwort *Marchantia polymorpha* L. *Scientific Reports, 3*, 1532.

Ishizaki, K., Mizutani, M., Shimamura, M., Masuda, A., Nishihama, R., et al. (2013). Essential role of the E3 ubiquitin ligase NOPPERABO1 in Schizogenous intercellular space formation in the liverwort *Marchantia polymorpha. Plant Cell, 25*, 4075–4084.

Ishizaki, K., Nishihama, R., Ueda, M., Inoue, K., Ishida, S., et al. (2015). Development of gateway binary vector series with four different selection markers for the liverwort *Marchantia polymorpha. PLoS One, 10*, e0138876.

Ishizaki, K., Nishihama, R., Yamato, K. T., & Kohchi, T. (2016). Molecular genetic tools and techniques for *Marchantia polymorpha* research. *Plant and Cell Physiology, 57*, 262–270.

Iwasaki, M., Kajiwara, T., Yasui, Y., Yoshitake, Y., Miyazaki, M., Kawamura, S., et al. (2021). Sex determination factor and sex chromosome evolution in a haploid system. *Current Biology, 31*, 5522–5532.

Johnson, D. S. (1904). The development and relationship of *Monoclea. Botanical Gazette, 38*, 185–205.

Kamerling, Z. (1897). Zur Biologie und Physiologie der Marchantiaceen. *Flora, 84*, 1–68.

Kanazawa, T., Morinaka, H., Ebine, K., Shimada, T. L., Ishida, S., et al. (2020). The liverwort oil body is formed by redirection of the secretory pathway. *Nature Communications, 11*, 6152.

Kato, H., Ishizaki, K., Kouno, M., Shirakawa, M., Bowman, J. L., et al. (2015). Auxin-mediated transcriptional system with a minimal set of components is critical for morphogenesis through the life cycle in *Marchantia polymorpha. PLoS Genetics, 11*, e1005365.

Kato, H., Kouno, M., Takeda, M., Suzuki, H., Ishizaki, K., et al. (2017). The roles of the sole activator-type auxin response factor in pattern formation of *Marchantia polymorpha. Plant and Cell Physiology, 58*, 1642–1651.

Kato, H., Mutte, S. K., Suzuki, H., Crespo, I., Das, S., et al. (2020). Design principles of a minimal auxin response system. *Nature Plants, 6*, 473–482.

Kelley, C. (1969). Wall projections in the sporophyte and gametophyte of *Sphaerocarpos. Journal of Cell Biology, 41*, 910–914.

Kenrick, P., & Crane, P. R. (1997). *The origin and early diversification of land plants: A cladistic study*. Washington: Smithsonian Institution Press.

Kienitz-Gerloff, F. (1874). Vergleichende Untersuchungen über die Entwickelungsgeschichte des Lebermoos—Sporogoniums. *Botanische Zeitung, 32*, 161–172. 193-204; 209-221; 225-238.

Knapp, E. (1935). Untersuchungen über die Wirkung von Röntgenstrahlen an dem Lebermoos *Sphaerocarpus*, mit Hilfe der tetraden-Analyse. I. *Zeitschrift für Induktive Abstammungs- und Vererbungslehre, 70*, 309–349.

Kny, L. (1865). Beiträge zur Entwickelungs geschichte der laubigen Lebermoose. *Jahrbucher fur Wissenschaftliche Botanik, 4*, 64–100.

Kohchi, T., Yamato, K. T., Ishizaki, K., Yamaoka, S., & Nishihama, R. (2021). Development and molecular genetics of *Marchantia polymorpha. Annual Review of Plant Biology, 72*.

Koi, S., Hisanaga, T., Sato, K., Shimamura, M., Yamato, K. T., et al. (2016). An evolutionarily conserved plant RKD factor controls germ cell differentiation. *Current Biology, 26*, 1775–1781.

Kreh, W. (1909). Über die Regeneration der Lebermoose. Nova Acta Abh. der Kaiserl. Leop.-Carol. *Deutschen Akademie der Naturforscher, 90*, 216–301.

Kubo, H., Nozawa, S., Hiwatashi, T., Kondou, Y., Nakabayashi, R., et al. (2018). Biosynthesis of riccionidins and marchantins is regulated by R2R3- MYB transcription factors in *Marchantia polymorpha. Journal of Plant Research, 131*, 849–864.

Kubota, A., Ishizaki, K., Hosaka, M., & Kohchi, T. (2013). Efficient *Agrobacterium*-mediated transformation of the liverwort *Marchantia polymorpha* using regenerating thalli. *Bioscience Biotechnology and Biochemistry, 77*, 167–172.

Kubota, A., Kita, S., Ishizaki, K., Nishihama, R., Yamato, K. T., et al. (2014). Co-option of a photoperiodic growth-phase transition system during land plant evolution. *Nature Communications, 5*, 3668.

Lagerkrantz, U., Billhardt, A., Rousku, S. N., Ljung, K., & Eklund, D. M. (2020). Nyctinastic thallus movement in the liverwort *Marchantia polymorpha* is regulated by a circadian clock. *Scientific Reports, 10*, 8658.

LaRue, C. D., & Narayanaswami, S. (1957). Auxin inhibition in the liverwort *Lunularia*. *New Phytologist, 56*, 61–70.

Lee, J.-H., Lin, H., Joo, S., & Goodenough, U. (2008). Early sexual origins of Homeoprotein Heterodimerization and evolution of the plant KNOX/BELL family. *Cell, 133*, 829–840.

Leitgeb, H. (1874). *Untersuchungen über die Lebermoose. Heft I. Blasia pusilla*. Jena: O. Deistung's Buchhandlung.

Leitgeb, H. (1877a). Die Keimung der Lebermoossporen in ihrer Beziehung zum Lichte. *Sitzb Akad Wiss Wien, 74*, 425–436. Tafel.

Leitgeb, H. (1877b). *Untersuchungen ueber die Lebermoose III. Heft. Dir Frondosen Jungermannieen*. Jena: O. Deistung's Buchhandlung (Hermann Dabis).

Leitgeb, H. (1879). *Untersuchungen über die Lebermoose. Heft IV. Die Riccieen*. Graz: Leuschner & Lubensky.

Leitgeb, H. (1880a). Die Athemöffnungen der Marchantiaceen Sitzungsberichte der Kaiserlichen Akademie der Wissenschaften. *Wien, 81*, 40–54.

Leitgeb, H. (1880b). Die Inflorescenzen der Marchantiaceen Sitzungsberichte der Kaiserlichen Akademie der Wissenschaften. *Wien, 81*, 123–143.

Leitgeb, H. (1881). *Untersuchungen über die Lebermoose Heft VI. Die Marchantieen*. Jena: Leuschner & Lubensky.

Lidforss, B. (1904-5). Über die Reizbewegungen der Marchantia-Spermatozoiden. *Jahrbucher fur Wissenschaftliche Botanik, 41*, 65–87.

Ligrone, R., Duckett, J. G., & Renzaglia, K. S. (1993). The gametophyte-sporophyte junction in land plants. *Advances in Botanical Research, 19*, 231–317.

Linde, A.-M., Eklund, D. M., Kubota, A., Pederson, E. R. A., Holm, K., et al. (2017). Early evolution of the land plant circadian clock. *New Phytologist, 216*, 576–590.

Lorbeer, G. (1936). Die Umwandlung eines haploiden, genotypisch weiblichen Gametophyten von *Sphaerocarpus Donnellii* in einen männlichen mit Hilfe von Röntgenstrahlen. *Planta, 25*, 70–83.

Maravolo, N. C. (1976). Polarity and localization of auxin movement in hepatic, *Marchantia polymorpha*. *American Journal of Botany, 63*, 526–531.

Marchant, J. (1713). *Nouvelle Découverte des Fleurs et des Graines D'une Plante rangée par les Botanistes fous le genre du Lichen* (pp. 229–234). Année: Histoire de l'Académie royale des sciences, avec les mémoires de mathématique et de physique, pour la meme cette Académie. MDCCXIII.

McConaha, M. (1941). Ventral structures effecting capillarity in the Marchantiales. *American Journal of Botany, 28*, 301–306.

Menge, F. (1930). Die Entwicklung der Keimpflanzen von *Marchantia polymorpha* L. und Plasgiochasma rupestre. *Flora, 124*, 423–478.

Mirbel, C.-F. (1835). Researches anatomiques et physiologiques sur le *Marchantia polymorpha*, pour servir a l'histoire du tissu cellulaire, de l'épiderme et des stomates. *Mémoires de l'Académie Royale Societie de l'Institut de France, 13*, 337–436.

Molisch, H. (1884). Untersuchungen über den Hydrotropismus. *Sitzb Akad Wiss Wien, 88*, 897–942.

Molisch, H. (1922). *Pflanzenphysiologie als Theorie der Gärtnerei*. Jena: Gustav Fischer.

Montgomery, S. A., Tanizawa, Y., Galik, B., Wang, N., Ito, T., et al. (2020). Chromatin organization in early land plants reveals an ancestral association between H3K27me3, transposons, and constitutive heterochromatin. *Current Biology*, *30*, 573–588.

Morris, J. L., Puttick, M. N., Clark, J. W., Edwards, D., Kenrick, P., et al. (2018). The timescale of early land plant evolution. *Proceeding of the National Academy of Sciences USA*, *115*, E2274–E2283.

Mottier, D. M. (1891). Notes on the apical growth of liverworts. *Botanical Gazette*, *16*, 141–143.

Muggoch, H., & Walton, J. (1942). On the dehiscence of the antheridium and the part played by surface tension in the dispersal of spermatocytes in Bryophyta. *Proceedings of the Royal Society of London. Series B*, *130*, 448–461. plate 414.

Nagai, I. (1915). Ueber roten Pigmentbildung bei einigen *Marchantia*-Arten. *Botanical Magazine Tokyo*, *29*, 90–98.

Nagai, I. (1919). Induced adventitious growth in the Gemmae of *Marchantia*. *The Botanical Magazine*, *33*, 99–109.

Nägeli, K. (1842). Botanische Beiträge. *Linnea*, *16*, 237–285.

Nägeli, K. (1845). Wachsthumgeschichte der Laub- und Lebermoose. *Zeitschrift fur Wissenschaftliche Botanik*, *2*, 138–210.

Naramoto, S., Jones, V. A. S., Trozzi, N., Sato, M., Toyooka, K., et al. (2019). A conserved regulatory mechanism mediates the convergent evolution of plant shoot lateral organs. *PLoS Biology*, *17*, 3000560.

Nishihama, R., Ishida, S., Urawa, H., Kamei, Y., & Kohchi, T. (2016). Conditional gene expression/deletion systems for *Marchantia polymorpha* using its own heat-shock promoter and Cre/loxP-mediated site-specific recombination. *Plant and Cell Physiology*, *57*, 271–280.

Nishihama, R., Ishizaki, K., Hosaka, M., Matsuda, Y., Kubota, A., et al. (2015). Phytochrome-mediated regulation of cell division and growth during regeneration and sporeling development in the liverwort *Marchantia polymorpha*. *Journal of Plant Research*, *128*, 407–421.

Nishiyama, T., Wolf, P. G., Kugita, M., Sinclair, R. B., Sugita, M., et al. (2004). Chloroplast phylogeny indicates that bryophytes are monophyletic. *Molecular Biology and Evolution*, *21*, 1813–1819.

Oda, K., Yamato, K., Ohta, E., Nakamura, Y., Takemura, M., et al. (1992). Gene Organization deduced from the complete sequence of liverwort *Marchantia polymorpha* mitochondrial-DNA - a primitive form of plant mitochondrial genome. *Journal of Molecular Biology*, *223*, 1–7.

O'Hanlon, M. E. (1926). Germination of spores and early stages in development of gametophyte of *Marchantia polymorpha*. *Botanical Gazette*, *82*, 215–222.

Ohno, S. (1970). *Evolution by gene duplication*. New York: Springer-Verlag.

Ohyama, K., Fukuzawa, H., Kohchi, T., Shirai, H., Sano, T., et al. (1986). Chloroplast Gene Organization deduced from complete sequence of liverwort *Marchantia polymorpha* chloroplast DNA. *Nature*, *322*, 572–574.

Oppenheimer, H. (1922). Das Unterbleiben der Keimung in den Behältern der Mutterpflanze. *Sitzungsberichte der Kaiserlichen Akademie der Wissenschaften*, *131*, 279–312.

Peirce, G. J. (1902). Forcible discharge of the Antherozoids in *Asterella californica*. *Bulletin of the Torrey Botanical Club*, *29*, 374–382.

Pettersson, B. (1940). Experimentelle untersuchungen über die euanemochore verbreitung der sporenpflanzen. *Acta Botanica Fennica*, *25*, 1–102.

Pfeffer, W. (1871). Studien über Symmetric und spezifische Wachstumsursachen. *Arbeiten Des Botanischen Instituts in Würzburg*, *1*, 77–98.

Pressel, S., & Duckett, J. G. (2019). Do motile spermatozoids limit the effectiveness of sexual reproduction in bryophytes? Not in the liverwort *Marchantia polymorpha*. *Journal of Systematics and Evolution*, *57*, 371–381.

Proust, H., Honkanen, S., Jones, V. A. S., Morieri, G., Prescott, H., et al. (2016). RSL class I genes controlled the development of epidermal structures in the common ancestor of land plants. *Current Biology*, *26*, 93–99.

Puttick, M. N., Morris, J. L., Williams, T. A., Cox, C. J., Edwards, D., et al. (2018). The interrelationships of Land plants and the nature of the ancestral Embryophyte. *Current Biology*, *28*, 733–745.

Qiu, Y. L., Li, L. B., Wang, B., Chen, Z. D., Knoop, V., et al. (2006). The deepest divergences in land plants inferred from phylogenomic evidence. *Proceedings of the National Academy of Sciences of the United States of America*, *103*, 15511–15516.

Radhakrishnan, G. V., Keller, J., Rich, M. K., Vernié, T., Mbadinga, D. L. M., et al. (2020). An ancestral signalling pathway is conserved in intracellular symbioses-forming plant lineages. *Nature Plants*, *6*, 280–289.

Rimington, W. R., Duckett, J. G., Field, K. J., Bidartondo, M. I., & Pressel, S. (2020). The distribution and evolution of fungal symbioses in ancient lineages of land plants. *Mycorrhiza*, *30*, 23–39.

Robinson, W. (1910). *Alpine flowers for gardens—Rock, wall, marsh plants, and mountain shrubs*. London: John Murray.

Romani, F., Baniç, E., Florent, S. N., Kanazawa, T., Goodger, J. Q. D., et al. (2020). Oil body formation in *Marchantia polymorpha* is controlled by MpC1HDZ and serves as a defense against arthropod herbivores. *Current Biology*, *30*, 2815–2828.

Romani, F., & Moreno, J. E. (2021). Molecular mechanisms involved in functional macroevolution of plant transcription factors. *New Phytologist*, *230*, 1345–1353.

Rövekamp, M., Bowman, J. L., & Grossniklaus, U. (2016). *Marchantia* MpRKD regulates the gametophyte- sporophyte transition by keeping egg cells quiescent in the absence of fertilization. *Current Biology*, *26*, 1782–1789.

Ruge, G. (1893). Beiträge zur Kenntniss der vegetationsorgane der lebermoose. *Flora*, *77*, 279–312.

Sachs, J. V. (1882). Ueber orthotrope und plagiotrope pflanzentheile. *Arbeiten des botan. Institut Würzburg*, *2*, 1004–1064.

Sakakibara, K., Ando, S., Yip, H. K., Tamada, Y., Hiwatashi, Y., et al. (2013). KNOX2 genes regulate the haploid-to-diploid morphological transition in land plants. *Science*, *339*, 1067–1070.

Sakakibara, K., Nishiyama, T., Deguchi, H., & Hasebe, M. (2008). Class 1 KNOX genes are not involved in shoot development in the moss *Physcomitrella patens* but do function in sporophyte development. *Evolution & Development*, *10*, 555–566.

Sauret-Güeto, S., Frangedakis, E., Silvestri, L., Rebmann, M., Tomaselli, M., et al. (2020). Systematic tools for reprogramming plant gene expression in a simple model, *Marchantia polymorpha*. *ACS Synthetic Biology*, *9*, 864–882.

Schmidel, C. C. (1762). *Icones Plantarum et analyses partium aeri incisae atque vivis coloribus insgnitae adiectis indicibus nominum necessariis figurarum explicationibus et brevibus animaduersionibus*. Nürnberh: Johann Christoph Keller.

Schönherr, J., & Ziegler, H. (1975). Hydrophobic Cuticular ledges prevent water entering the air pores of liverwort thalli. *Planta*, *124*, 51–60.

Schostakowitsch, W. (1894). Über die reproduction und regenerationserscheinungen bei den Lebermoosen. *Flora*, *79*, 350–384.

Schuster, R. M. (1992). *The Hepaticae and Anthocerotae of North America. Vol. VI*. New York: Columbia University Press.

Schwann, T., & Schleiden, M. J. (1847). *Microscopical researches into the accordance in the structure and growth of plants and animals*. London: Sydenham Society.

Shimamura, M. (2016). *Marchantia polymorpha*: Taxonomy, phylogeny and morphology of a model system. *Plant and Cell Physiology*, *57*, 230–256.

Skutch, A. F. (1929). Early stages of plant succession following forest fires. *Ecology, 10*, 177–190.

Smee, A. (1872). *My garden, its plan and culture together with a general description of its geology, botany, and natural history*. London: Bell and Daldy.

Stahl, E. (1882). Ueber der Einfluss des sonnigen oder schattigen Standortes auf die Ausbildung der Laubbätter. *Jenaische Zeitschrift für Naturwissenschaft, 16*, 162–200.

Stahl, E. (1888). Pflanzen und Schnecken: Biologische Studie über die Schutzmittel der Pflanzen gegen Schneckenfrass. *Jenaische Zeitschrift für Naturwissenschaft, 22*, 557–684.

Strasburger, E. (1869). Die Geschlechtsorgane und die Befruchtung bei *Marchantia polymorpha* L. *Jahrbucher fur Wissenschaftliche Botanik, 7*, 409–422.

Strother, P. K., & Taylor, W. A. (2018). The evolutionary origin of the plant spore in relation to the antithetic origin of the plant sporophyte. In M. Krings, C. J. Harper, N. R. Cúneo, & G. W. Rothwell (Eds.), *Transformative paleobotany: Papers to commemorate the life and legacy of Thomas N. Taylor* (pp. 3–20). Amsterdam, Netherlands: Elsevier.

Strullu-Derrien, C., Selosse, M.-A., Kenrick, P., & Martin, F. M. (2018). The origin and evolution of mycorrhizal symbioses—From palaeomycology to phylogenomics. *New Phytologist, 220*, 1012–1030.

Sugano, S. S., Nishihama, R., Shirakawa, M., Takagi, J., Matsuda, Y., et al. (2018). Efficient CRISPR/Cas9-based genome editing and its application to conditional genetic analysis in *Marchantia polymorpha*. *PLoS One, 13*, e0205117.

Sugano, S. S., Shirakawa, M., Takagi, J., Matsuda, Y., Shimada, T., et al. (2014). CRISPR/Cas9-mediated targeted mutagenesis in the liverwort *Marchantia polymorpha* L. *Plant and Cell Physiology, 55*, 475–481.

Suire, C., Bouvier, F., Backhaus, R. A., Begu, D., Bonneu, M., et al. (2000). Cellular localization of isoprenoid biosynthetic enzymes in *Marchantia polymorpha*. Uncovering a new role of oil bodies. *Plant Physiology, 124*, 971–978.

Suzuki, H., Harrison, C. J., Shimamura, M., Kohchi, T., & Nishihama, R. (2020). Positional cues regulate dorsal organ formation in the liverwort *Marchantia polymorpha*. *Journal of Plant Research, 133*, 311–321.

Suzuki, H., Kohchi, T., & Nishihama, R. (2021). Auxin biology in Bryophyta: A simple platform with versatile functions. *Cold Spring Harbor Perspectives in Biology*, a040055.

Takizawa, R., Hatada, M., Moriwaki, Y., Abe, S., Yamashita, Y., et al. (2021). Fungal-type terpene synthases in *Marchantia polymorpha* are involved in sesquiterpene biosynthesis in oil body cells. *Plant & Cell Physiology*. https://doi.org/10.1093/pcp/pcaa175.

Tanaka, M., Esaki, T., Kenmoku, H., Koeduka, T., Kiyoyama, Y., et al. (2016). Direct evidence of specific localization of sesquiterpenes and marchantin a in oil body cells of *Marchantia polymorpha* L. *Phytochemistry, 130*, 77–84.

Tarén, N. (1958). Regulating the initial development of Gemmae in *Marchantia polymorpha*. *The Bryologist, 61*, 191–204.

Taylor, T. (1837). De Marchantieis. *Transactions of the Linnean Society, 17*, 375–395.

Téodoresco, E. C. (1899). Influence de l'acide carbonique sur la forme et la structure des plantes. *Revue Générale de Botanique, 11*, 445–470.

Téodoresco, E. C. (1929). Observations sur la croissance des plantes aux lumiéres de diverses longeurs d'ondé. *Annales des sciences naturelles Botanique, 10*(11), 201–335.

Thamm, A., Saunders, T. S., & Dolan, L. (2020). MpFEW RHIZOIDS1 miRNA-mediated lateral inhibition controls rhizoid cell patterning in *Marchantia polymorpha*. *Current Biology, 30*, 1905–1915.

Thomas, R. J. (1980). Cell elongation in hepatics: The seta system. *Bulletin of the Torrey Botanical Club, 107*, 339–345.

Thuret, G. (1851). Recherches sur les Zoospores des Algues et les Anthéridies des Cryptogames. *Annales de Sciences Naturelles Botanique, 3.16*, 5–39, Planches 31-15.

Torrey, R. H. (1932). *Marchantia polymorpha* after forest fires. *Torreya, 32*(9–10), 128–129.

Tsuboyama, S., Nonaka, S., Exura, H., & Kodama, Y. (2018). Improved G-AgarTrap—A highly efficient transformation method for intact gemmalings of the liverwort *Marchantia polymorpha. Scientific Reports, 8*, 10800.

Tsuzuki, M., Futagami, K., Shimamura, M., Inoue, C., Kunimoto, K., et al. (2019). An early arising role of the MicroRNA156-529-SPL module in reproductive development revealed by the liverwort *Marchantia polymorpha. Current Biology, 29*, 3307–3314.

Unger, F. (1837). Weitere Beobachtungen über die Samenthiere der Pflanzen. *Nova Acta physico medica A C L, 18*, 786–796.

Vöchting, H. (1885). Über die regeneration der Marchantieen. *Jahrbucher fur Wissenschaftliche Botanik, 16*, 367–414.

Wann, F. B. (1925). Some of the factors involved in the sexual reproduction of *Marchantia polymorpha. American Journal of Botany, 12*, 307–318.

Wickett, N. J., Mirarab, S., Nguyen, N., Warnow, T., Carpenter, E., et al. (2014). Phylotranscriptomic analysis of the origin and early diversification of land plants. *Proceedings of the National Academy of Sciences of the United States of America, 111*, E4859–E4868.

Xu, B., Taylor, L., Pucker, B., Feng, T., Glover, B. J., et al. (2021). The land plant-specific MIXTA-MYB lineage is implicated in the early evolution of the plant cuticle and the colonization of land. *New Phytologist, 229*, 2324–2338.

Yamaoka, S., Nishihama, R., Yoshitake, Y., Ishida, S., Inoue, K., et al. (2018). Generative cell specification requires transcription factors evolutionarily conserved in land plants. *Current Biology, 28*, 479–486.

Yamaoka, S., Takenaka, M., Hanajiri, T., Shimizu-Ueda, Y., Nishida, H., et al. (2004). A mutant with constitutive sexual organ development in *Marchantia polymorpha* L. *Sexual Plant Reproduction, 16*, 253–257.

Yamato, K. T., Ishizaki, K., Fujisawa, M., Okada, S., Nakayama, S., et al. (2007). Gene organization of the liverwort Y chromosome reveals distinct sex chromosome evolution in a haploid system. *Proceedings of the National Academy of Sciences of the United States of America, 104*, 6472–6477.

Yasui, Y., Tsukamoto, S., Sugaya, T., Nishihama, R., Wang, Q.-X., et al. (2019). GEMMA CUP-ASSOCIATED MYB1, an Ortholog of axillary meristem regulators, is essential in vegetative reproduction in *Marchantia polymorpha. Current Biology, 29*, 3987–3995.

Young, S. B., & Kläy, J.-R. (1971). Bryophytes in the 1969 crater of Deception Island, Antarctica: An apparent case of rapid long-distance dispersal. *Ohio Journal of Science, 71*, 358–362.

Leclerc du Sablon M., (1885). Recherches sur le développement du sporogone des hépatiques. Bulletin de la Société Botanique de France, 7, t. 2: 126–180.

CHAPTER TWO

The wild grass *Brachypodium distachyon* as a developmental model system

Michael T. Raissig[a,b,†,*] and Daniel P. Woods[c,d,†,*]
[a]Centre for Organismal Studies Heidelberg, Heidelberg University, Heidelberg, Germany
[b]Institute of Plant Sciences, University of Bern, Bern, Switzerland
[c]Department of Plant Sciences, University of California, Davis, CA, United States
[d]Howard Hughes Medical Institute, Chevy Chase, MD, United States
*Corresponding authors: e-mail address: michael.raissig@ips.unibe.ch; dpwoods@ucdavis.edu

Contents

† These authors contributed equally to this work and authors are listed in alphabetical order.

Current Topics in Developmental Biology, Volume 147
ISSN 0070-2153
https://doi.org/10.1016/bs.ctdb.2021.12.012

Abstract

The arrival of cheap and high-throughput sequencing paired with efficient gene editing technologies allows us to use non-traditional model systems and mechanistically approach biological phenomena beyond what was conceivable just a decade ago. Venturing into different model systems enables us to explore for example clade-specific environmental responses to changing climates or the genetics and development of clade-specific organs, tissues and cell types. We—both early career researchers working with the wild grass model *Brachypodium distachyon*—want to use this review to (1) highlight why we think *B. distachyon* is a fantastic grass developmental model system, (2) summarize the tools and resources that have enabled discoveries made in *B. distachyon*, and (3) discuss a handful of developmental biology vignettes made possible by using *B. distachyon* as a model system. Finally, we want to conclude by (4) relating our personal stories with this emerging model system and (5) share what we think is important to consider before starting work with an emerging model system.

1. An evolutionary perspective on *Brachypodium distachyon* as a model in the grass family (Poaceae)

For decades there have been just a few model plant organisms. For example, *Arabidopsis thaliana* for many years has been the plant of choice as a representative to model flowering plants (Koornneef & Meinke, 2010). Certainly, *A. thaliana* is useful for studying a number of different aspects of plant biology that are shared among flowering plants and five decades of *A. thaliana* research has provided great insights into how flowering plants develop and function (Provart et al., 2016). However, there is a great diversity of biology found within plants and *A. thaliana* is less useful for exploring specific phenomena unique to a given clade of plants or traits that have evolved independently multiple times (Kellogg, 2006). For example, *A. thaliana* does not have floral nectar spurs, which have evolved independently multiple times and are a defining characteristic of the beautiful columbine flower (Ballerini, Min, Edwards, Kramer, & Hodges, 2020).

The grasses (Poaceae) are a large family of about 10,000 recognized species that have radiated to span the entire globe (Kellogg, 2001). Within this group of plants are cereal crops that provide the bulk of human caloric content world wide either directly through consuming grain or indirectly

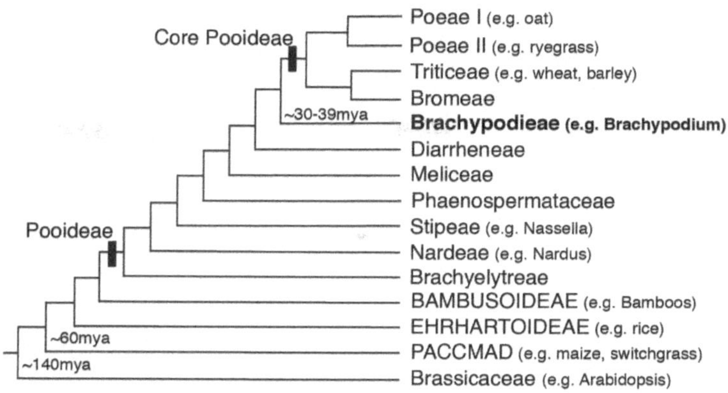

Fig. 1 Pooideae phylogeny showing the 11 major tribes and delimitation of the core pooids based on Schneider, Döring, Hilu, and Röser (2009) and Grass Phylogeny Working Group II (2012). Outgroups are the closely related grass subfamilies, Bambusoideae and Ehrhartoideae, which together with Pooideae form the BEP clade. Sister to the BEP clade is the PACMAD clade that contains tropical cereals such as maize and switchgrass. Brachypodieae sister to the core pooid clade is in bold. Stem node age of *B. distachyon* based on studies from Díaz-Pérez, López-Álvarez, Sancho, and Catalán (2018) and Sancho et al. (2018). Mya = million years ago.

through feed for livestock (FAO Statistical Pocketbook, 2015). However, grasses are morphologically, physiologically and developmentally distinct from *A. thaliana* and many aspects of their biology require a model more closely related than *A. thaliana*, which last shared a common ancestor with the grasses around 140 million years ago (Fig. 1; Kellogg, 2001). Maize has historically been utilized as a grass model and more recently so has rice; however, while they are useful in some respects, in many other regards they are not the ideal model organism due to their larger stature, more complex genomes and difficult growth requirements. Furthermore, despite being grasses, maize and rice are quite different from the pooid grass clade that contains the important crops wheat, oats, barley and rye (Fig. 1). Pooid crops indeed are some of the more challenging crops for genetic studies, as many of the pooid cereals we eat are polyploid and have some of the largest genomes in the grass family (Kellogg, 2015). Additionally, the grasses originated as tropical forest understory and many of the extant species remain tropical including rice (Kellogg, 2001). However, Pooideae (temperate grasses) contain some 3850 species, which have radiated into higher latitudes and thus have unique characteristics not found in other grasses that facilitated their radiation into temperate regions such as cold tolerance (Fig. 1; Grass Phylogeny Working Group II, 2012; Humphreys & Linder, 2013; Preston & Sandve, 2013).

The focus of this review will be on the emerging Pooid "model" grass *B. distachyon*. *B. distachyon* is placed phylogenetically within the Pooideae sub-family, sister to the agronomically important "crown pooids"—wheat, oat, barley and rye (Fig. 1; Grass Phylogeny Working Group II, 2012). Furthermore, *B. distachyon* is a wild grass that has not undergone a number of genetic bottlenecks as a domesticated species and thus can provide additional allelic variation in genes that impact different developmental processes.

2. Genomics enabled developmental biology

B. distachyon has a number of attributes that make it an attractive developmental model system. While sounding simplistic, one of the key aspects for a successful model organism is that it is easy and cheap to grow. *B. distachyon* is a fraction of the size of its domesticated relatives barley and wheat (Fig. 2) resulting in less required growth space enabling more replicates and experiments per growth area, which is notoriously sparse. Important aspects for plant models are that the species can self-fertilize to bulk and maintain stocks and that basic breeding protocols like cross-fertilization are established (see DOE-JGI resources website https://jgi.doe.gov/our-science/science-programs/plant-genomics/brachypodium/).

Fig. 2 Growth habits of *B. distachyon* compared to *A. thaliana* and other grasses. (A) A.t. *Arabidopsis thaliana*, B.d. *Brachypodium distachyon*, L.t. *Lolium temulentum*, T.m. *Triticum monococcum*, A.s. *Avena sativa*. (B) A flat of one hundred *B. distachyon* plants next to a single *Zea mays* (Z.m.) plant. Scale bar, 36 cm.

Indeed, most of the *B. distachyon* accessions isolated to date are already highly inbred making it an attractive developmental genetics model system (Gordon et al., 2017). For functional genetic analyses it is essential to have efficient transformation protocols available. Grasses are notoriously hard to transform yet the main accessions used like Bd21-3 are readily transformable compared to other model grasses (Scholthof, Irigoyen, Catalan, & Mandadi, 2018; https://jgi.doe.gov/our-science/science-programs/plant-genomics/brachypodium/). This allows for the generation of large collections of reporter lines and straightforward gene editing using CRISPR/Cas9. A recent high-efficiency transformation system for CRISPR/Cas9-mediated gene editing containing the wheat growth factors *TaGRF4* and *TaGIF1* (Debernardi et al., 2020) works efficiently in *B. distachyon* (D.P. Woods, Personal communication). Examples of vectors and cloning systems that work well to generate reporter lines in *B. distachyon* include the pIPK and pANIC vector series based on Gateway cloning (Himmelbach et al., 2007; Mann et al., 2012) and the GreenGate system based on GoldenGate cloning (Lampropoulos et al., 2013).

Apart from horticultural and biotechnological protocols, the availability of genomic and genetic resources is fundamental to study developmental mechanisms. In 2010, the first fully assembled and annotated draft genome was published (Initiative, International Brachypodium, 2010). *B. distachyon* has a relatively small and simple diploid genome (272 Mb) that is orders of magnitudes smaller in size compared with its close grass relatives such as hexaploid wheat (>15 Gb) and diploid barley (>5 Gb; Initiative, International Brachypodium, 2010; International Barley Genome Sequencing Consortium et al., 2012; International Wheat Genome Sequencing Consortium IWGSC et al., 2018). Since then pan-genome efforts have added *de-novo* assembled and fully annotated genomes of 54 additional accessions (Gordon et al., 2017; https://brachypan.jgi.doe.gov/). In addition, the diploid relative *B. stacei,* the perennial *B. sylvaticum* and the allotetraploid *B. hybridum* (hybrid of *B. stacei* and *B. distachyon;* Catalán et al., 2012) have been fully sequenced, assembled and annotated and are accessible as early release genomes in Phytozome (https://phytozome.jgi.doe.gov/pz/portal.html) and more perennial species will soon be sequenced (Community Science Program-Joint Genome Institute (CSP-JGI) Project ID 503006).

Finally, a growing repository of genetic resources and datasets facilitate functional genomic studies. There is a carefully curated diversity panel covering Spain, Italy and Turkey among which 54 accessions are fully sequenced, *de-novo* assembled and annotated (Gordon et al., 2017; https://brachypan.jgi.doe.gov/). A gene expression atlas has been built and integrated into the eFP browser (Sibout et al., 2017; http://bar.utoronto.ca/efp_brachypodium/cgi-bin/efpWeb.cgi). More than 23,000 T-DNA insertion lines are available (Bragg et al., 2012; Hsia et al., 2017) and searchable on Phytozome's J-Browse (https://phytozome.jgi.doe.gov/jbrowse/index.html). In addition, a heroic CSP-JGI (Project ID 1670) project whole-genome sequenced over 2000 mutant families mutagenized with sodium azide (NaN_3, hence the NaN population). Approximately 1.2 million variants were identified and accurately mapped to the reference genome Bd21-3 v1.1, which are annotated and searchable in Phytozome's J-Browse in the early release genome of Bd21-3 (https://phytozome.jgi.doe.gov/jbrowse/index.html). Finally, there are currently two additional CSP-JGI projects underway; one aims to generate extensive DNA Affinity Purification sequencing (DAP-seq) profiles of thousands of transcription factors (Project ID 504343) and another to use whole-genome sequence data to genotype 16 Recombinant Inbred Line (RIL) populations (Project ID 505702), which includes accessions that span the species range. For a detailed list of available resources see Scholthof et al. (2018).

3. *Brachypodium distachyon* as a model for Pooid development

B. distachyon as a grass model has and will accelerate our understanding of the genetic and molecular basis of a number of Pooid grass developmental traits. Here, we will focus on the progress that has been made in *B. distachyon* to expand our understanding of the transition to flowering, organ initiation, development of stomatal pores, leaf epidermal patterning and root development (Fig. 3). Particular focus will be on studies for which the use of *A. thaliana* or other grasses would not have been suitable to explore a given phenomena specific to Pooid grasses. Importantly, most of the work we will describe would not have been possible without the resources described above. There are a number of additional developmental processes being studied in *B. distachyon*; important contributions include for example, cell wall biology (see Coomey, Sibout, & Hazen, 2020 and references therein), vascular development (Sakai et al., 2021; Smertenko et al., 2019), pleiotropic genes like the NOOT-BOP-COCH-LIKE genes (Magne et al., 2020), seed germination (Wolny et al., 2018), species-specific

Fig. 3 The wild grass *B. distachyon* as a developmental model system. On the left panel, the shoot and the root system of a 6 week old, flowering *B. distachyon* plant is shown. Highlighted are the four developmental systems this review primarily covers; flowering transition, inflorescence development, stomatal development, and root development.

aspects of abscission zone formation (Yu, Hu, Doust, & Kellogg, 2020), and even thigmomorphogenesis—the developmental response to touch (Coomey et al., 2021). We apologize for not being able to cover all the developmental work done with *B. distachyon* due to space limitations.

4. Transition to flowering

The transition from vegetative to reproductive development is a key decision for which the timing is often directly influenced by the environment (Amasino, 2010). This critical life history trait has been shaped over evolutionary time to maximize the ability to flower at a time that optimizes reproductive success. In crops, such as wheat, the timing of this transition has been modified multiple times as wheat expanded through multiple environments across the globe (Distelfeld, Li, & Dubcovsky, 2009; Dubcovsky & Dvorak, 2007). Furthermore, the timing of flowering is one of many traits that have been manipulated by humans for increased crop productivity (Greenup, Peacock, Dennis, & Trevaskis, 2009). Thus, the study of flowering-time is not only a fascinating topic from a basic research standpoint, but also has relevance for crop improvement particularly in a changing global climate and a growing human population.

In many plant species, flowering occurs at a particular time of year in response to the sensing of seasonal cues such as changes in day-length and temperature. Photoperiodism is the process by which the sensing of changes in day-length leads to flowering (Song, Shim, Kinmonth-Schultz, & Imaizumi, 2015). Plants such as *B. distachyon* and other pooid grasses that flower in the spring or early summer are often responding to lengthening photoperiods and are referred to as long-day (LD) plants (Ream et al., 2014). Many plants adapted to temperate climates have a winter annual or biennial life history strategy (Bouché, Woods, & Amasino, 2017). These plants become established in the fall, overwinter, and flower rapidly in the spring. Essential to this adaptive strategy is that flowering does not occur prior to winter, during which flowering would not lead to successful reproduction. Thus, plants have evolved ways to prevent fall flowering and sense the passing of winter to establish competence to flower (Chouard, 1960). The block to flowering can be alleviated by exposure to two hallmarks of winter, either prolonged exposure to cold temperatures and/or the exposure to a prolonged period of short days (Chouard, 1960; Heide, 1994; Woods, Ream, & Amasino, 2014). The process by which flowering is promoted by cold exposure is known as vernalization and the process by which shorter photoperiods followed by growth in longer photoperiods promotes flowering is known as short-day (SD) vernalization (Fig. 4; Woods, Ream, & Amasino, 2014; Woods et al., 2019). A hallmark of cold-mediated vernalization is acquisition of *competence* to flower, not flowering *per se* (Chouard, 1960), and the SD vernalization phenomenon is similar: exposure to short days leads to competence, but plants must still be exposed to inductive long days to flower (Woods et al., 2019). These two processes exist in a diverse range of flowering plants, yet relatively little is known about the molecular nature of vernalization outside the eudicot model *Arabidopsis thaliana*, and only recently have molecular details been discovered about short-day vernalization in plants (*A. thaliana* does not

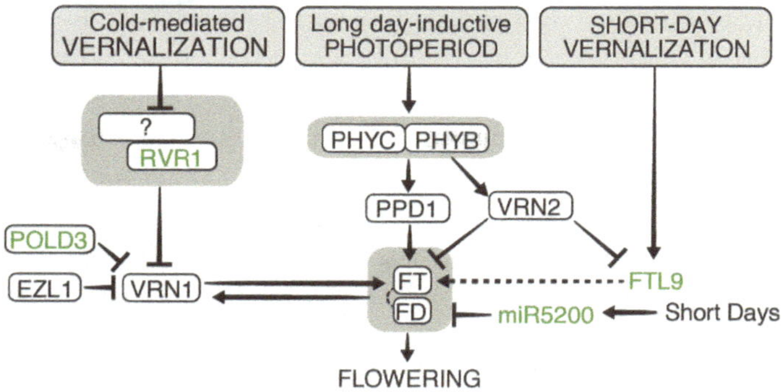

Fig. 4 Current working model of *B. distachyon* flowering pathways. Genes in green were identified in *B. distachyon*. Cold mediated vernalization results in the induction of *VRN1* which promotes flowering by activating *FT1*. Before cold there are a number of genes involved in establishing a vernalization requirement such as *RVR1*, *EZL1* and *POLD3*. Also before cold, genes such as *VRN2* repress *FT-like* genes. Essential to long day flowering are the photoreceptors PHYC and PHYB. Long days increase *PPD1* expression that has downstream consequences on the activation of *FT1* and flowering. During short days miR5200 expression is elevated which keeps *FT1* from functioning under these environmental conditions. Also SD triggers the expression of *FTL9* whose expression in SD provides competency to flower allowing plants to flower once shifted into LD (SD vernalization) where *FT1* can become expressed. Dotted arrow indicates *FTL9* is involved in providing competency to flower by effecting *FT1* expression but this is likely indirect and requires *VRN1*.

have the short-day vernalization phenomenon; Bouché et al., 2017; Woods et al., 2019). The characterization of vernalization in other species has revealed that different genes are involved in different plant groups, suggesting vernalization likely evolved independently multiple times throughout the diversification of flowering plants (Bouché et al., 2017; Ream et al., 2014; Zhong, Robbett, Poire, & Preston, 2017); thus, the rich information gained in *A. thaliana* cannot be directly applied to the grasses and studying vernalization in Pooid grasses provides an opportunity to explore the molecular basis of a classic example of convergent evolution. Furthermore, like vernalization, SD vernalization also likely evolved independently multiple times because a key gene required for SD vernalization is grass specific (Woods et al., 2019).

4.1 Molecular understanding of flowering in temperate grasses from natural variation work in wheat and barley

To date, information about the vernalization response in grasses has largely been the result of studies of existing allelic variation in wheat and barley (Fjellheim, Boden, & Trevaskis, 2014). Wheat and barley varieties have been broadly classified into spring and winter varieties (Distelfeld et al., 2009). Spring varieties do not require vernalization in order to flower rapidly in inductive LD, whereas winter varieties have a vernalization requirement that must be met before LD can promote flowering (Distelfeld et al., 2009; Greenup et al., 2009).

The current molecular model of vernalization in grasses, which is based on studies in wheat and barley, consists of the genes, *VERNALIZATION1* (*VRN1*), *VRN2* and *VRN3*, forming a regulatory loop (Bouché et al., 2017; Distelfeld et al., 2009; Greenup et al., 2009). *VRN3* is orthologous to *FT1* which is a small mobile floral signal often referred to as florigen and from here on will be referred to as *FT1* (Yan et al., 2006). The regulatory loop is thought to exist in leaves where all three genes are expressed (Sasani et al., 2009; Yan et al., 2006, 2004). In the fall, prior to cold exposure, high levels of VRN2 repress *FT1* to prevent flowering (Fig. 4; Yan et al., 2004). VRN2 is a CCT domain containing transcription factor and is part of the type VI *CO-like* family of genes (Griffiths, Dunford, Coupland, & Laurie, 2003; Yan et al., 2004). Prior to cold exposure, both *VRN1* and *FT1* are expressed at low levels (Greenup et al., 2009). However, during the cold of winter, *VRN1* is up-regulated proportionately to the length of cold experienced, and in turn, higher *VRN1* levels are thought to cause a decrease of *VRN2* expression (Dubcovsky et al., 2006; Sasani et al., 2009). *VRN1* is related to the *APETALA1/FRUITFULL* (*AP1/FUL*)-like class of MADS box transcription factors in *A. thaliana* (Yan et al., 2003) and arose from a duplication event at the base of the grass family (Preston & Kellogg, 2006). Upon repression of *VRN2* and in the presence of LD, *FT1* is activated by *Photoperiod-H1* (*PPD1*) a member of the pseudoresponse regulator gene family (Song et al., 2015; Turner, Beales, Faure, Dunford, & Laurie, 2005; Yan et al., 2006). Once activated *FT1* up-regulates *VRN1* expression in leaves creating a positive feedback loop of *VRN2* repression and *FT1* activation (Fig. 4; Distelfeld et al., 2009; Shimada et al., 2009; Yan et al., 2006). It is worth noting that while VRN1 can directly bind to the promoter region of *FT1* (Deng et al., 2015) this is not sufficient to cause *FT1* expression because during the cold *FT1* levels do not increase despite elevated levels of *VRN1* (Chen & Dubcovsky, 2012). As noted above a LD mediated photoperiod pathway via *PPD1* is required for *FT1* activation, however even if cold is provided under LD conditions *FT1* is still not activated perhaps because the photoperiod pathway does not work well in cold conditions.

4.2 Are the flowering pathways described in wheat and barley conserved in *B. distachyon*?

B. distachyon contains orthologs of all of the major flowering-time genes described so far in wheat and barley (Higgins, Bailey, & Laurie, 2010; Woods, McKeown, Dong, Preston, & Amasino, 2016). To a large extent these genes appear to be functionally conserved in *B. distachyon* (Higgins et al., 2010; Lv et al., 2014; Ream, Woods, & Amasino, 2012; Ream et al., 2014; Woods et al., 2016). For example, consistent with studies from wheat and barley, *BdFT1* and *BdVRN1* are promoters of flowering (Ream et al., 2014). Overexpression of *BdFT1* and *BdVRN1* results in rapid flowering (Ream et al., 2014) and knockdown of *BdFT1* and *BdVRN1* results in extremely delayed flowering (Lv et al., 2014; Woods et al., 2016). *BdVRN1* expression is induced in leaves proportional to the amount of time in cold temperatures and expression remains high post cold treatment, which correlates with higher *BdFT1* levels and rapid flowering (Ream et al., 2014). Also, we observe high expression levels of *BdFT1* in lines overexpressing *BdVRN1* and high expression levels of *BdVRN1* in lines overexpressing *BdFT1* consistent with the positive feedback loop suggested from flowering studies in wheat and barley (Ream et al., 2014). Also, as in wheat and barley, *BdVRN2* acts as a floral inhibitor, but a key difference is that in *B. distachyon BdVRN2* is not repressed by cold or by *BdVRN1*; rather, *BdVRN2* is constitutively expressed (Woods et al., 2016). This suggests that *BdVRN2* came under control of *VRN1* after *B. distachyon* split from wheat and barley and that the regulation of *BdVRN2* by *BdVRN1* is a derived trait that evolved within Triticeae (McKeown, Schubert, Marcussen, Fjellheim, & Preston, 2016; Woods et al., 2016). Furthermore, allelic variation at these genes probably contributes to natural variation in flowering responses in *B. distachyon* because they colocalize under quantitative trait loci controlling flowering-time variation (Bettgenhaeuser et al., 2017; Woods et al., 2017).

4.3 *B. distachyon* as a model plant for flowering time gene discovery in temperate grasses

Although there has been great progress in understanding photoperiodic flowering and vernalization at a molecular level in grasses, much remains much to be learned including identifying additional components of the vernalization response and the extent to which vernalization pathways are conserved throughout grasses (Fjellheim & Preston, 2018). *B. distachyon* has many attributes that make it an attractive model to pursue flowering research (Ream et al., 2012). There are hundreds of inbred accessions that have been collected throughout the Middle-East and Europe that display a wide range of flowering behaviors that can be used to map genes underlying that flowering variation (Brkljacic et al., 2011; Gordon et al., 2017; Ream et al., 2014; Schwartz et al., 2010; Tyler et al., 2014). Another route to gene discovery is the use of *B. distachyon* to conduct forward genetic screens to identify novel genes which is not possible in polyploid crops such as wheat (Ream et al., 2012). We have streamlined mapping *B. distachyon* mutants and have identified the causative lesion of many EMS-induced flowering time mutants using a mapping by sequencing pipeline (Minevich, Park, Blankenberg, Poole, & Hobert, 2012; Woods & Amasino, 2015; Woods, Bednarek, et al., 2017; Woods, Ream, et al., 2017; Woods, Ream, Minevich, Hobert, & Amasino, 2014; Woods et al., 2020). Interestingly, most of the genes we have mapped are in novel flowering-time genes—i.e., these genes have not been described before in other plants (Fig. 4) or are in genes which, in other plants have a minor role impacting flowering time but play a major role in *B. distachyon* (e.g., D.P. Woods & R. Amasino, Unpublished; Woods, Bednarek, et al., 2017; Woods, Ream, Minevich, et al., 2014).

4.4 Sensing photoperiod

One of the first mutants isolated and mapped was an extremely delayed flowering mutant that is due to a lesion in the light receptor *PHYTOCHROME C* (*PHYC*; Woods, Ream, Minevich, et al., 2014). We isolated three mutant alleles of *PHYC* all of which only flower sporadically after a year of growth in LD of 20h, which is a highly inductive photoperiod for flowering (wild type Bd21-3 flowers around 30 days after germination in 20h days; Ream et al., 2014). We found that *PHYC* is required for the transcriptional activation of all of the known genes within the photoperiod pathway including *FT1*, *CO* and *PPD1*, and ectopic expression of *FT1* in *phyC* results in rapid flowering (Woods, Ream, Minevich, et al., 2014). Additionally expression of genes that comprise the circadian clock were also altered in the *phyC* mutant background (Woods, Ream, Minevich, et al., 2014). Similar results were also obtained in studies of wheat and barley (Chen et al., 2014; Pankin et al., 2014). Thus, it appears that *phyC* is essential for photoperiodic flowering broadly across

Pooideae. However, the essential role of PHYC in photoperiodic flowering is not universal as loss of *phyC* function in *A. thaliana* and rice has only small effects on flowering (Monte et al., 2003; Takano et al., 2005).

The interconversion rates of the active Pfr and inactive Pr forms of PHYC is important in floral regulation. Specifically, phytochromes detect the levels and ratio of red (R) and far-red (FR) light in the environment (e.g., Quail, 2002). Upon absorption of red light, phytochromes photoconvert to an active form that initiates a signal transduction cascade. Recently a *tour de force* of genomic and biochemical work in *B. distachyon* revealed that the dark reversion rate of the Pfr form of PHYC is likely to be important in the measuring of the night length, which has downstream consequences in modulating the accumulation of the ELF3 clock component, which is a direct transcriptional repressor of *PPD1* (Gao et al., 2019). This suggests that the hourglass model first described decades ago (Borthwick & Hendricks, 1960) to explain how plants sense the length of the day may in fact be how temperate grasses sense photoperiod, although more work is needed to determine to what extent this is true (Gao et al., 2019). This appears distinct from *A. thaliana* where evidence supports the external coincidence model of CO in photoperiod sensing (e.g., Andrés & Coupland, 2012). It is worth noting, however, that in wheat *CO* plays a minor role in the activation of *FT1* especially in the presence of *PPD1* (Shaw et al., 2020), thus, the CO-FT1 regulon critical to flowering in *A. thaliana* appears to play a lesser role in temperate grasses (Ballerini & Kramer, 2011). However, we cannot rule out if both models are occurring in temperate grasses, which likely depends on the genetic background and environment. In summary, PHYC appears to be the main light receptor required for photoperiodic flowering in temperate grasses and may have played a role in the evolution of LD flowering in temperate grasses.

4.5 Sensing the cold of winter

A key feature in the evolution of all vernalization systems is a system to perceive the length of cold exposure. In pooid grasses *VRN1* induction by prolonged cold is a key feature of vernalization (McKeown et al., 2016). We have isolated a number of mutants from forward genetic screens looking for mutants where *VRN1* is expressed and flowering occurs without cold exposure in conditions in which the wild type requires cold for flowering and *VRN1* induction (e.g., Lomax et al., 2018; Woods & Amasino, 2015; Woods, Ream, et al., 2017; Woods et al., 2020).

Consistent with the key role of *VRN1* expression in achieving competence to flower upon prolonged cold exposure, these mutants no longer have a vernalization requirement for rapid flowering. These mutants are, in a sense, constitutively vernalized, and the role of the functional alleles is to repress *VRN1* prior to cold exposure (Woods, Ream, et al., 2017). Indeed, the transcriptomes (via RNAseq) of several of our mutants overlap substantially with vernalized plants (Woods et al., 2020; Woods, Ream, et al., 2017; D. Woods, Unpublished.). We found that one of the mutations was in a gene with a Bromo-Adjacent Homology and a TFIIS domain; we refer to the gene as *REPRESSOR OF VERNALIZATION1* (*RVR1*; Woods, Ream, et al., 2017). *RVR1* homologs are present in all plants, but *RVR1*-like genes are not found outside of the plant kingdom. Interestingly, no mutant phenotypes resulting from a *RVR1* lesion have been described in any other plant system and null mutants in *A. thaliana* do not have any obvious phenotype (Woods, Ream, et al., 2017). In other organisms, proteins containing BAH domains are often repressors that "read" chromatin states, and as noted above the role of RVR1 is to repress *VRN1* (Yang & Xu, 2013). Indeed, we observe loss of the repressive chromatin marks H3K27me3 in the *rvr1* mutants around *VRN1* similar to what has been observed in the Polycomb Repressive Complex 2 (PRC2) mutant *ENHANCER OF ZESTE1/CURLY LEAF* (*ezl1*; Goodrich et al., 1997; Lomax et al., 2018), which is a histone methyltransferase and the catalytic subunit of PRC2 responsible for the deposition of H3K27me3 marks in plants (De Lucia, Crevillen, Jones, Greb, & Dean, 2008). At present, it is unclear how RVR1 mediated repression of *VRN1* is overcome by vernalization because *RVR1* is constitutively expressed (Woods, Ream, et al., 2017). Perhaps, *RVR1* is abolished via a post transcriptional mechanism and/or *RVR1* could be part of a general repressive complex that has several subtypes with different target gene specificities, and loss of a flowering specific component of a *VRN1* repressing subtype after prolonged cold exposure could abolish *VRN1* repression (Fig. 4).

4.6 Sensing the short days of winter

Another route to gene discovery is exploiting natural variation in flowering responses. With regards to the SD vernalization response, some *B. distachyon* accessions exhibit SD vernalization whereas others do not (Woods et al., 2019). From crosses between such accessions, we found that SD vernalization segregates as a single locus and the responsible gene is a paralog of *FT1* referred to as *FT-LIKE 9* (*FTL9*) that is expressed in SD (Woods et al., 2019). However, unlike *FT1*, *FTL9* is unable to directly cause flowering; rather, its expression in SD enables flowering once plants are shifted into LD conditions (Woods et al., 2019). Accessions with a functional allele of *FTL9* have both the SD and cold vernalization responses, whereas accessions with loss-of-function alleles can only respond to cold vernalization (Woods et al., 2019). The adaptive value of an active allele of *FTL9* and the SD vernalization response is that in areas with a milder climate, SD is a more reliable indicator of winter than cold temperatures alone. In contrast, the adaptive value of loss of *FTL9* and the SD vernalization response might be to enable these *B. distachyon* accessions to grow in regions with more variable spring conditions. In these environmental conditions having a robust SD vernalization response may lead to premature flowering when there's still a chance of a hard freeze, which could damage the more sensitive floral organs. Consistent with this trend, accessions that lost the SD vernalization response also require longer periods of cold treatment in order to flower (Woods et al., 2019). Grasses have undergone an expansion of the *FT* gene family, and it is fascinating that a florigen-like gene acquired a role in creating *competence* to flower rather than directly inducing flowering as florigen (*FT1*) does. A recent study suggests that *FTL9* in some *B. distachyon* accessions might also be involved in actual SD flowering rather than SD vernalization, and thus, *FTL9* might have different functional roles depending on the *B. distachyon* accession (Qin et al., 2019). Taken together, the use of *B. distachyon* allowed the discovery of the first example of a gene required for SD vernalization—a phenomenon first described in the 1930s (Purvis & Gregory, 1937).

In addition to the progress that can be made using genetics, transcriptome studies have already begun to shed light on additional layers of flowering-time regulation such as the post-transcriptional regulation of *FT1* (Wu et al., 2013). An endogenous microRNA in *B. distachyon*, miR5200, has been shown to target *FT1* in SD and repress its activity (Wu et al., 2013). Consistent with miR5200's ability to repress *FT1*, miR5200 over-expressing plants have very low *FT1* levels even in LD conditions and delays flowering (Wu et al., 2013). Interestingly, miR5200 appears to be pooid specific as miR5200 is not detectable in other grasses or monocot species (McKeown, Schubert, Preston, & Fjellheim, 2017). However, miR5200's expression pattern is conserved across pooid grasses with miR5200 more highly expressed in SD than in LD (Wu et al., 2013). That miR5200 is specific to the pooid lineage indicates that miR5200 represents a fairly recent evolutionary path to regulate *FT1* expression.

5. Flower development

The vast majority of grasses are characterized by having their flowers arranged in novel structures called spikelets that are absent from the closest grass relatives such as Joinvillea (Malcomber, Preston, Reinheimer, Kossuth, & Kellogg, 2006). Similarly, grass flowers with lodicules (homologous to petals), a palea and a lemma surrounding the stamens and pistil, with the overall spikelet being subtended by empty bracts called glumes are substantially different from the two alternating whorls of tepals surrounding the stamen and pistil typical of monocot flowers in the immediate grass relatives (Whipple & Schmidt, 2006). *B. distachyon* has an unbranched inflorescence with the spikelets arising directly on the inflorescence axis, called a spike (Kellogg et al., 2013). There are a number of different inflorescence architecture genes that have been identified in grasses (e.g., Kellogg, 2007; Whipple, 2017); however, much less is known of the genes regulating unbranched inflorescence development.

Forward genetic screens have revealed mutants with more spikelets in *B. distachyon*. These mutants (called *more spikelets1*) had an increase in the number of axillary meristems and the causative gene for one of the more spikelets mutant phenotype is in an ethylene response factor *APETELA2* transcription factor orthologous to the *branched silkless* mutant in maize and *FRIZZY PANICLE* in rice (Derbyshire & Byrne, 2013). Therefore, it appears that the gene has a conserved role in determining spikelet meristem fate broadly in grasses. However, in *B. distachyon* it appears to have an additional role in the determinacy of the inflorescence meristem. This is potentially through the timing of the initiation of the terminal spikelet, which may be unique in grasses with a spike rather than a panicle for example (Derbyshire & Byrne, 2013). More recently it has been shown that in wheat and barley this gene affects the spike in a similar way and modifying the function of this gene could potentially be used to increase yield in these crops (Li et al., 2020; Poursarebani et al., 2015).

There is a great diversity in grass inflorescence architecture however, very few cross species functional comparisons between grasses have been done. Recently, a grass specific **T**eosinte branched1/**C**incinnata/**P**roliferating cell factor (TCP) transcription factor COMPOSITUM1 (COM1) was mapped in barley which plays a role inhibiting branch formation in the spikelet (Poursarebani et al., 2020). Interestingly, a role in inhibiting branches appears to be Triticeae specific because in non-triticeae grasses such as *B. distachyon* and sorghum, COM1 actually promotes branch formation as well as cell differentiation in the palea (Poursarebani et al., 2020). The different functional roles of COM1 between grasses appear to be due to variation in coding sequence within *COM1*. Therefore, coding variation in genes such as *COM1* may contribute to the diverse inflorescence architectures found in the grass family.

Recently, auxin transport has been implicated in inflorescence development. In most plants, development requires the growth hormone auxin (e.g., Leyser, 2010; Mcsteen, 2009). Auxin is actively transported to build sites with local auxin maxima, which are required for organ initiation (Mcsteen, 2010). Directional efflux of auxin is mediated by polarly localized PIN-FORMED (PIN) auxin efflux carriers (Gälweiler et al., 1998), which direct the flow of auxin to specific locations thereby orchestrating new organ formation. Thus, *pin* loss of function mutants in *A. thaliana* fail to initiate organs resulting in a "pin" like inflorescence phenotype (Reinhardt, Mandel, & Kuhlemeier, 2000; Reinhardt et al., 2003). A phylogenetic analysis revealed that a duplication of *PIN1* in grasses produced the paralogs *PIN1a* and *PIN1b*, and, in addition, another *PIN-like* clade sister to *PIN1*, aptly named *Sister-of-PIN1* (*SoPIN1*; O'Connor et al., 2014). *SoPIN1* is present in all angiosperms except Brassicaceae, which contains *A. thaliana,* thus it was not functionally characterized before. To dissect the functional relevance of the PIN1 and SoPIN1 proteins, reporter lines and an auxin signalling output reporter (DR5) were generated in *B. distachyon*. It was found that SoPIN1 defines auxin convergence points in the inflorescence meristem whereas the PIN1a and PIN1b are implicated in leaf vein patterning (O'Connor et al., 2014). Consistent with the difference in the localization of the PIN1 and SoPIN1 proteins, double loss of function of *bdpin1a/b* have short internodes but no defects were observed in the grass spikelet whereas *sopin1* loss of function mutants displayed the canonical "*pin*" phenotype and loss of organs (O'Connor et al., 2017). Likely, we would have missed the central role of *SoPIN1* in organ initiation if we relied on *A. thaliana* as the sole model for flowering plants. Some of the auxin-related reporter lines have recently been used to analyze auxin transport and signaling during grass embryogenesis in *B. distachyon* (Hao et al., 2021). In addition, stage-specific embryonic transcriptomes have been generated and identified common and diverged genetic programs during embryogenesis in *B. distachyon* and *A. thaliana* (Hao et al., 2021).

6. Development of stomatal pores and leaf epidermal patterning

The longitudinal grass leaf is morphologically and ontogenetically distinct from broad leaves typical for eudicots such as *A. thaliana* (Conklin, Strable, Li, & Scanlon, 2019). In grasses, leaf development follows a strict base-to-tip developmental gradient and patterning is highly stereotyped along the mediolateral leaf axis. Venation in grass leaves, for example, is parallel rather than reticulate (Conklin et al., 2019) and epidermal patterning is organized in semi-clonal longitudinal cell files that form different epidermal cell types like stomata, prickle hair cells, silica cells or pavement cells (Fig. 3; Stebbins & Shah, 1960).

Stomata are tiny breathing pores on the leaf's surface that can open and close to enable carbon dioxide uptake for photosynthesis while simultaneously balancing water vapor loss. The graminoid stomatal complex of grasses shows a highly innovative, derived morphology consisting of two central, dumbbell-shaped guard cells (GCs) that are flanked by two lateral subsidiary cells (SCs; Fig. 5A; Nunes, Zhang, & Raissig, 2020). *A. thaliana*, on the other hand, features ancestral two-celled stomata, where two kidney-shaped GCs surround the central pore (Fig. 5B; Nunes et al., 2020). Interestingly, the graminoid morphology is associated with faster stomatal opening and closing kinetics, which contribute to more water-use efficient photosynthesis (Franks & Farquhar, 2007; Lawson & Blatt, 2014; Lawson & Vialet-Chabrand, 2019; McAusland et al., 2016).

In *B. distachyon*, specific epidermal cell files adopt the capacity to form stomatal complexes (Fig. 5C; McKown & Bergmann, 2020; Nunes et al., 2020; Raissig, Abrash, Bettadapur, Vogel, & Bergmann, 2016; Stebbins & Shah, 1960). Stomatal files always run parallel and next to leaf veins and are usually composed of two adjacent cell files. Non-stomatal cell files make other epidermal cell types like prickle hair cells or adaxial bulliform cells (Kellogg, 2015). At the base of the growing leaf stomatal rows transversely divide more rapidly than adjacent prickle hair rows (McKown & Bergmann, 2020; Nunes et al., 2020; Stebbins & Shah, 1960). Eventually, a transverse asymmetric division in both the stomatal and the prickle hair cell rows in the *B. distachyon* epidermis generates a smaller daughter cell, which becomes a guard mother cell (GMC) or a prickle hair cell precursor, and a larger daughter cell that is destined to be a pavement cell (Fig. 5C; McKown & Bergmann, 2020 ; Nunes et al., 2020). While the prickle hair cell precursor

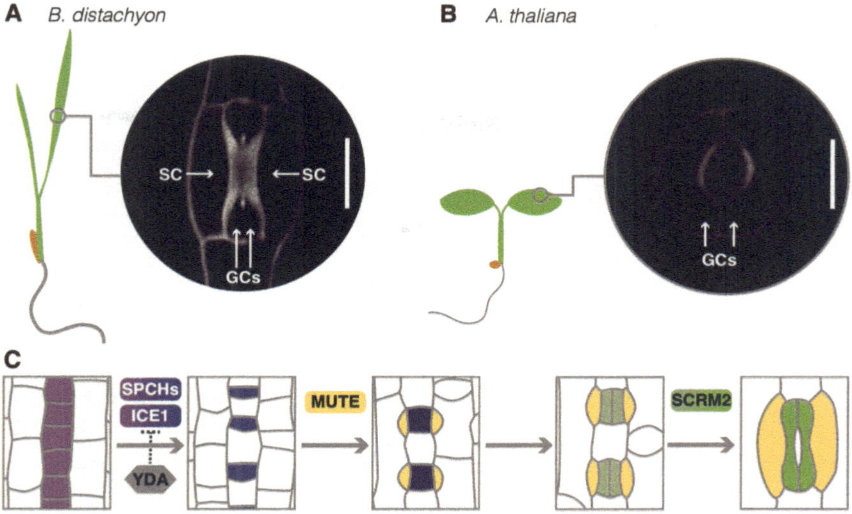

Fig. 5 Stomatal development in *B. distachyon*. (A) Graminoid stomata of *B. distachyon;* the central dumbbell-shaped guard cells (GCs) are flanked by lateral subsidiary cells (SCs); scale bar, 10 μm. (B) Anomocytic stomata of *A. thaliana*; two kidney-shaped guard cells (GCs) surround the central pore; scale bar, 10 μm. (C) Current working model of stomatal development in *B. distachyon*; the *BdSPCHs* and *BdICE1* together determine stomatal cell files (purple) and guide the smaller daughter cell toward GMC fate (blue). *BdMUTE* establishes the SC lineage and initiates polarization and asymmetric division that yields lateral SCs (yellow). After the symmetric division of the GMC, *BdSCRM2* orchestrates the differentiation and morphogenesis of the dumbbell-shaped, graminoid GCs (green).

will grow out and form the typical shark tooth shape (Fig. 3), the GMC induces its lateral neighbor cells to become subsidiary mother cells (SMCs), which polarize and asymmetrically divide longitudinally to contribute flanking SCs (Fig. 5C; Gray, Liu, & Facette, 2020; McKown & Bergmann, 2020; Nunes et al., 2020). Then the GMCs will longitudinally divide to form two equally sized GC precursors (Fig. 5C; McKown & Bergmann, 2020; Nunes et al., 2020; Stebbins & Shah, 1960). Finally, pore formation separates the GCs, which then undergo dramatic morphogenesis to build the dumbbell-shaped GCs unique to graminoid stomata (Fig. 5A, C; Galatis & Apostolakos, 2004; Spiegelhalder & Raissig, 2021).

Both the recruitment of lateral SCs and the morphogenetic events forming dumbbell-shaped GCs do not exist in *A. thaliana*. A forward genetic screen in *B. distachyon* instead yielded several players required for grass stomatal development including factors that determine stomatal identity, recruit the SC lineage, differentiate GCs and pattern stomata (McKown & Bergmann, 2020; Nunes et al., 2020). Surprisingly, many of the genes identified also have a role during stomatal development in *A. thaliana*, however, their function, their genetic interactions, and their regulation have been changed to generate novel morphological features and accommodate for the diverse grass leaf ontogeny and form (McKown & Bergmann, 2020; Nunes et al., 2020).

6.1 Establishing and differentiating the guard cell and subsidiary cell lineages

One of the key mutants identified in a forward genetic screen using chemically mutagenized *B. distachyon* families was a mutant called *stomataless* (*stl*) that lacked stomatal complexes (Raissig et al., 2016). It resembled the *speechless (spch)* mutant in *A. thaliana,* which disrupts the bHLH transcription factor *AtSPCH* that is required for entry into the stomatal lineage (MacAlister, Ohashi-Ito, & Bergmann, 2007). Surprisingly the *B. distachyon stl* mutation affected the *INDUCER OF CBF EXPRESSION1 (ICE1)* ortholog (Raissig et al., 2016), which in *A. thaliana* acts as AtSPCH's heterodimerization partner and does not have a mutant phenotype due to its redundancy with *AtSCREAM2* (*AtSCRM2*; Kanaoka et al., 2008). Notably, a loss of function gene-edited *bdscrm2* allele revealed a role for *BdSCRM2* in differentiation and pore formation of the four-celled complex (Raissig et al., 2016). Thus unlike in *A. thaliana*, the paralogs *ICE1/SCRM1* and *SCRM2* are non-redundant in grasses with distinct roles in controlling stomatal identity establishment and stomatal differentiation, respectively (Fig. 5C). In contrast, gene editing of the duplicated *SPCH* homologs *BdSPCH1* and *BdSPCH2* revealed these factors to be acting redundantly in establishing stomatal fate (Fig. 5C; Raissig et al., 2016).

The same screen also identified the *subsidiary cell identity defective* (*sid*) mutant, which failed to recruit SCs and produced ancestral, two-celled stomata without SCs (Raissig et al., 2017). The *sid* mutation caused a 5bp deletion in the *B. distachyon MUTE* homolog (Raissig et al., 2017). *A. thaliana MUTE* commits stomatal meristemoids to become guard mother cells (GMCs) and *atmute* plants arrest stomatal development before the symmetric GMC division causing seedling lethality (Pillitteri, Sloan, Bogenschutz, & Torii, 2007). In *bdmute*, however, 70% of the stomatal complexes still formed functional GCs yet failed to recruit SCs, suggesting that *BdMUTE* is primarily responsible to establish the SC lineage (Fig. 5C; Raissig et al., 2017). Accordingly, overexpressing *BdMUTE* in *B. distachyon* leaves caused excessive asymmetric SC-like divisions yielding multiple subsidiary cells around stomatal pores (Raissig et al., 2017). In *A. thaliana*, however, overexpressing *AtMUTE* reprogrammed all epidermal cells to GMCs, which divided and formed two-celled stomata (Pillitteri et al., 2007). Much like AtMUTE, BdMUTE is initially expressed in GMCs, but unlike the *A. thaliana* homolog, acquired cell-to-cell mobility and moves laterally into neighboring, non-stomatal cells to establish SMC identity (Raissig et al., 2017). Finally, the availability of a SC-less genotype allowed for testing the functional relevance of SCs to the extremely fast stomatal kinetics observed for graminoid stomata. In fact, SC-less *bdmute* stomata were not only much slower in opening and closing, they also failed to open and close to the same degree as wild-type stomata (Raissig et al., 2017). Interestingly, this work was only possible using the wild grass *B. distachyon* since mutations in *MUTE* in the domesticated cereals maize and rice not only lack SCs but also fail to make functional GC complexes leading to seedling lethality (Wang et al., 2019; Wu et al., 2019).

6.2 Patterning cell types in the leaf epidermis

Finally, a mutation that caused stomata to cluster within the stomatal cell file ("pearls on a string" phenotype) affected the mitogen activated protein (MAP) triple kinase *BdYODA1* (*BdYDA1*) (Abrash, Anleu Gil, Matos, & Bergmann, 2018). Much like *yda* in *A. thaliana*, the mutant was also dwarfed and intensiley green (hence the gene name *YODA*; Lukowitz, Roeder, Parmenter, & Somerville, 2004). Yet, while in *atyda* only stomatal spacing was affected (Bergmann, Lukowitz, & Somerville, 2004), in *bdyda1* not only spacing of stomata but also spacing of prickle hair cells and silica cells was affected (Abrash et al., 2018). This suggests a more general role for the grass MAPKKK *YDA1* homolog in patterning epidermal cell types. In *A. thaliana*, the YDA MAPK pathway directly targets and represses AtSPCH (Lampard, MacAlister, & Bergmann, 2008) to inhibit stomatal lineage entry. Therefore, overexpressing *AtSPCH* without mutating the MAPK phosphorylation sites does not have an effect on stomatal density or number of entry divisions (Lampard et al., 2008). In *B. distachyon*, on the other hand, *BdSPCH2-YFP* could be efficiently overexpressed and was able to reprogram prickle hair cell precursor cells to generate ectopic stomatal complexes (Raissig et al., 2016). Indeed, peptide sequence analysis of SPCH homologs in *B. distachyon* and *A. thaliana* revealed differences in MAPK target sites and protein degradation domains

(Raissig et al., 2016). Acquisition of high fidelity MAPK target sites in the protein degradation domain of BdICE1, on the other hand, together with the observation that BdICE1 was restricted to stomatal cell files even when overexpressed suggests that BdICE1 rather than the BdSPCH homologs might be targeted by the BdYDA1 MAPK pathway (Fig. 5C; Raissig et al., 2016).

Recently, quantitative polarization measurements of a BdYDA1 reporter revealed polarized localization at the base of the larger daughter cell, where stomatal fate needs to be repressed (Gong et al., 2021). This suggests that much like in *A. thaliana* a polarity scaffold in the larger, non-stomatal daughter cell might concentrate MAPK signaling components to efficiently downregulate key stomatal identity genes (Gong et al., 2021; Zhang, Wang, Shao, Zhu, & Dong, 2015).

In conclusion, a forward genetic screen in *B. distachyon* facilitated by its small size and small genome for efficient mapping by next-generation sequencing have significantly contributed to our understanding of how graminoid stomata form. Importantly, studies on *MUTE*'s function in SC development and SC function would not have been possible in domesticated grass models since *mute* mutants in both rice and maize are seedling-lethal (Wang et al., 2019; Wu et al., 2019). Taken together, the relatively efficient *B. distachyon* transformation protocols, which facilitated the generation of dozens of stomatal reporter and overexpression lines, coupled with stomatal developmental genetics will continue to reveal mechanistic insights into the development of graminoid stomata in the grasses.

7. Development of monocotyledonous root systems and root hair patterning

Monocotyledonous root systems are different from dicotyledonous root systems; in eudicots, primary tap roots and their lateral branches compile the root system. In monocots on the other hand, root systems are more complex and are dominated by adventitious, post-embryonic roots (Fig. 3). *B. distachyon* forms a seminal primary root and at least two adventitious roots from the coleoptile node below ground, which all persist throughout the whole life cycle (Hardtke & Pacheco-Villalobos, 2016). Depending on environmental conditions, the genetic background of specific accessions and tiller proximity to the soil, the above-ground leaf node can form adventitious roots, too (Hardtke & Pacheco-Villalobos, 2016). In fact, wind-stimulated *B. distachyon* shoots formed leaf node roots at tiller positions in proximity to soil and this process could be significantly inhibited when auxin transport was inhibited in wind-stimulated plants (Nam et al., 2020). Both seminal and adventitious roots can form lateral roots and root hairs (Hardtke & Pacheco-Villalobos, 2016). Therefore, *B. distachyon* is an ideal model to study monocotyledonous root system formation in grasses as it possesses all root types of domesticated cereals but tends to be less complex and smaller allowing efficient imaging and root growth analysis. Furthermore, *B. distachyon* is able to functionally associate with arbuscular mycorrhiza, which requires substantial transcriptional reprogramming in root cortex cells and the development of symbiotic interfaces between plant and fungi (Buendia, Ribeyre, Bensmihen, & Lefebvre, 2019; Hong et al., 2012; Müller et al., 2020; Torabi, Varshney, Villaécija-Aguilar, Keymer, & Gutjahr, 2021). It was shown, for example, that microbial symbiotic signals like Lipo-chitooligosaccharides stimulate lateral root formation in *B. distachyon* much like in legumes (Buendia, Maillet, et al., 2019).

7.1 Auxin and root growth regulation

Due to distinct redundancy levels, whole genome duplications and diversification of life histories, mutant phenotypes of homologous genes can be substantially different between *A. thaliana* and grasses. A mutation in the auxin biosynthesis gene *TRYPTOPHAN AMINOTRANSFERASE OF ARABIDOPSIS 1-RELATED 2-LIKE* (*BdTAR2L*) and the ethylene signaling gene *ETHYLENE INSENSITIVE-2-LIKE-1* (*BdEIN2L1*) display a longer root phenotype (Pacheco-Villalobos, Sankar, Ljung, & Hardtke, 2013). In *A. thaliana* the respective mutants show either no phenotype due to redundancy or shorter roots, respectively (Stepanova et al., 2008; Tao et al., 2008). The longer root phenotype results from enhanced cell elongation and cell anisotropy in the root, and, counterintuitively, is accompanied by higher rather than lower auxin levels (Pacheco-Villalobos et al., 2013). This is caused by an alternative wiring of ethylene-auxin crosstalk and by the common dependency of ethylene and auxin biosynthesis on the product of Tryptophan aminotransferases, IPA. In *A. thaliana*, ethylene positively regulates *TAA1* and the *YUCCA* genes and, consequently, auxin production, whereas in *B. distachyon*, ethylene signaling represses the *YUCCA* genes (Pacheco-Villalobos et al., 2013). Therefore, the *bdtar2l* mutant leads to lower IPA, and thus, lower ethylene, which hyperactivates the YUCCAs and leads to higher auxin levels (Pacheco-Villalobos et al., 2013). Similarly, *bdein2l1* mutants fail to repress YUCCAs in an ethylene-dependent manner resulting in higher auxin levels (Pacheco-Villalobos et al., 2013). Together, both mutations lead to increased auxin levels, exaggerated cell elongation and, thus, longer roots (Pacheco-Villalobos et al., 2013). The isolation of these mutants also facilitated testing the effect of high cellular auxin levels on transcriptional changes and whether auxin indeed stimulates cell wall acidification to promote cell elongation via expansin-mediated cell wall loosening. Transcriptionally, high cellular auxin upregulated cell wall remodeling factors including expansins and it was shown that the arabinogalactan content was reduced in the mutants (Pacheco-Villalobos et al., 2016). However, proton extrusion was not increased due to high auxin levels and, much like in *A. thaliana*, root elongation is very well buffered against external pH changes (Pacheco-Villalobos et al., 2016).

Similar to the auxin biosynthesis mutant *bdtar2l*, mutants in the auxin importer *BdAUX1* also caused roots to elongate more. In addition, *bdaux1* showed shoot dwarfism, flower developmental defects and reduced fertility (van der Schuren et al., 2018). While in *A. thaliana*, *AtAUX1* and its three *LIKE-AUX1* (*LAX*) homologs are partially redundant, *BdAUX1* seems to play a more central role (van der Schuren et al., 2018). Importantly, *bdaux1* mutant phenotypes are also more severe than corresponding *AUX1* single mutants in other grasses like *osaux1* in rice, *zmaux1* in maize or *svaux1* in *Setaria viridis*, where much milder shoot and inflorescence defects were observed (Huang et al., 2017; Zhao et al., 2015). This might be due to the presence of five *AUX1* homologs in rice, maize and *S. viridis* rather than just the three in *B. distachyon* (van der Schuren et al., 2018).

7.2 Root hair patterning

Much like the leaf epidermal cell types, root hairs alternate with non-hair cells within a longitudinal epidermal cell file (Dolan, 2017). In *A. thaliana*, only specific root epidermal cell files can form root hairs and the root hair forming files are flanked by at least two hairless cell files (Dolan, 2017). In grasses, however, all root epidermal cell files make root hairs (Dolan, 2017).

In *B. distachyon* a terminal, physically asymmetric division generates a smaller daughter cell that will develop into a root hair and a larger daughter cell making a non-hair cell (Kim & Dolan, 2011). In contrast, the domesticated cereals rice and barely execute a physically symmetric terminal division and differential elongation creates the size dimorphism between root hair and non-hair cells (Kim & Dolan, 2011; Marzec, Melzer, & Szarejko, 2013). In *A. thaliana*, *ROOT HAIRLESS SIX LIKE* (RSL) class I basic helix-loop-helix transcription factors generate root hair formation competency in specific cell files and activate RSL class II genes, which are necessary and sufficient for root hair outgrowth and elongation (Dolan, 2017). In *B. distachyon*, the three RSL class I genes are preferentially expressed in the smaller daughter cells and future hair cells and double knock-down lines showed shorter root hairs (Kim & Dolan, 2016). Unlike in *A. thaliana*, overexpression of RSL class I genes was sufficient to induce root hair outgrowth suggesting that root hair morphogenesis might not require RSL class II genes in *B. distachyon* (Kim & Dolan, 2016). Overexpression lines showed threefold root hair elongation and could increase phosphorus uptake, yet showed decreased overall biomass (Zhang et al., 2018).

Taken together, *B. distachyon* is well suited to answer fundamental questions specific to root development in grasses. Again, the small stature and the reduced gene redundancy space compared to other, domesticated grasses and even *A. thaliana* paired with a simplified yet complete monocotyledonous root system will allow for the identification of additional developmental mechanisms shaping root systems in grasses.

8. Joys and pains of working with a new model system: A personal perspective

8.1 *B. distachyon* & Daniel

Like many biologists, Daniel is interested in the diversity of life. He began investigating this diversity in Simon Malcomber's plant evolutionary developmental laboratory. Here, he explored the evolution of genes involved in the regulation of auxin transport and how this affects axillary meristem formation (Woods, Hope, & Malcomber, 2011). It was here that he became interested in the evolution of the novel grass spikelet. At the time the evolution of grass spikelet formation was mainly investigated by doing comparative expression work in candidate genes that affect spikelet formation across a variety of grasses and near grass relatives. Particular focus was paid to genes that were recently duplicated in the grasses, which could potentially lead to neo/subfunctionalization of the gene. Yet, while documenting changes in expression of a given candidate gene in a specific tissue is a proxy for a change in gene function, functional work was not possible outside of maize and rice. This prompted the move during his PhD studies in joining Richard Amasino's lab and the Brachypodium community. He was particularly excited at the prospect of using genetics to accelerate gene discovery and to not rely on a candidate gene approach identified in other model organisms. The lab traditionally had worked in *A. thaliana* but a small group in the laboratory began working in *B. distachyon*. At that time there was little to no molecular understanding of flowering in *B. distachyon*, but it was clear this provided the opportunity to discover new genes previously not identified in other plants. When very little is known about a given phenomenon genetics offers a powerful tool to identify the important genes controlling a given trait. In particular doing forward genetics screens to identify mutants with altered flowering times and exploiting natural variation in flowering behaviors across *B. distachyon* accessions was an exciting avenue for gene discovery (Fig. 6).

There are a lot of unknowns at the beginning of any research endeavor; however with a new model system this is amplified. For example, it was unclear which might be the best mapping partner to generate mapping populations or if EMS polymorphisms would need to be used for mapping in backcrossed populations. Fortunately, they identified a mapping partner in an accession that has a similar flowering behavior as the mutant

Fig. 6 Two plant developmental geneticists in the field looking for their favorite plant model during the 4th International Brachypodium Conference 2019 in Huesca, Spain. The authors Michael (A) and Daniel (C) were surprised to see how small the growth habitus of *B. distachyon* is, when it grows in its natural habitat in Northern Spain (B). *Photographs by Daniel P. Woods (A), David B. Lowry (B) and Michael T. Raissig (C).*

background but also enough sequence variation that the density of markers would not limit mapping resolution. This required the community resources of a collection of diverse inbred germplasm and also whole genome sequence information across those accessions. Despite the many unknowns, too many to list here, the excitement of identifying new genes involved in flowering pushed the research forward.

8.2 *B. distachyon* & Michael

During his PhD studies Michael investigated how the maternal and paternal genomes differentially affect embryo formation and seed development in *A. thaliana* with a particular focus on genomic imprinting and zygotic genome activation. After having spent half of his PhD sitting in front of a custom-made embryo isolation microscope (Raissig, Gagliardini, Jaenisch, Grossniklaus, & Baroux, 2013) to harvest minuscule embryos deeply embedded within developing seed tissue, he decided to change to a developmental system that is more readily accessible. He thus joined Dominique Bergmann's lab to study how stomata—tiny breathing pores on the plant's surface—develop and mediate gas exchange between the plant and the atmosphere. While the Bergmann lab was and is mostly focused on stomatal development in *A. thaliana*, a small subgroup (Team Brachy) led by the fantastic PhD student Emily Abrash was establishing *B. distachyon* as a model system to study grass stomatal development. Michael was immediately hooked by the unique morphology and cellular composition of grass stomata, their linear, cell-file specific patterning, and the strict developmental gradient that grass leaves present, so that he was willing to take the risk to work with a novel model system. He was extremely fortunate that the *B. distachyon* biotechnological (i.e., transformation), molecular and horticultural protocols were already up and running in the lab and very lucky that a handful of mutant phenotypes had already been isolated in a forward genetic screen. So upon arrival Michael could immediately start "doing the biology" and applying his developmental genetics background to map mutants and characterize gene functions while learning new techniques in cell biology, imaging and gas exchange physiology.

Together, it was rather accidental that Michael started working with *B. distachyon* and his reasoning to do so was mostly motivated by the compelling fascination of studying a derived stomatal morphology with improved functionality rather than an active choice of an alternate model system. In other words, the biological question, the phenomenon of graminoid stomatal morphology motivated Michael to embark on a journey with a novel model system. Clearly, the prospect of entering a research field that was wide open and joining the *B. distachyon* community, which is known to be open and supportive and which freely shared reagents, resources and datasets way ahead of publication were additional motivations.

9. Trials & tribulations

Working with a novel model system is risky and—like experimental research in general—requires an absurd amount of luck, timing and, well, luck again to succeed. And it strongly depends on the people around you, your mentors, your collaborators and your colleagues. We were both extremely fortunate to have had incredible mentors that created a supportive and vibrant research environment and fantastic collaborators within the Brachypodium community. We both want to stress how important it is to be part of an open, sharing and caring research community that puts the progress of the field before personal interests. We are very fortunate and grateful that it has always been at the core of the International Brachypodium Initiative to grant free access to genomic and genetic resources, where datasets and protocols are shared ahead of publication. In addition, the availability of molecular, genetic, genomic and biotechnological resources facilitate deciphering developmental mechanisms in any species, so ideally you do not have to start from scratch.

With emerging model systems it can be difficult to fit in with established communities; we keep claiming that *B. distachyon* is the "Arabidopsis of grasses," yet we are lagging at least one if not two decades behind the *A. thaliana* research community in terms of technologies, reagents, datasets and the sheer size of the community. On the other hand, we are not working on a domesticated grass relevant to agriculture, which can hinder us from tapping into some of the funding opportunities that support cereal research and can evoke comments regarding why we did not work on a "real grass" or with a crop "that matters." We hope this review provides some reasons for the utility of using an emerging model over either an established model system or domesticated crops with complex genomes and growth requirements.

Novel model systems can be frustrating at times; frustrating because required reagents and technologies have not been developed yet (and frankly, usually you have to develop these things yourself, because nobody else is going to do it). And of course it can be frustrating because some things are just harder in your favorite model system than in the established ones. Crossing *B. distachyon*, for example, is a complete pain in the back compared to crossing *A. thaliana*. In *A. thaliana*, dehisced pollen is viable for several days and one cross-pollinated flower produces approximately 50 seeds in a silique. In *B. distachyon*, however, pollen is only viable for about 30 minutes after another dehiscence and one cross-pollinated floret produces only one seed. Michael's crossing success rate is around 20% so he gets one seed per five crosses. Furthermore, while Michael was able to cross 50 Arabidopsis flowers in an hour (so 2.5k seeds!), he manages only to cross approximately 10 florets in *B. distachyon* (so 2 seeds at a success rate of 20%), which means crossing *B. distachyon* is a thousand times less efficient than crossing *A. thaliana*. And we do not even want to get started comparing the benefits of plant transformation via floral dipping with the tediousness of plant transformation via embryonic tissue cultures.

That being said, it can also be extremely rewarding to work with an emerging model system. First of all, we both are convinced that this is the right time to start thinking and researching questions that go beyond the classical model systems. Recent biotechnological progress like the identification of precise gene editing tools such as CRISPR/Cas9 and affordable, accurate and efficient next-generation sequencing technologies means we can read the stored and expressed genetic information of any given species, and, if it is transformable, can mutate any given gene. This allows for a purely phenomenon-driven research approach, where unique biological aspects can be investigated that cannot be studied in established model systems because they simply do not have this cell type (i.e., stomatal SCs) or environmental response (i.e., short-day vernalization flowering). Studying unique biological aspects also means that there are so-called low-hanging fruits—significant discoveries that can define novel new research directions. Finally, it is extremely rewarding to be embedded within, be supported by, contribute to and help shape a pioneering research community. Both authors irrespective of their junior positions serve on the International Brachypodium Research Steering committee since 2020.

In summary, we are both strong advocates of curiosity-driven, phenomenon-oriented and somewhat untargeted fundamental basic research, because we think it is both fun and has the potential for real breakthrough discoveries. The research community studying bacterial immunity would never have anticipated that their findings would create a multi-billion dollar patent war over who owns the CRISPR/Cas9 gene editing technology (Ledford, 2017). Also in 1928, Andrew Fleming did not intend to alter modern medicine by sorting through old plates containing Staphylococcus colonies and by chance finding one plate with the antibiotic-secreting fungus *Penicillium notatum* (Aldridge, 1999). Therefore, if you are fascinated by an unusual biological structure, an unexpected environmental response, or a unique biotic interaction, you should try to find an appropriate model species with decent growth requirements and let the adventure begin.

Acknowledgements

The authors would like to thank the International Brachypodium Initiative for being an open, sharing and caring community. Furthermore, we would like to thank our mentors Dominique C. Bergmann and Rick Amasino for their constant encouragement and support. D.P.W. would like to thank Thomas Ream and Frederic Bouche along with the many undergraduate colleagues that were essential in establishing *B. distachyon* as a flowering model while in the Amasino lab. D.P.W. is also grateful to Jorge Dubcovsky for allowing him to continue work with *B. distachyon* in his laboratory. M.T.R. would like to thank Emily Abrash for her pioneering and foundational work on establishing *B. distachyon* as a model for grass stomatal development. Furthermore, M.T.R. would like to thank Maria Ximena Anleu Xil, his past and present students, and, of course, his Co-PI Heike Lindner for sharing his passion for *B. distachyon* developmental genetics. D.P.W. is a Howard Hughes Medical Institute fellow of the Life Sciences Research Foundation (http://www.lsrf.org/). M.T.R. is supported by the German Research Foundation (DFG) through Emmy Noether Programme Grant RA 3117/1-1.

References

Abrash, E., Anleu Gil, M. X., Matos, J. L., & Bergmann, D. C. (2018). Conservation and divergence of YODA MAPKKK function in regulation of grass epidermal patterning. *Development, 145*, 1–9. https://doi.org/10.1242/dev.165860.

Aldridge, S. (1999). *The discovery and development of penicillin 1928-1945: An international historic chemical landmark*. American Chemical Society.

Amasino, R. (2010). Seasonal and developmental timing of flowering. *The Plant Journal, 61*, 1001–1013.

Andrés, F., & Coupland, G. (2012). The genetic basis of flowering responses to seasonal cues. *Nature Reviews. Genetics, 13*, 627–639.

Ballerini, E. S., & Kramer, E. M. (2011). In the light of evolution: A reevaluation of conservation in the CO?FT regulon and its role in photoperiodic regulation of flowering time. *Frontiers in Plant Science, 2*, 1–13.

Ballerini, E. S., Min, Y., Edwards, M. B., Kramer, E. M., & Hodges, S. A. (2020). POPOVICH, encoding a C2H2 zinc-finger transcription factor, plays a central role in the development of a key innovation, floral nectar spurs, in Aquilegia. *Proceedings of the National Academy of Sciences of the United States of America, 117*, 22552–22560. https://doi.org/10.1073/pnas.2006912117.

Bergmann, D. C., Lukowitz, W., & Somerville, C. R. (2004). Stomatal development and pattern controlled by a MAPKK kinase. *Science, 304*, 1494–1497.

Bettgenhaeuser, J., Corke, F. M. K., Opanowicz, M., Green, P., Hernández-Pinzón, I., Doonan, J. H., et al. (2017). Natural variation in brachypodium links vernalization and flowering time loci as major flowering determinants. *Plant Physiology*, *173*, 256–268.

Borthwick, H. A., & Hendricks, S. B. (1960). Photoperiodism in plants. *Science*, *132*, 1223–1228.

Bouché, F., Woods, D. P., & Amasino, R. M. (2017). Winter memory throughout the plant kingdom: Different paths to flowering. *Plant Physiology*, *173*, 27–35. https://doi.org/10.1104/pp.16.01322.

Bragg, J. N., Wu, J., Gordon, S. P., Guttman, M. E., Thilmony, R., Lazo, G. R., et al. (2012). Generation and characterization of the Western Regional Research Center Brachypodium T-DNA insertional mutant collection. *PLoS One*, 7, e41916.

Brkljacic, J., Grotewold, E., Scholl, R., Mockler, T., Garvin, D. F., Vain, P., et al. (2011). Brachypodium as a model for the grasses: Today and the future. *Plant Physiology*, *157*, 3–13.

Buendia, L., Maillet, F., O'Connor, D., van de- Kerkhove, Q., Danoun, S., Gough, C., et al. (2019). Lipo-chitooligosaccharides promote lateral root formation and modify auxin homeostasis in Brachypodium distachyon. *The New Phytologist*, *221*, 2190–2202. https://doi.org/10.1111/nph.15551.

Buendia, L., Ribeyre, C., Bensmihen, S., & Lefebvre, B. (2019). Brachypodium distachyon tar2lhypo mutant shows reduced root developmental response to symbiotic signal but increased arbuscular mycorrhiza. *Plant Signaling & Behavior*, *14*, e1651608. https://doi.org/10.1080/15592324.2019.1651608.

Catalán, P., Müller, J., Hasterok, R., Jenkins, G., Mur, L. A. J., Langdon, T., et al. (2012). Evolution and taxonomic split of the model grass Brachypodium distachyon. *Annals of Botany*, *109*, 385–405.

Chen, A., & Dubcovsky, J. (2012). Wheat tilling mutants show that the vernalization gene VRN1 down-regulates the flowering repressor VRN2 in leaves but is not essential for flowering. *PLoS Genetics*, *8*, e1003134.

Chen, A., Li, C., Hu, W., Lau, M. Y., Lin, H., Rockwell, N. C., et al. (2014). Phytochrome C plays a major role in the acceleration of wheat flowering under long-day photoperiod. *Proceedings of the National Academy of Sciences of the United States of America*, *111*, 10037–10044. https://doi.org/10.1073/pnas.1409795111.

Chouard, P. (1960). Vernalization and its relations to dormancy. *Annual Review of Plant Physiology*, *11*, 191–238.

Conklin, P. A., Strable, J., Li, S., & Scanlon, M. J. (2019). On the mechanisms of development in monocot and eudicot leaves. *The New Phytologist*, *221*, 706–724. https://doi.org/10.1111/nph.15371.

Coomey, J. H., MacKinnon, K. J.-M., Handakumbura, P. P., McCahill, I. W., Trabucco, G. M., Mazzola, J., et al. (2021). Touch-triggered bZIP translocation regulates elongation and secondary wall biosynthesis. *bioRxiv*, 1–34. https://doi.org/10.1101/2021.02.03.429573.

Coomey, J. H., Sibout, R., & Hazen, S. P. (2020). Grass secondary cell walls, Brachypodium distachyon as a model for discovery. *The New Phytologist*, 1649–1667. https://doi.org/10.1111/nph.16603.

De Lucia, F., Crevillen, P., Jones, A. M. E., Greb, T., & Dean, C. (2008). A PHD-polycomb repressive complex 2 triggers the epigenetic silencing of FLC during vernalization. *Proceedings of the National Academy of Sciences of the United States of America*, *105*, 16831–16836.

Debernardi, J. M., Tricoli, D. M., Ercoli, M. F., Hayta, S., Ronald, P., Palatnik, J. F., et al. (2020). A GRF–GIF chimeric protein improves the regeneration efficiency of transgenic plants. *Nature Biotechnology*, *38*, 1274–1279. https://doi.org/10.1038/s41587-020-0703-0.

Deng, W., Casao, M. C., Wang, P., Sato, K., Hayes, P. M., Finnegan, E. J., et al. (2015). Direct links between the vernalization response and other key traits of cereal crops. *Nature Communications*, *6*, 5882. https://doi.org/10.1038/ncomms6882.

Derbyshire, P., & Byrne, M. E. (2013). MORE SPIKELETS1 is required for spikelet fate in the inflorescence of brachypodium. *Plant Physiology*, *161*, 1291–1302.

Díaz-Pérez, A., López-Álvarez, D., Sancho, R., & Catalán, P. (2018). Reconstructing the origins and the biogeography of species' genomes in the highly reticulate allopolyploid-rich model grass genus Brachypodium using minimum evolution, coalescence and maximum likelihood approaches. *Molecular Phylogenetics and Evolution*, *127*, 256–271. https://doi.org/10.1016/j.ympev.2018.06.003.

Distelfeld, A., Li, C., & Dubcovsky, J. (2009). Regulation of flowering in temperate cereals. *Current Opinion in Plant Biology*, *12*, 178–184. https://doi.org/10.1016/j.pbi.2008.12.010.

Dolan, L. (2017). Root hair development in grasses and cereals (Poaceae). *Current Opinion in Genetics & Development*, *45*, 76–81. https://doi.org/10.1016/j.gde.2017.03.009.

Dubcovsky, J., & Dvorak, J. (2007). Genome plasticity a key factor in the success of polyploid wheat under domestication. *Science*, *316*, 1862–1866.

Dubcovsky, J., Loukoianov, A., Fu, D., Valarik, M., Sanchez, A., & Yan, L. (2006). Effect of photoperiod on the regulation of wheat vernalization genes VRN1 and VRN2. *Plant Molecular Biology*, *60*, 469–480.

FAO Statistical Pocketbook, F.A.O. (2015). *World Food and Agriculture (2015)*. Food and Agriculture Organization of the United Nations.

Fjellheim, S., Boden, S., & Trevaskis, B. (2014). The role of seasonal flowering responses in adaptation of grasses to temperate climates. *Frontiers in Plant Science*, *5*, 431.

Fjellheim, S., & Preston, J. C. (2018). Shedding light on the role of seasonal flowering in plant niche transitions. *American Journal of Botany*, *105*, 136–138.

Franks, P. J., & Farquhar, G. D. (2007). The mechanical diversity of stomata and its significance in gas-exchange control. *Plant Physiology*, *143*, 78–87. https://doi.org/10.1104/pp.106.089367.

Galatis, B., & Apostolakos, P. (2004). The role of the cytoskeleton in the morphogenesis and function of stomatal complexes. *The New Phytologist*, *161*, 613–639.

Gälweiler, L., Guan, C., Müller, A., Wisman, E., Mendgen, K., Yephremov, A., et al. (1998). Regulation of polar auxin transport by AtPIN1 in Arabidopsis vascular tissue. *Science*, *282*, 2226–2230.

Gao, M., Geng, F., Klose, C., Staudt, A.-M., Huang, H., Nguyen, D., et al. (2019). Phytochromes measure photoperiod in Brachypodium. *bioRxiv*, *16*, 1–53.

Gong, Y., Varnau, R., Wallner, E.-S., Acharya, R., Bergmann, D. C., & Cheung, L. S. (2021). Quantitative and dynamic cell polarity tracking in plant cells. *The New Phytologist*, *230*, 867–877. https://doi.org/10.1111/nph.17165.

Goodrich, J., Puangsomlee, P., Martin, M., Long, D., Meyerowitz, E. M., & Coupland, G. (1997). A polycomb-group gene regulates homeotic gene expression in Arabidopsis. *Nature*, *386*, 44–51.

Gordon, S. P., Contreras-Moreira, B., Woods, D. P., Des Marais, D. L., Burgess, D., Shu, S., et al. (2017). Extensive gene content variation in the Brachypodium distachyon pangenome correlates with population structure. *Nature Communications*, *8*, 2184. https://doi.org/10.1038/s41467-017-02292-8.

Grass Phylogeny Working Group II. (2012). New grass phylogeny resolves deep evolutionary relationships and discovers C4 origins. *The New Phytologist*, *193*, 304–312.

Gray, A., Liu, L., & Facette, M. (2020). Flanking support: How subsidiary cells contribute to stomatal form and function. *Frontiers in Plant Science*, *11*, 881. https://doi.org/10.3389/fpls.2020.00881.

Greenup, A., Peacock, W. J., Dennis, E. S., & Trevaskis, B. (2009). The molecular biology of seasonal flowering-responses in Arabidopsis and the cereals. *Annals of Botany*, *103*, 1165–1172.
Griffiths, S., Dunford, R. P., Coupland, G., & Laurie, D. A. (2003). The evolution of CONSTANS-like gene families in barley, rice, and Arabidopsis. *Plant Physiology*, *131*, 1855–1867.
Hao, Z., Zhang, Z., Xiang, D., Venglat, P., Chen, J., Gao, P., et al. (2021). Conserved, divergent and heterochronic gene expression during Brachypodium and Arabidopsis embryo development. *Plant Reproduction*, 207–224. https://doi.org/10.1007/s00497-021-00413-4.
Hardtke, C. S., & Pacheco-Villalobos, D. (2016). The Brachypodium distachyon root system: A tractable model to investigate grass roots. In J. P. Vogel (Ed.), *Genetics and Genomics of Brachypodium* (pp. 245–258). Cham: Springer International Publishing. https://doi.org/10.1007/7397_2015_6.
Heide, O. M. (1994). Control of flowering and reproduction in temperate grasses. *The New Phytologist*, *128*, 347–362.
Higgins, J. A., Bailey, P. C., & Laurie, D. A. (2010). Comparative genomics of flowering time pathways using brachypodium distachyon as a model for the temperate grasses. *PLoS One*, *5*, e10065.
Himmelbach, A., Zierold, U., Hensel, G., Riechen, J., Douchkov, D., Schweizer, P., et al. (2007). A set of modular binary vectors for transformation of cereals. *Plant Physiology*, *145*, 1192–1200.
Hong, J. J., Park, Y.-S., Bravo, A., Bhattarai, K. K., Daniels, D. A., & Harrison, M. J. (2012). Diversity of morphology and function in arbuscular mycorrhizal symbioses in Brachypodium distachyon. *Planta*, *236*, 851–865.
Hsia, M. M., O'Malley, R., Cartwright, A., Nieu, R., Gordon, S. P., Kelly, S., et al. (2017). Sequencing and functional validation of the JGI Brachypodium distachyon T-DNA collection. *The Plant Journal*, *91*, 361–370. https://doi.org/10.1111/tpj.13582.
Huang, P., Jiang, H., Zhu, C., Barry, K., Jenkins, J., Sandor, L., et al. (2017). Sparse panicle1 is required for inflorescence development in Setaria viridis and maize. *Nature Plants*, *3*, 17054. https://doi.org/10.1038/nplants.2017.54.
Humphreys, A. M., & Linder, H. P. (2013). Evidence for recent evolution of cold tolerance in grasses suggests current distribution is not limited by (low) temperature. *The New Phytologist*, *198*, 1261–1273.
Initiative, International Brachypodium. (2010). Genome sequencing and analysis of the model grass Brachypodium distachyon. *Nature*, *463*, 763–768.
International Barley Genome Sequencing Consortium, Mayer, K. F. X., Waugh, R., Brown, J. W. S., Schulman, A., Langridge, P., et al. (2012). A physical, genetic and functional sequence assembly of the barley genome. *Nature*, *491*, 711–716.
International Wheat Genome Sequencing Consortium (IWGSC), IWGSC RefSeq Principal investigators, Appels, R., Eversole, K., Feuillet, C., Keller, B., et al. (2018). Shifting the limits in wheat research and breeding using a fully annotated reference genome. *Science*, *361*, 1–13. https://doi.org/10.1126/science.aar7191.
Kanaoka, M. M., Pillitteri, L. J., Fujii, H., Yoshida, Y., Bogenschutz, N. L., Takabayashi, J., et al. (2008). SCREAM/ICE1 and SCREAM2 specify three cell-state transitional steps leading to Arabidopsis stomatal differentiation. *Plant Cell*, *20*, 1775–1785.
Kellogg, E. A. (2001). Evolutionary history of the grasses. *Plant Physiology*, *125*, 1198–1205.
Kellogg, E. (2007). Floral displays: Genetic control of grass inflorescences. *Current Opinion in Plant Biology*, *10*, 26–31.
Kellogg, E. A. (2006). Progress and challenges in studies of the evolution of development. *Journal of Experimental Botany*, *57*, 3505–3516. https://doi.org/10.1093/jxb/erl132.

Kellogg, E. A. (2015). *Flowering plants. Monocots: Poaceae*. Springer International Publishing.

Kellogg, E. A., Camara, P. E. A. S., Rudall, P. J., Ladd, P., Malcomber, S. T., Whipple, C. J., et al. (2013). Early inflorescence development in the grasses (Poaceae). *Frontiers in Plant Science*, *4*, 250. https://doi.org/10.3389/fpls.2013.00250.

Kim, C. M., & Dolan, L. (2011). Root hair development involves asymmetric cell division in Brachypodium distachyon and symmetric division in Oryza sativa. *The New Phytologist*, *192*, 601–610. https://doi.org/10.1111/j.1469-8137.2011.03839.x.

Kim, C. M., & Dolan, L. (2016). Root hair defective six-like class I genes promote root hair development in the grass brachypodium distachyon. *PLoS Genetics*, *12*, e1006211.

Koornneef, M., & Meinke, D. (2010). The development of Arabidopsis as a model plant. *The Plant Journal*, *61*, 909–921.

Lampard, G. R., MacAlister, C. A., & Bergmann, D. C. (2008). Arabidopsis stomatal initiation is controlled by MAPK-mediated regulation of the bHLH SPEECHLESS. *Science*, *322*, 1113–1116.

Lampropoulos, A., Sutikovic, Z., Wenzl, C., Maegele, I., Lohmann, J. U., & Forner, J. (2013). GreenGate—A novel, versatile, and efficient cloning system for plant transgenesis. *PLoS One*, *8*, e83043.

Lawson, T., & Blatt, M. R. (2014). Stomatal size, speed, and responsiveness impact on photosynthesis and water use efficiency. *Plant Physiology*, *164*, 1556–1570. https://doi.org/10.1104/pp.114.237107.

Lawson, T., & Vialet-Chabrand, S. (2019). Speedy stomata, photosynthesis and plant water use efficiency. *The New Phytologist*, *221*, 93–98. https://doi.org/10.1111/nph.15330.

Ledford, H. (2017). Bitter CRISPR patent war intensifies. *Nature*, 1–4. https://doi.org/10.1038/nature.2017.22892.

Leyser, O. (2010). The power of auxin in plants. *Plant Physiology*, *154*, 501–505.

Li, Y., Li, L., Zhao, M., Guo, L., Guo, X., Zhao, D., et al. (2020). Wheat FRIZZY PANICLE activates VERNALIZATION1-A and HOMEOBOX4-A to regulate spike development in wheat. *Plant Biotechnology Journal*, 1141–1154. https://doi.org/10.1111/pbi.13535.

Lomax, A., Woods, D. P., Dong, Y., Bouché, F., Rong, Y., Mayer, K. S., et al. (2018). An ortholog of CURLY LEAF/ENHANCER OF ZESTE like-1is required for proper flowering in Brachypodium distachyon. *The Plant Journal*, *93*, 871–882.

Lukowitz, W., Roeder, A., Parmenter, D., & Somerville, C. (2004). A MAPKK kinase gene regulates extra-embryonic cell fate in Arabidopsis. *Cell*, *116*, 109–119.

Lv, B., Nitcher, R., Han, X., Wang, S., Ni, F., Li, K., et al. (2014). Characterization of FLOWERING LOCUS T1 (FT1) Gene in Brachypodium and Wheat. *PLoS One*, *9*, e94171.

MacAlister, C. A., Ohashi-Ito, K., & Bergmann, D. C. (2007). Transcription factor control of asymmetric cell divisions that establish the stomatal lineage. *Nature*, *445*, 537–540.

Magne, K., Liu, S., Massot, S., Dalmais, M., Morin, H., Sibout, R., et al. (2020). Roles of BdUNICULME4 and BdLAXATUM-A in the non-domesticated grass Brachypodium distachyon. *The Plant Journal*, *103*, 645–659. https://doi.org/10.1111/tpj.14758.

Malcomber, S. T., Preston, J. C., Reinheimer, R., Kossuth, J., & Kellogg, E. A. (2006). *Advances in botanical research* (p. 00652296). Elsevier.

Mann, D. G. J., Lafayette, P. R., Abercrombie, L. L., King, Z. R., Mazarei, M., Halter, M. C., et al. (2012). Gateway-compatible vectors for high-throughput gene functional analysis in switchgrass (Panicum virgatum L.) and other monocot species. *Plant Biotechnology Journal*, *10*, 226–236. https://doi.org/10.1111/j.1467-7652.2011.00658.x.

Marzec, M., Melzer, M., & Szarejko, I. (2013). Asymmetric growth of root epidermal cells is related to the differentiation of root hair cells in Hordeum vulgare (L.). *Journal of Experimental Botany*, *64*, 5145–5155. https://doi.org/10.1093/jxb/ert300.

McAusland, L., Vialet-Chabrand, S., Davey, P., Baker, N. R., Brendel, O., & Lawson, T. (2016). Effects of kinetics of light-induced stomatal responses on photosynthesis and water-use efficiency. *The New Phytologist, 211*, 1209–1220. https://doi.org/10.1111/nph.14000.

McKeown, M., Schubert, M., Marcussen, T., Fjellheim, S., & Preston, J. C. (2016). Evidence for an early origin of vernalization responsiveness in temperate pooideae grasses. *Plant Physiology, 172*, 416–426.

McKeown, M., Schubert, M., Preston, J. C., & Fjellheim, S. (2017). Evolution of the miR5200-FLOWERING LOCUS T flowering time regulon in the temperate grass subfamily Pooideae. *Molecular Phylogenetics and Evolution, 114*, 1–36.

McKown, K. H., & Bergmann, D. C. (2020). Stomatal development in the grasses: Lessons from models and crops (and crop models). *The New Phytologist, 227*, 1636–1648. https://doi.org/10.1111/nph.16450.

Mcsteen, P. (2009). Hormonal regulation of branching in grasses. *Plant Physiology, 149*, 46–55.

Mcsteen, P. (2010). Auxin and monocot development. *Cold Spring Harbor Perspectives in Biology, 2*, a001479.

Minevich, G., Park, D. S., Blankenberg, D., Poole, R. J., & Hobert, O. (2012). CloudMap: A cloud-based pipeline for analysis of mutant genome sequences. *Genetics, 192*, 1249–1269. https://doi.org/10.1534/genetics.112.144204.

Monte, E., Alonso, J. M., Ecker, J. R., Zhang, Y., Li, X., Young, J., et al. (2003). Isolation and characterization of phyC mutants in Arabidopsis reveals complex crosstalk between phytochrome signaling pathways. *Plant Cell, 15*, 1962–1980.

Müller, L. M., Campos-Soriano, L., Levesque-Tremblay, V., Bravo, A., Daniels, D. A., Pathak, S., et al. (2020). Constitutive overexpression of RAM1 leads to an increase in arbuscule density in Brachypodium distachyon. *Plant Physiology, 184*, 1263–1272. https://doi.org/10.1104/pp.20.00997.

Nam, B. E., Park, Y.-J., Gil, K.-E., Kim, J.-H., Kim, J. G., & Park, C.-M. (2020). Auxin mediates the touch-induced mechanical stimulation of adventitious root formation under windy conditions in Brachypodium distachyon. *BMC Plant Biology, 20*, 335. https://doi.org/10.1186/s12870-020-02544-8.

Nunes, T. D. G., Zhang, D., & Raissig, M. T. (2020). Form, development and function of grass stomata. *The Plant Journal, 101*, 780–799. https://doi.org/10.1111/tpj.14552.

O'Connor, D. L., Elton, S., Ticchiarelli, F., Hsia, M. M., Vogel, J. P., & Leyser, O. (2017). Cross-species functional diversity within the PIN auxin efflux protein family. *eLife, 6*, 1–29. https://doi.org/10.7554/eLife.31804.

O'Connor, D. L., Runions, A., Sluis, A., Bragg, J., Vogel, J. P., Prusinkiewicz, P., et al. (2014). A division in PIN-mediated auxin patterning during organ initiation in grasses. *PLoS Computational Biology, 10*, e1003447. https://doi.org/10.1371/journal.pcbi.1003447.

Pacheco-Villalobos, D., Díaz-Moreno, S. M., van der Schuren, A., Tamaki, T., Kang, Y. H., Gujas, B., et al. (2016). The effects of high steady state auxin levels on root cell elongation in Brachypodium. *Plant Cell, 28*. https://doi.org/10.1105/tpc.15.01057. tpc.01057.2016.

Pacheco-Villalobos, D., Sankar, M., Ljung, K., & Hardtke, C. S. (2013). Disturbed local auxin homeostasis enhances cellular anisotropy and reveals alternative wiring of auxin-ethylene crosstalk in Brachypodium distachyon seminal roots. *PLoS Genetics, 9*, e1003564.

Pankin, A., Campoli, C., Dong, X., Kilian, B., Sharma, R., Himmelbach, A., et al. (2014). Mapping-by-sequencing identifies HvPHYTOCHROME C as a candidate gene for the early maturity 5 locus modulating the circadian clock and photoperiodic flowering in barley. *Genetics, 198*, 383–396. https://doi.org/10.1534/genetics.114.165613.

Pillitteri, L. J., Sloan, D. B., Bogenschutz, N. L., & Torii, K. U. (2007). Termination of asymmetric cell division and differentiation of stomata. *Nature*, *445*, 501–505.

Poursarebani, N., Seidensticker, T., Koppolu, R., Trautewig, C., Gawroński, P., Bini, F., et al. (2015). The genetic basis of composite spike form in barley and "miracle-wheat". *Genetics*, *201*, 155–165. https://doi.org/10.1534/genetics.115.176628.

Poursarebani, N., Trautewig, C., Melzer, M., Nussbaumer, T., Lundqvist, U., Rutten, T., et al. (2020). COMPOSITUM 1 contributes to the architectural simplification of barley inflorescence via meristem identity signals. *Nature Communications*, *1:5138*, 1–16.

Preston, J. C., & Kellogg, E. A. (2006). Reconstructing the evolutionary history of paralogous APETALA1/FRUITFULL-like genes in grasses (Poaceae). *Genetics*, *174*, 421–437.

Preston, J. C., & Sandve, S. R. (2013). Adaptation to seasonality and the winter freeze. *Frontiers in Plant Science*, *4*, 167.

Provart, N. J., Alonso, J., Assmann, S. M., Bergmann, D., Brady, S. M., Brkljacic, J., et al. (2016). 50 years of Arabidopsis research: Highlights and future directions. *The New Phytologist*, *209*, 921–944. https://doi.org/10.1111/nph.13687.

Purvis, O. N., & Gregory, F. G. (1937). Studies in vernalisation of cereals. *Annals of Botany*, *1*, 1–26.

Qin, Z., Bai, Y., Muhammad, S., Wu, X., Deng, P., Wu, J., et al. (2019). Divergent roles of FT-like 9 in flowering transition under different day lengths in Brachypodium distachyon. *Nature Communications*, *10*, 812. https://doi.org/10.1038/s41467-019-08785-y.

Quail, P. H. (2002). Phytochrome photosensory signalling networks. *Nature Reviews. Molecular Cell Biology*, *3*, 85–93.

Raissig, M. T., Abrash, E., Bettadapur, A., Vogel, J. P., & Bergmann, D. C. (2016). Grasses use an alternatively wired bHLH transcription factor network to establish stomatal identity. *Proceedings of the National Academy of Sciences of the United States of America*, *113*, 8326–8331. https://doi.org/10.1073/pnas.1606728113.

Raissig, M. T., Gagliardini, V., Jaenisch, J., Grossniklaus, U., & Baroux, C. (2013). Efficient and rapid isolation of early-stage embryos from Arabidopsis thaliana seeds. *Journal of Visualized Experiments*, e50371.

Raissig, M. T., Matos, J. L., Anleu Gil, M. X., Kornfeld, A., Bettadapur, A., Abrash, E., et al. (2017). Mobile MUTE specifies subsidiary cells to build physiologically improved grass stomata. *Science*, *355*, 1215–1218. https://doi.org/10.1126/science.aal3254.

Ream, T. S., Woods, D. P., & Amasino, R. M. (2012). The molecular basis of vernalization in different plant groups. *Cold Spring Harbor Symposia on Quantitative Biology*, 77, 105–115. https://doi.org/10.1101/sqb.2013.77.014449.

Ream, T. S., Woods, D. P., Schwartz, C. J., Sanabria, C. P., Mahoy, J. A., Walters, E. M., et al. (2014). Interaction of photoperiod and vernalization determines flowering time of Brachypodium distachyon. *Plant Physiology*, *164*, 694–709. https://doi.org/10.1104/pp.113.232678.

Reinhardt, D., Mandel, T., & Kuhlemeier, C. (2000). Auxin regulates the initiation and radial position of plant lateral organs. *Plant Cell*, *12*, 507–518. https://doi.org/10.1105/tpc.12.4.507.

Reinhardt, D., Pesce, E.-R., Stieger, P., Mandel, T., Baltensperger, K., Bennett, M., et al. (2003). Regulation of phyllotaxis by polar auxin transport. *Nature*, *426*, 255–260.

Sakai, K., Citerne, S., Antelme, S., Le Bris, P., Daniel, S., Bouder, A., et al. (2021). BdERECTA controls vasculature patterning and phloem-xylem organization in Brachypodium distachyon. *BMC Plant Biology*, *21*, 196. https://doi.org/10.1186/s12870-021-02970-2.

Sancho, R., Cantalapiedra, C. P., López-Alvarez, D., Gordon, S. P., Vogel, J. P., Catalán, P., et al. (2018). Comparative plastome genomics and phylogenomics of Brachypodium: Flowering time signatures, introgression and recombination in recently diverged ecotypes. *The New Phytologist*, *218*, 1631–1644. https://doi.org/10.1111/nph.14926.

Sasani, S., Hemming, M. N., Oliver, S. N., Greenup, A., Tavakkol-Afshari, R., Mahfoozi, S., et al. (2009). The influence of vernalization and daylength on expression of flowering-time genes in the shoot apex and leaves of barley (Hordeum vulgare). *Journal of Experimental Botany*, *60*, 2169–2178.

Schneider, J., Döring, E., Hilu, K. W., & Röser, M. (2009). Phylogenetic structure of the grass subfamily Pooideae based on comparison of plastid matK gene-3′trnK exon and nuclear ITS sequences. *Taxon*, *58*, 405–424. https://doi.org/10.1002/tax.582008.

Scholthof, K.-B. G., Irigoyen, S., Catalan, P., & Mandadi, K. K. (2018). Brachypodium: A monocot grass model genus for plant biology. *Plant Cell*, *30*, 1673–1694. https://doi.org/10.1105/tpc.18.00083.

Schwartz, C. J., Doyle, M. R., Manzaneda, A. J., Rey, P. J., Mitchell-Olds, T., & Amasino, R. M. (2010). Natural variation of flowering time and vernalization responsiveness in Brachypodium distachyon. *Bioenergy Research*, *3*, 38–46.

Shaw, L. M., Li, C., Woods, D. P., Alvarez, M. A., Lin, H., Lau, M. Y., et al. (2020). Epistatic interactions between PHOTOPERIOD1, CONSTANS1 and CONSTANS2 modulate the photoperiodic response in wheat. *PLoS Genetics*, *16*, e1008812. https://doi.org/10.1371/journal.pgen.1008812.

Shimada, S., Ogawa, T., Kitagawa, S., Suzuki, T., Ikari, C., Shitsukawa, N., et al. (2009). A genetic network of flowering-time genes in wheat leaves, in which an APETALA1/FRUITFULL-like gene, VRN1, is upstream of FLOWERING LOCUS T. *The Plant Journal*, *58*, 668–681.

Sibout, R., Proost, S., Hansen, B. O., Vaid, N., Giorgi, F. M., Ho-Yue-Kuang, S., et al. (2017). Expression atlas and comparative coexpression network analyses reveal important genes involved in the formation of lignified cell wall in Brachypodium distachyon. *The New Phytologist*, *215*, 1009–1025. https://doi.org/10.1111/nph.14635.

Smertenko, T., Turner, G., Fahy, D., Brew-Appiah, R. A. T., Alfaro-Aco, R., de Almeida Engler, J., et al. (2019). Brachypodium distachyon MAP20 functions in metaxylem pit development and contributes to drought recovery. *The New Phytologist*, *227*, 1587–1590. https://doi.org/10.1111/nph.16383.

Song, Y. H., Shim, J. S., Kinmonth-Schultz, H. A., & Imaizumi, T. (2015). Photoperiodic flowering: Time measurement mechanisms in leaves. *Annual Review of Plant Biology*, *66*, 441–464. https://doi.org/10.1146/annurev-arplant-043014-115555.

Spiegelhalder, R. P., & Raissig, M. T. (2021). Morphology made for movement: Formation of diverse stomatal guard cells. *Current Opinion in Plant Biology*, *63*, 102090. https://doi.org/10.1016/j.pbi.2021.102090.

Stebbins, G. L., & Shah, S. S. (1960). Developmental studies of cell differentiation in the epidermis of monocotyledons. II. Cytological features of stomatal development in the Gramineae. *Developmental Biology*, *2*, 477–500.

Stepanova, A. N., Robertson-Hoyt, J., Yun, J., Benavente, L. M., Xie, D.-Y., Dolezal, K., et al. (2008). TAA1-mediated auxin biosynthesis is essential for hormone crosstalk and plant development. *Cell*, *133*, 177–191. https://doi.org/10.1016/j.cell.2008.01.047.

Takano, M., Inagaki, N., Xie, X., Yuzurihara, N., Hihara, F., Ishizuka, T., et al. (2005). Distinct and cooperative functions of phytochromes A, B, and C in the control of deetiolation and flowering in rice. *Plant Cell*, *17*, 3311–3325.

Tao, Y., Ferrer, J.-L., Ljung, K., Pojer, F., Hong, F., Long, J. A., et al. (2008). Rapid synthesis of auxin via a new tryptophan-dependent pathway is required for shade avoidance in plants. *Cell*, *133*, 164–176. https://doi.org/10.1016/j.cell.2008.01.049.

Torabi, S., Varshney, K., Villaécija-Aguilar, J. A., Keymer, A., & Gutjahr, C. (2021). Controlled assays for phenotyping the effects of strigolactone-like molecules on arbuscular mycorrhiza development. *Methods in Molecular Biology*, *2309*, 157–177. https://doi.org/10.1007/978-1-0716-1429-7_13.

Turner, A., Beales, J., Faure, S., Dunford, R. P., & Laurie, D. A. (2005). The pseudo-response regulator Ppd-H1 provides adaptation to photoperiod in barley. *Science*, *310*, 1031–1034.

Tyler, L., Fangel, J. U., Fagerström, D., Steinwand, M. A., Raab, T. K., Willats, W. G. T., et al. (2014). Selection and phenotypic characterization of a core collection of Brachypodium distachyon inbredlines. *BMC Plant Biology*, *14*, 1–15.

van der Schuren, A., Voiniciuc, C., Bragg, J., Ljung, K., Vogel, J., Pauly, M., et al. (2018). Broad spectrum developmental role of Brachypodium AUX1. *The New Phytologist*, *219*, 1216–1223. https://doi.org/10.1111/nph.15332.

Wang, H., Guo, S., Qiao, X., Guo, J., Li, Z., Zhou, Y., et al. (2019). BZU2/ZmMUTE controls symmetrical division of guard mother cell and specifies neighbor cell fate in maize. *PLoS Genetics*, *15*, e1008377. https://doi.org/10.1371/journal.pgen.1008377.

Whipple, C. J. (2017). Grass inflorescence architecture and evolution: The origin of novel signaling centers. *The New Phytologist*, *216*, 367–372.

Whipple, C. J., & Schmidt, R. J. (2006). Genetics of grass flower development. *Advances in Botanical Research: Incorporating Advances in Plant Pathology*, *44*(44), 385–424.

Wolny, E., Betekhtin, A., Rojek, M., Braszewska-Zalewska, A., Lusinska, J., & Hasterok, R. (2018). Germination and the early stages of seedling development in Brachypodium distachyon. *International Journal of Molecular Sciences*, *19*, 2916. https://doi.org/10.3390/ijms19102916.

Woods, D. P., & Amasino, R. M. (2015). Dissecting the control of flowering time in grasses using Brachypodium distachyon. In *Genetics and Genomics of Brachypodium* (pp. 259–273). Cham: Springer International Publishing. https://doi.org/10.1007/7397_2015_10.

Woods, D. P., Bednarek, R., Bouché, F., Gordon, S. P., Vogel, J. P., Garvin, D. F., et al. (2017). Genetic architecture of flowering-time variation in Brachypodium distachyon. *Plant Physiology*, *173*, 269–279. https://doi.org/10.1104/pp.16.01178.

Woods, D., Dong, Y., Bouche, F., Bednarek, R., Rowe, M., Ream, T., et al. (2019). A florigen paralog is required for short-day vernalization in a pooid grass. *eLife*, *8*, 27. https://doi.org/10.7554/eLife.42153.

Woods, D. P., Dong, Y., Bouché, F., Mayer, K., Varner, L., Ream, T. S., et al. (2020). Mutations in the predicted DNA polymerase subunit POLD3 result in more rapid flowering of Brachypodium distachyon. *The New Phytologist*, *227*, 1725–1735. https://doi.org/10.1111/nph.16546.

Woods, D. P., Hope, C. L., & Malcomber, S. T. (2011). Phylogenomic analyses of the BARREN STALK1/LAX PANICLE1 (BA1/LAX1) genes and evidence for their roles during axillary meristem development. *Molecular Biology and Evolution*, *28*, 2147–2159. https://doi.org/10.1093/molbev/msr036.

Woods, D. P., McKeown, M. A., Dong, Y., Preston, J. C., & Amasino, R. M. (2016). Evolution of VRN2/Ghd7-like genes in vernalization-mediated repression of grass flowering. *Plant Physiology*, *170*, 2124–2135. https://doi.org/10.1104/pp.15.01279.

Woods, D. P., Ream, T. S., & Amasino, R. M. (2014). Memory of the vernalized state in plants including the model grass Brachypodium distachyon. *Frontiers in Plant Science*, *5*, 99. https://doi.org/10.3389/fpls.2014.00099.

Woods, D. P., Ream, T. S., Bouché, F., Lee, J., Thrower, N., Wilkerson, C., et al. (2017). Establishment of a vernalization requirement in Brachypodium distachyon requires REPRESSOR OF VERNALIZATION1. *Proceedings of the National Academy of Sciences of the United States of America*, *114*, 6623–6628. https://doi.org/10.1073/pnas.1700536114.

Woods, D. P., Ream, T. S., Minevich, G., Hobert, O., & Amasino, R. M. (2014). PHYTOCHROME C is an essential light receptor for photoperiodic flowering in the temperate grass, Brachypodium distachyon. *Genetics*, *198*, 397–408. https://doi.org/10.1534/genetics.114.166785.

Wu, Z., Chen, L., Yu, Q., Zhou, W., Gou, X., Li, J., et al. (2019). Multiple transcriptional factors control stomata development in rice. *The New Phytologist, 223*, 220–232. https://doi.org/10.1111/nph.15766.

Wu, L., Liu, D., Wu, J., Zhang, R., Qin, Z., Liu, D., et al. (2013). Regulation of FLOWERING LOCUS T by a microRNA in Brachypodium distachyon. *Plant Cell, 25*, 4363–4377.

Yan, L., Fu, D., Li, C., Blechl, A., Tranquilli, G., Bonafede, M., et al. (2006). The wheat and barley vernalization gene VRN3 is an orthologue of FT. *Proceedings of the National Academy of Sciences of the United States of America, 103*, 19581–19586.

Yan, L., Loukoianov, A., Blechl, A., Tranquilli, G., Ramakrishna, W., SanMiguel, P., et al. (2004). The wheat VRN2 gene is a flowering repressor down-regulated by vernalization. *Science, 303*, 1640–1644.

Yan, L., Loukoianov, A., Tranquilli, G., Helguera, M., Fahima, T., & Dubcovsky, J. (2003). Positional cloning of the wheat vernalization gene VRN1. *Proceedings of the National Academy of Sciences of the United States of America, 100*, 6263–6268.

Yang, N., & Xu, R.-M. (2013). Structure and function of the BAH domain in chromatin biology. *Critical Reviews in Biochemistry and Molecular Biology, 48*, 211–221.

Yu, Y., Hu, H., Doust, A. N., & Kellogg, E. A. (2020). Divergent gene expression networks underlie morphological diversity of abscission zones in grasses. *The New Phytologist, 225*, 1799–1815. https://doi.org/10.1111/nph.16087.

Zhang, C., Simpson, R. J., Kim, C. M., Warthmann, N., Delhaize, E., Dolan, L., et al. (2018). Do longer root hairs improve phosphorus uptake? Testing the hypothesis with transgenic Brachypodium distachyon lines overexpressing endogenous RSL genes. *The New Phytologist, 217*, 1654–1666. https://doi.org/10.1111/nph.14980.

Zhang, Y., Wang, P., Shao, W., Zhu, J.-K., & Dong, J. (2015). The BASL polarity protein controls a MAPK signaling feedback loop in asymmetric cell division. *Developmental Cell, 33*, 136–149.

Zhao, H., Ma, T., Wang, X., Deng, Y., Ma, H., Zhang, R., et al. (2015). OsAUX1 controls lateral root initiation in rice (Oryza sativa L.). *Plant, Cell & Environment, 38*, 2208–2222. https://doi.org/10.1111/pce.12467.

Zhong, J., Robbett, M., Poire, A., & Preston, J. C. (2017). Successive evolutionary steps drove Pooideae grasses from tropical to temperate regions. *The New Phytologist, 217*, 925–938.

CHAPTER THREE

The history of *Salpingoeca rosetta* as a model for reconstructing animal origins

David S. Booth[a,*] and Nicole King[b,*]

[a]Chan Zuckerberg Biohub and Department of Biochemistry and Biophysics, University of California, San Francisco, CA, United States
[b]Howard Hughes Medical Institute and Department of Molecular and Cell Biology, University of California, Berkeley, CA, United States
*Corresponding authors: e-mail address: David.Booth@ucsf.edu; nking@berkeley.edu

Contents

Abstract

Choanoflagellates, the closest living relatives of animals, have the potential to reveal the genetic and cell biological foundations of complex multicellular development in animals. Here we describe the history of research on the choanoflagellate *Salpingoeca rosetta*. From its original isolation in 2000 to the establishment of CRISPR-mediated genome editing in 2020, *S. rosetta* provides an instructive case study in the establishment of a new model organism.

1. Introduction

By virtue of their close relationship with animals (Fig. 1A) and the many animal-like features of their cell physiology and genomes, choanoflagellates have great potential as experimental models for reconstructing animal origins and core mechanisms underlying animal cell biology. Interest in choanoflagellates began following their first observations in the mid-to-late nineteenth century (Leadbeater, 2015). Early microscopists documented

Current Topics in Developmental Biology, Volume 147
ISSN 0070-2153
https://doi.org/10.1016/bs.ctdb.2022.01.001

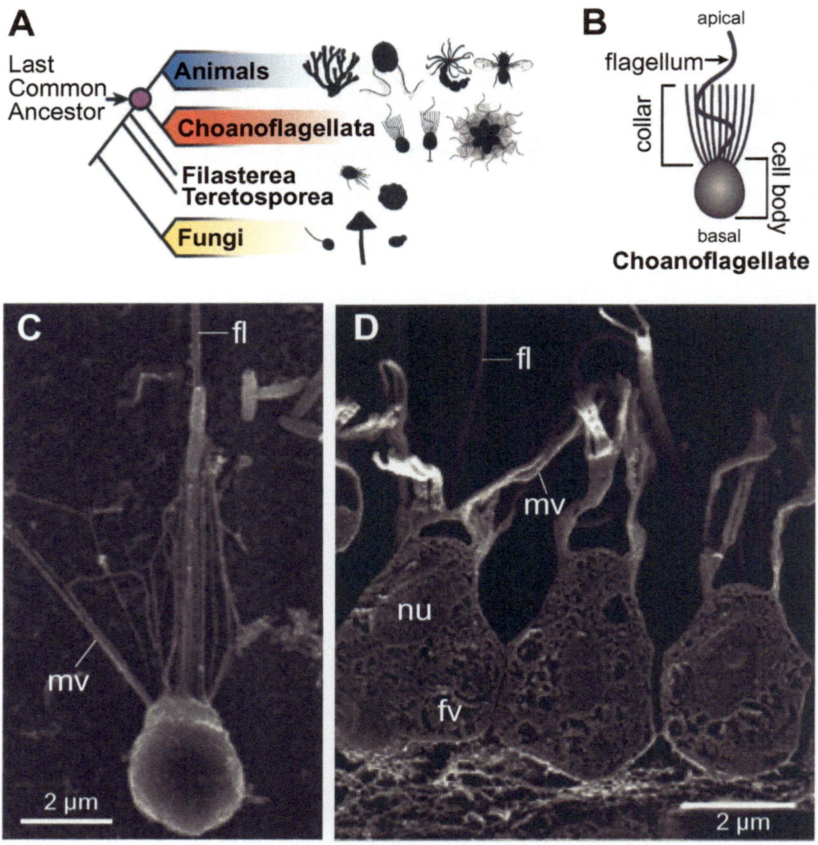

Fig. 1 Choanoflagellates and their closest living relatives, the animals, possess cells with a collar complex. (A) Choanoflagellates are the closest living relatives of animals and diverged from their last common ancestor over 600 million years ago (Parfrey, Lahr, Knoll, & Katz, 2011). Fungi, the closest living group of organisms with complex multicellularity, last shared a common ancestor with animals and choanoflagellates ~1.5 billion years ago. (B) A schematic of a choanoflagellate cell highlights their distinctive trait: the collar complex. At the apical pole of the cell, the collar forms from actin-filled microvilli encircling the flagellum. (C) A scanning electron micrograph of *S. rosetta* (from Dayel et al., 2011), showing the apical flagellum (fl) surrounded by the collar of microvilli (mv). (D) Scanning electron micrograph of a cross-section through sponge choanocytes (Leys & Hill, 2012) showing the collar complex: the apical flagellum (fl) surrounded by microvilli (mv). Also indicated are the nucleus (nu) and food vacuole (fv).

the fascinating collar complex, a diagnostic feature of choanoflagellates in which a picket fence of actin-filled microvilli encircles an apical flagellum (Fig. 1B and C; Karpov & Leadbeater, 1998). Cells in the feeding chambers of sponges (choanocytes; Fig. 1D) also contain a collar complex and these similarities in cellular architecture led Elie Metchnikoff in 1886 to hypothesize that these features were homologous and that animals and choanoflagellates shared a recent common ancestor (Brunet & King, 2021; Metchnikoff, 1886). The collar complex is now recognized as a widespread feature of animal cell biology, consistent with the results from multiple independent phylogenetic and phylogenomic comparisons that firmly established choanoflagellates as the sister group of animals (Fig. 1A; Wainright, Hinkle, Sogin, & Stickel, 1993; Lang, O'Kelly, Nerad, Gray, & Burger, 2002; Burger, Forget, Zhu, Gray, & Lang, 2003; Philippe et al., 2004; Ruiz-Trillo, Roger, Burger, Gray, & Lang, 2008). Subsequent forays into comparative genomics and transcriptomics revealed that many mediators of intercellular signaling, adhesion, cell differentiation, extracellular matrix and immunity in animals are also encoded in and transcribed from choanoflagellate genomes (Fairclough et al., 2013; King et al., 2008; King & Carroll, 2001; King, Hittinger, & Carroll, 2003; López-Escardó et al., 2019; Richter, Fozouni, Eisen, & King, 2018; Snell et al., 2006). These shared cellular and genetic traits suggested that mechanistic studies in choanoflagellates could prove valuable for core features of animal cell and molecular biology.

Nonetheless, the value of choanoflagellates to molecular and cell biologists was historically limited by the absence of key molecular genetic approaches that would allow mechanistic investigation of the functions of conserved genes. Moreover, understanding the transition between unicellular and colonial (i.e. multicellular) life history stages in choanoflagellates was poorly understood because of the absence of multicellular development in laboratory cultures.

Here we describe the history of molecular genetic approaches in choanoflagellates and the establishment of the colony-forming choanoflagellate *Salpingoeca rosetta* as a simple model for investigating the ancestry of animal gene function and cell biology. We focus on rosette colony development to show how discoveries from the study of this process not only revealed the biology of *S. rosetta* but also provided crucial insights that led to the establishment of genetic tools. We emphasize that a deep appreciation of choanoflagellate biology gained through the application of (relatively) inexpensive and general tools provided foundational knowledge required to establish methods for gene delivery, transgene expression, and genome editing.

2. *S. rosetta* growth and life history in the laboratory

When one of us (NK) began to study choanoflagellates in 2000, choanoflagellates were known to exist in diverse unicellular and colonial forms in nature (Leadbeater, 1983, 2015), but there were no readily available colony-forming species in culture. Fortunately, a visit with Dr. Tom Nerad at the ATCC inspired him to isolate a colony-forming choanoflagellate from brackish water samples collected near Hog Island, Virginia. This species, initially named *Proterospongia* sp. 50818 (Carr, Leadbeater, Hassan, Nelson, & Baldauf, 2008; King et al., 2003; Ruiz-Trillo et al., 2007) and later renamed *Salpingoeca rosetta* (Fairclough, Dayel, & King, 2010), was isolated as a single rosette colony (Fig. 2A) that was then propagated in cereal grass media with environmental bacteria (as food) to establish a stable culture.

Within a small number of generations in the laboratory, *S. rosetta* cultures transitioned into a stably unicellular state (Fig. 2B) and the rosette colonies that were apparent in the first environmental samples were rarely detected, preventing their study. Diverse environmental perturbations, including changes in media content, temperature, pH, and salinity, failed to re-stimulate rosette formation. However, a serendipitous observation made during the preparation of *S. rosetta* for genome sequencing revealed how rosette development could be induced (Fairclough et al., 2013).

Like all other choanoflagellates, *S. rosetta* must be co-cultured with bacteria as a food source. The diversity and abundance of bacteria in choanoflagellate cultures can interfere with eukaryotic genome sequencing and assembly, so then-undergraduate Rick Zuzow treated the cultures with different cocktails of antibiotics (Fairclough et al., 2010). One antibiotic cocktail blocked all rosette development while another unexpectedly yielded a culture in which rosettes developed readily and reproducibly. Experiments with these cultures performed by then-post-doc Rosie Alegado and members of the laboratory of Prof. Jon Clardy (Harvard Medical School) revealed that a key factor in rosette induction was the presence of a previously-uncharacterized bacterial species, which we named *Algoriphagus machipongonensis* (Alegado et al., 2012, 2013). Addition of *A. machipongonensis* to a culture of single cells was sufficient to induce rosette development through serial rounds of cell division.

Subsequent improvements to *S. rosetta* culture conditions, including co-culture with the feeder bacterium *Echinicola pacifica* and the development of new semi-defined media formulations (Levin & King, 2013; Woznica et al., 2016), increased the efficiency of rosette induction and later aided the establishment of transfection and genome editing methods (Booth & King, 2020; Booth, Szmidt-Middleton, & King, 2018). These advances provided the first evidence of bacterial regulation of choanoflagellate cell state transitions and provided a means by which to control rosette development in the laboratory.

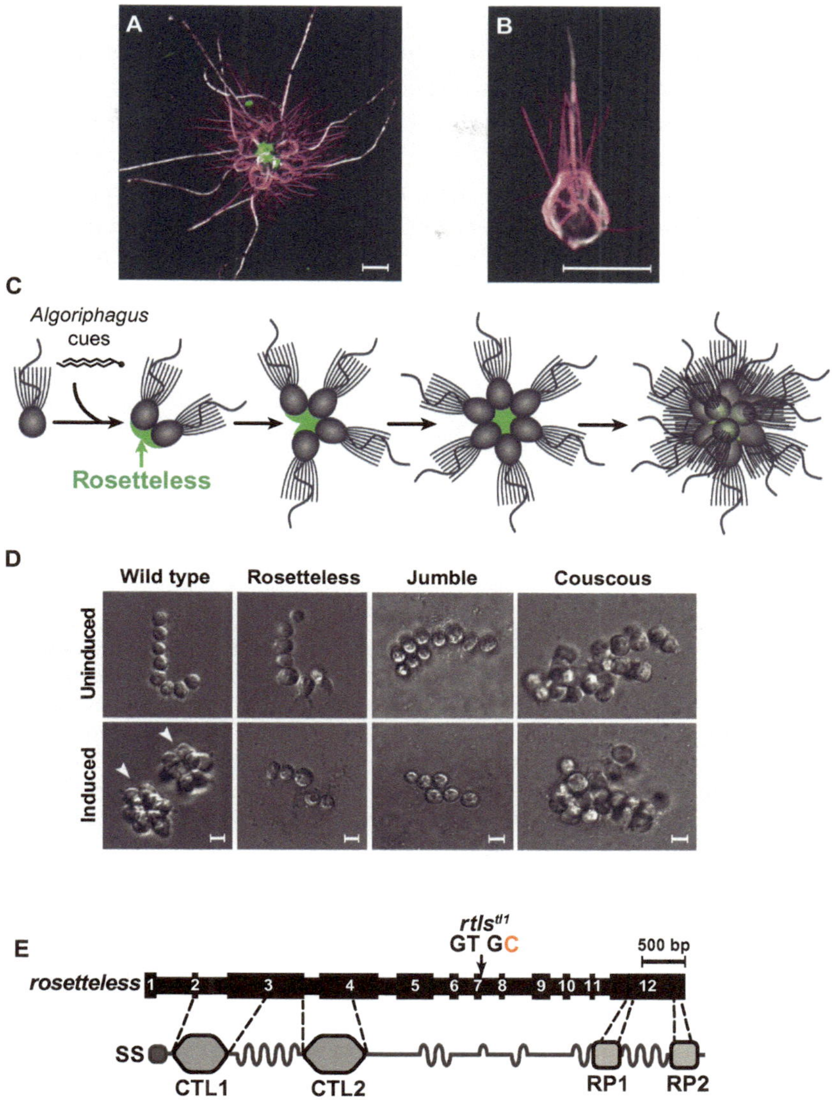

Fig. 2 Environmental cues and a C-type lectin together regulate multicellular development in *S. rosetta*. The dynamic life history of *S. rosetta* includes a multicellular "rosette" stage (A) and a single celled "slow swimmer" stage (B). The phalloidin-stained microvilli are highlighted in magenta, microtubules in white, and the extracellular matrix protein Rosetteless in green. (C) The transition from the slow swimmer stage to the development of the multicellular rosette proceeds through multiple rounds of incomplete cytokinesis (Fairclough et al., 2010). Thus, every rosette is clonal. Rosette development can

(Continued)

Multicellular development was not the only *S. rosetta* life history transition regulated by environmental cues. In a separate line of research, then-graduate student Tera Levin found that prolonged starvation (i.e., over 2 weeks) induced mating in *S. rosetta* (Levin & King, 2013). Four years later, then-graduate student Arielle Woznica found that treatment of *S. rosetta* cultures with the bacterium *Vibrio fischeri* induced *S. rosetta* to swarm and then mate, in this case within an hour after treatment (Woznica, Gerdt, Hulett, Clardy, & King, 2017). These findings demonstrated the essentiality of bacterial cues to the regulation of choanoflagellate biology and provided two independent methods for inducing mating, which later proved critical for the establishment of genetic screens. This period of *S. rosetta* research demonstrated that simple alterations to growth conditions, including altering nutrient content or changing the bacterial composition, could reveal novel biology and provide levers for controlling life history transitions in the laboratory.

Fig. 2—Cont'd be induced by secreted cues from *Algoriphagus* bacteria (Alegado et al., 2012; Woznica et al., 2016) and requires the activity of the C-type lectin protein Rosetteless (*green*; Levin, Greaney, Wetzel, & King, 2014), that contributes to the extracellular matrix at the basal poles of cells in rosettes. (D) The Rosetteless mutant was uncovered, along with rosette-defective mutants Jumble and Couscous (previously called Branched), as part of a forward genetic screen. Wild type cells typically exist as single cells or simple chains in the absence of inducing factors produced by *Algoriphagus*; whereas, the presence of *Algoriphagus* cues induces the development of compact and robust rosettes (Alegado et al., 2012; Levin et al., 2014). In contrast, Rosetteless mutants never develop into rosettes, even in the presence of *Algoriphagus* conditioned medium (Levin et al., 2014). Jumble and Couscous (previously called Branched in Levin et al., 2014) mutant cells never develop into rosettes, and instead aggregate into large amorphous clumps of cells (Wetzel et al., 2018). (E) The 12-exon *rosetteless* gene encodes a protein with an N-terminal signal sequence (SS), two C-type lectin domains (CTL1 and 2) and a pair of repeats (RP1 and 2) near the C-terminus. The mutant allele of *rosetteless* ($rtls^{tl1}$) in the original Rosetteless mutant contained a GT to GC transition at the splice junction between exon 7 and intron 7, which disrupted splicing and protein expression. All scale bars are 5 μm. *(Panels A and E) Adapted from Booth, D. S., & King, N. (2020). Genome editing enables reverse genetics of multicellular development in the choanoflagellate* Salpingoeca rosetta. eLife, 9, *e56193. doi:10.7554/eLife.56193, (Panel B) Adapted from Booth, D. S., Szmidt-Middleton, H., & King, N. (2018). Choanoflagellate transfection illuminates their cell biology and the ancestry of animal septins.* Molecular Biology of the Cell *mbcE18080514. doi:10.1091/mbc.E18-08-0514, and (Panel D) From Levin, T. C., Greaney, A. J., Wetzel, L., King. N. (2014). The rosetteless gene controls development in the choanoflagellate* S. rosetta. eLife, 3. *doi:10.7554/eLife.04070.*

3. Forward genetics and the hunt for rosette defective mutants

Although rosette development in *S. rosetta* could now be readily induced by treatment with *A. machipongonensis*, nothing was known about the genes that regulated the process. The gold standard for discovering the genes underlying a biological process is to perform an unbiased mutant screen, but such screens faced multiple barriers in choanoflagellates. Methods for mutagenesis had not been established (and there were reasons to think that it might be challenging), no genetic markers had yet been identified and the cells are grown in liquid media, meaning mutants could not be isolated, propagated, and phenotyped on plates.

The same Tera Levin who previously uncovered a method for inducing mating (Levin & King, 2013; see above) systematically overcame each of these barriers, allowing the establishment of forward genetics in *S. rosetta* (Levin et al., 2014). She first developed methods for mutagenesis of *S. rosetta* with both EMS and X-rays and identified effective doses using "death curves." Next, she established a phenotypically wild type mapping strain containing over 39,000 genetic variants distributed throughout the genome. The mapping strain was clonally propagated from an isolated cell that had been exposed to high doses of mutagen, found to have no obvious mutant phenotypes (e.g., growth defects or loss of rosette development), and then sequenced to identify genetic variants across the genome that could be used for mapping mutations through crosses.

These tools set up the stage for the first genetic screen in a choanoflagellate. Over 37,000 clones were isolated by serial dilution and screened by eye to identify 16 mutants that fell into seven phenotypic classes. One mutant, Rosetteless, was of particular interest; while it lacked the ability to form rosettes, all other aspects of its life history and cell biology seemed to be wild type. By crossing the Rosetteless mutant to the mapping strain, Levin identified a genetic lesion that co-segregated with the loss of rosette development. The lesion fell in a C-type lectin gene that we named "*rosetteless*" after its phenotype. Rosetteless protein is secreted into the extracellular matrix (ECM) at the basal ends of cells in rosettes (Fig. 2A and C), and the mutant allele abrogates Rosetteless protein expression, resulting in cells that grow normally except for their inability to form rosettes (Fig. 2D and C; Levin et al., 2014).

In two other rosette-defect mutants, Jumble and Couscous, single cells aggregated into large amorphous clumps and true rosettes were never observed to develop (Wetzel et al., 2018). Then-graduate student Laura Wetzel used crosses between these mutants and the mapping strain, coupled with bulk segregant analysis, to identify genetic lesions in two different glycosyltransferases that co-segregated with the clumping phenotype. Using new transgenic approaches (described in more detail below), Wetzel corroborated the mapping cross results and demonstrated that Jumble localizes to the Golgi, whereas Couscous localizes to puncta that may represent the endoplasmic reticulum.

Taken together, these studies demonstrated the efficacy of forward genetics for identifying genes required for *S. rosetta* rosette development. Moreover, the unbiased genetics approach emphasized the importance of the extracellular matrix during the development of rosettes. The criticality of the ECM for rosette development was shown independently through a recent study of biophysical constraints on rosette morphogenesis (Larson et al., 2020). Finally, the connection of genotype to phenotype in the context of the Rosetteless mutant provided a toehold against which different genome editing approaches could later be tested.

4. Gene delivery and genome editing in *S. rosetta*

While genomic, biochemical, and forward genetic approaches provided key early insights into rosette development, the absence of transgenic approaches in *S. rosetta* limited its experimental tractability and potential relevance as a laboratory model. From its isolation in 2000 until 2018, repeated efforts to establish transgenics failed, in large part because the field lacked methods for robustly delivering macromolecules into *S. rosetta* cells. Choanoflagellate cells are too small (~5 μm) for microinjection and standard electroporation methods had thus far failed, both because of the necessity of culturing *S. rosetta* in sea water and because we lacked a positive control for transfection. Key advances in our understanding of *S. rosetta's* genetics and cell physiology that arose through our basic research into rosette development proved to be critical for overcoming these barriers.

In many single-celled organisms, electroporation with nucleic acids or proteins is physically blocked by an organic layer (e.g., a cell wall or extracellular matrix) surrounding the cell (Benatuil, Perez, Belk, & Hsieh, 2010; Kawai, Hashimoto, & Murata, 2010; Meilhoc, Masson, & Teissié, 1990; Shimogawara, Fujiwara, Grossman, & Usuda, 1998; Thompson, Register, Curotto, Kurtz, & Kelly, 1998). However, because choanoflagellates lack a cell wall and prior electron microscopy studies of unicellular life stages suggested that they might be "naked," we admittedly paid little heed to the extracellular matrix in our early forays into establishing transgenics. This changed with the discovery of *rosetteless* (Levin et al., 2014) and improvements in electron microscopy that revealed a layer of extracellular matrix encapsulating *S. rosetta* cells (Fig. 3A; Dayel et al., 2011). Following lessons learned in other microeukaryotes, we therefore developed a method (Fig. 3B) to prime *S. rosetta* for transfection by bathing single cells in a cocktail of papain (a protease) and specific salts that gently removes extracellular proteins (dx.doi.org/10.17504/protocols.io.h68b9hw), leaving the cell membrane vulnerable to pore-forming electrical pulses delivered through electroporation, but healthy enough for the pores to seal and the cells to recover.

The next step was to identify the best buffering conditions and electrical pulses to create pores in the *S. rosetta* membrane. We chose to use nucleofection, a specialized form of electroporation, as it has proven effective for delivering DNA and protein into difficult-to-transfect cells (Gresch et al., 2004; Hamm, Krott, Breibach, Blindt, & Bosserhoff, 2002). Although the exact details of nucleofection remain proprietary, an excellent review of the biophysical principles that underlie a variety of gene delivery methods can be found in (Stewart, Langer, & Jensen, 2018). A major strength of nucleofection, particularly when used in conjunction with a plate-based platform, is that the combination of pulses and diverse transfection buffers creates a large parameter space to explore for testing a wide range of nucleofection conditions. By first removing the extracellular matrix and then applying the most effective nucleofection conditions, we achieved transfection efficiencies of 1%. (An important downside for nucleofection is the expense of the machine itself and the proprietary buffers. Therefore, we have recently established a second transfection protocol for *S. rosetta* using an inexpensive, square-pulse electroporator (Fig. 3C; dx.doi.org/10.17504/protocols.io.bx5vpq66). We hope that these initial parameters will make transfection accessible to a broader community of choanoflagellate researchers and serve as the start of further optimization.)

Our second big challenge was the detection of successful gene delivery events. Until everything worked—gene delivery, gene expression, and detection—we would not know if any part of the transfection protocol was working and therefore were operating in the dark. To improve our chances of success, we cloned promoters and regulatory elements from highly-expressed *S. rosetta* genes that lacked introns and replaced their open reading frames with genes, e.g., *luciferase* or fluorescent proteins, that would

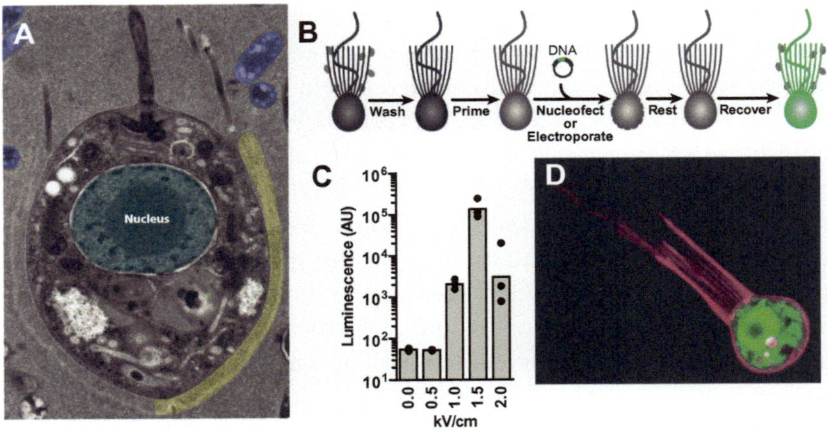

E

Strain	Locus (supercontig:nucleotide)	
	rtls exon 4 (8:429,354)	*rtls* intron 7 (8:427,804)
Wild Type	GAT------------------TGACGG	TCGTGAG
$rtls^{tl1}$	GAT------------------TGACGG	TCGCGAG
$rtls^{PTS1}rpl36a^{P56Q}$	GATTTTATTTAATTAAATAAATGACGG	TCGTGAG

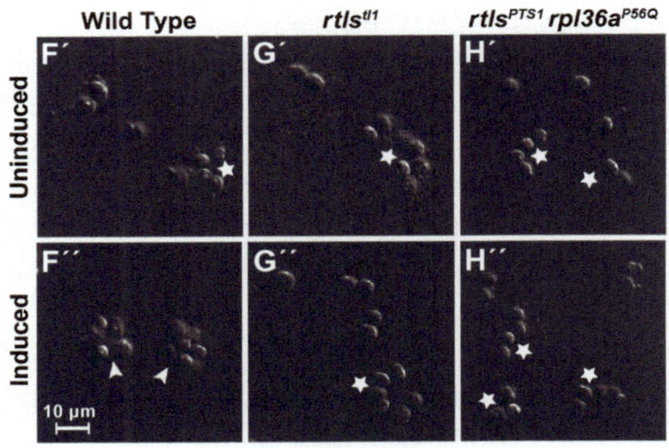

Fig. 3 Establishment of transfection and genome editing in *S. rosetta.* (A) Transmission electron microscopy reveals a layer of extracellular matrix (ECM; highlighted in *yellow* on one half of the cell, leaving the ECM on the other half unlabeled for easier viewing). Also labeled with pseudocolor: nucleus (*teal*) and representative bacteria (*blue*). (B) A step-wise procedure to deliver plasmids and other macromolecules into *S. rosetta* (Booth et al., 2018). Cells are prepared for gene delivery by washing away feeder bacteria—through centrifugation—from *S. rosetta* that has been grown to its mid-logarithmic phase. Afterward, *S. rosetta* is primed for gene delivery by removing a proteinaceous ECM with a buffer comprised of a chaotropic cation (Lithium), a gentle chelator (Citrate), a reducing agent (Cysteine), and a gentle protease (Papain).

The primed cells and cargoes (expression plasmids or gene editing materials) are then combined in a nucleofection or electroporation buffer. After applying an electrical pulse to deliver cargoes inside of *S. rosetta*, a rest buffer, which has low ionic strength and high osmolarity, is added to cells to promote the closure of holes that persist in membranes long after the application of the electrical pulse. Finally, cells are transferred to growth media to recover before being assayed for gene activity or cultured further to establish transgenic strains. An optimized protocol for transfection was developed using a nucleofector and proprietary buffers (Booth & King, 2020 and https://dx.doi.org/10.17504/protocols.io.bx58pq9w). (C) More recently we have established a transformation protocol using an inexpensive, square-pulse electroporator (https://dx.doi.org/10.17504/protocols.io.bx5vpq66), which can be used to deliver plasmids for exogenous gene expression in *S. rosetta*. By electroporating cells with an expression plasmid pEFL-*nanoluc* (Addgene NK606), containing the EFL promoter from *S. rosetta* fused in-frame to the *nanoluc* luciferase gene and varying the field strength (kV/cm), we found that voltages between 1 and 1.5 kV/cm transfected cells most effectively and yielded the highest nanoluc activities. Each dot represents one of three independent transfections, and the gray bar indicates the average of those experiments. Because the nanoluc activity in one electroporation reaction uses 10-fold more material to achieve a similar level of activity as one nucleofection reaction, we estimate that these electroporation conditions transfect cells with 10-fold lower efficiency than nucleofection. (D) Transfection of *S. rosetta* with plasmids expressing fluorescent proteins can be used to highlight cellular architecture in live cells. The collar and cell membranes here are visualized through the expression of mCherry fused to an *S. rosetta* geranyl-geranylation sequence (magenta; Addgene NK644) and the cell body, illuminated by the expression of cytoplasmic mTFP (green; Addgene NK591). (E and F) Genome editing of the *rosetteless* gene recapitulates the phenotype of the previously isolated rosetteless mutant. (E) Genotypes of *S. rosetta* strains show two distinct alleles in *rosetteless* that were either isolated from a forward genetic screen (*rtls*^{tl}; Levin et al., 2014) or engineered through CRISPR-Cas9 mediated genome editing (*rtls*^{PTS1} *rpl36a*^{P56Q}; Booth & King, 2020). The *rtls*^{tl1} allele is a T to C transition in the 5′-splice site of intron 7 that abrogates the expression of *rosetteless* through aberrant splicing (also see Fig. 2). The *rtls*^{PTS1} *rpl36a*^{P56Q} strain was engineered to bear a cycloheximide resistance allele (*rpl36*^{P56Q}) and a *rosetteless* allele with a palindromic premature termination sequence (PTS: TTTATTTAATTAAATAAA) inserted into the fourth exon that stops the transcription and translation of *rosetteless* products. (F) Independent mutations show the necessity of *rosetteless* for multicellular development in *S. rosetta*. In the absence of inducing cues from *Algoriphagus*, all strains (F′-H′) grew as chains (stars) or single cells. Upon the addition of inducing cues, only the wild-type (F″) strain formed rosettes (arrowheads) while strains with *rtls*^{tl1} (E″) or *rtls*^{PTS1} *rpl36*^{P56Q} (F″) alleles failed to form rosettes. *(Panel A) Courtesy of Kent McDonald and adapted from Booth, D. S., Szmidt-Middleton, H., & King, N. (2018). Choanoflagellate transfection illuminates their cell biology and the ancestry of animal septins.* Molecular Biology of the Cell, *mbcE18080514. doi:10.1091/mbc.E18-08-0514; (Panels E–H) Adapted from Booth, D. S., & King, N. (2020). Genome editing enables reverse genetics of multicellular development in the choanoflagellate* Salpingoeca rosetta. eLife, 9, *e56193. doi:10.7554/eLife.56193.*

report successful transfection. Although genes that confer resistance to a toxic drug can aid in selection for rare drug-resistant transformants, we were discouraged by the variable and unpredictable responses of *S. rosetta* cultures to drug treatments. Fluorescent protein expression in *S. rosetta* also presented problems because, as we later learned, green fluorescent protein (GFP) and its derivatives did not fold properly in *S. rosetta*.

Therefore, we settled on a reporter gene that encodes a small, sensitive, and stable protein, Nanoluc (Hall et al., 2012), to detect expression from transfected plasmids. In other experimental models (Vinayak et al., 2015), the use of *nanoluc* provided exceptional signal to noise, allowing clear distinction between successful and unsuccessful transfection experiments (Fig. 3C). Moreover, the small size of *nanoluc* (a 519 bp open reading frame that encodes a ~19 kDa protein), helped reduce the overall size of our transgenic constructs. After testing a broad array of transfection parameters with *nanoluc* constructs, we identified a small combination of nucleofection buffers and pulse programs that yielded luciferase activity. We subsequently optimized around these parameters to develop a robust and reproducible transfection protocol (dx.doi.org/10.17504/protocols.io.h68b9hw).

The establishment of this protocol almost immediately transformed the study of choanoflagellates. The expression of fluorescent markers brought the cellular architecture of *S. rosetta* into living color in a way not possible previously (Fig. 3D). The ability to express and perform live imaging of fluorescently tagged proteins also helped to validate the mapping of the *jumble* and *couscous* glycosyltransferases (Wetzel et al., 2018), revealed that septins localize to the basal pole of *S. rosetta* cells (Booth et al., 2018), and revealed the cytoskeletal dynamics that accompany the transition from flagellated swimming cells to crawling amoeboid cells upon confinement (Brunet et al., 2021). In addition, this protocol served as the foundation for transfection of the choanoflagellate *Monosiga brevicollis* (Woznica et al., 2021). Finally, and most importantly for the development of *S. rosetta* as an experimental model, it opened the possibility for delivering the machinery for genome editing.

CRISPR/Cas9-mediated genome editing has become a robust method for performing reverse genetics in established and emerging model systems. An easily designed synthetic guide RNA (gRNA) associates with the Cas9 endonuclease, and together they drive cleavage at specific targeted sequences in the genome (Cong et al., 2013; Jinek et al., 2012; Jinek et al., 2013). After the Cas9 endonuclease introduces double-stranded breaks in the genome, the incorporation of mutations ultimately relies on errors introduced during

DNA repair (Bibikova et al., 2001; Choulika, Perrin, Dujon, & Nicolas, 1995; Cong et al., 2013; Jinek et al., 2013; Rouet, Smih, & Jasin, 1994). The nature of those errors depends on the type of DNA repair mechanism implemented by an organism—either homology-directed repair (HDR) in which repair of the double stranded break incorporates mutations from a DNA template that shares homology with the region surrounding the cut site or non-homologous end-joining (NHEJ) in which repair of the double stranded break can lead to untemplated insertions or deletions as the DNA ends rejoin. Thus, a snag for implementing Cas9-mediated genome editing in emerging model systems can be the paucity of information on the DNA repair mechanisms, which can widely vary between organisms and cell types (Yeh, Richardson, & Corn, 2019). For example, *S. rosetta* preferentially incorporates mutations from HDR templates at a frequency 10-fold higher than untemplated insertions and deletions (Booth & King, 2020). In contrast, *M. brevicollis* seems to prefer the incorporation of untemplated mutations, as evidenced by the incorporation of insertions and deletions in a targeted gene despite the co-delivery of HDR templates (Woznica et al., 2021).

To establish genome editing in *S. rosetta*, we first focused on converting the ribosomal protein gene *rpl36a* to $rpl36a^{P56Q}$, a cycloheximide resistance allele (Dehoux, Davies, & Cannon, 1993; Kawai et al., 1992; Kim et al., 1998; Kondo, Saito, Kajiwara, Takagi, & Misawa, 1995; Stevens, Atteia, Franzén, & Purton, 2001). This relatively simple edit allowed us to select edited cells due to their cycloheximide resistance, thereby simplifying one stage of the optimization process. By iterating around the editing of *rpl36a*, we identified key parameters that promoted genome editing success, including the need to use least 20 pmol of Cas9 ribonucleoprotein (RNP) and more than 200 nmol of a single-stranded DNA repair template that had 50 bases of homology flanking the desired mutation (Booth & King, 2020).

Afterward, we targeted *rosetteless*, both because a successful edit should yield a predictable phenotype (loss of rosette development) and, more broadly, to demonstrate the utility of genome editing for investigating gene function in *S. rosetta*. The original *rosetteless* allele was likely a hypomorph whose phenotype manifested through aberrant splicing from a transition mutation in a 5′ splice site. By introducing an independent knockout of *rosetteless* by genome editing, we corroborated the necessity of *rosetteless* for rosette development (Fig. 3E–H). Thus, genome editing in combination with mating crosses are essential tools for the verification of mutations identified in forward genetic screens. Most importantly, genome editing now provides a method to directly test hypotheses about gene function in *S. rosetta*.

The major barrier that remains for genome engineering in *S. rosetta* is the accurate and precise introduction of large DNA fragments. Even with small insertions (~20 bases), the efficiency of genome editing (~1%) requires screening through hundreds of clonal strains to find one that bears a genome edit. Engineering larger insertions, which will likely integrate into the genome with even lower frequencies, will require modifications to improve gene delivery, to increase integration frequencies, and/or to select for stable transgenes. A puromycin resistance gene has already been used for the selection of transgenic cassettes (Brunet et al., 2021; Wetzel et al., 2018; Woznica et al., 2021), but puromycin also affects the feeder bacteria that support choanoflagellate growth, leading to variable selection conditions. The development of more selectable markers that confer resistance to drugs that selectively target eukaryotes would help advance genetics in choanoflagellates and other heterotrophic organisms that require bacterial feeding.

5. The door is now open to study the molecular biology of *S. rosetta* and other relatives of animals

The lessons learned from *S. rosetta* have aided efforts to establish genetics in other emerging microeukaryotic models. Of particular note is the recent development of molecular methods for *M. brevicollis* (Woznica et al., 2021). The establishment of gene delivery and genome editing in *M. brevicollis* helped to uncover potential ancestral functions of the innate immunity gene STING. Although the exact protocols from *S. rosetta* were not readily exported to *M. brevicollis*, variations on the general principles used during the development of *S. rosetta* transfection facilitated the construction of transgenic lines and the generation of a STING knockout with genome editing. This work clearly demonstrates the maturation of choanoflagellates as model systems to investigate core process conserved in both choanoflagellates and animals; the Woznica et al., 2021 study drew on comparative genomics (Richter et al., 2018), reported the discovery of a bacterial pathogen of *M. brevicollis*, and developed methods for molecular genetics to show that STING likely functioned as an innate immunity gene in the last common ancestor of choanoflagellates and animals. Similar advances in the development of genetic methods in other protistan relatives of choanoflagellates and animals (Faktorová et al., 2020; Kożyczkowska et al., 2021; Li, Neundorf, & Nitsche, 2018; Parra-Acero et al., 2018; Suga & Ruiz-Trillo, 2013) now provide a menagerie of microeukaryotes to reconstruct how evolutionary changes prior to the emergence of animals laid the foundations for complex, multicellular development.

We hope that this review serves as an invitation for the "choano curious" to pursue their research in these fascinating organisms. Rosette development is just one dimension of the rich biology that is now amenable to molecular investigation in *S. rosetta*; others include cell differentiation, interspecies interactions, and evolutionary cell biology. As we move forward with a new era of molecular genetics in choanoflagellates, we highlight the important roles that classical methods from biochemistry, cytology, and genetics played in developing modern tools for transforming *S. rosetta*, editing its genome, and illuminating its cell biology (Supplementary Table 1 in the online version at https://doi.org/10.1016/bs.ctdb.2022.01.001).

Acknowledgments

No amount of thanks can properly convey our deep gratitude to the many scientists—those from the King Lab and those in the larger community of holozoan researchers—for the passion, discoveries and insights that fostered the emergence of choanoflagellates as model systems, only a fraction of whom could be highlighted here. It really takes dedication from a community to found new model systems. We also thank the Gordon and Betty Moore Foundation Marine Microbiology and HHMI for their generous support of work that established fundamental biology and genetic methods in choanoflagellates. Finally, we thank Matt Welch and laboratory for use of the square-wave electroporator.

References

Alegado, R. A., Brown, L. W., Cao, S., Dermenjian, R. K., Zuzow, R., Fairclough, S. R., et al. (2012). A bacterial sulfonolipid triggers multicellular development in the closest living relatives of animals. *eLife*, *1*, e00013. https://doi.org/10.7554/eLife.00013.

Alegado, R. A., Grabenstatter, J. D., Zuzow, R., Morris, A., Huang, S. Y., Summons, R. E., et al. (2013). Algoriphagus machipongonensis sp. nov., co-isolated with a colonial choanoflagellate. *International Journal of Systematic and Evolutionary Microbiology*, *63*, 163–168. https://doi.org/10.1099/ijs.0.038646-0.

Benatuil, L., Perez, J. M., Belk, J., & Hsieh, C.-M. (2010). An improved yeast transformation method for the generation of very large human antibody libraries. *Protein Engineering, Design & Selection*, *23*, 155–159. https://doi.org/10.1093/protein/gzq002.

Bibikova, M., Carroll, D., Segal, D. J., Trautman, J. K., Smith, J., Kim, Y.-G., et al. (2001). Stimulation of homologous recombination through targeted cleavage by chimeric nucleases. *Molecular and Cellular Biology*, *21*, 289–297. https://doi.org/10.1128/MCB.21.1.289-297.2001.

Booth, D. S., & King, N. (2020). Genome editing enables reverse genetics of multicellular development in the choanoflagellate *Salpingoeca rosetta*. *eLife*, *9*, e56193. https://doi.org/10.7554/eLife.56193.

Booth, D. S., Szmidt-Middleton, H., & King, N. (2018). Choanoflagellate transfection illuminates their cell biology and the ancestry of animal septins. *Molecular Biology of the Cell*, mbcE18080514. https://doi.org/10.1091/mbc.E18-08-0514.

Brunet, T., Albert, M., Roman, W., Coyle, M. C., Spitzer, D. C., & King, N. (2021). A flagellate-to-amoeboid switch in the closest living relatives of animals. *eLife*, *10*. https://doi.org/10.7554/eLife.61037.

Brunet, T., & King, N. (2021). The single-celled ancestors of animals: A history of hypotheses. In M. Herron, W. Ratcliff, & P. Conlin (Eds.), *The Evolution of Multicellularity* CRC Press. In press.

Burger, G., Forget, L., Zhu, Y., Gray, M. W., & Lang, B. F. (2003). Unique mitochondrial genome architecture in unicellular relatives of animals. *Proceedings of the National Academy of Sciences*, *100*, 892–897. https://doi.org/10.1073/pnas.0336115100.

Carr, M., Leadbeater, B. S. C., Hassan, R., Nelson, M., & Baldauf, S. L. (2008). Molecular phylogeny of choanoflagellates, the sister group to Metazoa. *Proceedings of the National Academy of Sciences of the United States of America*, *105*, 16641–16646. https://doi.org/10.1073/pnas.0801667105.

Choulika, A., Perrin, A., Dujon, B., & Nicolas, J. F. (1995). Induction of homologous recombination in mammalian chromosomes by using the I-SceI system of *Saccharomyces cerevisiae*. *Molecular and Cellular Biology*, *15*, 1968–1973.

Cong, L., Ran, F. A., Cox, D., Lin, S., Barretto, R., Habib, N., et al. (2013). Multiplex genome engineering using CRISPR/Cas systems. *Science*, *339*, 819–823. https://doi.org/10.1126/science.1231143.

Dayel, M. J., Alegado, R. A., Fairclough, S. R., Levin, T. C., Nichols, S. A., McDonald, K., et al. (2011). Cell differentiation and morphogenesis in the colony-forming choanoflagellate *Salpingoeca rosetta*. *Developmental Biology*, *357*, 73–82. https://doi.org/10.1016/j.ydbio.2011.06.003.

Dehoux, P., Davies, J., & Cannon, M. (1993). Natural cycloheximide resistance in yeast. The role of ribosomal protein L41. *European Journal of Biochemistry*, *213*, 841–848.

Fairclough, S. R., Chen, Z., Kramer, E., Zeng, Q., Young, S., Robertson, H. M., et al. (2013). Premetazoan genome evolution and the regulation of cell differentiation in the choanoflagellate *Salpingoeca rosetta*. *Genome Biology*, *14*, R15. https://doi.org/10.1186/gb-2013-14-2-r15.

Fairclough, S. R., Dayel, M. J., & King, N. (2010). Multicellular development in a choanoflagellate. *Current Biology*, *20*, R875–R876. https://doi.org/10.1016/j.cub.2010.09.014.

Faktorová, D., Nisbet, R. E. R., Fernández Robledo, J. A., Casacuberta, E., Sudek, L., Allen, A. E., et al. (2020). Genetic tool development in marine protists: Emerging model organisms for experimental cell biology. *Nature Methods*, *17*, 481–494. https://doi.org/10.1038/s41592-020-0796-x.

Gresch, O., Engel, F. B., Nesic, D., Tran, T. T., England, H. M., Hickman, E. S., et al. (2004). New non-viral method for gene transfer into primary cells. *Methods: Transfection of Mammalian Cells*, *33*, 151–163. https://doi.org/10.1016/j.ymeth.2003.11.009.

Hall, M. P., Unch, J., Binkowski, B. F., Valley, M. P., Butler, B. L., Wood, M. G., et al. (2012). Engineered luciferase reporter from a deep sea shrimp utilizing a novel Imidazopyrazinone substrate. *ACS Chemical Biology*, *7*, 1848–1857. https://doi.org/10.1021/cb3002478.

Hamm, A., Krott, N., Breibach, I., Blindt, R., & Bosserhoff, A. K. (2002). Efficient transfection method for primary cells. *Tissue Engineering*, *8*, 235–245. https://doi.org/10.1089/107632702753725003.

Jinek, M., Chylinski, K., Fonfara, I., Hauer, M., Doudna, J. A., & Charpentier, E. (2012). A programmable dual-RNA-guided DNA endonuclease in adaptive bacterial immunity. *Science*, *337*, 816–821. https://doi.org/10.1126/science.1225829.

Jinek, M., East, A., Cheng, A., Lin, S., Ma, E., & Doudna, J. (2013). RNA-programmed genome editing in human cells. *eLife, 2*, e00471. https://doi.org/10.7554/eLife.00471.
Karpov, S. A., & Leadbeater, B. S. C. (1998). Cytoskeleton structure and composition in choanoflagellates. *The Journal of Eukaryotic Microbiology, 45*, 361–367. https://doi.org/10.1111/j.1550-7408.1998.tb04550.x.
Kawai, S., Hashimoto, W., & Murata, K. (2010). Transformation of *Saccharomyces cerevisiae* and other fungi. *Bioengineering Bugs, 1*, 395–403. https://doi.org/10.4161/bbug.1.6.13257.
Kawai, S., Murao, S., Mochizuki, M., Shibuya, I., Yano, K., & Takagi, M. (1992). Drastic alteration of cycloheximide sensitivity by substitution of one amino acid in the L41 ribosomal protein of yeasts. *Journal of Bacteriology, 174*, 254–262.
Kim, I. G., Nam, S. K., Sohn, J. H., Rhee, S. K., An, G. H., Lee, S. H., et al. (1998). Cloning of the ribosomal protein L41 gene of *Phaffia rhodozyma* and its use a drug resistance marker for transformation. *Applied and Environmental Microbiology, 64*, 1947–1949.
King, N., & Carroll, S. B. (2001). A receptor tyrosine kinase from choanoflagellates: Molecular insights into early animal evolution. *Proceedings of the National Academy of Sciences, 98*, 15032–15037. https://doi.org/10.1073/pnas.261477698.
King, N., Hittinger, C. T., & Carroll, S. B. (2003). Evolution of key cell signaling and adhesion protein families predates animal origins. *Science, 301*, 361–363. https://doi.org/10.1126/science.1083853.
King, N., Westbrook, M. J., Young, S. L., Kuo, A., Abedin, M., Chapman, J., et al. (2008). The genome of the choanoflagellate *Monosiga brevicollis* and the origin of metazoans. *Nature, 451*, 783–788. https://doi.org/10.1038/nature06617.
Kondo, K., Saito, T., Kajiwara, S., Takagi, M., & Misawa, N. (1995). A transformation system for the yeast *Candida utilis*: Use of a modified endogenous ribosomal protein gene as a drug-resistant marker and ribosomal DNA as an integration target for vector DNA. *Journal of Bacteriology, 177*, 7171–7177.
Kożyczkowska, A., Najle, S. R., Ocaña-Pallarès, E., Aresté, C., Shabardina, V., Ara, P. S., et al. (2021). Stable transfection in protist *Corallochytrium limacisporum* identifies novel cellular features among unicellular animals relatives. *Current Biology*. https://doi.org/10.1016/j.cub.2021.06.061.
Lang, B. F., O'Kelly, C., Nerad, T., Gray, M. W., & Burger, G. (2002). The closest unicellular relatives of animals. *Current Biology, 12*, 1773–1778. https://doi.org/10.1016/S0960-9822(02)01187-9.
Larson, B. T., Ruiz-Herrero, T., Lee, S., Kumar, S., Mahadevan, L., & King, N. (2020). Biophysical principles of choanoflagellate self-organization. *Proceedings of the National Academy of Sciences, 117*, 1303–1311. https://doi.org/10.1073/pnas.1909447117.
Leadbeater, B. S. C. (1983). Life-history and ultrastructure of a new marine species of Proterospongia (Choanoflagellida). *Journal of the Marine Biological Association of the United Kingdom, 63*, 135–160. https://doi.org/10.1017/S0025315400049857.
Leadbeater, B. S. C. (2015). *The choanoflagellates: Evolution, biology, and ecology*. Cambridge, United Kingdom: Cambridge University Press.
Levin, T. C., Greaney, A. J., Wetzel, L., & King, N. (2014). The Rosetteless gene controls development in the choanoflagellate S. rosetta. *eLife*, 3. https://doi.org/10.7554/eLife.04070.
Levin, T. C., & King, N. (2013). Evidence for sex and recombination in the Choanoflagellate *Salpingoeca rosetta*. *Current Biology, 23*, 2176–2180. https://doi.org/10.1016/j.cub.2013.08.061.
Leys, S. P., & Hill, A. (2012). Chapter one - the physiology and molecular biology of sponge tissues. In M. A. Becerro, M. J. Uriz, M. Maldonado, & X. Turon (Eds.), *Advances in marine biology, advances in sponge science: Physiology, chemical and microbial diversity, biotechnology* (pp. 1–56). Academic Press. https://doi.org/10.1016/B978-0-12-394283-8.00001-1.

Li, R., Neundorf, I., & Nitsche, F. (2018). First efficient transfection in Choanoflagellates using cell-penetrating peptides. *bioRxiv*, 260190. https://doi.org/10.1101/260190.

López-Escardó, D., Grau-Bové, X., Guillaumet-Adkins, A., Gut, M., Sieracki, M. E., & Ruiz-Trillo, I. (2019). Reconstruction of protein domain evolution using single-cell amplified genomes of uncultured choanoflagellates sheds light on the origin of animals. *Philosophical Transactions of the Royal Society, B: Biological Sciences, 374*, 20190088. https://doi.org/10.1098/rstb.2019.0088.

Meilhoc, E., Masson, J. M., & Teissié, J. (1990). *High efficiency transformation of intact yeast cells by electric field pulses. Vol. 8* (pp. 223–227). Biotechnol Nat Publ Co.

Metchnikoff, E. (1886). *Embryologische Studien an Medusen: Ein Beitrag Zur Genealogie Der Primitiv-Organe*. Vienna: A. Hölder.

Parfrey, L. W., Lahr, D. J. G., Knoll, A. H., & Katz, L. A. (2011). Estimating the timing of early eukaryotic diversification with multigene molecular clocks. *Proceedings of the National Academy of Sciences of the United States of America, 108*, 13624–13629. https://doi.org/10.1073/pnas.1110633108.

Parra-Acero, H., Ros-Rocher, N., Perez-Posada, A., Kożyczkowska, A., Sánchez-Pons, N., Nakata, A., et al. (2018). Transfection of *Capsaspora owczarzaki*, a close unicellular relative of animals. *Development (Cambridge, England), 145*. https://doi.org/10.1242/dev.162107.

Philippe, H., Snell, E. A., Bapteste, E., Lopez, P., Holland, P. W. H., & Casane, D. (2004). Phylogenomics of eukaryotes: Impact of missing data on large alignments. *Molecular Biology and Evolution, 21*, 1740–1752. https://doi.org/10.1093/molbev/msh182.

Richter, D. J., Fozouni, P., Eisen, M. B., & King, N. (2018). Gene family innovation, conservation and loss on the animal stem lineage. *eLife*, 7, e34226. https://doi.org/10.7554/eLife.34226.

Rouet, P., Smih, F., & Jasin, M. (1994). Introduction of double-strand breaks into the genome of mouse cells by expression of a rare-cutting endonuclease. *Molecular and Cellular Biology, 14*, 8096–8106. https://doi.org/10.1128/MCB.14.12.8096.

Ruiz-Trillo, I., Burger, G., Holland, P. W. H., King, N., Lang, B. F., Roger, A. J., et al. (2007). The origins of multicellularity: A multi-taxon genome initiative. *Trends in Genetics: TIG, 23*, 113–118. https://doi.org/10.1016/j.tig.2007.01.005.

Ruiz-Trillo, I., Roger, A. J., Burger, G., Gray, M. W., & Lang, B. F. (2008). A Phylogenomic investigation into the origin of Metazoa. *Molecular Biology and Evolution, 25*, 664–672. https://doi.org/10.1093/molbev/msn006.

Shimogawara, K., Fujiwara, S., Grossman, A., & Usuda, H. (1998). High-efficiency transformation of *Chlamydomonas reinhardtii* by electroporation. *Genetics, 148*, 1821–1828.

Snell, E. A., Brooke, N. M., Taylor, W. R., Casane, D., Philippe, H., & Holland, P. W. H. (2006). An unusual choanoflagellate protein released by hedgehog autocatalytic processing. *Proceedings of the Royal Society B: Biological Sciences, 273*, 401–407. https://doi.org/10.1098/rspb.2005.3263.

Stevens, D. R., Atteia, A., Franzén, L. G., & Purton, S. (2001). Cycloheximide resistance conferred by novel mutations in ribosomal protein L41 of *Chlamydomonas reinhardtii*. *Molecular and General Genetics MGG, 264*, 790–795.

Stewart, M. P., Langer, R., & Jensen, K. F. (2018). Intracellular delivery by membrane disruption: Mechanisms, strategies, and concepts. *Chemical Reviews, 118*, 7409–7531. https://doi.org/10.1021/acs.chemrev.7b00678.

Suga, H., & Ruiz-Trillo, I. (2013). Development of ichthyosporeans sheds light on the origin of metazoan multicellularity. *Developmental Biology, 377*, 284–292. https://doi.org/10.1016/j.ydbio.2013.01.009.

Thompson, J. R., Register, E., Curotto, J., Kurtz, M., & Kelly, R. (1998). An improved protocol for the preparation of yeast cells for transformation by electroporation. *Yeast, 14*, 565–571. https://doi.org/10.1002/(SICI)1097-0061(19980430)14:6<565::AID-YEA251>3.0.CO;2-B.

Vinayak, S., Pawlowic, M. C., Sateriale, A., Brooks, C. F., Studstill, C. J., Bar-Peled, Y., et al. (2015). Genetic modification of the diarrhoeal pathogen *Cryptosporidium parvum*. *Nature*, *523*, 477–480. https://doi.org/10.1038/nature14651.

Wainright, P. O., Hinkle, G., Sogin, M. L., & Stickel, S. K. (1993). Monophyletic origins of the Metazoa: An evolutionary link with fungi. *Science*, *260*, 340–342.

Wetzel, L. A., Levin, T. C., Hulett, R. E., Chan, D., King, G. A., Aldayafleh, R., et al. (2018). Predicted glycosyltransferases promote development and prevent spurious cell clumping in the choanoflagellate *S. rosetta*. *eLife*, *7*. https://doi.org/10.7554/eLife.41482.

Woznica, A., Cantley, A. M., Beemelmanns, C., Freinkman, E., Clardy, J., & King, N. (2016). Bacterial lipids activate, synergize, and inhibit a developmental switch in choanoflagellates. *Proceedings of the National Academy of Sciences of the United States of America*, *113*, 7894–7899. https://doi.org/10.1073/pnas.1605015113.

Woznica, A., Gerdt, J. P., Hulett, R. E., Clardy, J., & King, N. (2017). Mating in the closest living relatives of animals is induced by a bacterial Chondroitinase. *Cell*, *170*, 1175–1183. e11. https://doi.org/10.1016/j.cell.2017.08.005.

Woznica, A., Kumar, A., Sturge, C. R., Xing, C., King, N., & Pfeiffer, J. K. (2021). STING mediates immune responses in the closest living relatives of animals. *eLife*, *10*, e70436. https://doi.org/10.7554/eLife.70436.

Yeh, C. D., Richardson, C. D., & Corn, J. E. (2019). Advances in genome editing through control of DNA repair pathways. *Nature Cell Biology*, *21*, 1468–1478. https://doi.org/10.1038/s41556-019-0425-z.

CHAPTER FOUR

Emerging models: The "development" of the ctenophore *Mnemiopsis leidyi* and the cnidarian *Nematostella vectensis* as useful experimental models

Mark Q. Martindale*
Whitney Lab for Marine Bioscience, University of Florida, St. Augustine, FL, United States
*Corresponding author: e-mail address: mqmartin@whitney.ufl.edu

Contents

Abstract

The goal of this chapter is to explain the reasoning for developing two understudied invertebrate animal species for asking specific biological questions. The first is the ctenophore (comb jelly) *Mnemiopsis leidyi* and the second is the anthozoan cnidarian (starlet sea anemone) *Nematostella vectensis*. Although these two taxa belong to some of the earliest branching extant metazoan clades, their developmental features could hardly be more different from one another. This should serve as a general warning to be careful when extrapolating comparisons of one species to another. Two-taxon comparisons are especially flawed; and to interpret features in a phylogenetic context one must sample carefully within a given taxon to determine how representative certain features are before comparing with other clades. The other benefit of this comparison is to identify key practical factors when attempting to develop new species for experimental investigation.

Current Topics in Developmental Biology, Volume 147
ISSN 0070-2153
https://doi.org/10.1016/bs.ctdb.2022.02.002

1. Introduction

I initially became interested in cell and developmental biology as an undergraduate at New College, a small liberal arts college in Sarasota, Florida where we primarily worked on sea urchin (*Lytechinus variegatus)* and pulmonate snail (*Lymnaea palustris)* embryos. Once I got to graduate school at the University of Texas Austin, (in the "pre-genomic era," 1981–85), the predominant trend of my contemporaries was to focus their research on genetic model systems (reminds me of the old saying, geneticists are to Developmental Biology what surgeons are to Medicine). At that time, genetic systems were essentially limited to *Drosophila* and mice, as *C. elegans* was the new kid on the block and virtually no one outside of the aquarium trade had ever heard of zebrafish. Indeed, I published a paper on the transient metamerism of head paraxial mesoderm, but used Medaka of zebrafish (Martindale, Meier, & Jacobson, 1987). Instead, we were taught to find the right species for the question being asked. This approach requires two important components. First of all, you had to have some good, novel ideas. Second, you had to know a lot (or at least a little) about a variety of different species so that you knew what your options might be. One of the best ways of learning about biodiversity is to work at different marine labs and field stations that give endless opportunities to sample, explore, watch, poke, tug and overall play with different animal species to determine which, if any, have the properties you can exploit for your own research questions. Working on different kinds of embryos also fosters a natural sense of curiosity, provides the satisfaction of collecting, nurturing, spawning, and rearing delicate animals and, if you are willing to work hard and make detailed observations, create opportunities to stumble upon new and unanticipated important questions. There are almost always species that exhibit common biological processes and provide favorable features (optical properties, speed of development, etc.) that can be used if one has the knowledge and the prepared mind.

My graduate advisor at the University of Texas at Austin, Gary Freeman, was well aware of both of these things and provided opportunities for all his students to critically explore idea space and to be exposed to biodiversity. He challenged us to strive to have our research change the way that people think about our field while emphasizing the difference between a significant contribution to Science, and a mere compilation to Science. He pressured us to think about what we wanted to do, our methodological approach and, of course, what species was the most powerful system in which to invest.

In the early 1980s, at the dawn of the field now called EvoDevo, there emerged a renewed interest in exploiting various aspects of comparative developmental studies in a phylogenetic context. For example, there is a large group of animals including mollusks, annelids, nemerteans, and polyclad flatworms that display a highly stereotyped pattern of embryogenesis known as spiral cleavage. The group was a center of attention by comparative embryologists (especially at the Marine Biological Lab in Woods Hole, MA and the Stazione Zoologica de Napoli, Italy) in the late 20th and early 21st century. The temporal and spatial features of their cleavage patterns were described in detail. However, virtually no comparative experimental studies were performed, despite apparent variations among species. In the late 1950–60s, there were a handful of experiments performed on spiralian embryos that made their way into all of the developmental biology textbooks, but all of these centered around species whose first two cleavages were asymmetric, generating embryos with a larger cell at the four-cell stage (called the D blastomere). These species were instrumental in the formulation of ideas about how embryos worked, as manipulative experimentation demonstrated that these differences in their development were associated with the asymmetric localization of developmental potential. For example, equalization of cleavage by the removal of an anuclear cytoplasmic protuberance at the vegetal pole (called the polar lobe) caused the failure of the formation of dorso-ventral polarity and the radialization of the operated embryos (Clement, 1952; Henry & Martindale, 1999). These same species whose embryos undergo unequal cleavage went on to provide many additional insights into the timing of particular cell fate decisions, and the role of descendants of the D cell on inductive cell-cell interactions that form the so-called embryonic organizer (Clement, 1962).

At the time, textbooks gave the impression that all spiral-cleaving embryos would utilize the same mechanism of precocious, cytoplasmically localized, dorso-ventral polarity formation. However, there are many species of spirally cleaving embryos belonging to different animal clades that instead, undergo equal cleavages at the first two cell divisions, resulting in four equal-sized blastomeres all appearing identical to one another. It was assumed that there was also a precocious specification of quadrant fate in these embryos. However, my college roommate, Chris Q. Doe, and I followed up on some previous observations to experimentally show that that this was not always true. We showed that in pond snail (*L. palustris*) embryos that give rise to four equal-sized blastomeres during the first two rounds of cleavage, dorso-ventral polarity is determined by cell-cell interactions at the

24 cell stage, not by cytoplasmic segregation of morphogenetic determinants (Arnolds, van den Biggelaar, & Verdonk, 1983; Martindale, 1986, 1987). Although many of the differences in developmental patterning between species within a clade were attributed to life history evolution, for example, direct vs indirect development (Henry & Martindale, 1994; Raff, 1992) or long vs short germ band insect development (Scholtz & Wolff, 2013), these early experiments were an early wake up call to scientists about making assumptions about the developmental mechanisms behind cell fate or body plan determination based on descriptive studies of a small number of animals as well as relying on the phylogenetic conservation of developmental mechanisms.

The problem with these comparisons is that these species have been separated by long evolutionary time periods (tens to hundreds of millions of years) allowing for independent genome evolution that may or may not be causally related to the changes in developmental programming. My graduate career was well before even the hope of total genome sequencing (especially of a variety of different "non-canonical model" species), so being able to find animal groups in which one could compare experimentally induced differences in cellular pattering in the same species or genomic context was one of my central goals.

1.1 Ctenophores

As a graduate student, I wanted to use a single organism to examine its differential developmental responses to the same experimental perturbation. This would allow me to determine the response of this specific perturbation under a unified genomic regulation and evolutionary history. This would essentially control for evolutionary differences between species as the same genomic regulation would be involved at two distinct life history stages. After desperately looking for a favorable example that met these criteria, I ended up choosing to work on a group of animals called ctenophores (Figs. 1 and 2). The fundamental feature of virtually all members of this group is that their embryos and their adults respond much differently to surgical manipulations. In 1880, Charles Chun working at the Naples Marine Station in Italy that noticed that after a storm there were two complete "half animals" (Fig. 2) in the same fertilization membrane of a local species of ctenophore (Chun, 1880). He hypothesized that the first two blastomeres had separated and developed independently. He was able to reproduce this result experimentally in the lab, and hence, was among the first to show the phenomenon of "mosaic" development in animals. Notably, this was in

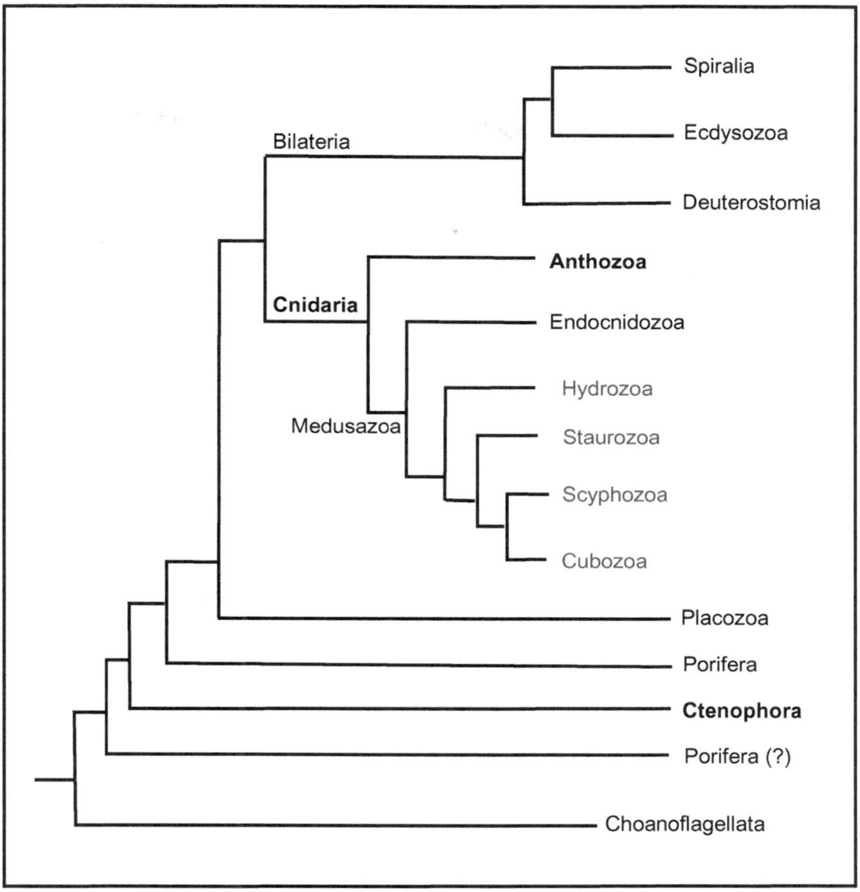

Fig. 1 Simplified phylogram of the relationship of ctenophores and cnidarians to other animal groups. Ctenophores such as *Mnemiopsis* have been proposed to be the earliest branching group of extant metazoans although sponges (poriferans) have traditionally held this position. Cnidarians are the sister group to Bilateria. *Nematostella vectensis* is a member of the direct developing anthozoans, the sister group to the highly diverse group of medusozoans.

contrast to results from similar experiments on sea urchin embryos in the same era by Hans Driesch (1891) who obtained the exact opposite result of embryonic regulation. Although there was personal and professional tension between Chun and other workers at the time (e.g., Driesch & Morgan, 1895) Chun's results were largely confirmed by a variety of workers through the next century (for reviews of ctenophore development see Martindale & Henry, 2015).

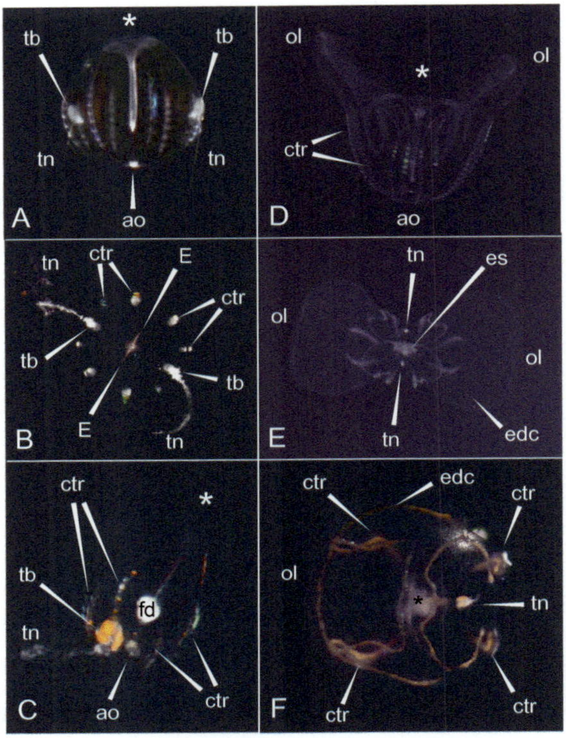

Fig. 2 Comparison of the body plan of the lobate ctenophore *Mnemiopsis leidyi* (A, B, D, E) to its metastabile "half animal" form (C, F). (A) Lateral view of a 3 mm diameter cydippid stage animal with the oral pole (*) up. The apical organ (ao) is located at the aboral pole. The two tentacles (tn) grow out of the tentacle bulbs (tb) and define the tentacular body axis. (B) Aboral view of a *Mnemiopsis* cydippid. Four of the eight ctene rows (ctr) are labeled. The flattened plane of the esophagus defines a body axis (E) orthogonal to the tentacular axis. (C) A "half animal" at the cydippid stage derived from an embryo bisected prior to the switchpoint (see text). Note the presence of only four ctene rows and one tentacle. These animals are functional and are able to capture food (fd). (D) Fully grown *Mnemiopsis* at the lobate stage. Note the growth of the oral lobes (ol). Only two of the eight comb rows are labeled. (E) Oral view of a lobate stage *Mnemiopsis* showing the adult symmetry properties. Note the endodermal canal system (edc) that distributes nutrients to the periphery of the animal. (F). High magnification oral view of a lobate stage "half animal" generated by bisecting an embryo during early development. Note the presence of a single oral lobe, four ctene rows, and a single tentacle bulb.

In contrast to the inability of ctenophore embryos to generate structures derived from missing blastomeres, adult ctenophores have an amazing capacity to regenerate (Coonfield, 1936, 1937; Martindale, 2016; Ramon-Mateu, Edgar, Mitchell, & Martindale, 2021). Adult organisms in the handful of species that have been tested (excluding species in the atentaculate genus *Beröe*) can replace all missing structures that have been surgically removed at all locations along their major longitudinal axis (the oral-aboral axis), including being cut in to two, four, or even eight pieces. In fact, there is even a highly derived group of ctenophores, the platyctenes, that regularly undergo asexual fission (Freeman, 1967). Ctenophores, therefore, had the differences I was looking for with respect to their response to similar operative procedures at different life history stages!

The problem then became which species of ctenophore to focus on. Because there were no genomic, molecular, or biochemical resources of any kind available for any ctenophore species, these were obviously not a factor. One of the hotspots for marine "jelloplankton," including ctenophores, is the University of Washington's Friday Harbor Lab on San Juan Island in Puget Sound. I spent a summer collecting animals and performing pilot experiments on a couple of different species there, but had limited success. This was primarily due to the length of time it took for animals to develop or regenerate at local temperatures (the seawater temperatures averaged 10–12 °C) and I discovered that one of the predominant types of ctenophore there (in the genus Beröe), had a limited ability to regenerate at all! The next summer I spent time at the University of Texas's marine lab in Port Aransas. Unfortunately, the lab was situated on a shipping canal and it was difficult to find healthy animals on a regular basis.

Finally, I was able to work in the lab of a researcher at the Univ. of Miami's Rosensteil School for Marine and Atmospheric Sciences (RSMAS). I had first met Dr. Ken King when we were both working at Friday Harbor and by the time I could arrange the opportunity to work at RSMAS Ken had decided to leave academia (he was taking his contractor's license test) and was in the process of shutting down his lab. Luckily, Ken allowed me to utilize his lab, microscopes, plankton nets, etc. I rode my motorcycle from Austin to Miami, slept on a rubber mat on the floor in the lab and cooked over a Bunsen burner for 3 months because I could not afford housing on the affluent Key Biscayne. Why RSMAS? A particular species of lobate ctenophore called *Mnemiopsis mccradyi* (recently renamed *Mnemiopsis leidyi*) was prevalent there in the wintertime and there was a group of outstanding physiological ecologists (Drs. Reeve, Walter, and King) on site at RSMAS that had experience with the care and handling of adult ctenophores. In addition, I had access to a dock with outstanding water quality where I could do plankton tows daily to provide an optimum natural food source for adults and growing juveniles, which was critical for working with holopelagic carnivorous animals. I was able to set up spawnings, perform embryonic and adult surgeries, and score experiments literally daily and around the clock. Furthermore, the warmer water temperatures at that latitude proved instrumental in promoting more rapid wound healing and adult regeneration, enabling experiments that were not possible at marine labs such as Friday Harbor and Woods Hole. After a couple of years of failure and frustration working at other locations, every piece of data for my dissertation was collected during those 3 months at RSMAS.

1.2 What have we learned from working on *Mnemiopsis*??

Without including too many of the gory details, the initial work at RSMAS generated two important findings. The first was to identify the "switch point" in which regenerative potential manifested itself during ctenophore development. By bisecting embryos at later and later stages of development, I determined that regenerative potential begins at 12 h after fertilization (Martindale, 1986). Embryos cut before this time generated two "half-animals," while those cut later generated two smaller, morphologically complete animals. This time point was well after gastrulation (4–5 h) and after most differentiated cell types were already apparent. This window appeared to correspond to the time in which locomotory ability (through comb plate beating) became coordinated throughout the animal, suggesting that there may be some type of "physiological integration," perhaps mediated by the developing nervous system? No further insight into what is going on at this "switch point" is known at this time (despite numerous rejected grant proposals focused on this important biological phenomenon).

The second important finding from these experiments has to do with experiments on "half animals" generated by cutting before the "switch point." If these "half-animals" are grown for a day or 2 and then themselves are cut in half (to make "1/4 pieces"), obviously well past the adult "switch point," how will they respond? If one thinks that the role of development is to produce the adult body plan, and the "role" of the adult action system is to maintain and functionally adapt to what development has handed it, then the adult regenerative response should faithfully reproduce what it was given. After all, these "half animals" have never been "whole animals" at any time on their life. Indeed, when this experiment is performed, the majority of animals reform "half animals." However, if the surgeries were performed such that different combinations of structures were juxtaposed, the rates for "whole animal" formation increased to nearly 40% (Martindale, 1986). Furthermore, we now know that the nutritional status of "half animals" affects their reproductive potential (Bading et al., 2017). These results indicate that the genomic regulation of an organism appears to know what normal is, and, if a robust adult regenerative response can be instigated, normal adult patterning can be achieved. While the cellular and molecular underpinnings of this phenomenon are currently unknown, these experiments provided some of the first indications that congenital or accidental defects in animals that do not normally regenerate (e.g., mammals), could be corrected if adult regenerative programs can be initiated. People, most notably people on NIH grant panels, argue that nothing useful can be learned from studying the regenerative properties of animals like ctenophores. However, I argue that a comparative approach is the only way to provide insight into whether there are very deep ancestral properties of the regenerative process (e.g., controlling the stability cell fate), or conversely, whether regenerative processes have evolved independently numerous times (Bely & Nyberg, 2010; Zattara, Fernández-Álvarez, Hiebert, Bely, & Norenburg, 2019). However, it is difficult to determine which of these two extreme views is more interesting and exciting. If some components of regeneration are ancient, then we know where to look in species that have lost the ability to regenerate. If regeneration is easy in the sense that there are multiple ways to accomplish regenerative gains, then we can attempt to implement one or more of these evolutionarily successful approaches in other species.

This early work helped establish ctenophores as an interesting new experimental system for a variety of different questions that my long term friend and colleague (and fellow grad student) Jonathan Q. Henry and I exploited over several summers at the Marine Biological Lab in Woods Hole, and other labs around the world. Among the interesting findings that ctenophores have provided include:

1. Most ctenophores are normally polyspermic (Carré & Sardet, 1984). Multiple sperms enter eggs indicating that there are no fast or slow blocks to polyspermy. Instead, the female pronucleus chooses which sperm nucleus to fuse with, forming an interesting case of female sexual selection at the subcellular level.
2. The first intracellular fate map for the first 60 cells of the *Mnemiopsis* embryo was generated (Martindale & Henry, 1999).
3. All of the mesodermal cell types (definitive muscle cells and mesenchyme) in *Mnemiopsis* are derived from endomesodermal sources, the

so called oral micromeres derived from the animal pole (it had previously been assumed that mesoderm was derived from ectodermal sources (Metschnikoff (1882, 1885).

4. The first cleavage defines the sagittal plane (esophageal axis) (Martindale & Henry, 1999).
5. The second cleavage defines the tentacular axis and the anal axis (Martindale & Henry, 1995).
6. Third cleavage separates developmental potential for the E vs M lineages (Freeman & Reynolds, 1973; Martindale & Henry, 1999).
7. Fourth cleavage segregates ectodermal comb plate fates from endomesodermal fates (Martindale & Henry, 1999).
8. The discovery of the first inductive interactions in ctenophore development were made (Martindale & Henry, 1997a, 1997b; Henry and Martindale, 2000). While it might seem surprising that the development of any animal could happen in the absence of inductive interactions, these results had profound implications for the interpretation of some historical cell deletion experiments (Farfaglio, 1963; Martindale, 1986).
9. Evidence that ctenophores are truly direct developers and do not go through a larval period of reproduction (Martindale, 1987). We have shown nutritional requirements required to maintain continuous reproductive output from cydippid "larval" stages (Edgar et al., 2022).

Another advantage of having *Mnemiopsis* material around during the summertime at the MBL is that it provided study material for the Embryology course students. A few years back, a couple of embryology students helped discover a "cleavage clock" in ctenophore development. Ctenophores develop incredibly rapidly with embryonic cell cycles occurring every 12–15 min at ambient temperatures. These students inhibited cytokinesis using the drug cytochalasin-B and found that although cell divisions did not occur, nuclear divisions remained unaffected and proceeded at the same rate as untreated controls. Rather than washing these "sick" embryos down the drain, the students raised these treated embryos and discovered that differentiated cell type-specific organelles (motile comb plate cilia and bioluminescent photoproteins) appeared at the precise time in cleavage-arrested embryos as they did in untreated control embryos (Fischer, Pang, Henry, & Martindale, 2014). Furthermore, by inhibiting cleavage at later and later stages (e.g., 8- or 16-cell stage), they were able to show that the developmental potential to form these structures segregated to their normal cell lineages (i.e., comb plate cilia to the e_1 lineage and photoproteins to the M lineage), demonstrating that the cleavage program had an active role in the segregation of embryonic "determinants." These students proposed that ctenophore embryos "counted" their cell divisions and that the regular increase in DNA content titrated out some kind of hypothetical inhibitor of differentiation, and when it reached a critical threshold, cells would turn on cell type-specific gene expression and differentiation. The time point in which these structures appeared (approximately 9h) preceded the "switch point" when adult regenerative ability appears. We currently have no idea if these developmental phenomena are related to one another.

Ctenophores joined the ranks of taxa with sequenced genomes in 2013 with the sequencing of the *Mnemiopsis* genome (Ryan et al., 2013). The *Mnemiopsis* genome is relatively small (approximately 150MB) and includes about 16,500 genes (Ryan et al., 2013). Although the analysis of its genomic content is still ongoing, one of the most interesting aspects is the apparent lack of most bilaterian cell-cell signaling pathways. While canonical Wnt, TGFβ, and MAPK/MEK/ERK pathways appear relatively intact (although lacking most extracellular antagonists), others, including Wnt-PCP, hedgehog, retinoic acid, JAK/STAT, and parts of the FGF and Delta-Notch pathways appear to be absent (Babonis & Martindale, 2016; Moroz et al., 2014; Ryan et al., 2013). Furthermore, key transcription factors involved in mesoderm and neural development (Moroz et al., 2014; Ryan et al., 2013), and structural components of proteins involved in epithelial organization (Salinas-Saavedra & Martindale, 2020; Salinas-Saavedra, Rock, & Martindale, 2018; Salinas-Saavedra, Wikramanayake, & Martindale, 2018) in bilaterians also appear to be absent, throwing into question the homology of these cell/tissue types between ctenophores and bilaterians.

The phylogenetic position of ctenophores has been highly contested in recent years (Fig. 1). Early phylogenomic studies placed them as the sister group to all other extant metazoans (Dunn et al., 2008; Hejnol et al., 2009; Whelan, Kocot, Moroz, & Halanych, 2015; Whelan et al., 2017). This result has been highly scrutinized by some workers who can't believe that sponges are not the most ancient existing group of animals based almost exclusively on morphological criteria. However, the most rigorous recent phylogenomic analyses appear to continue to support the early branching status of ctenophores (Hernandez & Ryan, 2021; Li, Shen, Evans, Dunn, & Rokas, 2021; Li, Zhang, et al., 2020). The maddening thing about the comparison of sponges to ctenophores is that they have a strikingly similar gene set complement, even though their morphological output/life history are so strikingly different (Ryan et al., 2007; Ryan, Pang, Mullikin, Martindale, & Baxevanis, 2010; Pang & Martindale, 2008;

Pang, Ryan, Baxevanis, & Martindale, 2011). For example, ctenophores have functional guts, muscles and nerve cells and an active carnivorous pelagic lifestyle while sponges are benthic filter feeding adults and do not have a single cell type that can be easily homologized with any other metazoan. It will be interesting to follow this controversy in the future due to its profound implications for animal evolution.

Regardless of the true evolutionary position of ctenophores, this group has many conceptual and experimental advantages over current sponge "model" systems. As such, ctenophores have much left to tell us about the development and the evolution of animals. Ctenophores continue to be a promising system for future work. Their outstanding optical properties, rapid development, and small genome all provide unique opportunities to learn about one of the most ancient groups of animals. We have reinitiated work on ctenophore regeneration (Martindale, 1986; Ramon, Ellison, Angelini, & Martindale, 2019), developed new techniques and procedures for fixation (Mitchell, Edgar, & Martindale, 2021), RNAseq (Babonis et al., 2018), ATACseq, morpholino knock-down (Yamada, Martindale, Fukui, & Tochinai, 2010), whole mount single and double fluorescent in situ hybridization (Pang, Ryan, Baxevanis, & Martindale, 2010; Pang, Ryan, Mullikin, Baxevanis, & Martindale, 2010; Ryan et al., 2010; Schnitzler et al., 2012; Schnitzler, Simmons, Pang, Martindale, & Baxevanis, 2014), single embryo (Levin, 2016), and single cell sequencing (Sebé-Pedrós et al., 2018). We have made significant advances in understanding the nutritional requirements of young adults such that we can grow them through their entire life cycle in 2–3 weeks in standing sea water (Edgar et al., 2022). These technical and animal handling advances open-up opportunities to a multitude of new researchers, and provide a realistic basis for maintaining transgenic lines, etc. The most challenging aspects of working on this group is the fragility of the adult body, their rapid embryonic development (which limits how much you can work on any particular developmental stage) and the difficulty of maintaining these animals in a land-locked laboratory.

1.3 *Nematostella vectensis*

Another experimental system growing in popularity is the "starlet sea anemone," *N. vectensis*. My graduate advisor, Gary Freeman was an outstanding experimental embryologist and one of the major groups he worked on were cnidarians (among lots of other things). One aspect of his work in the early 1980s was focused on variation in the developmental patterning in a large group of cnidarians called medusozoans (Fig. 1). Most medusozoans, including the hydrozoans, scyphozoans, and cubozoans, have a metagenic life cycle that consists of embryogenesis giving rise to a benthic polyp stage, which later gives rise to the sexual "adult" phase called a medusa. The medusa is also normally a pelagic (swimming) phase that are commonly referred to as "jellyfish." There are numerous well-known deviations to this metagenic life history (e.g., the medusa remaining attached to the polyp and becoming mobile, to the complete loss of the medusa phase such as Hydra), however one aspect of cnidarian developmental evolution that intrigued Gary were a couple of groups of direct-developing hydrozoans (the trachylines and siphonophores) that skipped the benthic phase altogether with embryogenesis resulting in the adult pelagic medusa. Gary showed that the direct developing species had a precocious specification of cell fate during early embryogenesis relative to other hydrozoan species such as *Phialidium* (now called *Clytia*) that maintained the ancestral hydrozoan life history.

I had always been interested in the "other" group of cnidarians, the anthozoans, a group that includes corals, sea fans, sea whips, and sea anemones. This clade of cnidarians likely represents the ancestral cnidarian life history in which embryogenesis results in the formation of the polyp and the polyp is the sexual phase adult (no "jellyfish" is produced). These would then in effect be direct-developers, and one might suppose that there would be a precocious segregation of cell fates similar to direct developing medusozoans. The problem with anthozoans is that it is difficult to obtain embryonic material. Most corals, for example, spawn on a single full moon night in the summertime, and others brood embryos (that are highly asynchronous in their staging and hard to collect) or produce asexual propagules that resemble sexual planula. Most sea anemones do not reproduce on a regular schedule, and thus anthozoans had been completely, and I mean completely, unexplored by experimental embryologists.

In 1991 I was invited to give a seminar at the Bodega Marine lab (Univ. California Davis). During my faculty interview with the late "cnidariologist" Cadet Hand and his graduate student Kevin Uhlinger, I noticed several bowls of sea anemones (Fig. 3) on the counter. I remember asking Cadet, "Are those what I think they are"? Sure enough, there were a number of egg masses that had been spontaneously spawned in the dishes. Kevin and Cadet were kind enough to send us some animals and we started our own colony at the University of Chicago. Kevin and Cadet went on to write a short paper on the care and handling of the burrowing starlet sea anemone,

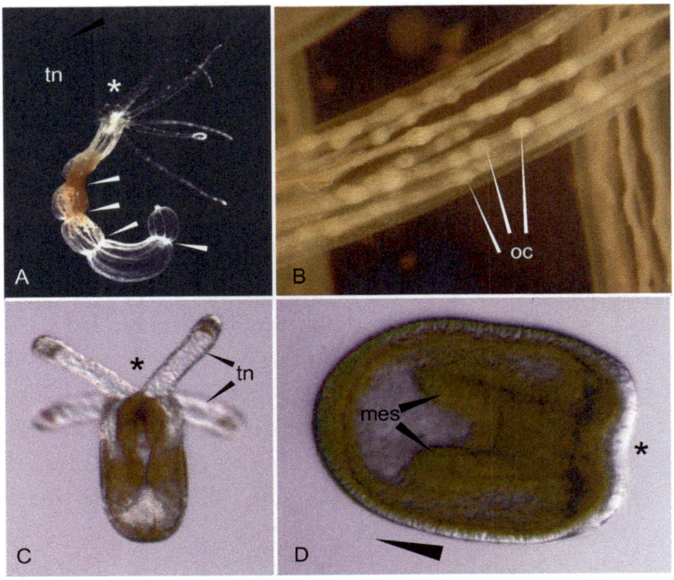

Fig. 3 Images of the body plan of the starlet sea anemone *Nematostella vectensis*. (A) a fully grown 3 cm adult (oral aboral length). Note the tentacles (tn) surrounding the oral opening (*) and the myoepithelial driven peristaltic constrictions (arrows). (B) High magnification of the mesenteries of a spawning adult *Nematostella*. The arrows indicate oocytes (oc) prior to spawning. (C) An 8 day-old *Nematostella* polyp with its first four feeding tentacles. (D) The swimming planula stage of *Nematostella* development. The future mouth (*) is located at the trailing end of the direction of swimming (arrowhead), a phenomenon that has confused some evolutionary biologists about the evolution of the metazoan anterior-posterior axis. Note the first two mesenteries (mes) that are the first morphological indications of anthozoan bilateral symmetry.

N. vectensis (Hand & Uhlinger, 1992). A star was born! Just to give you an idea of how long it took for this system to become established, it should be noted that it took until 1998 for the NSF to fund its first grant on *Nematostella* and, for many excellent reasons, *Nematostella* was the first cnidarian genome sequenced (Putnum et al., 2007). Now, thousands of *Nematostella* embryos can be obtained 365 days of the year and dozens of labs all over the globe are making important discoveries based on this species.

Having an ideal anthozoan species to study led to a number of important findings. The embyros are incredibly easy to cut, manipulate and microinject, and a number of functional molecular techniques have been developed including gene misexpression, knockdowns and genome editing (Ikmi, McKinney, Delventhal, & Gibson, 2014; Layden, Röttinger, Wolenski,

Gilmore, & Martindale, 2013; Nakanishi & Martindale, 2018; Servetnick et al., 2017). Some of the earliest experimental findings in the early stages of *Nematostella* investigations include the following:

1. The unipolar site of first cleavage occurs at the animal pole just as it does in all other experimentally tested cnidarians (Freeman, 1980; Lee, Kumburegama, Marlow, Martindale, & Wikramanayake, 2007).
2. There is an asymmetric distribution of developmental potential along the animal-vegetal axis in *Nematostella*, with only oral cell fates determined prior to gastrulation (Lee et al., 2007).
3. Gastrulation in Nematostella occurs by epithelial invagination at the animal pole, just as in many other anthozoan (but not medusozoan) cnidarians (Byrum & Martindale, 2004; Freeman, 1980; Magie, Daly, & Martindale, 2007).
4. The site of gastrulation occurs at the animal pole in cnidarians (and ctenophores) but occurs at the vegetal pole in bilaterians (Freeman, 1977; Lee et al., 2007; Martindale, 2005; Martindale & Hejnol, 2009). There is currently no reliable information on other early branching metazoans (e.g., sponges and placozoans), therefore, the most parsimonious conclusion is that the site of gastrulation moved to the vegetal pole prior to the evolution of the bilataria, and we have a pretty good idea of how this occurred (Lee et al., 2007; Martindale & Hejnol, 2009).
5. The *Nematostella* gastrodermis expresses many of the same genes as bilaterian endomesoderm (Martindale, Pang, & Finnerty, 2004; Röttinger, Dahlin, & Martindale, 2012) providing compelling evidence that a single bifunctional gastrodermal epithelium of diploblastic cnidarians (the sister group to bilaterians) was the precursor of both endoderm and mesoderm in triploblastic animals.
6. The nuclear localization of β-catenin is one of the initial steps to the specification of endomesoderm at the animal pole (Lee et al., 2007; Röttinger et al., 2012; Wikramanayake et al., 2003) and that the morphological invagination of endomesoderm is regulated by the Wnt-PCP pathway (Kumburegama, Wijesena, Xu, & Wikramanayake, 2011).
7. A new oral-aboral axis can be induced in *Nematostella* embryos by the ectopic injection of β-catenin and its cofactor Tcf mRNA (Wijesena & Martindale, 2018). This provides a plausible explanation for the translocation of gastrulation to the vegetal pole in bilaterians.
8. The mechanisms responsible for β-catenin stabilization/nuclearization at the vegetal pole in bilaterians results in an entire gene regulatory

network "module" being translocated to the vegetal pole. The fate map of *Nematostella* is essentially the mirror image of that seen in echinoids and helps explain why some bilaterian developmental biologists describe these genes as being "posterior" genes (Martindale & Hejnol, 2009).

9. The complex molecular inputs into individual gene expression explains the "bipolar" expression of genes in derivatives of both animal and vegetal pole cells (Amiel et al., 2017), a situation also seen in other embryos such as sea urchins (Gross & McClay, 2001; Oliveri, Walton, Davidson, & McClay, 2006).
10. *Nematostella* has the full complement of PAR complex cell polarity genes. However, these genes are only deployed during epithelial tissue formation, and not during early embryogenesis indicating that the PAR system was heterochronically co-opted for early embryonic polarity in bilaterian taxa (Salinas-Saavedra, Rock, & Martindale, 2018; Salinas-Saavedra, Stephenson, Dunn, & Martindale, 2015; Salinas-Saavedra, Wikramanayake, & Martindale, 2018).
11. That *Nematostella*, and perhaps all other anthozoans, displays clear aspects of bilateral symmetry at the molecular level that are not apparent morphologically (Finnerty, Pang, Burton, Paulson, & Martindale, 2004; Matus, Pang, Marlow, et al., 2006; Matus, Thomsen, & Martindale, 2006), whether this is precisely homologous to the bilateral symmetry seen in bilaterians is still controversial.
12. The cnidarian fate map and molecular understanding of *Nematostella* reveal that the oral-aboral axis is homologous to the bilaterian anterior-posterior axis, and that the oral pole of cnidarians is homologous to the anterior pole in bilaterians (Martindale and Hejnol, 2009).
13. Hox genes help pattern the oral-aboral axis through Wnt signaling (Dubuc, Stephenson, Rock, & Martindale, 2018), particularly at the cells of the blastopore lip that secretes multiple Wnt ligands and serves as an "embryonic organizer" (Kraus, Aman, Technau, & Genikhovich, 2016; Kraus, Fritzenwanker, Genikhovich, & Technau, 2007; Kusserow et al., 2005).

2. Lessons from the genome

Due to the many favorable properties this species possesses, such as the ability to easily maintain breeding colonies in the lab, gain access to and manipulate embryological material, being able to study the robust regenerative ability of adults, their ability to generate asexually, in combination with its relatively small compact genome (approximately 400 Mb), it is of little surprise that *Nematostella* was the first cnidarian genome sequenced. The technology of the time was Sanger sequencing performed at the Joint Genome Institute of the U.S. Department of Energy in Walnut Creek, CA (Putnum et al., 2007) at an estimated cost of $2 million. Since that time, the field has made extraordinarily important improvements to genomic/molecular resources by a number of labs around the world, including Uli Technau (Sars Institute, Norway and now at the Univ. of Vienna), Matt Gibson (Stowers Institute in Kansas City, Missouri), Casey Dunn (formerly at Brown University now at Yale), John Finnerty (Boston University), Adam Reitzel (Univ. of North Carolina. Charlotte), and Yehu Moran and Uri Gat's groups in Israel, among others.

The sequenced genome paid immediate benefits in the sense that it gave numerous workers in the field the chance to check the complete genomic repertoire of distinct molecular pathways, many of which had already been explored using degenerate PCR approaches (e.g., Finnerty et al., 2004; Finnerty, Paulson, Burton, Pang, & Martindale, 2003; Kusserow et al., 2005; Ryan et al., 2010). But the most mind-blowing revelation from the *Nematostella* genome project (Putnum et al., 2007) was that the common ancestor of cnidarians and bilaterians possessed more genes in common with human beings than either of the most common developmental model systems of fruit flies (Drosophila) or nematodes (*C. elegans*)! The common (and quite reasonable) assumption at the time was that genomic complexity evolved over time, and that more and more new genes would be added over evolutionary time. While this certainly happens, the *Nematostella* genome project revealed that many bilaterian clades have lost genes that were present in the common ancestor of cnidarians and bilaterians. This disconnect between genomic "complexity" and morphological/behavioral "complexity" was completely unexpected, yet was reinforced by simple gene expression studies. Cnidarians used to be thought of as simple, radially symmetrical animals with a dozen or so cell distinct cell types (Frank & Bleakney, 1976). What was *Nematostella* doing with all these genes?? If they had no function, they should have been lost; if they were really simple radially symmetrical animals, perhaps all these genes would be ubiquitously expressed everywhere. A multitude of whole mount in situ hybridization studies revealed complicated and unexpected patterns of cell and tissue specific gene expression, including clear patterns of bilateral symmetry and dorso-ventral polarity.

One striking example of how the molecular complexity of the *Nematostella* genome impacts our understanding of organismal and cell type complexity is shown by studies of ion channel family composition (Jegla et al., 2012; Li, Liu, et al., 2015; Li, Martinson, et al., 2015; Martinson et al., 2014). Shaker potassium channels are composed of four individual subunits. Flies and human beings have one or a small number of Shaker channel proteins that makes up all four subunits (Jegla et al., 2012; Jegla, Zmasek, Batalov, & Nayak, 2009). *Nematostella* has 20 distinct Shaker potassium channel subunits, about half of which have distinct regulatory properties. If each *Nematostella* neuron only expressed homotetramers of each Shaker channel, it would indicate that there would be 20 different "kinds" of neurons, all of which look morphologically identical to one another. However, in situ hybridization experiments have shown that any one cell can express two or more of these channel subunits. Furthermore, in vitro expression and electrophysiological characterization of *Nematostella* channels indicates that each subunit has unique functional attributes (gating, latency, speed of recovery, etc.) which affects cellular physiology (Jegla et al., 2012). Thus, the independent expression of channel components in *Nematostella* has the potential to generate a dizzying number of potentially functionally distinct neural "cell types" (4 raised to the 20th power, more potentially distinct neurons than are in the human brain)! Now who has the more "complex" nervous system?

The use of *Nematostella* as a powerful experimental system and EvoDevo model continues to grow. Numerous examples of its importance for addressing a diverse set of questions, including topics such as the evolution of gene expression (e.g., Ikmi et al., 2014; Karabulut, He, Chen, McKinney, & Gibson, 2019; Li, Shen, et al., 2021; Li, Zhang, et al., 2020; Moran et al., 2014; Praher et al., 2017; Renfer, Amon-Hassenzahl, Steinmetz, & Technau, 2010; Schwaiger et al., 2014; Xu et al., 2019) to the origins of novel cell types (Babonis & Martindale, 2017; Babonis, Martindale, & Ryan, 2016; Babonis, Ryan, Enjolras, & Martindale, 2019), to an understanding of the evolution of venoms (Moran et al., 2013; Sachkova et al., 2019) or neural cell types (Jegla et al., 2012; Li, Liu, et al., 2015; Li, Martinson, et al., 2015; Martinson et al., 2014; Moran, Barzilai, Liebeskind, & Zakon, 2015) .

One of the aspects my lab is really excited about now is to use the growing functional techniques in *Nematostella* embryos for "cellular engineering." We have become interested in the phenomenon of biomineralization which has evolved independently numerous times on Earth, and indeed none of the proteins throught to be required to initiate the PILP biomineralization process (Gower & Odom, 2000) from distinct taxa show any sequence homology with one another. Instead, proteins involved in biomineralization are categorized as "intrinsically disordered proteins" (IDPs) that provides unique opportunities to understand structure-function properties of these diverse proteins. We have been expressing a variety of different biomineralization inducing proteins from organisms as diverse as corals, sea urchins, mollusks, and even human tooth proteins in *Nematostella* embryos, driving them in distinct cell types and tissue layers using transgenic reporters. IDPs are post-translationally modified so performing site-directed mutagenesis and expression in *Nematostella* provides the opportunity to functionally characterize the structural details promoting biomineralization. One of our main long term goals is to understand the process of biomineralization to the point where we can effectively turn *Nematostella* into a biomineralizing "coral." This is not as easy as it might seem because there are complex interactions with other cellular components, (e.g., carbonic anhydrases and anion transporters) that regulate distinct biomineralizing "microenvironments," but our ability to functionally manipulate gene expression in *Nematostella* embryos provides exciting opportunities for future work. We also hope to use similar functional approaches to gain a better understanding of cnidarian dinoflagellate symbiosis.

3. Final thoughts on the comparison of *Mnemiopsis* and *Nematostella* development

The contrast between *Mnemiopsis* and *Nematostella* is even more extreme when considering their developmental programs. Although very few species of ctenophores have been experimentally manipulated, all species appear to have the exact same cleavage program and cell lineage. Therefore, the study of *Mnemiopsis* would appear to be a good model for many aspects of ctenophore development. Cnidarians are extremely diverse in their developmental programs, for example, displaying all known patterns of gastrulation (Byrum & Martindale, 2004). The greatest amount of diversity in developmental pattering in cnidarians occurs in the medusozoans that have complex metagenic life histories. Anthozoans, in contrast, are direct developers and appear to have a much more conserved developmental program. This is not to say that *Nematostella* is a "living fossil" (there is, of course, no such thing), but this species is likely to provide considerable insight into the evolution of cnidarian development.

The details of development between *Mnemiopsis* and *Nematostella* could not be more different from one another. *Mnemiopsis* has crystal clear, centrolecithal embryos that have a highly stereotyped phylum-specific cleavage

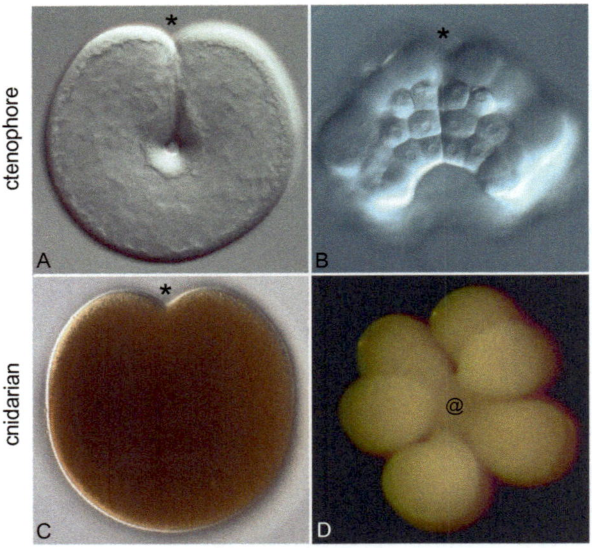

Fig. 4 Simplified comparison of the development of *Mnemiopsis* to *Nematostella*. (A, C) Although both embryos undergo unipolar first cleavage (*) that marks the site of gastrulation and the location of the oral pole, ctenophores (A, B) have crystal clear centrolecithal embryos and a highly stereotyped cleavage program in which every cell can be identified up until the 60 cell stage (B). *Nematostella* on the other hand has yolky embryos and the unipolar first cleavage regresses while nuclear divisions continue. Cytoplasmic connections (@) connect blastomeres until the 4–8 cell stage (D).

program (Fig. 4A, B) and rapid development, generating a miniature adult in less than 24h. Experimental studies have shown that the cleavage program itself is causally involved with the asymmetric segregation of developmental potential at each of the early cleavage divisions and these embryos have a remarkable lack of ability to regulate following removal of embryonic blastomeres (Martindale & Henry, 1997a, 1997b, 2015). In contrast, *Nematostella* embryos are yolky and undergo "chaotic" cleavage (Fig. 4C, D), where no two embryos even in the same clutch undergo the same series of cell divisions. In fact, *Nematostella* rarely undergoes complete cytokinesis at the first two cell divisions. The first two cleavage furrows normally regress to give rise to an early syncytium that formally cellularizes at the 4–8 cell stage (Fig. 4D). Cnidarian embryos in general, including *Nematostella,* are highly regulative and will develop normally despite removal of various embryonic domains.

So what could possibly unite these two very different developmental programs? Fantastic experimental work by Gary Freeman has revealed important insight into the development of early-branching metazoans.

Both ctenophores and cnidarians undergo what is called "unipolar" cell division (Fig. 4A,C). The cleavage furrow begins at the animal pole and then progresses to the vegetal pole (in contrast to most cleavage programs in most bilaterians in which the cleavage furrow is organized around the entire circumference of the zygote). By using vital cell staining and mild centrifugation experiments to translocate the zygotic nucleus, Gary was able to show (Freeman, 1977, 1983) that the site of first cleavage was causally involved in the establishment of the primary developmental and organismal axis (e.g., the site of gastrulation and the formation of the oral pole). These experiments showed that there is not a rigid maternally derived animal-vegetal axis (as we see in most bilaterian embryos) and that the cleavage program itself is responsible for organizing all aspects of subsequent development! Furthermore, experiments in *Nematostella* (an embryo Gary never had the chance to work on) have revealed the molecular explanation for his experimental results. Lee et al. (2007) showed that a protein called disheveled (Dsh) is localized to the membrane of the egg and zygotic nucleus. Dsh is causally involved with the stabilization of the protein β-catenin, that has been shown to be required for endomesoderm development in *Nematostella* (Wikramanayake, Huang, & Klein, 1998). Thus, the translocation of the zygotic nucleus to a new location results in a shift of the site of β-catenin stabilization and the formation of the site of gastrulation and the formation of the oral-aboral axis. More recently, Amiel and Houliston (2009) showed that a maternal RNA, CheFz1 RNA, also involved in the Wnt signaling pathway, was translocated along with the female pronucleus when displaced by centrifugal force, supporting Freeman's original experiments (in the same species Gary worked on). These are extremely satisfying molecular explanations for the formation of the oral-aboral axis in cnidarians (and presumably ctenophores?) that is akin to the discovery of the molecular basis of vertebrate limb development initially descried by John Saunders and his colleagues (reviewed by Verheyden & Sun, 2017). These types of experiments reveal that there is much to be learned about the origins of developmental pattering form early branching metazoan lineages.

References

Amiel, A., & Houliston, E. (2009). Three distinct RNA localization mechanisms contribute to oocyte polarity establishment in the cnidarian Clytia hemisphaerica. *Developmental Biology*, *327*(1), 191–203.

Amiel, A. R., Johnston, H., Chock, T., Dahlin, P., Iglesias, M., Layden, M., et al. (2017). A bipolar role of the transcription factor ERG for cnidarian germ layer formation and apical domain patterning. *Developmental Biology*, *430*(2), 346–361.

Arnolds, W. J., van den Biggelaar, J. A., & Verdonk, N. H. (1983). Spatial aspects of cell interactions involved in the determination of dorsoventral polarity in equally cleaving gastropods and regulative abilities of their embryos, as studied by micromere deletions in Lymnaea and Patella. *Wilhelm Roux's Archives of Developmental Biology*, *192*(2), 75–85.

Babonis, L. S., DeBiasse, M. B., Francis, W. R., Christianson, L. M., Haddock, S. H. D., Martindale, M. Q., et al. (2018). Integrating embryonic development and evolutionary history to characterize tentacle-specific cell types in a ctenophore. *Molecular Biology and Evolution*, *35*(12), 2940–2956.

Babonis, L. S., & Martindale, M. Q. (2016). Phylogenetic evidence for the modular evolution of metazoan signalling pathways. *Philosophical Transactions of the Royal Society B*, *372*, 20150477. https://doi.org/10.1098/rstb.2015.0477.

Babonis, L. S., & Martindale, M. Q. (2017). PaxA, but not PaxC, is required for cnidocyte development in the sea anemone *Nematostella vectensis*. *EvoDevo*, *8*(1), 14.

Babonis, L., Martindale, M. Q., & Ryan, J. (2016). Do novel genes drive morphological novelty? An investigation of the nematosomes in the sea anemone *Nematostella vectensis*. *BMC Evolutionary Biology*, *16*, 114. https://doi.org/10.1186/s12862-016-0683-3.

Babonis, L. S., Ryan, J. F., Enjolras, C., & Martindale, M. Q. (2019). Genomic analysis of the tryptome reveals molecular mechanisms of gland cell evolution. *EvoDevo*, *10*, 23. https://doi.org/10.1186/s13227-019-0138-1.

Bading, K. T., Kaehlert, S., Chi, X., Jaspers, C., Martindale, M. Q., & Javidpour, J. (2017). Food availability drives plastic self-repair response in a basal metazoan-case study on the ctenophore Mnemiopsis Leidyi A. Agassiz. *Scientific Reports*, *7*, 1–9. https://doi.org/10.1038/s41598-017-16346-w.

Bely, A. E., & Nyberg, K. G. (2010). Evolution of animal regeneration: Re-emergence of a field. *Trends in Ecology & Evolution*, *25*, 161–170.

Byrum, C., & Martindale, M. Q. (2004). Cnidarians and ctenophores. In C. Stern (Ed.), *Gastrulation: From cells to embryos* (pp. 33–50). Cold Spring Harbor Press.

Carré, D., & Sardet, C. (1984). Fertilization and early development in Beroe ovata. *Developmental Biology*, *105*(1), 188–195. https://doi.org/10.1016/0012-1606(84)90274-4.

Chun, C. (1880). *Die Ctenophoren des Golfes von Neapel, und der angrenzenden Meeres-Abschnitte* (p. 1880). Leipzig, Germany: W. Engelmann.

Clement, A. C. (1952). Experimental studies on germinal localization in Ilyanassa. I. the role of the polar lobe in determination of the cleavage pattern and its influence in later development. *The Journal of Experimental Zoology*, *121*(3), 593–625.

Clement, A. C. (1962). Development of Ilyanassa following removal of the D macromere at successive cleavage stages. *Journal of Experimental Zoology*, *149*(3), 193–215.

Coonfield, B. R. (1936). Regeneration in Mnemiopsis. *The Biological Bulletin*, *71*, 421–428.

Coonfield, B. R. (1937). Symmetry and regulation in Mnemiopsis Leidyi, Agassiz. *The Biological Bulletin*, *72*, 299–310. https://doi.org/10.2307/1537689.

Driesch, H., Willier, B., & Oppenheimer, J. M. (Eds.). (1891–1974). Entwicklungsmechanische studien,I. Der werth der beiden ersten furchungszellen in der echinodermentwicklung. Experimentelle erzeugen von theil—und doppelbildung. Zeitschr. Wissenschaft. Zool. 53:166–178. Abridged English translation. In *Foundations of experimental embryology*. New York: Hafner Press.

Driesch, H., & Morgan, T. H. (1895). Zur Analysis Der Ersten Entwickelungsstadien Des Ctenophoreneies. *Archiv für Entwicklungsmechanik der Organismen*, *2*, 216–224. https://doi.org/10.1007/bf02084248.

Dubuc, T. Q., Stephenson, T. B., Rock, A. Q., & Martindale, M. Q. (2018). Hox and Wnt pattern the primary body Axis of an Anthozoan cnidarian before gastrulation. *Nature Communications*, *9*(1), 1–12. https://doi.org/10.1038/s41467-018-04184-x.

Dunn, C. W., Hejnol, A., Matus, D. Q., Pang, K., Browne, W. E., Smith, S. A., et al. (2008). Broad phylogenomic sampling improves resolution of the animal tree of life. *Nature*, *452*, 745–749. https://doi.org/10.1038/nature06614.

Edgar, A., Ponciano, J., and Martindale, M. Q. 2022. Ctenophores are direct developers that reproduce continuously beginning very early after hatching. PNAS, In Press.

Farfaglio, G. (1963). Experiments on the formation of the ciliated plates in ctenophores. *Acta Embryologiae et Morphologiae Experimentalis*, *6*, 191–203.

Finnerty, J. R., Pang, K., Burton, P., Paulson, D., & Martindale, M. Q. (2004). Deep origins for bilateral symmetry: Hox and Dpp expression in a sea anemone. *Science*, *304*, 1335–1337.

Finnerty, J. R., Paulson, D., Burton, P., Pang, K., & Martindale, M. Q. (2003). Early evolution of a Homeobox gene: The ParaHox gene Gsx in the Cnidaria and the Bilateria. *Evolution & Development*, *5*, 331–345.

Fischer, A., Pang, K., Henry, J. Q., & Martindale, M. Q. (2014). A cleavage clock regulates features of lineage-specific differentiation in the development in a basal branching metazoan, the ctenophore *Mnemiopsis leidyi*. *EvoDevo*, *5*, 4. https://doi.org/10.1186/2041-9139-5-4.

Frank, P., & Bleakney, J. S. (1976). Histology and sexual reproduction of the anemone Nematostella vectensis Stephenson 1935. *Journal of Natural History*, *10*(4), 441–449. https://doi.org/10.1080/00222937600770331.

Freeman, G. (1967). Studies on regeneration in the creeping ctenophore, *Vallicula multiformis*. *Journal of Morphology*, *123*, 71–783. https://doi.org/10.1002/jmor.1051230107.

Freeman, G. (1977). The establishment of the oral–aboral axis in the ctenophore embryo. *Journal of Embryology and Experimental Morphology*, *42*, 237–260.

Freeman, G. (1980). The role of cleavage in the establishment of the anterior–posterior axis of the hydrozoan embryo. In P. Tardent, & R. Tardent (Eds.), *Development and cellular biology of coelenterates* (pp. 97–108). North Holland, Amsterdam: Elsevier.

Freeman, G. (1983). Experimental studies on embryogenesis in hydrozoans (Trachylina and Siphonophora) with direct development. *The Biological Bulletin*, *165*(3), 591–618.

Freeman, G., & Reynolds, G. T. (1973). The development of bioluminescence in the ctenophore *M. leidyi*. *Developmental Biology*, *31*(1), 61–100. https://doi.org/10.1016/0012-1606(73)90321-7. 4150750.

Gower, L. B., & Odom, D. J. (2000). Deposition of calcium carbonate films by a polymer-induced liquid-precursor (PILP) process. *Journal of Crystal Growth*, *210*(4), 719–734.

Gross, J. M., & McClay, D. R. (2001). The role of Brachyury (T) during gastrulation movements in the sea urchin L. *variegatus*. *Developmental Biology*, *239*(1), 132–147. https://doi.org/10.1006/dbio.2001.0426. 11784024.

Hand, C., & Uhlinger, K. R. (1992). The culture, sexual and asexual reproduction, and growth of the sea anemone *N. vectensis*. *The Biological Bulletin*, *182*(2), 169–176.

Hejnol, A., Obst, M., Stamatakis, A., Ott, M., Rouse, G., Edgecombe, G., et al. (2009). Rooting the Bilaterian Tree with Scalable Phylogenomic and Supercomputing Tools. *Proceedings of the Royal Society B*, *276*, 4261–4270. https://doi.org/10.1098/rspb.2009.0896.

Henry, J. J., & Martindale, M. Q. (1999). Conservation and innovation in spiralian development. *Hydrobiologia*, *402*, 255–265.

Henry, J. Q., & Martindale, M. Q. (1994). Establishment of the dorsoventral axis in nemertean embryos: Evolutionary considerations of spiralian development. *Developmental Genetics*, *15(1)*, 64–78. https://doi.org/10.1002/dvg.1020150108.

Henry, J. Q., & Martindale, M. Q. (2000). Regulation and regeneration in the ctenophore *Mnemiopsis*. *Developmental Biology*, *227*, 720–733.

Hernandez, A. M., & Ryan, J. F. (2021). Six-state amino acid recoding is not an effective strategy to offset compositional heterogeneity and saturation in phylogenetic analyses. *Systematic Biology*, 1200–1212. https://doi.org/10.1093/sysbio/syab027. syab027.

Ikmi, A., McKinney, S. A., Delventhal, K. M., & Gibson, M. C. (2014). TALEN and CRISPR/Cas9-mediated genome editing in the early-branching metazoan. *Nature Communications*, *5*(1), 1–8.

Jegla, T., Marlow, H. Q., Chen, B., Simmons, D. K., Jacobo, S. M., & Martindale, M. Q. (2012). Expanded functional diversity of shaker K+ channels in cnidarians is driven by gene expansion. *PLoS One*, 7(12), e51366.

Jegla, T. J., Zmasek, C. M., Batalov, S., & Nayak, S. K. (2009). Evolution of the human ion channel set. *Combinatorial Chemistry & High Throughput Screening*, *12*(1), 2–23.

Karabulut, A., He, S., Chen, C. Y., McKinney, S. A., & Gibson, M. C. (2019). Electroporation of short hairpin RNAs for rapid and efficient gene knockdown in the starlet sea anemone, *N. vectensis*. *Developmental Biology*, *448*(1), 7–15.

Kraus, Y., Aman, A., Technau, U., & Genikhovich, G. (2016). Pre-bilaterian origin of the blastoporal axial organizer. *Nature Communications*, 7(1), 1–9.

Kraus, Y., Fritzenwanker, J. H., Genikhovich, G., & Technau, U. (2007). The blastoporal organizer of a sea anemone. *Current Biology*, *17*(20), R874–R876.

Kumburegama, S., Wijesena, N., Xu, R., & Wikramanayake, A. H. (2011). Strabismus-mediated primary archenteron invagination is uncoupled from Wnt/β-catenin-dependent endoderm cell fate specification in *N. vectensis* (Anthozoa, Cnidaria): Implications for the evolution of gastrulation. *EvoDevo*, *2*(1), 1–15.

Kusserow, A., Pang, K., Sturm, C., Hrouda, M., Lentfer, J., Technau, U., et al. (2005). Unexpected complexity of Wnt gene family in a sea anemone. *Nature*, *433*, 156–160.

Layden, M. J., Röttinger, E., Wolenski, F. S., Gilmore, T. D., & Martindale, M. Q. (2013). Microinjection of mRNA or morpholinos for reverse genetic analysis in the starlet sea anemone, *Nematostella Vectensis*. *Nature Protocols*, *8*, 924. https://doi.org/10.1038/nprot.2013.009.

Lee, P. N., Kumburegama, S., Marlow, H. Q., Martindale, M. Q., & Wikramanayake, A. H. (2007). Asymmetric developmental potential along the animal–vegetal axis in the anthozoan cnidarian, *N. vectensis*, is mediated by Disheveled. *Developmental Biology*, *310*(1), 169–186.

Li, X., Liu, H., Luo, J. C., Rhodes, S. A., Trigg, L. M., van Rossum, D. B., et al. (2015). A major diversification of voltage-gated K^+ channels occurred in ancestral parahoxozoans. *PNAS*, *112*, E1010–E1019.

Li, X., Martinson, A. S., Layden, M. J., Diatta, F. H., Sberna, A. P., Simmons, D. K., et al. (2015). Ether-a-go-go family voltage-gated K^+ channels evolved in an ancestral metazoan and functionally diversified in a Cnidarian/Bilaterian ancestor. *The Journal of Experimental Biology*, *218*, 526–536.

Levin, M., et al. (2016). The mid-developmental transition and the evolution of animal body plans. *Nature*, *532*(7596), 637–641. https://doi.org/10.1038/nature16994.

Li, Q., Zhang, P., Li, J., Yu, H., Zhan, X., Zhu, Y., et al. (2020). On the origin and evolution of RNA editing in metazoans. *bioRxiv*. https://doi.org/10.1101/2020.01.19.911685.

Li, Y., Shen, X. X., Evans, B., Dunn, A., & Rokas, A. (2021). Rooting the animal tree of life. *Molecular biology and evolution*, *38*(10), 4322–4333.

Magie, C. R., Daly, M., & Martindale, M. Q. (2007). Gastrulation in the cnidarian *N. vectensis* occurs via invagination not ingression. *Developmental Biology*, *305*(2), 483–497.

Martindale, M. Q. (1986). The expression and maintenance of adult symmetry properties in the ctenophore, *M. mccradyi*. *Developmental Biology*, *118*, 556–576.

Martindale, M. Q. (1987). Larval reproduction in the ctenophore *M. mccradyi* (order Lobata). *Marine Biology*, *94*, 409–414. https://doi.org/10.1007/BF00428247.

Martindale, M. Q. (2005). The evolution of metazoan axial properties. *Nature Reviews. Genetics*, *6*(12), 917–927.

Martindale, M. Q. (2016). The onset of regenerative properties in Ctenophores. In E. Tanaka, & P. Reddien (Eds.), *Vol. 40. Current opinion in genetics & development* (pp. 113–119). https://doi.org/10.1016/j.gde.2016.06.017.

Martindale, M. Q., & Hejnol, A. (2009). A developmental perspective: Changes in the position of the blastopore during bilaterian evolution. *Developmental Cell*, *17*(2), 162–174.
Martindale, M. Q., & Henry, J. Q. (1995). Diagonal development: Establishment of the anal axis in the ctenophore *M. leidyi*. *The Biological Bulletin*, *189*, 190–192.
Martindale, M. Q., & Henry, J. Q. (1997a). Reassessing embryogenesis in the Ctenophora: The inductive role of e_1 micromeres in organizing ctene row formation in the "mosaic" embryo, *M. leidyi*. *Development*, *124*, 1999–2006.
Martindale, M. Q., & Henry, J. Q. (1997b). The Ctenophora. In S. Gilbert, & A. Raunio (Eds.), *Embryology, the construction of life* (pp. 87–111). Sinauer Press.
Martindale, M. Q., & Henry, J. Q. (1999). Intracellular fate mapping in a basal metazoan, the ctenophore *M. leidyi*, reveals the origins of mesoderm and the existence of indeterminate cell lineages. *Developmental Biology*, *214*, 243–257.
Martindale, M. Q., & Henry, J. Q. (2015). The embryology of the Ctenophora (chapter 1). In A. Wanninger (Ed.), *Evolutionary developmental biology of invertebrates* (pp. 179–201). Germany: Springer.
Martindale, M. Q., Meier, S., & Jacobson, A. G. (1987). Mesodermal metamerism in the teleost, *Oryzias latipes* (the Medaka). *Journal of Morphology*, *193*(3), 241–252.
Martindale, M. Q., Pang, K., & Finnerty, J. R. (2004). Investigating the origins of triploblasty: "Mesodermal" gene expression in a diploblastic animal, the sea anemone, *N. vectensis* (phylum, Cnidaria; Class Anthozoa). *Development*, *131*, 2463–2474.
Martinson, A. S., van Rossuma, D. B., Laydenb, M. J., Rhodesa, S. A., Martindale, M. Q., & Jegla, T. (2014). Functional evolution of Erg potassium channel gating reveals an ancient origin for IKr. *PNAS*, *111*(15), 5712–5717. https://doi.org/10.1073/pnas.1321716111.
Matus, D. Q., Pang, K., Marlow, H., Dunn, C., Thomsen, G. H., & Martindale, M. Q. (2006). Molecular evidence for deep evolutionary roots of bilaterality in animal development. *PNAS*, *103*, 11195–11200.
Matus, D. Q., Thomsen, G. H., & Martindale, M. Q. (2006). "Dorso-ventral" genes are asymmetrically expressed and involved in germ layer demarcation during cnidarian gastrulation. *Current Biology*, *16*, 499–505.
Metschnikoff, E. (1882). Vergleichend-embryologische studien. *Zeitschrift für wissenschaftliche Zoologie*, *36*, 433–444.
Metschnikoff, E. (1885). Vergleichend Embryologische Studien. IV. Ueber die gastrulation und mesodermbildung der Ctenophoren. *Zeitschrift für wissenschaftliche Zoologie*, *42*, 648–656.
Mitchell, D. G., Edgar, A., & Martindale, M. Q. (2021). Improved histological fixation of gelatinous marine invertebrates. *Frontiers in Zoology*, *18*, 29. https://doi.org/10.1186/s12983-021-00414-z.
Moran, Y., Barzilai, M. G., Liebeskind, B. J., & Zakon, H. H. (2015). Evolution of voltage-gated ion channels at the emergence of Metazoa. *Journal of Experimental Biology*, *218*(4), 515–525.
Moran, Y., Fredman, D., Praher, D., Li, X. Z., Wee, L. M., Rentzsch, F., et al. (2014). Cnidarian microRNAs frequently regulate targets by cleavage. *Genome Research*, *24*(4), 651–663.
Moran, Y., Praher, D., Schlesinger, A., Ayalon, A., Tal, Y., & Technau, U. (2013). Analysis of soluble protein contents from the nematocysts of a model sea anemone sheds light on venom evolution. *Marine Biotechnology*, *15*(3), 329–339.
Moroz, L. L., Kocot, K. M., Citarella, M. R., Dosung, S., Norekian, T. P., Povolotskaya, I. S., et al. (2014). The Ctenophore genome and the evolutionary origins of neural systems. *Nature*, *510*, 109–114. https://doi.org/10.1038/nature13400.
Nakanishi, N., & Martindale, M. Q. (2018). Crispr knockouts reveal an endogenous role for ancient neuropeptides in regulating developmental timing in a sea anemone. *eLife*, *7*, e39742. https://doi.org/10.7554/eLife.39742.

Oliveri, P., Walton, K. D., Davidson, E. H., & McClay, D. R. (2006). Repression of mesodermal fate by foxa, a key endoderm regulator of the sea urchin embryo. *Development, 133*(21), 4173–4181. https://doi.org/10.1242/dev.02577. 17038513.

Pang, K., & Martindale, M. Q. (2008). Developmental expression of homeobox genes in the ctenophore, *M. leidyi. Development Genes and Evolution, 218*, 307–319.

Pang, K., Ryan, J. F., Baxevanis, A. D., & Martindale, M. Q. (2010). Genomic insights into Wnt signaling in an early diverging metazoan, the ctenophore *M. leidyi. EvoDevo, 1*(10), 1–15. https://doi.org/10.1186/2041-9139-1-10.

Pang, K., Ryan, J. F., Baxevanis, A. D., & Martindale, M. Q. (2011). Evolution of the TGF-β signaling pathway and its potential role in the Ctenophore, *M. leidyi. PLoS One, 6*(9), e24152. https://doi.org/10.1371/journal.pone.0024152.

Pang, K., Ryan, J. F., Mullikin, J. C., Baxevanis, A. D., & Martindale, M. Q. (2010). Genomic insights into Wnt signaling in an early diverging metazoan, the ctenophore *M. leidyi. EvoDevo, 1*(1), 1–15.

Praher, D., Zimmermann, B., Genikhovich, G., Columbus-Shenkar, Y., Modepalli, V., Aharoni, R., et al. (2017). Characterization of the piRNA pathway during development of the sea anemone *N. vectensis. RNA Biology, 14*(12), 1727–1741.

Putnum, N., Srivastava, M., Hellsten, U., Dirks, B., Chapman, J., Salamov, A., et al. (2007). Sea anemone genome reveals ancestral eumetazoan gene repertoire and genomic organization. 2007. *Science, 317*, 86–94. https://doi.org/10.1126/science.1139158.

Raff, R. R. (1992). Direct-developing sea urchins and the evolutionary reorganization of early development. *BioEssays, 14*(4), 211–218. https://doi.org/10.1002/bies.950140403.

Ramon, J. M., Ellison, T., Angelini, T., & Martindale, M. Q. (2019). Regeneration in the ctenophore *M. leidyi* occurs in the absence of a blastema, requires cell division, and is temporally separable from wound healing. *BMC Biology, 17*(10), 1–25.

Ramon-Mateu, J., Edgar, A., Mitchell, D., & Martindale, M. Q. (2021). Experimental methods to study ctenophore whole-body regeneration. In S. Blanchoud, & B. Galliot (Eds.), *Whole body regeneration: Methods and protocols* EDS.

Renfer, E., Amon-Hassenzahl, A., Steinmetz, P. R., & Technau, U. (2010). A muscle-specific transgenic reporter line of the sea anemone, *N. vectensis. PNAS, 107*(1), 104–108.

Röttinger, E., Dahlin, P., & Martindale, M. Q. (2012). A provisional cnidarian gene regulatory network for "endomesoderm" specification: The inputs of Wnt/β-catenin signaling. *PLoS Genetics, 2012.* http://www.plosgenetics.org/doi/pgen.1003164.

Ryan, J. F., Mazza, M. E., Pang, K., Matus, D. Q., Baxevanis, A. D., Martindale, M. Q., et al. (2007). Pre-bilaterian origins of the hox cluster and the hox code: Evidence from the sea anemone, *N. vectensis. PLoS One, 2*(1), e153. https://doi.org/10.1371/journal.pone.0000153.

Ryan, J. F., Pang, K., Mullikin, J. C., Martindale, M. Q., & Baxevanis, A. D. (2010). The homeodomain complement of the ctenophore *M. leidyi* suggests that Ctenophora and Porifera diverged prior to the ParaHoxozoa. *EvoDevo, 1*(1), 1–18.

Ryan, J. F., Pang, K., Schnitzler, C. E., Nguyen, A., Moreland, R. T., Simmons, D. K., et al. (2013). The Genome of the Ctenophore *M. leidyi* and Its Implications for Cell Type Evolution. *Science, 342*, 1336. https://doi.org/10.1126/science.1242592.

Sachkova, M. Y., Singer, S. A., Macrander, J., Reitzel, A. M., Peigneur, S., Tytgat, J., et al. (2019). The birth and death of toxins with distinct functions: A case study in the sea anemone Nematostella. *Molecular Biology and Evolution, 36*(9), 2001–2012.

Salinas-Saavedra, M., & Martindale, M. Q. (2020). Par protein localization during the early development of *M. leidyi* suggests different modes of epithelial organization in the Metazoa. *eLife, 9*, e54927. https://doi.org/10.7554/eLife.54927.

Salinas-Saavedra, M., Rock, A. Q., & Martindale, M. Q. (2018). Germ layer specific regulation of cell adhesion: Insight in to the evolution of mesoderm. *eLife*, 7, e36740. https://doi.org/10.7554/eLife.36740.

Salinas-Saavedra, M., Stephenson, T. Q., Dunn, C. W., & Martindale, M. Q. (2015). Par system components are asymmetrically localized in ectodermal epithelia, but not during early development in the sea anemone *N. vectensis*. *EvoDevo*, *6*(1), 1–20. https://doi.org/10.1186/s13227-015-0014-6.

Salinas-Saavedra, M., Wikramanayake, A., & Martindale, M. Q. (2018). β-catenin has an ancestral role in cell fate specification but not cell adhesion. *bioRxiv*, 520957. https://doi.org/10.1101/520957.

Schnitzler, C., Pang, K., Powers, M., Reitzel, A. M., Ryan, J. F., Simmons, D., et al. (2012). Genomic organization, evolution, and expression of photoprotein and opsin genes in *Mnemiopsis Leidyi*: A new view of ctenophore photocytes. *BMC Biology*, *10*, 107. https://doi.org/10.1186/1741-7007-10-107.

Schnitzler, C. E., Simmons, D. K., Pang, K., Martindale, M. Q., & Baxevanis, A. D. (2014). Expression of multiple Sox genes through embryonic development in the ctenophore *M. leidyi* is spatially restricted to zones of cell proliferation. *EvoDevo*, *5*(1), 1–17.

Scholtz, G., & Wolff, C. (2013). Arthropod embryology: Cleavage and germ band development. In A. Minelli, G. Boxshall, & G. Fusco (Eds.), *Arthropod biology and evolution* (pp. 63–89). Berlin, Heidelberg: Springer.

Schwaiger, M., Schönauer, A., Rendeiro, A. F., Pribitzer, C., Schauer, A., Gilles, A. F., et al. (2014). Evolutionary conservation of the eumetazoan gene regulatory landscape. *Genome Research*, *24*(4), 639–650.

Sebé-Pedrós, A., Chomsky, E., Pang, K., Lara-Astiaso, D., Gaiti, F., Mukamel, Z., et al. (2018). Early metazoan cell type diversity and the evolution of multicellular gene regulation. *Nature Ecology & Evolution*, *2*(7), 1176–1188. https://doi.org/10.1038/s41559-018-0575-6.

Servetnick, M., Steinworth, B., Babonis, L. S., Simmons, D., Salinas-Saavedra, M., & Martindale, M. Q. (2017). Cas9-mediated excision of Nematostella brachyury disrupts endoderm development, pharynx formation and oral-aboral patterning. *Development*, *144*, 2951–2960. https://doi.org/10.1242/dev.145839.

Verheyden, J. M., & Sun, X. (2017). Embryology meets molecular biology: Deciphering the apical ectodermal ridge. *Developmental Biology*, *429*(2), 387–390.

Whelan, N. V., Kocot, K. M., Moroz, L. L., & Halanych, K. M. (2015). Error, signal, and the placement of Ctenophora sister to all other animals. *PNAS*, *112*(18), 5773–5778.

Whelan, N. V., Kocot, K. M., Moroz, T. P., Mukherjee, K., Williams, P., Paulay, G., et al. (2017). Ctenophore relationships and their placement as the sister group to all other animals. *Nature Ecology & Evolution*, *1*, 1737–1746.

Wijesena, N., & Martindale, M. Q. (2018). Reengineering the primary body axis by ectopic embryonic cWnt signaling. *Current Biology*, *28*(5), R206–R207.

Wikramanayake, A. H., Hong, M., Lee, P. N., Pang, K., Byrum, C. A., Bince, J. M., et al. (2003). An ancient role for nuclear beta-catenin in the evolution of axial polarity and germ layer segregation. *Nature*, *426*, 446–450.

Wikramanayake, A. H., Huang, L., & Klein, W. H. (1998). beta-Catenin is essential for patterning the maternally specified animal–vegetal axis in the sea urchin embryo. *PNAS*, *95*, 9343–9348.

Xu, X., Li, G., Li, C., Zhang, J., Wang, Q., Simmons, D. K., et al. (2019). Evolutionary transition between invertebrates and vertebrates via methylation reprogramming in embryogenesis. *National Science Review*, *6*(5), 993–1003. https://doi.org/10.1093/nsr/nwz064.

Yamada, A., Martindale, M. Q., Fukui, A., & Tochinai, S. (2010). Highly conserved functions of the Brachyury gene on morphogenetic movements: Insight from the early-diverging phylum Ctenophora. *Developmental Biology*, *339*, 212–222.

Zattara, E. E., Fernández-Álvarez, F. A., Hiebert, T. C., Bely, A. E., & Norenburg, J. L. (2019). A phylum-wide survey reveals multiple independent gains of head regeneration in Nemertea. *Proceedings of the Royal Society B*, *286*, 20182524. https://doi.org/10.1098/rspb.2018.2524.

CHAPTER FIVE

Past, present and future of *Clytia hemisphaerica* as a laboratory jellyfish

Evelyn Houliston[a,*], Lucas Leclère[a], Catriona Munro[a,b], Richard R. Copley[a], and Tsuyoshi Momose[a]

[a]Sorbonne Université, CNRS, Laboratoire de Biologie du Développement de Villefranche-sur-mer (LBDV), France
[b]Center for Interdisciplinary Research in Biology, Collège de France, PSL Research University, Paris, France
*Corresponding author: e-mail address: houliston@imev-mer.fr

Contents

Abstract

The hydrozoan species *Clytia hemisphaerica* was selected in the mid-2000s to address the cellular and molecular basis of body axis specification in a cnidarian, providing a reliable daily source of gametes and building on a rich foundation of experimental embryology. The many practical advantages of this species include genetic uniformity

Current Topics in Developmental Biology, Volume 147
ISSN 0070-2153
https://doi.org/10.1016/bs.ctdb.2021.12.014

of laboratory jellyfish, derived clonally from easily-propagated polyp colonies. Phylogenetic distance from other laboratory models adds value in providing an evolutionary perspective on many biological questions. Here we outline the current state of the art regarding available experimental approaches and *in silico* resources, and illustrate the contributions of *Clytia* to understanding embryo patterning mechanisms, oogenesis and regeneration. Looking forward, the recent establishment of transgenesis methods is now allowing gene function and imaging studies at adult stages, making *Clytia* particularly attractive for whole organism biology studies across fields and extending its scientific impact far beyond the original question of interest.

1. Introduction and historical background

Our original, pragmatic, motivation to select *Clytia hemisphaerica* as a laboratory model was because of unique features suited for studying embryonic axis specification. Specifically, we were seeking an experimentally tractable species in which embryo polarity was established de novo following fertilization, rather than being directed by positional cues ("maternal determinants") in the egg. Such determinants, often as localized mRNAs, are responsible for axis specification in almost all traditional model species, but the key events of symmetry breaking are challenging to image and analyse because they occur over protracted periods of oogenesis within the mother's ovary (Bashirullah, Cooperstock, & Lipshitz, 1998). An important body of previous experimental embryology indicated at the time (mid-2000s) that animals from the clades Ctenophora or Cnidaria might offer uniquely accessible material for studying embryo polarity establishment. Dye tracing and egg manipulation experiments on many species from these taxa, including the cnidarian *Clytia* (a.k.a. *Phialidium*) hydrozoan jellyfish species, had indicated that the orientation of their single principle body axis was determined de novo in the fertilized egg, the position of the zygote nucleus dictating that of the future "oral" pole (Freeman, 1976, 1977, 1981a). Thus embryo polarity establishment mechanisms in cnidarian and ctenophore species could potentially be analyzed in embryos developing in sea water on the bench.

Against expectations it turned out that the embryonic axis in *Clytia* is positioned by maternal localized mRNAs that cause polarized Wnt pathway activation, revealing close and unexpected parallels with axis determination in traditional model species from Bilateria and allowing the historical experiments to be reinterpreted in a new light (see Section 4). By then, *Clytia* had proved its worth as an experimental model, and the initial investment in techniques and resources provided a platform for further studies on many topics. This included early realization that the ovary of female *Clytia* has particular properties that make it an exceptional model for studying the regulation of oogenesis, oocyte maturation, and spawning, as detailed in Section 3. Furthermore, *Clytia's* phylogenetic distance from other laboratory models, including the increasingly popular cnidarian model *Nematostella vectensis* (Layden, Rentzsch, & Röttinger, 2016), proved increasingly to have significant added value in providing an evolutionary perspective on many processes, as illustrated across different sections of this chapter. In parallel the *Clytia* model has been exploited for more explicit "EvoDevo" studies to map the genomic changes accompanying animal evolution, notably in the early years by Michael Manuel and colleagues in Paris, with possibilities augmented more recently by availability of the whole genome sequence as outlined in Section 6.

The choice of *Clytia hemisphaerica* among many candidate cnidarian species available in Villefranche was essentially based on ease of culture and suitability for embryo experimentation. *Clytia* jellyfish reliably provide a daily supply of large (180 μm diameter) transparent eggs, devoid of extracellular envelopes or jelly, which are robust to manipulate and develop reliably. Crucially, previous studies on the site had shown that *C. hemisphaerica* could be cultured as "immortal" vegetatively propagating polyp colonies supplying medusae by constant budding (Fig. 1; Carré & Carré, 2000). The adults are a convenient size to be maintained in modestly-dimensioned beaker or tank based aquarium systems, and survive on a simple diet of *Artemia* nauplii (Lechable et al., 2020). Other unanticipated practical advantages became apparent as the model developed. The clonal origin of jellyfish by vegetative budding from stable polyp colonies was a huge boon, not only increasing reproducibility between experiments but meaning that when ESTs (Chevalier, Martin, Leclère, Amiel, & Houliston, 2006; Forêt et al., 2010) and later full transcriptome and genome resources (Leclère et al., 2019) were established, the sequences exactly matched those of the laboratory animals. This alleviated the polymorphism problems experienced for many marine animal models collected from wild populations, facilitating successful selection of sequence targets for gene knockdown approaches such as Morpholino antisense and later CRISPR/Cas9 (Section 2). The partial temperature dependence of colony sex determination (Carré & Carré, 2000) allowed a set of highly genetically-related core colonies to be established via self-crossing (Leclère et al., 2019). These colonies can be conveniently distributed to colleagues around the world as cuttings on glass slides (Lechable et al., 2020).

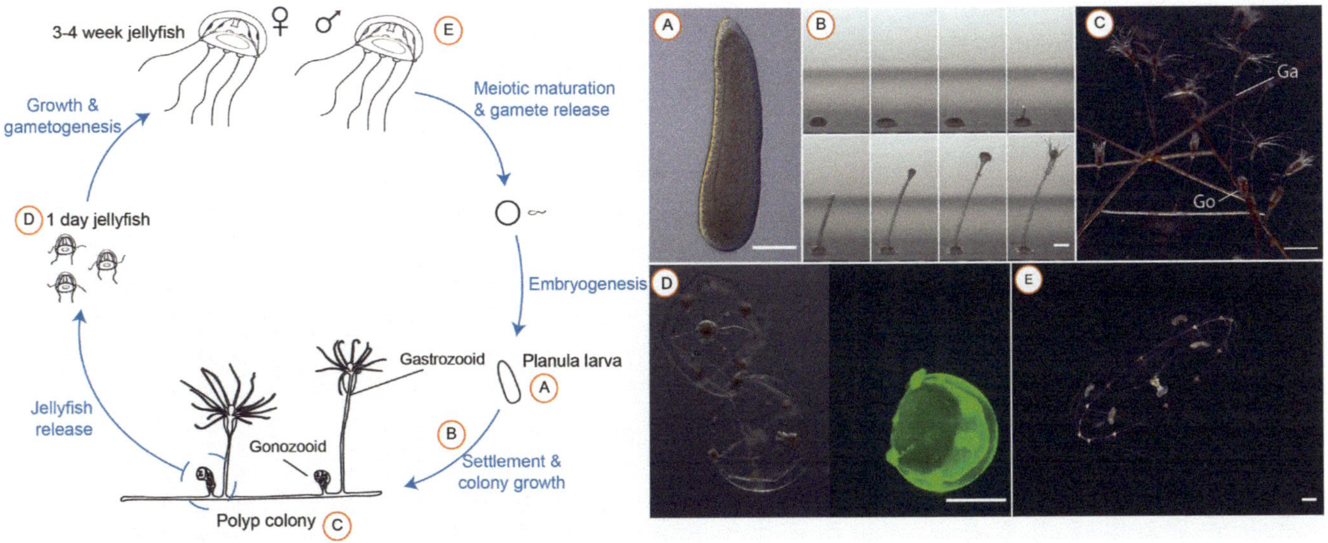

Fig. 1 *Clytia hemisphaerica* life cycle. On the left is a schematic of the full cycle, as maintained in the laboratory. Images on the right illustrate different stages. (A) 2-day planula larva. Its simple organization comprises two tissue layers: epidermis (ectoderm) and gastrodermis (endoderm) and an elongated "oral-aboral" axis (oral at the top). (B) Images from a time-lapse recording of planula metamorphosis recorded over 2 days following settlement on a substrate. The form drastically changes to generate the primary feeding polyp (gastrozooid). (C) Extension of a stolon from the base of the primary polyp and subsequent emergence of many polyps, both gastrozooids and gonozooids (budding polyps), leads to formation of a colony. Ga, gastrozooid; Go, gonozooid. (D) Bright field (left) and fluorescence (right) images of wildtype (bottom) and *GFP1/GFP2* knockout (top) jellyfish shortly following budding. Multiple GFP gene loci were simultaneously mutated by CRISPR/Cas9. (E) Adult jellyfish. Scale bars 50 μm for A, 200 μm for B, 500 μm for D, and 1 mm for C and E. *Lifecycle diagram adapted from https://commons.wikimedia.org/wiki/File:Clytia_hemisphaerica_Lifecycle.svg. (A–D) Adapted from https://commons.wikimedia.org/wiki/File:Clytia_hemisphaerica_polyp_colony.jpg, https://commons.wikimedia.org/wiki/File:Clytia_hemisphaerica_day2_planula.jpg, https://commons.wikimedia.org/wiki/File:Metamorphosis_of_Clytia_hemisphaerica.jpg, https://commons.wikimedia.org/wiki/File:Baby-medusae.jpg. (E) Adapted from Lechable, M., Jan, A., Duchene, A., Uveira, J., Weissbourd, B., & Gissat, L. (2020). An improved whole life cycle culture protocol for the hydrozoan*

Since the first "molecular era" papers on *Clytia* (Chevalier et al., 2006; Jager, Quéinnec, Houliston, & Manuel, 2006; Momose, Derelle, & Houliston, 2008; Momose & Houliston, 2007) which introduced ESTs, in situ hybridization and gene function analysis via Morpholino/mRNA microinjection, technical possibilities have steadily widened and techniques refined. These advances, along with the progressive accumulation of *in silico* resources, are detailed in Section 2 and Table 1. In parallel, attention has extended from embryos and larvae to the adult medusa stage. This one-centimeter diameter transparent, actively swimming and sexually reproducing animal (Fig. 1E) is already showing great promise as a model in the fields of regeneration and neural systems biology (see Section 5). *Clytia* thus illustrates how the impact of a new model organism can extend far beyond the original field of interest.

Table 1 Methods and resources.
A. Methods

Method	References—first citation and updated methods	Target life cycle stage	Application
Animal culture	Lechable et al. (2020)	Whole life cycle	Planula, polyp and medusa culture
Morpholino antisense oligo	Momose and Houliston (2007); Amiel, Leclère, Robert, Chevalier, and Houliston (2009)	Oocyte to planula	Loss of function
mRNA microinjection	Momose and Houliston (2007); Amiel et al. (2009)	Oocyte to planula	Gain of function, protein localization
CRISPR/Cas9 gene KO	Momose et al. (2018); Quiroga Artigas et al., (2018), Quiroga Artigas et al. (2020)	Whole life cycle	Loss of function
Transgenesis	Weissbourd et al. (2021)	Whole life cycle	Gain of function
Cell type specific ablation	Weissbourd et al. (2021)	Whole life cycle	Specific elimination of cells
Pharmacological treatments (NB target specificity is not guaranteed)	Momose and Houliston (2007); Amiel et al. (2009); Sinigaglia et al. (2020)	Whole life cycle	Loss- or Gain-of-function

Continued

Table 1 Methods and resources.—cont'd

A. Methods

Method	References—first citation and updated methods	Target life cycle stage	Application
In situ hybridization	Chevalier et al. (2006); Sinigaglia, Thiel, Hejnol, Houliston, and Leclère (2018)	Whole life cycle	Visualizing gene expression patterns
Immuno/phalloidin staining	Denker, Manuel, Leclère, Le Guyader, and Rabet (2008); Fourrage, Chevalier, and Houliston (2010), Sinigaglia et al. (2020)	Whole life cycle	Visualizing protein localization
Electron microscopy	Thomas, Freeman, and Martin (1987); Carré and Carré (2000); Kraus, Chevalier, and Houliston (2020); Roosen-Runge (1962)	Whole life cycle	Ultrastructural descriptions of cell types and tissue organization
Medusa dissection/ grafting	Kamran et al. (2017); Sinigaglia et al. (2020)	Medusa	Regeneration experiments
DIC microscopy for live imaging	Malamy and Shribak (2018)	Medusa	Live imaging of epithelial cells

B. *In silico* resources

Description	References	Database	Details
Whole genome sequence	Leclère et al. (2019)	marimba.obs-vlfr.fr NCBI/EBI	Improved genome assembly (N50 > 1 Mb)
Genome browser	Leclère et al. (2019)	marimba.obs-vlfr.fr ensembl.org	Marimba combines genomic and gene expression data
Transcriptome assembly	Lapébie et al. (2014); Quiroga Artigas et al. (2018); Leclère et al. (2019)	marimba.obs-vlfr.fr NCBI/EBI	Genome-guided transcriptome available

Table 1 Methods and resources.—cont'd
B. *In silico* resources

Description	References	Database	Details
Expression profiling (RNAseq)	Lapébie et al. (2014); Quiroga Artigas et al. (2018); Condamine et al. (2019); Leclère et al. (2019); Sinigaglia et al. (2020)	NCBI/EBI	Wnt-manipulated Gastrulae; Ovary tissues; Tentacle bulb regions; Life-cycle stages; Regenerating manubrium time course
Expressed Sequence Tags (EST)	Chevalier et al. (2006); Forêt et al. (2010)	NCBI/EBI	Mixed stage libraries
Single cell RNA seq: Data and analysis pipelines	Chari et al. (2021)	github.com/pachterlab/CWGFLHGCCHAP_2021	Cell types from adult female jellyfish, control and starved
Ontology terms for anatomy and developmental stages	Manuscript in preparation	www.ebi.ac.uk/ols/ontologies/clyh	Reviewed by the OBO Foundry

2. Practical state of the art

2.1 Animal culture and life cycle

Clytia hemisphaerica culture methods have been refined progressively over the last 15 years, and a detailed update is available (Lechable et al., 2020). A specific challenge compared to polyp-only cnidarian model species (notably *Hydra, Nematostella* and *Hydractinia*) is maintenance of the fragile and planktonic jellyfish (medusa) stage (Fig. 1D and E), which requires constant gentle water flow. A convenient solution is to use adapted "Kreisel" tanks for juvenile and adult stages. *Clytia* jellyfish reach sexual maturity 2–3 weeks after budding from specialized polyps called gonozooids (Fig. 1C). They then release gametes for the rest of their lifespan, typically around 2 months total. Between 400 and 1000 eggs can be collected every day from a batch of 30–40 females in one tank, and successfully fertilized within one hour of spawning by mixing with sperm released in parallel from a few males. Embryogenesis (see Section 3) results in formation of a planula larva (Fig. 1A), described in detail in Section 4. After metamorphosis of the planula into a primary feeding polyp (gastrozooid) on glass slides (Fig. 1B), extension of the stolon network creates a polyp colony, which is nourished by prey captured by the gastrozooids. Individual gastrozooids and gonozooids periodically atrophy, but colony growth allows their indefinite renewal. The colony is thus an extremely convenient form to amplify, maintain and distribute genetically uniform strains including mutants without need for crossing.

2.2 Functional genomics

Functional approaches, notably gene loss- and gain-of-function, enable a range of molecular and genetic studies. For *C. hemisphaerica* as a model, the first successful approach was microinjection of Morpholino antisense oligonucleotide (MO) and synthetic mRNAs into eggs before fertilization. MOs designed to block mRNA splicing or translation generated reproducible loss-of polarity phenotypes for developmental regulators, starting with the Wnt receptors Fz1 and Fz3 (Momose & Houliston, 2007). Other examples are illustrated in Sections 3 and 4 below. Injection of in vitro synthesized mRNA allows gene overexpression, verification of MO phenotype specificity, and expression of fluorescent protein markers to visualize protein localization (Momose, Kraus, & Houliston, 2012).

MO and mRNA egg microinjection can address biological events only between early blastula stage and metamorphosis, so other gene manipulation methods are required for studying processes in the polyp and jellyfish stages. Efforts to achieve reliable and effective targeted gene knockout started in the early 2010s, initially exploring the use of TALEN (Zhang et al., 2011) before switching to the CRISPR/Cas9 system (Momose et al., 2018). We were able to establish a protocol that consistently generates gene knock-out *Clytia* strains (i.e., F0 polyp colonies) with very low mosaicism in a single generation, involving prediction of the dominant mutation that will be induced by each sgRNA (Momose et al., 2018). This method exploits the observation that the Microhomology Mediated End-Joining (MMEJ) pathway is the principal pathway for DNA double strand break repair active in the early *Clytia* embryo. It has so far been used to knock-out endogenous GFPs (Fig. 1D) as well as to identify key genes mediating oocyte maturation and spawning (see Section 3).

Development of high-efficiency transgenesis and gene knock-in approaches has been more difficult. The MMEJ pathway likely favors rapid and error-prone DNA repair over homology-mediated insertion, explaining the relatively low incorporation of transgenes achieved in trials with CRISPR/Cas9 methods using various template DNA types. Observed accumulation of injected DNA around nuclei (T. Momose and B. Weissbourd, unpublished data) suggests that effective DNA delivery to the genomic target is also an issue. A recent breakthrough is the successful establishment of a transposon-mediated transgenesis protocol (manuscript in preparation), through extensive collaboration between Caltech (B. Weissbourd and D. J. Anderson) and Villefranche teams. This achieves highly efficient random transgenesis using the Tol2 transposon vector system, co-injected with purified Tol2 protein (Weissbourd et al., 2021). A first study successfully employed this method to create a stable F1 transgenic strain expressing the calcium indicator GCaMP in RFamide-expressing neurons in jellyfish using the promoter of the RFamide precursor gene *ChePP5* (Weissbourd et al., 2021). The use of polycistronic expression with 2A-peptide (Liu et al., 2017), further allowed metronidazole-inducible specific ablation of RFamide positive neurons. So far, strong and uniform transgene expression has been obtained using promoters from a non-muscle actin gene and from the RFamide precursor gene *ChePP5*. Methods to predict and select promoter sequences are currently not fully established. Further development will be aided by more information concerning the gene regulatory landscape in *Clytia*.

2.3 Medusa cell type atlas and single cell sequencing

The medusa (jellyfish) stage is the sexual phase of the life cycle, and has much more complex anatomy than the polyp stage. *Clytia* jellyfish are tetraradial, with 4 gonads each connected via radial canals to the manubrium (mouth) in the center and the circular canal around the periphery (Fig. 2A). The circular canal bears the tentacle bulbs, which continually replenish the tentacles. Adjacent to the circular canal, positioned between each tentacle bulb, are sensory balance organs called statocysts (Singla, 1975). The nervous system consists of various nerve nets and two nerve rings running around the margin of the bell, connected to the statocysts. The jellyfish has both smooth muscle (radially orientated) and striated muscle (circularly orientated). The nervous system and muscles enable complex behaviors such as swimming and feeding (Houliston, Momose, & Manuel, 2010; Leclère & Röttinger, 2016; Weissbourd et al., 2021). Until recently, the diversity of different cell types comprising the *Clytia* jellyfish was underexplored. A significant addition to available resources is a cell type atlas documenting the presence, distribution

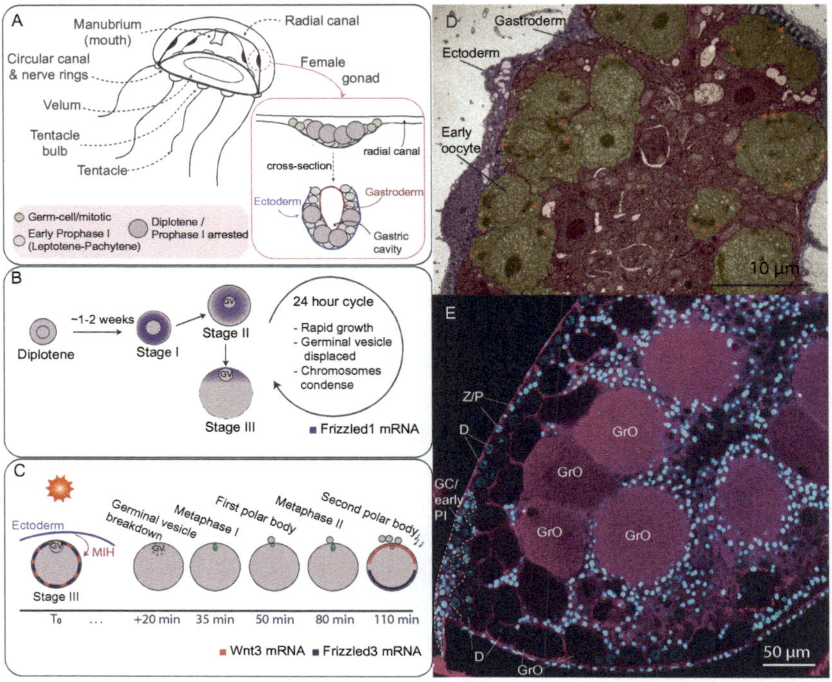

Fig. 2 Jellyfish anatomy and gametogenesis. (A) Schematic of a female jellyfish, with the organization of the female gonad highlighted. (B) Schematic of oocyte growth in a prophase I arrested oocyte. In fully grown jellyfish, cohorts of stage I oocytes (approx 50 μm diameter) grow daily to stage III (approx 180 μm diameter) prior to spawning. Fz1 mRNA becomes concentrated in the animal cytoplasm during this growth phase. (C) Schematic of oocyte maturation. Light triggers the release of oocyte maturation hormone (MIH) from cells in the gonad epidermis, triggering oocyte maturation and spawning. Wnt3 and Fz3 mRNAs are localized around the periphery of the stage III oocyte. During meiotic maturation, Fz3 becomes localized to the vegetal cortex and then Wnt3 to the animal cortex. (D) False coloured electron micrograph from the ovary of a 1 week-old jellyfish. Ectoderm is highlighted in blue, gastroderm in red and early oocytes in green (arrows indicate the perinuclear "nuage" characteristic of the germ cells). (E) Confocal microscopy image of a two week ovary by Anna Ferraioli. Nuclei are stained with hoechst (cyan), and actin is stained with phalloidin (magenta). GC/Early PI, Germ cell, Early Prophase I; Z/P, Zygotene/Pachytene; D, Diplotene; GrO, Growing oocyte, GVBD, Germinal vesicle breakdown. *Part A: Adapted from https://commons.wikimedia.org/wiki/File:Clytia_hemisphaerica_Lifecycle.svg, https://commons.wikimedia.org/wiki/File:Clytia_hemisphaerica_female_gonad.svg.*

and transcriptomes of 36 distinct cell types (Chari et al., 2021). This cell atlas was derived from analysis of single cell transcriptome data (scRNAseq) combined with in situ hybridization to localize the expression of marker genes.

3. Gametogenesis

The *Clytia* gonad is one of the simplest organs in the animal kingdom. Its basic structure comprises the two canonical cnidarian body layers, with the basal surfaces of the epidermal and gastrodermal epithelial cells facing each other across a common, loosely-structured extracellular matrix (ECM). Within this ECM layer, gametogenic cells proliferate and differentiate. In female medusae, daily batches of small oocytes grow from around 50 μm diameter to the full size of around 180 μm ready for the spawning cue (Amiel & Houliston, 2009; Jessus, Munro, & Houliston, 2020; Fig. 2A). The four gonads hang as pouches beneath the medusa bell, with the gonad gastrodermis derived from localized thickening of the radial canals and made up of specialized digestive-absorptive and secretory cell types (Fig. 2A and D) (Chari et al., 2021).

Two principal features render *Clytia* gonads particularly suited to gametogenesis research compared to the most studied animal species in this field (notably *Xenopus*, mouse and starfish; see Jessus et al., 2020). Firstly, they are highly transparent, allowing the entire progression of events to be observed in whole animals, live or fixed. Secondly, they function largely autonomously, such that gonads separated by simple dissection from adult mature jellyfish support repeated cycles of gamete production and release for several days until nutritional resources (including recycled somatic tissue) are exhausted (Freeman, 1987; Takeda et al., 2018). So far, molecular studies of oogenesis have focussed mainly on females. This reflects the historical interest in maternal determinants for axis specification, as outlined in Sections 1 and 3. Identification of the maternal determinant mRNAs Wnt3, Fz1 and Fz3 allowed the cellular mechanisms responsible for their localization to be linked to different oogenesis events: Fz1 mRNA localizes to the animal hemisphere (future oral) cytoplasm by a microtubule-based mechanism during the major daily oocyte growth phase (Fig. 2B), while Wnt3 and Fz3mRNA segregate respectively to oral and aboral domains of the oocyte cortex during the two hour process of light-induced oocyte maturation (Fig. 2C; Amiel & Houliston, 2009).

3.1 Oocyte maturation

Oocyte meiotic maturation transforms a fully grown, tetraploid, diplotene oocyte into a haploid gamete ready to be released from the gonad for fertilization. Completion of the first meiotic division terminates in emission of a first polar body containing one set of replicated homologous chromosomes, the second meiotic division separates sister chromatids and expels one set in the second polar body. In *Clytia*, meiotic maturation ends in an unfertilized egg containing a haploid female pronucleus (Fig. 2C). These events can be followed in isolated gonads exposed to light after darkness, or in isolated oocytes treated with oocyte maturation hormone (MIH) or with cell-permeable cAMP analogs (Amiel et al., 2009; Takeda et al., 2018). As extensively discussed recently in a recent review (Jessus et al., 2020) they are regulated by core kinase systems highly conserved across the animal kingdom: the Cdk1/Cyclin B complex (also known as MPF) active during each M phase, MAP-kinase which progressively activates as maturation proceeds, and cAMP-dependent kinase (PKA) activation during the initial triggering mechanism (Freeman & Ridgway, 1988; Takeda, Kyozuka, & Deguchi, 2006). A study focussed on *Clytia* orthologues of the animal oocyte cytoplasmic kinase Mos, which initiates MAP-kinase activation by phosphorylating the MAP-kinase kinase MEK in animal oocytes, illustrates how *Clytia* can reveal both core and novel features of this regulation (Amiel et al., 2009). Mos activity is regulated by the timing of its translation during meiotic maturation, as extensively analysed in the *Xenopus* and mouse vertebrate models (Gebauer & Richter, 1997). In *Clytia*, hydrozoan Mos gene family expansion has generated several paralogs, including two expressed in oocytes. Morpholino-mediated inhibition of their translation revealed a conserved role for *Clytia CheMos1* in the oocyte-specific processes of polar body formation and post-meiotic arrest, but suggest that *CheMos2* has evolved a distinct function in preparing oocytes for maturation. Since the kinase activities of the two proteins appear equivalent, this difference appears to be mediated by evolution of UTR sequence mediating translation timing (Amiel et al., 2009). It will be of interest to explore in *Clytia* the role and evolution of other oocytes regulatory kinases such as PKA and Greatwall, to help understand how oocyte maturation regulation has adapted to different species-specific requirements to synchronize gamete release and thus optimize reproductive success.

3.2 Triggering oocyte maturation and spawning

Studies exploiting CRISPR/Cas9-mediated gene knockdown have dissected the initiation mechanism for oocyte meiotic maturation in *Clytia*. This process takes about two hours and ends with the release of unfertilized eggs via localized rupturing of the gonad epidermis. This spawning process in females, and also the parallel process of sperm activation and release in males, is initiated by a light cue after darkness (Freeman, 1987; Miller, 1979; Roosen-Runge, 1962; Takeda et al., 2018). Specialized cells with neural-like morphology in the gonad ectoderm co-express one of 10 *Clytia* opsin proteins, Opsin9, and precursors of the oocyte maturation-inducing neuropeptide hormone MIH (Quiroga Artigas et al., 2018; Takeda et al., 2018). Upon light stimulation, Opsin9 mediates the release of MIH both inside and outside the gonad. Subsequent binding of MIH to a specific G-protein coupled receptor in the oocyte membrane initiates activation of meiotic maturation via GalphaS, cytoplasmic cAMP rise and PKA activation (Fig. 2C) (Quiroga Artigas et al., 2020; Takeda et al., 2006). Thus medusae from F0 female polyp colonies carrying mutations in the *Opsin9* or MIH receptor (*MIHR*) genes accumulate fully grown oocytes but fail to spawn. In both cases the failure of light-induced oocyte maturation can be overridden by treatment with cell-permeable cAMP analogs.

3.3 Origins of the germ cells

In both male and female medusae, the germ cells originate from a population of cells positioned within the gastrodermis close to the canal. These are characterized by small cytoplasm: nuclear ratio, expression of a "Germ line multipotency" gene set (Juliano, Swartz, & Wessel, 2010) including Piwi and Nanos genes (Leclère et al., 2012), and perinuclear clumps of "nuage" material (Fig. 2E). Knowledge from other hydrozoans, notably *Hydra*, suggests that this cell population likely includes both i-cells ("interstitial cells"), the classic hydrozoan multipotent stem cells (Hobmayer et al., 2012), and their derivatives that have embarked towards a germ cell fate. Whether a separate germ-line stem cell population derived from i-cells undergoes self renewal in the adult, as suspected in *Hydra* (Nishimiya-Fujisawa & Kobayashi, 2012), remains to be addressed.

3.4 Entry into meiosis

Clytia gonads, through their transparency and accessibility within the living animal, have excellent potential for studying the very early events of meiosis including the transition from mitotic to meiotic cycles, synapsis, and recombination. In female jellyfish oocytes are produced continuously, and early-stage meiotic oocytes are present in the gonads of female medusae throughout its lifespan. Meiosis progresses through leptotene, zygotene and pachytene before arresting in diplotene during oocyte growth (Jessus et al., 2020). The early meiotic stages are most accessible for study inside young jellyfish within the first 1 to 2 weeks following their release from the gonozooid. During this period they dominate the population compared to growing and fully grown oocytes. On the first day of jellyfish release, a handful of germ cells are found in a patch along the radial canal at the future site of the gonad. After several days, the number of cells has increased and oocytes at zygotene and pachytene stages can be recognized. In one week old female jellyfish, the gonad is full of early oocytes at all stages of prophase I (Fig. 2D). After 2 weeks, diplotene stage oocytes have already started to grow (Fig. 2B and E), while oocytes of earlier stages are still present in the periphery of the gonad (Fig. 2E) (Amiel, Chang, Momose, & Houliston, 2010; Amiel & Houliston, 2009; Jessus et al., 2020). In a collaboration between the Villefranche lab and Jean-René Huynh at Collège de France, we have recently initiated studies of entry into meiosis and the early meiotic events in prophase I. CRISPR/Cas9 is being used to target key genes, including the highly conserved meiotic gene Spo11, which is responsible for initiating meiotic double stranded breaks.

Recent technical advances open up exciting possibilities for future studies of gametogenesis in *Clytia* based on live imaging, particularly of the cellular mechanisms of early meiotic entry where cells are inaccessible for injection. As well as studying meiosis in females, many issues concerning spermatogenesis could also be addressed using the transparent and highly structured male gonads (Carré & Carré, 1992; Roosen-Runge & Szollosi, 1965; Szollosi, 1964).

4. Embryogenesis

The daily spawning from laboratory cultures of hundreds of transparent eggs provides ideal material for studies of embryogenesis, allowing the comparison of mechanisms within Cnidaria as well as more widely across animals. The cnidarian larval form, the planula, is typically elongated along the principle "oral-aboral" body axis, and made up of the two basic cnidarian tissue layers: epidermis and gastrodermis. Unlike as in some other cnidarian larvae, the oral end of the *Clytia* planula is not marked by a blastopore or later by development of a mouth opening, but it transforms into the feeding "head" of the polyp following the drastic morphological transformations of metamorphosis (Freeman, 2009). Coordinated ciliary beating propels the larva with the aboral pole in front. It becomes competent to metamorphose after about three days of development, and at that time will settle on a suitable substrate and transform into a primary polyp (Fig. 1B). The different cell types of the planula have not yet been fully characterized. Nematocytes (cnidocytes) are scattered through the lateral epidermis and concentrated at the oral pole, while specialized secretory cells involved in substrate adhesion are concentrated at the aboral pole (Bodo & Bouillon, 1968). Nematocytes are the stinging cells that characterize Cnidaria, whose pressurized stinging capsules include products of genes acquired by horizontal transfer from bacteria (Denker, Bapteste, Le Guyader, Manuel, & Rabet, 2008). Various neural cell types classed as sensory and ganglionic are distributed within the epidermis and in the underlying mesoglea (Thomas et al., 1987), with an aboral plexus likely involved in coordinating sensory inputs. Neural cells in the larva, including the nematocytes (Rentzsch, Layden, & Manuel, 2017), are likely generated principally from hydrozoan-specific multipotent stem cells called i-cells occupying the endodermal region, although neurosensory and secretory cells likely derive from the aboral ectoderm during gastrulation (Kraus et al., 2020; Thomas et al., 1987).

4.1 Axis establishment

As introduced in Section 1, historically important studies (Freeman 1981a, 1981b) demonstrated that polarity of the fertilized egg presages the larval oral-aboral body axis, and that endoderm derives from the egg animal/future oral pole. In the early 2000s, experimental manipulation of embryos from the hydrozoan *Podocoryne* (Momose & Schmid, 2006) and demonstration of localized Wnt/β-catenin pathway activation in relation to endoderm formation during *Nematostella* embryogenesis (Wikramanayake et al., 2003) together suggested that cnidarian eggs likely contain axis determinants at the animal (future oral) pole that initiate Wnt/β-catenin signaling. Such determinants were identified for the first time in *Clytia*, in the form of localized mRNAs for the Frizzled family receptors, CheFz1 and CheFz3 and the Wnt ligand CheWnt3 (Fig. 2B and C, Fig. 3) (Momose et al., 2008; Momose & Houliston, 2007). Experimental reorientation of the oral-aboral axis achieved by low-speed centrifugation of fertilized eggs (Freeman, 1981a), can be explained by the co-displacement of Fz1 mRNA with the zygote nucleus in these experiments (Amiel & Houliston, 2009). Morpholino and mRNA injection experiments showed that Fz1, Wnt3 and Fz3 proteins produced from the localized mRNAs act together to stabilize β-catenin in the oral domain and thus set up distinct programmes of gene expression along the axis (Fig. 3A and B). Zygotic gene expression, which starts from mid/late blastula stage, is globally controlled

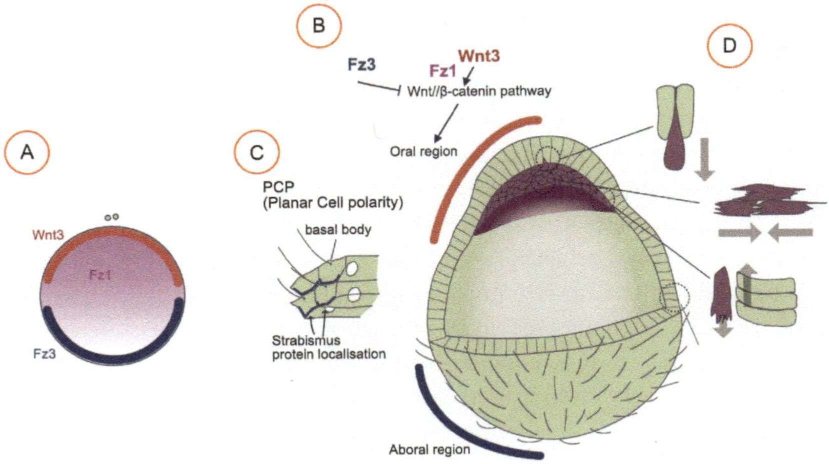

Fig. 3 Establishment of oral-aboral polarity in the embryo. (A) Schematic of a unfertilized egg showing the localization of Wnt/β-catenin pathway component mRNAs. (B) The proteins synthesized from these mRNAs locally activate the pathway and initiate transcriptional programmes that define oral and aboral regional characters, including presumptive endoderm fate at the oral pole. (C) In parallel, conserved planar cell polarity (PCP) pathway proteins coordinate tissue polarity in the ciliated ectoderm. (D) Individual ingression of presumptive endoderm cells from the oral pole, and PCP-mediated intercalation of cells in both layers underlie the morphogenetic changes during gastrulation that generate the elongated form of the planula.

by Wnt3. Two distinct transcriptional programmes activated downstream of Wnt3 can be distinguished in the oral ectoderm vs ingressing presumptive endoderm on the basis of sensitivity to PCP-mediated cellular interactions (Lapébie et al., 2014).

4.2 Morphogenesis directed by global polarity

Based on the extensive ability of embryo fragments, and even dissociated cells, to reestablish polarity, Gary Freeman proposed that the embryonic axis was determined by a global, i.e., non-localized, polarity property (Freeman, 1981b). We can now equate this property with Planar Cell Polarity (PCP), the common alignment of cells within epithelia or other coherent tissues mediated by specific interactions between neighboring cells (Butler & Wallingford, 2017). *Clytia* possesses a full set of proteins for the conserved 'Wnt-PCP" pathway, including Strabismus, Flamingo and Prickle orthologs (Momose et al., 2012). Morpholino-mediated knockdown of *Clytia* Strabismus (*CheStbm*) revealed a requirement for PCP both for cell alignment in the ciliated ectoderm and for planula elongation (Fig. 3C). The role in elongation likely involves intercalation between the ectodermal epithelial cells perpendicular to the axis of elongation (Byrum, 2001), but also between ingressing presumptive endoderm cells (Fig. 3D) (Kraus et al., 2020; van der Sande, Kraus, Houliston, & Kaandorp, 2020).

As first described at the end of the 19th century, gastrulation in *Clytia* involves ingression of individual cells from the future oral pole of a single cell-layered blastula (Byrum, 2001; Kraus et al., 2020; Metchnikoff, 1886). As they ingress, the presumptive endoderm cells crawl along the blastocoel wall and against each other, simultaneously narrowing and elongating the embryo along the oral-aboral axis (Kraus et al., 2020). Computational simulations of gastrulation require lateral interactions between ingressing cells to be included, simulating PCP, to account for full embryo elongation (van der Sande et al., 2020). Thus PCP acts in the *Clytia* embryo to generate the main morphological features of the larva: the axially elongated form and oriented cilia beating. It can be considered as a non-localized axial determinant acting in parallel with Wnt/β-catenin signaling to direct gene expression programmes across the embryo (Lapébie et al., 2014).

Modes of embryogenesis between different species of Cnidaria vary widely, with the two body layers forming by ingression in *Clytia*, but by cell sheet invagination ahead of embryo elongation in *Nematostella,* and partitioning during the cleavage divisions in *Hydractinia* (Kraus & Markov, 2016; Technau, 2020). Cnidarian embryos are thus a very valuable source of comparative material for addressing how morphogenetic processes function and evolve. In this context it was revealing to find that simulations of gastrulation in *Clytia* could be derived from simulations set up for *Nematostella* by changing parameters representing adhesion between cells, allowing apical cell contraction to occur in all cells rather than being orally restricted, and adding lateral interactions ("PCP") between ingressing cells (van der Sande et al., 2020).

5. Medusa formation and regeneration

Hydrozoan jellyfish have provided a rich source of material for biologists since the late nineteenth century. Early workers documented for example their development, nervous systems, musculature and regeneration. Molecular studies to explore this rich heritage have only just begun

(Leclère, Copley, Momose, & Houliston, 2016; Peron, Houliston, & Leclère, 2021). Here we will provide an overview of recent findings in each of these areas.

5.1 Medusa development

Clytia jellyfish are produced clonally by lateral budding of specialized polyps, the gonozooids (Fig. 1C). This process appears highly similar to that described in other hydrozoan species (e.g. Boelsterli, 1977). Jellyfish budding starts with bulging of the gonozooid body wall. A specialized cell layer called the entocodon, generates smooth and striated muscle layers of the subumbrella and manubrium (Kraus, Fredman, Wang, Khalturin, & Technau, 2015). The epidermis of the bud gives rise to the exumbrella, the tentacle epidermis and the outer layer of the velum of the medusa, while the bud gastrodermis provides the gastrovascular system of the medusa. The precise ontogeny of medusa formation as well as the molecular processes underlying budding, growth and aging of *Clytia* jellyfish are still very largely unexplored.

5.2 Nervous system

The nervous system of the *Clytia* jellyfish is still poorly characterized. As in other hydrozoan jellyfish, two nerve rings running around the bell margin play a central role in integrating sensory inputs to coordinate behavioral responses, while subpopulations of neural cells also occupy epidermal and gastrodermal layers of the manubrium and gonads. Analysis of scRNAseq data (Chari et al., 2021) revealed at least 12 neural cell types expressing different neuropeptides, whose precise morphologies, locations and roles mainly remain to be determined. A recent pioneering functional study showed that RFamide-expressing neurons in the subumbrella mediate the oriented flexing of the bell that brings prey captured by the tentacle to the mouth (Weissbourd et al., 2021). GCaMP imaging revealed these RFamide-positive neurons to be organized in distinct and largely independent modules under the control of the nerve ring.

Nerve cells and nematocytes likely derive principally from i-cells in the *Clytia* medusa, as shown in the polyp-only hydrozoans *Hydra* (Hobmayer et al., 2012) and *Hydractinia* (Gahan, Bradshaw, Flici, & Frank, 2016). In the *Clytia* medusa, i-cell pools are located not only in the gonads (see Section 3) but also at the base of the manubrium and, conspicuously, in the ectoderm at the base of each tentacle bulb (Condamine et al., 2019;

Denker, Manuel, et al., 2008; Leclère et al., 2012). Given the continuous growth of the tentacles and the constant renewal of nematocytes, the *Clytia* jellyfish tentacle bulb is an attractive system to investigate neurosensory cell differentiation. It can be viewed as a cellular conveyor belt, with stem cells close to the bell margin and differentiation into nematocytes progressing along the longitudinal axis of the bulb and into the tentacle (Denker, Manuel, et al., 2008; Chari et al., 2021) (Fig. 4A). Transcriptome comparisons between the base, middle and distal parts of the bulb uncovered sets of genes differentially expressed along the bulb (Condamine et al., 2019), confirming previous in situ hybridization studies of particular regulatory gene families (Denker, Manuel, et al., 2008; Hwang et al., 2010; Jager, Quéinnec, Le Guyader, & Manuel, 2011; Leclère et al., 2012; Steinmetz et al., 2012). Regulation of this differentiation likely involves both the Hippo (Coste, Jager, Chambon, & Manuel, 2016) and Wnt (Condamine et al., 2019) pathways. Wnt related genes are expressed differentially both along the proximo distal axis of the bulb and in relation to its perpendicular oral-aboral axis, likely reflecting complex spatial regulation of neural differentiation within the bulb.

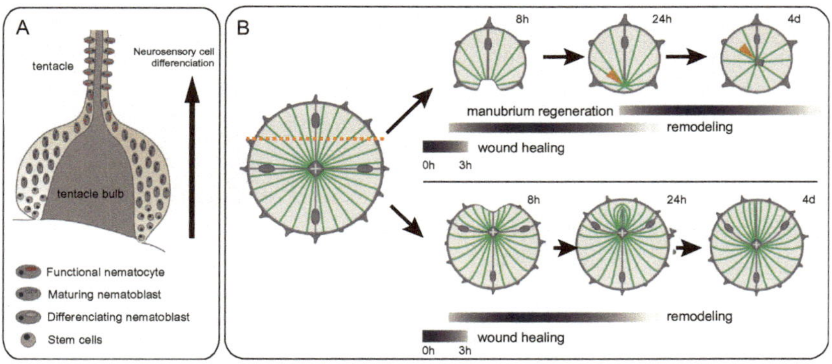

Fig. 4 Nematogenesis and regeneration in *Clytia* medusae. (A) Nematocyte differentiation ordered along the proximo-distal axis of the *Clytia* medusa tentacle bulb (Condamine et al., 2019; Denker, Manuel, et al., 2008). (B) Successive stages of wound healing, remodeling and regeneration in *Clytia* medusa following an asymmetric bisection (orange dotted line; Sinigaglia et al., 2020). Only the small fragment missing the manubrium (top) will reform one. The muscle fibers (green) merge into a hub (orange arrows), which becomes the regeneration site of a new manubrium 4 days after dissection. In the larger fragment, bearing a manubrium (bottom), re-circularization also causes the formation of a muscle hub, but this later disaggregates, and no new manubrium forms.

5.3 Muscle systems and regeneration

The *Clytia* jellyfish umbrella consists of exumbrella and subumbrella cell layers, separated by a thick ECM structure called the mesoglea (the "jelly"). The exumbrella comprises flattened, non-contractile polygonal cells. The sub-umbrella is made up of three layers of cells: (i) the endodermal plate of polygonal non contractile epithelial cells, covered by two contractile epithelial layers; (ii) a muscle layer displaying smooth radial muscle fibers oriented radially and responsible for the crumpling of the bell, and between these (iii) a layer of circular striated muscle layer responsible for the pulsing bell contractions used for swimming. *Clytia* striated muscles strongly resemble those of bilaterian animals in structure, but molecular phylogeny and gene expression studies of the proteins comprising the typical periodic actomyosin arrays suggest that this similarity arose through convergent evolution (Steinmetz et al., 2012). The radial smooth muscles have been recently studied for their role in food transfer from the bell margin to the mouth, stimulated by the action of associated networks of RFamide positive neurons (Weissbourd et al., 2021). Intriguingly, reorganization of these radial muscle fibers upon injury driven by actomyosin-driven wound closure may also have a patterning role in the jellyfish, determining the site where a new manubrium will form (Sinigaglia et al., 2020). These observations on the jellyfish muscle systems open an exciting era of exploration of the molecular basis of their development and function, comparisons of the composition and structure of the contractile fibers with canonical bilaterian muscles, and understanding of their role in regeneration and body patterning (Sinigaglia et al., 2020).

Cnidarians are well known for their extensive regenerative capacities, which have been well characterized in polyp-only species such as *Hydra, Nematostella* and *Hydractinia,* but less so in medusae. Building on a body of historical work (Neppi, 1918; Schmid, 1974; Schmid, Schmid, Schneider, Stidwill, & Baker, 1976; Schmid & Tardent, 1971), recent cellular and molecular analyses have shed a new light on regeneration processes in *Clytia.* Studies of wound healing exploited the optical clarity of the exumbrella monolayered epithelium for high resolution imaging and revealed the extreme rapidity and simplicity of this process, which occurs much faster than for cultured cells or bilaterian embryos (Kamran et al., 2017; Malamy & Shribak, 2018). Both purse-string-like supra-cellular actomyosin contractions and lamellipodia-dependent cell crawling contribute to wound healing (Kamran et al., 2017; Malamy & Shribak, 2018). The balance between these mechanisms depends on the depth of the wound and the extent of damage to the basement membrane/ECM.

Remarkably, the *Clytia* medusa can also cope with much more extreme damage. Even small fragments are able to rapidly restore shape, functionality and even missing organs (Sinigaglia et al., 2020). This occurs in three steps: initial wound healing is followed by a purse-string mediated remodeling process to re-establish the circular shape of the medusa. Finally, reformation of missing organs occurs, prioritizing the manubrium to allow feeding to resume (Fig. 4B). Manubrium reformation requires both cell proliferation and cell migration from other parts of the medusa, notably the gonads. Grafting experiments revealed that the position of the new manubrium is not directed by a morphogen based system of coordinates within the subumbrella (Schmid et al., 1976; Sinigaglia et al., 2020). It seems rather that patterns of reorganization of the radial smooth muscles of the sub-umbrella during remodeling define the site of the new manubrium. These mechanical contraction cues lead to activation of Wnt signaling at the remodeling site, which promotes manubrium regeneration or not, depending on the organization of the muscle fibers in the regenerating fragment (Fig. 4B). A similar patterning function for muscle is also emerging from studies of regeneration in *Hydra* (Livshits, Shani-Zerbib, Maroudas-Sacks, Braun, & Keren, 2017, Livshits et al., 2021; Maroudas-Sacks et al., 2021). Understanding of how this influence of muscle integrates with the well established role of Wnt in *Hydra* head regeneration (Cazet, Cho, & Juliano, 2021; Wang, Steele, & Collins, 2020) awaits further studies. Another important player during manubrium regeneration in *Clytia* is the radial canal system, which both transports construction material to the blastema site and acts locally to direct manubrium morphogenesis.

Much remains to be understood about the extreme and fascinating regeneration capacity of *Clytia* medusae. Open questions to be addressed notably include (i) the role of cell migration in regeneration, (ii) the integration of mechanical cues and signaling pathways during regeneration, (iii) the link between signaling, ECM, wound healing and mechanical forces, and (iv) the nature of the local cues that direct regeneration and morphogenesis.

6. Genomics of the jellyfish

Cnidarians are the closest living relatives of the bilaterians, the major group of bilaterally symmetric animals that includes most familiar species including humans. Cnidaria itself comprises two great clades, Anthozoa, consisting of corals and anemones (including the laboratory model *Nematostella vectensis*), and Medusozoa, most simply understood as those that have a jellyfish-like stage in their life-cycle. The divergence between

Anthozoa and Medusozoa is ancient (dos Reis et al., 2015), and both are large and diverse clades. The four medusozoan classes are Hydrozoa (including *Clytia* and *Hydra*) Staurozoa (stalked jellyfish), Scyphozoa (true jellyfish) and Cubozoa (box jellyfish). The last few years have seen the publication of high quality genomes of several Medusozoa, *Nemopilema nomurai* (Kim et al., 2019); *Rhopilema esculentum* (Li et al., 2020; Nong et al., 2020); *Sanderia malayensis* (Nong et al., 2020); *Chrysaora quinquecirrha* (Xia et al., 2020, 2021); *Aurelia aurita* (Gold et al., 2019; Khalturin et al., 2019); *Morbakka virulenta* (Khalturin et al., 2019). All of these, except the box jellyfish *Morbakka*, are Scyphozoa. Another box jellyfish, *Alatina alata* and a staurozoan *Calvadosia cruxmelitensis* are available as drafts (Ohdera et al., 2019).

After *Hydra* (Chapman et al., 2010; Hamada, Satoh, & Khalturin, 2020), *Clytia* was the second hydrozoan species for which the genome has been sequenced, and the first with planula and medusa stages (Leclère et al., 2019). The 445 Mb assembled genome of *Clytia* is typical for hydrozoans, with moderate GC and repeat contents (35% and 39%, respectively). As in other hydrozoans, messenger RNAs in *Clytia* are frequently subject to transplicing, involving addition of short RNA leader sequences to their 5' ends (Derelle et al., 2010). Work on improving the contiguity of the *Clytia hemisphaerica* genome continues; an improved assembly with N50 >1 Mb is now available at Ensembl (link in Table 1). A central result of the published *Clytia* genome analysis was that gene order in *Clytia* and *Hydra* is disrupted relative to the last common ancestor of cnidarians and bilaterians. Comparisons with the new scyphozoan genomes suggest that this disruption is indeed a hydrozoan phenomenon, with better gene order conservation between Anthozoa and Scyphozoa than between Anthozoa and *Clytia* or *Hydra*. These results will need to be confirmed as more contiguous hydrozoan genomes become available (Zimmermann et al., 2021). In this context, it is interesting that *Clytia* appears to have the same number of chromosomes (15) as the ancestral Cnidarian (Leclère et al., 2019; Zimmermann et al., 2021). Gene order disruption of Hydrozoa does not, therefore, appear to have been an inevitable consequence of genomic restructuring necessary to form another life-cycle stage.

The accessibility of the planula, polyp and medusa in the laboratory environment has enabled detailed study of gene use in the different stages of *Clytia*. The likelihood that the jellyfish is an evolutionary novelty (i.e. "invented" in the medusozoan lineage), raised the possibility that it might be enriched in lineage specific genes. All studies that have looked at this question, however, have concluded that this is not the case

(Gold et al., 2019;Khalturin et al., 2019; Leclère et al., 2019). Instead, genes showing differential expression between life cycle stages are more likely to be taxonomically restricted (i.e. more recently evolved), but they are not preferentially associated with the medusa stage (Khalturin et al., 2019; Leclère et al., 2019). In contrast, a careful analysis of transcription factors showing medusa specific expression showed that several had been lost in *Hydra*, despite being ancestral to Medusozoa and present in the closest medusoid-producing relatives of *Hydra* for which data were available (Leclère et al., 2019). These cases of gene loss *are* thus associated with loss of a life-cycle stage. Analysis of life-cycle stage gene use also suggested that the *Clytia* planula and polyp stages have been secondarily simplified relative to the Anthozoa, as genes with apparent conserved roles in patterning the apical organ and directive axis formation were lost or are not expressed. In this context it is notable that Scyphozoa similarly lacks many of these genes (*Gbx*, *Hox2* etc.). These analyses also suggested that, of all the stages, the medusa expresses the most complex set of transcriptional regulators. Many of these appear to associate with specific cell types, notably within the nervous system (Chiori et al., 2009; Leclère et al., 2019). A current research focus is understanding this phenomenon within the context of the medusa cell-types and their evolution (Chari et al., 2021).

More widely, the accumulation of genomic and sequence resources for *Clytia* (see Table 1) is providing a wealth of information valuable for the understanding of the evolution of animals and their genomes. Improvements to genome annotation and epigenomic studies in *Clytia* and other jellyfish species will fuel further insights into life cycle evolution and development in medusozoans, and the evolution of cell types in these different stages.

7. Perspectives

This review has illustrated how practical advantages combined with phylogenetic position have enabled *Clytia hemisphaerica* to contribute a number of insights in the fields of cell and developmental biology, and animal genome evolution. Existing *in silico* resources and well established methods like egg MO/mRNA injection and CRISPR/Cas9-mediated gene knockout already allow efficient functional testing of genes expressed during larval development or specifically in the adult, respectively. The recent emergence of an efficient transposon-mediated transgenesis method opens significant further perspectives to the community, notably widening the scope of gene function exploration in adult (polyp and medusa) stages via the use of inducible promoters. A detailed protocol including workflows to select and maintain strains is being prepared (B. Weissbourd and T. Momose, in preparation). Transgenesis will also allow researchers to tease out relationships between cell lineages via *in vivo* tracking, ablation experiments, imaging of particular cellular components etc. We thus anticipate that use of transgenesis in *Clytia* will advance hand in hand with the development and wider application of live imaging approaches. Recent work focusing on jellyfish wound healing (Kamran et al., 2017), regeneration (Sinigaglia et al., 2020), neurobiology (Weissbourd et al., 2021) and cell type characterization via scRNAseq (Chari et al., 2021) represent valuable steps forward in this direction, opening many new questions. For instance, what are the functions and organization of the different neural subtypes of the jellyfish in relation to physiology and behavior such as swimming, diel migrations, spawning and feeding? How do mechanical forces and transcriptional processes cooperate during wound healing and regeneration in *Clytia* jellyfish? The power of genomics, imaging, transgenesis and ablation approaches coupled with the transparency, small size and relative complexity of the jellyfish stage, place *Clytia* in an ideal position to tackle these fundamental questions.

The time is also ripe now for *Clytia* to be used also for studies at the level of the whole organism. One notable application in the context of climate change concern is to dissect biological responses to changing environmental parameters such as temperature and UV irradiation and food availability. Transparent planktonic organisms are especially vulnerable to UV damage, especially when they rise to the ocean surface to spawn. One hypothesized adaptation for UV protection is its absorption by a set of endogenous GFPs, highly expressed in the epidermis of the gonad and tentacle bulbs as well as in the planula ectoderm and egg mitochondria (Fourrage, Swann, Gonzalez Garcia, Campbell, & Houliston, 2014). This idea can now be tested using stable CRISPR/Cas9 Knockout strains lacking different GFP subtypes (Fig. 1D). Population genomics approaches to correlate the GFP repertoires of natural populations with their environment would also be informative. Another possible adaptation to UV exposure currently being explored is the differential deployment of alternative DNA repair pathways between life-cycle stages and cell types. This idea stems from analyses of repair responses to Cas9 mediated DNA breaks, which revealed microhomology mediated end-joining (MMEJ) pathway to be dominant over better known double strand break repair pathways such as non-homologous end-joining

(NHEJ) or homology-dependent repair pathway (HR) in the cells from the larva that found the polyp colony (Momose et al., 2018). Another aspect of *Clytia* biology clearly impacted by environmental factors is life-cycle transitions. Metamorphosis of hydrozoan planulae is promoted by bacterial biofilms by largely unknown mechanisms (Guo, Rischer, Westermann, & Beemelmanns, 2021). The temperature at which the polyp colonies are established then influences the sex of the medusa which bud from it (Carré & Carré, 2000). The molecular and cellular basis of this intriguing observation is now open to be addressed.

Clytia also could be a useful model for investigating the physiological relationships between food processing and regulation of starvation and gametogenesis. Study of the impact of food availability on the medusa has recently been leveraged by a proof of concept Whole Animal Multiplexed Single-Cell RNA-Seq ("WHAM-seq") study in which cells from control and starved *Clytia* jellyfish were simultaneously processed (Chari et al., 2021). This provided a rich body of information concerning the cells and genes affected by starvation, and revealed certain gastrodermal cell types to be particularly responsive. These likely include the "mobilising gastrodermal digestive cells" ("MGD cells"; Sinigaglia et al., 2020) that relocate through the gastrovascular canal system during starving and regeneration, hypothesized to mediate redistribution of nutrient resources between the different organs of the medusa. The functional specializations of the many distinct secretory and absorptive cell types and physiological states of the jellyfish digestive system are now open for investigation.

To conclude, *Clytia* is now fully established as a laboratory animal, offering many opportunities to a community of researchers progressively expanding across many fields, as well as ideal material for educators. As well as *in silico* resources (Table 1) animal resources can be shared as "cuttings" of polyp colonies established on glass slides (Lechable et al., 2020). *Clytia* wildtype strains are currently available through the Marine Biological Resource Centre (CRBM, request to be submitted from https://www.embrc-france.fr/) We hope that community expansion will continue, allowing further accelerated development of approaches and resources, and fueling scientific discovery across domains for years to come.

Acknowledgments

We thank all the past and present members of our groups who have contributed to the success of *Clytia* in the laboratory, as well as past and current collaborators including Michael Manuel, Muriel Jager and Eric Quéinnec (Sorbonne University), Ulrich Technau (University of

Vienna), Noriyo Takeda (Hiroshima University), Ryusaku Deguchi (Miyagi University of Education), Gáspár Jékely (University of Exeter), Yulia Kraus (Lomonosov Moscow State University), Jaap Kaandorp (Univ. Amsterdam), Jean-René Huynh (Collège de France) and Brady Weissbourd, David Anderson, Tara Chari, Jase Gehring and Lior Pachter (Caltech). We thank Anna Ferraoili for Fig. 2E. Original funding for work from our labs was from the Agence Nationale de la Recherche ANR-17-CE13-0016 (i-MMEJ), ANR-19-CE13-0003 (MYODEVO), the EU Training networks NEPTUNE and EvoCELL, as well as core CNRS and Sorbonne University funding to the LBDV. CM received funding from the European Union's Horizon 2020 research and innovation programme under the Marie Skłodowska-Curie grant agreement No 841433 (JOLI), as well as from the "Diversity of Biological Mechanisms" program of INSB CNRS. We thank the Marine Resources Centre (CRBM and PIV imaging platform) of Institut de la Mer de Villefranche (IMEV), supported by EMBRC-France whose French state funds are managed by the ANR within the Investments of the Future program under reference ANR-10-INBS-02. *Clytia* resource development was supported by the European Union's Horizon 2020 research and innovation programme under grant agreement No 730984 ASSEMBLE-Plus Joint Research Activity (JRA3) and the CORBEL European Research Infrastructure cluster project.

References

Amiel, A., Chang, P., Momose, T., & Houliston, E. (2010). *Clytia hemisphaerica*: A cnidarian model for studying oogenesis. In M.-H. Verlhac, & A. Villeneuve (Eds.), *Oogenesis: the Universal process* (pp. 81–101). Wiley. https://doi.org/10.1002/9780470687970.ch3.

Amiel, A., & Houliston, E. (2009). Three distinct RNA localization mechanisms contribute to oocyte polarity establishment in the cnidarian Clytia hemisphaerica. *Developmental Biology*, *327*(1), 191–203.

Amiel, A., Leclère, L., Robert, L., Chevalier, S., & Houliston, E. (2009). Conserved functions for Mos in eumetazoan oocyte maturation revealed by studies in a cnidarian. *Current Biology*, *19*(4), 305–311.

Bashirullah, A., Cooperstock, R. L., & Lipshitz, H. D. (1998). RNA localization in development. *Annual Review of Biochemistry*, *67*, 335–394.

Bodo, F., & Bouillon, J. (1968). Étude histologique du développement embryonnaire de quelques hydroméduses de Roscoff: *Phialidium hemisphaericum* (L.), *Obelia* sp. Péron et Lesueur, *Sarsia eximia* (Allman), *Podocoryne carnea* (Sars), Gonionemus vertens Agassiz. *Cahiers De Biologie Marine*, *9*(1), 69–104.

Boelsterli, U. (1977). An electron microscopic study of early developmental stages, myogenesis, oogenesis and cnidogenesis in the anthomedusa *Podocoryne carnea* M. Sars. *Journal of Morphology*, *154*(2), 259–289.

Butler, M. T., & Wallingford, J. B. (2017). Planar cell polarity in development and disease. *Nature Reviews. Molecular Cell Biology*, *18*(6), 375–388.

Byrum, C. A. (2001). An analysis of hydrozoan gastrulation by unipolar ingression. *Developmental Biology*, *240*(2), 627–640.

Carré, D., & Carré, C. (1992). Biologie et étude ultrastructurale des spermatozoïdes du cnidaire *Clytia hemispherica* (Leptomedusae): mise en évidence d'un processus acrosomal. *Canadian Journal of Zoology*, *70*(5), 866–873. https://doi.org/10.1139/z92-123.

Carré, D., & Carré, C. (2000). Origin of germ cells, sex determination, and sex inversion in medusae of the genus Clytia (Hydrozoa, leptomedusae): The influence of temperature. *The Journal of Experimental Zoology*, *287*(3), 233–242.

Cazet, J. F., Cho, A., & Juliano, C. E. (2021). Generic injuries are sufficient to induce ectopic Wnt organizers. *eLife*, *10*, e60562. https://doi.org/10.7554/eLife.60562.
Chapman, J. A., Kirkness, E. F., Simakov, O., Hampson, S. E., Mitros, T., Weinmaier, T., et al. (2010). The dynamic genome of Hydra. *Nature*, *464*(7288), 592–596.
Chari, T., Weissbourd, B., Gehring, J., Ferraioli, A., Leclère, L., Herl, M., et al. (2021). Whole-animal multiplexed single-cell RNA-seq reveals transcriptional shifts across *Clytia* medusa cell types. *Science Advances*, 7(48). https://doi.org/10.1126/sciadv.abh1683.
Chevalier, S., Martin, A., Leclère, L., Amiel, A., & Houliston, E. (2006). Polarised expression of FoxB and FoxQ2 genes during development of the hydrozoan *Clytia hemisphaerica*. *Development Genes and Evolution*, *216*(11), 709–720.
Chiori, R., Jager, E., Denker, P., Wincker, C. D., Silva, H. L., Guyader, M. M., et al. (2009). Are Hox genes ancestrally involved in axial patterning? evidence from the hydrozoan *Clytia hemisphaerica* (Cnidaria). *PLoS One*, *4*, e4231.
Condamine, T., Jager, M., Leclère, L., Blugeon, C., Lemoine, S., Copley, R. R., et al. (2019). Molecular characterisation of a cellular conveyor belt in *Clytia* medusae. *Developmental Biology*, *456*(2), 212–225.
Coste, A., Jager, M., Chambon, J.-P., & Manuel, M. (2016). Comparative study of Hippo pathway genes in cellular conveyor belts of a ctenophore and a cnidarian. *EvoDevo*, 7, 4.
Denker, E., Bapteste, E., Le Guyader, H., Manuel, M., & Rabet, N. (2008). Horizontal gene transfer and the evolution of cnidarian stinging cells. *Current Biology*, *18*(18), R858–R859.
Denker, E., Manuel, M., Leclère, L., Le Guyader, H., & Rabet, N. (2008). Ordered progression of nematogenesis from stem cells through differentiation stages in the tentacle bulb of *Clytia hemisphaerica* (Hydrozoa, Cnidaria). *Developmental Biology*, *315*(1), 99–113.
Derelle, R., Momose, T., Manuel, M., Da Silva, C., Wincker, P., & Houliston, E. (2010). Convergent origins and rapid evolution of spliced leader trans-splicing in metazoa: Insights from the ctenophora and hydrozoa. *RNA*, *16*(4), 696–707.
dos Reis, M., Thawornwattana, Y., Angelis, K., Telford, M. J., Donoghue, P. C. J., & Yang, Z. (2015). Uncertainty in the timing of origin of animals and the limits of precision in molecular timescales. *Current Biology*, *25*(22), 2939–2950.
Forêt, S., Knack, B., Houliston, E., Momose, T., Manuel, M., Quéinnec, E., et al. (2010). New tricks with old genes: The genetic bases of novel cnidarian traits. *Trends in Genetics*, *26*(4), 154–158.
Fourrage, C., Chevalier, S., & Houliston, E. (2010). A highly conserved Poc1 protein characterized in embryos of the hydrozoan *Clytia hemisphaerica*: Localization and functional studies. *PLoS One*, *5*(11), e13994.
Fourrage, C., Swann, K., Gonzalez Garcia, J. R., Campbell, A. K., & Houliston, E. (2014). An endogenous green fluorescent protein-photoprotein pair in *Clytia hemisphaerica* eggs shows co-targeting to mitochondria and efficient bioluminescence energy transfer. *Open Biology*, *4*, 130206.
Freeman, G. (1976). The role of cleavage in the localization of developmental potential in the ctenophore *Mnemiopsis leidyi*. *Developmental Biology*, *49*(1), 143–177.
Freeman, G. (1977). The establishment of the oral-aboral axis in the ctenophore embryo. *Development*, *42*(1), 237–260.
Freeman, G. (1981a). The cleavage initiation site establishes the posterior pole of the hydrozoan embryo. *Wilhelm Roux's Archives of Developmental Biology*, *190*(2), 123–125.
Freeman, G. (1981b). The role of polarity in the development of the hydrozoan planula larva. *Wilhelm Roux's Archives of Developmental Biology*, *190*(3), 168–184.
Freeman, G. (1987). The role of oocyte maturation in the ontogeny of the fertilization site in the hydrozoan *Hydractinia echinata*. *Roux's Archives of Developmental Biology*, *196*(2), 83–92. https://doi.org/10.1007/bf00402029.

Freeman, G. (2009). Is the remodeling of the polyp prepattern in a hydrozoan planula a function of larval age or size and is it of "adaptive" significance? *Zoology, 112*(3), 169–184.

Freeman, G., & Ridgway, E. B. (1988). The role of cAMP in oocyte maturation and the role of the germinal vesicle contents in mediating maturation and subsequent developmental events in hydrozoans. *Roux's Archives of Developmental Biology, 197*(1), 197–211.

Gahan, J. M., Bradshaw, B., Flici, H., & Frank, U. (2016). The interstitial stem cells in *Hydractinia* and their role in regeneration. *Current Opinion in Genetics & Development, 40*, 65–73. https://doi.org/10.1016/j.gde.2016.06.006.

Gebauer, F., & Richter, J. D. (1997). Synthesis and function of Mos: The control switch of vertebrate oocyte meiosis. *BioEssays: News and Reviews in Molecular, Cellular and Developmental Biology, 19*(1), 23–28.

Gold, D. A., Katsuki, T., Li, Y., Yan, X., Regulski, M., Ibberson, D., et al. (2019). The genome of the jellyfish *Aurelia* and the evolution of animal complexity. *Nature Ecology & Evolution, 3*(1), 96–104.

Guo, H., Rischer, M., Westermann, M., & Beemelmanns, C. (2021). Two distinct bacterial biofilm components trigger metamorphosis in the colonial hydrozoan *Hydractinia echinata. mBio, 12*(3), e0040121. https://doi.org/10.1101/2019.12.23.887182.

Hamada, M., Satoh, N., & Khalturin, K. (2020). A reference genome from the symbiotic hydrozoan, *Hydra viridissima. G3 Genes|Genomes|Genetics, 10*(11), 3883–3895.

Hobmayer, B., Jenewein, M., Eder, D., Eder, M.-K., Glasauer, S., Gufler, S., et al. (2012). Stemness in Hydra—A current perspective. *The International Journal of Developmental Biology, 56*, 509–517. https://doi.org/10.1387/ijdb.113426bh.

Houliston, E., Momose, T., & Manuel, M. (2010). *Clytia hemisphaerica*: A jellyfish cousin joins the laboratory. *Trends in Genetics, 26*(4), 159–167.

Hwang, J. S., Takaku, Y., Momose, T., Adamczyk, P., Özbek, S., Ikeo, K., et al. (2010). Nematogalectin, a nematocyst protein with GlyXY and galectin domains, demonstrates nematocyte-specific alternative splicing in *Hydra. Proceedings of the National Academy of Sciences of the United States of America, 107*(43), 18539–18544.

Jager, M., Quéinnec, E., Houliston, E., & Manuel, M. (2006). Expansion of the SOX gene family predated the emergence of the Bilateria. *Molecular Phylogenetics and Evolution, 39*(2), 468–477.

Jager, M., Quéinnec, E., Le Guyader, H., & Manuel, M. (2011). Multiple Sox genes are expressed in stem cells or in differentiating neuro-sensory cells in the hydrozoan *Clytia hemisphaerica. EvoDevo, 2*, 12.

Jessus, C., Munro, C., & Houliston, E. (2020). Managing the oocyte meiotic arrest-lessons from frogs and jellyfish. *Cells, 9*(5), 1150. https://doi.org/10.3390/cells9051150.

Juliano, C. E., Swartz, S. Z., & Wessel, G. M. (2010). A conserved germline multipotency program. *Development, 137*(24), 4113–4126.

Kamran, Z., Zellner, K., Kyriazes, H., Kraus, C. M., Reynier, J.-B., & Malamy, J. E. (2017). In vivo imaging of epithelial wound healing in the cnidarian *Clytia hemisphaerica* demonstrates early evolution of purse string and cell crawling closure mechanisms. *BMC Developmental Biology, 17*(1), 17.

Khalturin, K., Shinzato, C., Khalturina, M., Hamada, M., Fujie, M., Koyanagi, R., et al. (2019). Medusozoan genomes inform the evolution of the jellyfish body plan. *Nature Ecology & Evolution, 3*(5), 811–822.

Kim, H.-M., Weber, J. A., Lee, N., Park, S. G., Cho, Y. S., Bhak, Y., et al. (2019). The genome of the giant Nomura's jellyfish sheds light on the early evolution of active predation. *BMC Biology, 17*(1), 1–12.

Kraus, Y., Chevalier, S., & Houliston, E. (2020). Cell shape changes during larval body plan development in *Clytia hemisphaerica. Developmental Biology, 468*(1-2), 59–79.

Kraus, J. E. M., Fredman, D., Wang, W., Khalturin, K., & Technau, U. (2015). Adoption of conserved developmental genes in development and origin of the medusa body plan. *EvoDevo*, *6*, 23.

Kraus, Y. A., & Markov, A. V. (2016). The gastrulation in Cnidaria: A key to understanding phylogeny or the chaos of secondary modifications? *Zhurnal obshchei biologii*, 77(2), 83–105.

Lapébie, P., Ruggiero, A., Barreau, C., Chevalier, S., Chang, P., Dru, P., et al. (2014). Differential responses to Wnt and PCP disruption predict expression and developmental function of conserved and novel genes in a cnidarian. *PLoS Genetics*, *10*(9), e1004590.

Layden, M. J., Rentzsch, F., & Röttinger, E. (2016). The rise of the starlet sea anemone *Nematostella vectensis* as a model system to investigate development and regeneration. *Wiley Interdisciplinary Reviews. Developmental Biology*, *5*(4), 408–428.

Lechable, M., Jan, A., Duchene, A., Uveira, J., Weissbourd, B., Gissat, L., et al. (2020). An improved whole life cycle culture protocol for the hydrozoan genetic model *Clytia hemisphaerica. Biology Open*, *9*(11), bio051268. https://doi.org/10.1242/bio.051268.

Leclère, L., Copley, R. R., Momose, T., & Houliston, E. (2016). Hydrozoan insights in animal development and evolution. *Current Opinion in Genetics & Development*, *39*, 157–167. https://doi.org/10.1016/j.gde.2016.07.006.

Leclère, L., Horin, C., Chevalier, S., Lapébie, P., Dru, P., Peron, S., et al. (2019). The genome of the jellyfish Clytia hemisphaerica and the evolution of the cnidarian life-cycle. *Nature Ecology & Evolution*, *3*(5), 801–810.

Leclère, L., Jager, M., Barreau, C., Chang, P., Le Guyader, H., Manuel, M., et al. (2012). Maternally localized germ plasm mRNAs and germ cell/stem cell formation in the cnidarian Clytia. *Developmental Biology*, *364*(2), 236–248.

Leclère, L., & Röttinger, E. (2016). Diversity of cnidarian muscles: Function, anatomy, development and regeneration. *Frontiers in Cell and Developmental Biology*, *4*, 157.

Li, Y., Gao, L., Pan, Y., Tian, M., Li, Y., He, C., et al. (2020). Chromosome-level reference genome of the jellyfish *Rhopilema esculentum*. *GigaScience*, *9*(4), giaa036. https://doi.org/10.1093/gigascience/giaa036.

Liu, Z., Chen, O., Blake Joseph Wall, J., Zheng, M., Zhou, Y., Wang, L., et al. (2017). Systematic comparison of 2A peptides for cloning multi-genes in a polycistronic vector. *Scientific Reports*, 7(1), 2193. https://doi.org/10.1038/s41598-017-02460-2.

Livshits, A., Garion, L., Maroudas-Sacks, Y., Shani-Zerbib, L., Keren, K., & Braun, E. (2021). Plasticity of body axis polarity in Hydra regeneration under constraints. *BioRxiv*. https://doi.org/10.1101/2021.02.04.429818.

Livshits, A., Shani-Zerbib, L., Maroudas-Sacks, Y., Braun, E., & Keren, K. (2017). Structural inheritance of the actin cytoskeletal organization determines the body axis in regenerating hydra. *Cell Reports*, *18*(6), 1410–1421. https://doi.org/10.1016/j.celrep.2017.01.036.

Malamy, J. E., & Shribak, M. (2018). An orientation-independent DIC microscope allows high resolution imaging of epithelial cell migration and wound healing in a cnidarian model. *Journal of Microscopy*, *270*(3), 290–301.

Maroudas-Sacks, Y., Garion, L., Shani-Zerbib, L., Livshits, A., Braun, E., & Keren, K. (2021). Topological defects in the nematic order of actin fibers as organization centers of *Hydra* morphogenesis. *Nature Physics*, *17*(2), 251–259. https://doi.org/10.1038/s41567-020-01083-1.

Metchnikoff, E. (1886). *Embryologische Studien an Medusen: Ein Beitrag zur Genealogie der Primitiv-organe*. Wien: A. Hölder.

Miller, R. L. (1979). Sperm chemotaxis in the hydromedusae. I. Species-specificity and sperm behavior. *Marine Biology*, *53*(2), 99–113. https://doi.org/10.1007/bf00389182.

Momose, T., De Cian, A., Shiba, K., Inaba, K., Giovannangeli, C., & Concordet, J.-P. (2018). High doses of CRISPR/Cas9 ribonucleoprotein efficiently induce gene knockout with low mosaicism in the hydrozoan *Clytia hemisphaerica* through microhomology-mediated deletion. *Scientific Reports*, *8*(1), 11734.

Momose, T., Derelle, R., & Houliston, E. (2008). A maternally localised Wnt ligand required for axial patterning in the cnidarian *Clytia hemisphaerica*. *Development*, *135*(12), 2105–2113.

Momose, T., & Houliston, E. (2007). Two oppositely localised frizzled RNAs as axis determinants in a cnidarian embryo. *PLoS Biology*, *5*(4), e70.

Momose, T., Kraus, Y., & Houliston, E. (2012). A conserved function for Strabismus in establishing planar cell polarity in the ciliated ectoderm during cnidarian larval development. *Development*, *139*(23), 4374–4382.

Momose, T., & Schmid, V. (2006). Animal pole determinants define oral-aboral axis polarity and endodermal cell-fate in hydrozoan jellyfish *Podocoryne carnea*. *Developmental Biology*, *292*(2), 371–380.

Neppi, V. (1918). Sulla rigenerazione nelle idromeduse. *Pubblicazioni Della Stazione Zoologica Di Napoli*, *2*, 191–207.

Nishimiya-Fujisawa, C., & Kobayashi, S. (2012). Germline stem cells and sex determination in *Hydra*. *The International Journal of Developmental Biology*, *56*(6–8), 499–508.

Nong, W., Cao, J., Li, Y., Qu, Z., Sun, J., Swale, T., et al. (2020). Jellyfish genomes reveal distinct homeobox gene clusters and conservation of small RNA processing. *Nature Communications*, *11*(1), 3051.

Ohdera, A., Ames, C. L., Dikow, R. B., Kayal, E., Chiodin, M., Busby, B., et al. (2019). Box, stalked, and upside-down? Draft genomes from diverse jellyfish (Cnidaria, Acraspeda) lineages: *Alatina alata* (Cubozoa), *Calvadosia cruxmelitensis* (Staurozoa), and *Cassiopea xamachana* (Scyphozoa). *GigaScience*, *8*(7), giz069. https://doi.org/10.1093/gigascience/giz069.

Peron, S., Houliston, E., & Leclère, L. (2021). The marine jellyfish model *Clytia hemisphaerica*. In A. Boutet, & B. Schierwater (Eds.), *Established and emerging marine model organisms in experimental biology* (pp. 129–147). CRC Press.

Quiroga Artigas, G., Lapébie, P., Leclère, L., Bauknecht, P., Uveira, J., Chevalier, S., et al. (2020). A G protein-coupled receptor mediates neuropeptide-induced oocyte maturation in the jellyfish Clytia. *PLoS Biology*, *18*(3), e3000614.

Quiroga Artigas, G., Lapébie, P., Leclère, L., Takeda, N., Deguchi, R., Jékely, G., et al. (2018). A gonad-expressed opsin mediates light-induced spawning in the jellyfish. *eLife*, 7, e29555. https://doi.org/10.7554/eLife.29555.

Rentzsch, F., Layden, M., & Manuel, M. (2017). The cellular and molecular basis of cnidarian neurogenesis. *Wiley Interdisciplinary Reviews. Developmental Biology*, *6*(1), e257. https://doi.org/10.1002/wdev.257.

Roosen-Runge, E. (1962). On the biology of sexual reproduction of hydromedusae, genus *Phialidium Leuckhart*. *Pacific Science*, *16*(1), 15–24.

Roosen-Runge, E. C., & Szollosi, D. (1965). On biology and structure of the testis of *Phialidium Leuckhart* (Leptomedusae). *Zeitschrift für Zellforschung und Mikroskopische Anatomie*, *68*(5), 597–610. https://doi.org/10.1007/bf00340088.

Schmid, V. (1974). Regeneration in medusa buds and medusae of Hydrozoa. *American Zoologist*, *14*(2), 773–781. https://doi.org/10.1093/icb/14.2.773.

Schmid, V., Schmid, B., Schneider, B., Stidwill, R., & Baker, G. (1976). Factors effecting manubrium-regeneration in hydromedusae (Coelenterata). *Wilhelm Roux's Archives of Developmental Biology*, *179*(1), 41–56.

Schmid, V., & Tardent, P. (1971). The reconstitutional performances of the Leptomedusa *Campanularia jonstoni*. *Marine Biology*, *8*(2), 99–104. https://doi.org/10.1007/bf00350924.

Singla, C. L. (1975). Statocysts of hydromedusae. *Cell and Tissue Research, 158*(3), 391–407. https://doi.org/10.1007/bf00223835.

Sinigaglia, C., Peron, S., Eichelbrenner, J., Chevalier, S., Steger, J., Barreau, C., et al. (2020). Pattern regulation in a regenerating jellyfish. *eLife, 9*, e54868. https://doi.org/10.7554/eLife.54868.

Sinigaglia, C., Thiel, D., Hejnol, A., Houliston, E., & Leclère, L. (2018). A safer, urea-based in situ hybridization method improves detection of gene expression in diverse animal species. *Developmental Biology, 434*(1), 15–23.

Steinmetz, P. R. H., Kraus, J. E. M., Larroux, C., Hammel, J. U., Amon-Hassenzahl, A., Houliston, E., et al. (2012). Independent evolution of striated muscles in cnidarians and bilaterians. *Nature, 487*(7406), 231–234.

Szollosi, D. (1964). The Structure and function of centrioles and their satellites in the jellyfish *Phialidium gregarium. Journal of Cell Biology, 21*(3), 465–479. https://doi.org/10.1083/jcb.21.3.465.

Takeda, N., Kon, Y., Quiroga Artigas, G., Lapébie, P., Barreau, C., Koizumi, O., et al. (2018). Identification of jellyfish neuropeptides that act directly as oocyte maturation-inducing hormones. *Development, 145*(2), dev156786. https://doi.org/10.1242/dev.156786.

Takeda, N., Kyozuka, K., & Deguchi, R. (2006). Increase in intracellular cAMP is a prerequisite signal for initiation of physiological oocyte meiotic maturation in the hydrozoan *Cytaeis uchidae. Developmental Biology, 298*(1), 248–258.

Technau, U. (2020). Gastrulation and germ layer formation in the sea anemone *Nematostella vectensis* and other cnidarians. *Mechanisms of Development, 163*, 103628.

Thomas, M. B., Freeman, G., & Martin, V. J. (1987). The embryonic origin of neurosensory cells and the role of nerve cells in metamorphosis in Phialidium gregarium (Cnidaria, Hydrozoa). *International Journal of Invertebrate Reproduction and Development, 11*(3), 265–285. https://doi.org/10.1080/01688170.1987.10510286.

van der Sande, M., Kraus, Y., Houliston, E., & Kaandorp, J. (2020). A cell-based boundary model of gastrulation by unipolar ingression in the hydrozoan cnidarian *Clytia hemisphaerica. Developmental Biology, 460*(2), 176–186.

Wang, R., Steele, R. E., & Collins, E.-M. S. (2020). Wnt signaling determines body axis polarity in regenerating Hydra tissue fragments. *Developmental Biology, 467*(1-2), 88–94.

Weissbourd, B., Momose, T., Nair, A., Kennedy, A., Hunt, B., & Anderson, D. J. (2021). A genetically tractable jellyfish model for systems and evolutionary neuroscience. *Cell, 184*(24), 5854–5868. https://doi.org/10.1016/j.cell.2021.10.021.

Wikramanayake, A. H., Hong, M., Lee, P. N., Pang, K., Byrum, C. A., Bince, J. M., et al. (2003). An ancient role for nuclear beta-catenin in the evolution of axial polarity and germ layer segregation. *Nature, 426*(6965), 446–450.

Xia, W., Li, H., Cheng, W., Li, H., Mi, Y., Gou, X., et al. (2020). High-quality genome assembly of Chrysaora quinquecirrha provides insights into the adaptive evolution of jellyfish. *Frontiers in Genetics, 11*, 535. https://doi.org/10.3389/fgene.2020.00535.

Xia, W.-X., Li, H.-R., Ge, J.-H., Liu, Y.-W., Li, H.-H., Su, Y.-H., et al. (2021). High-continuity genome assembly of the jellyfish. *Zoological Research, 42*(1), 130–134.

Zhang, F., Cong, L., Lodato, S., Kosuri, S., Church, G. M., & Arlotta, P. (2011). Efficient construction of sequence-specific TAL effectors for modulating mammalian transcription. *Nature Biotechnology, 29*(2), 149–153. https://doi.org/10.1038/nbt.1775.

Zimmermann, B., Robb, S. M. C., Genikhovich, G., Fropf, W. J., Weilguny, L., He, S., et al. (2021). Sea anemone genomes reveal ancestral metazoan chromosomal macrosynteny. *bioRxiv*, 2020.10.30.359448. https://doi.org/10.1101/2020.10.30.359448.

CHAPTER SIX

Studying development, regeneration, stem cells, and more in the acoel *Hofstenia miamia*

Mansi Srivastava*
Department of Organismic and Evolutionary Biology, Museum of Comparative Zoology, Harvard University, Cambridge, MA, United States
*Corresponding author: e-mail address: mansi@oeb.harvard.edu

Contents

Abstract

Acoel worms represent an enigmatic lineage of animals (Acoelomorpha) that has danced around the tree of animal life. Morphology-based classification placed them as flatworms (Phylum Platyhelminthes), with much of their biology being interpreted as a variation on what is observed in better-studied members of that phylum. However, molecular phylogenies suggest that acoels belong to a clade (Xenacoelomorpha) that could be a sister group to other animals with bilateral symmetry (Bilateria) or could belong within deuterostomes, closely related to a group that includes sea stars (Ambulacraria). This change in phylogenetic position has led to renewed interest in the biology of

Current Topics in Developmental Biology, Volume 147
ISSN 0070-2153
https://doi.org/10.1016/bs.ctdb.2022.01.003

acoels, which can now offer insights into the evolution of many bilaterian traits. The acoel *Hofstenia miamia* has emerged as a powerful new research organism that enables mechanistic studies of xenacoelomorph biology, especially of developmental and regenerative processes. This article explains the motivation for developing *Hofstenia* as a new model system, describes *Hofstenia* biology, highlights the tools and resources that make *Hofstenia* a good research organism, and considers the questions that *Hofstenia* is well-positioned to answer. Finally, looking to the future, this article serves as an invitation to new and established scientists to join the growing community of researchers studying this exciting model system.

1. Why was *Hofstenia* brought to the lab?

1.1 The original question

Wound repair and regeneration are fundamental features of animal biology (Brockes & Kumar, 2008), and the capacity to replace all missing tissues ("whole-body regeneration") is widely distributed across animal phyla (Bely & Nyberg, 2010) (Fig. 1A). Genetic networks that mediate whole-body regeneration are beginning to be elucidated; but little is known about how these pathways compare across animal lineages. Studying one model system can reveal mechanisms for regeneration in one species; to understand how the mechanisms underlying regeneration have evolved, functional studies of multiple species in phylogenetically informative positions are needed (Srivastava, 2021; Tiozzo & Copley, 2015). At the time *Hofstenia* was brought into laboratory culture, planarian worms (Platyhelminthes) had emerged as a major bilaterian model system for studying whole-body regeneration (also see Newmark & Alvarado, 2022). To make statements about the evolution of regeneration, the mechanisms of regeneration uncovered in planarians would need to be compared to those in other animal lineages. For example, regeneration processes in planarians could be uniquely-evolved in the platyhelminth lineage, or they could be similar to those found in other animal groups, possibly reflecting ancestral or convergently-evolved mechanisms. Studies in other animals would enable us to distinguish between these two scenarios.

I joined Dr. Peter Reddien's lab in 2009 as a postdoctoral researcher with a specific interest in studying regeneration from an evolutionary perspective. The Reddien Lab uses powerful tools, such as functional studies *via* systemic RNA interference (RNAi) in adult animals, to study molecular and cellular mechanisms of regeneration in the planarian flatworm *Schmidtea*

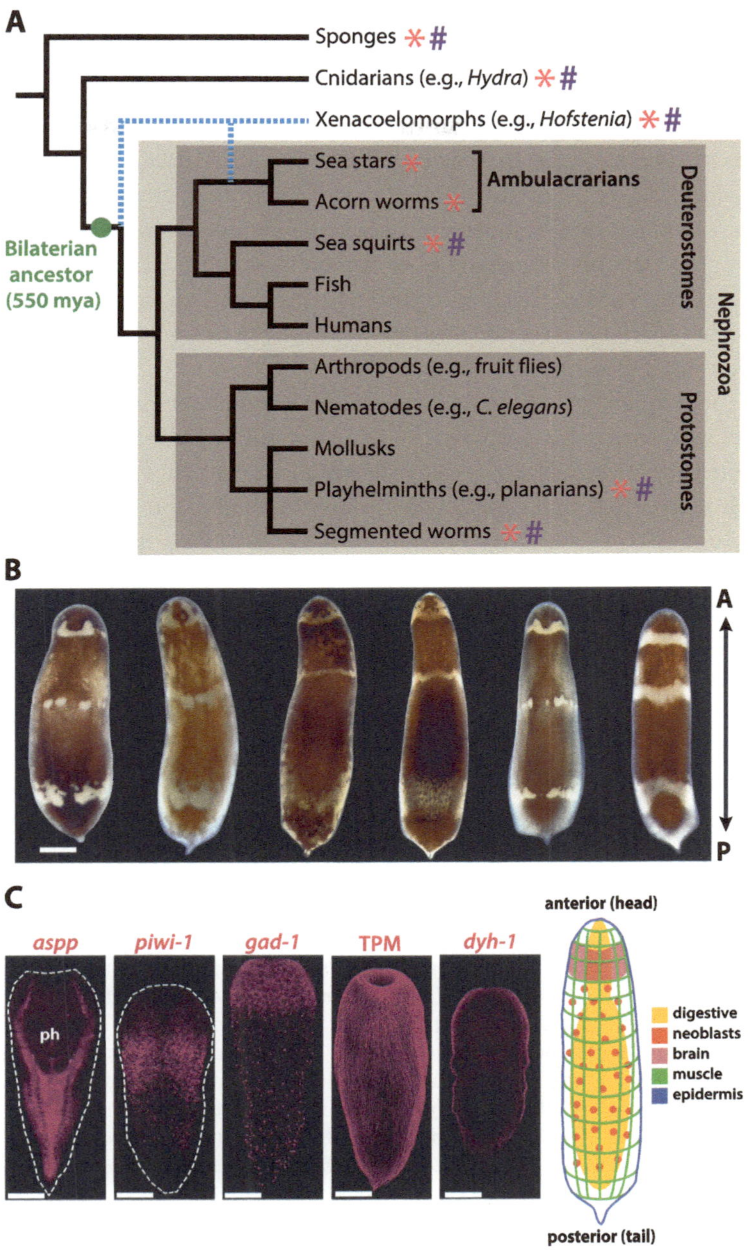

Fig. 1 Phylogenetic position and anatomy. (A) A phylogenetic tree depicting the relationships of major animal lineages including those where whole-body regeneration is studied (*pink* asterisks). Purple hash indicates presence of putatively pluripotent stem cells associated with regeneration. Blue dashed lines indicate the two alternative positions of xenacoelomorphs, including *Hofstenia*, based on molecular analyses. (B) Adult *Hofstenia* showing diversity of pigmentation patterns on the dorsal side;

(Continued)

mediterranea. To make comparative statements in a rigorous manner, we sought additional species that could live in the lab, could regenerate extensively, and were amenable to functional studies in adult animals. Strong evolutionary statements rely on studies of multiple taxa and no one other species can be the one perfect comparison to planarians, and therefore it was important to be open-minded to species from many animal lineages. The road to *Hofstenia* included efforts to develop regeneration assays in acorn worms (*Saccoglossus kowalevskii*), to apply RNAi in adult sea anemones (*Nematostella vectensis*), and to culture another platyhelminth (*Microstomum*) in the lab. *Hofsenia*, an acoel worm, just happens to be the system that worked out successfully.

Acoels, a group of marine worms that contains many regenerative species, are particularly attractive as a comparand to planarians because their bilaterian body plan bears similarity to planarians at the level of gross morphology, such that one could ask questions about the regeneration of homologous body parts, *e.g.*, heads and tails. Based on this morphological similarity, acoels were previously considered to be platyhelminths, belonging to the same phylum as planarian worms (Ax & Jeffries, 1987; Hyman, 1951; Smith, Teyler, & Rieger, 1986). However, molecular phylogenetic analyses at the time indicated that acoels likely belong to a lineage that forms a sister-group to all other animals with bilateral symmetry (Hejnol et al., 2009; Ruiz-Trillo et al., 2002; Ruiz-Trillo, Riutort, Fourcade, Baguñà, & Boore, 2004; Ruiz-Trillo, Riutort, Littlewood, Herniou, & Baguna, 1999; Sempere, Martinez, Cole, Baguñà, & Peterson, 2007) (Fig. 1A).

Fig. 1—Cont'd images oriented with anterior (A) to the top and posterior (P) to the bottom. Scale bar, 500 μm. (C) Fluorescence *in situ* hybridization labeling in juvenile worms of gut (*aspp*), stem cells (*piwi-1*), nervous system (*gad-1*), and epidermis (*dyh-1*). Fluorescent immunostaining of muscle using a custom anti-Tropomyosin antibody (TPM). All images and schematic oriented with anterior to the top and posterior to the bottom. *piwi-1* image provided by Ryan Hulett, *aspp* image provided by D. Marcela Bolaños. White dashed line indicates the boundary of the specimen. ph, pharynx. Scale bars, 100 μm. *Panel B: based on images from Srivastava, M., Mazza-Curll, K., van Wolfswinkel, J., & Reddien, P. W. (2014). Whole-body acoel regeneration is controlled by Wnt and Bmp-Admp signaling.* Current Biology, *24(10), 1107–1113; Panel C: images reproduced with permission from Hulett, R. E., Potter, D., & Srivastava, M. (2020). Neural architecture and regeneration in the acoel* Hofstenia miamia. Proceedings of the Biological Sciences, 287*(1931), 20201198. doi:10.1098/rspb.2020.1198 and Kimura, J. O., Ricci, L., & Srivastava, M. (2021). Embryonic development in the acoel Hofstenia miamia.* Development, 148*(13). doi:10.1242/dev.188656.*

In this position, comparisons of regeneration between acoels and planarians could generate hypotheses about mechanisms in the bilaterian ancestor that existed 550 million years ago.

1.2 A new acoel system for laboratory research

In 2009, laboratory research on acoel regeneration utilized various species of acoels including *Isodiametra pulchra*, *Symsagittifera roscoffensis*, and species in the genus *Convolutriloba* (Bourlat & Hejnol, 2009). These research organisms separately offered powerful biology (*e.g.* whole-body regeneration in *Convolutriloba*, but not RNAi) and tools (*e.g.* RNAi in *Isodiametra*, but not anterior regeneration) (Bely & Sikes, 2010; De Mulder et al., 2009; Sikes & Bely, 2008, 2010). We sought a system where whole-body regeneration, including regrowth of both anterior and posterior tissue, could be studied functionally in one species. The knowledge of experts in acoel biology and taxonomy played a crucial role. Matt Hooge and colleagues published a revision of the systematics of *Hofstenia* and ended their manuscript musing about the potential of three-banded panther worms, *Hofstenia* spp., as a research organism (Hooge et al., 2007) (Fig. 1B). Dr. Ulf Jondelius at the Swedish Museum of Natural History recommended *Hofstenia miamia* as a species that might be amenable to laboratory culture. Additionally, previous work by Steinbock suggested that the worms could regenerate (Steinbock, 1967).

Based on reports of where *Hofstenia* had been previously collected in the field (Hooge et al., 2007), we obtained a permit to collect 120 worms in Bermuda. With the guidance of Dr. Wolfgang Sterrer at the Bermuda Museum of Natural History, we identified the specific part of Walsingham Pond where researchers had previously been successful in finding *Hofstenia*. We collected decaying mangrove leaves at the bottom of the pond and algae growing on the mangrove roots in plastic bags. The Bermuda Museum of Natural History generously provided space and access to microscopes for us to work with this material. Sitting in large buckets, the material became anoxic over time, and many small organisms, including *Hofstenia*, began to crawl to the surface of the water. A quick look under a microscope revealed the characteristic banding pattern on the worms, confirming their identity as *Hofstenia*. Two days' worth of collecting yielded the 120 animals we sought, which flew to Boston, MA in 50 mL falcon tubes. After being cleared for import by a US Fish and Wildlife officer at Logan International airport, *Hofstenia* were ready to take the field of regeneration biology by storm.

2. Who is *Hofstenia*?

Although other hofsteniids had been described as early as 1923 (Bock, 1923), *Hofstenia miamia* was first described by Diva Diniz Corrêa in 1960 (Corrêa, 1960, 1963). The worms are about 4–10 mm in length, with an anteriorly-located mouth and a tail that forms a sharp point in the posterior. The dorsal surface of the worms carries the three characteristic cream-colored bands on an otherwise brown-pigmented animal, although individual worms show variations on this pattern (Fig. 1B). Internal anatomy has been described using traditional histological approaches, use of molecular markers, and transgenic lines (Hooge et al., 2007; Hulett, Potter, & Srivastava, 2020; Kimura, Ricci, & Srivastava, 2021; Ricci & Srivastava, 2021; Srivastava, Mazza-Curll, van Wolfswinkel, & Reddien, 2014) (Fig. 1C). The mouth leads to a muscular pharynx, which leads to a large syncytial gut. Several major muscle groups, including pharyngeal, parenchymal, body wall, and peripheral, provide structure to the body (Ricci & Srivastava, 2021). A subepidermal nerve net is found throughout the body, and many neuron types are present and organized into a brain-like anterior condensation (Hulett et al., 2020). The animal is covered in epidermal cells, but no basement membrane is observable.

Hofstenia have been found in mangrove habitats including in Bermuda, Bahamas, and Panama. Although two species occupying different niches in these habitats had been named (*H. giselae* living among algae and *H. miamia* living among decaying leaves), these species have been synonymized as *Hofstenia miamia* based on molecular studies (Hooge et al., 2007). Within acoels, *H. miamia* belongs to the family Hofstenidae, a clade that diverged before the radiation of the convolutid and isodiametrid acoels, which have been the focus of many developmental and regeneration studies (Bailly et al., 2014; De Mulder et al., 2009; Hejnol & Martindale, 2008a, 2008b, 2009; Semmler, Bailly, & Wanninger, 2008; Sikes & Bely, 2008, 2010).

Whereas progress has been made in resolving the relationships of acoel groups relative to each other (Jondelius, Wallberg, Hooge, & Raikova, 2011), the position of acoels relative to other bilaterians has been harder to pin down and has garnered much attention. The closest relatives of acoel worms are nemertodermatids, and together this group is referred to as the Acoelomorpha. Molecular analyses that include the enigmatic xenoturbellid worms suggest a close affinity to acoelomorphs, revealing a group referred to as Xenacoelomorpha (Philippe et al., 2011), albeit some data argue against this view (Mwinyi et al., 2010). Xenacoelomorphs finds themselves in different positions on the animal tree of life, with some analyses placing the lineage as the sister group to all other bilaterians and others placing it as a sister lineage to ambulacrarians (hemichordates and echinoderms) (Cannon et al., 2016; Hejnol et al., 2009; Kapli et al., 2021; Kapli & Telford, 2020; Philippe et al., 2011; Philippe, Brinkmann, Martinez, Riutort, & Baguna, 2007; Philippe et al., 2019; Ruiz-Trillo et al., 2002; Ruiz-Trillo et al., 2004; Ruiz-Trillo et al., 1999; Sempere et al., 2007; Telford, Lockyer, Cartwright-Finch, & Littlewood, 2003) (Fig. 1A). Resolving the position of xenacoelomorphs has been challenging likely because the lineage emerged around the time of the bilaterian radiation, which occurred a long time ago and occurred rapidly. This process may have led to low signal or systematic errors in sequence data, and currently available models for phylogenetic analyses may not be effective in analyzing these data. The disagreement in the placement of acoels has emerged from the utilization of different models of evolution and different approaches to addressing systematic error (Kapli, Flouri, & Telford, 2021). This uncertainty is fueling the generation of new datasets and the development of new methods in phylogenetics to obtain resolution on the placement of acoels. The last section of this review discusses approaches to studying a system with an unresolved phylogenetic position.

3. Why is *Hofstenia* a good research organism?

3.1 Easy laboratory culture

Hofstenia have thrived in the lab since their arrival in the United States. The worms can be cultured in artificially synthesized sea water in plastic boxes at room temperature (Srivastava et al., 2014). The worms are hermaphroditic and reproduce sexually, producing abundant embryos that can be raised in the lab and become sexually reproductive in 4–6 weeks (Kimura et al., 2021). The worms have a simple diet—adult worms are fed brine shrimp whereas juvenile worms eat rotifers. The current lab populations at MIT, Harvard, and other universities are derived from the progeny of the original 120 worms collected from Bermuda.

3.2 Opportunity to study both development and regeneration

Hofstenia regenerate extensively and enable mechanistic studies of this process (Fig. 2A–C). Upon transverse amputation, both head and tail fragments

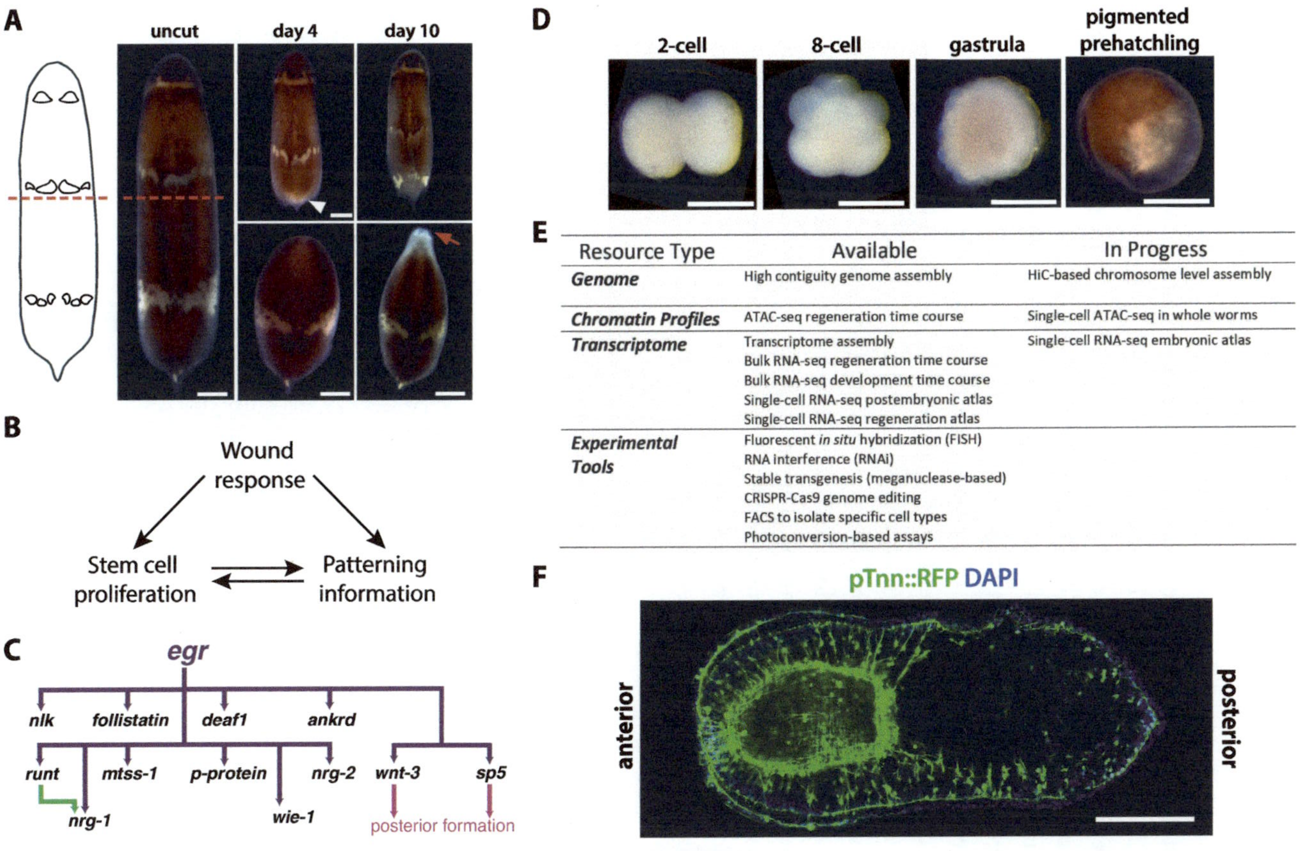

Fig. 2 See figure legend on opposite page.

replace all missing tissue to produce fully-functional worms, including a fully-functioning new brain (Hulett et al., 2020; Srivastava et al., 2014). Regeneration proceeds *via* the formation of new tissue, which is seen as a prominent unpigmented blastema in tail fragments, as well as changes in preexisting tissue, a process referred to as morphallaxis. In addition to transverse amputation into two fragments, worms will also undergo regeneration if amputated into more pieces or if cut sagittally or diagonally (unpublished data, Srivastava Lab).

Although the original objective in developing *Hofstenia* as a new system was to study regeneration, *Hofstenia* is now also facilitating studies of development because the worms also produce plentiful embryos in the lab. Embryonic development in *Hofstenia* has been characterized from zygote to hatching and occurs over 8–9 days (Kimura et al., 2021) (Fig. 2D). Broadly, embryos undergo a stereotyped cleavage program called duet cleavage, which results in the formation of small cells (micromeres) at the animal pole and two large cells (macromeres) at the vegetal pole. Upon gastrulation, macromeres become internalized, becoming enveloped by micromere progeny. Two days into development, a second cell internalization event results in the movement of cells at the surface to the interior of the embryo (Dimple stage). Major body axes and differentiated tissues such as gut, muscle, and the nervous system are detected after this second internalization event.

Fig. 2 *Hofstenia* as a model system for studying regeneration and development. (A) *Hofstenia* can regenerate any missing issue, with transverse amputations (dashed red lines) serving as a standard assay. Tails regenerate within a few days (white arrowhead) and head regeneration is first observable as an unpigmented outgrowth called a blastema (red arrow). images oriented with anterior to the top and posterior to the bottom. Scale bars, 500 μm. (B) A schema summarizing major aspects of regeneration that can be studied in *Hofstenia*. (C) A wound-induced gene regulatory network (GRN) controlled by the master regulator Egr, which is also likely a direct regulator of Wnt pathway components (*wnt-3* and *sp5*) (Gehrke et al., 2019; Ramirez, Loubet-Senear, & Srivastava, 2020). This GRN provides a direct link between generic wound-induced genes (*egr*) and patterning information (Wnt pathway). (D) Stages of embryonic development in *Hofstenia* showing stereotype early cleavage and direct development. Scale bars, 150 μm. (E) A table summarizing currently available and in progress tools and resources in *Hofstenia*. (F) A transgenic animal showing fluorescently labeled peripheral muscle where fluorescent protein expression is driven by a muscle-specific promoter *troponin* (*tnn*). Scale bar, 100 μm. *Panel (A): based on images from Srivastava, M., Mazza-Curll, K., van Wolfswinkel, J., & Reddien, P. W. (2014). Whole-body acoel regeneration is controlled by Wnt and Bmp-Admp signaling.* Current Biology, *24(10), 1107–1113. Panel F: image reproduced with permission from Ricci, L., & Srivastava, M. (2021). Transgenesis in the acoel* Hofstenia miamia. Development Cell, *56, 3160–3170.*

3.3 Tools for mechanistic studies

Many tools and resources are now available for in depth mechanistic studies of development and regeneration in *Hofstenia* (Fig. 1E). Bulk transcriptome data available for time courses of regeneration (in head and tail wound sites) and development can be mapped to a high quality reference transcriptome to identify genes that are putatively important for these processes (Gehrke et al., 2019; Kimura et al., 2021; Tewari, Owen, Petersen, Wagner, & Reddien, 2019). Single-cell transcriptome data from postembryonic stages and regenerating worms provide a cell type atlas and information of cell-type specific gene expression dynamics (Hulett et al., 2022). Gene expression can be studied in whole-mount embryos and worms *via* colorimetric or fluorescent *in situ* hybridization (FISH) (Kimura et al., 2021; Srivastava et al., 2014). Double-stranded RNA (dsRNA) administered *via* injection into the gut or *via* soaking results in RNAi, enabling the study of gene function during regeneration (Srivastava et al., 2014). Cells can be sorted and isolated with FACS (Raz, Srivastava, Salvamoser, & Reddien, 2017). A high-quality genome assembly and chromatin profiling data from regenerating worms enable functional genomics approaches for the identification of transcriptional gene regulatory networks. The availability of accessible embryos has facilitated the development of tools for lineage tracing in embryos and for transgenesis (Kimura et al., 2021; Ricci & Srivastava, 2021). Stable transgenic lines with fluorescently labeled epidermal, muscle, and gut tissue are available (Fig. 1F). Additionally, it has been shown that genome editing *via* CRISPR-Cas9 is feasible in this system. Progress in tool and resource development is ongoing, for example, HiC data have been applied to generate a draft chromosome-scale assembly and single-cell RNA-sequencing data have revealed cell types and their associated molecular profiles (unpublished data, Srivastava Lab).

4. What is *Hofstenia* helping us understand?

4.1 Regeneration and stem cell biology

Whole-body regeneration hinges upon three crucial steps: (1) signaling from the wound site, *i.e.*, the wound response, (2) proliferation of undifferentiated or dedifferentiated cells, and (3) patterning of the newly formed cells to correctly integrate new tissue into the existing animal body plan (Fig. 2B). *Hofstenia* offers an opportunity to study all of these aspects.

The wound response has been studied transcriptomically and epigenomically in *Hofstenia*, enabling a functional epigenomics approach for building gene regulatory networks (GRNs) for regeneration (Gehrke et al., 2019). We applied the assay for transposase-accessible chromatin using sequencing (ATAC-seq), which revealed thousands of chromatin regions that respond dynamically during regeneration. By looking for transcription factor (TF) binding motifs within these regions, we found that chromatin containing the Egr (Early Growth Response) motif was the most dynamic. Our analysis of ATAC-seq data from worms where Egr had been knocked down with RNAi allowed us to propose a gene regulatory network, with Egr as its top regulatory, that is launched upon wounding in *Hofstenia* (Fig. 2C). We were then able to link this GRN to the activation of the Wnt pathway, a key player needed for the regeneration of correctly patterned tissue along the anterior-posterior axis. Egr itself is a likely direct transcriptional regulator of the Wnt ligand, *wnt-3*, a patterning gene whose activation within hours of amputation, specifically at posterior-facing wounds, is indicative of the worm figuring out where it needs to make a head or a tail (Ramirez et al., 2020).

In addition to providing deep mechanistic insight into regulatory decision-making during the wound response and repatterning processes, this GRN provides a substrate for cross-species comparisons to inform the evolution of regeneration because it includes homologs of genes that are involved in regeneration in other species (Srivastava, 2021), including Wnt pathway components which appear to have similar roles during regeneration in acoels and planarians (Srivastava et al., 2014). Comparative work in acoels and planarians has revealed that Wnt pathway components are expressed in similar cell types (Raz et al., 2017) and that a core set of conserved genes are targeted by Wnt signaling in both systems (Tewari et al., 2019). A next major question GRN construction could address is about how the inputs into Wnt signaling compare between the two animals (Srivastava, 2021).

The wound response and patterning systems ultimately regulate the proliferation and differentiation of cells, which in the case of *Hofstenia*, are a large population of parenchymal stem cells called "neoblasts". Neoblasts appear to be the only proliferating cells in *Hofstenia*, are required for regeneration, and express homologs of *piwi* (Srivastava et al., 2014). Notably, similar populations of *piwi*-expressing stem cells are found in other species that regenerate extensively, *e.g.*, planarians (neoblasts), cnidarians (i-cells), sponges (archaeocytes), and sea squirts (Gehrke & Srivastava, 2016; Lai & Aboobaker, 2018) (Fig. 1A)). Studies of these species and *Hofstenia* can reveal the basic molecular and cellular principles that regulate pluripotent cells and their progeny in an *in vivo* context (Srivastava, 2021). Analysis of single-cell RNAseq data (Hulett et al., 2022) suggests that *Hofstenia* neoblasts, similarly to planarian ones, are a heterogeneous population that includes many lineage-primed progenitors for differentiated tissue. The ease of RNAi in this system is facilitating a thorough functional investigation of neoblast subpopulations.

4.2 Developmental biology

The abundant and accessible embryos of *Hofstenia* are ripe for many investigations (Kimura et al., 2021). The stereotyped cleavage program raises the question of whether early embryonic blastomeres are fated to produce specific tissues in hatched worms. *Hofstenia* blastomeres can be targeted for microinjection of dyes or nucleic acids to facilitate lineage tracing for fate mapping (Kimura et al., 2021). Fate mapping of these cells would enable comparisons of duet cleavage to spiral cleavage, a pattern of cell division seen in many animal lineages (such as annelids, mollusks, platyhelminths) that superficially resembles duet cleavage (Boyer, Henry, & Martindale, 1996; Boyer & Jonathan, 1998; Henry, Martindale, & Boyer, 2000; Martin-Duran & Marletaz, 2020). Comparisons of cleavage patterns and fate potentials across xenacoelomorphs and spiralians will yield answers to questions about how cellular mechanisms for germ layer and fate specification have evolved.

As a representative xenacoelomorph, *Hofstenia* and its embryos also provide opportunities to apply classical approaches in evolutionary developmental biology to study the evolution of many bilaterian traits, such as having three germ layers and a centralized nervous system (Bourlat & Hejnol, 2009; Hejnol & Pang, 2016). Consider for example the nervous system. In *Hofstenia*, neurons are organized as a subepidermal nerve net, similar to those observed in cnidarians and hemichordates. Additionally, a bi-layered condensation with a network of neurite bundles is present in the anterior, which could have a brain-like function in performing centralized computation (Hulett et al., 2020). If xenacoelomorphs are a sister-group to all other bilaterians, then studies of the *Hofstenia* brain, in terms of both patterning and function, would facilitate ancestral trait reconstruction for the urbilaterian brain. In the alternative position of acoels with ambulacrarians, these studies would inform the evolution of vertebrate brains much as investigations of hemichordate brains have (Hejnol & Lowe, 2015; Pani et al., 2012; Yao et al., 2016).

In the "xenacoelomorphs sister to other bilaterians" scenario, all other bilaterian lineages are contained within a monophyletic group called the "nephrozoa" (Fig. 3). This name indicates one possible shared derived character of this group - the presence of cells specialized for excretion. Although no distinct excretory organs have been identified in xenacoelomorphs thus far, mechanistic studies in *Hofstenia* could reveal cells that perform excretory function and their associated regulatory factors, furthering the study of the evolution of these cell types (Andrikou, Thiel, Ruiz-Santiesteban, & Hejnol, 2019; Gasiorowski et al., 2021). The question of excretory (or other) cell type evolution will be informed by the availability of single-cell RNA-seq data for *Hofstenia*, which provide molecular profiles for cell types in the animal.

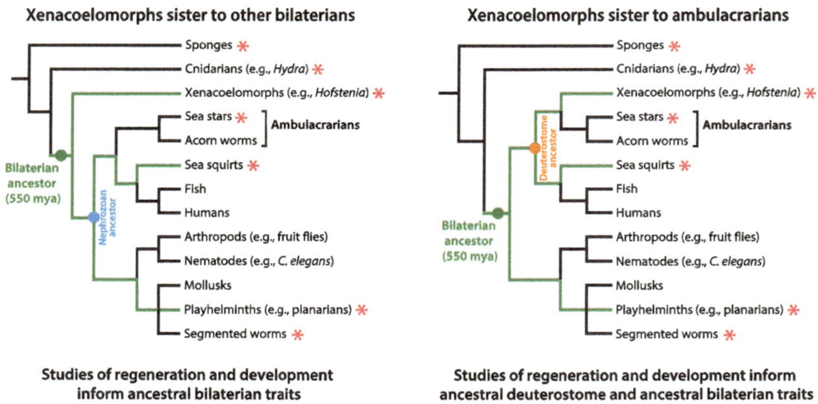

Fig. 3 Evolutionary inferences with uncertain phylogenetic relationships. *Left*: A phylogenetic tree depicting xenacoelomorphs, including *Hofstenia*, as the sister lineage to all other bilaterians; blue dot marks the last common ancestor of all other bilaterians, the nephrozoan ancestor. *Right*: A phylogenetic tree depicting xenacoelomorphs, including *Hofstenia*, as the sister lineage to ambulacrarians; orange dot marks the last common ancestor of xenacoelomorphs and vertebrates, the deuterostome ancestor. Green dot marks the ancestor of all bilaterians, including *Hofstenia*. Green color highlights relationships of xenacoelomorphs, platyhelminths, and sea squirts. In both scenarios, studies of regeneration and development in *Hofstenia* can reconstruct ancestral bilaterian traits when comparative statements include multiple taxa, *e.g.,* planarians and sea squirts for studying stem cells. In the scenario on the right, studies of *Hofstenia* are also important for understanding deuterostome evolution, *e.g.,* they can inform the evolution of the vertebrate nervous system.

4.3 Novel cell biology

Acoels present striking anatomical features that are unique relative to other bilaterians. For example, they lack basement membranes, and many species contain syncytial guts (Bourlat & Hejnol, 2009; Gavilan, Sprecher, Hartenstein, & Martinez, 2019). The biological significance of these traits is not known, and *Hofstenia* provides a system for determining how these tissues work. Photoconversion-based studies of the *Hofstenia* gut showed that it is a functional syncytium, for example (Ricci & Srivastava, 2021). Transgenic lines with fluorescently labeled epidermal, gut, and muscle tissue have revealed a high-resolution view of cellular anatomy in *Hofstenia*. Notably, terminal ramifications of muscle cells create an interlocking grid of longitudinal, circumferential, and parenchymal muscle. Epidermal cells project finger-like projections basally toward the interior of the animal and muscle fibers align closely with these projections. Additionally, muscle fibers also align with thin extensions of the gut. The structural or signaling roles of these closely associated cell types and other features of acoel cells and tissues could be probed functionally in *Hofstenia*.

5. What's the outlook?

5.1 Mechanistic depth into many questions

With the number of tools that *Hofstenia* offers as a system, the big next steps will be in deepening the level of mechanistic insight that can be gained. Current studies have pointed to putative enhancers in the genome, which will now be tested functionally in reporter assays. GRNs for regeneration will be inferred with cellular resolution. Genes required for neoblast biology have been identified, and now transgenic worms can be used to visualize and trace the fates of individual stem cells.

In addition to adding power to existing questions, *Hofstenia* is also open to new investigations, ones we did not originally have in mind when collecting the animals. The surprise of finding that *Hofstenia* produce embryos in the lab is now an opportunity to understand how regeneration and development compare, a long-standing pursuit in the field of regeneration biology. The worms can now be used to study any question, including those that don't rely on the significance of the phylogenetic position of the worms, or those that don't rely on the worms' ability to regenerate. For example, a neuroscientist might find *Hofstenia* to be a good system for studying the brain and behavior, and a cell biologist might seek to understand epithelial cell morphology.

5.2 Embracing uncertainty: Evo-devo with an unclear phylogenetic placement

Given the bright outlook for *Hofstenia* as a system, some may wonder how the uncertainty in its phylogenetic position will impact inferences in the long term. For many evolutionary questions, the precise answer to whether *Hofstenia* is a deuterostome or whether it represents a sister group to other bilaterians does not matter. In either case, comparing mechanisms for regeneration in *Hofstenia* to other regenerative species would be insightful because *Hofstenia* would still be distantly related to other major lineages where these questions are being studied (Fig. 3). For example, the last common ancestor of *Hofstenia*, planarians, and sea squirts would be the bilaterian ancestor (550mya) in both scenarios. *Hofstenia* developmental mechanisms would also be important to infer in either scenario. With increasing interest in resolving the relationships of animal groups and more investment in method development, we can expect the position of acoels to be resolved in the coming years. In the meantime, work in *Hofstenia* and other similarly unplaced animal lineages can build an arsenal of mechanistic inferences about animal biology, particularly for processes such as whole-body regeneration that cannot be studied in well-established model systems.

In the context of evolutionary comparisons, it is also important to note that powerful inferences from comparative approaches come from multi-taxon statements and not two-taxon statements. The reason that comparative work in Evo-Devo often relies on two or three species is that few systems for investigations at a mechanistic level are available. This has been changing substantially in the last 10–20 years, and this special issue highlights the diversity of species that can now be studied. *Hofstenia* is just one of many species that regenerate or offer similar other biology, and the best long-term outcome would be to study and compare many more species. Continued work on many acoel model systems such as *Isodiametra*, *Convolutriloba*, and *Symsagittifera* will be important for addressing a diversity of questions, including the evolution of developmental, regenerative, and stem cell mechanisms within and beyond acoels (Bailly et al., 2014; De Mulder et al., 2009; Perea-Atienza et al., 2013; Sikes & Bely, 2008, 2010; Sprecher et al., 2015).

5.3 Growing community

Hofstenia started out as a laboratory research organism at the Whitehead Institute in Cambridge, MA. They moved up the road two miles to Harvard University when I started my lab, and since then the worms have traveled to many other labs, including those in Illinois, Norway, Switzerland, the UK, the Netherlands, and China. Being adopted by an increasing number of research groups is the best a new research organism can hope to achieve. So I invite you to consider *Hofstenia* as a potential next research organism. To new scientists it would deliver the thrill of discovering foundational knowledge in a system, and to established researchers it would bring a new perspective, it may offer an evolutionary insight or a new tool for your questions. The worms are charming, easy to rear, and ready to ship.

Acknowledgments

Hofstenia could not have become a laboratory research organism without the efforts of many people, primary among whom is Peter Reddien, who saw the value in studying acoels and provided the mentoring and research support to get *Hofstenia* started as a system. Ulf Jondelius pointed us to the potential of *Hofstenia* as a laboratory research organism and Wolfgang Sterrer and the Bermuda Museum of Natural History guided the collection of the animals. Stan Rachootin provided key input on culturing methods and Kathleen Mazza-Curll worked tirelessly to keep the worms happy. Thank you also to the many scientists who have fearlessly worked on this new system over the years, always advancing knowledge and tools: Amelie Raz, Aneesha Tewari, Emily Neverett, Andrew Gehrke, Lorenzo Ricci, D. Marcela Bolaños, Julian Kimura, Ryan Hulett, Alyson Ramirez, Vanessa Poirier, Yi-Jyun Luo, Hafsa Sadiq, Katy Loubet-Senear, Amber Rock, Vikram Chandra, Sarah Arnold, Carlos Rivera-Lopez, Catriona Breen, Andre Franco-Vasquez, Sylvia Golumbeanu, Annika Gompers, Juliet Kim, Deirdre Potter, Elizabeth D'Haiti, Justin Duffy, Núria Ros i Rocher, James Sikes, Jan Engelhardt, and Mark Martindale. MS is supported by the National Science Foundation (1652104), the National Institutes of Health (1R35GM128817), and the Richard and Susan Smith Family Foundation.

References

Andrikou, C., Thiel, D., Ruiz-Santiesteban, J. A., & Hejnol, A. (2019). Active mode of excretion across digestive tissues predates the origin of excretory organs. *PLoS Biology*, *17*(7), e3000408. https://doi.org/10.1371/journal.pbio.3000408.

Ax, P., & Jeffries, R. P. S. (1987). *The phylogenetic system: The systematization of organisms on the basis of their phylogenies*. Chichester and New York: John Wiley and Sons.

Bailly, X., Laguerre, L., Correc, G., Dupont, S., Kurth, T., Pfannkuchen, A., et al. (2014). The chimerical and multifaceted marine acoel Symsagittifera roscoffensis: From photosymbiosis to brain regeneration. *Frontiers in Microbiology*, *5*, 498. https://doi.org/10.3389/fmicb.2014.00498.

Bely, A. E., & Nyberg, K. G. (2010). Evolution of animal regeneration: Re-emergence of a field. *Trends in Ecology & Evolution*, *25*(3), 161–170. https://doi.org/10.1016/j.tree.2009.08.005.

Bely, A. E., & Sikes, J. M. (2010). Acoel and platyhelminth models for stem-cell research. *Journal of Biology*, *9*(2), 14. https://doi.org/10.1186/jbiol223.

Bock, S. (1923). Eine neue marine Turbellariengattung aus Japan. *Uppsala Universitets Årsskrift, Mathematikoch Naturvetenskap*, *1*, 1–52.

Bourlat, S. J., & Hejnol, A. (2009). Acoels. *Current Biology, 19*(7), R279–R280. https://doi.org/10.1016/j.cub.2009.02.045.

Boyer, B. C., Henry, J. Q., & Martindale, M. Q. (1996). Modified spiral cleavage: The duet cleavage pattern and early Blastomer fates in the Acoel Turbellarian Neochildia fusca. *The Biological Bulletin, 191*(2), 285–286. https://doi.org/10.1086/BBLv191n2p285.

Boyer, B. C., & Jonathan, Q. H. (1998). Evolutionary modifications of the spiralian developmental program. *American Zoologist, 38*(4), 621–633. https://doi.org/10.1093/icb/38.4.621.

Brockes, J. P., & Kumar, A. (2008). Comparative aspects of animal regeneration. *Annual Review of Cell and Developmental Biology, 24*, 525–549. https://doi.org/10.1146/annurev.cellbio.24.110707.175336.

Cannon, J. T., Vellutini, B. C., Smith, J., 3rd, Ronquist, F., Jondelius, U., & Hejnol, A. (2016). Xenacoelomorpha is the sister group to Nephrozoa. *Nature, 530*(7588), 89–93. https://doi.org/10.1038/nature16520.

Corrêa, D. D. (1960). Two new marine Turbellaria from Florida. *Bulletin of Marine Science of the Gulf and Caribbean, 10*, 208–216.

Corrêa, D. D. (1963). The turbellarian Hofstenia miamia in the Caribbean Sea. *Studies Fauna Curaçao, 17*, 38–40.

De Mulder, K., Kuales, G., Pfister, D., Willems, M., Egger, B., Salvenmoser, W., et al. (2009). Characterization of the stem cell system of the acoel Isodiametra pulchra. *BMC Developmental Biology, 9*, 69. https://doi.org/10.1186/1471-213X-9-69.

Gasiorowski, L., Andrikou, C., Janssen, R., Bump, P., Budd, G. E., Lowe, C. J., et al. (2021). Molecular evidence for a single origin of ultrafiltration-based excretory organs. *Current Biology, 31*(16), 3629–3638 e3622. https://doi.org/10.1016/j.cub.2021.05.057.

Gavilan, B., Sprecher, S. G., Hartenstein, V., & Martinez, P. (2019). The digestive system of xenacoelomorphs. *Cell and Tissue Research, 377*(3), 369–382. https://doi.org/10.1007/s00441-019-03038-2.

Gehrke, A. R., Neverett, E., Luo, Y. J., Brandt, A., Ricci, L., Hulett, R. E., et al. (2019). Acoel genome reveals the regulatory landscape of whole-body regeneration. *Science, 363*(6432). https://doi.org/10.1126/science.aau6173.

Gehrke, A. R., & Srivastava, M. (2016). Neoblasts and the evolution of whole-body regeneration. *Current Opinion in Genetics & Development, 40*, 131–137. https://doi.org/10.1016/j.gde.2016.07.009.

Hejnol, A., & Lowe, C. J. (2015). Embracing the comparative approach: How robust phylogenies and broader developmental sampling impacts the understanding of nervous system evolution. *Philosophical Transactions of the Royal Society of London. Series B, Biological Sciences, 370*(1684). https://doi.org/10.1098/rstb.2015.0045.

Hejnol, A., & Martindale, M. Q. (2008a). Acoel development indicates the independent evolution of the bilaterian mouth and anus. *Nature, 456*(7220), 382–386. https://doi.org/10.1038/nature07309.

Hejnol, A., & Martindale, M. Q. (2008b). Acoel development supports a simple planula-like urbilaterian. *Philosophical Transactions of the Royal Society of London. Series B, Biological Sciences, 363*(1496), 1493–1501. https://doi.org/10.1098/rstb.2007.2239.

Hejnol, A., & Martindale, M. Q. (2009). Coordinated spatial and temporal expression of Hox genes during embryogenesis in the acoel Convolutriloba longifissura. *BMC Biology, 7*, 65. https://doi.org/10.1186/1741-7007-7-65.

Hejnol, A., Obst, M., Stamatakis, A., Ott, M., Rouse, G. W., Edgecombe, G. D., et al. (2009). Assessing the root of bilaterian animals with scalable phylogenomic methods. *Proceedings of the Royal Society B: Biological Sciences, 276*(1677), 4261–4270. https://doi.org/10.1098/rspb.2009.0896.

Hejnol, A., & Pang, K. (2016). Xenacoelomorpha's significance for understanding bilaterian evolution. *Current Opinion in Genetics & Development, 39*, 48–54. https://doi.org/10.1016/j.gde.2016.05.019.

Henry, J. Q., Martindale, M. Q., & Boyer, B. C. (2000). The unique developmental program of the acoel flatworm, *Neochildia fusca*. *Developmental Biology*, *220*(2), 285–295. https://doi.org/10.1006/dbio.2000.9628.

Hooge, M., Wallberg, A., Todt, C., Maloy, A., Jondelius, U., & Tyler, S. (2007). A revision of the systematics of panther worms (*Hofstenia* spp., Acoela), with notes on color variation and genetic variation within the genus. *Hydrobiologia*, *592*(1), 439–454.

Hulett, R. E., Kimura, J. O., Bolaños Rodreiguez, D. M., Luo, Y.-J., Ricci, L., & Srivastava, M. (2022). Acoel single-cell atlas reveals expression dynamics and heterogeneity of a pluripotent stem cell population. *BioRxiv*. https://doi.org/10.1101/2022.02.10.479464.

Hulett, R. E., Potter, D., & Srivastava, M. (2020). Neural architecture and regeneration in the acoel Hofstenia miamia. *Proceedings of the Biological Sciences*, *287*(1931), 20201198. https://doi.org/10.1098/rspb.2020.1198.

Hyman, L. (1951). *The invertebrates, Vol II: Platyhelminthes and Rhynchocoela*. New York: McGraw-Hill Book Company, Inc.

Jondelius, U., Wallberg, A., Hooge, M., & Raikova, O. I. (2011). How the worm got its pharynx: Phylogeny, classification and Bayesian assessment of character evolution in Acoela. *Systematic Biology*, *60*(6), 845–871. https://doi.org/10.1093/sysbio/syr073.

Kapli, P., Flouri, T., & Telford, M. J. (2021). Systematic errors in phylogenetic trees. *Current Biology*, *31*(2), R59–R64. https://doi.org/10.1016/j.cub.2020.11.043.

Kapli, P., Natsidis, P., Leite, D. J., Fursman, M., Jeffrie, N., Rahman, I. A., et al. (2021). Lack of support for Deuterostomia prompts reinterpretation of the first Bilateria. *Science Advances*, 7(12). https://doi.org/10.1126/sciadv.abe2741.

Kapli, P., & Telford, M. J. (2020). Topology-dependent asymmetry in systematic errors affects phylogenetic placement of Ctenophora and Xenacoelomorpha. *Science Advances*, *6*(50). https://doi.org/10.1126/sciadv.abc5162.

Kimura, J. O., Ricci, L., & Srivastava, M. (2021). Embryonic development in the acoel Hofstenia miamia. *Development*, *148*(13). https://doi.org/10.1242/dev.188656.

Lai, A. G., & Aboobaker, A. A. (2018). EvoRegen in animals: Time to uncover deep conservation or convergence of adult stem cell evolution and regenerative processes. *Developmental Biology*, *433*(2), 118–131. https://doi.org/10.1016/j.ydbio.2017.10.010.

Martin-Duran, J. M., & Marletaz, F. (2020). Unravelling spiral cleavage. *Development*, *147*(1). https://doi.org/10.1242/dev.181081.

Mwinyi, A., Bailly, X., Bourlat, S. J., Jondelius, U., Littlewood, D. T., & Podsiadlowski, L. (2010). The phylogenetic position of Acoela as revealed by the complete mitochondrial genome of *Symsagittifera roscoffensis*. *BMC Evolutionary Biology*, *10*, 309. https://doi.org/10.1186/1471-2148-10-309.

Newmark, P. A., & Alvarado, A. S. (2022). *Schmidtea* happens: Re-establishing the planarian as a model for studying the mechanisms of regeneration. *Current Topics in Developmental Biology*, *147*, 307–342.

Pani, A. M., Mullarkey, E. E., Aronowicz, J., Assimacopoulos, S., Grove, E. A., & Lowe, C. J. (2012). Ancient deuterostome origins of vertebrate brain signalling centres. *Nature*, *483*(7389), 289–294. https://doi.org/10.1038/nature10838.

Perea-Atienza, E., Botta, M., Salvenmoser, W., Gschwentner, R., Egger, B., Kristof, A., et al. (2013). Posterior regeneration in Isodiametra pulchra (Acoela, Acoelomorpha). *Frontiers in Zoology*, *10*(1), 64. https://doi.org/10.1186/1742-9994-10-64.

Philippe, H., Brinkmann, H., Copley, R. R., Moroz, L. L., Nakano, H., Poustka, A. J., et al. (2011). Acoelomorph flatworms are deuterostomes related to *Xenoturbella*. *Nature*, *470*(7333), 255–258. https://doi.org/10.1038/nature09676.

Philippe, H., Brinkmann, H., Martinez, P., Riutort, M., & Baguna, J. (2007). Acoel flatworms are not platyhelminthes: Evidence from phylogenomics. *PLoS One*, *2*(8), e717. https://doi.org/10.1371/journal.pone.0000717.

Philippe, H., Poustka, A. J., Chiodin, M., Hoff, K. J., Dessimoz, C., Tomiczek, B., et al. (2019). Mitigating anticipated effects of systematic errors supports sister-group relationship between Xenacoelomorpha and Ambulacraria. *Current Biology*, *29*(11), 1818–1826 e1816. https://doi.org/10.1016/j.cub.2019.04.009.

Ramirez, A. N., Loubet-Senear, K., & Srivastava, M. (2020). A regulatory program for initiation of Wnt signaling during posterior regeneration in acoels. *Cell Reports*, *32*, 108098.

Raz, A. A., Srivastava, M., Salvamoser, R., & Reddien, P. W. (2017). Acoel regeneration mechanisms indicate an ancient role for muscle in regenerative patterning. *Nature Communications*, *8*(1), 1260. https://doi.org/10.1038/s41467-017-01148-5.

Ricci, L., & Srivastava, M. (2021). Transgenesis in the acoel *Hofstenia miamia*. *Developmental Cell*, *56*, 3160–3170.

Ruiz-Trillo, I., Paps, J., Loukota, M., Ribera, C., Jondelius, U., Baguna, J., et al. (2002). A phylogenetic analysis of myosin heavy chain type II sequences corroborates that Acoela and Nemertodermatida are basal bilaterians. *Proceedings of the National Academy of Sciences of the United States of America*, *99*(17), 11246–11251. https://doi.org/10.1073/pnas.172390199.

Ruiz-Trillo, I., Riutort, M., Fourcade, H. M., Baguñà, J., & Boore, J. L. (2004). Mitochondrial genome data support the basal position of Acoelomorpha and the polyphyly of the Platyhelminthes. *Molecular Phylogenetics and Evolution*, *33*(2), 321–332. https://doi.org/10.1016/j.ympev.2004.06.002.

Ruiz-Trillo, I., Riutort, M., Littlewood, D. T., Herniou, E. A., & Baguna, J. (1999). Acoel flatworms: Earliest extant bilaterian metazoans, not members of Platyhelminthes. *Science*, *283*(5409), 1919–1923.

Semmler, H., Bailly, X., & Wanninger, A. (2008). Myogenesis in the basal bilaterian Symsagittifera roscoffensis (Acoela). *Frontiers in Zoology*, *5*, 14. https://doi.org/10.1186/1742-9994-5-14.

Sempere, L. F., Martinez, P., Cole, C., Baguñà, J., & Peterson, K. J. (2007). Phylogenetic distribution of microRNAs supports the basal position of acoel flatworms and the polyphyly of Platyhelminthes. *Evolution & Development*, *9*(5), 409–415. https://doi.org/10.1111/j.1525-142X.2007.00180.x.

Sikes, J. M., & Bely, A. E. (2008). Radical modification of the A-P axis and the evolution of asexual reproduction in Convolutriloba acoels. *Evolution & Development*, *10*(5), 619–631. https://doi.org/10.1111/j.1525-142X.2008.00276.x.

Sikes, J. M., & Bely, A. E. (2010). Making heads from tails: Development of a reversed anterior-posterior axis during budding in an acoel. *Developmental Biology*, *338*(1), 86–97. https://doi.org/10.1016/j.ydbio.2009.10.033.

Smith, J. P. S., Teyler, S., & Rieger, R. M. (1986). Is the Turbellaria polyphyletic? *Hydrobiologia*, *132*(1), 13–21. https://doi.org/10.1007/bf00046223.

Sprecher, S. G., Bernardo-Garcia, F. J., van Giesen, L., Hartenstein, V., Reichert, H., Neves, R., et al. (2015). Functional brain regeneration in the acoel worm Symsagittifera roscoffensis. *Biology Open*, *4*(12), 1688–1695. https://doi.org/10.1242/bio.014266.

Srivastava, M. (2021). Beyond casual resemblances: Rigorous frameworks for comparing regeneration across species. *Annual Review of Cell and Developmental Biology*. https://doi.org/10.1146/annurev-cellbio-120319-114716.

Srivastava, M., Mazza-Curll, K., van Wolfswinkel, J., & Reddien, P. W. (2014). Whole-body acoel regeneration is controlled by Wnt and bmp-Admp signaling. *Current Biology*, *24*(10), 1107–1113.

Steinbock, O. (1967). Regenerationsversuche mit *Hofstenia giselae* Steinb. (Turbellaria Acoela). *Roux Archiv fur Entwicklungsmechanik*, *158*, 394–458.

Telford, M. J., Lockyer, A. E., Cartwright-Finch, C., & Littlewood, D. T. (2003). Combined large and small subunit ribosomal RNA phylogenies support a basal position of the acoelomorph flatworms. *Proceedings of the Biological Sciences*, *270*(1519), 1077–1083. https://doi.org/10.1098/rspb.2003.2342.

Tewari, A. G., Owen, J. H., Petersen, C. P., Wagner, D. E., & Reddien, P. W. (2019). A small set of conserved genes, including sp5 and Hox, are activated by Wnt signaling in the posterior of planarians and acoels. *PLoS Genetics*, *15*(10), e1008401. https://doi.org/10.1371/journal.pgen.1008401.

Tiozzo, S., & Copley, R. R. (2015). Reconsidering regeneration in metazoans: an evo-devo approach. *Frontiers in Ecology and Evolution*, 3.

Yao, Y., Minor, P. J., Zhao, Y. T., Jeong, Y., Pani, A. M., King, A. N., et al. (2016). Cis-regulatory architecture of a brain signaling center predates the origin of chordates. *Nature Genetics*, *48*(5), 575–580. https://doi.org/10.1038/ng.3542.

CHAPTER SEVEN

Tardigrades and their emergence as model organisms

Bob Goldstein*

Department of Biology and Lineberger Comprehensive Cancer Center, University of North Carolina at Chapel Hill, Chapel Hill, North Carolina, United States
*Corresponding author: e-mail address: bobg@email.unc.edu

Contents

Abstract

Experimentally tractable organisms like *C. elegans*, *Drosophila*, zebrafish, and mouse are popular models for addressing diverse questions in biology. In 1997, two of the most valuable invertebrate model organisms to date—*C. elegans* and *Drosophila*—were found to be much more closely related to each other than expected. *C. elegans* and *Drosophila* belong to the nematodes and arthropods, respectively, and these two phyla and six other phyla make up a clade of molting animals referred to as the Ecdysozoa. The other ecdysozoan phyla could be valuable models for comparative biology, taking advantage of the rich and continual sources of research findings as well as tools from both *C. elegans* and *Drosophila*. But when the Ecdysozoa was first recognized, few tools were available for laboratory studies in any of these six other ecdysozoan phyla. In 1999 I began an effort to develop tools for studying one such phylum, the tardigrades. Here, I describe how the tardigrade species *Hypsibius exemplaris* and tardigrades more generally have emerged over the past two decades as valuable new models for answering diverse questions. To date, these questions have included how animal body plans evolve and how biological materials can survive some remarkably extreme conditions.

Current Topics in Developmental Biology, Volume 147
ISSN 0070-2153
https://doi.org/10.1016/bs.ctdb.2021.12.008

1. Emergence of a model system: A personal history

My interest in tardigrades was sparked by a discovery made by others. In 1997, Anna Marie Aguinaldo and her colleagues reported an 18S rDNA molecular phylogeny of the metazoa, which revealed that nematodes and arthropods form a single clade together with other phyla of animals that periodically shed their exoskeletons (Aguinaldo et al., 1997). They dubbed this clade the Ecdysozoa, or molting animals. Broader phylogenomic studies later buttressed this finding (see Giribet & Edgecombe, 2017 for review), supporting the emerging realization that *C. elegans* (a nematode) and *Drosophila* (an arthropod) were much more closely related than had been thought. Before 1997, *Drosophila* was thought to be even more closely related to humans than to *C. elegans* (Sidow & Thomas, 1994). As a result, there were no reasonable options for studying groups of closely related animals that were also among the strongest genetic model systems unless one turned to the vertebrates, which were more cumbersome to study than invertebrate models like *C. elegans* and *Drosophila*.

In my PhD and first postdoc, I had used *C. elegans* as a model organism for uncovering fundamental mechanisms of development. Like many scientists with some curiosity about how developmental programs evolved, for nearly every developmental mechanism I thought about, I wondered if evolution had tinkered with that mechanism in ways that might have contributed to the evolution of animal body plans. When the Aguinaldo et al. (1997) paper was published, I was in a second postdoc, working on an evolution-of-development (evo-devo) project that was largely predicated on the old animal phylogeny. The 1997 paper forced me to rethink my plans.

Several labs at the time were already using arthropods other than *Drosophila*, or nematodes other than *C. elegans*, as comparative models in evo-devo, with success (for reviews see Carroll, Grenier, & Weatherbee, 2001; Félix, 1999; Patel, 1994; Sommer, 2000). Evo-devo was widely considered to encompass some of the greatest unsolved mysteries in developmental biology, but one for which future progress was expected to be challenging (Barinaga, 1994). Yet these labs were transforming evo-devo research, in two major ways: (1) by identifying key genes whose expression patterns could help identify homologous structures in extant organisms, and thus help trace paths of evolution, and (2) by identifying some of the developmental mechanisms that had been modified through evolution to diversify animal body plans. The genes and mechanisms studied by these labs were, for the most part, those that had conserved functions across a great breadth of animals. Choosing such highly conserved genes and mechanisms for lab study was in large part a practical issue: the likelihood was high that such genes and mechanisms would have conserved and interpretable functions in less-studied animals. As a result, a small number of genes with deeply conserved functions, such as Hox genes, were frequently recurring themes in evo-devo studies.

I imagined that using an ecdysozoan relative of *C. elegans* and *Drosophila*, but outside of the nematode and arthropod phyla, might allow future researchers to maximally make use of *both C. elegans* and *Drosophila* as comparative model systems when asking how developmental mechanisms evolve. I pictured a large and continually growing menu of genes and mechanisms that could be productively studied to address interesting questions in comparative biology in general, and in evo-devo in particular. *C. elegans* and *Drosophila* researchers would likely pour out new findings continually, serving as fountains for future comparative studies on related nonmodel organisms. Some of the methods developed for studies of *C. elegans* and *Drosophila* might feasibly work in a close ecdysozoan relative as well. Unfortunately, there had been almost no modern, molecular work on development in the other ecdysozoan phyla, with only rare exceptions (see Grenier, Garber, Warren, Whitington, & Carroll, 1997 for example). The other ecdysozoan phyla—the loriciferans, priapulids, onychophorans, nematomorphs, kinorhynchs, and tardigrades—were some of the least studied major groups of animals. Few tools existed for studying development in these animals. Even old descriptive literature on development in these phyla, dating from before much of developmental biology research had become focused on just a handful of genetic model organisms, was meager (Grassé, 1949, 1965; Hyman, 1951).

In 1999 I began a lab at UNC Chapel Hill. I had equipped the lab with two microscopes for automated, multiplane DIC imaging of living *C. elegans* embryos (Thomas, DeVries, Hardin, & White, 1996). These microscopes also seemed ideally suited to the kind of descriptive embryology that might be valuable for attempting some foundational work on a little-studied phylum. Lab members used these microscopes for *C. elegans* experiments, but the *C. elegans* films were generally short, leaving the microscopes rarely in use overnight. I decided to try getting my hands on a little-studied ecdysozoan to film their embryos overnight on some nights, to see if I could collect data on normal development. The list of nonnematode, nonarthropod ecdysozoan phyla included mostly organisms I had never seen, but I had encountered tardigrades in an undergraduate course on animal diversity, and I had collected them from moss before. I went outside the lab and found some tardigrades, and I bought some tardigrades from multiple biological supply companies. I contacted tardigrade biologists Diane Nelson, Harold Heatwole, and Jette Eibye-Jacobsen for advice, and they kindly offered me tips on finding animals and keeping them alive in the lab. Among the tardigrade species I gathered and filmed, I looked for ones with small, optically clear embryos, and lacking ornate embryonic envelopes, so that I could film development by DIC microscopy and record patterns of cell divisions. I also wanted a species that had embryos with small cells and fast cell division cycles, because these features correlate at least very roughly with small genome size (Gregory, 2001; Gregory & Hebert, 1999). I envisioned that a large and potentially highly duplicated genome would make future gene function studies more difficult than would a compact genome. Looking ahead to gene function studies in tardigrades at this stage was highly wishful, since at this stage, sequences from only six genes existed in GenBank for the entire tardigrade phylum.

A species of tardigrades sold by Bob McNuff appeared suitable (Fig. 1). Bob single-handedly ran a small biological supply company, Sciento, from his home or at times, from a lab that he had assembled in a shed behind his home near Manchester, England. Mark Blaxter's lab had started collecting some sequence data from this species and other tardigrade

Fig. 1 The tardigrade *Hypsibius exemplaris. Photo credit: Sinclair Stammers.*

species (Blaxter, Elsworth, & Daub, 2004; Goldstein & Blaxter, 2002). Embryos of this species were small, about the length of a *C. elegans* embryo (about 60 μm), and I could see cells dividing every ~50 min. The generation time was short—just 13–14 days—and I found that cultures could be frozen long-term and successfully revived. Although little work had been done with tardigrade embryos in the previous 60 years, near this time a second group published on the development of another tardigrade species, *Thulinius* (formerly *Thulinia*) *stephaniae*. The *Thulinius* embryos could replace ablated parts and lacked stereotyped patterns of cell divisions (Hejnol & Schnabel, 2005). In the species I was filming, I found invariant patterns of unequal cell divisions and nuclear migrations (these patterns are detailed further below). These invariant features made it possible to identify the same cells in each embryo, and hence to begin to recognize additional stereotyped features of early development, including stereotyped cell cycle periods in different cells. I sent some animals of the species I was filming to tardigrade biologist Roberto Bertolani, who kindly agreed to identify them: among previously-described tardigrade species, they most closely resembled *Hypsibius dujardini*. This species is composed primarily of females that can develop by parthenogenesis (in which eggs develop in the absence of mating or sperm; hence all-female cultures can be kept). Males also exist in this species but are rarely found (Ammermann, 1967; Bertolani, 2001). In our cultures, we have yet to recognize a male.

After we had published several papers on this species (discussed below), other scientists seeking to improve the characterization of *H. dujardini* collected animals from as close to their original 1840 collection site near Paris (Doyère, 1840) as they could find them and discovered that the Parisian species they observed and the species sold by Bob McNuff were distinct (Gąsiorek, Stec, Morek, & Michalczyk, 2018). They concluded that we, and by this time several other labs, had likely been working on a previously undescribed species—probably a close relative of *H. dujardini* rather than *H. dujardini* itself. They dubbed the species *Hypsibius exemplaris*, for "exemplar," or "model," in recognition of the by-then "wide use of the species as a laboratory model for various types of scientific studies" (Gąsiorek et al., 2018). For this reason, pre-2018 literature on *H. exemplaris* uses the name *H. dujardini*, often also referring to Bob McNuff or his company Sciento as the source of animals used, or using the Sciento catalog number Z151 as the strain designation (see Gąsiorek et al., 2018; Goldstein, 2018 for lists of the pre-2018 papers). Post-2018 literature uses the name

H. exemplaris. Literature from before 2004—before Bob McNuff's cultures began to enter the scientific literature—and that referred to *H. dujardini* (or with an earlier synonymous genus designation) may have used any of several possible species (for example Ammermann, 1967; Kaufmann, 1851; Marcus, 1929; see Gąsiorek et al., 2018 for a discussion of relevant clues).

2. Tardigrades

Tardigrades, also known as water bears, are a phylum of eight-legged microscopic animals. Although tardigrades are less well known to many scientists than the more widely-used model organisms, they have been known to science for almost as long as microscopic life has been known. Tardigrades were first described by several scientists in the 1770s (see Greven, 2018 for review).

1380 living tardigrade species have been recognized in the phylum to date (Degma, Bertolani, & Guidetti, 2021; Degma & Guidetti, 2007; Guidetti & Bertolani, 2005), and it is likely that at least this many species again have yet to be discovered and described (Bartels, Apodaca, Mora, & Nelson, 2016; Guil & Cabrero-Sañudo, 2007). Tardigrades are most often placed by phylogenetic methods as a sister taxon to the arthropods plus onychophorans (see Giribet & Edgecombe, 2020 for review). This placement on the tree of life implies that tardigrades must have first arisen more than 500 million years ago.

Tardigrades live nearly everywhere on earth, although they are often overlooked because they are small (typically less than 1 mm long) and often transparent. Some tardigrade species live in marine sediments, and many live in water films in soil. Clean biological substrates that harbor water films, like mosses and lichens, are easy places to find tardigrades, perhaps primarily because collecting from clean substrates obviates the need to search among sediments. On mosses and lichens, tardigrades are often found together with nematodes and rotifers—which also possess the rare ability to survive drying, and hence that might be carried onto these surfaces by wind (Nkem et al., 2006; Ptatscheck, Gansfort, & Traunspurger, 2018; Rivas, Schröder, Gill, Wallace, & Walsh, 2019). Many tardigrades feed on algae or on microscopic animals, using piercing mouthparts and a muscular pharynx to suck cytoplasm out of these organisms (see Møbjerg, Jørgensen, Kristensen, & Neves, 2018 for review).

Tardigrades have a simple body plan (Figs. 2 and 3, and Supplementary Video 1 in the online version at https://doi.org/10.1016/bs.ctdb.2021.12.008)

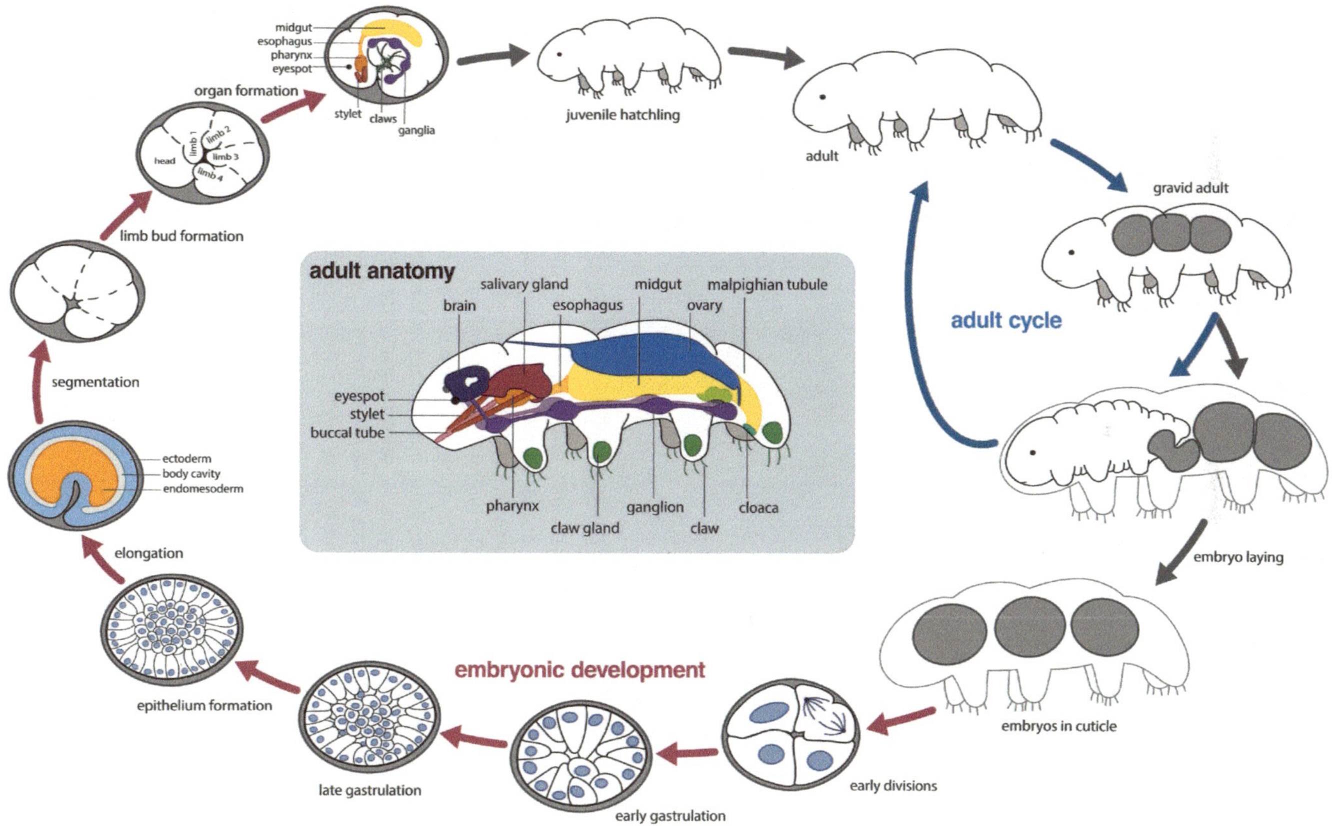

Fig. 2 The life cycle and development of *Hypsibius exemplaris*, and a diagram of adult anatomy. Illustration by Heather Barber.

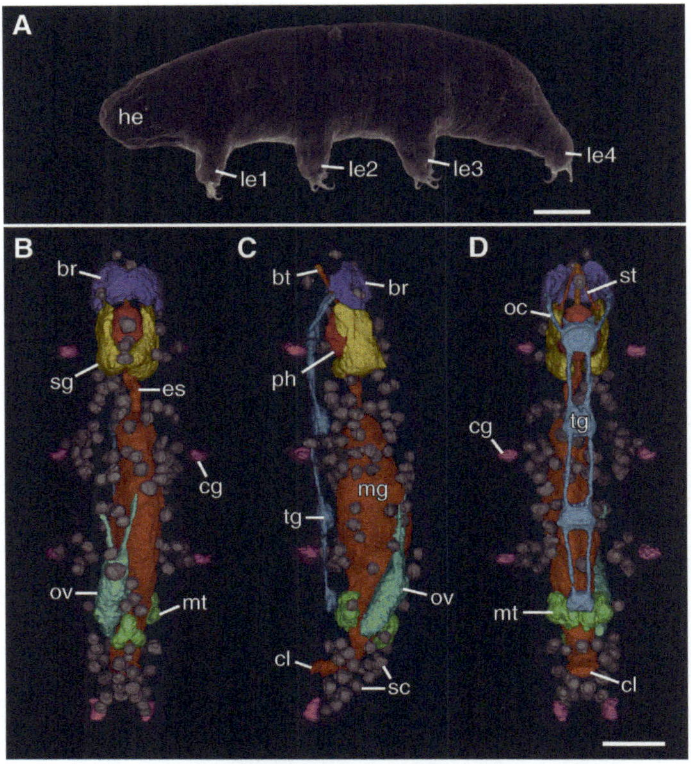

Fig. 3 Anatomy of *Hypsibius exemplaris*: Scanning electron micrograph and 3D renderings of X-ray nanoCT data. (A) Scanning electron micrograph of an adult specimen. Lateral view; anterior is left, dorsal is up. (B–D) 3D renderings of X-ray nanoCT data showing the body in dorsal (B), lateral (C), and ventral (D) views. Anterior is up. Abbreviations: br, brain; bt, buccal tube; cg, claw glands; cl, cloaca; es, esophagus; he, head; le1–le4, leg pairs one to four; mg, midgut; mt, malpighian tubules; oc, outer connectives; ov, ovary; ph, pharynx; sg, salivary glands; sc, storage cells; tg, trunk ganglion. Scale bars: 20 μm. *From Gross, V., Müller, M., Hehn, L., Ferstl, S., Allner, S., Dierolf, M., et al. (2019). X-ray imaging of a water bear offers a new look at tardigrade internal anatomy.* Zoological Letters, *5, 14, used with permission.*

and the same major tissue types as most other animals (Møbjerg et al., 2018; for details of *H. exemplaris* anatomy, see for example Gross & Mayer, 2019; Gross et al., 2019; Jezierska et al., 2021; Mayer et al., 2013; Richaud et al., 2020; Smith, Bartels, & Goldstein, 2017; Smith & Jockusch, 2014). The body has a tube-within-a-tube organization that is typical of most animals, in which a one-way digestive tract runs through the center of an elongated body. The gonad connects to the posterior digestive tract, along with malpighian tubules that function in osmoregulation. The fluid-filled body cavity that lies between the epidermis and digestive tract typically has hundreds of freely moving cells inside, called storage cells. Muscles criss-cross the body cavity, which functions as a hydrostatic skeleton. Each of the eight legs has several muscle cells extending into it from multiple directions, and connecting to the body wall, which dimples inward at each muscle attachment site. The nervous system includes a pair of eyespots, an anterior brain, and a ladder-like arrangement of paired segmental ganglia along the ventral side of the body. Major elements of the body plan become apparent during embryogenesis (Gabriel et al., 2007; Gross, Minich, & Mayer, 2017).

Tardigrades are famous for surviving extreme environments. When dried, many species of tardigrades lose nearly all of their intracellular water (Crowe, 1972; Richaud et al., 2020) and form a compact "tun" form that can remain desiccated for years and yet revive upon rehydration (Jørgensen, Møbjerg, & Kristensen, 2007; Richaud et al., 2020; Schill & Hengherr, 2018). In the desiccated state, the animals can survive extremes of pressure, temperature, and radiation, and they can survive extremes of radiation in the hydrated state as well (see Schill, 2018 for review).

3. Early descriptions of tardigrade development

Literature on development in tardigrades was scant when I began filming embryos. Four articles, all from before 1930, described embryos of different tardigrade species throughout much of embryonic development (Kaufmann, 1851; Marcus, 1929; von Erlanger, 1895; von Wenck, 1914). To my knowledge, the embryonic development of a tardigrade had never 8been described until the 1851 article, which included drawings of living embryos at 18 stages from early cell divisions through hatching (Kaufmann, 1851). The article also included a drawing of a tardigrade in the midst of releasing eggs into her own molt. Kaufmann's observations established some key facts for the species he observed that also hold for all tardigrade species observed to date. The embryos underwent apparently complete embryonic cell divisions (rather than producing a syncytium, as occurs in some arthropods). Early divisions produced daughter cells of at least approximately equal sizes. Development appeared to be direct, i.e., the animal that hatched at the end of embryonic development looked like a small tardigrade, rather than transitioning through a larva of a very different form before taking on the familiar form of a tardigrade. In 1895, the Baron

Raphael Slidell von Erlanger noted a more than 40 year gap since Kaufmann's description of embryos and produced a set of color illustrations of embryos of a different species, through germ layer and organ formation, based on fixed and cleared samples (von Erlanger, 1895). von Erlanger's illustrations were only diagrammatic, and he did not follow up on this work because he died of pneumonia 2 years later at age 32. In 1914, Wanda von Wenck reported on whole and sectioned embryos of what may have been a third species (von Wenck, 1914). And in 1929, Ernst Marcus described the development of *H. dujardini* or a closely related species (Gąsiorek et al., 2018) based in part on his sections of embryos (Marcus, 1929) (Marcus' slides bearing sectioned embryos later became part of the Smithsonian Natural History Museum's collection; the slides were deaccessioned by Smithsonian scientists in 2016 because of the slides' mixed quality by then, and they gave them to me).

4. Raising *H. exemplaris* cultures in a lab

Long-term culturing of tardigrades in labs has faced some challenges (Altiero & Rebecchi, 2001). Some tardigrade species had been maintained in labs before I began filming tardigrade embryos, but in at least some of the early papers it was not clear if long-term culturing had been used, or if animals had been recollected periodically from the wild. Some tardigrade species are still repeatedly collected from the wild for studies, and reliable long-term wild sources of animals make this a practical option for certain species (Rebecchi et al., 2009 for example). I was surprised when Bob McNuff told me by phone in 2001 that he did not recollect his animals from the wild periodically. Rather, he had been growing his cultures continuously for 14 years at the time, since collecting them in 1987. Bob's cultures may have represented the longest continuously maintained cultures of any tardigrade species to date, suggesting promise for indefinite lab maintenance of cultures—a valuable feature for any model organism. Bob kindly shared his methods for growing animals. Lab culture methods have been developed for other tardigrade species as well (Altiero, Giovannini, Guidetti, & Rebecchi, 2015; Horikawa et al., 2008; Suzuki, 2003; Tsujimoto, Suzuki, & Imura, 2015). We have continued to focus most of our studies on *H. exemplaris* because of several advantages they offer over other tardigrade species for the questions we have addressed, including their clear embryos, consistently successful embryonic development in the lab, and their short and consistent embryonic development period and generation time (Goldstein, 2018; Yoshida, Sugiura, Tomita, Matsumoto, & Arakawa, 2019). Still, some other tardigrade species offer some attractive features including even more compact genomes, a constitutive production of some protectants that contribute to survival in extremes, and an availability of males that might make classical genetics possible (Bemm et al., 2017; Boothby et al., 2017; Ramazzotti & Maucci, 1995; Suzuki, 2008; Yoshida et al., 2017).

H. exemplaris cultures can be maintained in labs at room temperature in Petri plates (or in shallow cultures in Erlenmeyer flasks) in commercial spring water, feeding the animals an immotile, unicellular alga. At 1–2-week intervals we replace the water and algae. This can be done by pipetting liquid in cultures up and down to disperse old algae. Then we let the tardigrades settle for about a minute, then pour off most of the liquid and the dispersed algae, and add back more spring water, repeating this a few times to remove most of the old algae. Then we add spring water and some of a healthy culture of unicellular algae as fresh food. Alternatively, animals can be simply swirled to the center of a Petri plate and pipetted into a plate with fresh spring water and algae. Bob McNuff is listed (together with his then-home address) as second author on one of our early papers based on his contributing methods for culturing animals long term (Gabriel et al., 2007; see McNuff, 2018 for a fuller description of Bob's collection site and culture methods). His work has highlighted that one can make an important contribution to science even without the myriad benefits of an institute or an institutional affiliation.

In cultures or on slides under microscopes, embryos develop to hatching in 4–4.5 days (Gabriel et al., 2007; Vasanthan & Stone, 2020). The resulting juvenile animals feed and grow for approximately 8–9 more days before they begin to lay embryos of their own. Gravid animals lay embryos into their own molts (often called an exuvia; "exuvia" is singular, "exuviae" is plural). The mothers exit their molts soon after laying, typically when the embryos are at the 1–2-cell stage, leaving the embryos behind in the molts. The resulting clutches of nearly-synchronous embryos encased in transparent molts are ideal for microscopy of multiple living embryos simultaneously (Gabriel et al., 2007; Heikes & Goldstein, 2018). In diverse animals, molting permits growth beyond the constraints of a molt's size, but because the embryos of *H. exemplaris* are a significant fraction of the mother's body volume, the mother that leaves the molt after laying is *smaller* than the molt. Growth occurs as the mothers feed in the days between each laying. This laying and molting cycle repeats every ~5 days. Clutch size varies, likely depending on algal quality and/or on other food added in some cases, and varying with the age of the mother: clutches of 2–5 embryos are common, but with a range from 1 embryo to as high as 10–11 embryos (Gabriel et al., 2007; Vasanthan & Stone, 2020) or even more (Gąsiorek et al., 2018).

5. Our early descriptive work on the animals and their development

I worked on tardigrades mostly alone in my lab among the graduate students and postdocs working on *C. elegans* until 2003, when one graduate student, Willow Gabriel, began to focus on tardigrades for her PhD project. Willow developed embryo immunostaining methods and reported the first immunolocalization data in the phylum, using cross-reactive antibodies from Nipam Patel (Gabriel & Goldstein, 2007). Willow used the data she collected to begin to address questions about segmentation mechanisms. She also produced a staging series for the embryos, describing embryonic development from DIC films and from fixed, DAPI-stained embryos. We published the staging series together with several other findings that helped establish a platform for further work on the system (Gabriel et al., 2007). For example, collaborator T. Ryan Gregory contributed a genome size estimate based on Feulgen densitometry and propidium iodide flow cytometry, suggesting a compact genome of about 75 Mb, which is at least roughly in line with the current estimate of 104 Mb in the best current genome assembly (Yoshida et al., 2017). We saw five pairs of chromosomes, the same number that had been reported previously for *H. dujardini* (Ammermann, 1967). We also reported on cell division patterns in the early embryo, and we related these patterns to the body axes. The first mitosis produced anterior and posterior daughter cells, although with no apparent indication at the two-cell stage of which cell is which (we can recognize these cells as anterior vs. posterior daughter cells only retrospectively, in films that last until anterior and posterior ends are apparent when the embryo begins to elongate within its eggshell). At the four-cell stage, I found that the nuclei in two of the four cells consistently migrated toward a common cell-cell boundary, and the two cells harboring these nuclei then divided unequally, producing a slightly smaller daughter cell near the cell-cell boundary toward which nuclear migration had occurred. This pattern of nuclear migrations and unequal divisions repeated for multiple rounds of division, and the site where this occurred became consistently the ventral side of the embryo. These observations contributed to an initial draft of an early-stage cell lineage, although without cell fates indicated, because we had no cell fate markers yet that we could relate reliably to the positions of lineaged cells (Gabriel et al., 2007). Current work in the lab is revising some features of the published cell lineage using observations of more embryos and assigning one of the major cell fates on the lineage by the molecular identification of the germline precursor cells.

6. The tardigrade toolbox: Resources for tardigrade research

To enable mechanistic studies using tardigrades as a model, we needed to develop several tools for the first time in the phylum. We developed methods for immunolocalization (Gabriel et al., 2007; Smith & Gabriel, 2018), *in situ* hybridization (Smith, 2018; Smith et al., 2016), live microscopy of embryos using DIC microscopy or fluorescent cell staining (Gabriel et al., 2007; Heikes & Goldstein, 2018; McGreevy, Heikes, Kult, Tharp, & Goldstein, 2018; Fig. 4), microinjection of animals

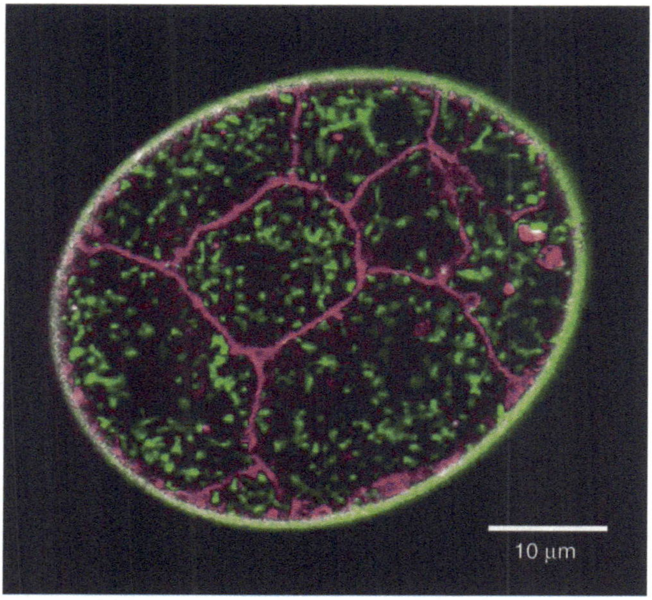

Fig. 4 *Hypsibius exemplaris* embryo with membranes labeled by FM 4–64 (*magenta*) and mitochondria labeled by MitoTracker Green (*green*). *From McGreevy, K. M., Heikes, K. L., Kult, S., Tharp, M. E., & Goldstein, B. (2018). Fluorescent cell staining methods for living Hypsibius exemplaris embryos.* Cold Spring Harbor Protocols, 2018, *878–884. doi:10.1101/pdb.prot106021, used with permission.*

(Tenlen, 2018; Tenlen, McCaskill, & Goldstein, 2013), RNA interference (Tenlen, 2018; Tenlen et al., 2013), and methods for identifying proteins that mediate tolerance to desiccation (Boothby et al., 2017). We share methods openly in order to help build a community of researchers that can answer key scientific questions more efficiently. We have shared protocols on our lab website, in publications, and as a volume of *H. exemplaris* protocols in the Cold Spring Harbor Emerging Model Systems series (Goldstein, 2018).

The best current *H. exemplaris* genome assembly is fairly complete in terms of genes found compared to benchmark gene sets, and partially assembled: half of the genome is represented in 85 scaffolds of 342 kb or larger, and 90% of the genome is represented in 343 scaffolds of 65 kb or larger (Yoshida et al., 2017). We published a draft genome in 2015 and presented evidence of extensive horizontal gene transfer, but our conclusion was soon disproven (Arakawa, 2016; Arakawa, Yoshida, & Tomita, 2016; Bemm, Weiß, Schultz, & Förster, 2016; Boothby et al., 2015; Delmont & Eren, 2016; Koutsovoulos et al., 2016; Yoshida et al., 2017). Yoshida and coauthors demonstrated that gene prediction software can produce a systematic elevation of horizontal gene transfer predictions, which likely accounted for some of the elevated rate we inferred (Yoshida et al., 2017). Because our assembly included contaminants, they incorporated our sequence data in the current assembly only where our reads mapped to their own initial assembly (Yoshida et al., 2017). Perhaps the most convincing data disproving our hypothesis of elevated horizontal gene transfer was the genome sequencing of individual tardigrades that had been starved to clear their intestinal contents for 48 h, in the presence of antibiotics, and without filtering resulting reads for predicted contaminants (Arakawa, 2016; Arakawa et al., 2016). These data together with Yoshida and coauthors' reassembly indicate that horizontal gene transfer has occurred in tardigrades only at a rate that is typical for animals, and the data provide a fairly complete and partially assembled genome that serve as a valuable platform for further studies using the organism (Yoshida et al., 2017). The genomes of two other tardigrade species have also been sequenced: *Ramazzottius varieornatus* (Hashimoto et al., 2016) and *Milnesium tardigradum* (Bemm et al., 2017).

Perhaps the methods most needed now for diverse kinds of further studies are transgenesis and CRISPR-mediated DNA editing. My lab's strategy for seeking to develop much-needed experimental tools, while maintaining a focus on our goal to answer scientific questions, has been to seek a balance: We have been opportunistic about addressing questions of deep interest that can be answered with the tools developed to date, while also seeking to spend a fraction of our time developing new methods. This strategy has afforded us a gradually improved understanding of the biology of the organism, and we hope that this improved understanding can make tool development increasingly efficient. We have also taken a 2-week period dedicated to methods development as a team, instigated and organized by lab member Kira Heikes with enthusiastic help from everyone involved. Our plan to repeat for 2 weeks each summer what we called "Camp Tardigrade" has been put on hold by the COVID-19 pandemic, temporarily.

7. Using tardigrades to contribute to understanding of how animal body plans evolve

Although our efforts with tardigrades were motivated by a vision of expanding the menu of genes and mechanisms that might be of interest beyond primarily the Hox genes and other deeply conserved developmental genes, once we had some tools in hand, we turned to Hox genes as an initial gene set of value for understanding homology of major body regions. Frank Smith visited the lab for a summer in the midst of his PhD work at the University of Connecticut to pursue his interest in developing an *in situ* hybridization method for tardigrades. He succeeded, and then he came back after his PhD, where he made use of the method for his postdoc. To understand how a common ancestor diverged to produce diverse organisms including tardigrades, Frank established segment homologies by examining Hox gene expression patterns in *H. exemplaris* (Smith et al., 2016). Addressing this issue had the potential to shed light on the path of evolution that led to the origin of the tardigrade body plan. Hox genes whose expression patterns define the heads of arthropods and other animals were expressed in a similar anterior-posterior order in tardigrades. But we were surprised to see that the ordered expression pattern boundaries covered not just the tardigrade head but most of the tardigrade body, suggesting that much of the tardigrade body might be homologous to only the heads of arthropods and other animals. Frank found that a posterior Hox gene of animals was expressed in the posterior end of the tardigrade, abutting an expression pattern of a head Hox gene homolog, suggesting that a large region of the body in between had been deleted in evolution—a region corresponding to the entire thorax and nearly the entire abdomen of *Drosophila*. These results and the expression patterns of other regional identity genes, together with a loss of certain Hox genes in diverse tardigrade species, suggested that the body plan of an animal phylum—the compact body plan of tardigrades—can arise by the loss of a much larger part of an ancestor's body than we had anticipated (Smith et al., 2016).

Frank used nervous system patterning genes to test a similar hypothesis about the evolution of nervous systems in tardigrades and their close relatives, the arthropods and onychophorans (Smith & Goldstein, 2016). The expression domains of nervous system patterning genes appeared to buttress the conclusion we reached on the basis of Hox gene expression: tardigrade homologs of genes that pattern arthropod brains are expressed in a similar anterior-to-posterior order in tardigrades as in arthropods, but with the three-segment brain of arthropods corresponding to a one-segment tardigrade brain plus the ganglia of the first two leg-bearing segments (Smith, Cumming, & Goldstein, 2018). Frank established his own lab at the University of North Florida, where he and his lab have found evidence from leg gap gene expression that limbs may have experienced a similar loss of a central region along with the loss of a major leg gap gene, *Dac* (Game & Smith, 2020). Together, these data build a working hypothesis that the tardigrade body plan may have arisen by the loss of major body parts of tardigrades' ancestors.

8. Using tardigrades to contribute to understanding how biological materials can survive extremes

Tardigrades have survived desiccation for decades, freezing to below 1 K (at which molecular motion is nearly stopped), heating to 151 °C, ionizing radiation at doses about 1000 times higher than humans or most animals can survive, and extremes of pressure including multiple-day exposure in the vacuum of space (see Schill, 2018 for review). Even in purified form, DNA, RNA, proteins, and other biological materials would be damaged or destroyed by some of these conditions—suggesting that tardigrades must harbor exceptional molecules that can protect them (referred to as protectants) and/or damage repair mechanisms. We are interested in understanding such protectants and repair mechanisms because of our fascination with how living organisms and biological materials can survive extremes and because of the practical potential for using biologically-derived substances as protectants for fragile drugs and medicines.

Until 2013, we had no way to disrupt gene functions in tardigrades, limiting our ability to identify roles for identified genes. Postdoc Jennifer Tenlen, with contributions from undergraduate student Shaina McCaskill, tried several methods for RNAi and found that injecting double-stranded RNA into *H. exemplaris* adults could result in the specific depletion of a targeted mRNA (Tenlen et al., 2013). Jennifer found that targeting different genes resulted in different phenotypes that depended on which gene was targeted, giving us some confidence in the specificity of RNAi in tardigrades. The method that Jennifer developed involves injecting double-stranded RNA into individual adults, which is labor intensive, but with practice, lab members have been able to inject up to ~100 animals in a day. We proposed that this development of an RNAi method for tardigrades could form a platform for discovering tardigrade gene functions, and that this ability to study gene functions could help advance our understanding of both the evolution of development and survival of extremes (Tenlen et al., 2013).

Postdoc Thomas Boothby used transcriptomics methods to identify *H. exemplaris* genes that become upregulated during desiccation (Boothby et al., 2017). *H. exemplaris* requires gradual drying to survive desiccation, and there is evidence that the animals mount a gene expression response to gradual drying that promotes their survival (Boothby et al., 2017; Kondo, Kubo, & Kunieda, 2015; Wright, 1989). Among the genes that Thomas found upregulated in response to drying was a set of *H. exemplaris* homologs of previously-identified intrinsically disordered tardigrade proteins, dubbed cytosolic-abundant, heat-stable proteins, or CAHS proteins (Yamaguchi et al., 2012). Thomas targeted CAHS-encoding genes by RNAi and found that this reduced the ability of animals to survive specifically when the animals were challenged with desiccation. As a step toward determining whether these proteins might be used to protect other biological materials, Thomas and collaborators expressed the genes encoding tardigrade CAHS proteins in bacteria or yeast and found that some of these genes could promote bacterial or yeast desiccation tolerance. Some of the CAHS proteins in buffer could also protect enzyme activities upon desiccation, freezing, or lyophilization (Boothby et al., 2017; Piszkiewicz et al., 2019). The proteins and the slow-dried tardigrades could form glassy solids—amorphous rather than crystalline solids—up to high temperatures, and the temperature above which a glassy state was lost roughly coincided with the maximum temperature that dried tardigrades, or yeast expressing these proteins, could survive. Other tardigrade species survive roughly up to the temperature at which they maintain a glassy state as well (Hengherr, Worland, Reuner, Brümmer, & Schill, 2009). Together, these results identified tardigrade proteins upregulated in response to desiccation and that contribute to desiccation tolerance, and they suggested a mechanism by which the proteins might work—by forming a glassy matrix that might prevent the disruption of other molecules during desiccation (Boothby & Pielak, 2017; Boothby et al., 2017).

Another group has searched for tardigrade proteins that can protect DNA from damage (Hashimoto et al., 2016). Hashimoto and colleagues biochemically purified tardigrade chromatin and identified some of the associated proteins. One such protein was found to localize to tardigrade DNA in immunostaining experiments and to colocalize with nuclear DNA when expressed in insect or human tissue culture cells. Interestingly, expression of this protein could suppress DNA damage in X-ray irradiated tissue culture cells and improve the viability of the irradiated cells. This work hence identified the first tardigrade-derived suppressor of DNA damage, and Hashimoto and colleagues dubbed the protein Dsup, for Damage suppressor. Dsup appears to be a tardigrade-specific protein, highlighting the value of turning to nontraditional model systems to identify proteins with extraordinary features (see Hashimoto & Kunieda, 2017 for review). Expressing Dsup in plants can also protect plant DNA from damage (Kirke, Jin, & Zhang, 2020). Understanding how Dsup suppresses DNA damage will be of interest. To date, it has been demonstrated that Dsup can bind nucleosomes, and it suppresses the potential for hydroxyl radicals—which can form when ionizing radiation splits water molecules—to mediate cleavage of DNA (Chavez, Cruz-Becerra, Fei, Kassavetis, & Kadonaga, 2019).

9. Unanswered questions and future prospects

The use of *H. exemplaris* as a model system for diverse studies has grown over the past 15 years (Fig. 5). Here, I address some unanswered questions of broad interest for which *H. exemplaris* can serve as a valuable model.

Tardigrades and a handful of other animal groups survive both ecologically-relevant extremes as well as several more remarkable extremes, but the mechanisms by which they do so are only incompletely understood. For the protectants identified to date from tardigrades, including the CAHS proteins and Dsup, there is much yet to be learned about the biochemical mechanisms by which they can suppress damage (Hashimoto & Kunieda, 2017; Hesgrove & Boothby, 2020; Hibshman, Clegg, & Goldstein, 2020; Janis, Belott, & Menze, 2018; Tanaka et al., 2015). And the potential for these protectants to be used to stabilize fragile protein-based medicines has only recently been explored. CAHS proteins, which have been perhaps the most explored in this regard, have not yet been found to generally outperform other proteins in stabilizing enzymes at least in conditions tested to

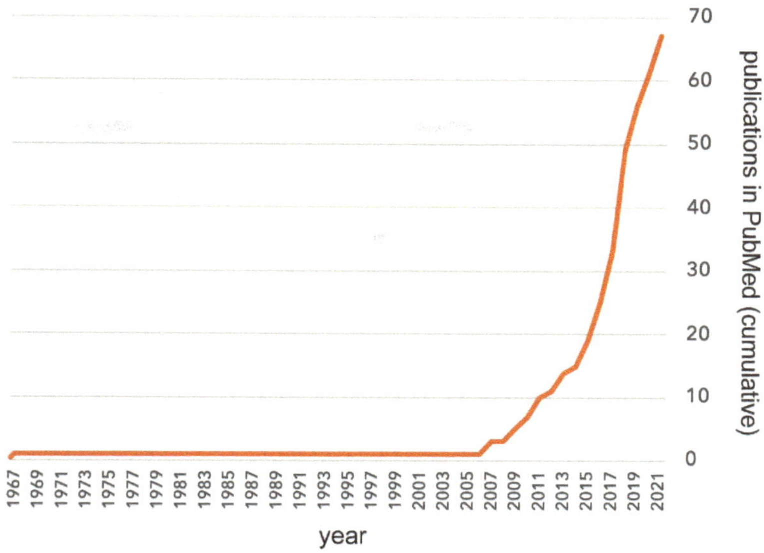

Fig. 5 Emergence of *Hypsibius exemplaris* as a model system. Cumulative number of publications over time in PubMed-indexed journals with *Hypsibius dujardini* or *Hypsibius exemplaris* in the title and/or abstract (to date as of 1 November 2021).

date (Piszkiewicz et al., 2019). Modifying CAHS proteins might be a route to increasing their potency as stabilizers, and this approach might be taken with other proteins that can function as stabilizers as well (Piszkiewicz et al., 2019). CAHS proteins or other protectants might also be used to stabilize fragile vaccines or cells. One CAHS protein has been injected in mice, and this did not result in harmful lysis of red blood cells above a low level caused by routine injection of Ringer's solution, and there were no signs of inflammation or toxicity, suggesting some promise in its use as a safe vaccine stabilizer (Esterly et al., 2020). Tardigrades are likely to harbor more protectants than are known to date, since different extreme conditions induce different gene expression responses, and protectants known to date are not known to protect against multiple extreme conditions (Boothby et al., 2017), nor to localize to every cellular compartment that would presumably need to be protected. As protectants are discovered, it will be fascinating to learn whether ancient protective mechanisms were retained in the rare animal groups that can survive extremes, or whether some protectants evolved such roles *de novo*.

Methods for transgenesis and CRISPR in tardigrades would be valuable for accelerating discovery of mechanisms used to survive extremes as well as a variety of other discoveries. RNA interference allows some studies of gene function, as discussed above, and chemical inhibitors have also been used to discover genes or pathways of interest from tardigrades (Kondo et al., 2015; Kondo, Mori, Tomita, & Arakawa, 2019, 2020; Wojciechowska et al., 2021). To date, forward genetic screening using mutagenesis has not been reported.

Cultured cell lines have not been isolated from tardigrades, although there are cells that continue to divide in the adults of *H. exemplaris*, raising the attractive possibility of developing cell lines even from adults (Gross, Bährle, & Mayer, 2018). If a cultured cell line that can survive extreme conditions could be developed from tardigrades, this could be valuable for studying cellular and molecular mechanisms of extreme tolerance, much as has been the desiccation-tolerant Pv11 cell line from the midge *Polypedilum vanderplanki* (Nakahara et al., 2010; Sogame & Kikawada, 2017).

How the body form of a tardigrade is generated during embryogenesis, by specification of cell fates and body regions and the generation of segmental-like patterns, is not well understood at all. And although tardigrades likely lost major body regions that were present in ancestors, the developmental and evolutionary mechanisms that led to such losses are incompletely understood. Tardigrades will likely continue to be useful along with other organisms to help trace evolutionary changes in the arthropods and nematodes, and more generally to aid in establishing the gene complement, development, and morphology of ancestors at various nodes of animal phylogeny (see for example Bavan, Straub, Blaxter, & Ennion, 2009; Gross & Mayer, 2015; Hering, Bouameur, Reichelt, Magin, & Mayer, 2016; Hering & Mayer, 2014; Koziol, 2018; Mapalo et al., 2020; Nelson, 2017; Smith et al., 2017; Smith & Goldstein, 2016; Thiruketheeswaran, Greven, & D'Haese, 2016).

Acknowledgments

I thank lab members to date who have made discoveries, developed novel ideas, and contributed to a stimulating and interactive environment for tardigrade work in the lab including Willow Gabriel, Sapna Patel, Jennifer Tenlen, Shaina McCaskill, Susan Clark, Thomas Boothby, Frank Smith, Kiera Patanella, Samantha Piszkiewicz, Sophia Tintori, Kristen McGreevy, Matt Niederhuber, Kira Heikes, Heather Barber, Emily Shaljian, Jonathan Hibshman, Courtney Clark-Hachtel, Sydney Lai, Noemi Gavino-Lopez, and Adriana Coke, many collaborators including Lorena Rebecchi and Bob McNuff, many MBL Physiology and Embryology course students, and Anna Marie Aguinaldo and colleagues for their clever use of animals with slow molecular clock estimates for reconstructing metazoan phylogeny. Our development of tardigrades as a model system has been supported by National Science Foundation grants IOS 0235658, IOS 1257320, IOS 1557432, and IOS 2028860.

References

Aguinaldo, A. M., Turbeville, J. M., Linford, L. S., Rivera, M. C., Garey, J. R., Raff, R. A., et al. (1997). Evidence for a clade of nematodes, arthropods and other moulting animals. *Nature, 387*, 489–493.

Altiero, T., Giovannini, I., Guidetti, R., & Rebecchi, L. (2015). Life history traits and reproductive mode of the tardigrade Acutuncus antarcticus under laboratory conditions: Strategies to colonize the Antarctic environment. *Hydrobiologia, 761*, 277–291.

Altiero, T., & Rebecchi, L. (2001). Rearing tardigrades: Results and problems. *Zoologischer Anzeiger—A Journal of Comparative Zoology, 240*, 217–221.

Ammermann, D. (1967). Die Cytologie der Parthenogenese bei dem Tardigraden Hypsibius dujardini. *Chromosoma, 23*, 203–213.

Arakawa, K. (2016). No evidence for extensive horizontal gene transfer from the draft genome of a tardigrade. *Proceedings of the National Academy of Sciences of the United States of America, 113*, E3057.

Arakawa, K., Yoshida, Y., & Tomita, M. (2016). Genome sequencing of a single tardigrade Hypsibius dujardini individual. *Scientific Data, 3*, 160063.

Barinaga, M. (1994). Frontiers in biology: Development. *Science, 266*, 561–614.

Bartels, P. J., Apodaca, J. J., Mora, C., & Nelson, D. R. (2016). A global biodiversity estimate of a poorly known taxon: Phylum Tardigrada. *Zoological Journal of the Linnean Society, 178*, 730–736.

Bavan, S., Straub, V. A., Blaxter, M. L., & Ennion, S. J. (2009). A P2X receptor from the tardigrade species Hypsibius dujardini with fast kinetics and sensitivity to zinc and copper. *BMC Evolutionary Biology, 9*, 17.

Bemm, F., Burleigh, L., Förster, F., Schmucki, R., Ebeling, M., Janzen, C. J., et al. (2017). Draft genome of the Eutardigrade Milnesium tardigradum sheds light on ecdysozoan evolution. *bioRxiv*. https://doi.org/10.1101/122309.

Bemm, F., Weiß, C. L., Schultz, J., & Förster, F. (2016). Genome of a tardigrade: Horizontal gene transfer or bacterial contamination? *Proceedings of the National Academy of Sciences of the United States of America, 113*, E3054–E3056.

Bertolani, R. (2001). Evolution of the reproductive mechanisms in tardigrades—A review. *Zoologischer Anzeiger—A Journal of Comparative Zoology, 240*, 247–252.

Blaxter, M., Elsworth, B., & Daub, J. (2004). DNA taxonomy of a neglected animal phylum: An unexpected diversity of tardigrades. *Proceedings of the Biological Sciences, 271*(Suppl. 4), S189–S192.

Boothby, T. C., & Pielak, G. J. (2017). Intrinsically disordered proteins and desiccation tolerance: Elucidating functional and mechanistic underpinnings of Anhydrobiosis. *BioEssays, 39*, 1700119. https://doi.org/10.1002/bies.201700119.

Boothby, T. C., Tapia, H., Brozena, A. H., Piszkiewicz, S., Smith, A. E., Giovannini, I., et al. (2017). Tardigrades use intrinsically disordered proteins to survive desiccation. *Molecular Cell, 65*. 975–984.e5.

Boothby, T. C., Tenlen, J. R., Smith, F. W., Wang, J. R., Patanella, K. A., Nishimura, E. O., et al. (2015). Evidence for extensive horizontal gene transfer from the draft genome of a tardigrade. *Proceedings of the National Academy of Sciences of the United States of America, 112*, 15976–15981.

Carroll, S. B., Grenier, J. K., & Weatherbee, S. D. (2001). *From DNA to diversity: Molecular genetics and the evolution of animal design*. Blackwell Science.

Chavez, C., Cruz-Becerra, G., Fei, J., Kassavetis, G. A., & Kadonaga, J. T. (2019). The tardigrade damage suppressor protein binds to nucleosomes and protects DNA from hydroxyl radicals. *eLife, 8*. https://doi.org/10.7554/eLife.47682.

Crowe, J. H. (1972). Evaporative water loss by tardigrades under controlled relative humidities. *The Biological Bulletin, 142*, 407–416.

Degma, P., Bertolani, R., & Guidetti, R. (2021). *Actual checklist of Tardigrada species (2009-2021: 19-07-2021) [WWW document].* https://doi.org/10.25431/11380_1178608 (accessed 10.26.21).

Degma, P., & Guidetti, R. (2007). Notes to the current checklist of Tardigrada. *Zootaxa, 1579*, 41–53.

Delmont, T. O., & Eren, A. M. (2016). Identifying contamination with advanced visualization and analysis practices: Metagenomic approaches for eukaryotic genome assemblies. *PeerJ, 4*, e1839.

Doyère, M. (1840). Mémoire sur les Tardigrades. *Annales des Sciences Naturelles*, 269–361.

Esterly, H. J., Crilly, C. J., Piszkiewicz, S., Shovlin, D. J., Pielak, G. J., & Christian, B. E. (2020). Toxicity and immunogenicity of a tardigrade cytosolic abundant heat soluble protein in mice. *Frontiers in Pharmacology*. https://doi.org/10.3389/fphar.2020.565969.

Félix, M. A. (1999). Evolution of developmental mechanisms in nematodes. *The Journal of Experimental Zoology, 285*, 3–18.

Gabriel, W. N., & Goldstein, B. (2007). Segmental expression of Pax3/7 and engrailed homologs in tardigrade development. *Development Genes and Evolution, 217*, 421–433.

Gabriel, W. N., McNuff, R., Patel, S. K., Gregory, T. R., Jeck, W. R., Jones, C. D., et al. (2007). The tardigrade Hypsibius dujardini, a new model for studying the evolution of development. *Developmental Biology, 312*, 545–559.

Game, M., & Smith, F. W. (2020). Loss of intermediate regions of perpendicular body axes contributed to miniaturization of tardigrades. *Proceedings of the Biological Sciences, 287*, 20201135.

Gąsiorek, P., Stec, D., Morek, W., & Michalczyk, Ł. (2018). An integrative redescription of Hypsibius dujardini (Doyère, 1840), the nominal taxon for Hypsibioidea (Tardigrada: Eutardigrada). *Zootaxa, 4415*, 45–75.

Giribet, G., & Edgecombe, G. D. (2017). Current understanding of Ecdysozoa and its internal phylogenetic relationships. *Integrative and Comparative Biology, 57*, 455–466.

Giribet, G., & Edgecombe, D. G. (2020). *The invertebrate tree of life*. Princeton University Press.

Goldstein, B. (2018). The emergence of the tardigrade Hypsibius exemplaris as a model system. *Cold Spring Harbor Protocols, 2018*, 859–866.

Goldstein, B., & Blaxter, M. (2002). Tardigrades. *Current Biology, 12*, R475.

Grassé, P. P. (1949). *Ed. Traite de zoologie. Anatomie, systématique, biologie. Tome VI. Onychophores, Tardigrados, Arthropodes, Trilobltomorphes, Gbélicérates. Ed. Traite de zoologie. Anatomie, systématique.*

Grassé, P.-P. (1965). *Traite de zoologie: anatomie, systematique, biologie Tome 4, Fascicle 3*. Paris: Masson.

Gregory, T. R. (2001). The bigger the C-value, the larger the cell: Genome size and red blood cell size in vertebrates. *Blood Cells, Molecules & Diseases, 27*, 830–843.

Gregory, T. R., & Hebert, P. D. (1999). The modulation of DNA content: Proximate causes and ultimate consequences. *Genome Research, 9*, 317–324.

Grenier, J. K., Garber, T. L., Warren, R., Whitington, P. M., & Carroll, S. (1997). Evolution of the entire arthropod Hox gene set predated the origin and radiation of the onychophoran/arthropod clade. *Current Biology, 7*, 547–553.

Greven, H. (2018). From Johann August Ephraim Goeze to Ernst Marcus: A ramble through the history of early tardigrade research (1773 until 1929). In R. O. Schill (Ed.), *Water bears: The biology of tardigrades* (pp. 1–55). Cham: Springer International Publishing.

Gross, V., Bährle, R., & Mayer, G. (2018). Detection of cell proliferation in adults of the water bear Hypsibius dujardini (Tardigrada) via incorporation of a thymidine analog. *Tissue & Cell, 51*, 77–83.

Gross, V., & Mayer, G. (2015). Neural development in the tardigrade Hypsibius dujardini based on anti-acetylated α-tubulin immunolabeling. *EvoDevo, 6*, 12.

Gross, V., & Mayer, G. (2019). Cellular morphology of leg musculature in the water bear Hypsibius exemplaris (Tardigrada) unravels serial homologies. *Royal Society Open Science*, *6*, 191159.
Gross, V., Minich, I., & Mayer, G. (2017). External morphogenesis of the tardigrade Hypsibius dujardini as revealed by scanning electron microscopy. *Journal of Morphology*, *278*, 563–573.
Gross, V., Müller, M., Hehn, L., Ferstl, S., Allner, S., Dierolf, M., et al. (2019). X-ray imaging of a water bear offers a new look at tardigrade internal anatomy. *Zoological Letters*, *5*, 14.
Guidetti, R., & Bertolani, R. (2005). Tardigrade taxonomy: An updated check list of the taxa and a list of characters for their identification. *Zootaxa*, *845*, 1–46.
Guil, N., & Cabrero-Sañudo, F. J. (2007). Analysis of the species description process for a little known invertebrate group: The limnoterrestrial tardigrades (Bilateria, Tardigrada). In D. L. Hawksworth, & A. T. Bull (Eds.), *Vertebrate conservation and biodiversity* (pp. 237–260). Netherlands, Dordrecht: Springer.
Hashimoto, T., Horikawa, D. D., Saito, Y., Kuwahara, H., Kozuka-Hata, H., Shin-I, T., et al. (2016). Extremotolerant tardigrade genome and improved radiotolerance of human cultured cells by tardigrade-unique protein. *Nature Communications*, 7, 12808.
Hashimoto, T., & Kunieda, T. (2017). DNA protection protein, a novel mechanism of radiation tolerance: Lessons from tardigrades. *Life*, 7. https://doi.org/10.3390/life7020026.
Heikes, K. L., & Goldstein, B. (2018). Live imaging of tardigrade embryonic development by differential interference contrast microscopy. *Cold Spring Harbor Protocols*, *2018*, 874–877. https://doi.org/10.1101/pdb.prot102335.
Hejnol, A., & Schnabel, R. (2005). The eutardigrade Thulinia stephaniae has an indeterminate development and the potential to regulate early blastomere ablations. *Development*, *132*, 1349–1361.
Hengherr, S., Worland, M. R., Reuner, A., Brümmer, F., & Schill, R. O. (2009). High-temperature tolerance in anhydrobiotic tardigrades is limited by glass transition. *Physiological and Biochemical Zoology*, *82*, 749–755. https://doi.org/10.1086/605954.
Hering, L., Bouameur, J.-E., Reichelt, J., Magin, T. M., & Mayer, G. (2016). Novel origin of lamin-derived cytoplasmic intermediate filaments in tardigrades. *eLife*, *5*, e11117.
Hering, L., & Mayer, G. (2014). Analysis of the opsin repertoire in the tardigrade Hypsibius dujardini provides insights into the evolution of opsin genes in panarthropoda. *Genome Biology and Evolution*, *6*, 2380–2391.
Hesgrove, C., & Boothby, T. C. (2020). The biology of tardigrade disordered proteins in extreme stress tolerance. *Cell Communication and Signaling*, *18*, 178. https://doi.org/10.1186/s12964-020-00670-2.
Hibshman, J. D., Clegg, J. S., & Goldstein, B. (2020). Mechanisms of desiccation tolerance: Themes and variations in brine shrimp, roundworms, and tardigrades. *Frontiers in Physiology*, *11*, 592016. https://doi.org/10.3389/fphys.2020.592016.
Horikawa, D. D., Kunieda, T., Abe, W., Watanabe, M., Nakahara, Y., Yukuhiro, F., et al. (2008). Establishment of a rearing system of the extremotolerant tardigrade Ramazzottius varieornatus: A new model animal for astrobiology. *Astrobiology*, *8*, 549–556.
Hyman, L. H. (1951). *Acanthocephala, aschelminthes, and entoprocta: The pseudocoelomate bilateria.* New York: McGraw-Hill.
Janis, B., Belott, C., & Menze, M. A. (2018). Role of intrinsic disorder in animal desiccation tolerance. *Proteomics*, *18*, e1800067.
Jezierska, M., Miernik, A., Sojka, J., Student, S., Śliwińska, M. A., Gross, V., et al. (2021). Oogenesis in the tardigrade Hypsibius exemplaris Gąsiorek, Stec, Morek & Michalczyk, 2018 (Eutardigrada, Hypsibiidae). *Micron*, *150*, 103126.
Jørgensen, A., Møbjerg, N., & Kristensen, R. M. (2007). A molecular study of the tardigrade Echiniscus testudo (Echiniscidae) reveals low DNA sequence diversity over a large geographical area. *Journal of Limnology*, *66*, 77–83.

Kaufmann, F. J. (1851). Ueber die Entwicklung und systematische Stellung der Tardigraden. *Zeitschrift für Wissenschaftliche Zoologie*, *3*, 220–232.

Kirke, J., Jin, X.-L., & Zhang, X.-H. (2020). Expression of a tardigrade Dsup gene enhances genome protection in plants. *Molecular Biotechnology*, *62*, 563–571. https://doi.org/10.1007/s12033-020-00273-9.

Kondo, K., Kubo, T., & Kunieda, T. (2015). Suggested involvement of PP1/PP2A activity and de novo gene expression in anhydrobiotic survival in a tardigrade, Hypsibius dujardini, by chemical genetic approach. *PLoS One*, *10*, e0144803.

Kondo, K., Mori, M., Tomita, M., & Arakawa, K. (2019). AMPK activity is required for the induction of anhydrobiosis in a tardigrade Hypsibius exemplaris, and its potential up-regulator is PP2A. *Genes to Cells*, *24*, 768–780.

Kondo, K., Mori, M., Tomita, M., & Arakawa, K. (2020). Pre-treatment with D942, a furancarboxylic acid derivative, increases desiccation tolerance in an anhydrobiotic tardigrade Hypsibius exemplaris. *FEBS Open Bio*, *10*, 1774–1781.

Koutsovoulos, G., Kumar, S., Laetsch, D. R., Stevens, L., Daub, J., Conlon, C., et al. (2016). No evidence for extensive horizontal gene transfer in the genome of the tardigrade Hypsibius dujardini. *Proceedings of the National Academy of Sciences of the United States of America*, *113*, 5053–5058.

Koziol, U. (2018). Precursors of neuropeptides and peptide hormones in the genomes of tardigrades. *General and Comparative Endocrinology*, *267*, 116–127.

Mapalo, M. A., Arakawa, K., Baker, C. M., Persson, D. K., Mirano-Bascos, D., & Giribet, G. (2020). The unique antimicrobial recognition and signaling pathways in tardigrades with a comparison across Ecdysozoa. *G3*, *10*, 1137–1148.

Marcus, E. (1929). Zur Embryologie der Tardigraden. *Zoologische Jahrbücher. Abteilung für Anatomie und Ontogenie der Tiere*, *50*, 333–384.

Mayer, G., Martin, C., Rüdiger, J., Kauschke, S., Stevenson, P. A., Poprawa, I., et al. (2013). Selective neuronal staining in tardigrades and onychophorans provides insights into the evolution of segmental ganglia in panarthropods. *BMC Evolutionary Biology*, *13*, 230.

McGreevy, K. M., Heikes, K. L., Kult, S., Tharp, M. E., & Goldstein, B. (2018). Fluorescent cell staining methods for living Hypsibius exemplaris embryos. *Cold Spring Harbor Protocols*, *2018*, 878–884. https://doi.org/10.1101/pdb.prot106021.

McNuff, R. (2018). Laboratory culture of Hypsibius exemplaris. *Cold Spring Harbor Protocols*, *2018*, 867–870. https://doi.org/10.1101/pdb.prot102319.

Møbjerg, N., Jørgensen, A., Kristensen, R. M., & Neves, R. C. (2018). Morphology and functional anatomy. In R. O. Schill (Ed.), *Water bears: The biology of tardigrades* (pp. 57–94). Cham: Springer.

Nakahara, Y., Imanishi, S., Mitsumasu, K., Kanamori, Y., Iwata, K.-I., Watanabe, M., et al. (2010). Cells from an anhydrobiotic chironomid survive almost complete desiccation. *Cryobiology*, *60*, 138–146.

Nelson, D. R. (2017). Cytochrome P450 diversity in the tree of life. *Biochimica et Biophysica Acta: Proteins and Proteomics*, *1866*, 141–154.

Nkem, J. N., Wall, D. H., Virginia, R. A., Barrett, J. E., Broos, E. J., Porazinska, D. L., et al. (2006). Wind dispersal of soil invertebrates in the McMurdo Dry Valleys, Antarctica. *Polar Biology*, *29*, 346–352.

Patel, N. H. (1994). Developmental evolution: Insights from studies of insect segmentation. *Science*, *266*, 581–590.

Piszkiewicz, S., Gunn, K. H., Warmuth, O., Propst, A., Mehta, A., Nguyen, K. H., et al. (2019). Protecting activity of desiccated enzymes. *Protein Science*, *28*, 941–951. https://doi.org/10.1002/pro.3604.

Ptatscheck, C., Gansfort, B., & Traunspurger, W. (2018). The extent of wind-mediated dispersal of small metazoans, focusing nematodes. *Scientific Reports*, *8*, 6814.

Ramazzotti, E., & Maucci, W. (1995). *The phylum Tardigrada.* (C. W. Beasley, Trans.).

Rebecchi, L., Altiero, T., Guidetti, R., Cesari, M., Bertolani, R., Negroni, M., et al. (2009). Tardigrade resistance to space effects: First results of experiments on the LIFE-TARSE mission on FOTON-M3 (September 2007). *Astrobiology*, *9*, 581–591. https://doi.org/10.1089/ast.2008.0305.

Richaud, M., Le Goff, E., Cazevielle, C., Ono, F., Mori, Y., Saini, N. L., et al. (2020). Ultrastructural analysis of the dehydrated tardigrade Hypsibius exemplaris unveils an anhydrobiotic-specific architecture. *Scientific Reports*, *10*, 4324.

Rivas, J. A., Jr., Schröder, T., Gill, T. E., Wallace, R. L., & Walsh, E. J. (2019). Anemochory of diapausing stages of microinvertebrates in North American drylands. *Freshwater Biology*, *64*, 1303–1314.

Schill, R. O. (2018). *Water bears: The biology of tardigrades*. Switzerland: Springer.

Schill, R. O., & Hengherr, S. (2018). Environmental adaptations: Desiccation tolerance. In R. O. Schill (Ed.), *Water bears: The biology of tardigrades* (pp. 273–293). Cham: Springer International Publishing.

Sidow, A., & Thomas, W. K. (1994). A molecular evolutionary framework for eukaryotic model organisms. *Current Biology*, *4*, 596–603.

Smith, F. W. (2018). Embryonic in situ hybridization for the tardigrade Hypsibius exemplaris. *Cold Spring Harbor Protocols*, *2018*, 891–899. https://doi.org/10.1101/pdb.prot102350.

Smith, F. W., Bartels, P. J., & Goldstein, B. (2017). A hypothesis for the composition of the tardigrade brain and its implications for panarthropod brain evolution. *Integrative and Comparative Biology*, *57*, 546–559.

Smith, F. W., Boothby, T. C., Giovannini, I., Rebecchi, L., Jockusch, E. L., & Goldstein, B. (2016). The compact body plan of tardigrades evolved by the loss of a large body region. *Current Biology*, *26*, 224–229.

Smith, F. W., Cumming, M., & Goldstein, B. (2018). Analyses of nervous system patterning genes in the tardigrade Hypsibius exemplaris illuminate the evolution of panarthropod brains. *EvoDevo*, *9*, 19.

Smith, F. W., & Gabriel, W. N. (2018). Embryonic immunostaining for the tardigrade Hypsibius exemplaris. *Cold Spring Harbor Protocols*, *2018*, 885–890. https://doi.org/10.1101/pdb.prot102343.

Smith, F. W., & Goldstein, B. (2016). Segmentation in Tardigrada and diversification of segmental patterns in Panarthropoda. *Arthropod Structure & Development*, *46*, 328–340.

Smith, F. W., & Jockusch, E. L. (2014). The metameric pattern of Hypsibius dujardini (Eutardigrada) and its relationship to that of other panarthropods. *Frontiers in Zoology*, *11*, 1–16.

Sogame, Y., & Kikawada, T. (2017). Current findings on the molecular mechanisms underlying anhydrobiosis in Polypedilum vanderplanki. *Current Opinion in Insect Science*, *19*, 16–21. https://doi.org/10.1016/j.cois.2016.10.008.

Sommer, R. J. (2000). Evolution of nematode development. *Current Opinion in Genetics & Development*, *10*, 443–448. https://doi.org/10.1016/s0959-437x(00)00110-6.

Suzuki, A. C. (2003). Life history of Milnesium tardigradum Doyère (Tardigrada) under a rearing environment. *Zoological Science*, *20*, 49–57. https://doi.org/10.2108/zsj.20.49.

Suzuki, A. C. (2008). Appearance of males in a thelytokous strain of Milnesium cf. tardigradum (Tardigrada). *Zoological Science*, *25*, 849–853.

Tanaka, S., Tanaka, J., Miwa, Y., Horikawa, D. D., Katayama, T., Arakawa, K., et al. (2015). Novel mitochondria-targeted heat-soluble proteins identified in the anhydrobiotic tardigrade improve osmotic tolerance of human cells. *PLoS One*, *10*, e0118272. https://doi.org/10.1371/journal.pone.0118272.

Tenlen, J. R. (2018). Microinjection of dsRNA in tardigrades. *Cold Spring Harbor Protocols*, *2018*, 900–904. https://doi.org/10.1101/pdb.prot102368.

Tenlen, J. R., McCaskill, S., & Goldstein, B. (2013). RNA interference can be used to disrupt gene function in tardigrades. *Development Genes and Evolution*, *223*, 171–181.

Thiruketheeswaran, P., Greven, H., & D'Haese, J. (2016). Gelsolin in Onychophora and Tardigrada with notes on its variability in the Ecdysozoa. *Comparative Biochemistry and Physiology. Part B, Biochemistry & Molecular Biology*, *203*, 47–52.

Thomas, C., DeVries, P., Hardin, J., & White, J. (1996). Four-dimensional imaging: Computer visualization of 3D movements in living specimens. *Science*, *273*, 603–607.

Tsujimoto, M., Suzuki, A. C., & Imura, S. (2015). Life history of the Antarctic tardigrade, Acutuncus antarcticus, under a constant laboratory environment. *Polar Biology*, *38*, 1575–1581.

Vasanthan, T., & Stone, J. (2020). Life history traits for the freshwater Tardigrade species Hypsibius exemplaris reared under laboratory conditions. *Journal of Wildlife and Biodiversity*, *4*, 65–72.

von Erlanger, R. (1895). Beiträge zur Morphologie der Tardigraden. I. Zur Embryologie eines Tardigraden: Macrobiotus macronyx Dujardin. *Morphologisches Jahrbuch*, *22*, 491–513.

von Wenck, W. (1914). Entwicklungsgeschichtliche Untersuchungen an Tardigraden (Macrobiotus lacustris Duj.). *Zoologische Jahrbücher. Abteilung für Anatomie und Ontogenie der Tiere*, *37*, 465–514.

Wojciechowska, D., Roszkowska, M., Kaczmarek, Ł., Jarmuszkiewicz, W., Karachitos, A., & Kmita, H. (2021). The tardigrade Hypsibius exemplaris has the active mitochondrial alternative oxidase that could be studied at animal organismal level. *PLoS One*, *16*, e0244260.

Wright, J. C. (1989). Desiccation tolerance and water-retentive mechanisms in tardigrades. *Journal of Experimental Biology*, *142*, 267–292. https://doi.org/10.1242/jeb.142.1.267.

Yamaguchi, A., Tanaka, S., Yamaguchi, S., Kuwahara, H., Takamura, C., Imajoh-Ohmi, S., et al. (2012). Two novel heat-soluble protein families abundantly expressed in an anhydrobiotic tardigrade. *PLoS One*, *7*, e44209. https://doi.org/10.1371/journal.pone.0044209.

Yoshida, Y., Koutsovoulos, G., Laetsch, D. R., Stevens, L., Kumar, S., Horikawa, D. D., et al. (2017). Comparative genomics of the tardigrades Hypsibius dujardini and Ramazzottius varieornatus. *PLoS Biology*, *15*, e2002266.

Yoshida, Y., Sugiura, K., Tomita, M., Matsumoto, M., & Arakawa, K. (2019). Comparison of the transcriptomes of two tardigrades with different hatching coordination. *BMC Developmental Biology*, *19*, 24.

CHAPTER EIGHT

The crustacean model *Parhyale hawaiensis*

Mathilde Paris[a,b], Carsten Wolff[c], Nipam H. Patel[c,d,*], and Michalis Averof[a,b,*]

[a]Institut de Génomique Fonctionnelle de Lyon, École Normale Supérieure de Lyon, Lyon, France
[b]Centre National de la Recherche Scientifique (CNRS), France
[c]Marine Biological Laboratory, Woods Hole, MA, United States
[d]Department of Organismal Biology and Anatomy, University of Chicago, Chicago, IL, United States
[*]Corresponding authors: e-mail address: npatel@mbl.edu; michalis.averof@ens-lyon.fr

Contents

Current Topics in Developmental Biology, Volume 147
ISSN 0070-2153
https://doi.org/10.1016/bs.ctdb.2022.02.001

Abstract

Arthropods are the most abundant and diverse animals on earth. Among them, pancrustaceans are an ancient and morphologically diverse group, comprising a wide range of aquatic and semi-aquatic crustaceans as well as the insects, which emerged from crustacean ancestors to colonize most terrestrial habitats. Within insects, *Drosophila* stands out as one of the most powerful animal models, making major contributions to our understanding of development, physiology and behavior. Given these attributes, crustaceans provide a fertile ground for exploring biological diversity through comparative studies. However, beyond insects, few crustaceans are developed sufficiently as experimental models to enable such studies. The marine amphipod *Parhyale hawaiensis* is currently the best established crustacean system, offering year-round accessibility to developmental stages, transgenic tools, genomic resources, and established genetics and imaging approaches. The *Parhyale* research community is small but diverse, investigating the evolution of development, regeneration, aspects of sensory biology, chronobiology, bioprocessing and ecotoxicology.

1. Natural habitat and lifecycle

1.1 Where *Parhyale* live

Parhyale hawaiensis (Dana, 1853) (Peracarida, Amphipoda, Hyalidae) are marine amphipod crustaceans, belonging to the talitrid superfamily (commonly known as beach-hoppers, sand-hoppers or sand-fleas) (Fig. 1). First described from the Hawaiian islands (Dana, 1853), *Parhyale hawaiensis* are cosmopolitan amphipods with a worldwide, circumtropical distribution (Barnard, 1965; Shoemaker, 1956) and might represent a species complex (Myers, 1985). They occur in intertidal habitats such as bays, estuaries and mangroves (Poovachiranon, Boto, & Duke, 1986; Shoemaker, 1956) and prefer rocky beaches and macroalgal fauna (Barnard, 1965). Occupying shallow intertidal waters, *Parhyale* can tolerate variations in salinity (5 to 40 ppt) and temperature (Poovachiranon et al., 1986). As effective detritus feeders with large population sizes (up to 7000 individuals per m^2) they can play an important role in ecosystems such as mangrove forests (Poovachiranon et al., 1986).

1.2 Morphology

Parhyale exhibit a pattern of segmental organization that is common to malacostracan crustaceans, including head, thoracic (pereon) and abdominal

Fig. 1 *Parhyale hawaiensis* couple. The male clasps the smaller female. (Photo credit: Vincent Mongorgé).

(pleon) units with conserved numbers of segments. The head bears sensory antennae and feeding appendages. The pereon is composed of eight segments: a first segment bearing modified feeding appendages (maxillipeds, T1), followed by two grasping appendages (gnathopods, T2–3) and five walking legs (T4–8). Maxillipeds are morphologically and functionally associated with the head segments. Walking legs T4–5 are oriented anteriorly, while T6–8 are oriented posteriorly—hence the name "amphipoda," which derives from the Greek *αμφί*/both and *πούς*/leg. The pleon is composed of six segments, each with a pair of branched (biramous) appendages. The first three pairs are swimmerets, the following three are smaller and more stiff uropods for holding onto or jumping from the substrate. All these various appendages with different morphologies are considered to be serially homologous (Liubicich et al., 2009; Martin et al., 2016; Pavlopoulos et al., 2009; Pavlopoulos & Wolff, 2020). Divakaran (1976) provides a detailed description of the morphology, organ systems and biology of *Parhyale hawaiensis*.

1.3 Lifecycle

Like most amphipod crustaceans, *Parhyale* shows sexual dimorphism. Mature males are usually larger than females of the same age and bear T3 gnathopods in which the two last leg segments (propodus, dactylus) are significantly enlarged. Males use this enlarged claw-like structure to clasp and hold on to a female in a mating position known as amplexus (Fig. 1). *Parhyale* adults pair a few days before females are ready to lay eggs. When the female moults, the male transfers sperm into the female's brood pouch and releases her, as she starts to release her eggs in the brood pouch (Borowsky & Borowsky, 1987; Divakaran, 1976). As in most peracarid crustaceans, the females carry the fertilized eggs in a ventral brood pouch (marsupium), which is formed by specialized side branches (oostegites) of thoracic appendages T3–6 (Wolff & Gerberding, 2015). Depending on the size and the age of the female, the marsupium can contain up to ~30 eggs. The eggs of each brood are developmentally synchronized (Browne, Price, Gerberding, & Patel, 2005). The embryonic development is direct, with no larval stages. After about 10 days at 26 °C, juveniles hatch as miniature versions of the adult (Browne et al., 2005) and stay for a few more days inside the marsupium. *Parhyale* grow through successive molts, in which they shed their cuticle and replace it by a new one, during their entire lifetime. Their body length increases from ~1 mm in hatchlings to >10 mm in adults. Secondary sexual characters, such as oostegites and male claspers differentiate during reproductive maturation, which takes at least 6–7 weeks at 26 °C.

2. From mangrove to lab

2.1 Historical info

While *Parhyale* are found in tropical waters around the world, the lab cultures of *Parhyale* are derived from animals collected by William Browne from the seawater filtration system of the Shedd Aquarium (Chicago, IL, USA) almost 25 years ago, in 1997. Thus, the geographical origin of the cultured animals remains a mystery, as the *Parhyale* presumably entered the system as hitchhikers from other material purposely introduced into the seawater tanks.

The selection of an amphipod crustacean was not a complete accident, however, as the extensive work of Wolfgang Dohle (reviewed in Dohle & Scholtz, 1988) had documented a remarkable pattern of orderly cell divisions that generates the germband of amphipod crustaceans, and at that time there was great interest in understanding the process of segmentation in arthropods that, unlike *Drosophila*, add segments sequentially in a cellularized rather than a syncytial context. Thus, *Parhyale* was an ideal organism for comparative evolutionary developmental studies of early pattern formation (phylogenetic context presented in Fig. 2). Once introduced into the lab, *Parhyale* were readily established in shallow seawater tanks and easily maintained with relatively little effort (Gerberding, Browne, & Patel, 2002). There were some issues with rotifers in the female brood pouches that led to high mortality of embryos, but these rotifers were eventually removed from the cultures.

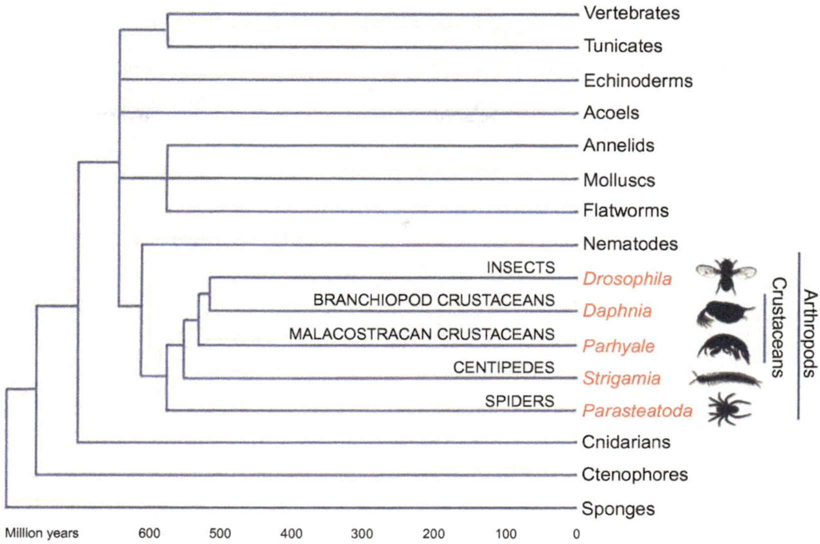

Fig. 2 *Parhyale* and other arthropod models, shown in the context of animal phylogeny, focusing on experimental and/or genomics models representing five major arthropod clades: insects, branchiopod and malacostracan crustaceans, centipedes and spiders. Crustaceans are paraphyletic with respect to the insects (Schwentner, Combosch, Pakes Nelson, & Giribet, 2017). The divergence times are tentative and in some cases controversial; unresolved nodes are shown as polytomies (see Kapli & Telford, 2020; Telford, Budd, & Philippe, 2015).

2.2 Lab culture

Parhyale are very hardy and do well in a variety of culture systems (Fig. 3). They are generally kept in shallow trays of artificial seawater with either an airstone or impeller pump to keep water circulating. They will readily accept a variety of food including fish flakes, kelp powder, carrots, and pellets formulated for feeding shrimp. A substrate composed of crushed coral or aragonite rock will help to buffer the pH and provide hiding spaces for juvenile animals. Water changes at two-week intervals help to maintain water quality, although animals will easily survive even under conditions of poor water quality provided that the water remains well oxygenated. Water salinity is kept at 24–32 ppt. Tanks with an active biological filtration system and strong circulation can support many hundreds of adult animals per liter. Under ideal conditions of 26–28 °C with frequent feeding, *Parhyale* can go through an entire generation in about 2 months. Individual adults can survive and remain reproductive for up to 2–3 years. *Parhyale* cultures are so robust that the species is being considered for aquaculture (Vargas-Abúndez, López-Vázquez, Mascaró, Martínez-Moreno, & Simões, 2021).

Fig. 3 A dense laboratory culture of *Parhyale* feeding on shrimp meal wafers.

2.3 Inbred line

Shortly after the *Parhyale* cultures were established, an inbred isofemale line, Chicago-F, was created through two rounds of sibling inbreeding, and then maintained as an isolated population since. This line was used for the creation of a BAC library (Parchem, Poulin, Stuart, Amemiya, & Patel, 2010, at the time the line was called Iso2) and for genome sequencing (Kao et al., 2016). Further attempts at inbreeding resulted in unhealthy animals which failed to survive (M. Vargas-Vila and N. Patel, unpublished observations). The Chicago-F line carries fewer polymorphisms than other laboratory populations, but still carries a high level of polymorphism in some parts of the genome (Kao et al., 2016).

3. Major interests and research questions

3.1 Embryonic development and germ layer formation

Embryogenesis takes ~11 days at 26 °C. Embryos undergo direct development, so hatch as miniature versions of the adults. Eggs can be removed from the female brood pouch without harming the female, and the eggs are easily maintained in petri dishes with sterile seawater.

Embryogenesis has been divided into a series of stages based on morphological criteria (Browne at el. 2005). Initial cell divisions are holoblastic, and at 8 h of development the embryo is composed of 8 cells: 4 macromeres and 4 micromeres. Initial lineage studies show that these cells subsequently give rise to defined germ layers of the embryo (Gerberding et al., 2002; Price, Modrell, Hannibal, & Patel, 2010). The four macromeres are called Ep, El, Er, and Mav, and the four micromeres are g, mr, ml, and en. Ep, Er, and El give rise to the ectoderm (posterior, right, and left, respectively). Mav gives rise to the visceral mesoderm and anterior somatic mesoderm. The somatic mesoderm on the left and right sides of the body are derived from ml and mr respectively, g gives rise to the germline. en was initially characterised as producing the endoderm, although subsequent studies suggest that it gives rise to extra-embryonic cells and the endoderm is derived from Mav (C. Chaw and N. Patel, unpublished). This lineage pattern is also observed in another amphipod, *Orchestia cavimana* (Wolff & Scholtz, 2002).

Ablation experiments of Ep, Er, El, ml, mr, and Mav reveal that compensation is possible when an individual macromere or micromere is ablated, but that compensation is restricted to within a given germ layer (Price et al., 2010). Thus, ablation of an ectodermal blastomere will be compensated, but only by another blastomere that would normally also give rise to ectoderm, and the same is true for the mesoderm. This compensation occurs if the ablation is done to a blastomere or its progeny prior to gastrulation. Once gastrulation has occurred, compensation either does not occur, or is very incomplete (Hannibal, Price, & Patel, 2012; Price et al., 2010). Studies of the g blastomere suggest that this cell contains germ plasm (Gupta & Extavour, 2013; Ozhan-Kizil, Havemann, & Gerberding, 2009) and other lineages cannot compensate for g cell ablation during embryogenesis (Extavour, 2005).

The ectoderm of the thorax and abdomen form by condensation of the ectodermal cells into an orderly array of rows and columns, which then follow a remarkably well-organized pattern of subsequent divisions, so that each subsequent parasegment of the embryo is derived from a single row of precursor cells, called a parasegment precursor row (PSPR) (Browne et al., 2005; Gerberding et al., 2002; Sun & Patel, 2019). This pattern of parasegment formation is also seen in other crustaceans (reviewed in Dohle & Scholtz, 1988), although amphipods are unique among malacostracan crustaceans in not possessing ectoteloblast stem cells that produce the PSPRs.

The somatic mesoderm of the thorax and abdomen of *Parhayle* are produced by eight mesotelobast stem cells (four per side) derived from ml and mr. These mesoteloblasts undergo a series of highly organized asymmetric divisions, which form rows of mesoblasts, with each row of mesoblasts then giving rise to a segmental unit of muscles (Gerberding et al., 2002; Hannibal, Price, Parchem, & Patel, 2012; Hannibal, Price, & Patel, 2012).

3.2 Body and limb patterning

Like most other crustaceans, *Parhyale* possess a pair of appendages on each segment of the body. Orthologs of the homeotic (*Hox*) genes have been identified in *Parhyale* and found to be expressed along the anterior-posterior axis of the embryo as in most other animals (Liubicich et al., 2009, Serano et al., 2016). They are present as a single complex spanning about 3.6 Mb (D. Sun and N. Patel, unpublished observations). The function in *Parhyale* of one of these genes, *Ultrabithorax* (*Ubx*), has been probed by ubiquitous misexpression using a transgenic approach (Pavlopoulos et al., 2009) and knockdown via RNAi (Liubicich et al., 2009). Both experiments confirmed that *Ubx* plays a role in establishing segmental identity, most obviously seen by transformations of appendage identity. Furthermore, these studies supported the previous hypothesis, based on comparison of Ubx protein expression across several crustacean lineages, that *Ubx* plays a role in distinguishing maxillipeds from more typical locomotory appendages in crustaceans (Averof & Patel, 1997).

More recently a systematic analysis of *Hox* gene function was carried out using CRISPR knockouts (Martin et al., 2016), which led to a detailed picture of how the identity of segments of the thorax and abdomen is specified. While the overall results are largely consistent with the role of *Hox* genes in specifying segmental identities, as seen in insects, many of the details suggest that there are important evolutionary differences both in the expression patterns and in the interaction between *Hox* genes in *Parhyale* and *Drosophila* (Martin et al., 2016; Serano et al., 2016). Most notably, the general pattern of posterior prevalence is not obeyed for *Ubx*, *abd-A*, and *Abd-B* (further supported through an analysis of double mutant combinations; E. Jarvis and N. Patel, unpublished data). These results provide a basis for analysing the potential role of *Hox* genes in evolutionary changes in body patterning across crustacean lineages.

Arthropods also display a clear regionalization along the proximal-distal axis of their limbs, which appears to be mediated by an evolutionarily conserved set of genes in all arthropods. CRISPR-based knockout of *Parhyale* orthologs of these leg patterning genes suggests a specific evolutionary relationship of individual leg segments between *Parhyale* and insects, which supports the hypothesis that proximal leg segments present in the last common ancestor of insects and crustaceans became incorporated into the body wall during insect evolution. This change may explain the evolutionary origin of insect wings from a proximal exite (Bruce & Patel, 2020). Studies on the orthologues of *Drosophila* wing patterning genes in *Parhyale* and in other crustaceans provide additional insights on the complex evolutionary origins of insect wings (Averof & Cohen, 1997; Clark-Hachtel & Tomoyasu, 2020; Shiga et al., 2017).

Cell lineage analyses of outgrowing thoracic limbs show that the early limb primordium becomes subdivided into anterior-posterior and dorsal-ventral compartments whose boundaries intersect at the distal tip of the growing limb. Limb formation is associated with a spatial modulation of cell proliferation, and limb elongation is driven by preferential orientation of cell divisions along the proximal-distal growth axis (Wolff et al., 2018).

3.3 Regeneration

Parhyale have the ability to regenerate their limbs, following limb injuries sustained by attacks from predators, cannibalism, or disease. They are able to regenerate any appendage (antennae, mouthparts, thoracic limbs, pleopods and uropods) and retain that ability throughout their lifetime.

Since their body is enclosed within a chitinous exoskeleton, arthropods depend on molting to fully restore the structure and function of their regenerating limbs. Live imaging using fluorescent nuclear markers has shown that the following events take place once a *Parhyale* leg has been amputated: (1) within minutes from the cut, haemocytes adhere to the wound surface and form a plug that stops the bleeding, (2) the wound surface becomes melanized within the next hours, a typical wounding reaction of arthropods that is mediated by the haemocytes at the wound, (3) within 1–2 days the wound epithelium stretches and closes over the wound surface, under the melanized scab, (4) extensive cell proliferation takes place in the region of the wound, (5) the epithelium at the wound site detaches from the cuticle and limb morphogenesis takes place below the scab, (6) the animal molts, the newly regenerated limb is released and becomes functional (Alwes, Enjolras, & Averof, 2016). Studies in other crustaceans suggest there is a complex interplay between the timing of regeneration and molting (Skinner, 1985), but this crosstalk has not yet been systematically investigated in *Parhyale*.

Many malacostracan crustaceans, including crabs and lobsters, can self-amputate limbs that are severely injured, a process known as autotomy. The injured limbs are broken off at specific sites near the base of the limb, under the control of a nervous reflex (Fredericq, 1882). We have not observed autotomy in *Parhyale*. Limbs can be amputated at any site along their proximo-distal axis, where they will heal and regenerate the missing parts.

Research in *Parhyale* has so far focused on the cellular basis of leg regeneration. Studies in mosaic animals, where individual blastomere lineages were marked by transposons carrying a fluorescent marker, established that *Parhyale* limb regeneration relies on distinct pools of progenitor cells for ectoderm and mesoderm, and that these progenitors reside locally, near the site of the regenerating limb (Konstantinides & Averof, 2014). There are no totipotent progenitors.

Cell transplantation and live imaging experiments have revealed the cellular sources for regeneration of muscle and epidermis: mesodermal cells resembling the satellite cells of vertebrates appear to act as progenitors for muscle (Konstantinides & Averof, 2014), while any epidermal cell near the site of amputation appears capable of regenerating the epidermis (Alwes et al., 2016).

A unique asset for studying leg regeneration in *Parhyale*, so far unparalleled in other experimental models, is the ability to image regeneration continuously, at single-cell resolution (Alwes et al., 2016; Sugawara, Çevrim, & Averof, 2022). This approach allows cell lineages to be tracked over the entire course of regeneration, from the time of amputation to the subsequent molt (~1 week), and provides a basis for studying the cell behaviors that underpin morphogenesis in regenerating legs. Detailed transcriptional profiling of leg regeneration has been performed in *Parhyale* (Sinigaglia et al., 2021), but to our knowledge no studies of gene function during regeneration have been undertaken so far. The impermeable cuticle that surrounds the limbs presents a significant challenge for in situ hybridization and immunochemical stainings.

3.4 Neurobiology, sensory systems and behavior

The central nervous system of *Parhyale* is composed of a brain and ventral nerve cord. The adult brain contains about 13,300 nuclei in total (Wittfoth, Harzsch, Wolff, & Sombke, 2019), a small number which renders this brain tractable for neuroanatomical studies. The ventral nerve cord comprises a fused subesophageal ganglion, seven segmental ganglia innervating the pereon, three segmental ganglia innervating pleon segments 1–3, and one fused ganglion innervating pleon segments 4–6 (Divakaran, 1982).

The brain of *Parhyale* is tripartite and formed during development through condensation of three segmental cephalic neuromeres. They represent three morphological regions with segmental identity: the protocerebrum, associated with the compound eyes, the deutocerebrum, associated with the first antenna, and the tritocerebrum, associated with the second antenna (Divakaran, 1982). The general architecture of all major neuropils is in agreement with the suggested ground-pattern of malacostracan crustaceans (see Kenning, Müller, Wirkner, & Harzsch, 2013), however, some of these neuropils are difficult to demarcate due to their close association and uniform appearance. For example, in the visual system there are uncertainties about the presence/absence of the lobula and the inner chiasma connecting the medulla (Ramos et al., 2019; Wittfoth et al., 2019). Studies of photoreceptor projections to the optic lobe neuropils have revealed some unexpected differences from a conserved pattern of photoreceptor projections found in other crustaceans and insects, but these differences remain poorly understood (Ramos et al., 2019).

Although there is significant anatomical conservation across amphipod brains, there is also a degree of variability in the construction and size of different neuropils (Hanström, 1932), which seems to correlate with ecology and life history (Ramm & Scholtz, 2017). The brain of *Parhyale* does not display any major modifications or bias towards one sensory modality, therefore it likely represents a common type of amphipod brain.

The visual system of *Parhyale* consists of a pair of eyes, with up to ~50 ommatidia per adult eye, and the underlying visual circuits (Ramos et al., 2019). Hatchlings start off with a small number of ommatidia, which then grow in size and in number, likely increasing visual sensitivity and resolution several fold as the animals grow to adulthood (Keskinen, Takaku, Meyer-Rochow, & Hariyama, 2002; Ramos et al., 2019).

Each ommatidium contains five photoreceptor cells, four with large rhabdomeres (R1–4) and one with a small rhabdomere (R5). These two types of photoreceptors express distinct opsins and project their axons to distinct regions of the first optic neuropil (Ramos et al., 2019). Photoreceptors R1–R4 have long, straight rhabdomeres that should make them intrinsically sensitive to the direction of light polarization. The rhabdomeres of R1 + R3 and R2 + R4 are arranged in perpendicular orientations, which would make them most sensitive to different directions of polarized light. Overall, this design gives *Parhyale* very low visual resolution (20° interommatidial angle and 50 resolved pixels per adult eye, compared with 5° and 700 pixels in *Drosophila*; Land, 1997), but the potential to detect color and polarized light (Ramos et al., 2019).

These results suggest that *Parhyale* do not use vision to perform tasks that require good spatial resolution, such as finding food or interacting with mates (e.g., see Holmes, 1903), but are likely to use vision to orient themselves within their habitat, e.g., by responding to gradients of light intensity and polarization. Behavioral experiments in other talitrid amphipods support this notion (Ciofini, Yamahama, Mercatelli, Hariyama, & Ugolini, 2020; Cohen, Cronin, Lessios, & Forward, 2010; Cohen & Putts, 2013; Ercolini & Scapini, 1976; Forward, Bourla, Lessios, & Cohen, 2009; Ugolini, 2014; Ugolini, Galanti, & Mercatelli, 2012). Simple behavioral experiments in *Parhyale* suggest that they have a phototactic response (Ramos et al., 2019).

Photoreceptors lacking overt pigmentation and a light-focusing apparatus have been found in the brain of other talitrid amphipods (Frelon-Raimond, Meyer-Rochow, Ugolini, & Martin, 2002). These photoreceptors could perform non-visual functions, such as entraining circadian rhythms. We speculate that in *Parhyale* these cells could be marked by the expression of the *3xP3* marker (see below).

Parhyale are likely to use chemosensory cues rather than vision to locate their food and mating partners. In laboratory conditions it is noticeable that males are more active in searching for females and probing whether they are receptive to mating (C. Wolff, personal observation). This observation is in agreement with studies on intertidal amphipod populations (Alegretti, Umbuzeiro, & Flynn, 2016) and suggests that waterborne attractants or contact pheromones might play an important role in reproductive behavior. There are studies on sexual dimorphisms of the peripheral sensory system, suggesting that the first antennae of male amphipods can be equipped with male-specific sensilla (Hallberg, Johansson, & Wallén, 1997; Lowry, 1986).

3.5 Circadian and tidal rhythms

Parhyale have recently attracted attention as an experimental model for studying biological clocks. Initial studies focused on probing circadian rhythms in locomotor activity, establishing a head transcriptome at 3-h intervals across the 24-h day/night cycle, and identifying core components of the circadian clock, including Bmal1/Cycle, Cry2 and Per (Hunt, 2016; Hunt, Mallon, & Rosato, 2019). The genome of *Parhyale* appears to lack orthologues of Cry1 and Timeless (Hunt et al., 2019). Preliminary observations on *Parhyale*'s locomotor behavior revealed 24-h and 12-h rhythms of activity, with highest activity in the dark hours of the circadian cycle (Hunt, 2016). As an organism living in intertidal habitats, *Parhyale* can also serve as a model for studying circatidal rhythms (E. Kwiatkowski, J. Rosenthal and P. Emery, unpublished observations).

3.6 Lignocellulose digestion

Insects such as termites and beetles have evolved complex digestive systems with the ability to digest wood and to use it as their main source of energy. To achieve this, they rely on populations of symbiotic microbes, resident in their digestive tract, which provide enzymes to digest lignocellulose (Watanabe & Tokuda, 2010). Within malacostracan crustaceans, some isopods and amphipods, including *Parhyale*, have evolved the capability to digest lignocellulose without relying on such microbes (Cragg et al., 2015; Zimmer et al., 2002). The *Parhyale* genome encodes GH7 family glycosyl hydrolase enzymes that are necessary for hydrolytically digesting lignocellulose (Kao et al., 2016). This offers an opportunity to understand the activity of glycosyl hydrolases in a digestive system, independently of gut microbes.

Lignocellulose (plant) biomass is the most abundant raw material on earth and exploiting this energy source on a larger scale would offer an exceptional chance to produce biofuels (Himmel et al., 2007). The genetic tools available in *Parhyale* could make significant contributions to this research.

3.7 Ecotoxicology

Amphipods are used as indicator or test species for environmental toxicology in aquatic habitats (e.g., Kunz, Kienle, & Gerhardt, 2010; Poynton et al., 2018). *Parhyale* are emerging as an attractive ecotoxicity test organism for coastal marine ecosystems, due to their worldwide distribution and abundance in tropical coastal and estuarine habitats, and their robust year-round reproduction in their natural habitats and in the laboratory (Alegretti et al., 2016; Artal, Dos Santos, Henry, & Umbuzeiro, 2018). Recent studies have tested the effects of ammonia, metals, the pesticide diflubenzuron and silver nanoparticles in this system (Artal et al., 2018, 2020; Diehl et al., 2021).

4. Experimental tools, approaches and resources

4.1 Experimental embryology

Cell ablation experiments can be carried out following a variety of methods in *Parhyale*. Injection of FITC-dextran into blastomeres is not harmful as long as embryos remain in subdued light, but exposure to intense blue light will kill cells containing the FITC-dextran. This method has been used to kill individual cells at the 4- and 8-cell stage, or entire lineages at gastrulation or even at germband stages (Hannibal, Price, & Patel, 2012; Price et al., 2010). Cell ablation can also be accomplished by injection of RNAse and DNAse (Chaw & Patel, 2012). Manual ablations are also possible by poking a hole into the target cell and removing the cell content (Alwes, Hinchen, & Extavour, 2011). Finally, individual cells, or small groups of cells, can also be efficiently killed by laser ablation; this technique has been used to examine the role of the *Parhyale* midline in establishing dorsal-ventral patterning of the germband (Vargas-Vila, Hannibal, Parchem, Liu, & Patel, 2010) and to probe the role of nerves in limb regeneration (C. Sinigaglia and M. Averof, unpublished).

Individual blastomeres can be separated and maintained in short term culture to probe their developmental potential in isolation from their neighbors (Extavour, 2005).

Robust protocols have been established for antibody staining (Rehm, Hannibal, Chaw, Vargas-Vila, & Patel, 2009a) and in situ hybridization (Rehm, Hannibal, Chaw, Vargas-Vila, & Patel, 2009b) of *Parhyale* embryos. A number of antibodies are available, including some made against *Parhyale* protein sequences, as well as cross-reacting antibodies raised against conserved epitopes. Recently, in situ HCR has been used to probe multiple transcripts per embryo (Bruce & Patel, 2020). Using HCR it is possible to simultaneously image five different transcripts at once, as well as to achieve robust results at hatchling stages when the cuticle has formed (Bruce et al., 2021).

4.2 Loss-of function approaches

Three approaches have been described for knockdown/knockout of gene expression in *Parhyale*: RNAi, morpholino, and CRISPR-Cas9. For RNAi, injection of Stealth siRNAs into 1–2 cell embryos was used to knockdown expression of *Distal-less* and *Ultrabithorax* (Liubicich et al., 2009) and *single-minded* (Vargas-Vila et al., 2010). Stealth siRNAs were found to be more efficient than dsRNA, possibly due to their increased stability, allowing them to persist to later stages of development. As is typical with RNAi experiments, this approach produced a phenotypic series, which could help to understand gene function. It also appears that RNAi acts cell-autonomously, as injection into a single blastomere at the 2-cell stage could create half-mutant embryos (D. Liubicich and N. Patel, unpublished observations). Integration of a *Minos*-based heat-inducible transgene generating a hairpin RNA of *abd-A* was also successfully used for knockdown and provides an approach for temporal control of RNAi in *Parhyale* (Martin et al., 2016).

The injection of morpholinos has been used successfully to inhibit the translation of *vasa*, which resulted in the death of germ cells (Ozhan-Kizil et al., 2009).

More recently, CRISPR-based approaches have been used to knock out a number of genes (Bruce & Patel, 2020; Clark-Hachtel & Tomoyasu, 2020; Kao et al., 2016; Martin et al., 2016). As in other systems, injections of Cas9 protein along with guide RNA is more effective than injections of Cas9 mRNA with guide RNA (Martin et al., 2016), and chemically modified sgRNAs (Synthego) have proven particularly efficient (Bruce & Patel, 2020; E. Jarvis and N. Patel, unpublished observations). CRISPR knock-outs allow for rapid phenotypic analysis as the gene edits occur early and (when using the most effective gRNAs) with such efficiency that fully mutant G0 animals can be readily generated. The technique can also be applied to analyze lineage specific gene function through injection at the 8-cell stage (E. Jarvis and N. Patel, unpublished).

4.3 Transgenesis

Transgenesis is routinely carried out in *Parhyale* using the *Minos* vector (Kontarakis & Pavlopoulos, 2014; Pavlopoulos & Averof, 2005), a versatile transposon that has been used with success in a wide variety of organisms (including insects, chordates and fungi; see Pavlopoulos, Oehler, Kapetanaki, & Savakis, 2007, Evangelinos et al., 2015). The vector typically carries a marker and a cargo that can be many kb in length. *Minos* plasmid constructs are co-injected with transposase mRNA in early embryos; the transposon jumps randomly in the *Parhyale* genome, giving rise to mosaic individuals (G0), carrying *Minos* insertions in different parts of the body. Once the transposase activity dies out, the insertions are fixed and thereafter inherited as stable genetic elements (Pavlopoulos & Averof, 2005).

As with RNAi and CRISPR, the slow holoblastic cleavage and stereotypic early cell lineage of *Parhyale* offer some unique advantages, such as the ability to test transgene constructs in G0s and the opportunity to generate mosaics where only half of the embryo or a specific germ layer is genetically marked/modified (see Konstantinides & Averof, 2014; Pavlopoulos et al., 2009).

Typically, about a third of embryos injected at the 1-cell stage give rise to mosaic G0s. Only a fraction of *Minos* insertions are present in the germline and will be carried through to subsequent generations (G1, G2, etc.). Transgenic individuals are identified through the marker carried in the transposon. For a long time transgenesis was based on *3xP3*-driven fluorescent markers (Pavlopoulos & Averof, 2005), previously shown to be active in the eyes of several insects (Berghammer, Klingler, & Wimmer, 1999; Horn & Wimmer, 2000; Uhlirova, Asahina, Riddiford, & Jindra, 2002). In *Parhyale*, *3xP3* markers are not expressed in the eyes, but in a pair of spots at the posterior of the head (Pavlopoulos & Averof, 2005). In recent years, a *Parhyale* opsin promoter (*PhOpsin1*) has been adopted for marker expression, which drives robust expression in the eyes and is much easier to score (Ramos et al., 2019).

Transgenic *Parhyale* lines often carry tens of genomic insertions of *Minos* (Pavlopoulos & Averof, 2005) and are maintained as "mixed stocks," in which individuals carry different sets of insertions. This speeds up the process of establishing transgenic lines, but results in heterogeneous stocks in which the frequency of individual insertions can drift over time.

Alternative transgenic approaches are available in *Parhyale*, in which the transgenes are targeted to specific loci rather than randomly integrated in the genome: (1) using the *phiC31* integrase to target transgenes to specific landing sites inserted in the genome (Kontarakis et al., 2011), and (2) using CRISPR knock-ins (Kao et al., 2016; Serano et al., 2016). Both could be useful for establishing transgenic lines carrying single, mapped insertions, but neither is yet routinely established. The *phiC31* approach requires establishing *Parhyale* lines with well characterized landing sites (ongoing work in the Pavlopoulos lab).

4.4 Genetic markers and drivers

Transgenesis in *Parhyale* has so far been used to fluorescently mark cells for live imaging and to build genetic tools for manipulating gene function. Both applications require the identification of functional *cis*-regulatory elements (CREs) to drive the expression of marker and effector proteins. Two types of CREs have been tested in *Parhyale* using reporter constructs: artificial elements consisting of multimerized binding sites of specific transcription factors, known to work in other species, and candidate CREs from *Parhyale* genes.

Among artificial elements, the Pax6-responsive *3xP3* element combined with the *Drosophila hsp70* core promoter (Berghammer et al., 1999; Horn & Wimmer, 2000) drives weak expression in a few cells at the posterior of the brain (Pavlopoulos & Averof, 2005). This weak expression as well as lack of expression in the eyes may be due to the use of the heterologous *Drosophila hsp70* core promoter. A second element, *DC5*, consisting of Pax and Sox2 binding sites (Kamachi, Uchikawa, Tanouchi, Sekido, & Kondoh, 2001; Blanco, Girard, Kamachi, Kondoh, & Gehring, 2005; combined with the *PhHS* promoter region), drives strong expression in the central nervous system. In the strongest lines, *DC5* also marks the motor neurons of the limbs (Alwes et al., 2016; Konstantinides & Averof, 2014). Reporters with multimerised Pax3/7 binding sites (Relaix et al., 2003) have also been tested but found to be nonfunctional in *Parhyale* (M. Grillo and M. Averof, unpublished).

So far, four robust endogenous CREs have been identified in *Parhyale*, two from hsp70 family genes and two from opsins. The first to be identified, *PhMS* (*MS* for muscle-specific), is a fragment cloned from a *Parhyale hsp70c* (*hsp70*-related) gene, which drives robust expression in muscles (Pavlopoulos & Averof, 2005, sequence accession FR749990) (Fig. 4). It includes upstream regions that contain an array of putative bHLH binding sites, the promoter and a 5' UTR with an intron (see Fig. S1 in Kontarakis et al., 2011). PhMS has been used to visualize muscle fibres and satellite-like muscle precursors in the context of leg regeneration (Konstantinides & Averof, 2014). The second, *PhHS* (*HS* for heat shock), is a heat inducible CRE cloned from a *Parhyale hsp70* gene (Pavlopoulos et al., 2009, sequence accession FM991730). The fragment includes upstream sequences with putative HSF binding sites, the promoter and a 5' UTR with an intron (Fig. S1 in Kontarakis et al., 2011). *PhHS* is routinely used to drive

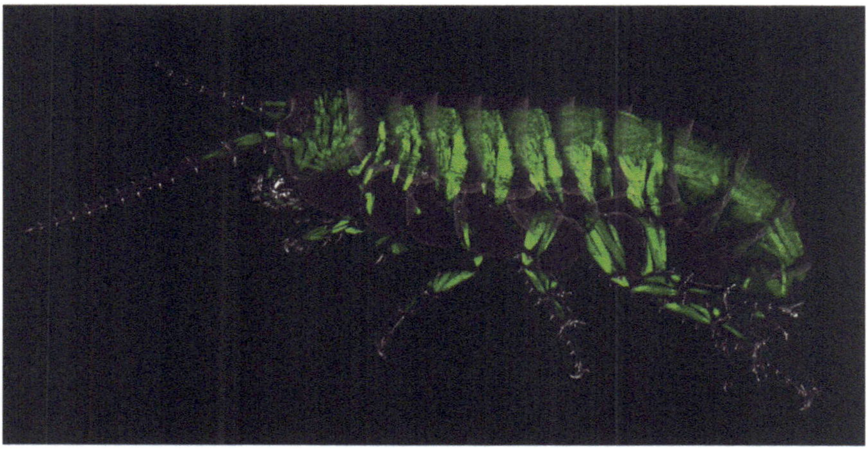

Fig. 4 Transgenic *Parhyale* juvenile expressing EGFP in muscles under the *PhMS* promoter. Confocal image of EGFP fluorescence (green) and cuticle autofluorescence (white).

ubiquitous and inducible gene expression (e.g., Alwes et al., 2016; Wolff et al., 2018). Finally, the *PhOpsin1* and *PhOpsin2* CREs, derived from the *Parhyale opsin 1* and *opsin 2* genes, drive expression in each of the two types of photoreceptor cells in the eyes of *Parhyale* (Ramos et al., 2019). *PhOpsin1* is now used to drive robust expression of transgenic markers, replacing the less reliable *3xP3* markers.

In addition to these robust CREs, numerous other gene fragments have been tested for CRE activity (Averof, Pavlopoulos and Patel labs, unpublished). These include gene fragments upstream of ubiquitously expressed housekeeping genes, developmental genes and viral promoters, in some cases selected after chromatin profiling via ATAC-seq (Sun & Patel, 2021). The failure to identify robust CREs in many cases could be due to poor genome annotation (e.g., mis-identification of promoter regions), the wide dispersal of CREs in *Parhyale*'s large genome, the size of the tested fragments (usually no larger than 5 kb), or a bad choice of core promoters.

Several core promoters have been tested in *Parhyale* (including the *Drosophila hsp70, Parhyale PhHS* and the artificial *SCP* core promoters), but none appear to work reliably with heterologous CREs (A. Pavlopoulos, A. Kiupakis, M. Grillo and M. Averof, unpublished).

An alternative approach for generating markers is generating enhancer or exon traps, in which the expression of a transgene is driven by endogenous *cis*-regulatory/transcriptional activity at the site of insertion. *Minos* constructs bearing a core promoter sequence upstream of a fluorescent marker have given putative enhancer traps (F. Alwes, M. Grillo, C. Cevrim and M. Averof, unpublished) and elements carrying a splice donor have been shown to mediate exon trapping in *Parhyale* (Kontarakis et al., 2011). Enhancer or exon traps can be repurposed into diverse genetic tools using the *phiC31* integrase (if they contain an *attP* site; Kontarakis et al., 2011) or by using CRISPR-mediated transgene replacement (Gilles, Schinko, & Averof, 2015).

The large average size of introns in *Parhyale* (Kao et al., 2016) may particularly facilitate exon trapping in this species. Exon trap screens have produced up to 2 trap lines per 100 injected embryos in *Parhyale* (Kontarakis et al., 2011, and F. Alwes unpublished). However it has been challenging to establish and maintain large numbers of exon trap lines.

A further alternative is to use CRISPR-mediated gene editing to knock-in marker or effector proteins in loci with the required pattern of gene expression (see Kao et al., 2016; Serano et al., 2016). The efficiency of CRISPR-mediated knock-ins can be very low (depending on target locus and gRNA) and careful consideration should be given to the knock-in strategy, to facilitate the selection of insertions and to avoid deleterious effects at the target locus.

4.5 Genetic tools based on transgenesis

Transgenesis provides opportunities to establish a wide range of genetic tools and approaches in *Parhyale*. First and foremost, the ability to express genes for gain-of-function studies. Since most genetic interventions are likely to have deleterious effects on survival, establishing conditional systems for gene expression has been a top priority. The *PhHS* regulatory element (described above) currently provides a robust tool for inducible gene expression in *Parhyale*, typically induced by a heat-shock of 30–60 min at 37 °C (Pavlopoulos et al., 2009). So far, besides the expression of fluorescent markers, *PhHS* has been used to mis-express different isoforms of the Hox protein Ubx, leading to homeotic transformations (Pavlopoulos et al., 2009). One flaw is that the *PhHS* element gives some variable/leaky expression in the absence of a heat shock, particularly in muscles.

Developing a binary system for mis-expressing genes, such as the *GAL4/UAS* system that has revolutionized studies in *Drosophila* (Brand & Perrimon, 1993), has been the focus of several unsuccessful efforts in *Parhyale* (A. Pavlopoulos, A. Kiupakis, V. Douris and M. Averof, unpublished; J. Serano and N. Patel, unpublished). Possible reasons for the lack of success so far may include the lack of an effective core promoter, or the variable/insufficient activity of GAL4 observed in some species (e.g., Asakawa et al., 2008; Schinko et al., 2010).

Some attempts have also been made to develop recombinase-mediated clonal approaches in *Parhyale* (N. Konstantinides, M. Grillo and M. Averof, unpublished). So far, only the *phiC31* integrase has been shown to work efficiently in *Parhyale* (Kontarakis et al., 2011) and a version of the *Raeppli* clonal marker which is based on this integrase (Kanca, Caussinus, Denes, Percival-Smith, & Affolter, 2013) has given promising early results (M. Grillo, P. Ramos and M. Averof, unpublished).

Various genetic approaches for inducible cell killing have been developed in different model organisms. One of the most versatile approaches combines transgene-mediated expression of the nitroreductase/NTR protein (providing spatial or cell-type specificity) with the application of the prodrug metronidazole to induce the ablation of NTR-expressing cells

(Curado, Stainier, & Anderson, 2008). Preliminary tests, using the *PhMS* and *DC5* regulatory elements to drive nitroreductase expression, suggest that this system provides an effective conditional approach for ablating specific cell types in *Parhyale* (I. Koltsaki and M. Averof, unpublished; C. Winchell and N. Patel, unpublished).

4.6 Live imaging and cell lineage analysis

The availability of fluorescent markers, the small size and the relative transparency *Parhyale* embryos, juveniles and adults, render *Parhyale* attractive for live imaging. The first live imaging and lineage analysis experiments in *Parhyale* were carried out by injection of fluorescently labelled high molecular weight dextrans or mRNA encoding fluorescent proteins between the 1- to 16-cell stage (Chaw & Patel, 2012; Gerberding et al., 2002; Hannibal, Price, Parchem, & Patel, 2012; Hannibal, Price, & Patel, 2012; Price et al., 2010; Price & Patel, 2007; Rehm, Hannibal, Chaw, Vargas-Vila, & Patel, 2009c). With these injection approaches, however, fluorescence signal fades with time due to a combination of dilution by cell division and degradation of the fluorescent reporter. With injections in individual blastomeres at the 8-cell stage, signal remains clearly visible until hatching for the g cell (germline) lineage, owing to the very few cell divisions in this lineage, while in the ectoderm or mesoderm reliable cell tracking is only possible until mid germband stages (approx. stage 19–21). Cell lineages up to the gastrulation stage have also been imaged and tracked in unlabelled embryos using Nomarski optics (Alwes et al., 2011).

More recently, live imaging experiments have used stable transgenic lines expressing histone-H2B-tagged fluorescent proteins to track cells during embryonic development and regeneration (Alwes et al., 2016; Sugawara et al., 2022; Wolff et al., 2018). The fusion with histone allows the nuclei of individual cells to be resolved and be tracked through mitoses. Imaging at 5–20 min time intervals is sufficient to detect mitoses and to track dividing cells during embryonic development and regeneration. Additionally, to visualize the cell cycle, a reporter of the S, G2 and M phases of the cell cycle has been developed by fusing EGFP with the N-terminus of *Parhyale* Geminin (Alwes et al., 2016).

Live imaging is still constrained by the limited number of markers available in *Parhyale*. In the absence of a characterized constitutive promoter, ubiquitous markers rely on the *PhHS* regulatory element, which requires repeated heat-shocks to sustain marker expression.

Continuous live imaging has been performed over long periods in *Parhyale* using fluorescence widefield microscopy (spanning embryonic development, Price & Patel, 2007, Hannibal, Price, & Patel, 2012), light-sheet microscopy (4–5 days on mid-late embryos, Wolff et al., 2018), and confocal microscopy (2–11 days on regenerating adult limbs, Alwes et al., 2016, Sugawara et al., 2022; C. Cevrim and M. Averof, unpublished). Over extended periods of imaging photodamage becomes a concern, especially when imaging on a confocal microscope. Steps to minimise light exposure include imaging strong fluorescent signals with low laser power, minimizing the number of imaged z planes and lengthening time lapse intervals, which constrain the spatial and temporal resolution of imaging. Under these conditions, *Parhyale* can be imaged over multiple days and give rise to fully developed embryos or regenerated limbs. Light-sheet microscopy reduces fluorophore bleaching and phototoxic effects to a large extent (Stelzer, 2014). *Parhyale* embryos can be easily mounted in agarose (Price & Patel, 2007; Wolff et al., 2018).

Cell tracking has been performed on time lapse recordings in early embryogenesis (Alwes et al., 2011; Chaw & Patel, 2012), segment formation (Hannibal, Price, & Patel, 2012; Price & Patel, 2007), limb development (Wolff et al., 2018) and adult limb regeneration (Alwes et al., 2016; Sugawara et al., 2022), in some cases yielding complete cell lineages that span several days of development and regeneration. Leading software for cell tracking—MaMuT/Mastodon (Wolff et al., 2018) and ELEPHANT (incorporating deep learning, Sugawara et al., 2022)—have been developed and tested on *Parhyale* light-sheet and confocal data.

Examples of live image recordings can be found at the following links, for ectoderm and mesoderm segmentation (https://ars.els-cdn.com/content/image/1-s2.0-S0012160611012772-mmc5.mov; Hannibal, Price, & Patel, 2012), limb development (https://elifesciences.org/articles/34410/figures#fig1video2; Wolff et al., 2018) and limb regeneration (https://zenodo.org/record/4557870; Sugawara et al., 2022).

4.7 Genetic crosses and forward genetics

Single-pair genetic crosses are performed routinely in *Parhyale* to establish transgenic lines and to combine transgenes. The females do not store sperm from previous matings, so there is no need to collect virgins. The lack of sperm transfer has been confirmed using fluorescent transgenic markers (G. Tsoumpekos and M. Averof, unpublished).

While establishing transgenic lines in *Parhyale*, we have in some cases identified heritable defects in embryonic patterning, eye morphology and eye color, likely to be caused by spontaneous mutations arising in our laboratory populations (P. Ramos and M. Averof, unpublished). These mutations were recessive and/or deleterious for survival, so we were not able to maintain them over long periods of time. The lack of easy solutions for maintaining stocks (such as cryopreservation), the slow reproductive rate (small brood size) and the long generation time of *Parhyale* make systematic forward genetics in this species a distant dream.

4.8 Genome sequence

The size of *Parhyale*'s genome is 3.6 Gb, similar to the size of the human genome and ~20 times larger than the genome of *Drosophila*.

The building of *Parhyale* genomic resources started in the early 2000s, at a time when sequencing gigabase-size genomes required huge investments. A BAC library of *Parhyale* genomic DNA, with ~5 × coverage of the genome, was generated, screened and partly sequenced (Parchem et al., 2010), allowing to identify important developmental genes (Hannibal, Price, Parchem, & Patel, 2012; Serano et al., 2016) and later to validate the full genome assembly (Kao et al., 2016).

The sequences of the *Parhyale* genome were obtained from a single male of the Chicago-F inbred line, using high throughput short read technology (Kao et al., 2016). The initial published assembly, named Phaw3.0, was one of the largest sequenced genomes at the time. It revealed high levels of heterozygosity, consistent with the high frequency of polymorphism found in lab populations. A large fraction of the genome was found to consist of repetitive sequences (57% of the assembled scaffolds), which at least partly explains the large size of the genome. High polymorphism and repeat content present serious challenges for genome assembly and caused this initial assembly to be somewhat fragmented (scaffold N50 of ~70 kb, scaffold L50 of ~14,000, https://www.ncbi.nlm.nih.gov/assembly/GCA_001587735.1/).

The genome assembly was subsequently updated by Dovetail Genomics using Chicago and Hi-C technologies (Burton et al., 2013; Putnam et al., 2016) and curated by D. Kao (sponsored by the Pavlopoulos, Patel, Averof, Extavour and Aboobaker teams). The current assembly, named Phaw_5.0, has much larger scaffolds (scaffold N50 of 20 Mb, scaffold L50 of 42), although contig length has only moderately increased (from N50 of 4 kb for Phaw3.0 to N50 of ~10 kb for Phaw_5.0). This genome assembly is publicly available at https://www.ncbi.nlm.nih.gov/assembly/GCA_001587735.2/.

Parhyale hawaiensis have 23 pairs of chromosomes (Kao et al., 2016). Although the latest genome assembly of *Parhyale* is of good quality, full contiguous chromosomal sequences are not available. With the advent of high throughput sequencing using long reads (tens or hundreds of kb long), chromosome-scale assemblies are quickly becoming the new standard, even for crustacean genomes (e.g., Baldwin-Brown, Weeks, & Long, 2018). It is likely that long reads will be necessary to cross through islands of repetitive sequences and to resolve haplotypes in the *Parhyale* genome, resulting in much longer contigs.

The closest relatives of *Parhyale* with assembled genomes are the amphipods *Hyalella azteca*, *Trinorchestia longiramus*, *Platorchestia hallaensis*, and *Orchestia grillus* (BioProject accessions PRJNA342675, PRJNA543945, PRJNA645242, and PRJNA557538 (Patra, Chung, Yoo, Baek, et al., 2020; Patra, Chung, Yoo, Kim, et al., 2020; Poynton et al., 2018). Of these, the best genome assembly is that of *Hyalella azteca*, which is significantly smaller (estimated genome size ~1 Gb, genome assembly 550 Mb) and more fragmented (N50 of ~200 kb for v2.0, Poynton et al., 2018) than *Parhyale*'s. It is estimated that these species have diverged from *Parhyale* for 40–90 million years (Copilas-Ciocianu, Borko, & Fiser, 2020), which means that non-functional sequences in these species are likely to have diverged extensively from those of *Parhyale*.

4.9 Gene identification and annotation

Before the genome sequencing, several transcriptomic datasets were produced in *Parhyale*, including EST data generated by Sanger sequencing (https://genome.jgi.doe.gov/portal/parha/parha.info.html) and de novo transcriptome assemblies generated by using early NGS sequencing technologies (454 pyrosequencing and SOLiD) on mRNA and small RNAs from maternal and embryonic tissues (Blythe et al., 2012; Zeng et al., 2011). DNA microarrays (Nestorov, Battke, Levesque, & Gerberding, 2013) and proteomic data (Trapp et al., 2016) were also produced to serve specific projects. Later, additional de novo transcriptomes were generated by Illumina sequencing to investigate embryonic development, limb regeneration, ecotoxicology and circadian rhythms in *Parhyale* (Artal et al., 2018; Calvo, Birgaoanu, Pettini, Ronshaugen, & Griffiths-Jones, 2022; Hunt et al., 2019; Sinigaglia et al., 2021; Sun & Patel, 2021).

When the *Parhyale* genome was sequenced, genes were annotated based on ab initio gene prediction (ORFs, splicing signals, etc.) and transcriptome data from diverse embryonic stages and adult regenerating limbs (Kao et al., 2016). In the Phaw3.0 assembly, many genes were split in unconnected contigs and the corresponding gene predictions were incomplete. In the Phaw_5.0 assembly, gene models were significantly improved and most manually curated genes in this dataset were found to be of good quality. However this assembly still contains errors, such as incorrect ordering or orientation of contigs within a scaffold (a known problem of Hi-C-scaffolded genomes) and tandem repetitions due to unresolved polymorphic repeats (M. Paris, unpublished observations). These problems have not been quantified genome-wide.

EST data and transcriptomes helped to reveal that a large fraction of mRNAs in *Parhyale* are subject to trans-splicing (Douris, Telford, & Averof, 2010), whereby a shared leader sequence is spliced to the 5′ end of diverse mRNAs (including mRNAs encoding the transcription factors Ultrabithorax, Twist and Mef2). The leader sequence comes from snRNAs that resemble spliceosomal snRNAs. Overall, we estimate that 10–20% of *Parhyale* mRNAs are trans-spliced (Douris et al., 2010, and M. Averof, unpublished observations).

Additional improvements in these gene models have been made by mapping a large number of new reads, obtained since the original gene models were assembled, and by long-read sequencing of cDNAs using Oxford Nanopore technology (Sinigaglia et al., 2021; Sun & Patel, 2021). Direct long-read sequencing of RNA would also allow the detection of modified bases in the RNA (Garalde et al., 2018).

Functional annotation of the *Parhyale* gene models is still limited. Gene orthologues of major classes of signaling molecules, transcription factors, non-coding RNAs and effectors of innate immunity were identified in the initial genome assembly (Kao et al., 2016). SID-1, implicated in systemic RNAi (Winston, Molodowitch, & Hunter, 2002; Maruekawong, Tirasophon, Panyim, & Attasart, 2018), was not found, which is consistent with the cell autonomous effects of RNAi described above.

Putative orthologies between *Parhyale*, *Drosophila* and human genes have been inferred on a genome-wide scale by reciprocal BLAST analysis (Sinigaglia et al., 2021) and by tree-based orthology reconstruction between *Parhyale* and 66 other metazoans (J. Huerta-Cepas and colleagues, unpublished data).

4.10 Expression and chromatin profiling

Genome-wide expression profiling by RNAseq has been applied in a number of projects to identify transcriptional dynamics and responses to treatments in *Parhyale*, including the timecourses of embryonic development (D. Sun,

H. Bruce and N. Patel, unpublished) and limb development and regeneration (Sinigaglia et al., 2021), the study of circadian rhythms (Hunt et al., 2019), and responses to environmental toxicants (Artal et al., 2020). Single-cell or single-nucleus RNAseq has been applied to embryos (L. Blondel and C. Extavour, unpublished) and adult limbs (Almazàn, Çevrim, Musser, Averof, & Paris, 2021). The large average intron size in *Parhyale* (median of >7 kb in *Parhyale* versus 1 kb in *Drosophila* and 18 kb in humans) favors capturing both spliced and unspliced/intronic reads, which can help to probe transcriptional dynamics at the single-cell level (La Manno et al., 2018) and to work on dissociated nuclei instead of whole cells.

Besides single-cell sequencing, cell- or tissue-specific transcriptional profiling can be carried out with the thiouracil/TU tagging approach, whereby transcripts expressed in specific cells over a chosen period of time can be tagged, isolated and sequenced (Gay, Karfilis, Miller, Doe, & Stankunas, 2014; Miller, Robinson, Cleary, & Doe, 2009). The approach is based on transgene-mediated expression of uracil-phosphoribosyltransferase/UPRT (providing spatial or cell type specificity) and the application of the nucleobase analogue 4-thiouracil (providing temporal control), which is converted by UPRT into a ribonucleotide that can be incorporated in RNA. Tests of the TU-tagging approach in *Parhyale*, using mosaic embryos that express UPRT in the mesoderm or the ectoderm, yielded 40- to 80-fold enrichments of mesodermal or ectodermal transcripts, respectively (N. Konstantinides and M. Averof, unpublished).

The study of chromatin states and epigenetic modifications is only just beginning in *Parhyale*. Bisulfite sequencing has revealed widespread CpG dinucleotide methylation in the *Parhyale* genome (Kao et al., 2016). Bulk and single-cell ATAC-seq data have been generated, revealing genome-wide patterns of chromatin accessibility in embryonic and adult tissues, providing a platform for the identification of novel *cis*-regulatory elements and many additional assays in the future (Sun & Patel, 2021; E. Skafida and M. Paris, unpublished).

5. Research community

The *Parhyale* research community is small, consisting of no more than 20–30 researchers at any time. Informal meetings, gathering most of the community, have been organised in Tübingen (2005), Lisbon (2012) and Janelia Research Campus (2017). The community welcomes new researchers, sharing genomic resources, transgenic lines and technical expertise.

Acknowledgments/Funding

We thank the *Parhyale* research community for sharing information and resources, and for allowing us to refer to unpublished information cited in this article. We acknowledge Vincent Moncorgé for the image shown in Fig. 1. *Parhyale* research in the Averof lab is supported by the European Research Council of the E.U., grant ERC-2015-AdG #694918. *Parhyale* research in the Patel lab has been supported by HHMI and NSF grant IOS-1257379.

References

Alegretti, L., Umbuzeiro, G.d. A., & Flynn, M. N. (2016). Population dynamics of *Parhyale hawaiensis* (Dana, 1853) (Amphipoda: Hyalidae) associated with an intertidal algal belt in southeastern Brazil. *Journal of Crustacean Biology, 36*, 785–791.

Almazàn, A., Çevrim, Ç., Musser, J. M., Averof, M., & Paris, M. (2021). Regenerated crustacean limbs are precise replicas. *bioRxiv*, 2021.12.13.472338.

Alwes, F., Enjolras, C., & Averof, M. (2016). Live imaging reveals the progenitors and cell dynamics of limb regeneration. *eLife, 5*, e19766.

Alwes, F., Hinchen, B., & Extavour, C. G. (2011). Patterns of cell lineage, movement, and migration from germ layer specification to gastrulation in the amphipod crustacean *Parhyale hawaiensis*. *Developmental Biology, 359*, 110–123.

Artal, M. C., Dos Santos, A., Henry, T. B., & Umbuzeiro, G. A. (2018). Development of an acute toxicity test with the tropical marine amphipod *Parhyale hawaiensis*. *Ecotoxicology, 27*, 103–108.

Artal, M. C., Pereira, K. D., Luchessi, A. D., Okura, V. K., Henry, T. B., Marques-Souza, H., et al. (2020). Transcriptome analysis in *Parhyale hawaiensis* reveal sex-specific responses to AgNP and AgCl exposure. *Environmental Pollution, 260*, 113963.

Asakawa, K., Suster, M. L., Mizusawa, K., Nagayoshi, S., Kotani, T., Urasaki, A., et al. (2008). Genetic dissection of neural circuits by Tol2 transposon-mediated Gal4 gene and enhancer trapping in zebrafish. *Proceedings of the National Academy of Sciences of the United States of America, 105*, 1255–1260.

Averof, M., & Cohen, S. M. (1997). Evolutionary origin of insect wings from ancestral gills. *Nature, 385*, 627–630.

Averof, M., & Patel, N. H. (1997). Crustacean appendage evolution associated with changes in hox gene expression. *Nature, 388*, 682–686.

Baldwin-Brown, J. G., Weeks, S. C., & Long, A. D. (2018). A new standard for crustacean genomes: The highly contiguous, annotated genome assembly of the clam shrimp *Eulimnadia texana* reveals HOX gene order and identifies the sex chromosome. *Genome Biology and Evolution, 10*, 143–156.

Barnard, J. L. (1965). Marine Amphipoda of atolls in Micronesia. *Proceedings of the United States National Museum, 117*, 459–552.

Berghammer, A. J., Klingler, M., & Wimmer, E. A. (1999). A universal marker for transgenic insects. *Nature, 402*, 370–371.

Blanco, J., Girard, F., Kamachi, Y., Kondoh, H., & Gehring, W. J. (2005). Functional analysis of the chicken delta1-crystallin enhancer activity in *Drosophila* reveals remarkable evolutionary conservation between chicken and fly. *Development, 132*, 1895–1905.

Blythe, M. J., Malla, S., Everall, R., Shih, Y.-H., Lemay, V., Moreton, J., et al. (2012). High throughput sequencing of the *Parhyale hawaiensis* mRNAs and microRNAs to aid comparative developmental studies. *PLoS One*, 7, e33784.

Borowsky, B., & Borowsky, R. (1987). The reproductive behaviors of the amphipod crustacean *Gammarus palustris* (Bousfield) and some insights into the nature of their stimuli. *Journal of Experimental Marine Biology and Ecology, 107*, 131–144.

Brand, A. H., & Perrimon, N. (1993). Targeted gene expression as a means of altering cell fates and generating dominant phenotypes. *Development*, *118*, 401–415.

Browne, W. E., Price, A. L., Gerberding, M., & Patel, N. H. (2005). Stages of embryonic development in the amphipod crustacean, *Parhyale hawaiensis*. *Genesis*, *42*, 124–149.

Bruce, H. S., Jerz, G., Kelly, S. R., McCarthy, J., Pomeranz, A., Senevirathne, G., & Patel, N. H. (2021). Hybridization chain reaction (HCR) in situ protocol. *protocols.io*. https://doi.org/10.17504/protocols.io.bunznvf6.

Bruce, H. S., & Patel, N. H. (2020). Knockout of crustacean leg patterning genes suggests that insect wings and body walls evolved from ancient leg segments. *Nature Ecology & Evolution*, *4*, 1703–1712.

Burton, J. N., Adey, A., Patwardhan, R. P., Qiu, R., Kitzman, J. O., & Shendure, J. (2013). Chromosome-scale scaffolding of de novo genome assemblies based on chromatin interactions. *Nature Biotechnology*, *31*, 1119–1125.

Calvo, L., Birgaoanu, M., Pettini, T., Ronshaugen, M., & Griffiths-Jones, S. (2022). The embryonic transcriptome of *Parhyale hawaiensis* reveals different dynamics of microRNAs and mRNAs during the maternal-zygotic transition. *Scientific Reports*, *12*, 174.

Chaw, R. C., & Patel, N. H. (2012). Independent migration of cell populations in the early gastrulation of the amphipod crustacean *Parhyale hawaiensis*. *Developmental Biology*, *371*, 94–109.

Ciofini, A., Yamahama, Y., Mercatelli, L., Hariyama, T., & Ugolini, A. (2020). Specializations in the compound eye of Talitrus saltator (Crustacea, Amphipoda). *Journal of Comparative Physiology. A*, *206*, 711–723.

Clark-Hachtel, C. M., & Tomoyasu, Y. (2020). Two sets of candidate crustacean wing homologues and their implication for the origin of insect wings. *Nature Ecology & Evolution*, *13*, 1–9.

Cohen, J. H., Cronin, T. W., Lessios, N., & Forward, R. B. (2010). Visual physiology underlying orientation and diel behavior in the sand beach amphipod Talorchestia longicornis. *The Journal of Experimental Biology*, *213*, 3843–3851.

Cohen, J. H., & Putts, M. R. (2013). Polarotaxis and scototaxis in the supratidal amphipod Platorchestia platensis. *Journal of Comparative Physiology. A*, *199*, 669–680.

Copilas-Ciocianu, D., Borko, S., & Fiser, C. (2020). The late blooming amphipods: Global change promoted post-Jurassic ecological radiation despite Palaeozoic origin. *Molecular Phylogenetics and Evolution*, *143*, 106664.

Cragg, S. M., Beckham, G. T., Bruce, N. C., Bugg, T. D., Distel, D. L., Dupree, P., et al. (2015). Lignocellulose degradation mechanisms across the tree of life. *Current Opinion in Chemical Biology*, *29*, 108–119.

Curado, S., Stainier, D. Y., & Anderson, R. M. (2008). Nitroreductase-mediated cell/tissue ablation in zebrafish: A spatially and temporally controlled ablation method with applications in developmental and regeneration studies. *Nature Protocols*, *3*, 948–954.

Dana, J. D. (1853). Crustacea. *Part II United States Exploring Expedition*, *14*, 900–901.

Diehl, O. J., Assano, P. K.d., Costa, T. R. G., Oliveira, R., Marques-Souza, H., & Umbuzeiro, G. A. (2021). Antenna regeneration as an ecotoxicological endpoint in a marine amphipod: A proof of concept using dimethyl sulfoxide and diflubenzuron. *Ecotoxicology*, *30*, 751–755.

Divakaran, O. (1976). *Studies on the biology of Parhyale hawaiensis (Crustacea: Amphipoda)*. PhD thesis (pp. 1–371). India: University of Kerala.

Divakaran, O. (1982). Nervous system of *Parhyale hawaiensis* Dana (Crustacea: Amphipoda). *Proceedings of the National Academy of Sciences India Section B - Biological Sciences*, *2*, 218–224.

Dohle, W., & Scholtz, G. (1988). Clonal analysis of the crustacean segment: The discordance between genealogical and segmental borders. *Development*, *104*(suppl), 147–160.

Douris, V., Telford, M. J., & Averof, M. (2010). Evidence for multiple independent origins of trans-splicing in Metazoa. *Molecular Biology and Evolution*, *27*, 684–693.

Ercolini, A., & Scapini, F. (1976). Sensitivity and response to light in the laboratory of the littoral amphipod Talitrus saltator Montagu. *The Italian Journal of Zoology*, *10*, 293–309.

Evangelinos, M., Anagnostopoulos, G., Karvela-Kalogeraki, I., Stathopoulou, P. M., Scazzocchio, C., & Diallinas, G. (2015). *Minos* as a novel Tc1/mariner-type transposable element for functional genomic analysis in *aspergillus nidulans*. *Fungal Genetics and Biology*, *81*, 1–11.

Extavour, C. G. (2005). The fate of isolated blastomeres with respect to germ cell formation in the amphipod crustacean *Parhyale hawaiensis*. *Developmental Biology*, *277*, 387–402.

Forward, R. B., Jr., Bourla, M. H., Lessios, N. N., & Cohen, J. H. (2009). Orientation to shorelines by the supratidal amphipod Talorchestia longicornis: Wavelength specific behavior during sun compass orientation. *Journal of Experimental Marine Biology and Ecology*, *376*, 102–109.

Fredericq, L. (1882). Amputation des pattes par mouvement réflexe chez le crabe. *Archives de Biologie III*, 235–240.

Frelon-Raimond, M., Meyer-Rochow, V. B., Ugolini, A., & Martin, G. (2002). Intracerebral ocelli in an amphipod: Extraretinal photoreceptors of the sandhopper *Talitrus saltator* (Crustacea, Amphipoda). *Invertebrate Biology*, *121*, 73–78.

Garalde, D. R., Snell, E. A., Jachimowicz, D., Sipos, B., Lloyd, J. H., Bruce, M., et al. (2018). Highly parallel direct RNA sequencing on an array of nanopores. *Nature Methods*, *10*, 57.

Gay, L., Karfilis, K. V., Miller, M. R., Doe, C. Q., & Stankunas, K. (2014). Applying thiouracil tagging to mouse transcriptome analysis. *Nature Protocols*, *9*, 410–420.

Gerberding, M., Browne, W. E., & Patel, N. H. (2002). Cell lineage analysis of the amphipod crustacean *Parhyale hawaiensis* reveals an early restriction of cell fates. *Development*, *129*, 5789–5801.

Gilles, A. F., Schinko, J. B., & Averof, M. (2015). Efficient CRISPR-mediated gene targeting and transgene replacement in the beetle *Tribolium castaneum*. *Development*, *142*, 2832–2839.

Gupta, T., & Extavour, C. G. (2013). Identification of a putative germ plasm in the amphipod *Parhyale hawaiensis*. *EvoDevo*, *4*, 34.

Hallberg, E., Johansson, K. U. I., & Wallén, R. (1997). Olfactory sensilla in crustaceans: Morphology, sexual dimorphism, and distribution patterns. *International Journal of Insect Morphology and Embryology*, *26*, 173–180.

Hannibal, R. L., Price, A. L., Parchem, R. J., & Patel, N. H. (2012). Analysis of *snail* genes in the crustacean *Parhyale hawaiensis*: Insight into *snail* gene family evolution. *Development Genes and Evolution*, *222*, 139–151.

Hannibal, R. L., Price, A. L., & Patel, N. H. (2012). The functional relationship between ectodermal and mesodermal segmentation in the crustacean, *Parhyale hawaiensis*. *Developmental Biology*, *361*, 427–438.

Hanström, B. (1932). Neue Untersuchungen über Sinnesorgane und Nervensysteme der Crustceen. II. *Zoologische Jahrbücher. Abteilung für Anatomie und Ontogenie der Tiere*, *56*, 387–520.

Himmel, M. E., Ding, S. Y., Johnson, D. K., Adney, W. S., Nimlos, M. R., Brady, J. W., et al. (2007). Biomass recalcitrance: Engineering plants and enzymes for biofuels production. *Science*, *315*, 804–807.

Holmes, S. J. (1903). Sex recognition among amphipods. *The Biological Bulletin*, *5*, 288–292.

Horn, C., & Wimmer, E. A. (2000). A versatile vector set for animal transgenesis. *Development Genes and Evolution*, *210*, 630–637.

Hunt, B. J. (2016). *Advancing molecular crustacean chronobiology through the characterisation of the circadian clock in two malacostracan species, Euphausia superba and Parhyale hawaiensis*. PhD thesis (pp. 1–240). U.K: University of Leicester.

Hunt, B. J., Mallon, E. B., & Rosato, E. (2019). In silico identification of a molecular circadian system with novel features in the crustacean model organism *Parhyale hawaiensis*. *Frontiers in Physiology*, *10*, 791.
Kamachi, Y., Uchikawa, M., Tanouchi, A., Sekido, R., & Kondoh, H. (2001). Pax6 and SOX2 form a co-DNA-binding partner complex that regulates initiation of lens development. *Genes and Development*, *15*, 1272–1286.
Kanca, O., Caussinus, E., Denes, A. S., Percival-Smith, A., & Affolter, M. (2013). Raeppli: A whole-tissue labeling tool for live imaging of *Drosophila* development. *Development*, *141*, 472–480.
Kao, D., Lai, A. G., Stamataki, E., Rosic, S., Konstantinides, N., Jarvis, E., et al. (2016). The genome of the crustacean *Parhyale hawaiensis*, a model for animal development, regeneration, immunity and lignocellulose digestion. *eLife*, *5*, e20062.
Kapli, P., & Telford, M. J. (2020). Topology-dependent asymmetry in systematic errors affects phylogenetic placement of Ctenophora and Xenacoelomorpha. *Science Advances*, *6*. eabc5162.
Kenning, M., Müller, C., Wirkner, C. S., & Harzsch, S. (2013). The Malacostraca (Crustacea) from a neurophylogenetic perspective: New insights from brain architecture in *Nebalia herbstii* leach, 1814 (Leptostraca, Phyllocarida). *Zoologischer Anzeiger - A Journal of Comparative Zoology*, *252*, 319–336.
Keskinen, E., Takaku, Y., Meyer-Rochow, V. B., & Hariyama, T. (2002). Postembryonic eye growth in the seashore isopod *Ligia exotica* (Crustacea, Isopoda). *The Biological Bulletin*, *202*, 223–231.
Konstantinides, N., & Averof, M. (2014). A common cellular basis for muscle regeneration in arthropods and vertebrates. *Science*, *343*, 788–791.
Kontarakis, Z., & Pavlopoulos, A. (2014). Transgenesis in non-model organisms: The case of *Parhyale*. *Methods in Molecular Biology*, *1196*, 145–181.
Kontarakis, Z., Pavlopoulos, A., Kiupakis, A., Konstantinides, N., Douris, V., & Averof, M. (2011). A versatile strategy for gene trapping and trap conversion in emerging model organisms. *Development*, *138*, 2625–2630.
Kunz, P. Y., Kienle, C., & Gerhardt, A. (2010). *Gammarus* spp. in aquatic ecotoxicology and water quality assessment: Toward integrated multilevel tests. *Reviews of Environmental Contamination and Toxicology*, *205*, 1–76.
La Manno, G., Soldatov, R., Zeisel, A., Braun, E., Hochgerner, H., Petukhov, V., et al. (2018). RNA velocity of single cells. *Nature*, *560*, 494–498.
Land, M. F. (1997). Visual acuity in insects. *Annual Review of Entomology*, *42*, 147–177.
Liubicich, D. M., Serano, J. M., Pavlopoulos, A., Kontarakis, Z., Protas, M. E., Kwan, E., et al. (2009). Knockdown of *Parhyale* ultrabithorax recapitulates evolutionary changes in crustacean appendage morphology. *Proceedings of the National Academy of Sciences of the United States of America*, *106*, 13892–13896.
Lowry, J. K. (1986). The callynophore, a eucaridan/peracaridan sensory organ prevalent among the Amphipoda (Crustacea). *Zoologica Scripta*, *15*, 333–349.
Martin, A., Serano, J. M., Jarvis, E., Bruce, H. S., Wang, J., Ray, S., et al. (2016). CRISPR/Cas9 mutagenesis reveals versatile roles of hox genes in crustacean limb specification and evolution. *Current Biology*, *26*, 14–26.
Maruekawong, K., Tirasophon, W., Panyim, S., & Attasart, P. (2018). Involvement of LvSID-1 in dsRNA uptake in *Litopenaeus vannamei*. *Aquaculture*, *482*, 65–72.
Miller, M. R., Robinson, K. J., Cleary, M. D., & Doe, C. Q. (2009). TU-tagging: Cell type-specific RNA isolation from intact complex tissues. *Nature Methods*, *6*, 439–441.
Myers, A. (1985). Shallow-water, coral reef and mangrove Amphipoda (Gammaridea) of Fiji. *Records of the Australian Museum*, *5*, 1–143.
Nestorov, P., Battke, F., Levesque, M. P., & Gerberding, M. (2013). The maternal transcriptome of the crustacean *Parhyale hawaiensis* is inherited asymmetrically to invariant cell lineages of the ectoderm and mesoderm. *PLoS One*, *8*, e56049.

Ozhan-Kizil, G., Havemann, J., & Gerberding, M. (2009). Germ cells in the crustacean *Parhyale hawaiensis* depend on vasa protein for their maintenance but not for their formation. *Developmental Biology*, *327*, 230–239.

Parchem, R. J., Poulin, F., Stuart, A. B., Amemiya, C. T., & Patel, N. H. (2010). BAC library for the amphipod crustacean *Parhyale hawaiensis*. *Genomics*, *95*, 261–267.

Patra, A. K., Chung, O., Yoo, J. Y., Baek, S. H., Jung, T. W., Kim, M. S., & Choi, J.-H. (2020). The draft genome sequence of a new land-hopper *Platorchestia hallaensis*. *Frontiers in Genetics*, *11*, 621301.

Patra, A. K., Chung, O., Yoo, J. Y., Kim, M. S., Yoon, M. G., Choi, J.-H., & Yang, Y. (2020). First draft genome for the sand-hopper *Trinorchestia longiramus*. *Scientific Data*, *7*, 85.

Pavlopoulos, A., & Averof, M. (2005). Establishing genetic transformation for comparative developmental studies in the crustacean *Parhyale hawaiensis*. *Proceedings of the National Academy of Sciences of the United States of America*, *102*, 7888–7893.

Pavlopoulos, A., Kontarakis, Z., Liubicich, D., Serano, J., Akam, M., Patel, N. H., et al. (2009). Probing the evolution of appendage specialization by hox gene mis-expression in an emerging model crustacean. *Proceedings of the National Academy of Sciences of the United States of America*, *106*, 13897–13902.

Pavlopoulos, A., Oehler, S., Kapetanaki, M. G., & Savakis, C. (2007). The DNA transposon *Minos* as a tool for transgenesis and functional genomic analysis in vertebrates and invertebrates. *Genome Biology*, *8*(Suppl 1), S2.

Pavlopoulos, A., & Wolff, C. (2020). Chapter 2. Crustacean limb morphogenesis during normal development and regeneration. In M. Thiel, K. Anger, & S. Harzsch (Eds.), *The natural history of the Crustacea* (pp. 45–78). Volume VII: Developmental Biology and Larval Ecology. Oxford University Press.

Poovachiranon, S., Boto, K., & Duke, N. (1986). Food preference studies and ingestion rate measurements of the mangrove amphipod *Parhyale hawaiensis* (Dana). *Journal of Experimental Marine Biology and Ecology*, *98*, 129–140.

Poynton, H. C., Hasenbein, S., Benoit, J. B., Sepulveda, M. S., Poelchau, M. F., Hughes, D. S. T., et al. (2018). The toxicogenome of *Hyalella azteca*: A model for sediment ecotoxicology and evolutionary toxicology. *Environmental Science & Technology*, *52*, 6009–6022.

Price, A. L., Modrell, M. S., Hannibal, R. L., & Patel, N. H. (2010). Mesoderm and ectoderm lineages in the crustacean *Parhyale hawaiensis* display intra-germ layer compensation. *Developmental Biology*, *341*, 256–266.

Price, A. L., & Patel, N. H. (2007). Investigating divergent mechanisms of mesoderm development in arthropods: The expression of *Ph-twist* and *Ph-mef2* in *Parhyale hawaiensis*. *The Journal of Experimental Zoology*, *310B*, 24–40.

Putnam, N. H., O'Connell, B. L., Stites, J. C., Rice, B. J., Blanchette, M., Calef, R., et al. (2016). Chromosome-scale shotgun assembly using an in vitro method for long-range linkage. *Genome Research*, *26*, 342–350.

Ramm, T., & Scholtz, G. (2017). No sight, no smell? - brain anatomy of two amphipod crustaceans with different lifestyles. *Arthropod Structure & Development*, *46*, 537–551.

Ramos, A. P., Gustafsson, O., Labert, N., Salecker, I., Nilsson, D. E., & Averof, M. (2019). Analysis of the genetically tractable crustacean *Parhyale hawaiensis* reveals the organisation of a sensory system for low-resolution vision. *BMC Biology*, *17*, 67.

Rehm, E. J., Hannibal, R. L., Chaw, R. C., Vargas-Vila, M. A., & Patel, N. H. (2009a). Antibody staining of *Parhyale hawaiensis* embryos. *Cold Spring Harbor Protocols*, *2009*(1). pdb.prot5129.

Rehm, E. J., Hannibal, R. L., Chaw, R. C., Vargas-Vila, M. A., & Patel, N. H. (2009b). In situ hybridization of labeled RNA probes to fixed *Parhyale hawaiensis* embryos. *Cold Spring Harbor Protocols*, *2009*(1). pdb.prot5130.

Rehm, E. J., Hannibal, R. L., Chaw, R. C., Vargas-Vila, M. A., & Patel, N. H. (2009c). Injection of *Parhyale hawaiensis* blastomeres with fluorescently labeled tracers. *Cold Spring Harbor Protocols, 2009*(1). pdb.prot5128.

Relaix, F., Polimeni, M., Rocancourt, D., Ponzetto, C., Schafer, B. W., & Buckingham, M. (2003). The transcriptional activator PAX3-FKHR rescues the defects of Pax3 mutant mice but induces a myogenic gain-of-function phenotype with ligand-independent activation of met signaling in vivo. *Genes and Development, 17*, 2950–2965.

Schinko, J. B., Weber, M., Viktorinova, I., Kiupakis, A., Averof, M., Klingler, M., et al. (2010). Functionality of the GAL4/UAS system in Tribolium requires the use of endogenous core promoters. *BMC Developmental Biology, 10*, 53.

Schwentner, M., Combosch, D. J., Pakes Nelson, J., & Giribet, G. (2017). A phylogenomic solution to the origin of insects by resolving crustacean-hexapod relationships. *Current Biology, 27*, 1818–1824.

Serano, J. M., Martin, A., Liubicich, D. M., Jarvis, E., Bruce, H. S., La, K., et al. (2016). Comprehensive analysis of hox gene expression in the amphipod crustacean *Parhyale hawaiensis*. *Developmental Biology, 409*, 297–309.

Shiga, Y., Kato, Y., Aragane-Nomura, Y., Haraguchi, T., Saridaki, T., Watanabe, H., et al. (2017). Repeated co-option of a conserved gene regulatory module underpins the evolution of the crustacean carapace, insect wings and other flat outgrowths. *BioRxiv*. https://doi.org/10.1101/160010.

Shoemaker, C. R. (1956). Observations on the amphipod genus *Parhyale*. *Proceedings of the United States National Museum, 106*, 345–358.

Sinigaglia, C., Almazàn, A., Sémon, M., Gillet, B., Hughes, S., Edsinger, E., & Paris, M. (2021). Distinct gene expression dynamics in developing and regenerating limbs. *bioRxiv*, 2021.06.14.448408.

Skinner, D. M. (1985). Molting and regeneration. In D. E. Bliss, & L. H. Mantel (Eds.), *The biology of Crustacea: Integument, pigments, and hormonal processes* (pp. 43–146). Academic Press.

Stelzer, E. H. K. (2014). Light-sheet fluorescence microscopy for quantitative biology. *Nature Methods, 12*, 23–26.

Sugawara, K., Çevrim, Ç., & Averof, M. (2022). Tracking cell lineages in 3D by incremental deep learning. *eLife, 11*, e69380.

Sun, D. A., & Patel, N. H. (2019). The amphipod crustacean *Parhyale hawaiensis*: An emerging comparative model of arthropod development, evolution, and regeneration. *WIREs Developmental Biology, 8*, e355.

Sun, D. A, & Patel, N. H. (2021). Identification and classification of *cis*-regulatory elements in the amphipod crustacean *Parhyale hawaiensis*. *bioRxiv*, 2021.09.16.460328.

Telford, M. J., Budd, G. E., & Philippe, H. (2015). Phylogenomic insights into animal evolution. *Current Biology, 25*, R876–R887.

Trapp, J., Almunia, C., Gaillard, J.-C., Pible, O., Chaumot, A., Geffard, O., et al. (2016). Proteogenomic insights into the core-proteome of female reproductive tissues from crustacean amphipods. *Journal of Proteomics, 135*, 51–61.

Ugolini, A. (2014). Optic flow and sea-land orientation in the sandhopper Talitrus saltator. *The Journal of Experimental Biology, 217*, 2041–2043.

Ugolini, A., Galanti, G., & Mercatelli, L. (2012). The skylight gradient of luminance helps sandhoppers in sun and moon identification. *Journal of Experimental Biology, 215*, 2814–2819.

Uhlirova, M., Asahina, M., Riddiford, L., & Jindra, M. (2002). Heat-inducible transgenic expression in the silkmoth *Bombyx mori*. *Development Genes and Evolution, 212*, 145–151.

Vargas-Abúndez, J. A., López-Vázquez, H. I., Mascaró, M., Martínez-Moreno, G. L., & Simões, N. (2021). Marine amphipods as a new live prey for ornamental aquaculture: Exploring the potential of *Parhyale hawaiensis* and *Elasmopus pectenicrus*. *PeerJ, 9*, e10840.

Vargas-Vila, M. A., Hannibal, R. L., Parchem, R. J., Liu, P. Z., & Patel, N. H. (2010). A prominent requirement for single-minded and the ventral midline in patterning the dorsoventral axis of the crustacean *Parhyale hawaiensis*. *Development*, *137*, 3469–3476.

Watanabe, H., & Tokuda, G. (2010). Cellulolytic systems in insects. *Annual Review of Entomology*, *55*, 609–632.

Winston, W. M., Molodowitch, C., & Hunter, C. P. (2002). Systemic RNAi in C. elegans requires the putative transmembrane protein SID-1. *Science*, *295*, 2456–2459.

Wittfoth, C., Harzsch, S., Wolff, C., & Sombke, A. (2019). The "amphi"-brains of amphipods: New insights from the neuroanatomy of *Parhyale hawaiensis* (Dana, 1853). *Frontiers in Zoology*, *16*, 30.

Wolff, C., & Gerberding, M. (2015). Crustacea: Comparative aspects of early development. In A. Wanninger (Ed.), *Vol. 4 Ecdysozoa II: Crustacea. Evolutionary developmental biology of invertebrates* (pp. 39–46). Springer Press.

Wolff, C., & Scholtz, G. (2002). Cell lineage, axis formation, and the origin of germ layers in the amphipod crustacean *Orchestia cavimana*. *Developmental Biology*, *250*, 44–58.

Wolff, C., Tinevez, J.-Y., Pietzsch, T., Stamataki, E., Harich, B., Guignard, L., et al. (2018). Multi-view light-sheet imaging and tracking with the MaMuT software reveals the cell lineage of a direct developing arthropod limb. *eLife*, 7, e34410.

Zeng, V., Villanueva, K. E., Ewen-Campen, B. S., Alwes, F., Browne, W. E., & Extavour, C. G. (2011). De novo assembly and characterization of a maternal and developmental transcriptome for the emerging model crustacean *Parhyale hawaiensis*. *BMC Genomics*, *12*, 581.

Zimmer, M., Danko, J., Pennings, S., Danford, A., Carefoot, T., Ziegler, A., et al. (2002). Cellulose digestion and phenol oxidation in coastal isopods (Crustacea: Isopoda). *Marine Biology*, *140*, 1207–1213.

CHAPTER NINE

My road to the ants: A model clade for eco-evo-devo

Ehab Abouheif*
Department of Biology, McGill University, Montreal, QC, Canada
*Corresponding author: e-mail address: ehab.abouheif@mcgill.ca

Contents

Abstract

This chapter is the story of how I pioneered ants as a system for studying eco-evo-devo, a field that integrates developmental biology with ecology and evolutionary biology. One aim of eco-evo-devo is to understand how the interactions between genes and their environments during development facilitates the origin and evolution of novel phenotypes. In a series of six parts, I review some of the key discoveries from my lab on how novel worker caste systems in ants––soldiers and supersoldiers––originated and evolved. I also discuss some of the ideas that emerged from these discoveries, including the role that polyphenisms, hidden developmental potentials, and rudimentary organs play in facilitating developmental and evolutionary change. As superorganisms, I argue that ants are uniquely positioned to reveal types of variation that are often difficult to observe in nature. In doing so, they have the potential to transform our view of biology and provide new perspectives in medicine, agriculture, and biodiversity conservation. With my story I hope to inspire the next generation of biologists to continue exploring the unknown regions of phenotypic space to solve some of our most pressing societal challenges.

Current Topics in Developmental Biology, Volume 147
ISSN 0070-2153
https://doi.org/10.1016/bs.ctdb.2022.01.005

1. Introduction

At present, there are ~2.3 million known species on the earth, with anywhere between 8 and 64 million yet to be described (Hinchliff et al., 2015; Lewin et al., 2018; Scheffers, Joppa, Pimm, & Laurance, 2012). I believe that each of these discovered and undiscovered species has a story to tell and that the secrets of nature and evolution are written in their genomes, their development, and their ecologies. Yet most of these species' secrets remain unexplored (Abzhanov et al., 2008; Goldstein & King, 2016; Marx, 2021; Russell et al., 2017; Zhao et al., 2021). Why should we care? Much of our knowledge of molecular and developmental mechanisms of animals has come from just a handful of model organisms—mainly the fruit fly *Drosophila melanogaster*, the worm *Caenorhabditis elegans*, the frog *Xenopus laevis*, the fish *Danio rerio*, and the mouse *Mus musculus* (Ankeny & Leonelli, 2011; Goldstein & King, 2016). This approach has been effective because the genes regulating developmental processes are remarkably conserved across these model organisms, as well as humans (Gilbert & Barresi, 2019). However, many cogent arguments have been made for the need to expand this current range of model organisms, including the fact that many of them cannot answer several fundamental biological questions, including pressing questions about the molecular and developmental processes underlying how organisms respond and have adapted to rapid environmental change (Gilbert & Epel, 2015; Goldstein & King, 2016; Russell et al., 2017). Here, I add to these important arguments that our knowledge of molecular and developmental processes obtained from these models represents but a small fraction of phenotypic space.

To study the genes that regulate development in model organisms, we most often standardize the temperature and humidity at which we raised them as well as the genetic background and developmental time at which we conduct our experiments (Bolker, 1995; Gilbert, 2001; Lewontin, 2002). However, the phenotypic effect of a mutation differs in response to variation in environment, genetic background, and developmental time (Dworkin et al., 2009; Gilbert & Epel, 2015; Lewontin, 2002; Paaby & Rockman, 2014; Schlichting, 2008). For example, the classic *Infrabar* mutation affecting the size and number of facets in the compound eye of a fruit fly has dramatically different phenotypic effects at 15 °C (~175 facets) than at 30 °C (~300 facets) (Lewontin, 2002). The phenotype of the classic *scalloped* mutation in the wings of a fruit fly differs depending on the genetic background in which the mutation occurs (Dworkin et al., 2009). And in water striders (insects that spend their entire life on the water surface), knocking down the expression of the developmental regulatory gene *distaless* early in development suppresses the growth of all limbs on the body, but if you knock it down later in development, you specifically affect the differentiation of the antenna (and only in males!) (Khila, Abouheif, & Rowe, 2012). These examples provide but a glimpse of unexplored phenotypic space in biology, which I predict is potentially immense. We will only know how immense it is, however, after we simultaneously account for the phenotypic variation arising in response to variation in environment, genetic background, and developmental timing across the millions of described and undescribed species on the earth. Organisms with extreme adaptations, like those able to tolerate extremely high or low temperatures, may take us to whole new regions of phenotypic space, to new frontiers yet to be discovered.

Conceiving the immensity of this unexplored phenotypic space is a challenge. One way to visualize it is as a hypervolume (a multi-dimensional space) of which we have explored but a small dot (Fig. 1). I have drawn this

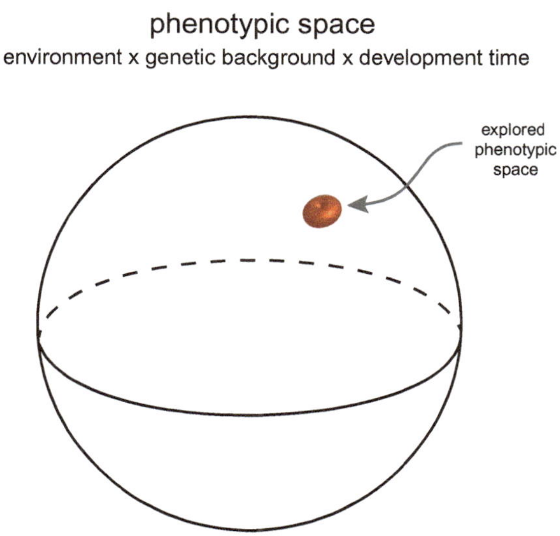

Fig. 1 Representation of unexplored versus explored phenotypic space. The three-dimensional sphere (black lines) represents a theoretical abstraction of the unexplored phenotypic space for organismal traits in response to variation in environment, genetic background, and developmental time, as well as the interactions between them. The red three-dimensional oval represents a prediction about the amount of phenotypic space that has actually been explored.

hypervolume as a sphere, although in principle it can assume any shape depending on the existence of constraints of biological systems, be they physical, historical, developmental, or ecological (Alba, Carthew, Carthew, & Mani, 2021). My hope is that the next generation of biologists are more successful at embracing and exploring phenotypic space before many of the discovered and undiscovered species disappear from the Earth. My dream is that diversity, both at the biological and societal level, is the rule rather than exception. When this happens, I predict that a new era of discovery in the life- and medical sciences will be ushered in. It is for this reason that I have dedicated my scientific career to revealing the development and evolution of 'non-model' species, no matter how challenging.

In this chapter, I recount my journey to the ants, which I pioneered as a system to integrate developmental biology with ecology and evolutionary biology almost 25 years ago. This synthesis of fields is now known as eco-evo-devo (Abouheif et al., 2014; Gilbert, 2016; Gilbert & Epel, 2015; Sultan, 2015). I hope my chapter achieves three things: (1) to make those wanting to explore and set up emerging models for eco-evo-devo aware of both the excitement and challenges involved; (2) to review some of the key discoveries my lab and I made along the way; and (3) to outline some of the major questions in biology that ants are poised to address in the coming years.

2. Serendipitous beginnings

I have often been asked "were you obsessed with ants as a kid?" To their surprise, my answer is "no." Although I did have a deep fascination with nature, and like any kid who grew up in a Canadian middle-class suburb, I played with insects a lot. But mostly, I was fascinated by the vastness of the ocean and all its creatures, especially sharks. I vividly recall going to the local library, borrowing 6 books on sharks, and flipping through them with both amazement and fear. So much so, I was afraid to swim in the deep end of the pool during swim practice. My fascination with the ocean and its creatures largely comes from my Egyptian uncle Abdellatif Abouheif, who according to the International Swiming Hall of Fame (Florida, USA) "became the greatest marathon swimmer in the history of the sport and set the standards for today's open water swimmers" (ISHOF, 2021) . As first-generation Arab-Muslim immigrants (Egyptian Father/Palestinian Mother), my family faced many challenges after immigrating to Canada in the 1960s, and I found myself continually looking to my Uncle for inspiration. Although I tried hard, swimming was not to be my calling, but I kept looking to the ocean for answers. So, my road to the ants was anything but fated from childhood, and in fact, my early fascination with the ocean and the sharks eventually led me to start a PhD in 1995 with Axel Meyer in the Department of Ecology and Evolution at Stonybrook University (New York, USA). Axel is a world leader in molecular phylogenetics and evolution of fish, and at the time I joined his lab, he had just published a series of exciting papers on the molecular phylogeny of whales, swordtail fish, and adaptive radiation of cichlid fish in the African Great Lakes (Meyer, Kocher, Basasibwaki, & Wilson, 1990; Meyer, Morrissey, & Schartl, 1994; Milinkovitch, Orti, & Meyer, 1993).

My time in Axel's lab was extremely productive. The field of evolutionary developmental biology (evo-devo) was, according to Günter Wagner, in its "enthusiastic" phase (Wagner, Chiu, & Laubichler, 2000), meaning that it was an era filled with unexpected and exciting discoveries. For example, Walter Gehring's lab had just discovered that the function of the *eyeless/ Pax-6* gene was highly conserved and regulates the development of the compound eyes of fruit flies and the camera eyes of mice—they expressed the mouse *Pax-6* gene in the developing wing tissue of fruit flies and showed that it induced the development of eyes on adult fruit fly wings (Halder, Callaerts, & Gehring, 1995; Quiring, Walldorf, Kloter, & Gehring, 1994). Sean B. Carroll's group had just discovered that the gene *distal-less* (*dll*), which regulates limb development in fruit flies, was co-opted during evolution to regulate the development of the eyespots on butterfly wings (Carroll et al., 1994). This was one of the first demonstrations of the co-option of a developmental regulatory gene to regulate a novel trait. It was in the context of these and many other exciting discoveries in evo-devo, that Axel and I set out to investigate the developmental basis of the second set of jaws in Cichlid fish, a key morphological and functional innovation thought to have facilitated their explosive speciation and radiation in the African Great Lakes (Liem, 1973). This second set of jaws, which evolved from the gill arches, freed the ancestral jaws to acquire novel functions, such as scale picking (Takeuchi, Hori, Tada, & Oda, 2016). So, I began my training in molecular biology in Axel's lab under the supervision of a talented postdoc by the name of Rafael Zardoya (now Director of the Natural History Museum in Madrid, Spain doing exciting work on the evolution of snail shells (Pardos-Blas et al., 2021). My training period resulted in two publications: one in the *Proceedings of the National Academy of the Sciences, USA*, and the other in *Trends in Genetics* on the molecular evolution of the *hedgehog* gene family, which plays key roles in limb development and other important functions (Zardoya, Abouheif, & Meyer, 1996a,b).

During the first 2 years of my PhD, I also became fascinated, almost obsessed, with the concept of homology, which is classically defined as any trait present in the most immediate common ancestor of two taxa (Hall, 1994). How then should we interpret the discovery from Gehring's lab that a homologous gene—*eyeless/Pax-6*—regulates the camera eyes of vertebrates and compound eyes of flies (Halder et al., 1995; Quiring et al., 1994), which according to the classical definition, are not homologous (the most immediate common ancestor of vertebrates and insects did not have image forming eyes)? Based on his lab's discovery, Gehring made the bold assertion that we should now interpret the fly and vertebrate eyes as homologous because the gene that regulates them—*eyeless/Pax-6*—is homologous (Gehring, 1996; Halder et al., 1995; Quiring et al., 1994). Dickinson and Seger (1996) wrote a reply to these papers, challenging this assertion in favor of the classical definition, but provided little in the way of a solution. And so, I tried to make sense of this conundrum by digging deep into the vast literature on homology.

I soon realized that there is a lot more to homology than how it is portrayed in textbooks. The writings or edited volumes by Brian Hall, Collin Patterson, Gunter Wagner, Louise Roth, David Wake, and George Lauder were particularly thought provoking (Hall, 1994, 1995b; Lauder, 1994; Patterson, 1988; Roth, 1984; Wagner, 1989; Wake, 1994, 1999; Wake, Wake, & Specht, 2011). They made me realize that the mechanisms through which traits are generated during development are just as important to consider as a trait's phylogenetic history for understanding homology. In response to the controversy raised by Pax-6/*eyeless*, Jessica Bolker and Rudy Raff first proposed the term 'deep homology' to describe the highly conserved function of *eyeless/Pax-6* in eye development across animals (Bolker & Raff, 1996). Shortly after, Neil Shubin, Sean B. Carroll and Cliff Tabin elaborated the concept of deep homology to refer to traits that are not homologous (evolved independently) at the level of morphology, but do share homology at deeper levels of organization, including homologous developmental genes and their expression, and in some cases, even homologous cell types (Shubin, Tabin, & Carroll, 2009) (Fig. 2A). Deep homology therefore places emphasis on the homology of the genetic and embryological underpinnings of novel morphologies, like animal eyes, to understand their origin and evolution.

My reading of the homology literature, however, led me to discover additional scenarios, such as traits that are homologous at all levels of

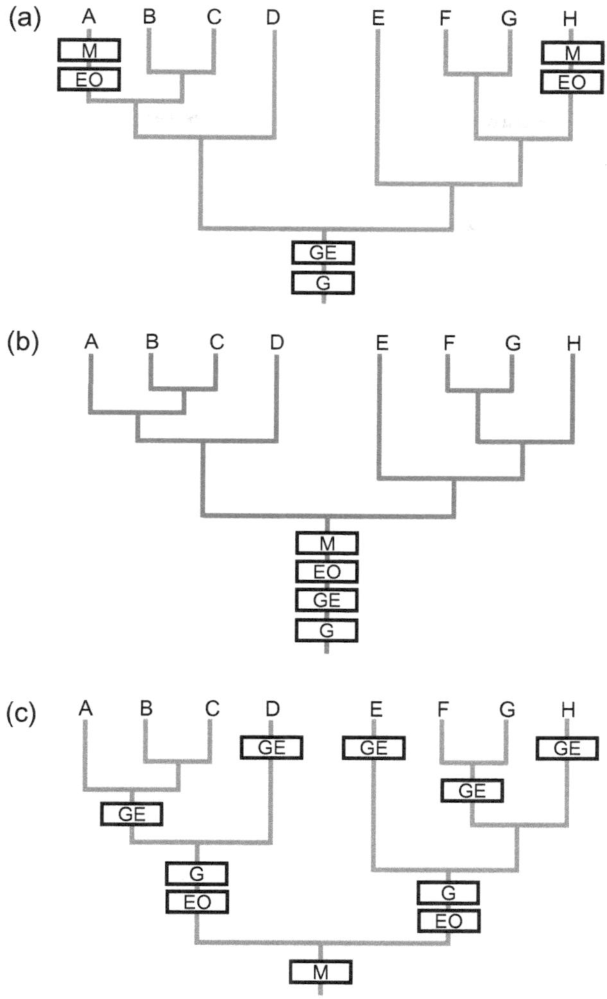

Fig. 2 The hierarchical approach to homology. The boxes represent traits at different levels of biological organization: genes (G), gene expression (GE), embryonic origins (EO), and morphology (M). These traits are mapped simultaneously on to a hypothetical phylogenetic tree with eight species (labeled A to H). *Scenarios adapted and modified from Abouheif, E., (1997). Developmental genetics and homology: A hierarchical approach.* Trends in Ecology & Evolution, 12 *(10), 405–408.*

biological organization (morphology embryonic origins, gene expression, and genes) (Abouheif, 1997) (Fig. 2B). I even began uncovering scenarios where a trait is homologous at the level of morphology, but the embryological origins, cells, or genes were not (Abouheif, 1997) (Fig. 2C).

For example, Ralf Sommer and Paul Sternberg had just discovered that the vulva (the female reproductive organ) in nematode worms is homologous at the level of morphology but is generated by different cell lineages during development in different species (Sommer, Carta, & Sternberg, 1994; Sommer & Sternberg, 1994). After uncovering these alternative scenarios, I felt that we needed an approach to homology that could reveal and account for each of these scenarios in an unbiased way. This led me to publish a series of four papers on homology arguing for a hierarchical approach, where levels of biological organization (genes, cells, embryonic origins, and morphology) are treated separately and as independent traits that are analyzed simultaneously in a phylogenetic framework (Abouheif, 1997, 1999a; Abouheif et al., 1997; Wray & Abouheif, 1998). This hierarchical approach allowed for scenarios where the levels are either associated (Fig. 2B) or dissociated (Fig. 2A and C) relative to one another, and I argued that each of these scenarios has important implications for understanding development and evolution (Abouheif, 1997). However, it was this third scenario (Fig. 2C) in Abouheif et al. (1997), where the morphology is homologous, but the underlying developmental and genetic basis is not, which was most controversial at the time.

After training with Axel for 2 years (1995–1997), he announced that he was leaving to take a position at the University of Konstanz in Germany and invited me to move with him. I was torn, I liked and worked well with Axel, and was excited about my research project on unraveling the developmental basis of the second set of jaws in cichlid fish. I had already started generating preliminary results and was gaining momentum—I cloned and visualized the expression pattern of *dll* in the developing second set of jaws. However, leaving the US would have meant leaving a very exciting milieu for evo-devo as well as being far away from my parents and girlfriend, and I wasn't quite ready for that. So, I made the difficult decision to stay and start all over again from scratch. Greg Wray, who was a young and rising star in evo-devo, had just started his lab at Stony Brook University a few years before I had arrived. His lab was addressing exciting evo-devo questions on how the radially symmetrical body plan of echinoderms (i.e., sea urchins, sea stars, sea cucumbers, and brittle stars) evolved from a bilaterally symmetrical ancestor (Lowe & Wray, 1997). Greg and I had insightful discussions on homology during my first 2 years as a PhD student, and together, we published two of my four papers on homology—what is now a classic commentary in *Trends in Genetics* "Homology and Developmental Genes" (Abouheif et al., 1997) and "When is Homology not Homology"

(Wray & Abouheif, 1998). After Axel's decision to move to the University of Konstanz, I approached Greg (in the bathroom of all places–my timing for such things was never good) to ask if I could join his lab. He said yes (and that was one of the best bathroom moments of my career)! But was I now going to work on echinoderms?

Fortunately, I was awarded a PhD fellowship from the Quebec Government, and they allowed me to use it in the USA. I will never forget our first formal meeting when Greg said: "you have a fellowship, so you can work on whatever you like, my only condition is that it can't have a backbone!" The brainstorming began. After thinking so hard about the homology concept, I told Greg that I wanted to steer clear of homology problems, and that I didn't want to face the challenges inherent in interpreting gene expression patterns between embryos of distantly related organisms. Greg proposed that we study a species that had dramatically different morphological forms within a species where there was no question or challenges posed by homology. We started naming several species from aphids, to termites, to bees, to marine colonial invertebrates ... and then ... the defining moment of my career came when I blurted out "ants!" and Greg and I erupted with excitement. Most ants show dramatic morphological differences between the queen caste and worker caste within a single colony, where the queen is substantially larger, lives for 30 years, and has fully functionally wings to engage in mating flights and reproduces thousands (and in some cases millions) of eggs during her lifetime (Hölldobler & Wilson, 1990). In contrast, the worker caste is smaller, completely wingless, lives for a few months, and has a greatly reduced reproductive capacity (Hölldobler & Wilson, 1990). Greg appreciated ants as a system because, as a PhD student, he had done some work on butterflies in Fred Nijhout's lab at Duke University (USA), which is also where Diana Wheeler did her pioneering work with Fred demonstrating that Juvenile Hormone (JH) is a key mediator of caste determination in ants (Nijhout & Wray, 1988; Nijhout, Wray, Kremen, & Teragawa, 1986; Wheeler & Nijhout, 1981b). And so, in 1997, my journey to the ants officially began.

If there is an important lesson up to this point in the story, it is that serendipity played a critical role in finding the ants—so many of the events leading up to this point could have played out much differently and might have directed me along other scientific paths. Serendipity, however, is only half the story. Equally important was my conceptual thinking on homology that had (unconsciously) guided me to the ants. As you will read, and looking back, I find it somewhat eerie how much my career has been (and continues to be) influenced by these first few baby steps. And finally, the mentorship I received along the way—the freedom, intellectual guidance, and respect for cultural and religious differences—was also key.

3. Finding the ants and the unsung heroes of science

I was excited to start exploring the evo-devo of ants, but I really knew nothing about them other than playing with them growing up. So, I went in search of ant colonies, saying to myself "ants are everywhere, how hard could this be?" But I needed to collect colonies where I could find both the queen and her workers and keep them alive in the lab, and this tuned out to be much harder than I expected! There are thousands of non-reproductive workers in the colony but only one (or a few) reproductive queens. Collecting only workers, which is much easier, would therefore not be much use for what I was trying to accomplish. The day after my key meeting with Greg, I grabbed a spoon and rushed out to the grounds in and around Stony Brook. I looked for small protruding mounds and dug them up, but one after the other I came up empty-handed. I found a few workers here and there but not a single queen. I searched for days with little success and started to think that while working with ants was exciting, it may not be as feasible as I hoped. I started questioning my decision to begin my PhD research from scratch. However, one day my fellow PhD student Alexa Bely (now Professor at the University Maryland, USA) returned from vacation and saw me digging at a far corner of the campus. I told her I was embarking on a new evo-devo journey with ants but was failing so far. That night, she told her partner (Leo Shapiro) about my project, and he recalled an old man on Long-Island—a dental technician that makes false teeth for a living—who once helped him collect ant colonies for one of their Introductory Biology labs. His name is Ray Sanwald. A few days later, Leo managed to get me the phone number for his dental studio. Though I was skeptical, I called out of desperation and Ray generously invited me to his home for a visit the next day. Little did I know that this visit would be one of the defining moments of my career.

Ray's home was in Medford, Long Island (NY, USA) about a 20 min drive from the University. When I arrived at his property, I discovered that he had three acres of land dedicated to studying ants. He took me on a tour of his property and explained that Long Island is one of the best places to collect ants because of the soft, sandy soil and mild climate. His property wasn't just any property, the landscape was especially welcoming for ants, consisting of

(a)

(b)

Fig. 3 Ray Sanwald's property, a living museum of ants. (A) Photo of Ray Sanwald's property in Long Island, NY (USA). (B) Photo of *Pheidole pallidula* minor worker (left) and soldier (right). *Photo by Hugo Durras.*

rolling mounds of green grass with a handful of trees for shade, and most importantly, lots of rocks for ants to colonize and nest beneath (Fig. 3A). His property was a living museum of ants! With a large shovel, we started digging up colony after colony, species after species, sifting through the soil looking for queens. Large, small, red, yellow, so many ants and so much variation. For the first time in my life, I felt like a naturalist. However, it was when I observed a colony of *Pheidole* ants for the first time that I truly fell in love. Unlike other ant species, whose workers are largely uniform in size, the worker caste in *Pheidole* colonies is made up of two kinds of workers: small-headed minor workers and big-headed soldiers (Fig. 3B). I will never forget the moment when I was lying flat on my stomach watching the minor workers and soldiers charismatically charging from their nest, like a legion of warriors ready to defend it.

Ray used part of his dental studio as an ant lab and showed me how he rears colonies. It's a very basic set up: all you need is plastic boxes, glass test tubes half-filled with water and plugged with a piece of cotton, and most importantly, you need "Fluon," a white liquid that reduces the friction of surfaces, so the ants cannot escape (Fig. 4). Most ants can be kept between

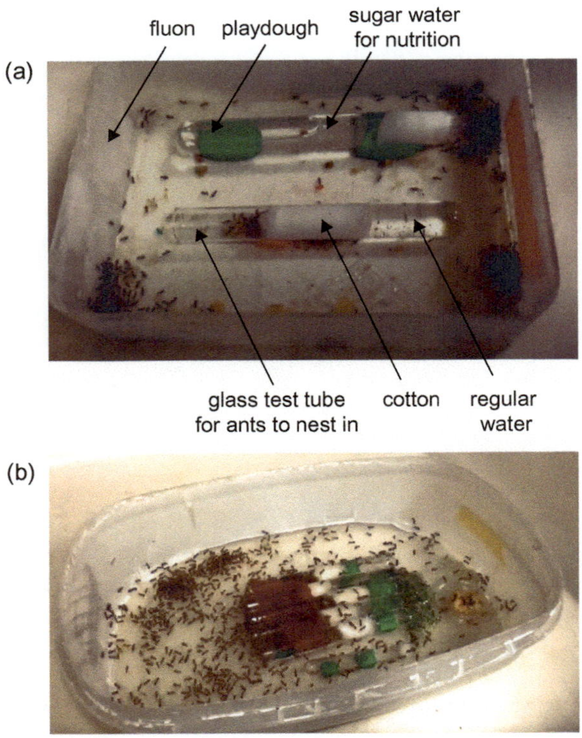

Fig. 4 Simple set up of an ant nest for evo-devo research. (A) A smaller ant nest in Tupperware plastic box showing one glass test tube (lower) for the ants to nest in with regular water filling it half-way and plugged with cotton and another glass test tube (upper) filled to the top with sugar water and plugged with cotton so the ants could feed. Both glass test tubes are held in place with playdough. (B) A larger ant nest in Tupperware plastic box. The glass test tubes for ants to nest in are larger but are prepared in the same way as in (A). Preparation of same size sugar water tubes as in (A). The red piece of plastic is to protect ants in their nests from direct light. The white chemical substance on the walls of the ant nests in (A) and (B) is called "Fluon," which prevents the ants from escaping by reducing friction of the surface making the ants literally slip down the wall.

23 and 27 °C and in a relatively humid environment (40–60% humidity). Although this setup will work for most ants, different species have different requirements depending on their life history and ecological environment. I left his property that day with 3–4 "queened colonies," meaning that these colonies had both the queen and her workers, which I could take back and set up in Greg's lab. That day was just the beginning of many shared adventures with Ray.

Ray and I collected ants all over Long Island; between power lines, in wooded lots, and in protected natural areas in the Pine Barrens (Fig. 5A).

Fig. 5 Collecting ants with Ray Sanwald, an unsung hero of science. (A) Photo of Ray and I collecting ants along the powerlines in Long Island, NY (USA). (B) Photo of Ray Sanwald with the infamous red carpet where he discovered a colony of *Crematogaster lineolata* (inset).

On one of these adventures, I told Ray that I wanted to look for the "acrobat ant" *Crematogaster lineolata*. He took me to a wooded lot at the back of a large "Big K" department store. We started walking, flipping rocks, breaking branches, and digging up mounds. Suddenly, Ray shouted, "here they are!" and he pointed to an abandoned car missing its hood and doors. As I approached closer, I found Ray pointing to an old and moist red carpet that was rolled up tightly and sitting in the middle of the car (Fig. 5B). We removed the carpet from the car and started slowly ripping off its top layer. Then, suddenly, the ants came pouring out of the carpet like water through a broken dam. The ants flowed frantically over my hands, and before long my body was crawling with them. We kept tearing away at the carpet until we finally found the queen. I stood in awe of this man, who had obviously searched long enough for the ants that he could find them in a rotting red carpet, in an abandoned car, in a wooded lot, behind a department store! Unfortunately, over the years, many of the places we used to collect have been destroyed and developed, and the ants that once thrived in these places are now gone.

What I find amazing about all this, is that Ray, a dental technician with a high school education, enabled a major research program on the evo-devo of ants that led to several breakthroughs in biology (as I will describe in the following parts of my journey). I think of Ray as one of the many unsung heroes of science that go without formal recognition. Impassioned lovers of nature, like Ray, can spend more time observing nature than scientists, who are often burdened by teaching, grant writing, and administration. And so, they can observe and potentially reveal details of the natural world that would otherwise remain hidden from scientists.

A case in point is Ray's passion for observing kidnapping behavior in ants (also known as dulosis, social parasitism, or slave-raiding behavior). Kidnapping ants will send out a few scouts, find nests of another ant species, and then conduct daily raids with thousands of individuals to enter their nests and kidnap their larvae (without killing the queen). These ants depend on these stolen larvae for their survival—these larvae acquire the chemical smell of their captors, and after developing into adults, believe they are part of the colony and work to sustain it. Remarkably, Ray has been taking detailed records on the kidnapping behavior of two ant species (*Polyergus lucidus* and *Polyergus sanwaldi*) every single day during the spring and summer for the last 60 years (from 1952 to 2012). He missed only a few days when his first wife died of cancer. His records include many variables, such as the temperature, humidity, how many and what time the scouts go out, and what time they tend to raid. After observing Ray do this for many years, I asked him what he plans to do with these notes, and his response was "Nothing! Since I am sitting here, I thought I might as well take note of what's happening, so I don't get bored." Eventually, Ray gave me all his notes, except that his wife threw out the first 12 years because they were cluttering his home! Luckily, I managed to salvage 48 years, and with the help of many undergraduate students at McGill we managed to enter all his notes into an electronic database. After we got all the data in and verified, we started looking for trends in the data. Amazingly enough, one of my undergraduate students (TingJia Lorigiano) found a clear signature of climate change on the kidnapping behavior of these ants over a 48-year period. We hope to publish these findings very soon to honor Ray's accomplishments as a citizen scientist.

Over my career, I have been fortunate to have encountered other unsung heroes of science—Bob Johnson, Lloyd Davis, and Stefan Cover—who like Ray have spent their lives observing nature (especially ants) out of sheer passion. I suspect that each of the emerging models described in this volume may have similar stories of people, who are not a formal part of the world of science but play significant roles in their discoveries, and like Ray, make significant discoveries of their own. My advice is to seek these heroes of science, and together with them, explore the species whose stories remain untold.

4. The first analysis of gene regulatory networks underlying polyphenic development (Abouheif & Wray, 2002)

Now that I was discovering the world of ants with Ray and able to culture ant colonies in Greg's lab, the next step was to put my molecular and developmental biology training into action. I found surprisingly little information about ant development in the literature. In his book *Ants: Their Structure, Development, and Behavior* (Wheeler, 1910), the renowned scientist William Morton Wheeler drew by hand a late-stage embryo (Fig. 6A) as well as the internal anatomy of an ant larva (Fig. 6B) from the genus *Formica*. Other than that, however, I couldn't find any comprehensive descriptions of ant development, from embryo to adult.

After intense discussions, Greg and I concluded that of all the dramatic morphological differences between the queen and worker caste, wing development was a good place to start. We reasoned that developmental geneticists had worked out the gene regulatory network (GRN) underlying wing development in the fruit fly, and most of the genes in this network were

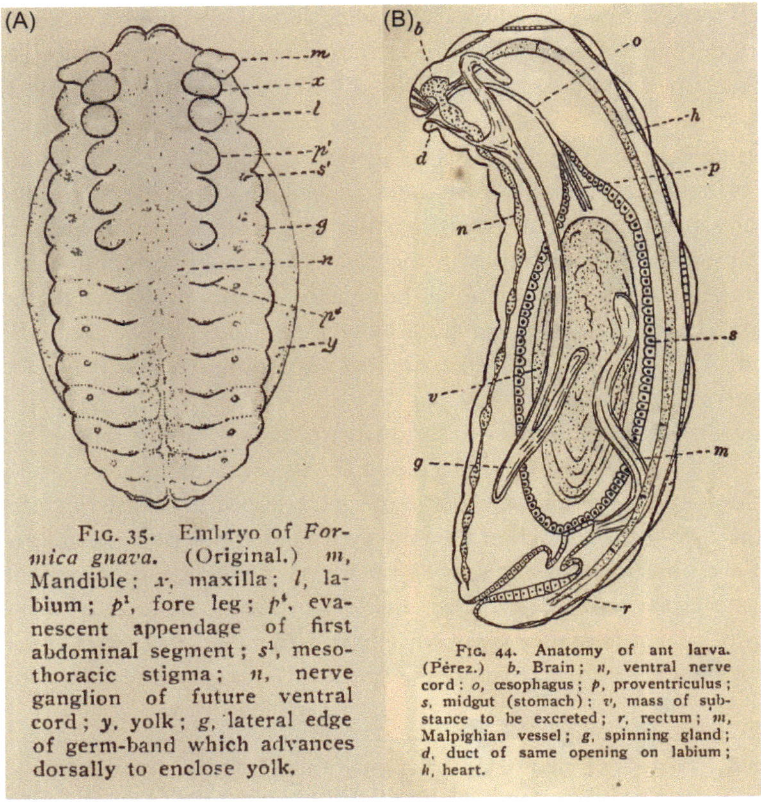

Fig. 6 Hand-drawn images of an ant embryo and of the internal anatomy of an ant larva from the genus *Formica. Drawings and accompanying figure legends from Wheeler, W. M. (1910). Ants: Their structure, development and behavior. Lancaster: Columbia University Press.*

known to be highly conserved between flies and vertebrates (Abouheif, 1999a; Brook, Diaz-Benjumea, & Cohen, 1996; Held, 2002) (Fig. 7A). So, there was a good chance that I would be able to clone some orthologs of these highly conserved genes from the ants. This was a key consideration in choosing the wings because there were no genomes sequenced at this time making it very challenging to clone orthologs in any non-model species. Furthermore, I would be able to compare the spatial expression of genes within the wing GRN between ants, flies, beetles, and butterflies because their wings develop from populations of cells called "imaginal discs," which are set aside during the larval phase and then evert during metamorphosis to give rise to the adult wings (Held, 2002; Truman & Riddiford, 1999)

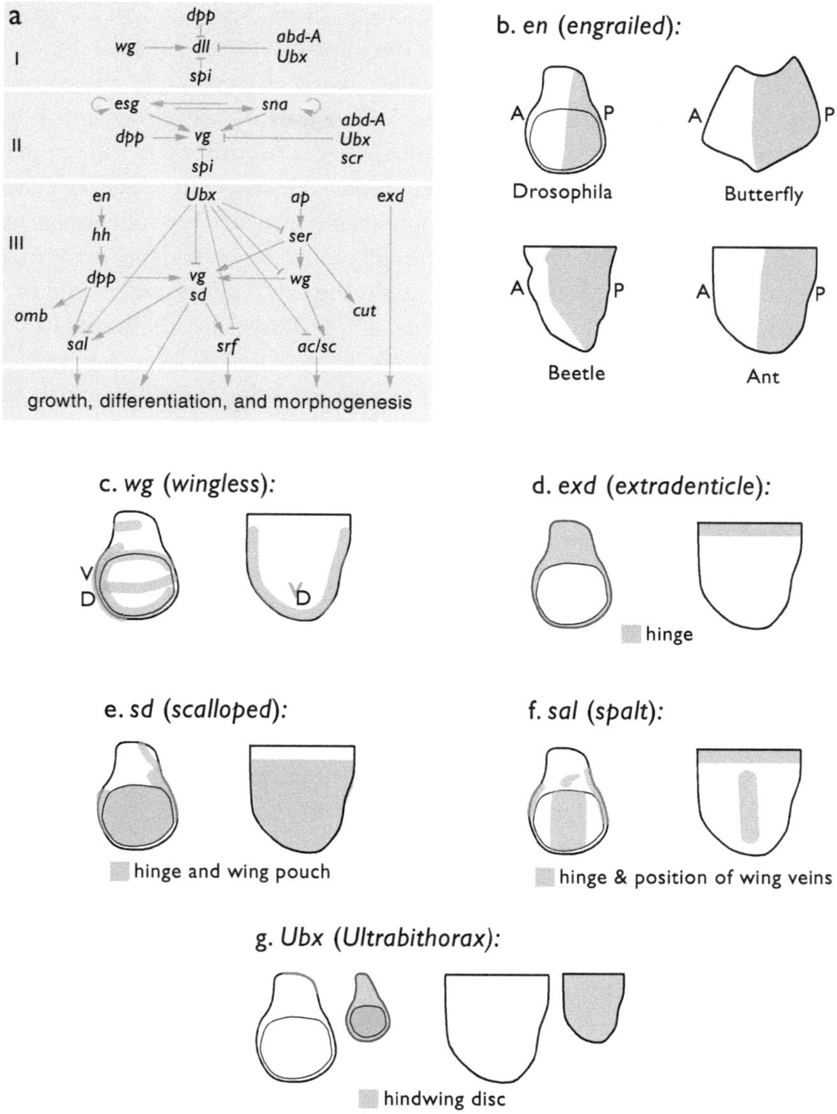

Fig. 7 Conservation of the wing-patterning network in holometabolous insects. (a) An overview of the wing GRN in *Drosophila melanogaster*, which regulates wing development in three phases (I–III). First (I), a set of transcription factors and signaling molecules establish a cluster of approximately 20 ectodermal cells as precursors of both the leg and the wing imaginal discs. Second (II), a set of interacting genes then divides these cells into separate clusters that give rise to three pairs of leg discs and two pairs of fully sized wing imaginal discs. Third (III), a set of interacting transcription factors and signaling

(Continued)

(Fig. 7B–G). Most importantly, these comparisons between species would allow me to compare the expression of the wing GRN between the imaginal discs of the winged queens and wingless workers in a single ant colony. Wing development in ants is called a "wing polyphenism" because wing development is activated in winged queens but is suppressed in the wingless workers in response to environmental cues such as temperature and/or nutrition (Hanna & Abouheif, 2021). While queen larvae develop fully functional wing imaginal discs, the workers—although completely wingless—have rudiments of wing imaginal discs that appear during larval development and then disappear before the adult stage (Dewitz, 1878; Wheeler & Nijhout, 1981a) (Fig. 8). Therefore, I would be able to ask if the wing GRN is conserved in the winged queens but is interrupted to suppress wing development in the wingless worker caste.

We were also excited by the possibility of being able to potentially infer developmental processes at play during the evolutionary origin of ants. Wing polyphenism evolved around the same time-period as eusociality in ants and is a nearly universal feature of the approximately 16,500 known ant species (Hanna & Abouheif, 2021). It is thought to have been a key innovation that allowed ants to colonize the ground, raising the possibility that wing polyphenism may have facilitated or reinforced the origin of eusociality (Hanna & Abouheif, 2021). Given that wing polyphenism is a nearly universal feature that evolved just once during the origin of ants and that expression of the wing GRN is generally highly conserved between flies and vertebrates, Greg and I predicted that the wing GRN was interrupted at a single point during the origin of ants and that this interruption point was conserved across the worker castes of all ant species.

In theory, all aspects of this new adventure seemed to be lining up and starting from scratch now seemed to have set me on an even more exciting

Fig. 7—Cont'd molecules pattern these wing discs, imparts to them a wing-specific identity, and activates downstream target genes that pattern the detailed structures of the wing. (b–g) Expression of several genes within this wing GRN is conserved in wing discs across holometabolous insects. (b) *engrailed* (*en*)expression is conserved in the posterior (P) compartment across *Drosophila*, butterfly, beetle, and ant wing imaginal discs. (c–f) Conservation, between *Drosophila* and ant wing discs of: (c) *wingless* (*wg*) expression along the dorsal (D) ventral (V) margin; (d) *extradenticle* (*exd*) expression in the hinge region; (e) *scalloped* (*sd*) expression in the wing pouch; (f) *spalt* (*sal*) expression along the AP axis marking the position of future wing veins; and (g) Ultrabithorax (*Ubx*) expression in the hindwing but not the forewing. *Reprinted with permission from Nahmad, M., Glass, L., & Abouheif, E. (2008).* Evolution & Development, 10, *360–74.*

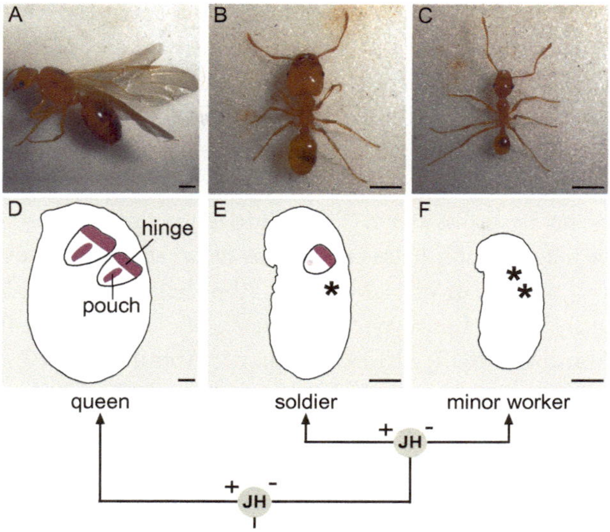

Fig. 8 Caste determination in *Pheidole* ants. Caste determination occurs at two JH-mediated switch points during development in response to environmental cues (temperature, nutrition, and social interactions) to produce either winged queens (A), wingless soldiers (B) or minor workers (C). Wing discs in queen larvae showing conserved hinge and pouch expression of the gene *spalt* (*sal*) (D). Rudimentary wing discs in soldier larvae showing a soldier-specific pattern of *sal* expression, where it is conserved in the hinge but downregulated in the pouch (E). Asterisks represent absence of visible wing discs and *sal* expression in (E) soldiers and minor worker larvae (F). Scale bars indicate relative size of queen, soldier, and minor worker larvae and adults. *Reprinted with permission from Rajakumar, R., San Mauro, D., Dijkstra, M. B., Huang, M. H., Wheeler, D. E., Hiou-Tim, F., et al. (2012). Ancestral developmental potential facilitates parallel evolution in ants.* Science, 335 *(6064), 79–82.*

journey in search of the key interruption point in the wingless worker caste of ants. Of course, no journey is without its obstacles. I started with *Pheidole morrisi*, the ant species that I first fell in love with. Diana Wheeler and Fred Nijhout had described the development of rudimentary wing discs in the minor worker and soldier larvae of *Pheidole bicarinata* in their now classic paper in 1981 (Wheeler & Nijhout, 1981a). Queen larvae have two pairs of fully functional wing discs. In contrast, soldier larvae have only one pair of rudimentary forewing discs while minor worker larvae have no visible rudimentary wing discs (Wheeler & Nijhout, 1981a). Although this was well described in Diana and Fred's paper, it took me more than 6 months to find them under the microscope. I was looking for them in younger larval stages, but I had missed a key point in Diana's paper, which is that rudimentary forewing discs become visible only during the last larval instar (Wheeler & Nijhout, 1981a). It was only when I stepped back and reconstructed a larval staging table that I was able to distinguish the different larval stages and castes, and finally, I spotted the rudimentary wing discs with my own eyes. This is one of the basic lessons I learned about establishing emerging model systems—it is critical to take the time to build a timetable of the major developmental events before starting. This approach is time consuming but was necessary for me to establish ants (as well as water striders and even snails) as eco-evo-devo models in my lab (Khila, Abouheif, & Rowe, 2009, 2014; Khila et al., 2012; Lesoway, Abouheif, & Collin, 2014; Lesoway, Collin, & Abouheif, 2017; Shbailat, Khila, & Abouheif, 2010). It still took me another 6 months to visualize the expression of genes in the wing imaginal discs in ants using immunohistochemistry and in situ hybridization. The main challenge for me to get these techniques to work was to learn how to properly dissect the larvae, which is now a key technique in my lab.

Just as the first gene expression results were rolling in and I was gaining momentum, Greg had announced he was moving his lab to Duke University in North Carolina (USA). Greg was a fantastic mentor, and I enjoyed our interactions very much, so I picked up my things and moved to Duke with Greg to help him set up his lab and continue my PhD. Yet again, I felt like I was starting from scratch. To save time, I tried to transport all my solutions with me in a truck down the East Coast, but this was another mistake that would cost me 6 months—all of my in situ hybridization experiments stopped working. While informally speaking to Nipam Patel at the University of Chicago (USA) about potential postdoctoral opportunities, he had mentioned that PH was a critical parameter in getting in situ hybridization experiments to work. Thanks to Nipam, I remade all of my solutions and my in situ experiments finally started working again. I was back on the road to ant evo-devo! Looking back at this move, I realized that the Department of Ecology and Evolution at Stony Brook University had given me a strong foundation in evolutionary biology, while Duke's Biology Department was strong in organismal biology. This combination in my training, as well as learning how to set-up a lab from scratch, served me well in my career. I have come to learn that there is real benefit to experiencing the culture of different academic institutions as a young scientist.

After 6 years (1997–2002), which felt more like failure sprinkled with success than success sprinkled with failure, I confirmed our first prediction—the expression of the wing GRN is conserved in the winged queen (and male) caste in ants relative to flies and butterflies (Abouheif & Wray, 2002)

(Fig. 7B–G). I thought of this result as a "natural positive control" and interruptions in the wing GRN of wingless workers as a "natural experiment." To our surprise, however, instead of finding a single interruption point in the wing GRN of wingless worker castes across ants, as Greg and I had predicted, we discovered that different species had different interruption points (Abouheif & Wray, 2002) (Fig. 9). In fact, within my favorite *Pheidole* species, interruption points were even different between the minor worker and soldier castes (Fig. 8D–F). Therefore, despite the fact that wing polyphenism evolved just once during the origin of ants, interruptions points within the wing GRN

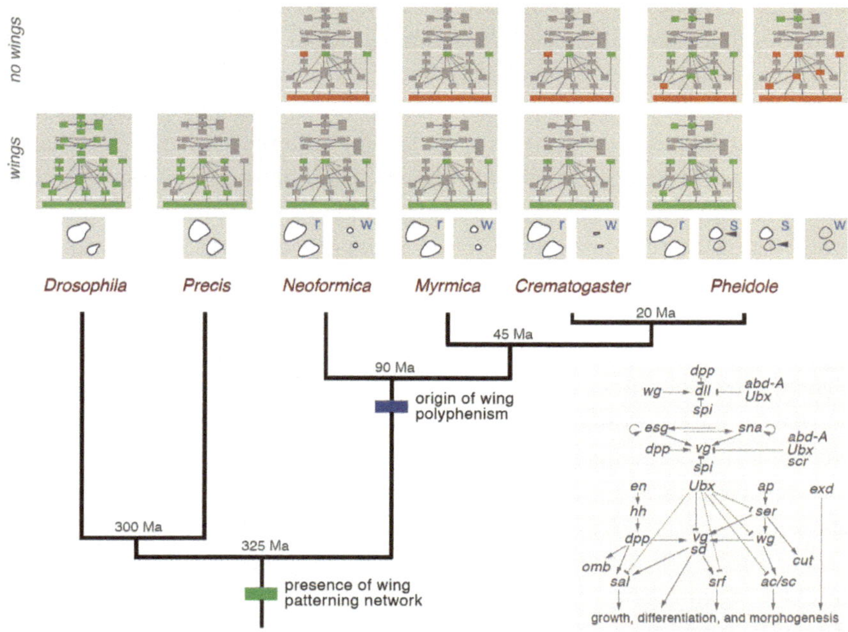

Fig. 9 Simultaneous conservation and lability of the GRN underlying wing polyphenism in ants. Wing polyphenism evolved just once, approximately 125 million years ago (blue bar). The wing GRN has been largely conserved across holometabolous insects for 325 million years (green bar). Diagrams of reproductive and worker disc morphology within each species are shown in the first row above the phylogeny (for *P. morrisi*, arrowheads refer to either the soldier forewing or hindwing). Network diagrams for winged reproductive castes are shown in the second row above the phylogeny; network diagrams in wingless worker castes are shown in the third row. Within each panel, conserved gene expression is indicated in green, and interrupted expression is indicated in red; genes not examined are shown in gray. *Reprinted with permission from Abouheif, E., & Wray, G. A. (2002).* Science, *297, 249–252.*

of workers were "labile," meaning that they had evolved on relatively short timescales (Abouheif & Wray, 2002).

Our discovery of the simultaneous conservation of the wing GRN in winged castes but lability of interruption points in wingless castes across ant species holds several fundamental implications for our understanding of eco-evo-devo. Abouheif and Wray (2002) was the first to reveal how the expression of a GRN changes in response to the polyphenic development of a trait, and the first to reveal how a GRN underlying a polyphenism evolves. Greg and I predicted that this simultaneous conservation and lability of GRNs was a general feature underlying polyphenism, which is widespread in nature and includes all cases where a single genome produces alternative phenotypes in response to internal or external cues. Polyphenism produces alternative phenotypes at multiple levels of biological organization. Examples include how pluripotent stem cells differentiate into different cell types in response to signaling molecules within an individual as well as how developing individuals (embryos or larvae) differentiate into alternative phenotypes in response to environmental cues within a colony or population (Arendt et al., 2016; West-Eberhard, 2003). Our prediction has since been confirmed in reptiles with temperature-dependent sex determination (Abouheif, 2004; Valenzuela, 2008), in beetles with alternative horned and hornless phenotypes (Moczek & Nagy, 2005; Moczek, Rose, Sewell, & Kesselring, 2006), in nematode worms with alternative active and dauer (diapause) phenotypes (Sommer & Mayer, 2015), and in cichlid fish with alternative jaw morphologies (Gunter, Schneider, Karner, Sturmbauer, & Meyer, 2017; Schneider, Li, Meyer, & Gunter, 2014).

Greg and I also proposed that the lability of interruption points in wingless worker castes across ants could have evolved either via neutral evolutionary processes, natural selection, or some combination of both. At the time, I favored the neutral hypothesis because I thought that, if wing development is suppressed or eliminated (perhaps by programmed cell death), it shouldn't matter which gene in the wing GRN was interrupted. However, I vividly remember having lunch with John H. Gillespie, a prominent molecular evolutionist who was invited to speak at Duke University. He wrote the popular book "*The Causes of Molecular Evolution*" in which he argued against the neutral theory of molecular evolution. After telling him about my results, we had a heated discussion about whether the lability of interruption points could be the result of neutral evolutionary processes or natural selection. Similarly, during a postdoc interview with evo-devo pioneer Michael Akam at Cambridge University (UK), Michael remarked "I would be very cautious about interpreting your results as neutral evolution." Nonetheless, I kept all possibilities on the table, mostly because I could not envision an alternative function that these wing rudiments may have.

Greg and I first submitted our results to *Nature*, and after 3 days, the Editor wrote a note saying that the lability of interruption points in the wingless workers would mean that wing polyphenism (as well as eusociality) must have evolved several times independently within ants, instead of just once. They concluded that such a result is too unbelievable without further evidence to confirm it. This interpretation of our results implied that natural selection was playing an active role in determining the most efficient route to halting wing development in the worker caste of each species and may operate directly on different genes in different species during the independent evolution of eusociality. However, this interpretation of our results did not fit with the fossil and phylogenetic evidence—there is not a single case within the ~16,500 known ant species where the worker caste has wings and the earliest known ant fossils had already evolved wing polyphenism (Hanna & Abouheif, 2021). So, we revised our manuscript making these points clearer and submitted it to *Science*.

While the manuscript was in review, I had applied for an Assistant Professorship Position in evolutionary biology at McGill University in my hometown Montreal (Quebec, Canada). I applied not expecting to get an interview while still in my PhD, but to my amazement, they called me for an interview! After seeing the high caliber of candidates invited to interview for this position, I thought my chances were quite slim. Then, the most unbelievable thing happened—at midnight, the very night before my interview, I received news from *Science* that my manuscript was accepted. This gave me a boost of confidence going into the interview the next morning. The interview process was intense and during the question period, as well as the "round table" discussion that followed, the search committee took aim at whether I would ever be able to distinguish whether interruption points in the wing GRN evolved through selection or neutral processes. Their skepticism came from the fact that molecular evolutionists had debated the neutral theory of molecular evolution for decades and struggled to distinguish between selection and neutral evolution of DNA sequences (Kreitman, 1996). What made me confident that I would be able to do so? Was there something distinct about GRNs and their evolution that would allow me to answer this question? I phoned Greg in a panic after my interview talk (but before the round table), and he gave me some key advice on how to respond to the critique that I would never be able to distinguish between natural selection and neutral evolution. At the end of the round table discussion, I looked the Committee members straight in the eyes and said: "if you are looking for someone to answer safe questions, then I am not the person you are looking for." To my surprise, I ranked second for the position, and fortunately for me, the person who ranked first took up a position at the University of Toronto (Canada). So, I was offered the Assistant Professor position as a Canada Research Chair in evolutionary developmental biology. After 20 years, I believe I have answered this question (see Section 6), but during that interview, I came out thinking that it might be impossible.

Landing this position at McGill was an important milestone in being able to set up ants as an emerging model for evo-devo because establishing such models takes vision and the sustained commitment from a team of researchers. The ants were finally going to get their shot at providing insights into the world of eco-evo-devo. I was still a PhD student when I received the offer and had already been offered a postdoc position in Nipam Patel's lab at the University of Chicago. I thought that if I was going to establish ants as an emerging model, then I really should learn the techniques for functionally manipulating non-model organisms that Nipam was pioneering in his lab. When I joined Nipam's lab, he was also setting up an emerging model himself—the amphipod crustacean *Parhyale hawaiensis* (Paris, Wolff, Patel, & Averof, 2022). Nipam was a master experimentalist and I had much to learn from him. I learned how to microinject embryos and larvae so I could experimentally manipulate the expression of genes as well as how to make cross-reactive antibodies that I could use to survey the expression of key developmental genes across ants.

However, after spending a year in Nipam's lab, he announced that he was moving to the University of California at Berkeley (USA), and so once again, I found myself packing up and helping to set-up yet another lab from scratch. Fortunately, Berkeley, like the University of Chicago, was an exciting place to be with many great scientists to talk to. I had illuminating discussions with David Wake on the concept of homology, which made me realize that the evolution of interruption points in the GRN underlying wing polyphenism in ants was in fact an excellent example of the third (controversial) scenario that I outlined in Abouheif et al. (1997) (Fig. 2C), where a homologous morphology (wingless worker caste) had evolved divergent underlying GRN expression across species (lability of interruption points in the wing GRN across ants). This scenario was later called "Developmental System Drift" (DSD) by John True and Eric Haig (True & Haag, 2001).

Throughout my PhD, my thinking was very much "bottom-up," meaning that I believed mutations in gene regulation are what led to (drove) the evolution of the GRN underlying ant wing polyphenism. However, while writing my thesis, I often debated this bottom-up view of biology with Louise Roth, a member of my PhD committee at Duke. During one of our meetings, Louise asked me: "how do you know it's not the other way around." What Louise meant was, how could I be so sure that it was changes in genetic regulation, and not changes in environmental factors, like nutrition and temperature, that drove the evolution of interruption points in the GRN underlying ant wing polyphenism? Although my immediate reaction was to dismiss Louise's "top-down" hypothesis, the seed for this top-down idea was planted, and it was only during my postdoc with Nipam that it started to grow and mature. Looking back now, Louise was right, and her question put me on the path towards "eco-evo-devo," making me realize just how important the "eco" in eco-evo-devo really is and how important the complex interaction between genes, development and environment is for understanding evolution.

5. Supersoldier ants and the storage and release of ancestral developmental potential in biological systems (Rajakumar et al., 2012)

While finishing up my postdoc at Berkeley, I spent time at a local cafe planning the first experiment I would do as soon as arrived at McGill. I was most attracted to the classic work on caste determination in *Pheidole* ants (Lillico-Ouachour & Abouheif, 2017). In 1937, Wilhelm Goetsch showed that the developmental switch between minor workers and soldiers is regulated by nutrition—he fed *Pheidole* colonies everything from solid meat chunks, to breadcrumbs soaked in sugar, to frog and rabbit blood (Metzl, Wheeler, & Abouheif, 2018). Goetsch observed that only those larvae fed with solid protein induced bipotential larvae (larvae that had not yet been determined as soldiers or minor workers) to develop into soldiers (Metzl et al., 2018). Almost 45 years later, Wheeler and Nijhout (1981b) showed that this nutrition-dependent developmental switch between minor workers and soldiers is mediated by JH (Fig. 8)—the majority of bipotential larvae treated with JH develop into soldiers. Meanwhile, Gregg (1942) and Passera (1974) showed that *Pheidole* colonies socially regulate the ratio of minor workers and soldiers in the colony—small-headed minor workers make up ~95% of the colony, while the big-headed soldiers make up only ~5%.

By experimentally increasing the percentage of soldiers in *Pheidole* colonies (in increments from 10% to 100%), they observed that adult soldiers inhibited the development of future soldiers until the ratio of minor workers to soldiers returned to their equilibrium ratio of ~95% minor workers and ~5% soldiers (Gregg, 1942; Passera, 1974). These classic experiments demonstrated that this equilibrium ratio is maintained by a balance between the activation of the soldier developmental program by JH and its inhibition by adult soldiers (Lillico-Ouachour & Abouheif, 2017).

Of all the classic experiments on *Pheidole* caste determination, however, I was most enamored by an experiment from Wheeler and Nijhout (1984), in which they simultaneously activated and inhibited the soldier developmental program. They applied JH to bipotential larvae to activate it, while raising these same bipotential larvae with 100% adult soldiers to inhibit it (Wheeler & Nijhout, 1984). The result? Most developed into regular minor workers, indicating that inhibition by adult soldiers can override and block the activation of the soldier development program by JH. However, a few of these larvae developed into minor workers that were almost as large as the soldiers (Fig. 10)! I spent days at that local cafe thinking about these "super minor workers"—did they develop rudimentary wing discs with the same or different interruption points in the wing GRN as those of similarly-sized soldiers? I realized that if I could answer this question, I could finally test whether the lability of interruption points in the wing GRN evolved either in a top-down or in a bottom-up manner. In other words, I could test whether combinations of social interactions, environmental cues, and JH (all top-down factors) induce variation in interruption points within the wing GRN within the worker caste. I reasoned that if variation in interruption points could be induced in this way, and this variation was advantageous for the colony, then it could be selected on and brought to fixation. I couldn't wait to begin this first experiment at McGill.

The time had finally come to start replicating Wheeler and Nijhout (1984)'s classic experiment in my new lab, which began humbly with two undergraduate students, a visiting PhD student from Spain, and an MSc student from the Physics Department. The visiting PhD student from Spain (Diego San Mauro) and one of the undergraduates (Michael Cornoyea) began replicating the Wheeler and Nijhout (1984) experiment. To simultaneously activate and inhibit the soldier developmental program, our experimental replicates contained 40 bipotential larvae treated with two different concentrations of JH (a low and high dose) for activation and were raised by 40 adult soldiers for inhibition. As one of our controls, we made

Fig. 10 Experimental induction of "super minor workers" in *P. morrisi*. Control soldier (left), control minor worker (right), and super minor worker (center), which is almost as large as the control soldier (left). Super minor workers were produced by simultaneously activating and inhibiting the soldier developmental program. We replicated the following experiment in Wheeler and Nijhout (1984) in *P. morrisi*—experimental replicates contained 40 bipotential larvae treated with JH to activate the soldier developmental program and raised by 40 adult soldiers to inhibit it. Control minor workers and soldiers were produced by treating 40 bipotential larvae with solvent only (acetone) and raised with 40 adult minor workers (no inhibition). For detailed methods see Wheeler and Nijhout (1984). The super minor workers we experimentally induced in *P. morrisi* are strikingly similar to the super minor workers that Wheeler and Nijhout (1984) experimentally induced in *Pheidole bicarinata* (Abouheif, 2021; Metzl et al., 2018), and both are similar to the anomalous super minor we discovered in a wild colony of *P. hyatti* (Rajakumar et al., 2018).

"no inhibition replicates" to make sure that our JH treatments were capable of activating the soldier program. These control replicates contained 40 bipotential larvae treated with two concentrations of JH (a low and high dose) raised by 40 minor workers (no inhibition). The first experiments in the Abouheif Lab were underway.

In the meantime, I had gone to Arizona to collect more *Pheidole* colonies and other ants with Bob Johnson. All 1167 currently described species in the genus *Pheidole* have a worker caste composed of two distinct subcastes: minor workers and soldiers. However, there are approximately 8 species known to have evolved a third type of worker called "supersoldiers," which are even larger in size and have disproportionately larger head than the

Fig. 11 In search of supersoldier ants in *Pheidole rhea*. Top row showing minor worker, soldier, and supersoldier subcastes in *P. rhea*. Bottom row showing myself and Robert Johnson searching for the queen of this large *P. rhea* colony, which typically nest in rock crevices, in the southwest of Arizona.

soldiers (Wilson, 2003) (Fig. 11). Bob and I drove down to the south of Arizona to collect a species that had evolved supersoldiers called *Pheidole rhea*, whose colonies were known to nest in the crevices of rocks. We lifted rock after rock looking for them until we finally found them! (Fig. 11). I brought back a colony of *P. rhea* containing minor workers, soldiers, and supersoldier ants back to McGill, mainly because I wanted to investigate what the interruption points in look like in the wing GRN of supersoldier ants. On his first visit to the Abouheif Lab, the second undergraduate student —Rajendhran (Rajee) Rajakumar—saw the supersoldiers of *P. rhea* for the first time. I saw his eyes grow big with amazement; it was love at first sight. I told him they hold the secret to understanding the complex interactions between genes, development, and the environment because a single genome was able to generate these dramatically different types of workers (minor workers, soldiers, and supersoldiers) within a single colony in response to nutrition and social interactions. I had no idea that soon they were about to change both our scientific paths.

Diego began by fixing and dissecting larvae from our no inhibition control replicates. Before long, he rushed into my office with excitement "Ehab,

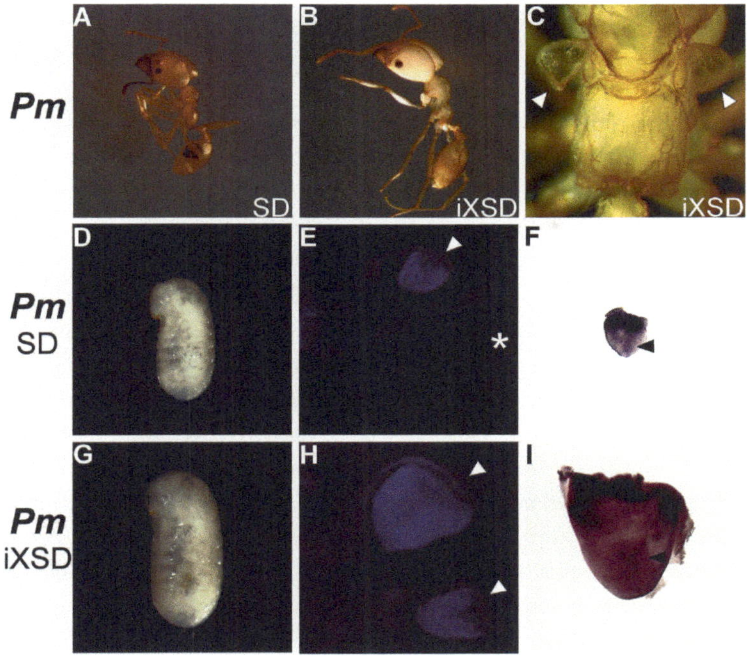

Fig. 12 Methoprene induced supersoldiers (iXSD) in *P. morrisi* (*Pm*). Comparison of (A) normal adult soldier (SD) with (B) induced supersoldier (iXSD). (C) Thorax of iXSD. Comparison of (D) SD and (G) iXSD larvae, and of vestigial wing discs stained for DAPI and expression of the gene *spalt* (*sal*) in (E and F) normal SD and (H and I) iXSD. White arrowheads indicate presence of mesothoracic vestigial wing buds or discs, whereas asterisks denote absence. Black arrowheads indicate *sal* expression in wing pouch. Adult, larval, and vestigial wing disc images are all to scale. *Reprinted with permission from Rajakumar, R., San Mauro, D., Dijkstra, M. B., Huang, M. H., Wheeler, D. E., Hiou-Tim, F., et al. (2012). Ancestral developmental potential facilitates parallel evolution in ants.* Science, 335 *(6064), 79–82.*

the no inhibition control larvae treated with the high dose of JH are much larger than those treated with the low dose, and not only that, those treated with the high dose have two pairs of very large rudimentary wing discs, while those treated with the low dose have just one pair like wildtype soldier larvae!" (Fig. 12D, E, G, and H). For each replicate, Diego fixed half of the larvae and let the other half develop into adults. Control larvae treated with the high dose of JH developed into very large adult soldiers that had significantly larger heads as compared to wildtype adult soldiers (Fig. 12A and B), and they had little wing buds on them too (Fig. 12C). These results took me by surprise, first because that was one of our "control" experiments, and second, because it supported Louise's top-down hypothesis for the evolution of interruption points.

The true significance of this finding, however, came soon after while brainstorming potential projects with Mischa Dijkstra—a newly arrived postdoc from the University of Copenhagen. We wanted to test whether supersoldiers in *Pheidole rhea* and *Pheidole obtusospinosa* have different interruption points in their wing GRN than their soldiers and minor workers. As I began summarizing Diego's results for Mischa, it suddenly struck me! Could the large adult soldiers with very big heads induced in our control replicates actually be supersoldier-like individuals? Had we accidentally induced the supersoldier developmental program in *Pheidole morrisi*, a species that completely lacked supersoldiers? If so, then we predicted that the larvae giving rise to supersoldiers in species that had evolved them should also be larger than soldiers and would also have two pairs of very large rudimentary wing discs, just like the control larvae we treated with a high dose of JH. Within a week, Mischa and I booked our flights down to Phoenix, Arizona (USA), and together with Bob Johnson, we drove down to Nogales, a town bordering the US-Mexican border in southern Arizona in search of *Pheidole* species that evolved supersoldiers.

I had always envisioned Arizona as a barren desert populated with cactus, but I never could have imagined it as a hotspot of biodiversity. It's spectacular in so many ways it's hard to describe; an incredible diversity of plant and animal life and its surrounding high elevation mountains. After a week of flipping rocks and digging, we successfully found and collected several queened colonies of the two species that had evolved supersoldiers—*P. rhea* and *P. obtusospinosa*—as well as of several other *Pheidole* species that had not. We started driving back towards Phoenix late in the evening to pack up the ants and head home. Unfortunately, Mischa forgot to fill the gas tank and before long we ran out of gas, leaving us stranded in the middle of the highway at midnight, with no service stations close by. This was just one of many mishaps and adventures while collecting ants with my students over the years.

Once back at McGill, Rajee and Mischa started testing our prediction on the development of supersoldier ants. Their results confirmed that, like the supersoldier-like individuals we accidentally induced with a high dose of JH (Fig. 12), the larvae that develop into evolved supersoldiers are also larger than those of soldiers and had two pairs of very large rudimentary wing discs! (Fig. 13E, F, G, I, J, and K). Furthermore, Rajee and postdoc Abderrahman

Khila found that the interruption points in the wing GRN in these large rudimentary wing discs changed (Fig. 13H and L). This confirmed that we really did activate a hidden or dormant supersoldier developmental program in a *Pheidole* species that lacks supersoldiers. Soon, we realized that larval size and the number and size of the rudimentary wing discs are the defining features of the *Pheidole* soldier and supersoldier developmental program—supersoldier larvae are larger in size than soldiers and develop two pairs of very large rudimentary wing discs, soldiers are larger than minor workers and develop one pair of large rudimentary wing discs, and minor worker larvae have none. No other differences are discernible between them (Figs. 8 and 13).

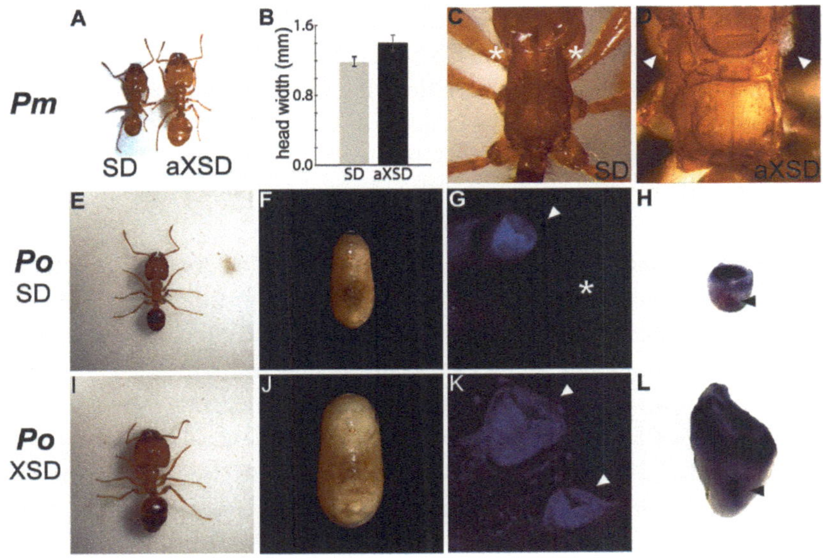

Fig. 13 Supersoldier anomalies in *P. morrisi* compared to the evolved supersoldier subcaste in *P. rhea*. Comparison of *P.morrisi* (*Pm*): (A left) normal soldier (SD) and (A right) anomalous supersoldier (aXSD), (B) mean and standard deviation of head width of normal SD (gray) and aXSD (black), thorax of (C) normal SD and (D) aXSD. Comparison of *P. obtusospinosa* (*Po*): adults of (E) SD and (I) XSD, larvae of (F) SD and (J) XSD, vestigial wing discs (DAPI) and *spalt* (*sal*) expression in SD (G, H) and XSD (K, L). White arrowheads indicate presence of mesothoracic vestigial wing buds or discs, whereas asterisks denote absence. Black arrowheads indicate *sal* expression in wing pouch. Adult, larval, and vestigial wing disc images are all to scale. *Reprinted with permission from Rajakumar, R., San Mauro, D., Dijkstra, M. B., Huang, M. H., Wheeler, D. E., Hiou-Tim, F., et al. (2012). Ancestral developmental potential facilitates parallel evolution in ants.* Science, 335 *(6064), 79–82.*

While these experiments were ongoing, we made another fundamental discovery in parallel, almost entirely by chance. Rajee and I had gone down to Long Island, NY to collect more *P. morrisi*, the same species in which Diego had accidentally activated the dormant supersoldier program. A couple of weeks after returning to McGill, Rajee and I were sitting outside McGill's Biological Sciences building collecting some sidewalk ants (*Tetramorium immigrans*), when he happened to mention to me (as a kind of "by the way") that there were a few large soldiers in one of the new colonies we just collected in Long Island. I leapt up with excitement, and we both ran to the growth chambers, only to find 8 supersoldier-like individuals running around the colony (Fig. 13A and B). These supersoldier-like individuals were rare anomalies that also had tiny wing-buds on them, and were strikingly similar to those we accidentally induced in the lab and to the supersoldiers that evolved in *P. obtusospinosa* and *P. rhea* (Fig. 13A–D). I use the word "rare" here to mean that I have been collecting *P. morrisi* for over 25 years, and unless you are actively looking for them with a specific search image, such anomalies are not obviously observed. This meant that the activation of the dormant supersoldier developmental program in species that lack supersoldiers was happening, albeit at rare frequency, in wild *Pheidole* colonies.

In light of these discoveries, we asked whether this dormant potential to produce supersoldiers is a unique feature of *P. morrisi* or was it present in other *Pheidole* species? By applying a high dose of JH to bipotential larvae in a closely related sister species *P. hyatti*, Rajee and Mischa were able to induce the dormant supersoldier developmental program (Fig. 14). Then, with the help of Diana Wheeler and her PhD student Ming Huang at the University of Arizona, we were able to induce the dormant supersoldier developmental program in *Pheidole spadonia*, a more distantly related *Pheidole* species that lacked supersoldiers (Fig. 14). These results showed that the dormant potential to produce supersoldiers is an ancestral and conserved feature of the genus *Pheidole*.

But where did this dormant and ancestral potential to produce supersoldiers come from? All of the described *Pheidole* species with a supersoldier subcaste are located in southwestern USA and Northern Mexico. Based on their common geographic origin, Wilson (2003) proposed that the supersoldier subcaste also had a common phylogenetic origin, meaning that they evolved only once in the genus *Pheidole* and subsequently diversified and speciated within southwestern USA and Northern Mexico. However, to our surprise, Rajee discovered that *P. obtusospinosa* and *P. rhea*, the two

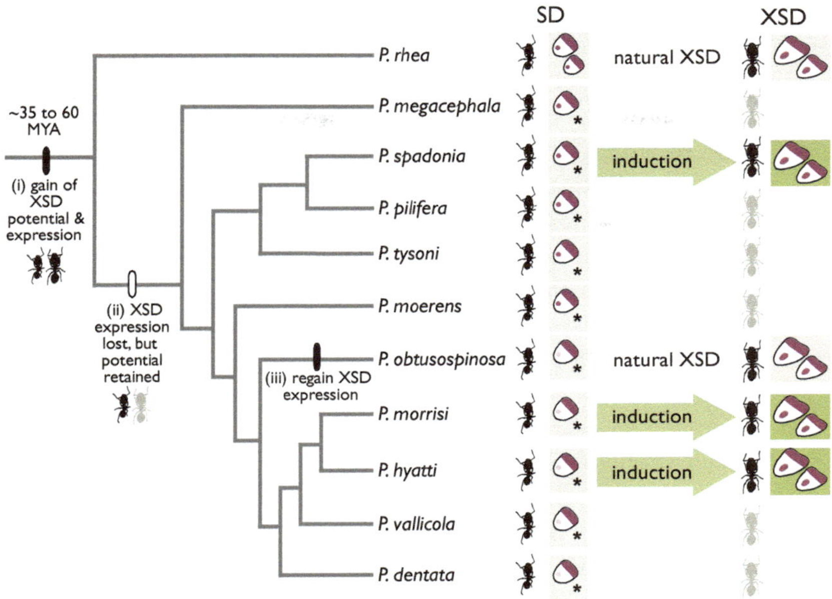

Fig. 14 Evolutionary history of ancestral developmental potential and phenotypic expression of supersoldiers (XSD). (i) indicates that the developmental potential and phenotypic expression of a novel supersoldier subcaste originated in an early common ancestor of *Pheidole*; (ii) indicates that the phenotypic expression of supersoldiers was subsequently lost, but the ancestral potential to produce them was retained; and (iii) indicates that this potential was then actualized in *P. obtusospinosa* leading to the re-evolution of a supersoldier subcaste. Purple represents pattern of *sal* expression, whereas asterisks indicate absence of rudimentary wing discs and *sal* expression. Green arrows and boxes represent induction of XSD potential. *Reprinted with permission from Rajakumar, R., San Mauro, D., Dijkstra, M. B., Huang, M. H., Wheeler, D. E., Hiou-Tim, F., et al. (2012). Ancestral developmental potential facilitates parallel evolution in ants.* Science, 335 *(6064), 79–82.*

species with a supersoldier subcaste, are not sister species, but instead, are on different parts of the *Pheidole* tree (Rajakumar et al., 2012) (Fig. 14). This means that a supersoldier subcaste evolved independently in these two species. In his phylogenetic reconstruction of 12 *Pheidole* species, *P. rhea* is the earliest diverging branch in *Pheidole*, while *P. obtusospinosa* is more derived (Fig. 14), a finding consistent with a much larger phylogenetic analysis of 140 *Pheidole* species by Moreau (2008).

The classical interpretation of this type of phylogenetic pattern would be that the supersoldier subcaste evolved completely independently of one another, meaning it evolved *de novo* in *P. rhea* and *de novo* in

P. obtusospinosa (Rajakumar et al., 2012). However, this classical interpretation fails to consider that one of the earliest diverging branches within the genus had evolved a supersoldier subcaste and that the dormant supersoldier developmental program could be induced in several closely and distantly related species within the genus (Rajakumar et al., 2012) (Fig. 14). Therefore, a more parsimonious explanation is that a supersoldier subcaste evolved early in the genus but was subsequently lost, such that the majority of the ~1167 species now lack a supersoldier subcaste (Rajakumar et al., 2012). However, even though supersoldiers were lost at the phenotypic level, the ancestral developmental potential to produce supersoldiers was retained in the genus for approximately 35–60 million years (Rajakumar et al., 2012) (Fig. 14).

By this point, I had started my year-long sabbatical at the KLI Institute for the Advanced Study of Life and Sustainability Sciences in Vienna, Austria, and was having a Viennese coffee with Thomas Flatt (now a Professor at the University of Lausanne) in front of the Franz-Joseph train station. As I was describing our results, he asked me a critical question: how is the ancestral developmental potential to produce supersoldiers retained for so long? A question that we had not yet focused on. After parting, he shared with me an insightful paper by Stephan Stearns on clade specific constraints in evolution (Stearns, 1994). However, his question got me thinking more mechanistically about how development could maintain the supersoldier potential for so long. As someone who has previously published on phylogenetic methods (Abouheif, 1998, 1999b), I was quite familiar with Dollo's Law, which posits that lost traits would decay too quickly for them to re-evolve after millions of years (Bull & Charnov, 1985; Gould, 1970). However, countless violations of Dollo's Law have been discovered (Bull & Charnov, 1985; Collin & Miglietta, 2008; Gould, 1970; Marshall, Raff, & Raff, 1994; Porter & Crandall, 2003; Seher et al., 2012; Walker-Larsen & Harder, 2007). We therefore reasoned that Dollo's Law is too simplistic and fails to consider the complexity of biological systems. In the case of the supersoldier ants in *P. obtusospinosa*, we argued that JH is redundantly used to mediate the developmental switch between minor workers and soldiers and used again shortly after to mediate the developmental switch between soldiers and supersoldiers (Rajakumar et al., 2012) (Fig. 15). We also argued that there is a redundancy in the developmental program that generates soldiers and supersoldiers as indicated by the activation of rudimentary wing imaginal disc development (one pair of rudimentary wing imaginal discs in soldiers and two pairs in supersoldiers) (Rajakumar et al., 2012)

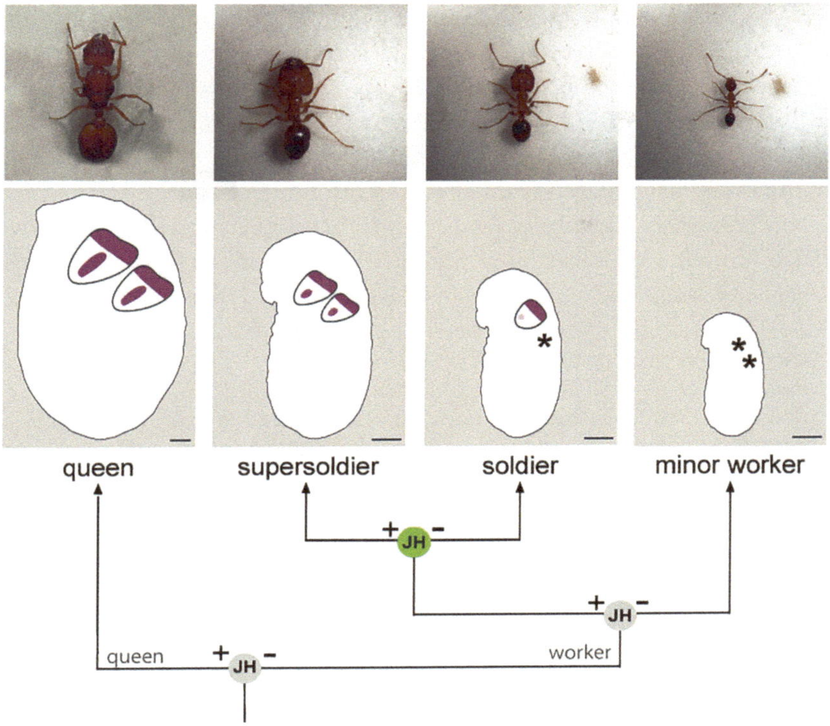

Fig. 15 Actualization of supersoldiers in *P. obtusospinosa* through the re-evolution of a second JH-sensitive period that mediates the switch (green circle) between soldiers and supersoldiers. Scale bars indicate the relative size of queen, soldier, and minor worker larvae. All adult images are to scale. Purple represents *sal* expression, whereas asterisks indicate absence of visible wing primordia and *sal* expression. *Reprinted with permission from Rajakumar, R., San Mauro, D., Dijkstra, M. B., Huang, M. H., Wheeler, D. E., Hiou-Tim, F., et al. (2012). Ancestral developmental potential facilitates parallel evolution in ants.* Science, 335 *(6064), 79–82.*

(Fig. 12E and H). Therefore, it may be not possible to lose the ancestral developmental potential to produce supersoldiers as long as the soldier developmental program is maintained and all *Pheidole* species have a soldier subcaste (Rajakumar et al., 2012).

Biological systems are replete with redundancies—the redundant use of hormones, signaling pathways, and developmental regulatory genes is a general feature of multicellular development (Gilbert & Barresi, 2019; Nijhout, 1994; Wray & Lowe, 2000). We therefore predicted that such redundancies should generally retain ancestral developmental potentials in multicellular organisms over the span of millions of years, despite mutations that accumulate in the genes that underlie these lost phenotypes (Rajakumar et al., 2012). This prediction is supported by the existence of "reversions" or "atavisms," which are the sporadic appearance of an ancestral trait in an individual that has evolutionarily lost it (Hall, 1995a, 2010; Stiassny, 2003). Reversions or atavisms reflect the induction of ancestral developmental potentials and have been documented in a vast number of organisms, including both animals and plants (Hall, 1984; West-Eberhard, 2003). Darwin (1868) himself recognized their ubiquity in populations in his book "*The Variation of Plants and Animals under Domestication*." Darwin (1868) wrote:

> "… I wish to guard the reader against supposing that reversion is due to some rare or accidental combination of circumstances. When a character, lost during hundreds of generations, suddenly reappears, no doubt some combination must occur; but reversions may constantly be observed, at least to the immediately preceding generations, in the offspring of most unions …"

Since Darwin, the sporadic appearance of atavism has been documented in natural populations, such as the induction of development of hind limbs in whales and dolphins, which lost their hindlimbs after the mammalian ancestor of whales and dolphins re-entered the water (Bejder & Hall, 2002; Hall, 1984). Atavisms have also been experimentally induced in the lab, such as the induction of teeth in chickens or the induction of hind wings in fruit flies (Gibson & Hogness, 1996; Hall, 1995a; Harris, Hasso, Ferguson, & Fallon, 2006; Stiassny, 2003). It is therefore likely that the retention and sporadic release of ancestral developmental potential, which we discovered in ants, is also a general property of multicellular development.

This generality of ancestral developmental potentials would be further reinforced during my sabbatical in Vienna, when I came across an old German paper written in 1939, which according to Gregg (1942) reported the appearance of soldier-like individuals in a genus of termites (*Anoplotermes*) that had lost its soldier caste (Metzl et al., 2018). My Family and I stayed at a charming apartment-hotel called the Mondial. The staff were extraordinarily friendly, and I became friends with one of the desk clerks, Christian Metzl, a biology student at the University of Vienna. We had long conversations about my research, and he became fascinated with our discoveries on the supersoldier ants. One day, I asked him to read and translate the paper I discovered by Goetsch (1939), which to my surprise, revealed a detailed discussion of ancestral developmental potentials 70 years earlier (Metzl et al., 2018). This discovery by Goetsch (1939) was remarkably prescient, as can be seen in this English summary of Goetsch's discovery by Gregg (1942):

> "In a recent paper Goetsch ('39) states that an intermediate nasute type soldier appeared in a captive colony of Anoplotermes (soldierless genus), and can be interpreted as furnishing evidence for genetic difference is in the eggs. It would mean that although soldiers do not appear in normal wild colonies and are supposed to have dropped out phylogenetically, their genes are in reality still retained by the species. This can be reinterpreted in favor of environmental control by assuming that, while genes are admittedly present, some change in the physiological thresholds in the particular colony has in part enabled them to come to expression. The anomaly might also be interpreted as the result of phylogenetically suppressed genes which cause the degeneration of certain characters (in this case the soldier caste), but which have not been entirely lost because of other vital influences in the termite colony."

Gregg (1942) not only provides hypotheses about the retention of ancestral developmental potentials to produce soldiers in this soldierless termite genus, but he also provides a hypothesis about the mechanism for their release—"some change in physiological thresholds"—through which this soldier anomaly in termites, like the supersoldier anomalies we found in *P. morrisi*, may have been induced. Since our serendipitous meeting leading to the translation of Goetsch (1939) on ancestral developmental potentials, Christian opened his own translation business (www.oldgermanpapers.com). We also joined forces with Diana Wheeler to unearth the roots of the field of eco-evo-devo in the old German literature, which suggests eco-evo-devo may be as old as the field of evo-devo itself (Metzl et al., 2018).

If atavisms and latent developmental capacities are so widespread in nature, why have their roles in evolution and development been largely underappreciated? Stiassny (2003) argues that one reason for this is the prevailing influence of the neoDarwinian argument put forth by Levinton (1983). He argued that atavisms and latent developmental capacities hold little evolutionary potential because they are freaks of nature or slips of the developmental system. Prior to our study on supersoldier ants in *Pheidole*, no single study had convincingly shown that the induction of an ancestral developmental potential in a natural population can lead to the re-evolution of a phenotypic trait in nature (Abouheif et al., 2014; Rajakumar et al., 2012). Rigorously demonstrating this requires uniting three types of evidence: (1) observing the sporadic release of ancestral potential for a trait in natural populations; (2) the ability to experimentally induce the release of ancestral potential for trait in the lab; and (3) phylogenetic evidence that the ancestral trait was lost and then re-evolved in the taxonomic group of interest (Abouheif et al., 2014). Rajakumar et al. (2012) was the first to unite all three types of evidence in a single study. First, we found evidence for the appearance of supersoldier-like anomalies in natural *Pheidole* colonies (Fig. 13A–D); second, by applying high doses of JH during the critical period, we could activate the supersoldier developmental program in the lab in species that have not evolved supersoldiers (Fig. 12); and finally, we showed that supersoldier subcaste in *P. obtusospinosa* re-evolved in the genus *Pheidole* through an evolutionary change in an ancestral, but cryptic, JH threshold through a process called genetic accommodation (Suzuki & Nijhout, 2006; West-Eberhard, 2003) (Figs. 8 and 15). By uniting these three findings in a single study (Rajakumar et al., 2012), the supersoldier ants provided evolutionary biology with the first solid evidence that the retention and sporadic release of ancestral developmental potential in natural populations can be realized by natural selection and can lead to the re-evolution of ancestral-like traits in extant organisms.

My Colleagues have often asked me whether our discovery is unique to ants because of their extreme plasticity and social life. As I commented earlier, every organism makes a particular biological process more evident. The existence of morphologically discrete worker subcastes in *Pheidole* colonies, combined with the fact that *Pheidole* colonies care for and keep anomalous individuals alive, made it possible for us to readily observe and characterize anomalies in wild colonies. In solitary organisms, most of these anomalies would have been wiped out by natural selection too quickly for biologists to observe them. Furthermore, ant colonies are "superorganisms," where the queen is analogous to the germline and workers are analogous to the soma of a solitary multicellular organism (Abouheif, 2004; Hölldobler & Wilson, 2009; Yang, 2007). The different worker subcastes (minor workers, soldiers, and supersoldiers) in a *Pheidole* colony are therefore analogous to differentiated cell types in a solitary multicellular organism, like those that give rise to the heart and immune system (Yang, 2007). In contrast, observing the sporadic appearance of anomalies that represent atavistic cell types within a solitary organism is challenging, even with our current imaging technologies. This has likely contributed to the general underappreciation of this type of variation in solitary organisms. Nonetheless, excellent examples that evolution has occurred through the retention and release of ancestral developmental potentials in solitary organisms is now accumulating (Foster & Baker, 2019; Foster, O'Neil, King, & Baker, 2019).

Our article was finally published in *Science* on January 6, 2012 (Rajakumar et al., 2012). We were excited but had no idea that our study would explode in the media. I was sitting in my office thinking it would be a relatively quiet day, when suddenly, my phone started ringing off the hook.

Before long, I was speaking to one reporter after another. One of these reporters, Ed Yong, had written a commentary called "Return of the Super Ants" for *Nature* (Yong, 2012), and our study was featured on the front page of *Nature*'s website as well as on hundreds of news networks, including the BBC, CBC, *Financial Times*, *Discover Magazine*, *Popular Mechanics*, and many others. Rajee and I gave multiple radio interviews and we appeared on the second page of the Saturday print edition of the *Montreal Gazette* and several other in-print newspapers throughout Canada and the world. And then, soon after, we appeared on the six o'clock nightly news on National Television (Global National Nightly News).

Why had the supersoldier ants become such a media sensation? My guess is that our paper was published at the height of the Marvel Cinematic Universe—both Captain America and The Hulk were supersoldiers that were created in the lab with supersoldier serum. Also, a bizarre event occurred just a week before our paper was published—images of an immensely tall soldier marching with the North Korean Army began to appear in the media. Headlines called him a "human supersoldier," like Gould (2011)'s article "*9-Foot-Tall Supersoldier Mourns Kim Jong Il*" in *The Atlantic* published on December 29, 2011. I started receiving several emails from the general public that were either volunteering themselves for clinical trials that would induce their innermost supersoldier potential or they were begging me not release the supersoldier ants outside my lab. All this media attention has largely been forgotten, but clearly, the idea that the power of our ancestors lies deep within us, and can be recalled, is an empowering feeling that captures the imagination.

One of the organizers of the 2013 Gordon Research Conference on Quantitative Genetics, Jack Dekkers, had heard my interview on CBC's *Quirks and Quarks* show and invited me to give a lecture on the supersoldier ants at the conference. Many animal and plant breeders attend this conference, and after my talk, I was invited by one of the largest Poultry Breeding Associations in the world—The National Breeders Round Table—to give a talk at their annual meeting in St. Louis. Why would chicken breeders be interested in an ancestral developmental potential to produce supersoldier ants? One of the online media websites *The A Register* that covered our *Science* paper wrote:

> "For example, the aurochs – the massive ancestor to modern cattle that was hunted to extinction by the 1600s – may be recreatable by examining a cow's genome and finding a way to activate the processes that would cause the much larger and more aggressive aurochs to develop. In the plant world too, crops could be subjected to environmental and chemical stressing to

> see if the dormant genotypes could be activated. This could usher in new crops that can better deal with current conditions – not to mention changing conditions as climate change wreaks its havoc."
>
> ***(Thomson, 2012)***

The National Breeders Round Table meeting inspired Rajee and I to write a basic protocol for how to release ancestral developmental potential in domesticated chickens, and once released, how to select on the most desirable characteristics (Rajakumar & Abouheif, 2013). Instead of resorting to genetic modification of chickens to enhance the breeding of desirable characteristics, our approach assumes that such desirable characteristics are already present, but dormant, in the genome. Our protocol still needs to be developed in further detail and tested in domesticated animals. It is also important to test this in domesticated crops, where it could potentially have a significant impact on enhancing global food security (Moczek et al., 2015). We are just at the beginning in terms of understanding all the ways we can harness ancestral developmental potentials to solve many critical societal challenges, including animal and plant breeding, medicine, and how to enhance our understanding of biodiversity conservation in the face of a rapidly changing climate (Rajakumar & Abouheif, 2013).

6. Rudimentary organs acquire novel functions and store ancestral developmental potential (Rajakumar et al., 2018)

After the publication of Rajakumar et al. (2012), many questions remained about supersoldier ants and ancestral developmental potentials, but one question kept coming up—why is the growth of rudimentary wing discs correlated with the soldier and supersoldier developmental programs in *Pheidole*? As you may recall, supersoldier larvae in *P. obtusospinosa* develop two pairs of very large rudimentary wing discs, soldiers develop just one pair of large rudimentary forewing discs, and minor worker larvae do not develop any (Fig. 15). Although I had continued to entertain several alternative evolutionary hypotheses—neutral evolution, natural selection, or a combination of both—to explain the lability of interruption points in the wing GRN of wingless worker castes across ants, I had always assumed that the wing rudiment itself had no function and was therefore evolving neutrally (Nahmad, Glass, & Abouheif, 2008; Shbailat & Abouheif, 2013; Shbailat et al., 2010). At best, I thought that the wing rudiment might have a minor function in providing cells to build a small part of the body wall.

In other words, I interpreted the growth of rudimentary wing imaginal discs to be a consequence, and not a cause, of activating the soldier or supersoldier developmental program.

Wheeler and Nijhout (1981a) proposed that the correlation between rudimentary wing disc growth and the soldier program in *Pheidole* occurs because the same hormone (JH) regulates the switch between queen or worker castes early in development and again between minor worker or soldier castes later in development (Fig. 8). Implicit in this idea is a "developmental constraint," meaning that parts of the queen program (wing discs) are thought to be activated as a secondary by-product of re-using JH to activate the soldier program later in development. Another possibility, which I proposed in Abouheif (2004), is that growth of the wing rudiment is functionless, but is pleiotropically linked to another developmental process necessary for soldier development, and therefore "hitchhikes" on that developmental process. These neutral evolution hypotheses felt intuitive because functional significance of rudiments and vestiges has largely been ambiguous (Darwin, 1859; Gould, 1977; Hall, 2003; Mayr, 1994). For example, the rudimentary tail in a human embryo has no known function, while the gill rudiments in a human embryo are known to go on to form part of the jaws and bones of the inner ear (Gould, 1977; Hall, 1994).

During one of our regular weekly lab meetings, however, all of these assumptions and ideas got turned on their head after reading about the classic wing disc transplant experiments in *Drosophila* by Ursprung and Hadorn (1962). They transplanted dissociated wing disc cells into late-stage wild type larvae and found that these larvae kept growing to a very large size until the transplanted wing disc cells stopped proliferating. This triggered another eureka moment! If a transplanted wing disc could signal to the rest of a fruit fly larva to delay its development until it stopped proliferating, could it be that the rudimentary wing discs in *Pheidole* soldier and supersoldier larvae also signal to larvae to regulate their size? Rajee and I rushed into my office after lab meeting to keep discussing this possibility. If this were true, that would mean that rudimentary wing discs in *Pheidole* are not functionless, but instead, play a causal role in regulating the size of soldier and supersoldier ants. We were so excited that we plunged head-first and began devising several ways that we could test this crazy idea.

What I find remarkable looking back is that I had already studied these classic transplant experiments when I first set up my lab at McGill in 2004 (Ephrussi & Beadle, 1936; Koch & Abouheif, 2019; Poodry & Woods, 1990; Ursprung & Hadorn, 1962). However, reading about the results of these transplant experiments with the correlation between rudimentary disc growth and soldier and supersoldier development in mind, gave them entirely new context and meaning. There are two brief lessons to be learned here: reading the older literature was key to helping make sense of our results, and second, it helps to re-read some of the key papers in your field because they take on new meanings depending on the data you have at hand.

Finally, Rajee started testing our idea that rudimentary wing discs play a functional role in regulating size of *Pheidole* soldier larvae. We enlisted the help of Marie-Julie Favé, another PhD student in my lab who was finishing her PhD thesis on wing GRN evolution in ants on the Sky Islands of Arizona (Favé et al., 2015). Rajee and Marie-Julie started by knocking down a wing-specific gene called *vestigial*, which was known to regulate the growth and patterning of wing discs in fruit flies. *Vestigial* is a master regulator of wing development—overexpressing *vestigial* in other imaginal discs, including the leg and head (eye-antennal) disc transforms them into wing discs such that wings grow in the place of legs and antenna (Halder et al., 1998; Kim et al., 1996). Rajee and Marie-Julie used RNA interference (RNAi) to knock down *vestigial* expression to halt the growth of the rudimentary forewing disc in soldier larvae. The first results came in, and they came into my office with a disappointed look on their face, announcing that, while making the individuals smaller, the statistical analyses showed that *vestigial* RNAi did not make them significantly smaller relative to control injections. I was disappointed, but as scientists we must be prepared for our ideas, no matter how elegant, to be slayed by the sword of experimental results. The next morning, however, Rajee and Marie-Julie ran into my office with great excitement! They had re-analyzed the results, but this time, they included the effect of *vestigial* RNAi on both head and body size together. Although the rudimentary wing disc affected both head and body size, it made the head much smaller relative to the body (Rajakumar et al., 2018) (Fig. 16A–C). We had envisioned that growth of rudimentary wing discs could function in regulating body size, but we never could have imagined that they functioned in regulating head-to-body scaling, one of the most variable traits across ants. This means that the rudimentary wing discs had been repurposed in evolution to generate novel big-headed soldier and supersoldier castes in *Pheidole* (Rajakumar et al., 2018).

Suddenly, that morning, more than 20 years of research on understanding the simultaneous conservation and lability of interruption points in the GRN underlying wing polyphenism in ants all started making sense. Because rudimentary wing discs function to generate the big-headed soldiers

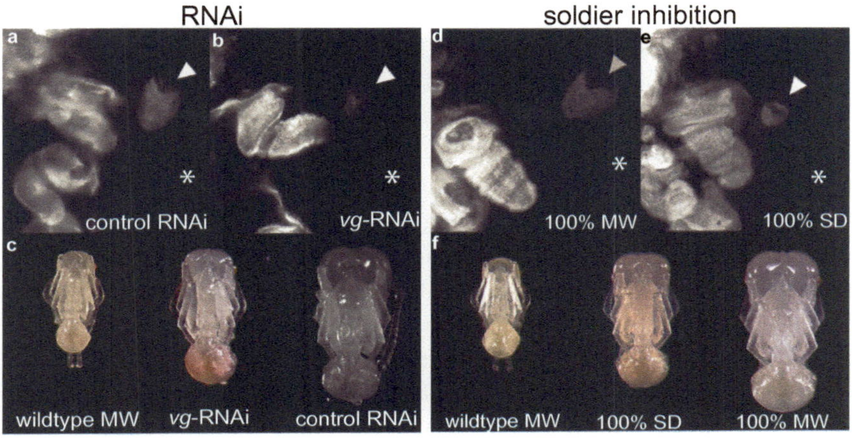

Fig. 16 Soldier rudimentary forewing discs regulate size and disproportionate head-to-body scaling, and in turn, are regulated by social inhibition. Comparing rudimentary forewing discs after RNAi: (A) *yfp*-RNAi (control RNAi); (B) *vg*-RNAi. (C) Comparing wildtype minor worker (MW) with individuals treated either with control RNAi (right) or *vg*-RNAi (center). Comparing rudimentary wing disc size after: (A) no-inhibition (100% MW), (B) high-inhibition (100% SD). (D) Comparing wildtype minor worker (MW) with individuals raised by 100% MW (right) or 100% SD (center). All manipulations (A–F) were on SD-destined larvae. Rudimentary wing disc presence (white arrowheads) or absence (white/black asterisks). Image comparisons are to scale. *Adapted with permission from Rajakumar, R., Koch, S., Couture, M., Fave, M. J., Lillico-Ouachour, A., Chen, T., et al. (2018).* Nature, 562, *574–577.*

in *Pheidole*, and perhaps in other species that had independently evolved a soldier caste, natural selection had modified the interruption points in these species to allow the rudimentary wing discs to perform their novel function. This discovery would also explain why the supersoldier potential was retained in the genus for 35–60 million years—because rudimentary wing discs function in generating soldiers (and all known *Pheidole* species have soldiers) the ancestral developmental potential to generate supersoldiers cannot be lost without first losing the function of rudimentary wing discs to generate the soldiers. Therefore, rudimentary wing discs store the ancestral developmental potential to produce supersoldiers (Rajakumar et al., 2018).

In the true spirit of eco-evo-devo, we began to think about how the rudimentary wing disc might be regulated in wild *Pheidole* colonies. We wondered whether social interactions between minor workers and soldiers (a colony-level property) could regulate the growth of the rudimentary wing disc in soldier larvae (an individual-level property). This brought us right back to the soldier inhibitory pheromone and its role in suppressing the development of soldiers when the number of soldiers exceeded ~5% (Gregg, 1942; Wheeler & Nijhout, 1984). Was the inhibitory pheromone capable of suppressing the soldier developmental program by halting the growth of the rudimentary wing disc? Rajee stayed on for a short postdoc to test this exciting possibility with the help of a talented team of students—Sophie Koch, Mélanie Couture, Angelica Lillico-Ouachour, Travis Chen, Giovanna De Blasis, Arjuna Rajakumar, and Dominic Ouellette. Together, they showed that, like *vestigial* RNAi, the soldier inhibitory pheromone halted growth of the rudimentary wing imaginal discs producing individuals that are smaller and with a modified head-to-body scaling (Rajakumar et al., 2018) (Fig. 16D–F). Therefore, *Pheidole* colonies in nature actively maintain a ratio of minor workers (~95%) and soldiers (~5%) by activating the growth of rudimentary wing discs through nutrition and JH to produce the big-headed soldiers, and when number of soldiers in the colony exceeds ~5%, the soldier inhibitory pheromone suppresses growth of the rudimentary wing discs thereby suppressing development of soldiers (Rajakumar et al., 2018) (Fig. 17). Therefore, the rudimentary wing discs not only function in generating the big-headed soldier castes, but are themselves regulated by the social interactions, hormones, and pheromones to maintain the ratio of minors and soldiers in the colony (Rajakumar et al., 2018). This direct and intricate feedback between regulation of the colony (ratio of minor workers and soldiers) and regulation of the individual (the rudimentary wing disc) unites these two levels of biological organization. In so doing, our discovery provides a unique model in biology to potentially connect, in a complex non-linear chain of events, changes in DNA nucleotides within genes, to genes within networks, to networks within organs, to organ systems within an individual, to individuals within colonies, to colonies within populations, and to populations within ecosystems, creating a high resolution genotype-phenotype map that traces the origin of changes in an organism's development, ecology, and evolution.

Finally published in 2018, our discovery extends beyond the ants, potentially holding far-reaching implications for biology (Rajakumar et al., 2018). The transient appearance of rudiments are a general feature of organismal development and evolution, such that Darwin (1859) in his *Origin of Species* was able to use rudiments as evidence of ancestry, and Haeckel was able to use them to reconstruct phylogenetic relationships between organisms (Gould, 1977). Their functional significance, on the other hand, has remained enigmatic and are generally thought to be functionless or have

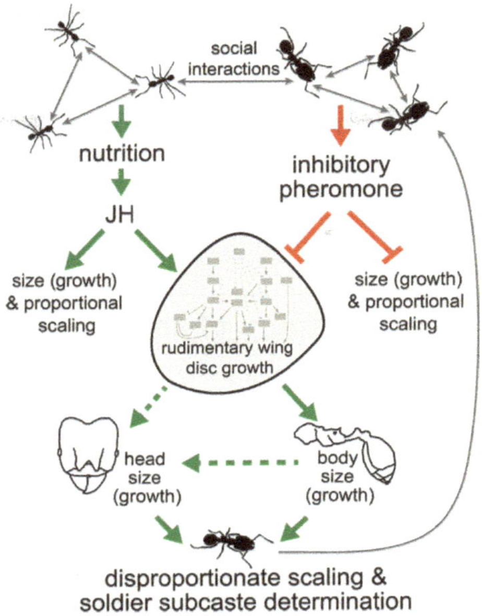

Fig. 17 Rudimentary wing discs: social regulation, development, and evolution of complex worker caste systems in ants. Interactions (arrows/lines) may be direct or indirect. Activation (green arrows); potential pathways to disproportionate scaling (dashed green arrows); inhibition (red arrows/lines); social interactions (gray arrows); rudimentary forewing discs (gray circle); wing gene network (gray boxes). *Reprinted with permission from Rajakumar, R., Koch, S., Couture, M., Fave, M. J., Lillico-Ouachour, A., Chen, T., et al. (2018).* Nature, 562, *574–577.*

minor functions during development (Hall, 2003). Research on the developmental and evolutionary significance of rudiments received little attention after Haeckel's approach to integrating ontogeny and phylogeny was largely abandoned (Gould, 1977; Mayr, 1994). Our discovery raises the possibility that the transient appearance of rudiments plays major regulatory roles during development, in which they regulate (and are themselves regulated) during critical developmental windows, most likely through highly conserved signaling pathways (Rajakumar et al., 2018). If this prediction is correct, then our discovery holds important implications for medicine—rudimentary traits during human development should be revisited and tested for possible signaling functions. Our discovery also shows that rudiments can store ancestral developmental potentials for millions of years, and when it is sporadically released, can serve as raw variation for natural selection to act upon (Rajakumar et al., 2018).

My scientific journey, as I partly document here through the discoveries in Abouheif and Wray (2002), Rajakumar et al. (2012), and Rajakumar et al. (2018), ends with the inescapable conclusion that rudimentary organs and the storage and release of ancestral developmental potentials is an underappreciated source of variation for adaptive evolution that can be harnessed to improve animal and plant breeding, biodiversity conservation, and medicine.

7. Conclusion: The future of ants as a model clade for eco-evo-devo

In his perspective piece on Abouheif and Wray (2002) in *Science* "*Sociogenomics Takes Flight*," Gene Robinson wrote:

> "These are good times for research on social insects. The social Hymenoptera—ants, bees, and wasps—are emerging as valuable models for the study of molecular processes ... Abouheif and Wray now formally welcome ants into the vigorous field of "evo-devo" with a fascinating description of changes in gene expression that are associated with the evolution of the wingless worker caste ... The heightened interest in social insects reflects the sense that the time has come to develop a comprehensive understanding in molecular terms of social life: how it evolved, how it is governed, and how it influences all aspects of genome structure, gene expression and organismal development, physiology and behavior."
>
> ***(Robinson, 2002)***

Fast forward 20 years, ant researchers have now sequenced over 100 ant genomes from almost every major subfamily (Boomsma et al., 2017; Gadau et al., 2012), they have developed cutting edge tools to manipulate gene expression during development (reviewed in Sieriebriennikov, Reinberg, and Desplan (2021), including CRISPR (Chiu, Hsu, Chang, Huang, & Wang, 2020; Trible et al., 2017; Yan et al., 2017), RNAi (Rajakumar et al., 2018; Ratzka, Gross, & Feldhaar, 2013), transgenics (Sieriebriennikov et al., 2021) and pharmacological inhibitors (Alvarado, Rajakumar, Abouheif, & Szyf, 2015; Fetter-Pruneda et al., 2021)). Researchers are increasingly combining these molecular tools with colony-level manipulations—what E.O. Wilson has called the "pseudo-mutant" approach (Wilson, 1985). For example, removing all the soldiers in a *Pheidole* colony to study its homeostatic properties is one way of making a pseudo-mutant, while brood-swapping experiments, where brood from one colony is swapped with the brood from another colony is another way (Lillico-Ouachour & Abouheif, 2017; Metzl et al., 2018; Teseo, Chaline, Jaisson, & Kronauer, 2014).

Unlike the community of researchers focused on a single model bee species—the honeybee *Apis mellifera*—there is no single model ant species. One reason for this is that all ~16,500 described species of ants are eusocial and show an extraordinary range of variation in their eusocial organization and caste structure, and therefore, there is no "standard" ant (Heinze, 2008). Ants should instead be thought of as a model clade for eco-evo-devo studies. Sanger and Rajakumar (2018) discuss the many insights that can be gained from adopting a model clade instead of model species approach in evo-devo, one of the most important being the ability to produce a high-resolution map of the molecular and developmental transitions underlying the origin and elaboration of novel phenotypes. In addition to using this model clade approach to gain insight into the origin and evolution of wing polyphenism and novel worker caste systems, my lab has used this approach to gain insight into the eco-evo-devo of reproductive division of labor and of obligate endosymbiosis in ants (Khila & Abouheif, 2008, 2010; Rafiqi, Rajakumar, & Abouheif, 2020; Ramsay, Lasko, & Abouheif, 2020). There is, however, a balance to be struck between the model clade versus model species approach, as I have discussed previously (Abouheif, 2014). In my view, the field should aim to have broad phylogenetic sampling across the ant tree of life with one "anchor" species in each of the major ant genera. Anchor species are species in which we explore the molecular and developmental processes in greater depth, allowing for more powerful mechanistic inferences from phylogenetic comparisons within the ants. Although I would not call these anchor species "model species" per se to avoid a singular focus on one specific species to the exclusion of others. Several research groups have started developing anchor species within specific ant genera, including the Florida carpenter ant *Camponotus floridanus*, the Indian jumping ant *Harpegnathos saltator*, the clonal raider ant *Ooceraea biroi*, and the tramp ant *Cardiocondyla obscurior* (Bonasio et al., 2010; Oettler et al., 2019; Oxley et al., 2014; Rafiqi et al., 2020; Schrader et al., 2021; Yan et al., 2017). My lab continues to search for a species of *Pheidole* to develop as an anchor with three species—*P. hyatti*, *P. dentata*, and *P. pallidula*—showing great promise.

In my opinion, however, although these advances in the molecular biology and genomics of social insects have been critical, the greatest advance positioning ants as a powerful model for eco-evo-devo is conceptual. For years, I have struggled to explain why ants can reveal general principles of biology. The missing link is the superorganism concept, which provides the conceptual tools to argue that we are in fact studying general principles of organismal development and evolution but at a higher-level of biological organization (Hölldobler & Wilson, 2009; Lüscher, 1953). In other words, what we discover in superorganisms applies to organisms, and in some cases, I would argue that it's even easier to observe specific types of variation in superorganisms, like the supersoldier anomalies we found in Long Island, New York (Rajakumar et al., 2012).

William Morton Wheeler first proposed that ant colonies are superorganisms during his 1911 lecture at the Marine Biological Labs at Woods hole (MA, USA). He argued that the male and queen castes are analogous to the germline while the worker caste is analogous to differentiated somatic cell types of a solitary organism (Wheeler, 1911). Since Wheeler (1911) proposed this concept, we are beginning to discover that the similarities between organisms and superorganisms run deep. In the same way that cells within a solitary organism differentiate from a single genome in response to signals from the internal environment, developing embryos or larvae within a colony differentiate from a single genome into different caste types in response to environmental cues (Abouheif, 2004; Hölldobler & Wilson, 2009; Lillico-Ouachour & Abouheif, 2017). Viewed in this way, the worker caste is not a degenerative form of the queen caste. Instead, like differentiated cell types, queen and worker caste types originated through the evolution of a novel regulatory switch or developmental threshold that produced two dramatically different phenotypes that divide labor by performing different functions in the superorganism. This difference in view is subtle but has fundamental implications for how we interpret data from ant eco-evo-devo studies. For example, if viewed at the level of the organism, the absence of wings in workers would be interpreted as the loss of a complex trait, like the loss of limbs in snakes or loss of eyes in blind cave fish. In fact, a colleague once told me that they rejected a grant I had written because they viewed the origin of winglessness in workers as a loss of complexity in evolution rather than a gain, which is what the granting body was aiming to fund. However, I have since managed to convince my colleague that, if viewed at the level of the superorganism, then the origin of a wingless worker caste represents a gain of complexity—the evolution of novel developmental threshold or condition sensitive switch that regulates the activation wing development in the queen to engage in mating flights or halts wing development in workers so they can work underground (Hanna & Abouheif, 2021). In other words, the origin, elaboration, and regulation of novel cell types at the organismal level is analogous to the origin, elaboration, and regulation of novel caste types at the superorganismal level.

Wheeler's (1911) superorganism concept, however, has a complex history and received little attention for most of the twentieth century (Boomsma & Gawne, 2018; Metzl et al., 2018; Wilson & Sober, 1989). It has only recently started to be embraced by social insect researchers (Kennedy et al., 2017; Yang, 2007). There are likely to be several reasons why the superorganism concept was largely ignored, and I will leave these complexities to the historians and philosophers of science. In my simplistic view, for much of its history, the superorganism concept has been equated with group selection, and therefore, has been excluded from the modern synthesis of evolutionary biology, mainly because it favors individual-level selection as the basis for evolutionary change (Wilson & Sober, 1989). In the early part of the 20th century, many famous evolutionary biologists and geneticists, including Fisher, Sturtevant, Weisman, Huxley, Spencer, and Haldane struggled to reconcile the biology of ant colonies and their differentiated castes with natural selection at the individual-level (Boomsma & Gawne, 2018; Chalmers Mitchell, 1894; Fisher, 1930; Sturtevant, 1938). It was not until 1964, that W.D. Hamilton proposed his theory of inclusive fitness—a theory that explains the origin of cooperative behavior based on the fitness of individuals, their genetic relatives, as well as the ecological costs and benefits incurred to these individuals (Hamilton, 1964a,b). Although Hamilton's theory laid the foundations for the field of Sociobiology in the 1970s (Wilson, 1975), it's focus on the population dynamics of individuals left out the mechanistic focus inherent in the superorganism concept (Metzl et al., 2018). I consider the longstanding debates over kin (individual-level) and group-level selection to be one of the principal factors holding back the promise and progress of ant eco-evo-devo, and the fundamental biological insights to be gained from it.

I am optimistic, however, that the tide is turning—over the last 20 years the frequency of use of the term "superorganism" in books has increased exponentially (Google N-grams; Fig. 18). Furthermore, the rise in the use of the term has been accompanied with key theoretical and empirical advances in evolutionary biology allowing superorganisms to be viewed as the product of multi-level selection (both individual and colony-level selection) (Canciani, Arnellos, & Moreno, 2019; Johnson & Linksvayer, 2010; Khila & Abouheif, 2008, 2010; Kramer & Meunier, 2016; Linksvayer, Fewell, Gadau, & Laubichler, 2012; Rajakumar et al., 2018; Ramsay et al., 2020; Taylor, Reuter, & Sumner, 2019; Wilson & Sober, 1989; Wilson & Wilson, 2007). Finally, superorganisms are now viewed as falling along a continuum of morphological and social complexity rather

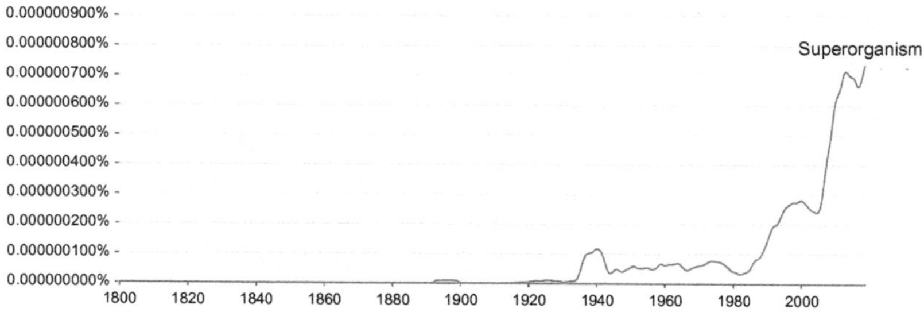

Fig. 18 Increasing frequency of use of the term "superorganism" in books from the year 1800 to 2021. Graph produced by Google Books Ngram Viewer (English, Case-Insensitive, smoothing of 3). X-axis is in years from 1800 to 2021, while y-axis is a measure of the percentage of all unigrams that are "superorganism."

than being categorized as "all or nothing" according to the criterion of a strict germ-soma divide between queens and workers (Mitchell, 2003; Ramsay et al., 2020; Reeve & Holldobler, 2007). For example, colonies of the Indian jumping ant *H. saltator* are thought to be relatively simple—they are made up of only 40–50 individuals and both queens and workers are relatively similar in reproductive ability, size, morphology. In these colonies, internal conflicts are high, and the queen actively polices her workers to stop them from reproducing (Hölldobler & Wilson, 2009). In contrast, colonies of the leaf cutting ant *Atta cephalotes* are thought to be relatively more complex—they are made up of millions of ants with dramatic differences in reproductive ability, size, and head-to-body scaling between the queen and worker castes, and even between individuals within the worker caste. In these colonies, workers have reduced reproductive capabilities and internal conflicts are consequently lower (Hölldobler & Wilson, 2009). This continuum of superorganism complexity to some degree parallels the complexity observed among solitary multicellular organisms (Bonner, 1993; Hölldobler & Wilson, 2009; Mitchell, 2003). For example, within animals, one can readily observe animals that range in morphological complexity, from sponges with only two cell layers and relatively few cell types to humans with three cell layers and over 130 cell types (Bell & Mooers, 1997; Bonner, 1993; Vickaryous & Hall, 2006; Wang et al., 2021).

I finally finish this chapter where I started—my dream is to use ants as emerging model systems for eco-evo-devo to significantly expand our exploration of phenotypic space. Together, we can harvest the power of superorganisms to connect levels of biological organization and unravel the molecular and developmental mechanisms underlying the origin and evolution of phenotypic novelty in complex biological systems. My hope is that some of the insights gained from superorganisms can address some major questions in biology as well as tackle some of society's most pressing challenges.

Acknowledgments

I dedicate this article to my Mentors (formal and informal) and Students (past and present) who made my journey to ant evo-devo possible. I thank R. Rajakumar and members of Abouheif lab for comments. This work was supported by Natural Sciences and Engineering Research Council (NSERC) Discovery Grant to E.A.

References

Abouheif, E. (1997). Developmental genetics and homology: A hierarchical approach. *Trends in Ecology & Evolution*, *12*(10), 405–408.

Abouheif, E. (1998). Random trees and the comparative method: A cautionary tale. *Evolution*, *52*, 1197–11204. https://doi.org/10.1111/j.1558-5646.1998.tb01845.x.

Abouheif, E. (1999a). Establishing homology criteria for regulatory gene networks: Prospects and challenges. In G. K. Bock, & G. Cardew (Eds.), *Homology* (pp. 207–225). John Wiley & Sons. https://doi.org/10.1002/9780470515655.ch14.

Abouheif, E. (1999b). A method for testing the assumption of phylogenetic independence in comparative data. *Evolutionary Ecology Research*, *1*(8), 895–909.

Abouheif, E. (2004). A framework for studying the evolution of gene networks underlying polyphenism: Insights from winged and wingless ant castes. In B. K. Hall, R. D. Pearson, & G. Müller (Eds.), *Environment, development, and evolution: Toward a synthesis* (pp. 125–137). Cambridge, MA: MIT Press.

Abouheif, E. (2014). Phylogenies and the evolution of development. In P. Vargas, & R. Zardoya (Eds.), *The tree of life* (p. 713). Sinauer Associates.

Abouheif, E. (2021). Ant caste evo-devo: It's not all about size. *Trends in Ecology & Evolution*, *36*(8), 668–670. https://doi.org/10.1016/j.tree.2021.04.002.

Abouheif, E., & Wray, G. A. (2002). Evolution of the gene network underlying wing polyphenism in ants. *Science*, *297*(5579), 249–252. https://doi.org/10.1126/science.1071468.

Abouheif, E., Akam, M., Dickinson, W. J., Holland, P. W. H., Meyer, A., Patel, N. H., et al. (1997). Homology and developmental genes. *Trends in Genetics*, *13*(11), 432–433. https://doi.org/10.1016/s0168-9525(97)01271-7.

Abouheif, E., Favé, M. J., Ibarraran-Viniegra, A. S., Lesoway, M. P., Rafiqi, A. M., & Rajakumar, R. (2014). Eco-evo-devo: The time has come. *Advances in Experimental Medicine and Biology*, *781*, 107–125. https://doi.org/10.1007/978-94-007-7347-9_6.

Abzhanov, A., Extavour, C. G., Groover, A., Hodges, S. A., Hoekstra, H. E., Kramer, E. M., et al. (2008). Are we there yet? Tracking the development of new model systems. *Trends in Genetics*, *24*(7), 353–360. https://doi.org/10.1016/j.tig.2008.04.002.

Alba, V., Carthew, J. E., Carthew, R. W., & Mani, M. (2021). Global constraints within the developmental program of the Drosophila wing. *eLife*, 10. https://doi.org/10.7554/eLife.66750.

Alvarado, S., Rajakumar, R., Abouheif, E., & Szyf, M. (2015). Epigenetic variation in the Egfr gene generates quantitative variation in a complex trait in ants. *Nature Communications*, *6*, 6513. https://doi.org/10.1038/ncomms7513.

Ankeny, R. A., & Leonelli, S. (2011). What's so special about model organisms? *Studies in History and Philosophy of Science Part A*, *42*(2), 313–323. https://doi.org/10.1016/j.shpsa.2010.11.039.

Arendt, D., Musser, J. M., Baker, C. V. H., Bergman, A., Cepko, C., Erwin, D. H., et al. (2016). The origin and evolution of cell types. *Nature Reviews. Genetics*, *17*(12), 744–757. https://doi.org/10.1038/nrg.2016.127.

Bejder, L., & Hall, B. K. (2002). Limbs in whales and limblessness in other vertebrates: Mechanisms of evolutionary and developmental transformation and loss. *Evolution & Development*, *4*(6), 445–458. https://doi.org/10.1046/j.1525-142x.2002.02033.x.

Bell, G., & Mooers, A. O. (1997). Size and complexity among multicellular organisms. *Biological Journal of the Linnean Society*, *60*(3), 345–363. https://doi.org/10.1111/j.1095-8312.1997.tb01500.x.

Bolker, J. A. (1995). Model systems in developmental biology. *BioEssays*, *17*(5), 451–455. https://doi.org/10.1002/bies.950170513.

Bolker, J. A., & Raff, R. A. (1996). Developmental genetics and traditional homology. *BioEssays*, *18*(6), 489–494. https://doi.org/10.1002/bies.950180611.

Bonasio, R., Zhang, G., Ye, C., Mutti, N. S., Fang, X., Qin, N., et al. (2010). Genomic comparison of the ants Camponotus floridanus and Harpegnathos saltator. *Science*, *329*(5995), 1068–1071. https://doi.org/10.1126/science.1192428.

Bonner, J. (1993). Dividing the labour in cells and societies. *Current Science*, *64*, 459–466.

Boomsma, J. J., & Gawne, R. (2018). Superorganismality and caste differentiation as points of no return: How the major evolutionary transitions were lost in translation. *Biological Reviews of the Cambridge Philosophical Society*, *93*(1), 28–54. https://doi.org/10.1111/brv.12330.

Boomsma, J. J., Brady, S. G., Dunn, R. R., Gadau, J., Heinze, J., Keller, L., et al. (2017). The Global Ant Genomics Alliance (GAGA). *Myrmecological News*, *25*, 61–66.

Brook, W. J., Diaz-Benjumea, F. J., & Cohen, S. M. (1996). Organizing spatial pattern in limb development. *Annual Review of Cell and Developmental Biology*, *12*, 161–180. https://doi.org/10.1146/annurev.cellbio.12.1.161.

Bull, J. J., & Charnov, E. L. (1985). On irreversible evolution. *Evolution*, *39*(5), 1149–1155. https://doi.org/10.1111/j.1558-5646.1985.tb00455.x.

Canciani, M., Arnellos, A., & Moreno, A. (2019). Revising the superorganism: An organizational approach to complex eusociality. *Frontiers in Psychology*, *10*, 2653. https://doi.org/10.3389/fpsyg.2019.02653.

Carroll, S. B., Gates, J., Keys, D. N., Paddock, S. W., Panganiban, G. E., Selegue, J. E., et al. (1994). Pattern formation and eyespot determination in butterfly wings. *Science*, *265*(5168), 109–114. https://doi.org/10.1126/science.7912449.

Chalmers Mitchell, P. (1894). The Spencer-Weismann controversy. *Nature*, *49*(1268).

Chiu, Y. K., Hsu, J. C., Chang, T., Huang, Y. C., & Wang, J. (2020). Mutagenesis mediated by CRISPR/Cas9 in the red imported fire ant, Solenopsis invicta. *Insectes Sociaux. Insectes Sociaux*, *67*(2), 317–326. https://doi.org/10.1007/s00040-020-00755-8.

Collin, R., & Miglietta, M. P. (2008). Reversing opinions on Dollo's Law. *Trends in Ecology & Evolution*, *23*(11), 602–609. https://doi.org/10.1016/j.tree.2008.06.013.

Darwin, C. (1859). *The origin of species*. Peguin Books.

Darwin, C. (1868). *The variation of animals and plants under domestication* (p. 922). New York: Cambridge University Press.

Dewitz, H. (1878). Beiträge zur postembryonalen Gliedmassenbildung bei denInsecten. *Zeitschrift für Wissenschaftliche Zoologie*, *30*, 78–105.

Dickinson, W. J., & Seger, J. (1996). Eye evolution. *Science*, *272*(5261), 467–469. https://doi.org/10.1126/science.272.5261.467b.

Dworkin, I., Kennerly, E., Tack, D., Hutchinson, J., Brown, J., Mahaffey, J., et al. (2009). Genomic consequences of background effects on scalloped mutant expressivity in the wing of Drosophila melanogaster. *Genetics, 181*(3), 1065–1076. https://doi.org/10.1534/genetics.108.096453.

Ephrussi, B., & Beadle, G. W. (1936). A technique of transplantation for Drosophila. *American Naturalist, 70*, 218–225.

Favé, M. J., Johnson, R. A., Cover, S., Handschuh, S., Metscher, B. D., Muller, G. B., et al. (2015). Past climate change on Sky Islands drives novelty in a core developmental gene network and its phenotype. *BMC Evolutionary Biology, 15*, 183. https://doi.org/10.1186/s12862-015-0448-4.

Fetter-Pruneda, I., Hart, T., Ulrich, Y., Gal, A., Oxley, P. R., Olivos-Cisneros, L., et al. (2021). An oxytocin/vasopressin-related neuropeptide modulates social foraging behavior in the clonal raider ant. *PLoS Biology, 19*(6), e3001305. https://doi.org/10.1371/journal.pbio.3001305.

Fisher, R. A. S. (1930). *The genetical theory of natural selection.* Oxford: Clarendon Press.

Foster, S. A., & Baker, J. A. (2019). Loss and re-emergence of plastic ancestral behavioural traits: Influences on phenotypic and evolutionary pattern. *Animal Behaviour, 155*, 271–277. https://doi.org/10.1016/j.anbehav.2019.06.013.

Foster, S. A., O'Neil, S., King, R. W., & Baker, J. A. (2019). Replicated evolutionary inhibition of a complex ancestral behaviour in an adaptive radiation. *Biology Letters, 15*(1), 20180647. https://doi.org/10.1098/rsbl.2018.0647.

Gadau, J., Helmkampf, M., Nygaard, S., Roux, J., Simola, D. F., Smith, C. R., et al. (2012). The genomic impact of 100 million years of social evolution in seven ant species. *Trends in Genetics, 28*(1), 14–21. https://doi.org/10.1016/j.tig.2011.08.005.

Gehring, W. J. (1996). Response: Eye evolution. *Science, 272*(5261), 468–469. https://doi.org/10.1126/science.272.5261.468.

Gibson, G., & Hogness, D. S. (1996). Effect of polymorphism in the Drosophila regulatory gene Ultrabithorax on homeotic stability. *Science, 271*(5246), 200–203.

Gilbert, S. F. (2001). Ecological developmental biology: Developmental biology meets the real world. *Developmental Biology, 233*(1), 1–12. https://doi.org/10.1006/dbio.2001.0210.

Gilbert, S. F. (2016). Developmental plasticity and developmental symbiosis: The return of eco-devo. *Current Topics in Developmental Biology, 116*, 415–433. https://doi.org/10.1016/bs.ctdb.2015.12.006.

Gilbert, S. F., & Barresi, M. J. F. (2019). *Developmental biology* (12th ed.). Oxford University Press.

Gilbert, S. F., & Epel, D. (2015). *Ecological developmental biology: Integrating epigenetics, medicine, and evolution, pp* (2nd ed.). Sunderland, MA: Sinauer Associates.

Goetsch, W. (1939). Neuartige Termitensoldaten aus Kunstnestern. *Zoologischer Anzeiger, 128*, 209–216.

Goldstein, B., & King, N. (2016). The future of cell biology: Emerging model organisms. *Trends in Cell Biology, 26*(11), 818–824. https://doi.org/10.1016/j.tcb.2016.08.005.

Gould, S. J. (1970). Dollo on Dollo's law: Irreversibility and the status of evolutionary laws. *Journal of the History of Biology, 3*(2), 189.

Gould, S. J. (1977). *Ontogeny and phylogeny.* Belknap Press of Harvard University Press.

Gould, J. J. (2011). *9-Foot-tall supersoldier mourns Kim Jong Il.* The Atlantic.

Gregg, R. E. (1942). The origin of castes in ants with special reference to *Pheidole morrisi* Forel. *Ecology, 23*(3), 295–308.

Gunter, H. M., Schneider, R. F., Karner, I., Sturmbauer, C., & Meyer, A. (2017). Molecular investigation of genetic assimilation during the rapid adaptive radiations of East African cichlid fishes. *Molecular Ecology, 26*(23), 6634–6653. https://doi.org/10.1111/mec.14405.

Halder, G., Polaczyk, P., Kraus, M. E., Hudson, A., Kim, J., Laughon, A., et al. (1998). The Vestigial and Scalloped proteins act together to directly regulate wing-specific gene expression in Drosophila. *Genes & Development, 12*(24), 3900–3909.

Halder, G., Callaerts, P., & Gehring, W. J. (1995). Induction of ectopic eyes by targeted expression of the eyeless gene in Drosophila. *Science*, *267*(5205), 1788–1792. https://doi.org/10.1126/science.7892602.

Hall, B. K. (1984). Development mechanisms underlying the formation of atavisms. *Biological Reviews of the Cambridge Philosophical Society*, *59*(1), 89–124. https://doi.org/10.1111/j.1469-185x.1984.tb00402.x.

Hall, B. K. (1994). *Homology: The hierarchical basis of comparative biology*. Academic Press.

Hall, B. K. (1995a). Atavisms and atavistic mutations. *Nature Genetics*, *10*(2), 126–127. https://doi.org/10.1038/ng0695-126.

Hall, B. K. (1995b). Homology and embryonic development. In M. K. Hecht, R. J. Macintyre, & M. T. Clegg (Eds.), *Evolutionary biology* (pp. 1–37). Boston, MA: Springer US.

Hall, B. K. (2003). Descent with modification: The unity underlying homology and homoplasy as seen through an analysis of development and evolution. *Biological Reviews*, *78*(3), 409–433. https://doi.org/10.1017/s1464793102006097.

Hall, B. K. (2010). Atavisms. *Current Biology*, *20*(20), R871. https://doi.org/10.1016/j.cub.2010.07.002.

Hamilton, W. D. (1964a). The genetical evolution of social behaviour II. *Journal of Theoretical Biology*, 7(1), 17–52.

Hamilton, W. D. (1964b). The genetical evolution of social behaviour I. *Journal of Theoretical Biology*, 7(1), 1–16.

Hanna, L., & Abouheif, E. (2021). The origin of wing polyphenism in ants: An eco-evo-devo perspective. *Current Topics in Developmental Biology*, *141*, 279–336. https://doi.org/10.1016/bs.ctdb.2020.12.004.

Harris, M. P., Hasso, S. M., Ferguson, M. W., & Fallon, J. F. (2006). The development of archosaurian first-generation teeth in a chicken mutant. *Current Biology*, *16*(4), 371–377.

Heinze, J. (2008). The demise of the standard ant (Hymenoptera: Formicidae). *Myrmecological News*, *11*, 9–20.

Held, L. I. (2002). *Imaginal Discs*. Cambridge University Press.

Hinchliff, C. E., Smith, S. A., Allman, J. F., Burleigh, J. G., Chaudhary, R., Coghill, L. M., et al. (2015). Synthesis of phylogeny and taxonomy into a comprehensive tree of life. *Proceedings of the National Academy of Sciences of the United States of America*, *112*(41), 12764–12769. https://doi.org/10.1073/pnas.1423041112.

Hölldobler, B., & Wilson, E. O. (1990). *The ants*. Cambridge: The Belknap Press of Harvard University Press.

Hölldobler, B., & Wilson, E. O. (2009). *The superorganism*. New York: W.W Norton & Company Inc.

ISHOF, 2021. Abdellatief Abouheif (EGY). Honor Pioneer Open Water Swimmer (1998). International Swimming Hall of Fame.

Johnson, B. R., & Linksvayer, T. A. (2010). Deconstructing the superorganism: Social physiology, groundplans, and sociogenomics. *The Quarterly Review of Biology*, *85*(1), 57–79. https://doi.org/10.1086/650290.

Kennedy, P., Baron, G., Qiu, B., Freitak, D., Helantera, H., Hunt, E. R., et al. (2017). Deconstructing superorganisms and societies to address big questions in biology. *Trends in Ecology & Evolution*, *32*(11), 861–872. https://doi.org/10.1016/j.tree.2017.08.004.

Khila, A., & Abouheif, E. (2008). Reproductive constraint is a developmental mechanism that maintains social harmony in advanced ant societies. *Proceedings of the National Academy of Sciences of the United States of America*, *105*(46), 17884–17889. https://doi.org/10.1073/pnas.0807351105.

Khila, A., & Abouheif, E. (2010). Evaluating the role of reproductive constraints in ant social evolution. *Philosophical Transactions of the Royal Society of London. Series B, Biological Sciences*, *365*(1540), 617–630. https://doi.org/10.1098/rstb.2009.0257.

Khila, A., Abouheif, E., & Rowe, L. (2009). Evolution of a novel appendage ground plan in water striders is driven by changes in the *Hox* gene *Ultrabithorax*. *PLoS Genetics*, *5*(7), e1000583. https://doi.org/10.1371/journal.pgen.1000583.

Khila, A., Abouheif, E., & Rowe, L. (2012). Function, developmental genetics, and fitness consequences of a sexually antagonistic trait. *Science*, *336*(6081), 585–589.

Khila, A., Abouheif, E., & Rowe, L. (2014). Comparative functional analyses of *Ultrabithorax* reveal multiple steps and paths to diversification of legs in the adaptive radiation of semi-aquatic insects. *Evolution*, *68*(8), 2159–2170. https://doi.org/10.1111/evo.12444.

Kim, J., Sebring, A., Esch, J. J., Kraus, M. E., Vorwerk, K., Magee, J., et al. (1996). Integration of positional signals and regulation of wing formation and identity by *Drosophila vestigial* gene. *Nature*, *382*(6587), 133–138. https://doi.org/10.1038/382133a0.

Koch, S., & Abouheif, E. (2019). The coordination of insect imaginal discs and the regulation and evolution of complex worker caste systems of ants. In C. Bishop & B. K. Hall (Eds.), Deferring Development (pp. 197–224). CRC Press.

Kramer, J., & Meunier, J. (2016). Kin and multilevel selection in social evolution: A never-ending controversy. *F1000Res*, *5*. https://doi.org/10.12688/f1000research.8018.1.

Kreitman, M. (1996). The neutral theory is dead. Long live the neutral theory. *BioEssays*, *18*(8), 678–683. https://doi.org/10.1002/bies.950180812.

Lauder, G. V. (1994). Homology, form, and function. In B. K. Hall (Ed.), *Homology: The hierarchical basis of comparative biology* Academic Press.

Lesoway, M. P., Abouheif, E., & Collin, R. (2014). The development of viable and nutritive embryos in the direct developing gastropod Crepidula navicella. *The International Journal of Developmental Biology*, *58*(6–8), 601–611. https://doi.org/10.1387/ijdb.140136rc.

Lesoway, M. P., Collin, R., & Abouheif, E. (2017). Early activation of MAPK and apoptosis in nutritive embryos of calyptraeid gastropods. *Journal of Experimental Zoology. Part B, Molecular and Developmental Evolution*, *328*(5), 449–461. https://doi.org/10.1002/jez.b.22745.

Levinton, J. S. (1983). Stasis in progress: The empirical basis of macroevolution. *Annual Review of Ecology and Systematics*, *14*(1), 103–137. https://doi.org/10.1146/annurev.es.14.110183.00053.

Lewin, H. A., Robinson, G. E., Kress, W. J., Baker, W. J., Coddington, J., Crandall, K. A., et al. (2018). Earth bioGenome project: Sequencing life for the future of life. *Proceedings of the National Academy of Sciences of the United States of America*, *115*(17), 4325–4333. https://doi.org/10.1073/pnas.1720115115.

Lewontin, R. C. (2002). *The triple helix*. Harvard University Press.

Liem, K. F. (1973). Evolutionary strategies and morphological innovations: Cichlid pharyngeal jaws. *Systematic Zoology*, *22*(4). https://doi.org/10.2307/2412950.

Lillico-Ouachour, A., & Abouheif, E. (2017). Regulation, development, and evolution of caste ratios in the hyperdiverse ant genus *Pheidole*. *Current Opinion in Insect Science*, *19*, 43–51. https://doi.org/10.1016/j.cois.2016.11.003.

Linksvayer, T. A., Fewell, J. H., Gadau, J., & Laubichler, M. D. (2012). Developmental evolution in social insects: Regulatory networks from genes to societies. *Journal of Experimental Zoology. Part B, Molecular and Developmental Evolution*, *318*(3), 159–169. https://doi.org/10.1002/jez.b.22001.

Lowe, C. J., & Wray, G. A. (1997). Radical alterations in the roles of homeobox genes during echinoderm evolution. *Nature*, *389*(6652), 718–721. https://doi.org/10.1038/39580.

Lüscher, M. (1953). The termite and the cell. *Scientific American*, *188*(5), 74–80.

Marshall, C. R., Raff, E. C., & Raff, R. A. (1994). Dollo's law and the death and resurrection of genes. *Proceedings of the National Academy of Sciences of the United States of America*, *91*(25), 12283–12287. https://doi.org/10.1073/pnas.91.25.12283.

Marx, V. (2021). Model organisms on roads less traveled. *Nature Methods*, *18*(3), 235–239. https://doi.org/10.1038/s41592-021-01086-7.

Mayr, E. (1994). Recapitulation reinterpreted: The somatic program. *The Quarterly Review of Biology*, *69*(2), 223–232.

Metzl, C., Wheeler, D. E., & Abouheif, E. (2018). Wilhelm Goetsch (1887-1960): pioneering studies on the development and evolution of the soldier caste in social insects. *Myrmecological News*, *26*, 81–95.

Meyer, A., Kocher, T. D., Basasibwaki, P., & Wilson, A. C. (1990). Monophyletic origin of Lake Victoria cichlid fishes suggested by mitochondrial DNA sequences. *Nature*, *347*(6293), 550–553. https://doi.org/10.1038/347550a0.

Meyer, A., Morrissey, J. M., & Schartl, M. (1994). Recurrent origin of a sexually selected trait in Xiphophorus fishes inferred from a molecular phylogeny. *Nature*, *368*(6471), 539–542. https://doi.org/10.1038/368539a0.

Milinkovitch, M. C., Orti, G., & Meyer, A. (1993). Revised phylogeny of whales suggested by mitochondrial ribosomal DNA sequences. *Nature*, *361*(6410), 346–348. https://doi.org/10.1038/361346a0.

Mitchell, S. D. (2003). Constitutive complexity. In S. D. Mitchell (Ed.), *Biological complexity and integrative pluralism* (pp. 13–37). Cambridge: Cambridge University Press.

Moczek, A. P., & Nagy, L. M. (2005). Diverse developmental mechanisms contribute to different levels of diversity in horned beetles. *Evolution & Development*, 7(3), 175–185. https://doi.org/10.1111/j.1525-142X.2005.05020.x.

Moczek, A. P., Sears, K. E., Stollewerk, A., Wittkopp, P. J., Diggle, P., Dworkin, I., et al. (2015). The significance and scope of evolutionary developmental biology: A vision for the 21st century. *Evolution & Development*, *17*(3), 198–219. https://doi.org/10.1111/ede.12125.

Moczek, A. P., Rose, D., Sewell, W., & Kesselring, B. R. (2006). Conservation, innovation, and the evolution of horned beetle diversity. *Development Genes and Evolution*, *216*(11), 655–665. https://doi.org/10.1007/s00427-006-0087-2.

Moreau, C. S. (2008). Unraveling the evolutionary history of the hyperdiverse ant genus Pheidole (Hymenoptera: Formicidae). *Molecular Phylogenetics and Evolution*, *48*(1), 224–239. https://doi.org/10.1016/j.ympev.2008.02.020.

Nahmad, M., Glass, L., & Abouheif, E. (2008). The dynamics of developmental system drift in the gene network underlying wing polyphenism in ants: A mathematical model. *Evolution & Development*, *10*(3), 360–374. https://doi.org/10.1111/j.1525-142X.2008.00244.x.

Nijhout, H. F. (1994). *Insect hormones*. pp. XI, 267 p. Princeton, NJ: Princeton University Press.

Nijhout, H. F., & Wray, G. A. (1988). Homologies in the colour patterns of the genus Heliconius (Lepidoptera: Nymphalidae). *Biological Journal of the Linnean Society*, *33*(4), 345.

Nijhout, H. F., Wray, G. A., Kremen, C., & Teragawa, C. K. (1986). Ontogeny, phylogeny and evolution of form: An algorithmic approach. *Systematic Zoology*, *35*(4). https://doi.org/10.2307/2413108.

Oettler, J., Platschek, T., Schmidt, C., Rajakumar, R., Fave, M. J., Khila, A., et al. (2019). Interruption points in the wing gene regulatory network underlying wing polyphenism evolved independently in male and female morphs in Cardiocondyla ants. *Journal of Experimental Zoology. Part B, Molecular and Developmental Evolution*, *332*(1–2), 7–16. https://doi.org/10.1002/jez.b.22834.

Oxley, P. R., Ji, L., Fetter-Pruneda, I., McKenzie, S. K., Li, C., Hu, H., et al. (2014). The genome of the clonal raider ant Cerapachys biroi. *Current Biology*, *24*(4), 451–458. https://doi.org/10.1016/j.cub.2014.01.018.

Paaby, A. B., & Rockman, M. V. (2014). Cryptic genetic variation: Evolution's hidden substrate. *Nature Reviews. Genetics*, *15*(4), 247–258. https://doi.org/10.1038/nrg3688.

Pardos-Blas, J. R., Irisarri, I., Abalde, S., Afonso, C. M. L., Tenorio, M. J., & Zardoya, R. (2021). The genome of the venomous snail Lautoconus ventricosus sheds light on the origin of conotoxin diversity. *Gigascience*, *10*(5). https://doi.org/10.1093/gigascience/giab037.

Paris, M., Wolff, C., Patel, N. H., & Averof, M. (2022). The crustacean model *Parhyale hawaiensis*. *Current Topics in Developmental Biology*, *147*, 199–230. https://doi.org/10.1016/bs.ctdb.2022.02.001.

Passera, L. (1974). Différenciation des soldats chez la Fourmi *Pheidole pallidula* Nyl. (Formicidae Myrmicinae). *Insectes Sociaux*, *21*(1), 71–86. https://doi.org/10.1007/bf02222980.

Patterson, C. (1988). Homology in classical and molecular biology. *Molecular Biology and Evolution*, *5*(6), 603–625. https://doi.org/10.1093/oxfordjournals.molbev.a040523.

Poodry, C. A., & Woods, D. F. (1990). Control of the developmental timer for Drosophila Pupariation. *Rouxs Archives of Developmental Biology*, *199*(4), 219–227. https://doi.org/10.1007/Bf01682081.

Porter, M. L., & Crandall, K. A. (2003). Lost along the way: The significance of evolution in reverse. *Trends in Ecology & Evolution*, *18*(10), 541–547. https://doi.org/10.1016/s0169-5347(03)00244-1.

Quiring, R., Walldorf, U., Kloter, U., & Gehring, W. J. (1994). Homology of the eyeless gene of Drosophila to the Small eye gene in mice and Aniridia in humans. *Science*, *265*(5173), 785–789. https://doi.org/10.1126/science.7914031.

Rafiqi, A. M., Rajakumar, A., & Abouheif, E. (2020). Origin and elaboration of a major evolutionary transition in individuality. *Nature*, *585*(7824), 239–244. https://doi.org/10.1038/s41586-020-2653-6.

Rajakumar, R., & Abouheif, E. (2013). Ancestral developmental potential: A new tool for animal breeding? (pp. 5–18). 62nd Annual National Breeders Roundtable.

Rajakumar, R., San Mauro, D., Dijkstra, M. B., Huang, M. H., Wheeler, D. E., Hiou-Tim, F., et al. (2012). Ancestral developmental potential facilitates parallel evolution in ants. *Science*, *335*(6064), 79–82. https://doi.org/10.1126/science.1211451.

Rajakumar, R., Koch, S., Couture, M., Fave, M. J., Lillico-Ouachour, A., Chen, T., et al. (2018). Social regulation of a rudimentary organ generates complex worker-caste systems in ants. *Nature*, *562*(7728), 574–577. https://doi.org/10.1038/s41586-018-0613-1.

Ramsay, C., Lasko, P., & Abouheif, E. (2020). Evo-devo lessons from the reproductive division of labor in eusocial Hymenoptera. In L. N. de la Rosa, & G. Müller (Eds.), *Evolutionary developmental biology* (pp. 1–14). Cham: Springer. https://doi.org/10.1007/978-3-319-33038-9_173-1.

Ratzka, C., Gross, R., & Feldhaar, H. (2013). Systemic gene knockdown in Camponotus floridanus workers by feeding of dsRNA. *Insectes Sociaux*, *60*(4), 475–484. https://doi.org/10.1007/s00040-013-0314-6.

Reeve, H. K., & Holldobler, B. (2007). The emergence of a superorganism through inter-group competition. *Proceedings of the National Academy of Sciences of the United States of America*, *104*(23), 9736–9740. https://doi.org/10.1073/pnas.0703466104.

Robinson, G. E. (2002). Sociogenomics takes flight. *Science*, *297*(5579), 204–205. https://doi.org/10.1126/science.1074493.

Roth, V. L. (1984). On Homology. *Biological Journal of the Linnean Society*, *22*(1), 13–29. https://doi.org/10.1111/j.1095-8312.1984.tb00796.x.

Russell, J. J., Theriot, J. A., Sood, P., Marshall, W. F., Landweber, L. F., Fritz-Laylin, L., et al. (2017). Non-model model organisms. *BMC Biology*, *15*(1), 55. https://doi.org/10.1186/s12915-017-0391-5.

Sanger, T. J., & Rajakumar, R. (2018). How a growing organismal perspective is adding new depth to integrative studies of morphological evolution. *Biological Reviews of the Cambridge Philosophical Society*, *94*, 184–198. https://doi.org/10.1111/brv.12442.

Scheffers, B. R., Joppa, L. N., Pimm, S. L., & Laurance, W. F. (2012). What we know and don't know about Earth's missing biodiversity. *Trends in Ecology & Evolution, 27*(9), 501–510. https://doi.org/10.1016/j.tree.2012.05.008.
Schlichting, C. D. (2008). Hidden reaction norms, cryptic genetic variation, and evolvability. *Annals of the New York Academy of Sciences, 1133*, 187–203. https://doi.org/10.1196/annals.1438.010.
Schneider, R. F., Li, Y., Meyer, A., & Gunter, H. M. (2014). Regulatory gene networks that shape the development of adaptive phenotypic plasticity in a cichlid fish. *Molecular Ecology, 23*(18), 4511–4526. https://doi.org/10.1111/mec.12851.
Schrader, L., Winter, M., Errbii, M., Delabie, J., Oettler, J., & Gadau, J. (2021). Inhibition of HSP90 causes morphological variation in the invasive ant Cardiocondyla obscurior. *Journal of Experimental Zoology Part B: Molecular and Developmental Evolution, 336*(4), 333–340. https://doi.org/10.1002/jez.b.23035.
Seher, T. D., Ng, C. S., Signor, S. A., Podlaha, O., Barmina, O., & Kopp, A. (2012). Genetic basis of a violation of Dollo's Law: Re-evolution of rotating sex combs in Drosophila bipectinata. *Genetics, 192*(4), 1465–1475. https://doi.org/10.1534/genetics.112.145524.
Shbailat, S. J., & Abouheif, E. (2013). The wing-patterning network in the wingless castes of Myrmicine and Formicine ant species is a mix of evolutionarily labile and non-labile genes. *Journal of Experimental Zoology. Part B, Molecular and Developmental Evolution, 320*(2), 74–83. https://doi.org/10.1002/jez.b.22482.
Shbailat, S., Khila, A., & Abouheif, E. (2010). Correlations between spatiotemporal changes in gene expression and apoptosis underlie wing polyphenism in the ant *Pheidole morrisi*. *Evolution & Development, 12*(6), 580–591. https://doi.org/10.1111/j.1525-142X.2010.00443.x.
Shubin, N., Tabin, C., & Carroll, S. (2009). Deep homology and the origins of evolutionary novelty. *Nature, 457*(7231), 818–823. https://doi.org/10.1038/nature07891.
Sieriebriennikov, B., Reinberg, D., & Desplan, C. (2021). A molecular toolkit for superorganisms. *Trends in Genetics, 37*(9), 846–859. https://doi.org/10.1016/j.tig.2021.05.005.
Sommer, R. J., & Mayer, M. G. (2015). Toward a synthesis of developmental biology with evolutionary theory and ecology. *Annual Review of Cell and Developmental Biology, 31*, 453–471. https://doi.org/10.1146/annurev-cellbio-102314-112451.
Sommer, R. J., & Sternberg, P. W. (1994). Changes of induction and competence during the evolution of vulva development in nematodes. *Science, 265*(5168), 114–118. https://doi.org/10.1126/science.8016644.
Sommer, R. J., Carta, L. K., & Sternberg, P. W. (1994). The evolution of cell lineage in nematodes. *Development, 1994*(Supplement), 85–95. https://doi.org/10.1242/dev.1994.Supplement.85.
Stearns, S. C. (1994). The evolutionary link between fixed and variable traits. *Acta Palaeontologica Polonica, 38*, 215.
Stiassny, M. L. J. (2003). Atavism. In *Keywords and concepts in evolutionary developmental biology* (pp. 10–14). Cambridge, MA: Harvard University Press.
Sturtevant, A. H. (1938). Essays on evolution. II. On the effects of selection on social insects. *The Quarterly Review of Biology, 13*(1), 74–76.
Sultan, S. E. (2015). *Organism and environment*. Oxford University Press. https://doi.org/10.1093/acprof:oso/9780199587070.001.0001.
Suzuki, Y., & Nijhout, H. F. (2006). Evolution of a polyphenism by genetic accommodation. *Science, 311*(5761), 650–652. https://doi.org/10.1126/science.1118888.
Takeuchi, Y., Hori, M., Tada, S., & Oda, Y. (2016). Acquisition of lateralized predation behavior associated with development of mouth asymmetry in a lake tanganyika scale-eating cichlid fish. *PLoS One, 11*(1), e0147476. https://doi.org/10.1371/journal.pone.0147476.

Taylor, B. A., Reuter, M., & Sumner, S. (2019). Patterns of reproductive differentiation and reproductive plasticity in the major evolutionary transition to superorganismality. *Current Opinion in Insect Science*, *34*, 40–47. https://doi.org/10.1016/j.cois.2019.02.007.

Teseo, S., Chaline, N., Jaisson, P., & Kronauer, D. J. (2014). Epistasis between adults and larvae underlies caste fate and fitness in a clonal ant. *Nature Communications*, *5*, 3363. https://doi.org/10.1038/ncomms4363.

Thomson, I. (2012). Boffins hack evolution, create supersoldier ants. In *The a register*.

Trible, W., Olivos-Cisneros, L., McKenzie, S. K., Saragosti, J., Chang, N. C., Matthews, B. J., et al. (2017). orco mutagenesis causes loss of antennal lobe glomeruli and impaired social behavior in ants. *Cell*, *170*(4), 727–735 e10. https://doi.org/10.1016/j.cell.2017.07.001.

True, J. R., & Haag, E. S. (2001). Developmental system drift and flexibility in evolutionary trajectories. *Evolution & Development*, *3*(2), 109–119. https://doi.org/10.1046/j.1525-142x.2001.003002109.x.

Truman, J. W., & Riddiford, L. M. (1999). The origins of insect metamorphosis. *Nature*, *401*(6752), 447–452. https://doi.org/10.1038/46737.

Ursprung, H., & Hadorn, E. (1962). Weitere Untersuchungen über Musterbildung in Kombinaten aus teilweise dissoziierten Flügel-Imaginalscheiben von Drosophila melanogaster. *Developmental Biology*, *4*(1), 40–66. https://doi.org/10.1016/0012-1606(62)90032-5.

Valenzuela, N. (2008). Evolution of the gene network underlying gonadogenesis in turtles with temperature-dependent and genotypic sex determination. *Integrative and Comparative Biology*, *48*(4), 476–485. https://doi.org/10.1093/icb/icn031.

Vickaryous, M. K., & Hall, B. K. (2006). Human cell type diversity, evolution, development, and classification with special reference to cells derived from the neural crest. *Biological Reviews of the Cambridge Philosophical Society*, *81*(3), 425–455. https://doi.org/10.1017/S1464793106007068.

Wagner, G. P. (1989). The biological homology concept. *Annual Review of Ecology and Systematics*, *20*(1), 51–69. https://doi.org/10.1146/annurev.es.20.110189.000411.

Wagner, G. P., Chiu, C.-H., & Laubichler, M. (2000). Developmental evolution as a mechanistic science: The inference from developmental mechanisms to evolutionary processes. *American Zoologist*, *40*(5), 819–831. https://doi.org/10.1093/icb/40.5.819.

Wake, D. B. (1994). Comparative terminology. *Science*, *265*(5169), 268–269. https://doi.org/10.1126/science.265.5169.268.

Wake, D. B. (1999). Homoplasy, homology and the problem of 'sameness' in biology. In B. K. Bock, & G. Cardew (Eds.), *Homology* (pp. 25–46). New York: John Wiley & Sons. https://doi.org/10.1002/9780470515655.ch3.

Wake, D. B., Wake, M. H., & Specht, C. D. (2011). Homoplasy: From detecting pattern to determining process and mechanism of evolution. *Science*, *331*(6020), 1032–1035. https://doi.org/10.1126/science.1188545.

Walker-Larsen, J., & Harder, L. D. (2007). Vestigial organs as opportunities for functional innovation: The example of the penstemon staminode. *Evolution*, *55*(3), 477–487. https://doi.org/10.1111/j.0014-3820.2001.tb00782.x.

Wang, J., Sun, H., Jiang, M., Li, J., Zhang, P., Chen, H., et al. (2021). Tracing cell-type evolution by cross-species comparison of cell atlases. *Cell Reports*, *34*(9), 108803. https://doi.org/10.1016/j.celrep.2021.108803.

West-Eberhard, M. J. (2003). *Developmental plasticity and evolution* (p. 794). Oxford/New York: Oxford University Press.

Wheeler, W. M. (1910). *Ants: Their structure, development and behavior*. Lancaster: Columbia University Press.

Wheeler, W. M. (1911). The ant colony as organism. *Journal of Morphology*, *22*, 307–325.

Wheeler, D. E., & Nijhout, H. F. (1981a). Imaginal wing discs in larvae of the soldier caste of Pheidole bicarinata vinelandica Forel (Hymenoptera : Formicidae). *International Journal of Insect Morphology and Embryology*, *10*(2), 131–139. https://doi.org/10.1016/s0020-7322(81)80018-9.

Wheeler, D. E., & Nijhout, H. F. (1981b). Soldier determination in ants: New role for juvenile hormone. *Science*, *213*(4505), 361–363. https://doi.org/10.1126/science.213.4505.361.

Wheeler, D. E., & Nijhout, H. F. (1984). Soldier determination in *Pheidole bicarinata*: Inhibition by adult soldiers. *Journal of Insect Physiology*, *30*(2), 127–135. https://doi.org/10.1016/0022-1910(84)90116-1.

Wilson, E. O. (1975). *Sociobiology: The new synthesis*. Cambridge/MA: Belknap Press of Harvard University Press.

Wilson, E. O. (1985). The sociogenesis of insect colonies. *Science*, *228*(4707), 1489–1495. https://doi.org/10.1126/science.228.4707.1489.

Wilson, E. O. (2003). *Pheidole in the new world: A dominant, hyperdiverse ant genus*. Cambridge/MA: Harvard University Press.

Wilson, D. S., & Sober, E. (1989). Reviving the superorganism. *Journal of Theoretical Biology*, *136*(3), 337–356. https://doi.org/10.1016/s0022-5193(89)80169-9.

Wilson, D. S., & Wilson, E. O. (2007). Rethinking the theoretical foundation of sociobiology. *The Quarterly Review of Biology*, *82*(4), 327–348. https://doi.org/10.1086/522809.

Wray, G. A., & Abouheif, E. (1998). When is homology not homology? *Current Opinion in Genetics & Development*, *8*(6), 675–680. https://doi.org/10.1016/s0959-437x(98)80036-1.

Wray, G. A., & Lowe, C. J. (2000). Developmental regulatory genes and echinoderm evolution. *Systematic Biology*, *49*(1), 28–51. https://doi.org/10.1080/10635150050207375.

Yan, H., Opachaloemphan, C., Mancini, G., Yang, H., Gallitto, M., Mlejnek, J., et al. (2017). An engineered orco mutation produces aberrant social behavior and defective neural development in ants. *Cell*, *170*(4), 736–747 e9. https://doi.org/10.1016/j.cell.2017.06.051.

Yang, A. S. (2007). Thinking outside the embryo: The superorganism as a model for evoDevo studies. *Biological Theory*, *2*(4), 398–408. https://doi.org/10.1162/biot.2007.2.4.398.

Yong, E. (2012). Return of the super ants. *Nature*. https://doi.org/10.1038/nature.2012.9746.

Zardoya, R., Abouheif, E., & Meyer, A. (1996a). Evolution and orthology of hedgehog genes. *Trends in Genetics*, *12*(12), 496–497. https://doi.org/10.1016/s0168-9525(96)20014-9.

Zardoya, R., Abouheif, E., & Meyer, A. (1996b). Evolutionary analyses of hedgehog and Hoxd-10 genes in fish species closely related to the zebrafish. *Proceedings of the National Academy of Sciences of the United States of America*, *93*(23), 13036–13041. https://doi.org/10.1073/pnas.93.23.13036.

Zhao, L., Gao, F., Gao, S., Liang, Y., Long, H., Lv, Z., et al. (2021). Biodiversity-based development and evolution: The emerging research systems in model and non-model organisms. *Science China Life Sciences*, *64*(8), 1236–1280. https://doi.org/10.1007/s11427-020-1915-y.

CHAPTER TEN

Cricket: The third domesticated insect

Taro Mito[a], Yoshiyasu Ishimaru[a], Takahito Watanabe[a], Taro Nakamura[b], Guillem Ylla[c,f], Sumihare Noji[a], and Cassandra G. Extavour[c,d,e,*]

[a]Division of Bioscience and Bioindustry, Graduate School of Technology, Industrial and Social Sciences, Tokushima University, Tokushima City, Tokushima, Japan
[b]Division of Evolutionary Developmental Biology, National Institute for Basic Biology, Okazaki, Aichi, Japan
[c]Department of Organismic and Evolutionary Biology, Harvard University, Cambridge, MA, United States
[d]Department of Molecular and Cellular Biology, Harvard University, Cambridge, MA, United States
[e]Howard Hughes Medical Institute, Chevy Chase, MD, United States
[f]Laboratory of Bioinformatics and Genome Biology, Faculty of Biochemistry, Biophysics and Biotechnology, Jagiellonian University, Krakow, Poland
*Corresponding authors: e-mail address: extavour@oeb.harvard.edu

Contents

Abstract

Many researchers are using crickets to conduct research on various topics related to development and regeneration in addition to brain function, behavior, and biological clocks, using advanced functional and perturbational technologies such as genome editing. Recently, crickets have also been attracting attention as a food source for the next generation of humans. In addition, crickets are increasingly being used as disease models and biological factories for pharmaceuticals. Cricket research has thus evolved over the last century from use primarily in highly important basic research, to use in a variety of applications and practical uses. These insects are now a state-of-the-art model animal that can be obtained and maintained in large quantities at low

Current Topics in Developmental Biology, Volume 147
ISSN 0070-2153
https://doi.org/10.1016/bs.ctdb.2022.02.003

cost. We therefore suggest that crickets are useful as a third domesticated insect for scientific research, after honeybees and silkworms, contributing to the achievement of global sustainable development goals.

1. Introduction

Crickets belong to the family Gryllidae (Insecta, Orthoptera). There are more than 2400 described species of crickets belonging to this family. The main cricket species used for research in the United States and Japan is *Gryllus bimaculatus*. The most commonly used wild type laboratory strain of this cricket originated from a population on Ishigaki Island in southern Japan, and was brought to Hiroshima City around 1970 by Midori Nishioka (Amphibian Research Center, Hiroshima University) as frog food. Today, these crickets are commonly sold for use as live food for amphibian and reptilian pets, and together with European crickets, are widely cultivated around the world.

In order to achieve Zero Hunger, the second of the 17 goals of the global sustainable development goals declared in 2015 (Atukunda, Eide, Kardel, Iversen, & Westerberg, 2021), namely efforts to improve the efficiency of food production, is gaining attention. Crickets, in particular, hold new potential for an edible insect as an alternative protein source to compensate for protein deficiencies in human food and livestock feed. In Japan, locusts are a representative of an edible insect. In Southeast Asia, including Thailand and Cambodia, crickets and other insects are a part of many people's daily diet. Recently, in the United States and Europe, a variety of foods using crickets have been developed. In recent years, products using crickets, such as "cricket rice crackers," have also been released in Japan.

It is noteworthy that many of these crickets, including *G. bimaculatus*, can lay eggs and breed all year round at a rearing temperature of about 28 °C. Their reproductive cycle is about 40–50 days at 28 °C, and each female produces more than 1000 eggs in her lifetime. The dry weight of one cricket is about 0.2 g, and current commercial cricket farming industries can harvest over 50 million tons per week. Because of their rapid growth and environmental friendliness, these crickets are very useful as a new source of protein in food and food products.

In addition to this important applied use of crickets, important findings have also been obtained in basic research. In particular, from the perspective of entomology, an outstanding problem is to understand the mechanism of morphogenesis of hemimetabolous insects, to elucidate the process by which insects evolved indirect development (holometabolous development), and crickets have been used as model insects for this purpose. RNA interference is an effective research tool for cricket gene function analysis, and has the advantage of being very simple and efficient. Genome editing technology can also be used, which makes cricket species very useful as research organisms.

In the future, it should be possible to use these genome editing technologies to develop a variety of health-enhancing properties in crickets, and to make crickets themselves into health foods and medicines that can be eaten and used in various types of production industries. Thus, we suggest crickets will become the third domesticated insect, following successful domestication of honeybees (*Apis mellifera*) and silkworms (*Bombyx mori*), thereby contributing to the fulfillment of global sustainable development goals for the betterment of humanity.

2. Advantages as a model system

The insect that has been used the most in genetic laboratory research is the fly *Drosophila melanogaster*. *Drosophila*-based work has been associated with five Nobel Prizes in Physiology or Medicine. This illustrates the utility of basic research on insects in elucidating basic biological principles of many organisms, including humans. Some major advantages of using crickets as insect model organisms, and examples of new biological and evolutionary insights contributed by research using crickets, are described below.

2.1 Availability and ease of husbandry

Crickets of multiple species are generally available in large quantities at low cost for use in research and teaching. In addition, for applications where uniformity of the genetic background is important, multiple laboratories maintain inbred strains of the cricket *G. bimaculatus*, using the white-eyed trait as a genetic marker.

2.2 Genome sequence and genetic manipulation

The number of chromosomes 2n in *G. bimaculatus* is 28 + XX (female)/XO (male). The only sex chromosome is the X chromosome (Yoshimura, Nakata, Mito, & Noji, 2006). The 1.66 billion base pair genome of the cricket has been sequenced, assembled and annotated (Ylla et al., 2021). For context, it is useful to consider the history of insect genome sequencing.

The *D. melanogaster* genome (Adams et al., 2000) was the first publicly available insect genome, and the second animal genome to be fully sequenced, preceded only by the genome of the nematode *Caenorhabditis elegans* (C. elegans Sequencing Consortium, 1998). After the *Drosophila* genome, the insect genomes that followed to be sequenced were those with human health or economic interests. These included the malaria-carrying mosquito *Anopheles gambiae* (Holt et al., 2002), the silkworm *B. mori* (Mita et al., 2004), the honeybee *A. mellifera* (Weinstock et al., 2006), and the pest beetle *Tribolium castaneum* (Richards et al., 2008). All of these insect species belong to the monophyletic Holometabola (also called Endopterygota), which undergo complete metamorphosis.

In 2010, nearly 10 years after the publication of the *D. melanogaster* genome, the genome of the pea aphid *Acyrthosiphon pisum* was published, representing the first fully sequenced genome of a hemimetabolous insect (International Aphid Genomics Consortium, 2010). This was closely followed by the genome of the hemimetabolous human parasite, the louse *Pediculus humanus corporis* (Kirkness et al., 2010). The pea aphid belongs to the order Hemiptera, and the louse belongs to the order Phthiraptera. These two orders are typically grouped under the superorder Paraneoptera, although it is not clear whether Paraneoptera is monophyletic or paraphyletic (Belles, 2020; Misof et al., 2014; Wang et al., 2016; Wipfler et al., 2019). In either case, paraneopteran species, despite displaying the ancestral state of hemimetabolous metamorphosis (direct development), are more closely related to the Holometabola than to the other 14 hemimetabolous orders, 10 of which belong to the superorder Polyneoptera.

In 2014 the first polyneopteran genome assembly was published. This was the genome of the dessert locust *Locusta migratoria*, which, at 6.5 Gb, represented the largest animal genome sequenced to date at the time of its release (Wang et al., 2014). The genomes of more Polyneoptera, including the termites *Zootermopsis nevadensis* and *Macrotermes natalensis* (Poulsen et al., 2014; Terrapon et al., 2014), quickly followed. The first cricket genome assembly was that of the Hawaiian cricket *Laupala kohalensis*, published in 2018 (Blankers, Oh, Bombarely, & Shaw, 2018), followed by the genome assemblies of the Australian field cricket *Teleogryllus oceanicus* (Pascoal et al., 2020), the Asian cricket *Teleogryllus occipitalis* (Kataoka et al., 2020), the house cricket *Acheta domesticus* (Gupta et al., 2020), and the Mediterranean field cricket *G. bimaculatus* (Ylla et al., 2021).

21 years after the publication of the first insect genome, at the time of writing NCBI contains the genome assemblies of 705 insect species. However, most of them belong to only a few specific groups of insects (Fig. 1A). 617 of them (87.52%) belong to holometabolous species, while there are only 87 (12.34%) genomes of hemimetabolous species and 1 (0.14%) is from an ametabolous species. Of the 87 hemimetabolous genomes, 54 belong to the Paraneoptera and only 27 belong to the Polyneoptera (Fig. 1A). Thus, despite the fact that Polyneoptera are one of the largest winged insect lineages with

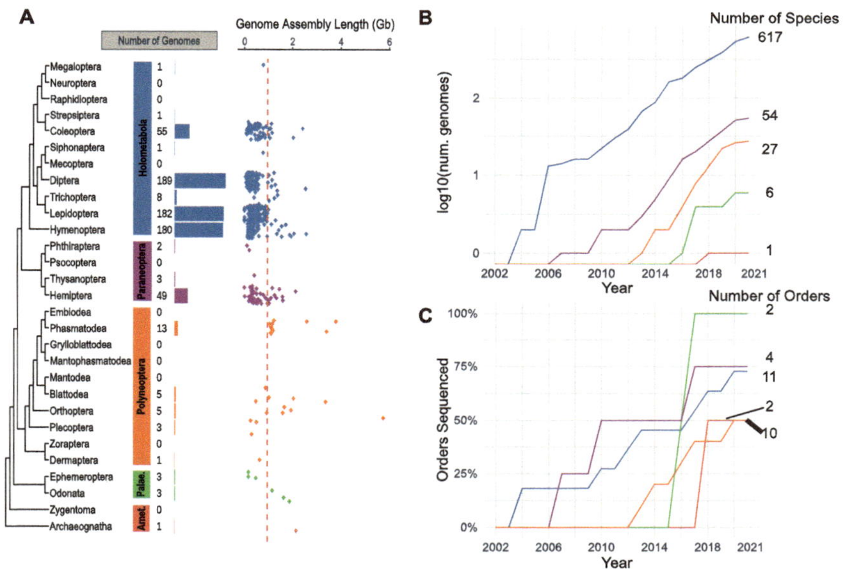

Fig. 1 (A) Phylogenetic tree of the main orders of insects based on Misof et al. (2014), modified to merge the former order Isoptera into Blattodea (Inward, Beccaloni, & Eggleton, 2007), maintaining monophyletic Paraneoptera (Wang et al., 2016) , and splitting Psocodea into Psocoptera and Phthiraptera, keeping NCBI-taxonomy ranks, showing the number of species with a publicly available NCBI genome assembly at the time of writing (bar plot with number shown at the base). Colored dots represent the reported lengths of each genome assembly in gigabases (Gb). The red-dashed vertical line indicates a genome length of 1 Gb. Colors indicate Holometabola (blue), Ametabola (red), Paraneoptera (purple), Polyneoptera (orange), and Palaeoptera (green). (B) Number of species with a publicly available NCBI genome assembly per year since 2002 in log10 scale for each of the five insect lineages shown in colors in (A). The number on the right indicates the number of insect species with a publicly available NCBI genome assembly available at the time of writing. (C) Proportion of insect orders with at least one publicly available NCBI genome assembly per year since 2002. The number on the right indicates the total number of insect orders in the given lineage; colors as in (A).

circa 40,000 described species (Belles, 2020; Wipfler et al., 2019), and their study being essential to fully understand insect evolution and biodiversity, fewer than 1% of species of this clade have genomes available.

Using the submission dates of the genome assemblies deposited in NCBI as a proxy for studying the progress on insect genome sequencing, we observe an exponential growth in the number of genomes per year for the major insect lineages (Fig. 1B). Although this exponential growth in the number of insect genomes is very exciting for the scientific community, the number of newly sequenced orders grows at a much slower pace (Fig. 1C). As result, we have a representative genome from only 72.71% of the Holometabola orders (8 out of 11), 50% of the Polyneoptera (5 out of 10), and 75% of the Paraneoptera (3 out of 4) (Fig. 1C).

This sampling bias might lead to inaccurate or erroneous conclusions regarding the evolution of various aspects of insect biology, given that holometabolous genomes appear to be quite different from those of polyneopteran species in many respects. For example, while most Holometabola do not perform genome-wide DNA methylation (the few known exceptions being primarily confined to the Hymenoptera), most polyneopteran genomes display signs of widespread DNA methylation (Bewick, Vogel, Moore, & Schmitz, 2017). Similarly, the genome sizes of the Holometabola and Paraneoptera are very different from those of Polyneoptera (Fig. 1A). Most of the genome assemblies in NCBI of Holometabola (589 out of 617) and Paraneoptera (45 out of 54) are under 1 Gb, and only 5 holometabolous species and 1 paraneopteran species have a genome larger than 2 Gb, the largest one being 2.6 Gb. By contrast, most of the polyneopteran genome assemblies (20 out of 27) exceed 1 Gb, and 6 of them exceed 2 Gb. The largest insect genome assembly available in NCBI to date belongs to the orthopteran *L. migratoria*, with an assembly length of 5.76 Gb (reported as 6.5 Gb in the original publication (Wang et al., 2014)). The genome sizes of some other Orthoptera, although not yet available in NCBI, are even larger, such as the recently assembled 8.55 Gb genome of the locust *Schistocerca gregaria* (Verlinden et al., 2020), and the estimated 16–18 Gb genome of the grasshopper *Podisma pedestris* (Bensasson, Petrov, Zhang, Hartl, & Hewitt, 2001; Camacho et al., 2015; Westerman, Barton, & Hewitt, 1987). This makes the Orthoptera one of the animal lineages with the widest range of genome sizes, and therefore one of the most interesting lineages in which to study genome size evolution and the mechanisms of genome size expansion and constraint.

At the time of writing, there are five cricket genome assemblies available, four of which have been annotated: *T. oceanicus*, *T. occipitalis*, *L. kohalensis*, and *G. bimaculatus* (Kataoka et al., 2020; Pascoal et al., 2020; Ylla et al., 2021). These four crickets have fairly similar genome sizes (ranging from 1.6 to 2 Gb), numbers of annotated protein-coding genes (ranging from 12,000 to 19,000), and proportions of reported repetitive content (ranging between 34% and 45%). These annotated cricket genomes have also provided new insights into cricket genomics, and allowed us to gain an understanding of the evolution of polyneopteran genomes. For example, comparing the genome annotations for two cricket species, *G. bimaculatus* and *L. kohalensis*, revealed that nearly 45% of annotated protein-coding genes display fingerprints of DNA methylation. Furthermore, these genes show significantly stronger signatures of purifying selection, and higher conservation across distant species, than genes predicted to be unmethylated, suggesting a putative biologically relevant role for DNA methylation in Polyneoptera that may have been lost in many species of Holometabola.

Genetic manipulation techniques achieved by injection of fertilized cricket eggs, including genome editing using CRISPR/Cas9, have been established (Watanabe, Noji, & Mito, 2017). In addition, the ReMOT control genome editing method, in which a complex of Cas9 and guide RNA injected into the abdominal hemolymph of an adult female insect is transferred to the developing eggs, wherein genome editing occurs, has been developed in mosquitoes (Chaverra-Rodriguez et al., 2018; Macias et al., 2020) and applied successfully in other arthropods (Chaverra-Rodriguez et al., 2020; Heu, McCullough, Luan, & Rasgon, 2020; Sharma et al., 2022). With this method, microinjection into fertilized eggs, which can be initially challenging for inexperienced researchers, becomes unnecessary for genome editing. This method should be applicable to crickets as well.

2.3 Germ cell formation and evolution

In animals that reproduce sexually, germ cells are the only cell lineage that transmits genetic information to the next generation. Elucidating the characteristics of germ cells and the mechanisms that control their formation is one of the most important issues in life science research. Differentiated cells specialized for reproductive ability are an evolutionary novelty typically acquired concomitantly with the evolution of multicellularity (Veit, 2019). Histological, embryological and molecular studies have shown that germ cell formation in animals can be broadly classified into two modes: (a) somatic induction by signals from somatic cells, and (b) cytoplasmic inheritance by localization of maternal effectors (germ plasm) during egg formation (Extavour, 2007; Extavour & Akam, 2003). It is hypothesized that the inheritance mode evolved from the somatic induction mode (Extavour, 2007; Extavour & Akam, 2003). However, the molecular mechanisms involved in the formation of primordial germ cells have only been elucidated in a few model organisms.

The cricket *G. bimaculatus* has played an important role in recent attempts to clarify the commonality and diversity of the molecular basis of the two modes of primordial germ cell formation from an evolutionary developmental perspective. The Extavour lab performed studies elucidating the molecular mechanisms of primordial germ cell formation in this cricket, which provided strong evidence for somatic cell-induced primordial germ cell formation by the hypothesized ancestral mode of induction, for the first time in an invertebrate animal. Interestingly, BMP signaling induces cricket primordial germ cells from somitic mesoderm by positively regulating the expression of the Blimp1 transcription factor, as is the case in mice (Donoughe et al., 2014; Nakamura & Extavour, 2016). In addition, the posterior Hox genes are required for correct cricket germ cell formation (Barnett, Nakamura, & Extavour, 2019).

2.4 Parental RNAi and the study of early development

Fertilized eggs of *G. bimaculatus* hatch in 10 days at 28 °C. 27 hours after fertilization, a large portion of the blastoderm cells assemble separately into dense condensations on the left and right sides of the ventral midline in the posterior third of the embryo. Later, they coalesce along the ventral midline to form a small embryonic rudiment in a posterior ventral position. At this stage both the anteroposterior and the dorsoventral axes of the embryo are determined, and are aligned with those of the egg. The rudiment tissues give rise to the head, thoracic, and telson primordia, whereas abdominal segments form at later stages, sequentially from anterior to posterior. Unlike the derivatives of the imaginal discs of Holometabola, limb buds in crickets are formed as lateral outgrowths from the thoracic segments, gradually elongate and eventually form legs, more similar to the developmental process that produces vertebrate limbs than to that involved in *Drosophila* leg formation.

Parental RNAi is efficient in male and female crickets. When double-stranded RNA is injected into the adult female abdominal cavity, typically more than 1000 eggs subsequently laid by the injected female gradually develop a strong phenotype (Ronco, Uda, Minelli, Noji, & Klingler, 2008). This is a very useful method for the analysis of genes involved in embryonic development.

2.5 Mechanisms of metamorphosis and its evolution

Insect development proceeds in one of two ways: full metamorphosis, seen in indirectly developing insects (e.g., flies, butterflies), or incomplete metamorphosis, seen in directly developing insects (e.g., dragonflies, crickets). Holometabolous insects undergo full metamorphosis, forming a pupa and then becoming an adult. As epitomized by the metamorphosis of butterflies, for example, the adult forms a completely different body plan from the larva. Holometaboly evolved from hemimetaboly, and hormones that control metamorphosis are present in both types of insects, but are deployed differently in the two groups. Studies in *G. bimaculatus* have been instructive in understanding how hormones regulating holometabolous metamorphosis operate in hemimetabolous insects, and thus in clarifying how insect metamorphosis evolved.

Juvenile hormone (JH) is a hormone that inhibits the metamorphosis of larvae into pupae in Holometabola, or of juveniles to adults in Hemimetabola. When JH activity is artificially and continuously increased by external treatment, larvae undergo additional growth and molt cycles, resulting in a dramatic transformation into giant-sized adults. On the other hand, when JH action is inhibited, the number of molts until metamorphosis is reduced, resulting in adults with smaller body size, referred to as precocious metamorphosis. In the final step of the juvenile hormone synthesis process, juvenile hormone acid methyltransferase (JHAMT) and epoxidase (Cytochrome P450: CYP15A1) are involved. JHAMT was first identified in the silkworm in 2003 by T. Shinoda and his colleagues (Shinoda & Itoyama, 2003). In the absence of this enzyme, JH is not activated. In early silkworm larvae, JHAMT is continuously expressed and produces juvenile hormones, but in final instar larvae, JH synthesis ceases due to a decrease in the amount of JHAMT, resulting in the induction of metamorphosis. Thus, the repression of JHAMT expression is an essential switch for insect metamorphosis. However, the mechanism of JHAMT repression has remained unclear. Elucidation of the molecular mechanism is very important for understanding the regulation of insect metamorphosis.

Studies in *G. bimaculatus* by T. Mito, Y. Ishimaru and their colleagues revealed that in this cricket, JH is regulated by *dpp* and *myoglianin*, which are members of the TGFb family of ligands (Ishimaru et al., 2016). Dpp signaling regulates the biosynthesis of active JH through the induction of JHAMT expression and suppresses molting to adulthood at early stages. At later stages, when myoglianin signaling represses JHAMT expression, the final molt to adulthood is triggered. Thus, the timing of metamorphosis is strictly regulated by the effects of both signals on JHAMT expression.

In the molecular mechanism of juvenile hormone action in holometabolous insect metamorphosis, the mutually inhibitory functions of *Krüppel homolog* 1 (*Kr-h*1), *Broad-Complex* (*Br*), and *E93* at two stages ((terminal larva → pupa) and (pupa → adult)) are also observed in the final transition from juvenile to adult in hemimetabolous insects. Based on the timing of action of this evolutionarily conserved molecular mechanism, Ishimaru and colleagues proposed the hypothesis that in the cricket, the preterminal → larval stage of hemimetabolous development corresponds to the pupal stage of Holometabola (Ishimaru, Tomonari, Watanabe, Noji, & Mito, 2019). Thus, although the metamorphosis process differs greatly between holometabolous and hemimetabolous insects in terms of relative degree of change in body plan organization, many elements of the molecular mechanisms controlling the process appear conserved.

2.6 Surgical experiments and RNAi elucidate leg regeneration mechanisms

The legs of the cricket larvae are capable of complete regeneration following amputation. Pioneering studies on the regeneration of the legs of hemimetabolous insects were primarily conducted by European researchers using cockroaches, but these studies had nearly ceased by around 1980. Performing similar experiments with crickets, T. Bando and colleagues hypothesized that the protocadherins Fat/Dachsous (Ft/Ds) might determine positional information in cricket legs during regeneration, and tested this hypothesis using RNAi. A related hypothesis had previously been proposed by P. Lawrence and colleagues based on his work on *Drosophila* (Lawrence, Struhl, & Casal, 2008). They suggested that these molecules may be responsible for positional information, and proposed the Ft/Ds gradient model as a mechanism to determine the length of leg segments. Bando and colleagues found that when cricket legs regenerate, if the function of Ft is lost, the regenerated legs are thicker and shorter than controls. This phenotype is similar to that of loss of function mutations of the *fat* gene in *Drosophila*, which results in thicker legs (hence the name of the gene). Bando and colleagues found that signaling by the protocadherins Ft/Ds is involved in the regeneration process via the Hippo signaling pathway, and proposed a novel gradient model for the determination of leg size (Bando, Mito, Nakamura, Ohuchi, & Noji, 2011). Elucidation of the regeneration mechanism at the molecular level is ongoing (Bando et al., 2018).

Recently, it was reported that during recovery from damage to multiple organ systems in larval and adult zebrafish, the innate immune response is involved in their regeneration (Iribarne, 2021). To investigate whether macrophages are similarly involved in the regeneration of cricket legs as an innate immune response, Bando and colleagues applied a pharmacological agent that kills macrophages to crickets subject to leg amputation, and observed no subsequent regeneration of the amputated leg (Bando et al., 2021). This suggests that the function of macrophages is necessary in early stages of regeneration, and that this function may be common to both insects and vertebrates. In addition, cricket regeneration requires a STAT protein that is also involved in the differentiation of human macrophages. A current model thus posits that when a cricket's leg is amputated, cells of the immune system, including macrophages, gather to seal the wound, and that substances secreted by these cells may determine the start of regeneration. Understanding how the innate immune response regulates regeneration in the cricket may provide important clues that could improve the therapeutic strategies for repairing injured mammalian tissues that do not have inherent regenerative capacity (Iribarne, 2021).

2.7 Internal clock and photoperiodism in crickets

Crickets, like most animals, have an internal circadian clock. Studies by Seymour Benzer and others of clock mutants in *Drosophila* (Konopka & Benzer, 1971; Price, 2005; Singh et al., 2019; Takahashi, 2004), and a detailed gene expression cascade (Claridge-Chang et al., 2001; Hardin, 2011) have laid important foundations for understanding the internal clock in this insect. While the regulation of insect physiological activities including reproduction and dormancy are regulated by photoperiodism (the response of an organism to the length of the light or dark period of the day), the mechanism of this regulation is still unclear. K. Tomioka and his group conducted experiments to address this problem using the cricket *Modicogryllus siamensis* (Miki, Shinohara, Chafino, Noji, & Tomioka, 2020). When the expression of myoglianin was suppressed by RNAi under long-day conditions, the number of molts increased, indicating short-day development. They also observed that the expression of myoglianin was regulated by day-length information through the brain. The question of what regulates the expression of myoglianin in the brain remains unclear, but its answer may be relevant to elucidation of the mechanism of hibernation in animals, given that the mammalian version of myoglianin is myostatin (GDF8), which may be involved in hibernation (Brooks, Myburgh, & Storey, 2011).

2.8 Brain and behavioral studies

The fact that crickets can learn and remember has been exploited by neurobiologists and neuroethologists to study the mechanisms of learning and memory. Researchers use genetic manipulation, RNAi and surgical procedures such as the introduction of electrodes, to pursue research into the higher functions of the brain. For example, the mechanisms underlying the formation and retention of memory of crickets has been under investigation by the Mizunami group (Matsumoto, Matsumoto, & Mizunami, 2018). Recently, they found that the cricket has the ability to learn, not only by previously known olfactory stimuli (Matsumoto & Mizunami, 2002, 2004), but also by observation (Ebina & Mizunami, 2020). In humans, "observational learning" refers to the ability to learn by observing the behavior of other individuals without having any experience of their own. Evidence for this ability comes from experiments wherein a "demonstrator" cricket is allowed to freely visit a water reward associated with a particular scent, or salt water (punishment) associated with a second scent. An "observer" cricket is allowed to observe the behavior of the "demonstrator" through a transparent partition, and then tested to see which of the two waters they prefer (Ebina & Mizunami, 2020). The neural substrate of this learning model in crickets is the subject of ongoing work.

2.9 Crickets as food

Around 2016, the phrase "entomophagy" began to appear with increased frequency in multiple scientific journals, often in the context of global sustainable development goals. Crickets, in particular, appear increasingly frequently as an example of edible insects. Since the wealth of existing developmental research in crickets should be useful to enhance the cultivation and breeding of edible crickets, crickets as an edible resource may contribute effectively to solving the planets' global human food shortage. Multiple commercial and nonprofit organizations, including Aspire Food Group and Gryllus Inc., are working on the development of automatic cricket breeding equipment, a stable supply of cricket powder, and genetic enhancements of cricket nutritional value.

3. Conclusion

Studies of crickets have made critical contributions to multiple areas of basic and applied scientific knowledge. As the number of insect genomes continues to grow in the future, we suggest that increased focus on incorporating species from currently underrepresented lineages will yield a more complete picture of the diversity of insect genomes. As for cricket genomes, 2020 was a great year, with the release of four annotated genomes and an unannotated fifth one. Hopefully, these genomes will help to establish crickets as even more widely used model organisms, promoting the development of enhanced research protocols, experimental techniques, and genome editing technologies.

References

Adams, M. D., Celniker, S. E., Holt, R. A., Evans, C. A., Gocayne, J. D., Amanatides, P. G., et al. (2000). The genome sequence of *Drosophila melanogaster*. *Science*, *287*, 2185–2195. https://doi.org/10.1126/science.287.5461.2185.

Atukunda, P., Eide, W. B., Kardel, K. R., Iversen, P. O., & Westerberg, A. C. (2021). Unlocking the potential for achievement of the UN Sustainable Development Goal 2—'Zero Hunger'—In Africa: Targets, strategies, synergies and challenges. *Food and Nutrition Research*, *65*. https://doi.org/10.29219/fnr.v65.7686.

Bando, T., Mito, T., Hamada, Y., Ishimaru, Y., Noji, S., & Ohuchi, H. (2018). Molecular mechanisms of limb regeneration: Insights from regenerating legs of the cricket *Gryllus bimaculatus*. *The International Journal of Developmental Biology*, *62*, 559–569. https://doi.org/10.1387/ijdb.180048ho.

Bando, T., Mito, T., Nakamura, T., Ohuchi, H., & Noji, S. (2011). Regulation of leg size and shape: Involvement of the Dachsous-fat signaling pathway. *Developmental Dynamics*, *240*, 1028–1041. https://doi.org/10.1002/dvdy.22590.

Bando, T., Okumura, M., Bando, Y., Hagiwara, M., Hamada, Y., Ishimaru, Y., & Ohuchi, H. (2021). Toll signalling promotes blastema cell proliferation during cricket leg regeneration via insect macrophages. *Development*, *149*, dev199916. https://doi.org/10.1242/dev.199916.

Barnett, A. A., Nakamura, T., & Extavour, C. G. (2019). Hox genes limit germ cell formation in the short germ insect *Gryllus bimaculatus*. *Proceedings of the National Academy of Sciences of the USA*, *116*, 16430–16435. https://doi.org/10.1073/pnas.1816024116.

Belles, X. (2020). *Insect metamorphosis: From natural history to regulation of development and evolution*. Elsevier. 60.

Bensasson, D., Petrov, D. A., Zhang, D.-X., Hartl, D. L., & Hewitt, G. M. (2001). Genomic gigantism: DNA loss is slow in mountain grasshoppers. *Molecular Biology and Evolution*, *18*, 246–253. https://doi.org/10.1093/oxfordjournals.molbev.a003798.

Bewick, A. J., Vogel, K. J., Moore, A. J., & Schmitz, R. J. (2017). Evolution of DNA methylation across insects. *Molecular Biology and Evolution*, *34*, 654–665. https://doi.org/10.1093/molbev/msw264.

Blankers, T., Oh, K. P., Bombarely, A., & Shaw, K. L. (2018). The genomic architecture of a rapid island radiation: Recombination rate variation, chromosome structure, and genome assembly of the Hawaiian cricket *Laupala*. *Genetics*, *209*, 1329–1344. https://doi.org/10.1534/genetics.118.300894.

Brooks, N. E., Myburgh, K. H., & Storey, K. B. (2011). Myostatin levels in skeletal muscle of hibernating ground squirrels. *The Journal of Experimental Biology*, *214*, 2522–2527. https://doi.org/10.1242/jeb.055764.

Camacho, J. P. M., Ruiz-Ruano, F. J., Martín-Blázquez, R., López-León, M. D., Cabrero, J., Lorite, P., et al. (2015). A step to the gigantic genome of the desert locust: Chromosome sizes and repeated DNAs. *Chromosoma*, *124*, 263–275. https://doi.org/10.1007/s00412-014-0499-0.

C. elegans Sequencing Consortium. (1998). Genome sequence of the nematode *C. elegans*: A platform for investigating biology. *Science*, *282*, 2012–2018. https://doi.org/10.1126/science.282.5396.2012.

Chaverra-Rodriguez, D., Benetta, E. D., Heu, C. C., Rasgon, J. L., Ferree, P. M., & Akbari, O. S. (2020). Germline mutagenesis of *Nasonia vitripennis* through ovarian delivery of CRISPR-Cas9 ribonucleoprotein. *Insect Molecular Biology*, *29*, 569–577. https://doi.org/10.1111/imb.12663.

Chaverra-Rodriguez, D., Macias, V. M., Hughes, G. L., Pujhari, S., Suzuki, Y., Peterson, D. R., et al. (2018). Targeted delivery of CRISPR-Cas9 ribonucleoprotein into arthropod ovaries for heritable germline gene editing. *Nature Communications*, *9*, 3008. https://doi.org/10.1038/s41467-018-05425-9.

Claridge-Chang, A., Wijnen, H., Naef, F., Boothroyd, C., Rajewsky, N., & Young, M. W. (2001). Circadian regulation of gene expression systems in the *Drosophila* head. *Neuron*, *32*, 657–671. https://doi.org/10.1016/s0896-6273(01)00515-3.

Donoughe, S., Nakamura, T., Ewen-Campen, B., Green, D. A., Henderson, L., & Extavour, C. G. (2014). BMP signaling is required for the generation of primordial germ cells in an insect. *Proceedings of the National Academy of Sciences of the United States of America*, *111*, 4133–4138. https://doi.org/10.1073/pnas.1400525111.

Ebina, H., & Mizunami, M. (2020). Appetitive and aversive social learning with living and dead conspecifics in crickets. *Scientific Reports*, *10*, 9340. https://doi.org/10.1038/s41598-020-66399-7.

Extavour, C. G. M. (2007). Evolution of the bilaterian germ line: Lineage origin and modulation of specification mechanisms. *Integrative and Comparative Biology*, *47*, 770–785. https://doi.org/10.1093/icb/icm027.

Extavour, C. G., & Akam, M. (2003). Mechanisms of germ cell specification across the metazoans: Epigenesis and preformation. *Development*, *130*, 5869–5884. https://doi.org/10.1242/dev.00804.

Gupta, Y. M., Tanasarnpaiboon, S., Buddhachat, K., Peyachoknagul, S., Inthim, P., & Homchan, S. (2020). Development of microsatellite markers for the house cricket, *Acheta domesticus* (Orthoptera: Gryllidae). *Biodiversitas Journal of Biological Diversity*, *21*. https://doi.org/10.13057/biodiv/d210921.

Hardin, P. E. (2011). Molecular genetic analysis of circadian timekeeping in *Drosophila*. *Advances in Genetics*, *74*, 141–173. https://doi.org/10.1016/b978-0-12-387690-4.00005-2.

Heu, C. C., McCullough, F. M., Luan, J., & Rasgon, J. L. (2020). CRISPR-Cas9-based genome editing in the silverleaf whitefly (Bemisia tabaci). *The CRISPR Journal*, *3*, 89–96. https://doi.org/10.1089/crispr.2019.0067.

Holt, R. A., Subramanian, G. M., Halpern, A., Sutton, G. G., Charlab, R., Nusskern, D. R., et al. (2002). The genome sequence of the malaria mosquito *Anopheles gambiae*. *Science*, *298*, 129–149. https://doi.org/10.1126/science.1076181.

International Aphid Genomics Consortium. (2010). Genome sequence of the pea aphid *Acyrthosiphon pisum*. *PLoS Biology*, *8*, e1000313. https://doi.org/10.1371/journal.pbio.1000313.

Inward, D., Beccaloni, G., & Eggleton, P. (2007). Death of an order: A comprehensive molecular phylogenetic study confirms that termites are eusocial cockroaches. *Biology Letters*, *3*, 331–335. https://doi.org/10.1098/rsbl.2007.0102.

Iribarne, M. (2021). Inflammation induces zebrafish regeneration. *Neural Regeneration Research*, *16*, 1693. https://doi.org/10.4103/1673-5374.306059.

Ishimaru, Y., Tomonari, S., Matsuoka, Y., Watanabe, T., Miyawaki, K., Bando, T., et al. (2016). TGF-β signaling in insects regulates metamorphosis via juvenile hormone biosynthesis. *Proceedings of the National Academy of Sciences of the United States of America*, *113*, 5634–5639. https://doi.org/10.1073/pnas.1600612113.

Ishimaru, Y., Tomonari, S., Watanabe, T., Noji, S., & Mito, T. (2019). Regulatory mechanisms underlying the specification of the pupal-homologous stage in a hemimetabolous insect. *Philosophical Transactions of the Royal Society of London. Series B, Biological Sciences*, *374*, 20190225. https://doi.org/10.1098/rstb.2019.0225.

Kataoka, K., Minei, R., Ide, K., Ogura, A., Takeyama, H., Takeda, M., et al. (2020). The draft genome dataset of the Asian cricket *Teleogryllus occipitalis* for molecular research toward entomophagy. *Frontiers in Genetics*, *11*, 470. https://doi.org/10.3389/fgene.2020.00470.

Kirkness, E. F., Haas, B. J., Sun, W., Braig, H. R., Perotti, M. A., Clark, J. M., et al. (2010). Genome sequences of the human body louse and its primary endosymbiont provide insights into the permanent parasitic lifestyle. *Proceedings of the National Academy of Sciences of the United States of America*, *107*, 12168–12173. https://doi.org/10.1073/pnas.1003379107.

Konopka, R. J., & Benzer, S. (1971). Clock mutants of *Drosophila melanogaster*. *Proceedings of the National Academy of Sciences of the United States of America*, *68*, 2112–2116. https://doi.org/10.1073/pnas.68.9.2112.

Lawrence, P. A., Struhl, G., & Casal, J. (2008). Do the protocadherins Fat and Dachsous link up to determine both planar cell polarity and the dimensions of organs? *Nature Cell Biology*, *10*, 1379–1382. https://doi.org/10.1038/ncb1208-1379.

Macias, V. M., McKeand, S., Chaverra-Rodriguez, D., Hughes, G. L., Fazekas, A., Pujhari, S., et al. (2020). Cas9-mediated gene-editing in the malaria mosquito *Anopheles stephensi* by ReMOT control. *G3: Genes, Genomes, Genetics*, *10*, 1353–1360. https://doi.org/10.1534/g3.120.401133.

Matsumoto, Y., Matsumoto, C. S., & Mizunami, M. (2018). Signaling pathways for long-term memory formation in the cricket. *Frontiers in Psychology*, *9*, 1014. https://doi.org/10.3389/fpsyg.2018.01014.

Matsumoto, Y., & Mizunami, M. (2002). Lifetime olfactory memory in the cricket *Gryllus bimaculatus*. *Journal of Comparative Physiology*, *188*, 295–299. https://doi.org/10.1007/s00359-002-0303-0.

Matsumoto, Y., & Mizunami, M. (2004). Context-dependent olfactory learning in an insect. *Learning and Memory*, *11*, 288–293. https://doi.org/10.1101/lm.72504.

Miki, T., Shinohara, T., Chafino, S., Noji, S., & Tomioka, K. (2020). Photoperiod and temperature separately regulate nymphal development through JH and insulin/TOR signaling pathways in an insect. *Proceedings of the National Academy of Sciences of the United States of America*, *117*, 5525–5531. https://doi.org/10.1073/pnas.1922747117.

Misof, B., Liu, S., Meusemann, K., Peters, R. S., Donath, A., Mayer, C., et al. (2014). Phylogenomics resolves the timing and pattern of insect evolution. *Science*, *346*, 763–767. https://doi.org/10.1126/science.1257570.

Mita, K., Kasahara, M., Sasaki, S., Nagayasu, Y., Yamada, T., Kanamori, H., et al. (2004). The genome sequence of silkworm, *Bombyx mori*. *DNA Research*, *11*, 27–35. https://doi.org/10.1093/dnares/11.1.27.

Nakamura, T., & Extavour, C. G. (2016). The transcriptional repressor Blimp-1 acts downstream of BMP signaling to generate primordial germ cells in the cricket *Gryllus bimaculatus*. *Development*, *143*, 255–263. https://doi.org/10.1242/dev.127563.

Pascoal, S., Risse, J. E., Zhang, X., Blaxter, M., Cezard, T., Challis, R. J., et al. (2020). Field cricket genome reveals the footprint of recent, abrupt adaptation in the wild. *Evolution Letters*, *4*, 19–33. https://doi.org/10.1002/evl3.148.

Poulsen, M., Hu, H., Li, C., Chen, Z., Xu, L., Otani, S., et al. (2014). Complementary symbiont contributions to plant decomposition in a fungus-farming termite. *Proceedings of the National Academy of Sciences of the United States of America*, *111*, 14500–14505. https://doi.org/10.1073/pnas.1319718111.

Price, J. L. (2005). Genetic screens for clock mutants in *Drosophila*. *Methods in Enzymology*, *393*, 35–60. https://doi.org/10.1016/s0076-6879(05)93003-6.

Richards, S., Gibbs, R. A., Weinstock, G. M., Brown, S. J., Denell, R., Beeman, R. W., et al. (2008). The genome of the model beetle and pest *Tribolium castaneum*. *Nature*, *452*, 949–955. https://doi.org/10.1038/nature06784.

Ronco, M., Uda, T., Minelli, A., Noji, S., & Klingler, M. (2008). Antenna and all gnathal appendages are similarly transformed by *homothorax* knock-down in the cricket *Gryllus bimaculatus*. *Developmental Biology*, *313*, 80–92. https://doi.org/10.1016/j.ydbio.2007.09.059.

Sharma, A., Pham, M. N., Reyes, J. B., Chana, R., Yim, W. C., Heu, C. C., … Gulia-Nuss, M. (2022). Cas9-mediated gene-editing in the black-legged tick, *Ixodes scapularis*, by embryo injection and ReMOT control. *iScience*. https://doi.org/10.1016/j.isci.2022.103781.

Shinoda, T., & Itoyama, K. (2003). Juvenile hormone acid methyltransferase: A key regulatory enzyme for insect metamorphosis. *Proceedings of the National Academy of Sciences of the United States of America*, *100*, 11986–11991. https://doi.org/10.1073/pnas.2134232100.

Singh, S., Giesecke, A., Damulewicz, M., Fexova, S., Mazzotta, G. M., Stanewsky, R., et al. (2019). New *Drosophila* circadian clock mutants affecting temperature compensation induced by targeted mutagenesis of timeless. *Frontiers in Physiology*, *10*, 1442. https://doi.org/10.3389/fphys.2019.01442.

Takahashi, J. S. (2004). Finding new clock components: Past and future. *Journal of Biological Rhythms*, *19*, 339–347. https://doi.org/10.1177/0748730404269151.

Terrapon, N., Li, C., Robertson, H. M., Ji, L., Meng, X., Booth, W., et al. (2014). Molecular traces of alternative social organization in a termite genome. *Nature Communications*, *5*, 3636. https://doi.org/10.1038/ncomms4636.

Veit, W. (2019). Evolution of multicellularity: Cheating done right. *Biology and Philosophy*, *34*, 34. https://doi.org/10.1007/s10539-019-9688-9.

Verlinden, H., Sterck, L., Li, J., Li, Z., Yssel, A., Gansemans, Y., et al. (2020). First draft genome assembly of the desert locust, *Schistocerca gregaria*. *F1000Research*, *9*, 775. https://doi.org/10.12688/f1000research.25148.1.

Wang, Y., Engel, M. S., Rafael, J. A., Wu, H., Rédei, D., Xie, Q., et al. (2016). Fossil record of stem groups employed in evaluating the chronogram of insects (Arthropoda: Hexapoda). *Scientific Reports*, *6*, 38939. https://doi.org/10.1038/srep38939.

Wang, X., Fang, X., Yang, P., Jiang, X., Jiang, F., Zhao, D., et al. (2014). The locust genome provides insight into swarm formation and long-distance flight. *Nature Communications*, *5*, 2957. https://doi.org/10.1038/ncomms3957.

Watanabe, T., Noji, S., & Mito, T. (2017). Genome editing in animals, methods and protocols. *Methods in Molecular Biology*, *1630*, 219–233. https://doi.org/10.1007/978-1-4939-7128-2_18.

Weinstock, G. M., Robinson, G. E., Gibbs, R. A., Weinstock, G. M., Weinstock, G. M., Robinson, G. E., et al. (2006). Insights into social insects from the genome of the honeybee *Apis mellifera*. *Nature*, *443*, 931–949. https://doi.org/10.1038/nature05260.

Westerman, M., Barton, N. H., & Hewitt, G. M. (1987). Differences in DNA content between two chromosomal races of the grasshopper *Podisma pedestris*. *Heredity*, *58*, 221–228. https://doi.org/10.1038/hdy.1987.36.

Wipfler, B., Letsch, H., Frandsen, P. B., Kapli, P., Mayer, C., Bartel, D., et al. (2019). Evolutionary history of Polyneoptera and its implications for our understanding of early winged insects. *Proceedings of the National Academy of Sciences of the United States of America*, *116*, 3024–3029. https://doi.org/10.1073/pnas.1817794116.

Ylla, G., Nakamura, T., Itoh, T., Kajitani, R., Toyoda, A., Tomonari, S., et al. (2021). Insights into the genomic evolution of insects from cricket genomes. *Communications Biology*, *4*, 733. https://doi.org/10.1038/s42003-021-02197-9.

Yoshimura, A., Nakata, A., Mito, T., & Noji, S. (2006). The characteristics of karyotype and telomeric satellite DNA sequences in the cricket, *Gryllus bimaculatus* (Orthoptera, Gryllidae). *Cytogenetic and Genome Research*, *112*, 329–336. https://doi.org/10.1159/000089889.

CHAPTER ELEVEN

Schmidtea happens: Re-establishing the planarian as a model for studying the mechanisms of regeneration

Phillip A. Newmark[a],* and Alejandro Sánchez Alvarado[b]
[a]Howard Hughes Medical Institute, Morgridge Institute for Research, Department of Integrative Biology, University of Wisconsin, Madison, WI, United States
[b]Howard Hughes Medical Institute, Stowers Institute for Medical Research, Kansas City, MO, United States
*Corresponding author: e-mail address: pnewmark@morgridge.org

Contents

Abstract

Understanding the remarkable regenerative abilities of freshwater planarians was a classic problem of developmental biology. These animals were widely studied until the late 1960s, when their use as experimental subjects declined precipitously after some infamous experiments on memory transfer. By the mid-1990s, only a handful of laboratories worldwide were investigating the mechanisms of planarian regeneration. Here, we provide the personal stories behind our work to reinvigorate studies of these fascinating animals. We recount many of the challenges that had to be overcome and reflect on some of the fortuitous events that helped launch the planarian *Schmidtea mediterranea* as a model organism for studying the molecular basis of regeneration.

1. Introduction

"What led you to start working on planarians?" is the most-often-asked question when we meet with trainees during seminar visits and scientific meetings. We each have our own distilled answer to that inquiry, but

Current Topics in Developmental Biology, Volume 147
ISSN 0070-2153
https://doi.org/10.1016/bs.ctdb.2022.01.002

because we try to focus such discussions on our current work, we never have enough time to flesh out the entire story. This chapter provides us with that opportunity. We'll begin by discussing each of our parallel pathways: how a *Drosophila* developmental geneticist and a molecular biologist/budding vertebrate embryologist independently wound up deciding that they would abandon well-established experimental systems to study the planarian, an organism whose scientific heyday had long since passed, at the risk of committing (according to some senior colleagues) "career suicide." We'll then cover the chance encounter that led to the convergence of our parallel pathways, bringing together our complementary skills and expertise in the right place at the right time to help revitalize studies on these fascinating animals.

2. PAN's planarian travelog: From flies in Boulder to flatworms in Barcelona

As a second-year graduate student in the Department of Molecular, Cellular, and Developmental Biology (MCDB) at the University of Colorado at Boulder, I participated in a developmental biology journal club run by my thesis advisor, Bob Boswell. The only rule was that we had to present papers outside of our individual research areas, which forced us to expand our thinking to other systems and questions. When my turn came to present, I found a fascinating paper that had just been published in *Development* on the role of stem cells in planarian regeneration (Baguñà, Saló, & Auladell, 1989). Decades of previous work had implicated proliferative cells known as neoblasts in planarian regeneration. After irradiation, neoblasts were eliminated and planarians lost their ability to regenerate; however, whether these cells represented true stem cells and the source of the newly regenerated tissue or whether dedifferentiation was responsible for regeneration remained open questions.

To directly address these issues, Jaume Baguñà, along with his colleagues Emili Saló and Carme Auladell at the University of Barcelona, injected neoblasts or differentiated cells into lethally irradiated planarians. They found that dissociated cell preparations enriched in neoblasts could rescue lethally irradiated planarians, whereas differentiated cells could not. Most strikingly, by transplanting neoblasts from a planarian species (*S. mediterranea*) in which sexually and asexually reproducing strains could be distinguished by a chromosomal translocation, they reported that neoblasts from the asexual strain could convert irradiated sexual planarians to an asexual mode of reproduction, whereas sexual neoblasts could convert irradiated asexual planarians to sexual reproduction. Thus, neoblasts could generate both somatic and germ cell lineages.

I had last encountered planarians in my freshman introductory biology class at Boston University and was completely fascinated by them back then. Now, seeing planarians in this context brought home how little we knew about the mechanisms of regeneration in any animal. Furthermore, as a graduate student studying one of the classic paradigms of cytoplasmic determinants, germ plasm localization to the posterior pole of the *Drosophila* oocyte, this example of inductive germ cell specification in an invertebrate was particularly striking.

This journal club presentation sparked a deep dive into the classic planarian literature (facilitated by H.V. Brøndsted's monograph, *Planarian Regeneration* (Brøndsted, 1969), which provided references to nearly all the classic papers on the topic) and the following year I wound up giving a talk on planarians in a graduate course on evolutionary developmental biology. The more I read, the more I became convinced that the problem of planarian regeneration merited being revisited using the tools of molecular biology. It occurred to me that because neoblasts could rescue lethally irradiated planarians, one should be able to introduce reporter genes randomly into the genomes of individual neoblasts, use these transfected cells to repopulate irradiated worms, and in this manner generate enhancer-trap lines—like those being used in *Drosophila* at the time (O'Kane & Gehring, 1987)—to identify genes upregulated during regeneration or in a tissue-specific manner. I now had a plan for my post-doc! Unfortunately, it seemed that there were no labs in the US working on planarians at the molecular level, and the number of such labs around the world could be counted on one hand, with room to spare. Also, I still had ~4 years left before finishing my Ph.D., so I had many other, far more pressing concerns than finding a post-doc lab.

My major concern was cloning *mago nashi*, a new posterior-group gene in *Drosophila* that Bob Boswell identified in screens for additional maternal-effect mutants resulting in sterile progeny (the grandchildless phenotype) (Boswell, Prout, & Steichen, 1991). While I was plugging away at the molecular biology, Bob patiently put up with me occasionally heading off to the library during long incubations, where I would dig through the stacks in search of old planarian papers to photocopy (google it). One of the joys of pulling out century-old journals in search of a specific paper was learning about other work that was being published contemporaneously; while digging for planarian papers by T.H. Morgan, in the same volume one could stumble across a paper on tadpole tail regeneration by Ross Harrison or on acoel regeneration by Nettie Stevens.

One day while skimming through the shelves containing books on regeneration, development, and related topics, I came across a 1990 monograph, "Experimental Embryology in Aquatic Plants and Animals," which resulted from a NATO Advanced Study Institute. Scanning the table of contents, I found a chapter by Baguñà and colleagues reviewing growth, degrowth, and regeneration in planarians (Baguñà et al., 1990). This chapter also described some of the group's ongoing work and future directions, including a figure outlining conceptually how to use transfected neoblasts to perform enhancer traps. I hadn't even finished my Ph.D. and my post-doc project was already going to be scooped (this worry is even funnier 30 years later)! Given that there were no labs in the US studying planarian regeneration with cellular and molecular tools, while the Barcelona group was already pursuing such approaches and had documented expertise in the system through a series of thoughtful, well-executed papers (Baguñà & Romero, 1981; Salò & Baguñà, 1984, 1985a, 1985b, to name a few), it seemed like the best course of action would be to apply for a post-doc position there. After consulting with Bob, my thesis committee, and various MCDB faculty members, and a warning from Bill Wood to avoid the infamous planarian RNA memory experiments (see Rilling (1996) for a history of this story), I was encouraged to follow my scientific passion. So, in early September 1991 I sent a letter to Jaume via air mail, applying for a post-doc position in his group (some readers may be surprised to know that e-mail was not universally available back then: some rudimentary Unix skills were required to send and edit e-mails; I used it for informal communication with friends, but not for professional correspondences).

Several months passed without receiving a response (they don't call it snail mail for nothing), so I became increasingly concerned that I would need to devise a plan B. However, at the end of January 1992 I heard back from Jaume, who expressed shock that anyone would consider leaving *Drosophila* to work on planarians (indeed, he was considering making the opposite switch!), but also welcomed the idea of me joining his lab. Future correspondences were facilitated somewhat by fax, as Jaume shared with me details of his group's ongoing work and we began outlining a project.

In the end, this plan would only work if I could come up with funding to support this project, so in the Summer of 1993, I began applying for various post-doc fellowships. I proposed two aims: (i) produce transgenic planarians using transfected neoblasts and (ii) identify genes expressed differentially in anterior or posterior regeneration using subtractive hybridization. Given the high-risk nature of this proposal, along with its focus on an organism that most scientists hadn't thought about since high school (if at all), it was unclear if it had any chance at all to succeed for these highly competitive awards. After a glimmer of hope on the first application (being invited to interview with Gerry Rubin for a Helen Hay Whitney Fellowship), a rather disheartening string of rejections followed (Whitney, Human Frontier Science Program, Life Sciences Research Foundation, Jane Coffin Childs Memorial Fund). Midway through June 1994, with my thesis defense scheduled for the end of August, the only applications still in the running were the Damon Runyon Cancer Research Foundation and the American Cancer Society, neither of which I was feeling particularly optimistic about. However, on June 17 I received a fax from Jaume informing me that I had been awarded a Damon Runyon Fellowship! The only cause for concern was the September 1 start date, which seemed unrealistic given the timing of my thesis defense and the scale of the trans-Atlantic move. Fortunately, the Foundation was flexible and postponed my start date to January 1, 1995.

So, at the end of December, my wife Sonia and I put most of our possessions in storage and brought whatever we could fit in our checked luggage and carry-on bags and moved to Barcelona. Jaume was an incredibly kind and gracious host, welcoming us at the airport, commandeering taxis to fit our belongings, booking us a hotel in the Gothic Quarter, and then taking us around to look at apartments, after a quick stop at the lab. While waiting for Jaume to finish his check-in at the lab, as we were seated in a plaza outside the Biology building and the Biology library, someone came up to us and asked in the worst Americanized Spanish accent imaginable, "Doenday esta la biblioetaycah dee beeolojeea?" Amazingly, since I didn't speak Spanish, I could understand his question; even more amazingly, it was one of the only two buildings whose locations I now knew in Barcelona. So, I could say to him, "It's ok, you can speak English, I'm an American, too. The Biology Library's right there." To this day, I still get asked for directions in places I'm visiting and have no idea where anything is (fortunately, advising graduate students has prepared me well for such encounters).

After all these years, I was finally in Barcelona and ready to begin working on planarians, but there was one major problem: there weren't any planarians in the lab when I arrived. At that time, planarians were not continuously cultured in the lab; rather, they were collected from various streams and rivers in northern Catalonia. These fun group collecting trips could yield anywhere from zero to thousands of planarians, which would be brought back to lab, and put in the cold room to be recovered and used as needed. Coming from a *Drosophila* background, this lack of consistent access to animals was a bit unsettling.

The other complication was that the planarian species that could be obtained in abundance was *Girardia* (formerly *Dugesia*) *tigrina*. Having access to large numbers of animals was critical when the group first began generating monoclonal antibodies and cloning genes before PCR, so it made sense initially to work on these animals. However, *G. tigrina* also had some sub-optimal characteristics from the perspective of a molecular biologist. *G. tigrina* was an invasive species from North America and these asexual invaders were mixoploid, meaning that they were a mosaic of cells with different ploidies, in this case diploid and triploid. Dealing with potentially three different alleles of a given gene was problematic, as were DNA content measurements suggesting that their haploid genome content was double that of several other planarian species. Also, large numbers of mariner transposable elements were scattered throughout the genome (Garcia-Fernàndez et al., 1995), which complicated chromosomal walking.

Concerned about the lack of easy access to animals, the individual variability of *G. tigrina*, and the long-term prospects of doing molecular biology on mixoploids, it seemed that it would be preferable to work on another species. Ironically, the asexual strain of *Schmidtea* (also formerly *Dugesia*) *mediterranea* could be found directly in Barcelona, in an artificial fountain near the Miró Foundation Museum. Since they were only a metro ride away, these worms were easily accessible to me. More importantly, they were stable diploids, with a smaller genome than *G. tigrina*. Jaume's Ph.D. research focused on this asexual population (Baguñà, 1973), which contained a chromosomal translocation distinguishing it from sexually reproducing individuals of the same species (Baguñà, Carranza, Pala, Ribas, & Riutort, 1999). The karyotypic difference between sexuals and asexuals was used effectively to monitor graft versus host cells in clever transplantation experiments (Salò & Baguñà, 1985a). Thinking longer term, this natural genetic variation would provide an opportunity to study the development and regeneration of germ cells and the reproductive system. So, I began collecting *S. mediterranea* and trying to rear them in the lab. Readers interested in learning more about the beginnings of Jaume's work on *S. mediterranea* in the late 1960s should consult Baguñà (2019) (this article also confirms my long-held, sneaking suspicion that my colleagues in Barcelona thought I "was nuts to change [from] flamboyant *Drosophila* to dull planarians").

My long run of experimental failures lent considerable support to the idea that my transition from *Drosophila* to planarians was more-than-somewhat crazy. I spent a great deal of time trying to culture neoblasts, playing with different matrix and culture conditions. Although I was able to keep cells alive in a dish for reasonable periods of time, they would not divide. On another front, several reports in the literature and lots of lab anecdotes suggested that planarians were refractory to nucleotide analogs. I contributed to this lore by my often-repeated, unsuccessful attempts to label neoblasts with the thymidine analog, BrdU.

Aside from these (and many other) failures, while familiarizing myself with the animals and the existing tool kit, I was trying to develop better methods for performing whole-mount immunostaining. Nearly all such work (along with *in situ* hybridization) was being performed on tissue sections, but I wanted to visualize what was happening in the context of the whole animal. It turns out that planarians are basically mucous-secreting machines; not only do they tend to curl up when exposed to many fixatives, but aldehyde-based fixatives tend to generate impermeable blobs that trap antibodies and lots of other labels non-specifically. So, I came up with ways to fix them so they would stay flat, and then to permeabilize them so I could visualize something other than their mucous. The University had an expertly staffed, excellent Leica confocal microscope, and although access to it was rather limited, we were able to generate some beautiful (and largely still unpublished!) images of a range of planarian tissues and cell types. Note that the royal "we" refers to me, Rafael (Rafa) Romero (in whose group and space I was embedded), and his student Francesc Cebrià (who would later join Alejandro and me for a short stint in Baltimore and then become my first post-doc after a productive post-doc in Kiyokazu Agata's laboratory). Most of our available time on the confocal was spent as a group, helping us to visualize how the various organs and tissues all fit together.

The kindness, generosity, and hospitality of so many colleagues in the Department made the time in Barcelona such a valuable, life-changing experience. Our daughter, Elena, was born during the first year of my post-doc, and prior to Elena's arrival, Roser Gonzàlez-Duarte, leader of the human molecular genetics group within the Department (and also Jaume's wife) organized members of the Department to lend us/donate their outgrown baby clothes and other necessities so we would have everything we needed as new parents. Jaume and Roser welcomed us (and our visiting family members) to their home for wonderful meals, while also occasionally serving as tour guides of Barcelona and its surroundings. Other faculty members of the Department (and their spouses) who displayed similar kindness and generosity outside of the lab included Rafa, Florenci Serras, Montse Corominas, Marta Riutort, David Bueno, and Emili Saló. The Department also had a remarkable group of graduate students, and I shared a bay with Francesc, Gonzalo Giribet, Salvador Carranza, and Iñaki Ruiz-Trillo, all of whom brought tremendous enthusiasm, dedication, and positive energy to their work.

In spite of the wonderful coworkers and living in one of the greatest cities in the world, my work was not progressing as I had hoped it would. Although I had learned a great deal about planarian biology, I hadn't yet produced anything worthy of publication and it seemed unlikely that I would have the molecular resources necessary to perform the proposed subtractive hybridizations. I had come up with this plan to bring molecular tools to study planarian regeneration and had gotten funding to pursue it, but I had not thought about how I would bring this work back to the United States. At the beginning of the second year of my fellowship, I applied to a few different labs with broad interests and expertise in a wide range of animals, describing my ideas and letting them know that I would come with my own funding; unfortunately, these queries were unsuccessful. It was beginning to look like this great planarian experiment was doomed to fail, and I began thinking about research areas to pursue in a second post-doc stint (parasitology wound up at the top of my list, but that's another story—see (Rozario, Collins, & Newmark, 2022)).

To help get the lay of the land I wound up attending (along with Jaume and Emili) the joint meeting of the British Societies for Developmental Biology and Cell Biology, which was held at the University of York in March 1996. One of the themes was Regeneration, Growth, and Pattern, which provided a great way to hear about recent work in a range of systems. During the first Regeneration session, one of the attendees sitting a few seats over from me was from the Carnegie Institution of Washington's Department of Embryology; his name was Alejandro Sánchez Alvarado, and although I knew about Carnegie and the great work going on there (I had written Joe Gall to see if I could bring planarians to his lab), I had not heard of Alejandro. I was hoping to run into him at some point so we could chat…

3. ASA's journey: From mice to frogs to planarians

In 1992, my last year as a graduate student working with mouse stem cells at the University of Cincinnati College of Medicine, my mentor Dr. Jeffrey Robbins told me that I should meet with a prominent scientist who would be visiting to give a seminar. Jeff had an hour to meet with our distinguished guest and he asked me to join the meeting so I could tell him about my work. The guest was none other than Donald D. Brown. Mind you, I had not heard of Don before, and as he was working with *Xenopus* I did not understand Jeff's insistence in getting us together. But it was clear to me that Don meant a lot to many people at the university. The day for Don's seminar came and I listened intently to his work on the thyroid hormone receptor and metamorphosis in *Xenopus*. Later, Jeff, Don, and I met in a small office to talk science. I shared with Don my work on growing mouse embryonic stem cells and how I was differentiating them into embryoid bodies with beating cardiomyocytes. He listened attentively, and asked a few questions, but suffice it to say that I could not read Don. I was still puzzled as to why Jeff would want me to meet with Don and I thought the matter closed.

A couple of weeks after our meeting, however, I received a letter from Don inviting me to visit the Carnegie Institution of Washington Department of Embryology. I flew to Baltimore to visit the Department and was struck by the fact that my taxi driver insisted that the address I had given him was a parking garage, and that I must have had the wrong address. I confirmed the address with Don's letter and insisted that the address was correct, and somewhat grudgingly my driver took me there. I have to admit that the simple, unassuming yet elegant lines of the old Embryology Department at 115 West University Parkway could have been easily confused with a parking structure, as its outward appearance did not provide any clues about the remarkable activities and contents of the building. However, not facing the street and discreetly emblazoned on a wall adjacent to the entrance stairs that led to the internal front entry were brass, sans-serif letters that read "Carnegie Institution of Washington Department of Embryology." I still remember the mixture of both relief and fear of being at the right place! Upon my arrival, Don greeted me warmly and described the events planned for the day. I met that day for the first time with Joe Gall, Doug Koshland, Andy Fire, Nena Fedoroff and Allan Spradling. I gave my seminar, which included videos I shot in VHS of beating embryoid bodies and was grilled by an audience of roughly 40 people for a good 30–45 min after I concluded my presentation. Sometime toward the end of my visit, Don offered me the opportunity of either joining his lab as a post-doctoral scientist or taking a position as an independent Staff Associate. As I knew little to no developmental biology at the time, I did not feel prepared to take the Staff Associate position and joined Don's lab instead.

I returned to Baltimore in 1994 to start my post-doctoral training. Between my two visits I pondered what problem I would tackle in Don's laboratory. Mouse stem cells intrigued me immensely, not as a tool to carry out homologous recombination per se, but because of what they could do. To have cells that could in principle generate a mouse fascinated me to no end and watching them proliferate *in vitro* and form beating embryoid bodies made me want to understand the mechanisms behind their remarkable potency. In the year between my first visit and eventual move to Baltimore, I read broadly and with abandon on the topic of potency and discovered a rich history of this concept that had remained up to that point wholly unknown to me. I searched for other reported and experimentally demonstrated examples of potency in the available literature and puzzled over the primarily embryonic nature of this property. Such was my thinking when in 1993 I became aware of two papers, one by Mohanty-Hejmadi (Mohanty-Hejmadi, Dutta, & Mahapatra, 1992) and the other by Malcolm Maden (Maden, 1993), reporting that the tails of amphibian tadpoles could be reprogrammed by retinoic acid (RA) to produce hindlimbs instead of regenerating tails after amputation. The papers called this phenomenon a true homeotic transformation in a vertebrate. I saw reprogramming of fates during regeneration. However, the amphibian biology that always caught my eye without fail during this time was always described in species other than *Xenopus*. In the case of the homeotic transformation, one frog was endemic to India (the marbled balloon frog *Uperodon systoma*), and the other was endemic to Europe (*Rana temporaria*). So the chances of getting this project off the ground in order to dissect it molecularly seemed improbable. However, I corresponded with Don about the possibility of giving this project a go and told him that this is what I wanted to do in his lab. I expected resistance, but instead received pointed criticisms and questions about the experimental design and approach, but most importantly encouragement to start and proceed with this project. In fact, Don insisted that I write a fellowship on this project, which I did and for which I would receive an NRSA from the NIH.

In Baltimore, I began to figure out a way to get the frogs. I gave up on the Indian species as my letters went unanswered. However, my correspondence with Malcolm Maden was more fruitful. He told me that he had collected his embryos and tadpoles near his house, but that perhaps I could find a commercial supplier, one in particular had been supplying frogs for dissection to schools across the United Kingom. Armed with this information in pre-internet times, I located the company in phone books (remember those?) only to find out it was no longer in business. I did find the name of the owner, Mr. Dennis McCarthy and later located his home phone number. It was March of 1994. It was spring. I had nothing to lose. So I picked up the phone and called him. I recall that his daughter answered the phone. I asked for Mr. McCarthy, and his daughter told me he was not available. Instead of hanging up and trying again later, I proceeded to tell her that I needed to get some tadpoles of *R. temporaria* and I was wondering if it could be possible for her father to collect some for me. She explained that her father was no longer in business, but I kindly insisted and as we were talking, she mentioned that they had recently brought some tadpoles from the field into an aquarium at the house and that it would not be very difficult to collect some for me, seeing how keen I was to secure some. I asked her if she could tell me whether the tadpoles they had collected had any limbs or limb buds and she said no, only a head and a tail. I knew the stage was young enough and that if I could get some I could try to reproduce Malcolm Maden's results in Baltimore. Mr. McCarthy's daughter promised to talk to her father, and a couple of hours later I received a call from Mr. McCarthy agreeing to help me. I learned that he had three daughters and five sons and that they would help him collect the tadpoles in the numbers I needed as soon as possible. And so began the shipping of *R. temporaria* from Mr. McCarthy's fields in Ireland to the laboratories at the Carnegie Institution.

After clearing customs (thanks to the customs agent hired by the Carnegie), I received my first batch of tadpoles in early April 1994. I began to treat the tadpoles in retinol palmitate and then proceeded to amputate their tails. I collected samples (blastemas) at different time points for RNA purification in order to carry out a differential display of expressed genes between treated and untreated animals using a methodology developed by Zhou (Joe) Wang and Don a year or so before I joined the lab (Wang & Brown, 1991). Using this method, I managed to identify genes that were up- and downregulated in both normal regeneration and in the tails undergoing RA-induced reprogramming from tail to hind limbs. I found a stark difference between the upregulated genes in the normal vs RA-treated samples, but no distinguishable difference in the genes down-regulated in both of these conditions. I cloned and sequenced hundreds of these genes, but their identity at the time remained unknown as many of the sequences represented 3′-UTRs (which diverge extensively in amphibians), and the collection of genes and open reading frames (ORFs) available at the time in GenBank was quite limited. I wished to learn more about these transcripts and needed more tadpoles for whole mount *in situ* hybridizations as well as earlier embryonic stages to determine what may be shared between regeneration, embryogenesis and the tail-to-hindlimb reprogramming induced by RA during regeneration. In October of 1994, I wrote to Mr. McCarthy once more, requesting not only tadpoles, but also gravid females and sexually mature males, both of which bury themselves in the mud during the fall to survive the winter to later emerge once the weather warms up in the spring to reproduce. The idea was to keep the females cold enough in the United States in one of our cold rooms and bring them to room temperature for ovulation when needed and thus obtain embryos and tadpoles year round at the laboratory. Remarkably, Mr. McCarthy agreed to do this for us, and in March of 1995 we received frogs and soon after tadpoles.

It is fair to say that without the help of Mr. McCarthy, I would have not been able to embark in this way on the pursuit to understand regeneration, to which I have now dedicated almost 30 years of my life. As I reminisce about my interactions with Mr. McCarthy–once or twice by phone, but mostly via post–I can distinctly recall the rising expectations at my end when he would write to tell me that the tadpoles were on their way to Baltimore via Aer Lingus. So much excitement surrounded the arrival of those parcels, so carefully and thoughtfully packaged. Without fail all animals were alive upon arrival, a testament to the care with which they were collected. On January 22, 2021, I received an e-mail from Pat McCarthy–one of Mr. McCarthy's sons–informing me that his dad had passed away in November of 2020. He wrote to me because he found in his late father's papers a letter I wrote to his father on October 28, 1994 thanking him for his help and updating him on our progress. I wrote that letter because I was immensely grateful to this man who lived an ocean away from me, whom I never had the chance to meet in person and who, nevertheless, was openly willing to help with our work. That Mr. McCarthy saved that letter for so many years moved me deeply, and it is yet another important addition to my list of constant reminders that scientists need the wills and minds of many to help them unclasp from the jaws of the unknown a measure of knowledge and understanding.

Far from being a commonplace tool, transgenesis in frogs was but a rumor at the time, and I grew frustrated with my inability to test the possible functions of the genes I had isolated from *R. temporaria*. I began to question the merit of pursuing my studies in amphibians. Again, I went to talk to Don to share with him my misgivings to which he asked: "So what do you propose to do?" To which I replied that perhaps the best move may be to identify an organism in which to study regeneration that may be amenable to molecular manipulations. Don said something along the lines: "Well, there you go. Why don't you do that?" Thus began my search for a system, a research organism that may be amenable to the molecular dissection of animal regeneration. I began by looking at all the organisms already available at the Carnegie. I played with zebrafish in Marnie Halpern's laboratory, where I learned that to study gene function in adults, the only way to do so at the time was by generating temperature-sensitive alleles, and given the low percentage of such mutants that can be generated in screens, I felt this was not the best approach. I did, however, manage to be an author on one of her papers describing the cloning and characterization of Chordin in zebrafish (Miller-Bertoglio, Fisher, Sánchez, Mullins, & Halpern, 1997).

In Andy Fire's lab, I also tried *C. elegans*, but quickly discovered that all of my attempts at physically injuring these transparent worms resulted not in healing or regeneration, but rather in producing very efficient methods for releasing zygotes and embryos from their entrails. I tried fly imaginal discs as well, and even considered plant regeneration. In the end, none of these systems appealed to me, and I decided to look broader and see what else may be out there to study regeneration. So I decided to apply for the Embryology course at the Marine Biological Laboratory in Woods Hole and began to explore the literature on regeneration at both the Library of Congress and the collections at the Smithsonian Institute. Around this time Allan Spradling, who had just become the new director of the Department, asked me to come to his office. I was worried because I had told Don that I thought amphibians were a dead end, so I was not particularly looking forward to what I thought Allan wanted to talk to me about. Instead of an uncomfortable conversation, Allan wanted to know if I wanted to have my own lab and take over the space that Nipam Patel had vacated when he moved to the University of Chicago to become faculty there. I asked Allan whether he knew that I was thinking of abandoning amphibians and that I wanted to look for a new animal to study and which I had yet to choose. He said he was aware of my intentions and told me that the position was ideal for this type of exploration as it came with a budget large enough to buy reagents and hire a technician. I said yes, and almost as soon as I took possession of the lab, I left for the Embryology course in Woods Hole.

My time in Woods Hole was transformative. Eric Davidson, Mike Levine and Andy Cameron were codirectors of the Embryology course that summer in 1995. In Woods Hole, I threw myself with abandon to learning about and doing experimental embryology. I, like most of the other students, bombarded all visiting faculty with questions, and played sometimes until sunrise with embryos that to me, at least, might as well have been the progeny of extraterrestrial organisms. I remember seeing under the dissecting scope a field of dozens of static, plain, unimpressive and diminutive translucent globes, only to watch them, as if on cue, split in half to produce blastomeres. I almost fell off my chair. I never saw a sea urchin again with the same eyes. Many wonders unfolded that summer under that dissecting microscope: the embryonic polar lobes of *Ilyanassa obsoleta*, the cytoplasmic streams of ascidian embryos, and the gastrulation movements of a host of organisms like nemertine worms, squids, the surf clam and *Chaetopterus,* to name just a few of the species that would simply take my breath away. I wanted to know more and that desire led me, of course, to the Marine Biological Laboratory library across the street from the embryology lab. To me the MBL Library is today, as it was back then in 1995, a living and breathing repository of freely accessible knowledge. So in between experiments in the lab, I would grab my notebook, cross the appropriately named MBL street and enter the library at any hour of the day or night and walk the stacks looking for information on both regeneration and the animals I was learning about. One afternoon, my undisciplined but relentless exploration of the library eventually led me to a book from 1901 appropriately entitled "Regeneration." On the spine of the book, under the title, the author's last name could be read: Morgan. Frankly, at the time, I did not think much of it, but I removed the book from the shelf, found a place to sit down and began to read its pages. I could not put it down. It was only the increased carrel activity and the sunrise that made me realize I had spent the whole night reading. I wanted to know who this Morgan was. I assumed perhaps a relative of the more famous geneticist associated with developing the fruit fly *Drosophila* as a research organism. Imagine my surprise when I figured out the author of the book I had just read was the one and the same Thomas Hunt Morgan! Nothing that I had been taught up to this point provided me with a hint that Morgan may have worked with other organisms. And yet, he had almost a dozen published papers from 1898 to the 1910s on the topic of planarian regeneration. Biographers, somehow, skipped this part of Morgan's life (Allen, 1978), an omission which has now been addressed by his biographers (Maienschein, 2016). In my own way, I thank Morgan for introducing me to planarians as his book was filled with references to the remarkable ability of these organisms to regenerate. In fact, I was so taken aback by the book that I convinced one of the librarians to make a photocopy of the entire book for me, a ream of paper that I treasure to this day.

This discovery, at least to me, of planarians having been used to study regeneration by none other than T.H. Morgan, plus the fact that he had done much of that work in Woods Hole at the MBL had a profound effect on me. I breathlessly tried to explain this to Eric Davidson, with whom I had numerous conversations about development, evolution, gene regulation among many other topics, and in typical heuristic fashion, Eric told me that Platyhelminthes were "the garbage can of evolution". I took that to mean one of two things: that he was right and I should desist entertaining any notions to study them or, alternatively, that we knew so little about these animals that it was easy to disregard them altogether. Time would show it was the latter.

After the embryology course, I continued to explore the platyhelminth and regeneration literature, an exploration that took me to both the Library of Congress in DC, where I secured a carrel for several months to read, and eventually to the Smithsonian Institute. I would collect the brunt of this exploration and distill my thoughts in a review that eventually appeared in BioEssays (Sánchez Alvarado, 2000).

4. Laying the foundation: From York to Carnegie Embryology

ASA: In 1996, I went to the United Kingdom to participate in three meetings that took place consecutively. Besides learning new science, I had a different and specific reason for attending each of these conferences. The first was the 15th Singer Symposium on Models of Regeneration at the Institute of Child Health, London, from 21st to 22nd of March. I wanted to learn about regeneration but chiefly wanted to have an opportunity to meet and talk science with Jeremy Brockes. The second meeting was the 5th International Conference on the Development and Regeneration of the Extremities, which took place at the University of York from the 24th to 27th of March, where I wanted to meet with Malcolm Maden and discuss my work on *R. temporaria*. And the third meeting and third reason for going to the United Kingdom was to attend the joint meeting of the British Societies for Developmental Biology and Cell Biology also at York from the 27th to 30th of March to announce at the meeting during my presentation that I was going to shift my attention from the study of amphibians to the study of regeneration in planarians. In retrospect, there was a fourth and more consequential reason plotted by fate and unknown to me, which was to meet Phil Newmark.

During my presentation, I discussed my ideas on regeneration and the approach I was using to exploit the ability of *R. temporaria* tadpoles to regenerate limbs in amputated tails treated with RA as a way to screen for genes associated with regeneration, and concluded that there were a number of gaps in the field that could be approached by developing an invertebrate research organism and proceeded to introduce planarians as prime candidates to help us do so. At the end of the session Emili Saló from the University of Barcelona introduced himself and invited me to attend his presentation on the role of HOX genes in planarians, which he was going to deliver at the meeting. During the course of our conversation, he mentioned that they had an American in their lab in Barcelona and that he would introduce him to me. Unfortunately, I did not run into Emili again and the promised introduction never happened. The last day, I went to the banquet mostly because of the setting at the National Railway Museum. I wanted to see the locomotives, monuments of the British Industrial Revolution, which I had only read about or seen photos of. I spent a great deal of time perambulating the museum and after a while, decided to join the banquet and sat down at the first table I ran into. That's when Phil and I met that fateful evening, without knowing we were establishing a scientific partnership and friendship that has endured for a quarter of a century.

PAN: The last evening of the York meeting featured a banquet at the National Railway Museum. Trying to save my fellowship funds and being generally uncomfortable in such situations anyway, I chose not to register for the banquet (a very bad decision, in retrospect). Jaume, who had registered for the banquet, decided to head back to Barcelona that evening, and he kindly gave me his banquet ticket. As I wandered around the museum looking at the trains, I wound up finally crossing paths with the guy from Carnegie Embryology who was seated near me at the beginning of the meeting. Unlike most of the people whom I met at that meeting, after looking at my nametag, seeing my affiliation, and learning that I was working on planarians, Alejandro did not treat me as if I had some weird communicable disease; rather, he told me that he was also interested in planarians and was actually considering working on them when he started his own lab at Carnegie. We hit it off immediately, and over dinner I told him about the work from the Barcelona group that had attracted me there, my research plans, and my hopes for bringing planarians back to the United States. Alejandro told me about his subtractive hybridization experiments to understand the tail-to-limb transformation in *R. temporaria* and how the hurdles associated with moving the work forward led to his own search for other regeneration models, which brought him to planarians, too.

This chance meeting led to months of e-mail correspondence and, finally, an opportunity for me to interview at the Carnegie that summer with the possibility of joining Alejandro's lab as a post-doc. Although I would be bringing my own funding for the final year of my fellowship, the major hurdle to this plan was that Alejandro's position (Staff Associate) did not allow him to have a post-doc.

ASA: After our conversation I resolved to have Phil join my lab back in Baltimore. As Phil wrote, the minor wrinkle in all of this was that Staff Associates were not allowed to have post-docs in their labs. So I figured I had to find a solution. Back at the Carnegie I began to run this idea by Don Brown, Joe Gall and Allan Spradling. It was neither disregarded nor warmly received, but it was also not rejected. Instead, I was asked a good number of hard questions that I did my best to answer. Some of these answers addressed and resolved the concerns and others merely managed to mitigate them. However, I was not told to discontinue this effort. Once I felt the time was right, I invited Phil to give a seminar at the Carnegie and share with us what he had been doing in Barcelona.

PAN: Given the opportunity to combine our complementary expertise to tackle this classic problem of developmental biology, Alejandro managed to convince the powers-that-be at the Carnegie to give us a chance to work together. Thus, I wound up flying to Baltimore during a family trip to the United States, giving a seminar on planarian regeneration to the entire Department, and then meeting individually with the Department's group leaders. Despite a few skeptical faces in the audience, the seminar and visit seemed to go well.

ASA: A few days after Phil left, I broached the topic again with Allan, and he agreed to make an exception and allow Phil to join my nascent lab as a post-doc. I promised myself that I would work as hard and as intelligently as possible to feel like I was deserving of this great opportunity: to start from scratch an entire research program aimed at a mechanistic dissection of a long-standing problem of biology.

PAN: With my pending return to the United States, there were still a few questions left to iron out: Which species would we study? How would we grow them in the lab? Based on the naturally occurring asexual and sexual populations that were distinguished by a chromosomal translocation, *S. mediterranea* remained my favored option. However, there were also arguments to be made for using *S. polychroa*, which had an even smaller genome and was more broadly distributed in European freshwaters. A series of classic papers, beginning with the work of Mario Benazzi and Giuseppina Benazzi Lentati examining the chromosomal mosaicism of some parthenogenetic strains (Benazzi & Benazzi Lentati, 1976) and by Vittorio Gremigni and colleagues using it to study transdifferentiation of the germ cell lineage into somatic cells (Gremigni & Miceli, 1980; Gremigni, Miceli, & Picano, 1980; Gremigni, Miceli, & Puccinelli, 1980; Gremigni, Nigro, & Puccinelli, 1982), made clear that there was also much interesting biology to uncover studying this species.

When discussing this issue with Jaume, he told me about the work of Nico Michiels, then a Group Leader at the former Max Planck Institute for Behavioral Physiology in Seewiesen, who was studying these parthenogenetic populations and had built aquatic systems for maintaining planarians and studying their behavior (Peters, Streng, & Michiels, 1996). Jaume suggested that I visit Nico's lab to learn about their work and their culture system. Nico kindly agreed to have me visit his lab, so in October I set off for Seewiesen, which is about an hour Southwest of Munich.

When I arrived at the Institute where Konrad Lorenz had performed his pioneering research on behavioral imprinting in geese, Nico showed me around the grounds and pointed out the incredible work of the Institute's workshop staff, who built the submersible used by the Institute's Hans Fricke to provide the first footage of living coelacanths swimming at their natural depth. These same remarkably skilled craftsmen built Nico's planarian culture and behavioral imaging facilities, recirculating systems in which planarians were maintained in constantly flowing water subjected to biological and ultraviolet filtration. Nico and his lab members generously shared all the details about their culture system and methods of planarian maintenance. I took copious notes and photographed the systems, passing along the information to Alejandro so he could begin having the Carnegie workshop staff build a similar culture system for our work (as a testament to his effectiveness, by the time I arrived in Baltimore, the system—dubbed the "Planarium" (Fig. 1)—was already built and ready for the planarians I had brought with me!).

Fig. 1 The authors in the "Planarium" at Carnegie Embryology (1997). The planarians were housed in a recirculating system in containers with mesh-covered openings that enabled water to flow in and out. It worked flawlessly for nearly 2 years before catastrophe struck and all the planarians died.

The work of Nico and his colleagues extended the findings of Benazzi and Benazzi Lentati and revealed the tremendous variability in ploidy that could be found in natural populations of *S. polychroa*: in addition to normal diploids, there were triploids, tetraploids, and pentaploids (Beukeboom, Weinzierl, Reed, & Michiels, 1996). Having to karyotype individual animals before beginning any molecular work and the lack of easily expandable asexually reproducing populations helped sway the decision of which species to work on in favor of *S. mediterranea*, nearly 1000 of which accompanied me and my family on our move to Baltimore, where we arrived on January 1, 1997.

ASA: Before I met Phil, I too grappled with the problem of what planarian species to use to develop a molecularly accessible research organism to study regeneration. I too wanted a species that was easy to rear in the lab, that was an exaggerator of the biological problem I wanted to study (i.e., regeneration), that was vulnerable to molecular manipulations, and which were diploid and with a relatively small genome size. I did my share of reading, and like Phil, found out very quickly about the extraordinary karyological acrobatics performed by freshwater planarians. Ploidy and mixoploidy, inversions and translocations. The list of chromosomal behaviors I was uncovering in the planarian literature was then and continues to be today quite astonishing. In late 1995, my reading eventually led me to the realization that the Smithsonian Institute had a department of worms, and so I reached out to them early in 1996 only to find out that Roman Kenk (a towering figure in the study of platyhelminthes) had recently passed away, and that the Smithsonian was preparing to box and store his extensive book, pre-print, and sample collections. I pleaded for time and access to consult these resources before they were to be carted away, and I received both. My experience at the Smithsonian was invaluable. Not only did I find a meticulously organized and comprehensive library and article re-print collection, but also hundreds of slide preparations of fixed specimens. Chiefly among these was Roman's 1974 "Index of the Genus and Species of the Freshwater Triclads (Turbellaria) of the World" (Kenk, 1974). This work and several others in Roman's collection helped me identify potential diploid species to go after, which included, among others, *Dugesia* (now *Schmidtea*) *polychroa*.

PAN: Upon my arrival in Baltimore, Alejandro and I immediately started a series of subtractive hybridization experiments to identify genes expressed differentially between anterior- and posterior-regenerating tissues as well as between heads and tails. We decided to focus on the earliest timepoints during which head and tail fates were determined; based on a series of grafting experiments by Emili, this occurred within the first 8–24 h after amputation, well before a blastema is visible. So, we set two dissecting scopes in Alejandro's office, turned on some music, and began amputating planarian heads and tails, immediately freezing them in liquid nitrogen, then reamputating the regenerating tissues and freezing them, too. This was followed by several months performing multiple rounds of subtractive hybridization to enrich for fragments of transcripts expressed differentially in heads, tails, head blastemas, or tail blastemas. Each candidate fragment then needed to be validated by northern blotting, at which point we would sequence it to see if its predicted product could provide a clue about its potential role in regeneration. After more than a year of assembling lists of validated, differentially expressed genes, we ran into two major roadblocks: the lack of robust whole-mount *in situ* hybridization protocols in planarians, and the inability to assess gene function in these animals.

Because we were at Carnegie Embryology, the solution to the gene-function problem turned out to be just one floor away. Weekly Progress Reports, at which trainees and PIs presented their research to the entire Department, were an important part of Embryology's unique culture. Since the expectation was that everyone would attend and present, the entire department knew what everybody else was doing. Thus, at some point after we had presented the results of our subtractive screen, Andy Fire wandered up from his first-floor lab to Alejandro's office on the second floor, where Alejandro and I were in the middle of a discussion, and asked us if our subtractions had yielded any genes whose phenotypes could be predicted if they were inhibited. At that point, we were still very much in the validation phase and unable even to say what cell types the genes were expressed in, so the answer was, "No, why do you ask?" At which point Andy began to tell us about his new results showing that double-stranded RNA (dsRNA) could inhibit gene expression. I recall both of us being incredulous, and as we asked Andy question after question, it was clear that he was onto something important.

I'd like to say that we immediately leapt into action to see whether dsRNA did the same thing in planarians, but we didn't. We were still busy with the hard work of validating candidate genes from the subtractive hybridizations, so testing gene function still seemed premature. Plus, this phenomenon could have been some weird *C. elegans* thing, so we went back to our subtraction clones and attempts at in situs. After another month or two, Andy returned to ask us about predicting phenotypes again, this time with the news that Rich Carthew told him that dsRNA also seemed to inhibit gene function in *Drosophila*. This wasn't a weird *C. elegans* thing after all, so this time we did leap into action: I would try soaking planarians in dsRNA, while Alejandro set out to inject them with it. In the course of our subtractive hybridizations, we had cloned gene fragments that encoded abundant proteins for which antibodies were available, including commercial anti-Tubulins and an anti-Myosin Heavy Chain monoclonal antibody generated in Rafa Romero's group (Cebrià, Vispo, Newmark, Bueno, & Romero, 1997). Thus, we could use immunofluorescence to assay whether or not delivery of dsRNA affected protein levels. I didn't get very far with soaking, but it didn't matter because Alejandro's injections wound up succeeding so spectacularly.

ASA: Carnegie's weekly Progress Reports were always memorable events because of the rigor and thoughtfulness invested by both the presenter and the audience. I distinctly remember Andy Fire's turn in 1997 during which he proceeded to lay out the evidence for dsRNA as a way of introducing genetic interference in *C. elegans*. While the approximately 40-min presentation was lucid, crisp, and at times humorous, it was the hour or so of questions that followed afterwards that I remember most. Skepticism, incredulity, and methodological questions were lobbed at Andy. Probing questions came from Allan Spradling, Joe Gall, Doug Koshland, and Don Brown among many others. It was as powerful a barrage as one can imagine. But Andy proceeded to swat each question down with logic, data, scholarship, and in the best of cases, all of them. As Phil has written above, Andy would drop by in the months leading to his progress report to ask whether we had genes we could test for function using dsRNA. But we had more practical matters to tackle, including purifying high-quality RNA from planarian tissues and purified cells to make cDNA libraries for our subtractive hybridizations. I was particularly concerned, because the first time I purified RNA from *S. mediterranea* instead of obtaining an opalescent white pellet, what I got was a dark, almost black pellet instead. I had purified RNA from many species by then (mice, chicks, nematodes, flies, surf clams, ascidians, echinoderms, even nemertine worms), and I had never encountered dark RNA pellets. I feared the worst: useless RNA. But the RNA was intact and we managed to prepare cDNA libraries, perform the subtractions and obtain a cohort of differentially expressed genes, so I was much more optimistic about making molecular inroads into the mysterious biology of planarians. Andy's progress report fully convinced me that dsRNA was something we needed to test thoroughly.

At the time, lore had it that injections in planarians were extremely difficult as they were acoelomates and it was believed that they had no room to accept any volumes of liquid introduced via injections. As unlikely as that sounds today, I took this potential limitation seriously and decided that I would try every microinjection rig I was familiar with. I began to make preparations like pulling needles of different geometries, and wondered whether slow delivery of very small volumes or fast-pulsed injections would do the trick. I had multiple parameters to test and began scheduling the use of the various rigs available at the Carnegie: Allan's *Drosophila* injection apparatus, Andy's nematode set up, Don's *Xenopus* injection station, and even Marnie Halpern's zebrafish injection rig. I decided to start from simple to complex and take it from there. As we had very few *S. mediterranea* left, I used planarians collected locally to test the rigs. My first attempt was with a Drummond piezoelectric injector capable of injecting nanoliter boluses of liquid. I was familiar with this injector as I had used it in Don's lab to inject *Xenopus* embryos. On July 30th, 1998, I pulled and backfilled the needed glass needles with saline and blue food dye, chilled the animals to slow them down and placed one of these animals under the dissecting microscope, steeled my resolve for what I had imagined would be the beginning of a long and arduous optimization process and proceeded to inject. Imagine my surprise when after the very first attempt I managed to deliver not one but three consecutive injections of ~32 nanoliters each... I was incredulous. So I tried it again on another chilled animal and same result. I knew right away I had found a way to directly deliver dsRNA to planarians, and after a few more tries, I remember walking to Phil's bench to ask him to come take a look at the blue worms.

I proceeded to use Andy's protocol to synthesize dsRNA from two of our cDNA clones: myosin (clone M71 in my notebook) and tubulin. Perhaps not the most exciting genes on the planet, but ones for which we had antibodies to detect loss-of-function effects by the injected dsRNA. Again, as we had very few *S. mediterranea* and predicting a high degree of sequence conservation, I injected the dsRNA into a local species in an effort to work out dosage, toxicity, and to see whether defects or phenotypes would become apparent after amputation. My laboratory notes at the time have drawings of changes I saw in the regenerating animals after three rounds of injections and amputation not seen in controls. I felt something was happening and wanted to now test the dsRNA in *S. mediterranea*. In early August I injected a few *S. mediterranea* with dsRNA made from the myosin and tubulin clones and showed Phil how to do it so we could inject more animals. Not only did we have a limited supply of *S. mediterranea*, but as Phil describes below, we also had an event in the "Planarium" that killed all our animals. At the time I felt like we embodied the proverbial saying of "being all dressed up and having no place to go." I remember being upset and proposing to Phil that we travel to Barcelona to collect animals ourselves. We went to my computer to open Netscape (remember that web browser?) and look for weather predictions for Barcelona. Rain was in the forecast, so we began to make preparations for our trip, which involved shopping for a cooler and getting all the necessary material to collect and transport the animals. Simultaneously, I worked the phones to let Carnegie's customs agents know of our impending collection trip so they could help us prepare the paperwork to bring the animals into the United States. In the midst of our preparations, I received a call from the US Fish and Wildlife office at Baltimore/Washington International (BWI) Airport. The inspector at the other end of the line wanted to know what these animals were that we wanted to bring to the lab from Europe. I remember explaining that they were non-parasitic, freshwater-dwelling animals, and the reasons why we needed this particular species. I remember him being very difficult and quite displeased with my answers to his questions. After he finished grilling me over the phone, he adamantly demanded that he had to be present on our arrival from Europe to personally inspect our documents and animals before we could bring them into the US. I had already made the necessary travel arrangements, so I shared with him our return flight number and arrival time. For the next couple of days, I prepared an extensive file with the forms prepared by our customs agents and a myriad of supporting material to have at the ready for my impending meeting with the inspector upon our return from Barcelona.

PAN: While exploring the potential inhibitory effects of dsRNA injections on expression of the corresponding proteins, we also needed to test the effects of dsRNA on transcript levels and validate our subtractive hybridization results. Fortunately, the planarian whole-mount *in situ* problem had recently been solved in the laboratory of Kiyokazu (Kiyo) Agata (then at the Himeji Institute of Technology), where graduate student Yoshihiko (Yoshi) Umesono and colleagues were systematically testing different fixation and permeabilization conditions. They published papers reporting beautiful whole-mount *in situ* hybridization patterns of neurally expressed genes at the end of 1997 (Umesono, Watanabe, & Agata, 1997) and in June 1998 (Agata et al., 1998). Counterintuitively, they found that acidic fixation conditions removed the mucous while still leaving the RNA intact.

Using the method outlined in these papers, we could now obtain reasonable *in situ* patterns for some of the differentially expressed genes from our subtractive hybridization screen; depressingly, several matrix metalloprotease-encoding transcripts, on which we had spent the first half of 1998, turned out not to be upregulated in the blastema, but were false positives from our screen based on their region-specific expression (in secretory cells surrounding the pharynx), a variable we had not considered when we started (these highly abundant transcripts later proved very useful when we were testing the ability of bacterially expressed dsRNA to inhibit gene expression in planarians (Newmark, Reddien, Cebrià, & Sánchez Alvarado, 2003). During these initial *in situ* experiments, non-specific epithelial background remained problematic; a problem we resolved by optimizing the hybridization conditions (which is why we phenol-chloroform extract and ethanol precipitate the yeast RNA to this day). It wasn't until Alejandro invited Kiyo to visit the Carnegie in 1999 that we learned the original published method had initially sufficed to remove the epithelium from *Dugesia japonica* (the species Kiyo's group was studying) and prevent non-specific background, but an unknown variable subsequently changed and they found that Proteinase K treatment was now required to minimize epithelial background and boost signal. With a protocol in hand, we could now begin validating our subtractive hybridization clones for cell type-specific expression. For example, our head-enriched library yielded a fragment corresponding to an *opsin* gene; we were able to detect its expression in photoreceptor cells, which would later enable us to examine the effects of injected dsRNA on transcript levels in a single cell type far from the injection site.

Our attention, however, was not focused solely on the effects of dsRNA: there was still the issue of how to label neoblasts. Carnegie Embryology provided a range of forums for discussing research and keeping abreast of new developments. In addition to weekly Progress Reports, Department-wide journal clubs, and external seminars, we participated in a weekly joint group meeting (Assorted Beasts) with the labs of Don Brown, Marnie Halpern, and another Staff Associate, Jimo Borjigin. This meeting was a great opportunity to follow the progress of each of these groups and to get constructive, critical feedback on our work. During Alejandro's first presentation to the group, Don asked why, if neoblasts are the only proliferating cells, we couldn't just label them with thymidine analogs. This led to a discussion of how such experiments had been tried unsuccessfully for decades. Don asked if planarians have thymidine kinase, and when I told him that they did, he said,

"Well, the labeling should work, so something else must be going on." He was challenging us to go back and try again, not accepting the negative lore from the field (or my past failed experiments).

While discussing the BrdU problem with Giovanni (Gio) Bosco, one of my old friends from BU who was then a post-doc in Terry Orr-Weaver's lab at the Whitehead Institute, Gio asked me if I had ever tried feeding the planarians BrdU, like one does to label *Drosophila* larvae. Here, I must note that my son, Carl, had been born 3–4 months before this conversation, and I was completely sleep deprived. I was so sleep deprived that I forgot about the paper by Best, Rosenvold, Souders, and Wade (1965), which showed that feeding planarians tritiated thymidine failed to label neoblasts; had I remembered the details from this paper, it is unlikely I would have pursued this approach. There is an important lesson here: just do the experiment!

At that time, we had a range of planarian species obtained from various biological supply companies, including some *Phagocata* sp. that were laying egg capsules. While feeding liver to some recently emerged hatchlings, I noticed that they appeared to transfer the yolk contained in their intestine into the mesenchyme as they were digesting the liver. Seeing this, I wondered if feeding these newly emerged hatchlings would enable them to take up BrdU. So, I added BrdU to the liver homogenate we used for feeding the planarians, testing several different concentrations. It turned out that it was surprisingly difficult to mix the BrdU solution into the liver, so there were pockets of liver with greater or lesser amounts of BrdU; even worse, when I added the BrdU/liver mix to the planarian water, it wound up dispersing rather than staying in one concentrated spot. The planarians ate, but there was no way of knowing if any of them were even exposed to BrdU. Nonetheless, I proceeded with the rest of the experiment, fully expecting it to fail, as it had every other time.

After the several-days-long protocol, I started the post-secondary antibody washes and after a relatively brief wash, I couldn't resist checking to see if there was any signal. I put the 24-well plate on our inverted scope, starting at the lowest of the BrdU concentrations. All the planarians were negative. The next well contained the planarians fed the mid-range concentration of BrdU: all negative. As I moved to the well of planarians fed the highest BrdU concentration, I started going through them one by one, and each was negative. After looking at five unlabeled planarians, it seemed as if this experiment was once again doomed to fail. At this point, I remember thinking, "Why do I keep trying this? Why am I wasting my time? This dumb experiment will never work..." While in mid-thought, I came to the sixth planarian in the well, and it was full of fluorescent nuclei, exactly where the neoblasts should be! This terribly executed experiment actually worked! Planarians could take up BrdU after all! My memory tells me that I called Alejandro in the middle of the night to share the good news; my lab notebook, however, tells me it was actually around 8:00 in the evening. With this newfound ability to label neoblasts, combined with *in situ* hybridization and the possibility of using RNAi, we finally had a viable experimental system in hand. Unfortunately, the "Culligan Man" had other ideas…

We maintained all our planarians in the Planarium (Fig. 1); the water source for this recirculating system was dechlorinated tap water to which we added extra salts. In addition to asexual *S. mediterranea*, we had thriving populations of sexual *S. mediterranea* (that were laying fertile egg capsules) and *Procotyla fluviatilis* that we had caught in a local stream (this latter species was of interest because of its limited regenerative abilities, a topic to which we eventually returned (Sikes & Newmark, 2013)). The week after the positive BrdU results, we went into the Planarium on a Thursday, and discovered that all of our planarians were dead. The populations had all been thriving, so we were mystified by what could have happened. Unbeknownst to us, the Culligan Man came to swap out our large dechlorination filter, and instead of using the pH-neutralized filter that was our standard order, he used a filter that had not been neutralized, so water passing through it would be acidic; as far as we could tell, low pH wound up killing the planarians.

Without planarians, there was nothing we could do. Since we had just dedicated nearly 2 years of intensive molecular biology to asexual *S. mediterranea* from Barcelona, continuing this project required a collecting trip to gather more. However, whether such a trip would have any chance of success depended on the weather: if it hadn't rained in Barcelona in quite some time, the fountain would be dry and it would be impossible to find any planarians. Luckily for us, the weather report told us that it had rained recently, so we started making arrangements to fly to Barcelona the very next day. We realized that if we left the United States on Friday, we could get to Barcelona Saturday morning, before the shops closed. That would enable us to buy fresh liver to use as bait for the planarians.

In preparation for our collecting expedition, we made a run to Wal-Mart to purchase a large cooler, many thermoses to fit inside the cooler for the trip home, Tupperware containers for sorting the planarians, as well as fishing hooks and fishing wire for the liver bait. Of course, we also needed razor blades for chopping up the liver, parafilm for sealing the thermoses, transfer pipettes and paintbrushes for transferring the planarians, as well as latex gloves for handling the liver and for protecting our hands from the disgusting water that the planarians lived in. So, on September 25th, 1998, this assortment of goodies all went into the cooler, which the two of us carried onto the train to the airport and then checked in as baggage on our British Airways flight (when asked what was in the cooler, we said "scientific materials," which apparently was a good enough answer).

Upon arrival in Barcelona, we went to the hotel to drop off our stuff, then headed straight to a local market to buy liver and spring water. There was no fresh calf liver to be found, but we were able to buy some sheep liver, which we brought back to the hotel. There we donned latex gloves and used razor blades to chop the liver into pieces for baiting the fishing hooks that we would use, if necessary. After disposing of our gloves and liver remnants in our hotel room's garbage can (which, in retrospect, may have caused some of the funny looks we got from hotel staff), we headed off in search of planarians.

The small, artificial fountain where I had routinely collected planarians during my time in Barcelona was on Montjuïc, a large hill with several notable landmarks, including the Olympic stadium, the National Museum of Catalan Art, and the Miró Foundation Museum. As we worked our way up the hill to the fountain by the Miró Foundation, we stopped at several others, just to see if they also had planarians (they did not). We finally made it to our destination (Fig. 2), and happily started picking up stones and turning over floating leaves in search of planarians. We continued patiently for several hours and, between the two of us, we found one small planarian. We were beyond discouraged at this point and began to wonder if this signaled the end of this great experiment. It was starting to get dark, so we had to end our unsuccessful outing; before leaving, we left our liver-baited fishing hooks at many places throughout the fountain, in hopes that they would attract planarians before we returned the next morning. On our way down the hill, we were treated to a full display of the Magic Fountain, built for the 1929 Universal Exhibition. This beautiful display helped ease some of the pain from our failed collecting run, as did a fantastic tapas dinner we shared with Rick Elinson, an expert on direct-developing frogs, who was on sabbatical in Don Brown's lab; we somehow ran into him entirely by chance in Barcelona.

First thing the next morning, we returned to the fountain and began pulling in our baited fishing lines. Fortunately for us, the baits were all swarming with planarians! We filled our thermoses with the planarians and brought them back to the hotel room to transfer them into clean water

Fig. 2 The authors at the Montjüic fountain as they started their planarian-collecting trip in Barcelona (September 1998). They are smiling because they thought they would be gathering lots of planarians that afternoon. A few hours later, with only a single planarian to show for their efforts, they were no longer smiling.

and count them. We wound up with 1544 planarians, so the trip had paid off after all. We had enough thermoses and spring water so the planarians wouldn't be overcrowded during the trip home, but we still had to figure out how to safely get the cooler through the airline's baggage-handling system. There was also a concern that the airline would not accept our package and it would have to be shipped another way. We decided that it would be best to box up the cooler so that it was completely sealed, so we set off in search of a store where we could buy a big enough box. Most stores were closed on Sundays, but we managed to find a store that sold packaging materials and bought two boxes that we duct-taped together in order to accommodate the cooler (Fig. 3). While we were packaging everything up that evening, we realized that we had not gotten a water sample from the fountain, so we stopped there on our way to the airport early the next morning.

At the airport, as we approached the check-in desk for our flight, we worried that they would not allow us to check as baggage our duct-taped box with a cooler containing fishing hooks, razor blades, etc. So, we began devising back-up plans that would have Alejandro staying behind to handle

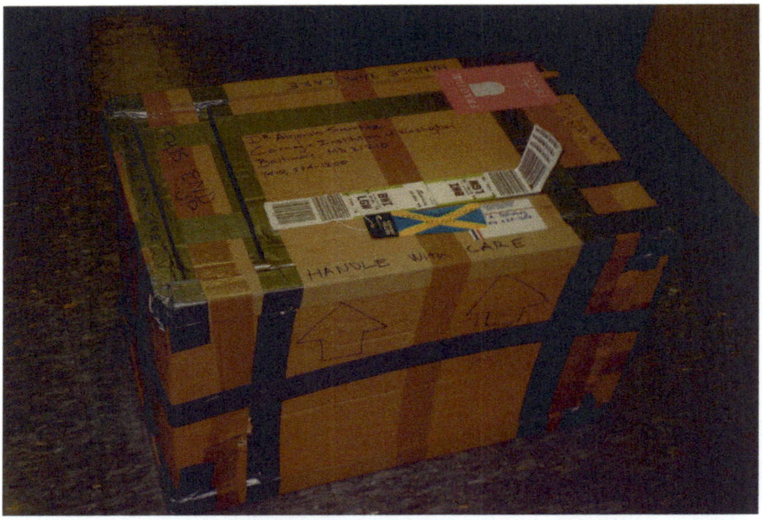

Fig. 3 The humble beginnings of *S. mediterranea* asexual clone CIW4 in British Airways cargo (September 1998). This image shows the jury-rigged cardboard box that the authors assembled to transport the cooler containing 1544 specimens of asexual *S. mediterranea* from Barcelona to Baltimore. One of those specimens grew up to be CIW4, whose descendants populate labs all over the world. Note the careful use of duct tape to form a single box from two boxes.

shipping the box to the United States. When we arrived at check-in and they asked us what was in the crate, Alejandro answered "materiales científicos," which, once again was apparently good enough, so we (and the planarians) were on our way back to Baltimore.

The final hurdle was to clear customs upon arrival in BWI. Before leaving for Barcelona, Alejandro had arranged for Carnegie staff to contact US Fish and Wildlife to obtain the appropriate permits for us to bring the planarians into the country. The Fish and Wildlife Inspector who was handling our request was quite a stickler, so we made sure that we had all the appropriate forms with all the appropriate signatures, and we were expecting him to be awaiting our arrival. With some trepidation we approached the customs agent and told them about our arrangement with the Fish and Wildlife officer, and as we started pulling out the required paperwork, the customs agent asked us what kind of animals we were bringing in. When we told him that we were bringing in planarians, he said, "We don't care about those," and waived us through without any further discussion.

Back at Carnegie, I vowed never to let the planarians out of my sight again, so the entire population took over most of the shelves and drawers around my bench. At feeding time, much to the consternation of my baymates, every available surface was occupied by planarians. To minimize experimental variability moving forward, I immediately started developing clonal lines: clone 4 wound up growing the best of the 12 individuals I started with, and we named it CIW4 to reflect its birthplace. Thanks to improvements in planarian culture systems by Shane Merriman at Stowers Institute, there are now millions of individuals derived from this single worm being reared in labs throughout the world.

As soon as we returned to Baltimore, Alejandro submitted for analysis the water sample from the asexuals' home fountain. Shortly thereafter, he began receiving panicked messages from the water-testing company, telling us that we absolutely should not drink that water due to fecal coliform contamination as well as iron, manganese, hardness, and total dissolved solids levels all exceeding maximum contaminant values. It's worth noting that this water analysis later helped resolve a major crisis in the first month of starting my own lab, when the planarians started crawling out of their water due to changes in the local water supply. That analysis from 1998 served as a guide for producing a planarian culture solution using milliQ water to avoid the variability of local tap water. It was originally meant to solve an immediate emergency, but my first graduate student, Tingxia Guo, found that the planarians were quite happy in it, and it is now known as Montjuïc water (Cebrià & Newmark, 2005). Given that it was created in the midst of a crisis and is rather a pain to make, there may be nobody more surprised than me that it has become the culture water of choice for the mass maintenance of *S. mediterranea* (Merryman, Sánchez Alvarado, & Jenkin, 2018).

ASA: Now that we had animals in hand again, we could return to the many promising lines of research that were underway. The RNAi experiments progressed rapidly. The many months of whole-mount immunofluorescence optimization paid off. On October 20th and 21st, 1998, I injected multiple animals with dsRNA targeting either myosin or tubulin and processed these for confocal microscopy analyses. Sometime after midnight on October 29th, I began to image the samples and saw for the first time two specific loss-of-function phenotypes caused by genetic interference in planarians: regenerated fragments lacking either muscle fibers or cilia. I called Phil and told him that he had to come to the lab to see something important. It is hard to describe the feeling of knowing that you are the first person ever to see this important result; we had now shared this experience twice over the course of just a few months! The time spent optimizing *in situ* hybridization also paid off and we went on to examine the effects of dsRNA on transcripts, showing that *opsin* mRNA in the photoreceptors was reduced below detectable levels less than a day after injection of dsRNA.

By December, we submitted to *Science* a manuscript showing that dsRNA inhibited gene expression in planarians, paving the way for a functional analysis of regeneration in these animals. In a pro forma letter dated 22 December 1998 we learned that the paper was rapidly rejected without review. Discussing this rejection with our senior colleagues at Carnegie, they encouraged us to appeal the decision, so on January 4, 1999 we sent a letter detailing why "the first demonstration of the targeted elimination of gene expression in a regenerating organism" merited review. The next day we received the following rejection of our request from the editor who handled the manuscript: "Although we find your work to be exciting for those studying regeneration in Planaria, the use of double stranded RNA in blocking gene expression was published in *C. elegans* previously. Thus, the conceptual novelty of this technique is lacking. Your submitted manuscript does not represent a significant advance in Planaria biology nor does it provide mechanistic insight into how the ds RNA interference works. Hence, we feel that the paper would be more appropriate for a specialized journal." Although we did not find it funny at the time, the idea that RNAi did not represent a significant advance in planarian biology is laughable.

PAN: As we finalized the proofs of the RNAi paper for *PNAS* in mid-March 1999 (Sánchez Alvarado & Newmark, 1999), we could return our attention to the BrdU experiments. Fortunately, with more animals and improved feeding conditions, we could reproducibly label planarians with BrdU. During the course of reading about cell-cycle kinetics, it occurred to me that we could update laborious fraction-of-labeled-mitoses approaches that used *in vivo* tritiated-thymidine labeling followed by autoradiographic detection on tissue sections, and instead use whole-mount, antibody-stained animals that could be visualized with confocal microscopy. Performing time courses after BrdU labeling, then double labeling with antibodies against BrdU and a (then) recently described marker of mitotic chromatin (phospho-histone H3, (Hendzel et al., 1997)), could allow us to define various cell-cycle stages in the animal.

After working out some of the final technical challenges, we set out to perform the critical experiments for the paper. Unfortunately, instead of finishing up the work for the paper, what followed is a section of my lab notebook entitled, "BrdU Hell Entrance." Everything stopped working.

For 3 full months we were unable to see any evidence of BrdU labeling. After months of troubleshooting, as unlikely as it sounds, we wound up being derailed by bad batches of primary antibodies and a bad lot of BrdU. Once our reagents were working again, we could complete the final experiments and by mid-September we submitted the manuscript to *Development*. We received the reviews at the end of October, with Referee #1's typewriter-written review recommending "Do not publish," marking the "Scientifically flawed" check box, and helpfully noting that "The authors (sic) haphazard approach to these regeneration experiments is a waste of journal space." This remains the snarkiest reviewer comment either of us has ever received. Unfortunately, our revised manuscript and response letter failed to convince this reviewer, who still selected the "Scientifically flawed" check box and suggested that we either delete the section of the manuscript dealing with measuring G2 "…or do the experiments the right way with suitable controls." Our rebuttal letter argued that we had, in fact, done the experiments the right way with suitable controls, but we failed to convince the editor handling the manuscript, so we reformatted the paper for submission to *Developmental Biology*. The reviews were much more positive this time around, and the manuscript was accepted by the end of January 2000 (Newmark & Sánchez Alvarado, 2000).

For planarians to be a viable experimental system, we realized that we would still need more tools and resources at our disposal, so while the BrdU paper was under review, we embarked on several parallel efforts to expand our tool kit. Lisa Timmons, a post-doc in Andy Fire's lab, had shown that *C. elegans* could be fed bacteria expressing dsRNA, leading to knock-down of the targeted gene (Timmons, Court, & Fire, 2001; Timmons & Fire, 1998). Relative to labor-intensive injections, dsRNA feeding could be scaled up in a manner that would allow screens to be performed. Lisa kindly shared her reagents with me and I set out to test whether planarians could also mount an RNAi response after ingesting bacterially expressed dsRNA, or if the bacteria would eat the planarians, instead. Several of the abundantly expressed transcripts that we had characterized during our first subtractive hybridization screens became useful targets for these experiments and allowed us to show the specificity and systemic nature of the response to dsRNA feeding.

To have access to more complete information on planarian gene expression, we also embarked on an expressed sequence tag (EST) project using cDNA libraries that we generated from our clonal planarian strain. We made cDNA libraries from heads and head blastemas (and later, enriched neoblasts), and in those pre-automation days, our excellent technicians, Sofia Robb and Rejeanne Juste performed the mini-preps and sequencing reactions for thousands of randomly picked clones.

As our list of expressed planarian genes grew, we hoped to scale up our *in situ* hybridization experiments accordingly. Chen-Ming Fan's laboratory was next door to Alejandro's, and Fan dropped by one day to see if we had any ideas about equipment that he could purchase with some surplus, end-of-year funds. Alejandro had recently returned from a conference, where he learned about an *in situ* hybridization robot that was being used to automate this tedious process. Alejandro told Fan about the *in situ* robot; Fan thought it sounded like a great resource, so he bought one. Fan then kindly allowed us to monopolize his machine for quite some time and we set out to adapt our *in situ* protocol for this apparatus. After a few rounds of trouble shooting, we had it working robustly.

By the start of the next academic year, both Alejandro and I went on the job market. To avoid competing for the same jobs, we split things geographically: I applied to Washington University in St. Louis and everything east of the Mississippi, while Alejandro applied to everything to the west. I wound up accepting an offer from the University of Illinois at Urbana-Champaign and Alejandro accepted an offer from the University of Utah. Before my move to Illinois, I was selected to give a talk on the opening night of the Developmental Biology Gordon Conference, an opportune place to share our work with the community. On the bus ride to the airport after the conference, I found myself seated next to Mark Patterson, then the Editor of *Nature Reviews Genetics*. Mark had missed my talk, but had heard about it, and asked me to give him a rundown of the work. Mark invited me to write a review discussing how genomic tools were being brought to bear on studies of planarian regeneration, while placing this work in a historical context. Such a review provided a great opportunity to introduce a wider audience to the advantages of *S. mediterranea* as a model organism and the fascinating open questions in planarian biology. Alejandro and I worked on this review while we were each getting settled into our new laboratories and submitted the manuscript at the beginning of 2002. The journal editors were unconvinced by the beginning of our title (This is not your father's planarian:); it took some negotiating, but in the end, they agreed to let us use most of it (Newmark & Sánchez Alvarado, 2002). At this point, we still had unpublished work to submit, so I made a trip to Salt Lake City so we could finish our paper on ESTs from the clonal asexual strain of *S. mediterranea*. To highlight the resources available for studying this animal, we included automated *in situ* hybridization to localize transcripts in various cell types, and the demonstration that bacterial feeding could be used for delivery of dsRNA. After review, we were requested to remove the entire section of the manuscript showing the efficacy of feeding bacterially expressed dsRNA, so our initial EST paper had only three figures (Sánchez Alvarado, Newmark, Robb, & Juste, 2002). We eventually published the bacterial feeding results as a collaboration between our two labs, with post-docs Peter Reddien (from Alejandro's lab) and Francesc Cebrià (from my lab) contributing experiments showing that comparable defects in regeneration could be generated by injecting or feeding dsRNA (Newmark et al., 2003).

The final missing piece to develop *S. mediterranea* as a model in which genomic approaches could be used to study regeneration and stem cell biology was the actual genome sequence. At the end of 2002, we learned about an opportunity to propose new organisms for sequencing by the National Human Genome Research Institute. Before the white paper application deadline, I again flew to Salt Lake City, where Alejandro, Peter Reddien, and I (figuratively) locked ourselves in a room for several days and drafted the sequencing proposal. During this time, we bounced ideas off each other and did everything we could to make the most convincing case for why the *S. mediterranea* genome should be sequenced. This remains the most fun I have ever had writing a grant application, which may sound like a dubious distinction, but it is still true. Even better was receiving the news that the proposal had been approved and that the Washington University Genome Sequencing Center would do the sequencing.

At the beginning, this journey to bring planarians into the molecular era felt a bit like a quixotic quest, and I had serious reservations about asking others to join me on this potentially career-killing adventure. With a reasonable set of tools now in hand, this concern no longer applied, and I was confident that we could make headway on many open questions in planarian biology.

5. Concluding thoughts

This chapter will be published 25 years after our first days cutting off planarian heads at Carnegie Embryology. Since then, we have both been fortunate to be joined by a fantastic group of colleagues interested in unlocking the many mysteries of planarian regeneration. As these colleagues have moved on to set up labs of their own and others have also joined the fray, the international community of planarian researchers has grown considerably, as has our knowledge about the mechanisms of regeneration in these animals. However, much work remains to be done and many technical challenges remain to be resolved, so we still encourage everybody to "grab a razor blade!" (Newmark, 2005) and join us.

Acknowledgments

We would like to thank past and present members of the Sánchez Alvarado and Newmark labs, along with the many colleagues around the world who have contributed to the re-emergence of planarians as experimental models. We have mentioned many in this review, and we extend our sincere apologies to those whose contributions and support we were unable to discuss. Unusually, beyond specific individuals, we also acknowledge the Carnegie Institution's Department of Embryology. Based on its unique scientific culture and the exceptional colleagues who worked there, it is hard to imagine a better place to develop a model system. We will always be grateful for our time there and the influence it had on the kind of science we strive to do. PAN would like to thank Melanie Issigonis for helpful comments on multiple drafts of this manuscript. We also gratefully acknowledge NIH (ASA, PAN) and the Damon Runyon Cancer Research Foundation (PAN) for their support of this work during the initial stages discussed here.

References

Agata, K., Soejima, Y., Kato, K., Kobayashi, C., Umesono, Y., & Watanabe, K. (1998). Structure of the planarian central nervous system (CNS) revealed by neuronal cell markers. *Zoological Science*, *15*, 433–440.

Allen, G. E. (1978). *Thomas Hunt Morgan: The man and his science*. Princeton, NJ: Princeton University Press.

Baguñà, J. (1973). *Estudios citotaxonómicos, ecológicos e histofisiología de la regulación morfogenética durante el crecimiento y la regeneración en la raza asexuada de la planaria Dugesia mediterranea n. sp. (Turbellaria, Tricladida, Paludicola)*. Ph.D. Thesis Univ. de Barcelona.

Baguñà, J. (2019). Planarian regeneration between 1960s and 1990s: From skilful baffled ancestors to bold integrative descendants. A personal account. *Seminars in Cell & Developmental Biology*, *87*, 3–12.

Baguñà, J., Carranza, S., Pala, M., Ribas, M., & Riutort, M. (1999). From morphology and karyology to molecules. New methods for taxonomical identification of asexual populations of freshwater planarians. A tribute to professor Mario Benazzi. *The Italian Journal of Zoology*, *66*, 207–214.

Baguñà, J., & Romero, R. (1981). Quantitative analysis of cell types during growth, degrowth and regeneration in the planarians *Dugesia mediterranea* and *Dugesia tigrina*. *Hydrobiologia*, *84*, 181–194.

Baguñà, J., Romero, R., Saló, E., Collet, J., Auladell, C., Ribas, M., et al. (1990). Growth, degrowth, and regeneration as developmental phenomena in adult freshwater planarians. In H.-J. Marthy (Ed.), *Experimental embryology in aquatic plants and animals* (pp. 129–162). New York: Plenum Press.

Baguñà, J., Saló, E., & Auladell, C. (1989). Regeneration and pattern-formation in planarians. III. Evidence that neoblasts are totipotent stem-cells and the source of blastema cells. *Development*, *107*, 77–86.

Benazzi, M., & Benazzi Lentati, G. (1976). *Animal Cytogenetics, Vol. 1: Platyhelminthes*. Berlin: Gebrüder Borntraeger.

Best, J. B., Rosenvold, R., Souders, J., & Wade, C. (1965). Studies on the incorporation of isotopically labeled nucleotides and amino acids in planaria. *The Journal of Experimental Zoology*, *159*, 397–403.

Beukeboom, L. W., Weinzierl, R. P., Reed, K. M., & Michiels, N. K. (1996). Distribution and origin of chromosomal races in the freshwater planarian *Dugesia polychroa* (turbellaria: Tricladida). *Hereditas*, *124*, 7–15.

Boswell, R. E., Prout, M. E., & Steichen, J. C. (1991). Mutations in a newly identified *Drosophila melanogaster* gene, *mago nashi*, disrupt germ cell formation and result in the formation of mirror-image symmetrical double abdomen embryos. *Development*, *113*, 373–384.

Brøndsted, H. V. (1969). *Planarian regeneration*. Oxford: Pergamon Press.

Cebrià, F., & Newmark, P. A. (2005). Planarian homologs of netrin and netrin receptor are required for proper regeneration of the central nervous system and the maintenance of nervous system architecture. *Development*, *132*, 3691–3703.

Cebrià, F., Vispo, M., Newmark, P., Bueno, D., & Romero, R. (1997). Myocyte differentiation and body wall muscle regeneration in the planarian *Girardia tigrina*. *Development Genes and Evolution*, *207*, 306–316.

Garcia-Fernàndez, J., Bayascas-Ramírez, J. R., Marfany, G., Muñoz-Mármol, A. M., Casali, A., Baguñà, J., et al. (1995). High copy number of highly similar mariner-like transposons in planarian (Platyhelminthe): Evidence for a trans-phyla horizontal transfer. *Molecular Biology and Evolution*, *12*, 421–431.

Gremigni, V., & Miceli, C. (1980). Cytophotometric evidence for cell 'transdifferentiation' in planarian regeneration. *Roux's Archives of Developmental Biology*, *188*, 107–113.

Gremigni, V., Miceli, C., & Picano, E. (1980). On the role of germ cells in planarian regeneration. II. Cytophotometric analysis of the nuclear Feulgen-DNA content in cells of regenerated somatic tissues. *Journal of Embryology and Experimental Morphology*, *55*, 65–76.

Gremigni, V., Miceli, C., & Puccinelli, I. (1980). On the role of germ cells in planarian regeneration. I. A karyological investigation. *Journal of Embryology and Experimental Morphology*, *55*, 53–63.

Gremigni, V., Nigro, M., & Puccinelli, I. (1982). Evidence of male germ cell redifferentiation into female germ cells in planarian regeneration. *Journal of Embryology and Experimental Morphology*, *70*, 29–36.

Hendzel, M. J., Wei, Y., Mancini, M. A., Van Hooser, A., Ranalli, T., Brinkley, B. R., et al. (1997). Mitosis-specific phosphorylation of histone H3 initiates primarily within pericentromeric heterochromatin during G2 and spreads in an ordered fashion coincident with mitotic chromosome condensation. *Chromosoma*, *106*, 348–360.

Kenk, R. (1974). *Index of the genera and species of the freshwater triclads (turbellaria) of the world*. Washington, D.C.: Smithsonian Institution Press.

Maden, M. (1993). The homeotic transformation of tails into limbs in *Rana temporaria* by retinoids. *Developmental Biology*, *159*, 379–391.

Maienschein, J. (2016). Garland Allen, Thomas Hunt Morgan, and development. *Journal of the History of Biology*, *49*, 587–601.

Merryman, M. S., Sánchez Alvarado, A., & Jenkin, J. C. (2018). Culturing planarians in the laboratory. In J. C. Rink (Ed.), *Planarian regeneration: Methods and protocols, methods in molecular biology* (pp. 241–258). Springer.

Miller-Bertoglio, V. E., Fisher, S., Sánchez, A., Mullins, M. C., & Halpern, M. E. (1997). Differential regulation of *chordin* expression domains in mutant zebrafish. *Developmental Biology*, *192*, 537–550.

Mohanty-Hejmadi, P., Dutta, S. K., & Mahapatra, P. (1992). Limbs generated at site of tail amputation in marbled balloon frog after vitamin A treatment. *Nature*, *355*, 352–353.

Newmark, P. A. (2005). Opening a new can of worms: A large-scale RNAi screen in planarians. *Developmental Cell*, *8*, 623–624.

Newmark, P. A., Reddien, P. W., Cebrià, F., & Sánchez Alvarado, A. (2003). Ingestion of bacterially expressed double-stranded RNA inhibits gene expression in planarians. *Proceedings of the National Academy of Sciences of the United States of America*, *100*(Suppl 1), 11861–11865.

Newmark, P. A., & Sánchez Alvarado, A. (2000). Bromodeoxyuridine specifically labels the regenerative stem cells of planarians. *Developmental Biology*, *220*, 142–153.

Newmark, P. A., & Sánchez Alvarado, A. (2002). Not your father's planarian: A classic model enters the era of functional genomics. *Nature Reviews. Genetics*, *3*, 210–219.

O'Kane, C. J., & Gehring, W. J. (1987). Detection *in situ* of genomic regulatory elements in *Drosophila*. *Proceedings of the National Academy of Sciences of the United States of America*, *84*, 9123–9127.

Peters, A., Streng, A., & Michiels, N. K. (1996). Mating behaviour in a hermaphrodite flatworm with reciprocal insemination: Do they assess their mates during copulation? *Ethology*, *102*, 236–251.

Rilling, M. (1996). The mystery of the vanished citations: James McConnell's forgotten 1960s quest for planarian learning, a biochemical engram, and celebrity. *The American Psychologist*, *51*, 589–598.

Rozario, T., Collins, J. J., 3rd, & Newmark, P. A. (2022). The good, the bad, and the ugly: From planarians to parasites. *Current Topics in Developmental Biology*, *147* (In press).

Salò, E., & Baguñà, J. (1984). Regeneration and pattern formation in planarians. I. The pattern of mitosis in anterior and posterior regeneration in *Dugesia (G) tigrina*, and a new proposal for blastema formation. *Journal of Embryology and Experimental Morphology*, *83*, 63–80.

Salò, E., & Baguñà, J. (1985a). Cell movement in intact and regenerating planarians. Quantitation using chromosomal, nuclear and cytoplasmic markers. *Journal of Embryology and Experimental Morphology*, *89*, 57–70.

Salò, E., & Baguñà, J. (1985b). Proximal and distal transformation during intercalary regeneration in the planarian *Dugesia (S) mediterranea*. *Roux's Archives of Developmental Biology*, *194*, 364–368.

Sánchez Alvarado, A. (2000). Regeneration in the metazoans: Why does it happen? *BioEssays*, *22*, 578–590.

Sánchez Alvarado, A., & Newmark, P. A. (1999). Double-stranded RNA specifically disrupts gene expression during planarian regeneration. *Proceedings of the National Academy of Sciences of the United States of America*, *96*, 5049–5054.

Sánchez Alvarado, A., Newmark, P. A., Robb, S. M., & Juste, R. (2002). The *Schmidtea mediterranea* database as a molecular resource for studying platyhelminthes, stem cells and regeneration. *Development*, *129*, 5659–5665.

Sikes, J. M., & Newmark, P. A. (2013). Restoration of anterior regeneration in a planarian with limited regenerative ability. *Nature*, *500*, 77–80.

Timmons, L., Court, D. L., & Fire, A. (2001). Ingestion of bacterially expressed dsRNAs can produce specific and potent genetic interference in *Caenorhabditis elegans*. *Gene*, *263*, 103–112.

Timmons, L., & Fire, A. (1998). Specific interference by ingested dsRNA. *Nature*, *395*, 854.

Umesono, Y., Watanabe, K., & Agata, K. (1997). A planarian *orthopedia* homolog is specifically expressed in the branch region of both the mature and regenerating brain. *Development, Growth & Differentiation*, *39*, 723–727.

Wang, Z., & Brown, D. D. (1991). A gene expression screen. *Proceedings of the National Academy of Sciences of the United States of America*, *88*, 11505–11509.

CHAPTER TWELVE

The good, the bad, and the ugly: From planarians to parasites

Tania Rozario[a,]*, James J. Collins III[b,]*, and Phillip A. Newmark[c,]*
[a]Center for Tropical and Emerging Global Diseases and Department of Genetics, University of Georgia, Athens, GA, United States
[b]Department of Pharmacology, UT Southwestern Medical Center, Dallas, TX, United States
[c]Howard Hughes Medical Institute, Morgridge Institute for Research, Department of Integrative Biology, University of Wisconsin, Madison, WI, United States
*Corresponding authors: e-mail address: Tania.Rozario@uga.edu; J.Collins@UTSouthwestern.edu; pnewmark@morgridge.org

Contents

Abstract

Platyhelminthes can perhaps rightly be described as a phylum of the good, the bad, and the ugly: remarkable free-living worms that colonize land, river, and sea, which are often rife with color and can display extraordinary regenerative ability; parasitic worms like schistosomes that cause devastating disease and suffering; and monstrous tapeworms that are the stuff of nightmares. In this chapter, we will explore how our research expanded beyond free-living planarians to their gruesome parasitic cousins. We start with *Schistosoma mansoni*, which is not a new model; however, approaching these parasites from a developmental perspective required a reinvention that may hold generalizable lessons to basic biologists interested in pivoting to disease models.

Current Topics in Developmental Biology, Volume 147
ISSN 0070-2153
https://doi.org/10.1016/bs.ctdb.2021.12.015

We then turn to our (re)establishment of the rat tapeworm *Hymenolepis diminuta*, a once-favorite model that had been largely forgotten by the molecular biology revolution. Here we tell our stories in three, first-person narratives in order to convey personal views of our experiences. Welcome to the dark side.

1. Seeing the parallels (PAN)

"What led a planarian lab to start working on parasites?" is the second-most-often-asked question when I meet with trainees during seminar visits and scientific meetings (see Chapter "*Schmidtea* happens: Re-establishing the planarian as a model for studying the mechanisms of regeneration" by Newmark and Sánchez Alvarado for the most-often-asked question). The seeds for our work on parasitic flatworms were planted when I was beginning my work on planarians as a post-doc in Barcelona. As I contemplated returning to the US, but was struggling to find any PIs interested in having me bring planarians to their labs (see Chapter "*Schmidtea* happens: Re-establishing the planarian as a model for studying the mechanisms of regeneration" by Newmark and Sánchez Alvarado), it was beginning to look as if I would need to come up with a plan for a second post-doc. As I started expanding my reading to think about other potential research topics to pursue, I read a small, paperback, introductory parasitology book (Whitfield, 1979) that was one of the few books I brought with me to Spain. This book came to me courtesy of Sean Eddy, a fellow grad student in Boulder, when he was shedding belongings in preparation for his move to the UK for a post-doc. Re-learning the complexities of many parasitic life cycles (which I hadn't thought about since freshman introductory biology) and thinking about the various ways in which parasites interacted with their hosts convinced me that if I had to abandon planarians, I would study parasitology, instead. Fortunately, I didn't have to give up on the planarian project, so I wound up not pursuing this plan "B" (at least not for another 15 years or so).

When Alejandro Sánchez Alvarado and I began generating thousands of expressed sequence tags (ESTs) from planarians, we noticed genes that were shared between planarians and parasitic flatworms, but not found elsewhere in the animal kingdom. Our manuscript reporting the initial EST database for asexual *Schmidtea mediterranea* and the development of high-throughput, automated in situ hybridization to identify cell-type-specific markers (Sánchez Alvarado, Newmark, Robb, & Juste, 2002) highlighted some aspects of the fascinating biology of parasitic flatworms, which infect hundreds of millions of people, and concluded by emphasizing the possibility of using planarians as free-living models for studying their parasitic counterparts. Indeed, Alejandro and I even submitted a patent invention disclosure based on this idea, but it was rejected by the Carnegie Institution's patent attorneys. My lab's paper characterizing ESTs from the sexual strain of *S. mediterranea* (Zayas, Hernández, et al., 2005) returned to this idea, noting that roughly half of the planarian ESTs had homologs in *Schistosoma mansoni*. We also pointed out that the identification of genes encoding signaling molecule receptors in the parasite's genome did not necessarily mean that those receptors were responding to host signals, particularly if those receptors were found in free-living flatworms; thus, it was important to compare the parasites to their free-living relatives before reaching such conclusions.

As the work on planarians continued to develop, I had hoped that researchers studying parasitic flatworms would begin to apply some of the tools we were developing and the lessons we were learning to their own experimental subjects. When it became clear that this wasn't really happening, I then hoped to recruit a post-doc with expertise in parasitic flatworms to pursue this idea. Such a recruit never materialized, so this plan lay dormant for several years.

Meanwhile, our work on planarian germ cell development was progressing well. Graduate student Yuying Wang showed that RNA interference (RNAi) knockdown of the conserved germ cell regulator, *nanos*, resulted in planarians without germ cells (Wang, Zayas, Guo, & Newmark, 2007); she used this knowledge to conduct a non-biased, gene expression-based, functional screen for genes required for germ cell development. Yuying compared gene expression profiles (via microarrays) of planarians with and without germ cells (*nanos* RNAi), then identified RNAi knockdowns leading to defects at distinct stages of male germ cell development. In addition to genes conserved between planarians and mammals, Yuying identified genes that were shared between planarians and schistosomes, but not found in other organisms (Wang, Stary, Wilhelm, & Newmark, 2010); thus, her work ascribed functions to flatworm-specific genes.

Concurrently, post-doc Jim Collins was characterizing the entire complement of peptide hormones/neuropeptides encoded by the planarian genome. These molecules arguably constitute the largest and most diverse class of signaling molecules in animals, and they regulate a wide range of physiological processes from neurotransmission and behavior to feeding and sexual development. Jim's work integrated genomic prediction of peptide hormones with biochemical/mass spectrophotometric analysis of peptides (in collaboration with Jonathan Sweedler's lab), followed by whole-mount in situ hybridization to map the expression patterns of each peptide hormone gene (Collins et al., 2010). Jim found that knocking down the expression of *prohormone convertase 2* (*pc2*), which encodes an enzyme critical for peptide hormone processing, led to testis regression in sexual planarians, resulting in testes that contained only the initial stages of germ cell development. This phenotype was reminiscent of the testis regression observed after head amputation (Fedecka-Bruner, 1967; Ghirardelli, 1965) and implicated peptide hormones in the systemic regulation of germ cell development. The question was: which peptide hormone(s) mediated this effect on the reproductive system?

Among his collection of 51 genes (which encoded over 200 peptides) Jim was able to find a single peptide hormone-encoding gene (*npy-8*) that was expressed robustly in the nervous system of sexual planarians but whose transcripts were not detectable in asexual worms. Surprisingly, *npy-8* RNAi in mature adults led to testis regression resembling that seen in *pc2* RNAi animals; knocking down *npy-8* during the course of juvenile growth blocked proper development of the reproductive system (Collins et al., 2010). Thus, a single neuropeptide was regulating the development and maintenance of the planarian reproductive system, suggesting an ancient origin for the neuroendocrine control of reproductive maturation (and that planarians also experience puberty!). Beyond its implications for understanding the systemic regulation of germ cell development, Jim identified planarian neuropeptides that had homologs in *S. mansoni* and/or *S. japonicum*, but had yet to be annotated, again demonstrating the utility of such comparative approaches.

As Jim was wrapping up this comprehensive work for publication (please note that I've omitted several other major conclusions of the paper to focus on the topic at hand), he came to an important choice point in his post-doc career: would he continue to study planarian neuropeptides or would he pursue a different biological problem? This question led him to his own deep dive into the literature, starting with Volume 2 of Libbie Hyman's remarkable series of monographs, "The Invertebrates" (Hyman, 1951), which included her overview of the Platyhelminthes. Jim became fascinated by the extreme biology exhibited by parasitic flatworms and by their parallels with planarians (see below). One of the most striking realizations to emerge from his readings was that the pathology of schistosomiasis, which impacts over 200 million people, was almost entirely due to the parasite's prolific reproductive output (Basch, 1991). Female schistosomes produce >300 eggs per day; blood flow away from the parasite's niche leads approximately half of these eggs to get trapped in host tissues, triggering massive inflammatory responses that can lead to tissue dysfunction and failure. Thus, schistosomiasis is essentially a disease of flatworm reproduction. Jim had just identified a single neuropeptide required for the development and maintenance of the planarian reproductive system; he found that the schistosome genome also encoded this neuropeptide. Could blocking this signaling pathway also switch off the parasite's reproductive system, thereby halting disease pathology and preventing parasite transmission? Here we had this fascinating convergence of a fundamental question of developmental biology with a public health problem of global significance.

As Jim and I discussed the possibility of pursuing this exciting idea ourselves, there were a few major barriers that dramatically increased the activation energy required to get started; all of these barriers were in some way related to the schistosome life cycle (Fig. 1). These blood flukes are obligate parasites of two different animal hosts: a mammalian definitive host (in which

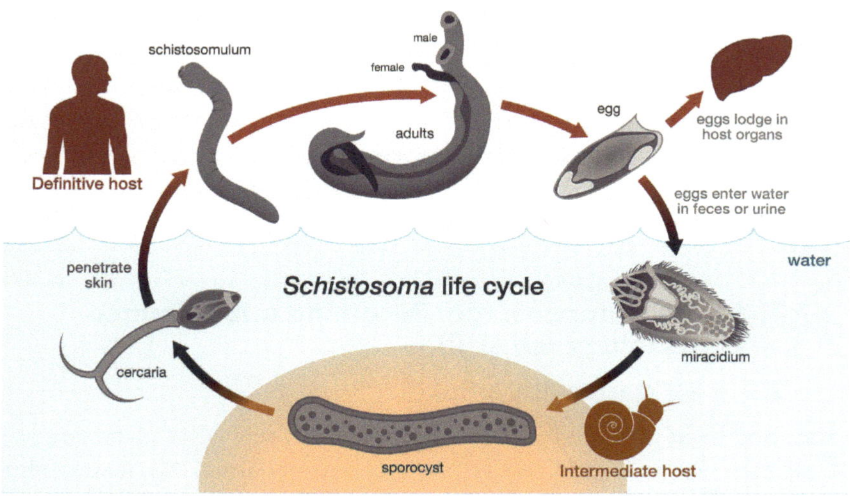

Fig. 1 The schistosome life cycle. Adult parasites lay eggs: roughly half of these eggs get stuck in host tissues, causing disease, whereas the other half are released via urine or feces. Eggs that reach freshwater hatch and liberate miracidia that infect a snail intermediate host. Inside the snail, the miracida become sporocysts that undergo clonal expansion, ultimately producing cercariae that emerge from the snail to infect a mammalian definitive host. After penetrating the skin of their host, cercariae become schistosomula that migrate through the tissue and into the bloodstream. Once in the portal circulation, the worms develop as either males or females and reproduce sexually.

sexual reproduction occurs) and a snail intermediate host (in which asexual reproduction occurs). Working with the mammalian host (*S. mansoni* can be maintained in the lab in mice) required us to dive into the world of animal care protocols, institutional animal care and use committees, and animal care facilities. Furthermore, working with the snails added the complexity of requiring biosafety level 2 (BSL-2) protocols, approvals, and appropriate lab space. As planarian researchers, we were blissfully unaware of such matters, and were daunted by all of these additional administrative, animal care, and safety burdens.

One day, while still in discussions about whether or how to proceed on the parasitic flatworm front, Jim came charging into my office with a *PLOS NTD* paper by Fred Lewis and colleagues, which described the Schistosomiasis Resource Center (SRC), a National Institutes of Health (NIH)-National Institute of Allergy and Infectious Diseases (NIAID)-funded resource dating back to the 1960s, which was established to facilitate research on these important parasites (Lewis, Liang, Raghavan, & Knight, 2008). It turned out that the SRC was actively maintaining several strains of the major schistosome species infecting humans, along with their different snail intermediate hosts. In addition to offering courses in maintaining the schistosome life cycle, at that time they provided, free-of-charge, schistosome-infected mice and snails to qualified laboratories around the world. After learning about this remarkable resource, within the hour we contacted Fred, who graciously arranged for us to visit his laboratory at the Biomedical Research Institute in Rockville, Maryland to learn how to handle schistosomes.

2. Schisto happens, too: Becoming a full-fledged parasitology lab (JJC)

2.1 It all starts with a fluke

I had long been inspired by Phil and Alejandro's revival of planarians as a model organism (see PAN and ASA chapter, this volume). So, it was exhilarating to have the opportunity to bring schistosomes to the lab. However, as a mid-stage post-doc with a career on the line, this exhilaration was tempered by a fair dose of apprehension. There were many unknowns and no guarantee of success. The biggest unknowns circled around two simple questions: as developmental biologists, with no background in parasitology, could we actually maintain these parasites in the lab and could we ask meaningful and experimentally tractable biological questions in an organism renowned for its experimental intractability?

My trepidation wasn't unwarranted when you consider the worm's life cycle (Fig. 1). This life cycle starts when the parasite's egg reaches freshwater, triggering the release of a ciliated larva, called a miracidium, which despite its simple appearance is quite sophisticated, possessing both an excretory system and a fairly elaborate nervous system. This short-lived larva then has less than 24 h to infect a snail intermediate host. It turns out each schistosome species can only infect specific genera of snails, and thus the parasite's geographic range is not dictated by its definitive vertebrate host but rather the range of the snail intermediate host.

Upon locating the appropriate snail, the miracidium transforms to another larval stage called a sporocyst. The sporocyst is arguably the most critical stage of the life cycle, as it serves as the conduit through which a single miracidium undergoes clonal expansion, thus amplifying the parasite's opportunity to pass to a mammal. This clonal expansion is driven by a small number of undifferentiated cells (germinal cells) that begin to proliferate once the miracidium sheds its cilia and transforms to a sporocyst inside the snail (Wang, Collins, & Newmark, 2013). The germinal cells then enter a phase of asexual embryogenesis, generating so-called daughter sporocysts that migrate to distal parts of the snail. Like their mothers, these daughter sporocysts also contain pools of undifferentiated cells that seed multiple rounds of clonal expansion, generating either new daughter sporocysts or infective larvae called cercariae. Triggered by light, these cercariae wriggle out of the snail tissue and are liberated into the water.

Cercariae use their muscular, forked tail to zip around the water, waiting for the opportunity to infect a vertebrate. Attracted by movement and fatty acids from the skin (Shiff, Cmelik, Ley, & Kriel, 1972), the worms seek out their host, deploy secretions that compromise the epidermis, and break off their tail as they wriggle into the skin. This penetration process occurs within a matter of minutes (McKerrow & Salter, 2002) and the parasites make their way through the tissue before entering the vasculature. The worms then make several passes through the circulation before taking up residence in either the portal vasculature or the venous plexus of the bladder. Once at their final destination, the worms mature to adulthood as either males or females. The male schistosome is large and muscular and possesses a groove on his ventral surface (gynecophoral canal) that he uses to grasp the long, slender female.

Not only do schistosomes have the distinction of being the only non-hermaphroditic, dioecious (i.e., separate male and female sexes) group of flatworms, but this close physical contact between the male and female worms is absolutely essential for female sexual development. Female worms grown in hosts without males are small and their reproductive organs are present as undeveloped primordia (LoVerde, 2002; Severinghaus, 1928). Upon pairing with a male, the female perceives a yet-to-be defined signal, initiating the development of the sexual organs and ultimately egg production. The eggs laid by the sexually mature female are then washed away into the blood with the goal of reaching the outside world. Roughly half of these eggs move from the blood to the intestine (or bladder for urogenital schistosomiasis) to be passed to the environment via feces (or urine). The other half, however, become lodged in host organs inducing inflammatory lesions that are the sole driver of morbidity and death due to schistosome infection (Pearce & MacDonald, 2002).

2.2 Getting exposed to the parasites

As Phil noted above, our journey to schistosomes moved from the "hypothetical" phase to the "ok, we are doing this" phase when we contacted Fred Lewis. Fred, a veteran of the schistosome field had spent his career optimizing every aspect of schistosome husbandry (Lewis, 2001). One of the main challenges of schistosome maintenance wasn't keeping the worms alive, but rather growing and maintaining the sometimes-fastidious snails. Since the SRC propagated the three major species that infect humans (*S. mansoni, S. japonicum, and S. haematobium*) that meant they had to grow three varieties of snails, each presenting their own challenges. In fact, the SRC is the only facility I am aware of that actively maintains the notoriously demanding *Oncomelania* snails that serve as the intermediate host of *S. japonicum*. These snails will not eat lettuce like other snail hosts and prefer rather to eat single-celled critters (e.g., diatoms and algae). Fred and his colleagues developed a methodology to grow *Nostoc* (cyanobacterium) on mud, lime and chicken manure and use it as a food source for *Oncomelania*. In fact, the SRC had a dedicated "mud room" that, well, was a room full of (you guessed it) mud they routinely collected from a specific locale in Maryland.

Fred had studied these snails so intently, that he even noted that these snails often harbor commensal rotifers and that some rotifer species could actually make a diffusible substance that paralyzes cercariae, preventing them from infecting mice (Stirewalt & Lewis, 1981). So as part of the SRC snail maintenance protocols, folks routinely use a Waterpik (yeah, the one for your teeth) to blast rotifers off of snail shells. (Side note: It turns out Fred did quite a bit of work in the late 1980s trying to find the rotifer-derived compound only to have funding for the work expire and his progress go unpublished. A few decades later, Fred shared his unpublished observations with Phil and me. This topic became the subject of Phil's 2011 summer "back to the bench" project, culminating in the elucidation of a novel, rotifer-derived alkaloid with schistosome paralytic activity (Gao et al., 2019).

After a few days with Fred at the SRC learning the ins and outs of snail maintenance, how to get cercariae from snails, and expose mice to the parasites, Phil and I got our first real "exposure" to parasitology when it came time to recover worms from mice. It was as impressive as it was gruesome. For a couple of guys who worked on invertebrates most of their careers, this was an eye-popping experience. It started with euthanasia of about 50 schistosome-infected mice and these expired mice made their way into a large sink full of water to mat down their fur. I don't recall in my career ever seeing a single dead mouse, much less a whole sink full. From here, the scissors came out, first for removing the skin and then for exposing the contents of the peritoneal cavity. By precisely severing the portal vein under the liver and plunging a needle into the aorta, the worms were perfused from the blood into a large bloody container beneath, with the worms often liberated as large boluses visible to the eye. Now collecting the worms wouldn't be enough to close the life cycle, since eggs and miracidia are required to infect snails. This requires the egg-riddled livers. So, Fred's team would retrieve the livers, puree them in a Waring Blender, pass the material through a series of mesh sieves, and then put what remained into freshwater. Once in freshwater, the miracidia are liberated from the eggs to infect snails, thus closing the circle.

2.3 The work begins

Following our brief visit to the SRC, we returned to Illinois and got to work. Unfortunately, before the parasite work could start we had to convert parts of a planarian developmental biology lab into a space suitable for both animal work and for handing BSL-2 pathogens. The investment in equipment was minimal, but the investment in time navigating both IACUC and the Institutional Biosafety Committee was considerable for uninitiated chaps like Phil and me. Fortunately, Fred was able to put us into contact with David Williams, a schistosome biochemist whose lab was just a couple of hours away at Rush University in Chicago. We established a collaboration with David, showing him the ropes on planarian husbandry while he provided us support for our nascent schistosome efforts, including sharing details of his animal and biosafety registrations.

In the intervening eight or so months between committing to working on schistosomes and actually receiving parasites in the lab, Phil and I spent a lot of time going through the literature. One thing that frustrated us in particular, was we had no idea what these parasites looked like on the inside. Certainly, there were volumes of beautiful ultrastructural studies from the 60s, but few studies using modern confocal microscopy to visualize the worm's tissues and organ systems. This is at a time when confocal images of planarians had made their way to countless journal covers, yet, we couldn't find a single image of a schistosome stained with something as simple as DAPI. So, our first priority once we finally got worms in the lab was to throw them in formaldehyde and then pull things out of the freezer to see what stuck. Together with another post-doc, Ryan King, who was interested in the parasite's protonephridial system, we threw basically every fluorescent stain, lectin, and antibody we could dig out of the freezer on various stages of the worms and then spent hours a day monopolizing the confocal microscope. These efforts, which spanned over just a couple of months, paid off, defining a series of beautiful markers to visualize the reproductive, protonephridial, and nervous systems as well as the worm's many types of secretory organs. As we wrote this "atlas" detailing the worm's organ systems (Collins, King, Cogswell, Williams, & Newmark, 2011), there was some trepidation about the reception the work would receive in the field. I have certainly heard of outsiders wandering into established fields and being met with a great deal of hostility. However, we were pleasantly surprised that not only did the paper fly through peer review, but we received a number of encouraging emails from people welcoming us to the field. In fact, we have always found folks in the schistosome community to be exceptionally collegial and welcoming to newcomers.

2.4 Worming our way into new biology

Armed with a collection of cell-type-specific markers and a newly adapted protocol for whole-mount in situ hybridization that we developed alongside David Williams' group (Cogswell, Collins, Newmark, & Williams, 2011; Collins et al., 2013), 2 years in we were finally ready to explore some biology. Based on Hyman's descriptions, it was clear that parasitic flatworms had all sorts of extreme biology that was strangely reminiscent of what occurs in their planarian relatives. Tapeworms can grow meters in length by perpetually adding segments to their body and as detailed above, schistosomes (and all other trematodes) asexually reproduce in their intermediate hosts by undergoing endless rounds of asexual embryogenesis. Importantly, all of this extreme biology appeared to be driven by cells that resembled the stem cells (neoblasts) that give the planarians their ability to regenerate. Given this, Phil and I were taken with the idea that during the course of evolution these parasites simply repurposed the tricks of their free-living, planarian-like ancestors to thrive as parasites.

One observation that Phil and I were interested in tackling first was the intriguing observation that schistosomes can live for decades inside their mammalian host. In the literature there is no shortage of medical case studies in which a person moves from a schistosome-endemic region to a non-endemic country as a child to find out as an adult that they have schistosomes alive in their bloodstream, laying eggs. Remember, the adult parasites that lay eggs do not replicate in the blood, thus, these adult parasites are the very same worms that have lived with these adult patients since childhood. The literature describes some extreme examples in which people are infected with schistosomes 30+ years after leaving an endemic region (Harris, Russell, & Charters, 1984; Hornstein et al., 1990; Payet et al., 2009)! So what is the mechanism underlying this longevity?

To Phil and me the answer was obvious: they must use neoblast-like cells to maintain their tissues in a manner similar to how the planarian ensures its immortality. However, examination of the literature found scant evidence that schistosomes (or any other trematode) possessed any proliferative cells outside their reproductive organs. It was also clear that no one had ever really looked for such cells systematically. So we did the most straightforward experiment we could imagine: culture the worms in vitro in tissue culture media and pulse them with the thymidine analog EdU to determine if they possessed any proliferative cells in their somatic tissues. Excitingly (but not totally unexpectedly) there were loads of EdU-labeled cells scattered throughout the worm's soma (Collins et al., 2013). We went on to show that these neoblast-like cells were true stem cells capable of self-renewal and differentiation. We also pulled a trick from the planarian biology grab bag to use irradiation to kill the stem cells and then perform transcriptional profiling to define neoblast-specific markers. Not only did this provide a description of the cells on a molecular level, it provided an avenue to begin conducting functional experiments to define genes required for stem cell function.

As of 2012 when we started exploring these neoblast-like cells, RNAi had already been shown years prior to function in in vitro-cultured schistosomes across a variety of life-cycle stages (Boyle, Wu, Shoemaker, & Yoshino, 2003; Skelly, Da'dara, & Harn, 2003). However, anecdotal observations and published reports (Krautz-Peterson, Bhardwaj, Faghiri, Tararam, & Skelly, 2010; Štefanić et al., 2010) suggested that the effects of RNAi were variable, with some genes totally non-susceptible to the effects of double-stranded RNA (dsRNA). This was unsurprising since this is the case in a variety of systems. What was more concerning was that there was little evidence that RNAi could generate reproducible phenotypes, particularly in the adult stage of the life cycle. If mRNA silencing was robust, then where were the phenotypes?

Working from our foundation in planarian biology, we reasoned the troubles were twofold. First, it was clear from many published reports that phenotypes were scored shortly after dsRNA treatment (usually 1 week). However, in planarians very few phenotypes could be seen in this timeframe. Since planarians have massive amounts of tissue turnover (and thus protein turnover) relative to schistosomes, it seemed unlikely that dramatic phenotypes could be observed in such a short interval. This was an easy fix: just wait longer. The second issue was that most studies only examined a single gene at a time, and since the effects of RNAi varied from gene to gene, one would likely have to examine multiple genes involved in a given process to yield a phenotype. What was the barrier to doing this? Well, to achieve optimal mRNA knockdown, worms had to be cultured in high concentrations of dsRNA in several mL of media (Krautz-Peterson et al., 2010; Krautz-Peterson, Radwanska, Ndegwa, Shoemaker, & Skelly, 2007; Štefanić et al., 2010). Since the field had relied on either expensive "high-yield" dsRNA synthesis kits (Štefanić et al., 2010) or commercially synthesized siRNAs (Krautz-Peterson et al., 2007) for their experiments, it was cost prohibitive to examine more than a few genes at a time.

So we took another trick from the planarian playbook. At the time, many in the Newmark lab had migrated from using bacterially expressed dsRNA for RNAi in planarians to in vitro-synthesized dsRNA made using a super-cheap, homemade "high-yield" recipe (Rouhana et al., 2013). This meant we could overcome the sampling error that in part obscured phenotypes and quickly screen tens, if not hundreds, of genes in a single experiment. So, with these tweaks in place, we cloned cDNAs from ~30 of our neoblast-expressed gene list, treated worms with dsRNAs, and examined the worms by EdU labeling about 2½ weeks later. Not only did we observe that loss of several canonical cell-cycle regulators caused stem cell maintenance defects, but loss of one of the worm's FGF receptors caused a dramatic reduction in stem cell number (Collins et al., 2013). I had been skeptical about whether we would be able to ask the types of developmental questions that drew us to the field in the first place (see question #2 above), but with the results of these RNAi experiments it was clear we had a real honest-to-goodness system to study new and interesting biology.

Buoyed by our studies on schistosome stem cells, we felt empowered and wanted to add another experimental arrow to our quiver. In particular, since these are obligate parasites, studies during in vitro culture can only elucidate so much of the parasite's biology. For instance, the reproductive biology of the parasite represents not only amazing biology but a key driver of disease pathology. Unfortunately, at the time culture conditions were insufficient to maintain sexual reproduction for more than a few days outside the host (we have since addressed this (Wang, Chen, & Collins, 2019)). However, there weren't just technical limitations to in vitro culture, there are simply many critical interactions between the worms and their host that cannot be replicated in a dish. For example, how is it that these worms thrive on the front lines of the host immune response for decades without being cleared? To this end, we wanted to develop a method that would allow us to experimentally manipulate worms in the context of their host; without the convenience of germline transgenics, we had to think outside the box a bit.

In the summer of 2013 when preparing materials for my upcoming job hunt, I came across a paper from 1976 by Donato Cioli, a schistosome researcher who had developed an approach to surgically transplant schistosomes from one host to another (Cioli, 1976). I reasoned that we could retrieve worms from a host, manipulate them in vitro (e.g., RNAi treatment), then transplant them to a new host and examine what happens to the worm. As luck would have it, Donato was still working on schistosomes and had come across our atlas paper. He reached out to us for our input with some of his ongoing work on the mechanism of action of the anti-schistosomal drug Praziquantel. So, I contacted Donato and inquired about the technique he developed almost 40 years prior. Within 24 h I received a response complete with a highly detailed protocol of the procedure, including catalog numbers of specific types of custom needles he used to inject worms and his preferred anesthetic. The protocol entailed anesthetizing the recipient mice, pulling the intestines through a small incision, injecting the worms into a specific vein (cecal vein), stopping the bleeding with hemostatic gauze, reinserting the intestines, and closing the incision with sutures.

Although Donato was a trained physician, the most complicated surgery I had ever done involved a planarian and a razor blade. So I was relieved that in the post-script of his email he offered to host us in his lab and teach us the technique. So I talked to Phil and, although he couldn't travel because of NIH study section duties, he gave me the green light to visit. Oh, did I mention that Donato lived in Rome? So, I packed my bags and traveled to sunny Italy for 3 days of training in Donato's lab at the Institute of Cell Biology and Neurobiology in Monterotondo outside of Rome. Aside from being an outstanding source of knowledge about schistosome biology he was an astounding teacher and I left Italy having successfully transplanted schistosomes into several mice. (Side note: Not only was Donato an amazing host and teacher, but perhaps the greatest tour guide of Italy anyone could imagine. He knew so much about Italian and Roman History and he actually spent his time outside of the lab as an archeologist exploring ancient Roman aqueducts. Also, I didn't even bring Phil a t-shirt, as he likes to remind me.)

My time at Illinois outfitting parts of Phil's lab for schistosome work set the stage for starting my own lab at UT Southwestern. With this relatively modest tool kit we've been able to dive deeper into schistosome stem cell biology (Collins, Wendt, Iyer, & Newmark, 2016; Wendt et al., 2018, 2020), define new regulators of sexual development (Wang et al., 2019; Wang & Collins, 2016), determine the functions of schistosome genes in vivo (Collins & Collins, 2016; Wang et al., 2020), and even leverage these tools to prioritize therapeutic targets (Wang et al., 2020).

3. Transitioning to tapeworms… (PAN)

In the summer of 2012, Jim and I went to the Gordon Conference on Host-Parasite Interactions, the first parasitology conference either of us had attended. I had the opportunity to present Jim's work showing that adult schistosomes have stem cells, just like their planarian cousins. I was feeling quite proud of how well the work had progressed and how the idea of approaching the study of these parasites from a developmental perspective had yielded new insights into their biology. When my talk ended, the first question came from Jayne Raper, who asked if we had also characterized such stem cells in tapeworms. I wound up saying something to the effect of, "I wish we had, but this was already quite a bit of work: it took us several years to get to this point, and as interesting as tapeworms are, we simply haven't looked at them yet." In our earlier discussions of the amazing biology of parasitic flatworms, Jim and I had talked about the seemingly limitless growth of tapeworms and how it was driven by stem cells, but we had our hands full with the schistosome work. Nonetheless, Jayne's question gnawed at me and I kept tapeworms at the back of my mind.

That November, I was invited to speak at the graduate student-organized departmental retreat for the Molecular Biology Program at the University of Colorado School of Medicine. Dick Davis was one of the faculty members attending the retreat; Dick and I had corresponded a decade or so earlier based on his previous work on trans-splicing in schistosomes (Davis, Hardwick, Tavernier, Hodgson, & Singh, 1995; Davis & Hodgson, 1997; Rajkovic, Davis, Simonsen, & Rottman, 1990) when my lab had found similar trans-splicing in planarians (Zayas, Bold, & Newmark, 2005), but we had never met in person. Dick was now doing fascinating work on chromatin diminution in *Ascaris*, bringing genomic tools to bear on the classic observations of Boveri, who described the loss of chromatin in somatic embryonic cells while cells of the germline retained their full chromosome complement (reviewed in Wilson, 1937). I wanted to learn more about this fascinating work, so made a point of tracking him down between sessions. During the course of our discussion, I learned that Dick studied the tapeworm, *Hymenolepis diminuta* for his Ph.D. When I told him that I was interested in looking at tapeworms to explore the similarities between their stem cells and the neoblasts from planarians and schistosomes, Dick graciously volunteered to help get us started with the life cycle. Soon enough, we would come knocking at his door.

4. Resurrecting a model tapeworm: *Hymenolepis diminuta* (TR)

4.1 An unexpected beginning

I arrived in the Newmark lab with every intention of being a good planarian biologist, ready to elucidate mechanisms of regeneration in these remarkable flatworms. Regeneration had long fascinated me and was the subject of my college admissions essay but I took somewhat of a detour in graduate school to study embryogenesis in *Xenopus*. I reasoned that a foundation in developmental biology would be a valuable asset in the study of regeneration and that has certainly remained true. As a newly minted Ph.D., I thought I was finally executing my master plan. Little did I know that there were more twists and turns ahead.

Regeneration is actually a common occurrence across most phyla, though the range of regenerative ability varies widely; many species can regenerate cells/tissues to a limited extent, while far fewer species exhibit whole-body regeneration (Bely & Nyberg, 2010). What brought me to the Newmark lab was its history working with both a champion specimen capable of regenerating both soma and germline—*S. mediterranea*, while also tapping the diversity of the flatworm phylum in which regenerative ability spans the gamut. Using a multi-species approach, one can probe why some animals can regenerate while others cannot. I was particularly struck by previous work in the lab from James Sikes who showed that *Procotyla fluviatilis*, which normally cannot regenerate a head from a tail fragment, could be coaxed to regenerate its head by simply reducing Wnt/β-catenin signaling levels (Sikes & Newmark, 2013). The Newmark lab was also making numerous discoveries about the parasitic blood fluke *S. mansoni*; while this species does not regenerate it does exhibit prolific stem cell-driven reproductive output with major implications for the pathology caused by these parasites, as described above. I was excited to work in this veritable zoo of a lab and yet I was unprepared to sit in Phil's office on my first day and hear him ask, "Have you ever thought about tapeworms?" I confessed, I had not. Without exerting any pressure, Phil suggested I just do some reading and see what I thought.

The flood gates opened. Why *hadn't* I thought about tapeworms? They are cousins to planarians that seem to do nothing but grow, often to enormous lengths! As part of their normal life cycle, many tapeworms have the ability to shed and regrow their body! Surely, these are the kinds of "monsters in our midst" that scientists should take advantage of to gain a richer understanding of stem cell and regenerative biology.

4.2 It takes a village to "raise" a tapeworm

As parasites that infect humans and livestock, causing disease and devastating economic burdens, tapeworms have long been a small but persistent subject of scientific research. There are international groups in pursuit of better diagnostic, treatment, and preventative options to eradicate the scourge of these parasites (e.g., The Cysticercosis Working Group in Peru). Could we as developmental biologists bring something to the table? Our approach was to try and uncover fundamental workings of tapeworms both to make basic biological findings and to expose vulnerabilities that can be exploited.

Naturally, most tapeworm research was focused on pathologically relevant species (primarily from the Taeniidae family) that cause cysticercoses and echinococcoses, devastating diseases in which tapeworm larval cysts invade host organs, including liver, lung, and brain (Brutto, 2012; Eckert & Deplazes, 2004). Recently, there has been growing recognition that like the adult tapeworms, the larval cystic forms maintain stem-cell-like populations that are sometimes called "germinative cells" (Koziol, Rauschendorfer, Rodríguez, Krohne, & Brehm, 2014; Orrego et al., 2021). Thus, understanding tapeworm stem cell regulation has potential for translational relevance (Brehm & Koziol, 2014). Despite all this, our understanding of both stem cells and regeneration in tapeworms was rudimentary at best. Furthermore, the field as a whole suffered from a decided dearth of molecular data, though that was beginning to change through the contributions of Klaus Brehm (Universität Würzburg, Germany), Estella Castillo and Uriel Koziol (Universidad de la República, Uruguay), Pete Olson (The Natural History Museum, United Kingdom) and others. Thus, we reasoned that we could leverage our experience with planarians to gain a molecular understanding of regeneration and stem cell regulation in tapeworms. What we needed was to establish a tapeworm laboratory model amenable to modern molecular experimentation.

How does one choose a model species? My experience teaches me that it is some combination of previous knowledge, intuition, and a healthy dose of luck. There are more than 4000 species of tapeworms and counting (Caira & Jensen, 2017). The majority of tapeworms require host species that would be daunting to work with (e.g., sharks, pigs, humans). For a laboratory model from which we could easily obtain adult worms, we prioritized rodent tapeworms. This quickly narrowed the list down to four popular species from the genus *Hymenolepis*: *H. diminuta*, *H. citelli*, *H. microstoma*, and *H. nana*. Each had their pros and cons. *H. nana* is the most common human tapeworm and thus has direct translational relevance but also posed additional biosafety concerns. *H. microstoma* had a published genome (Tsai et al., 2013), but is an atypical tapeworm that resides in the bile duct instead of the intestine. Furthermore, the Wellcome Sanger Institute was embarking on an ambitious project to sequence dozens of helminth genomes, making available a wealth of genomic data for many species (Holroyd & Sanchez-Flores, 2012). However, to be perfectly honest, the decision was not an agonizing one. As we wanted a good laboratory model, it made sense to choose the species that was most extensively studied in the lab and a clear winner emerged: *H. diminuta*. We were astonished to find a veritable tome numbering 747 pages called "The Biology of the tapeworm *Hymenolepis diminuta*" edited by Hisao P. Arai and published in 1980 (Arai, 1980). This book capped decades of beautiful biochemistry, light and electron microscopy, in vitro as well as in vivo manipulations, and more. How sad then that after 1980 almost all research on this and other rodent tapeworms seemed to cease. Why? Anything I say would be mere speculation from a newcomer. It was probably a combination of changing funding priorities, the difficulty of establishing molecular methods on such complicated animals, and the funneling of resources into an ever decreasing number of organisms. To me, I felt both humbled and excited by the chance to revitalize this work and bring *H. diminuta* back to the bench, having the advantage of previous work by outstanding experimentalists from a different era. I personally leaned heavily on the work of Clark Read, Marietta Voge, Chauncey Goodchild, John Ubelaker, Larry Roberts, Richard Lumsden, Robert Specian, and Teresa Sulgostowska, just to name a few.

With Phil's agreement, I cast my lot with *H. diminuta*. Phil then revealed that we were not alone, and relayed to me how Dick Davis had offered his expertise should we need it. As Phil described above, Dick had long since moved on from tapeworms to an even more disgusting parasite, *Ascaris* (Google image search, dear reader, at your peril). In an act of sheer generosity, Dick got his laboratory set up with rats and infected beetles, hosted me in his lab and home, and taught me how to propagate the life cycle and handle these worms. For this, he has my heartfelt and unending gratitude. Up to this point, all my knowledge was from the page and now I was able to touch and see my first tapeworm. I named her Aurora in honor of Aurora, Colorado, where we stood.

4.3 Good housekeeping matters, even for parasites

Back in the Newmark lab, I was now all set to get *H. diminuta* established. It was not difficult to source—infected beetles are actually sold by Carolina Biologicals as they are still used in parasitology classrooms across the country. The life cycle (Fig. 2A) is extraordinarily simple to maintain. The mealworm beetle *Tenebrio molitor* ingests embryonated eggs from infected rat feces and unwittingly plays host to the larval stages of *H. diminuta*. Larvae in the gut can burrow into the hemocoel, where larval development concludes in 2 weeks. In the hemocoel, a transformation occurs in which the anterior end forms structures that will persist in the adult (such as the suckers) and this region becomes encapsulated into a cysticercoid by the rest of the larval tissue. In essence, the cysticercoid forms a protective pouch around a dormant juvenile tapeworm, which can remain in suspended animation for the life of the beetle. Thus, batches of infected beetles can be easily maintained in the lab for months in a bed of oats with moisture from carrots as a water source.

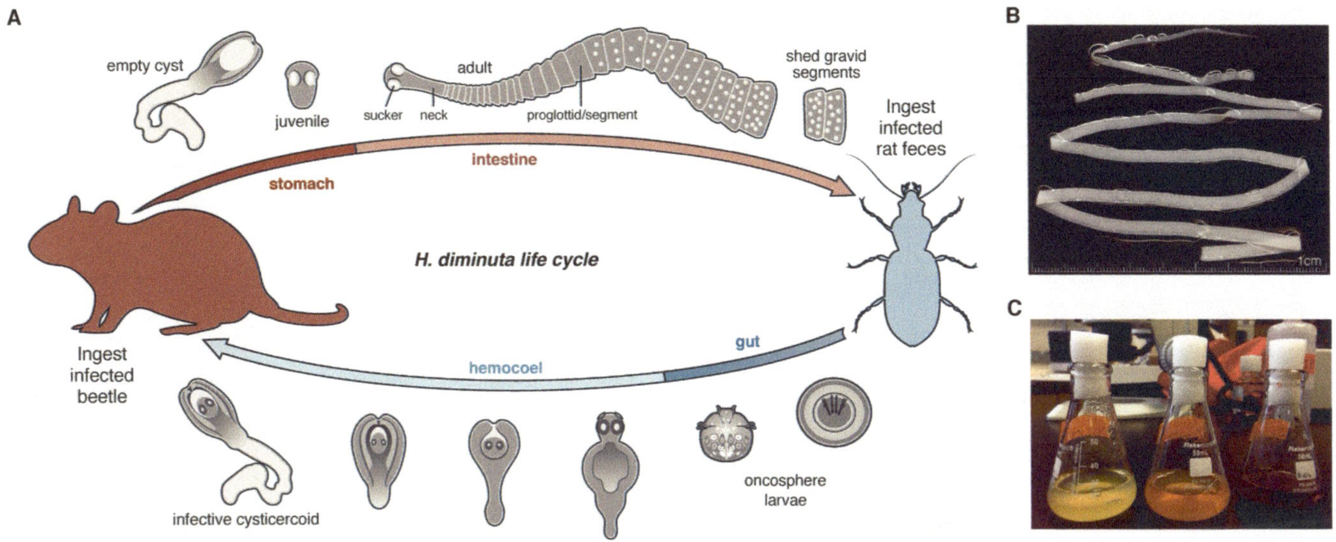

Fig. 2 In vivo and in vitro cultivation of *H. diminuta*. (A) Two hosts are used to complete the *H. diminuta* life cycle: mealworm beetles host larval development from oncosphere to infective cysticercoid and rats host the adult stage through sexual maturity. Each adult forms a head, neck, and strobilated body made of thousands of segments/proglottids. Each mature segment is hermaphroditic, can cross-fertilize, and develop to gravidity. Through apolysis, gravid segments are shed and excreted with the rat feces, which can be consumed by a beetle to complete the life cycle. (B) An adult *H. diminuta* at 20 days old arranged from anterior (top) to posterior (bottom). (C) Adult stages can be grown in vitro using biphasic cultures. Shown are flasks with agar that was enriched with either plasma, serum, or blood (preferred), overlain with Hanks Balanced Salt Solution. Amputated tapeworm fragments containing the neck can grow and regenerate using this in vitro culture system in a hypoxic chamber.

To infect rats with *H. diminuta*, I dissect open the infected beetles and administer 1–500 cysticercoids to rats via oral gavage. The rat's stomach acids and enzymes then permeabilize the cysticercoid and bile triggers excystment of the juvenile tapeworm. At this point, the tapeworm measures 100–200 μm in length and uses its suckers to latch onto intestinal microvilli where it will grow to reproductive maturity and reach an equilibrium length of ~60 cm in 38 days (Chandler, 1939; Roberts, 1980) (Fig. 2B). Adult *H. diminuta* has a head/scolex at the anterior-most end followed by an unsegmented neck (germinative region), which serves as a growth zone from which thousands of segments/proglottids bud (Fig. 2A). Each segment in these hermaphrodites is fated to form all the male and female reproductive structures. Within 2 weeks, *H. diminuta* reaches sexual maturity and can mate with itself or with another worm. The fertilized eggs become deposited in the uterus along with a yolky vitelline cell that forms extraembryonic structures to protect the developing embryo. However, development cannot proceed through larval stages in the rat host. In the wild, *H. diminuta* undergoes apolysis, in which posterior fragments containing gravid segments are released with the rat excrement and can be consumed by an unsuspecting beetle or other similar insects. In the laboratory, I can simply amputate 4–10 cm of adult posteriors, smash them with a little apple sauce on filter paper, and feed the embryo-laden mixture to starved beetles. And thus, the life cycle is completed. Adult *H. diminuta* causes no pathology to its rat hosts and up to 10 mature tapeworms can reside in the rat intestine for the life of the rat. As a former frog embryologist with no parasitology chops, the simple and affordable husbandry involved in this system was a great attraction.

With the guidance I received from Dick, I successfully established a colony of *H. diminuta* in the Newmark lab. I remember running through the hallway toward Phil's office exclaiming "I have tapeworms!" Phil rightfully noted that such a statement rarely came with the level of exuberance I was displaying. I decided to keep my first mature tapeworm as a demo worm that I could show to visitors. I named her Bathsheba (after the tenacious heroine of Thomas Hardy's "Far From the Madding Crowd": Bathsheba Everdene). She has been a useful communication aid that I have now shown to visiting speakers, prospective graduate students, school kids of all ages, and even my wedding guests. As exciting as all this was, the hard part had just begun.

4.4 Reflections on tool building

With no molecular tools established for *H. diminuta*, tool building was an immediate priority. From previous experience, we had learnt that fluorescently conjugated lectins and many common generic antibodies could preferentially label flatworm tissues and allow us to visualize features such as the osmoregulatory canals and tubules, flame cells, gonads, musculature, neurons, and domains of the tegument (skin) (Collins et al., 2011; Rozario & Newmark, 2015; Zayas, Cebrià, Guo, Feng, & Newmark, 2010). These observations confirmed many previous cell-biological findings that had been made by light and electron microscopy (Roberts, 1980) and gave us an appreciation for the complex structures that are continually formed throughout adult development. This was a good way to "get our hands wet" with something we had good reason to suspect would work and would give us a resource we could apply to many future experiments.

Though tool building is obviously necessary, one could get dragged down trying to perfect methods. It was really important that we forged ahead with the dual priorities of establishing methodologies and testing hypotheses. We immediately focused on elucidating regenerative competence in *H. diminuta* using amputation and to describe genetic regulators of stem cells that drive growth and regeneration. To this end, the tools we needed were clear. We prioritized assembling a transcriptome and establishing RNA in situ hybridization, as this would allow us to clone candidate genes of interest and assay gene-expression patterns. We collaborated with Jianbin Wang from Dick Davis's lab for the transcriptome assembly, which was a combination of de novo and map-based assemblies because at the time, the *H. diminuta* reference genome was highly fragmented. We decided against sequencing the genome ourselves reasoning that genome assemblies for helminths would continue to improve thanks to other researchers with more relevant expertise and indeed there is now a much improved *H. diminuta* genome assembly (Nowak et al., 2019).

For in situ hybridization, we relied on a wealth of previous experience in the Newmark lab but were also generously aided by Pete Olson who shared his working protocol for *H. microstoma*. These advances allowed us to perform RNA sequencing and identify putative stem cell markers by comparing transcriptomes of wild-type tapeworms to tapeworms that were first depleted of stem cells using irradiation (Rozario, Quinn, Wang, Davis, & Newmark, 2019). We were able to gain a molecular description of transcripts enriched in subpopulations of dividing cells, which include stem cells. It is worth noting that though our work benefits strongly from comparison with planarians, parasitic flatworm stem cells have unique characteristics: key stem cell genes like *piwi*, *tudor*, and *vasa* have been lost from the genomes of parasitic flatworms (Tsai et al., 2013). Thus, unbiased discovery of stem cell markers is also necessary and we are currently pursuing single-cell sequencing of stem cells in *H. diminuta* to capture transcriptional heterogeneity within the stem cell population.

To move this model away from being purely descriptive and into the realm of functional biology, it was crucial to establish loss-of-function methods. For this to be feasible, we first needed to grow these parasites in vitro. Admittedly, one of the main reasons we chose *H. diminuta* in the first place was that all stages had been grown in vitro before, though there were multiple different protocols, and no one had performed these experiments in over 30 years. We concentrated on the adult stage. Most previously described methods grew these worms in a hypoxic, biphasic culture composed of a supplemented nutrient agar that was overlain with a physiological saline in which the worms could reside. Working mostly from papers by Everett Schiller, Larry Roberts, Faith Mong, and Robert Turton (Roberts & Mong, 1973; Schiller, 1965; Turton, 1972), we tried supplementing with blood, serum, or plasma from numerous animal sources, as well as yeast and liver extracts (Fig. 2C). We also varied the saline solutions and glucose concentrations. In truth, many conditions worked to some extent though there was a great deal of variation.

I decided to set a "gold standard." I wanted a condition in which worms could grow to gravidity in vitro, at which point I would take the tissue with embryonated eggs and use that to complete one life cycle (in both beetle and rat hosts) and recover a fully mature adult tapeworm. The winning combination was defibrinated sheep blood-agar with 4 g/L glucose-supplemented Hanks Balanced Salt Solution. The mature tapeworm we recovered was a milestone indeed. I named her Cara (for "cara mia," or "my beloved" in Italian. I trust my naming pattern is emerging).

It is worth noting that while I was testing different culture conditions, I was simultaneously performing amputation experiments to test regeneration competence of *H. diminuta*. While this doubled or tripled the number of conditions tested, it was also highly valuable. First, we were able to gain experimental insights while building tools. We were able to uncover that *H. diminuta* regenerates in a regionally limited fashion: the neck is necessary and sufficient for segment regeneration, whereas the head and body are not regeneration competent. However, regeneration from the neck is finite: persistent regeneration requires the presence of the head (Rozario et al., 2019). Continuing studies with head amputations are revealing that the head either directly or indirectly regulates the microenvironmental signals that are necessary for segment regeneration to occur. Second, doing these kinds of amputation experiments concurrently with varying the culture conditions allowed us to identify the condition that was most permissive for the kinds of experiments we were interested in. Though many experiments then had to be repeated and standardized later, we gained a tremendous amount of confidence and excitement that *H. diminuta* was going to be a worthwhile and interesting model to pursue.

The next holy grail was RNAi. All parasitic flatworms belong to a monophyletic clade called the Neodermata, which refers to a "new skin": the tegument (Tyler & Tyler, 1997). The tegument is a single cell that encompasses the entire worm exterior and is connected to thousands of nuclei within cytons that make thin cytoplasmic bridges to the shared cytoplasm at the exterior. This unique structure is used for communication and nutrition as it forms the interface with the tapeworm's environment (Lumsden, 1975; Lumsden & Specian, 1980). Tapeworms have evolved to completely rely on the tegument for nourishment and have entirely dispensed with their digestive system; they have no mouths and no guts. Thus, one would think that they are tailor-made for RNAi and one could simply soak them in double-stranded RNA (dsRNA). This strategy works reasonably well in blood flukes (Wang et al., 2020), which was encouraging to us. However, unlike tapeworms, flukes have both a tegument and digestive tract, enabling two routes for dsRNA uptake. All efforts to soak *H. diminuta* in dsRNA (with carriers, with permeabilization agents, in different volumes and concentrations, using different sizes of dsRNA, etc.) have failed to produce robust gene knockdown. Thankfully, my years working on *Xenopus* came to my aid and I was able to microinject dsRNA into the rather finicky and narrow tapeworm necks to induce RNAi. Consequently, we were able to establish an RNAi paradigm to identify functional regulators of stems cells and regeneration (Rozario et al., 2019). Microinjecting is laborious, time consuming, and relies on "user ability" and thus requires more careful experimental design (additional controls, blinded experiments, and large sample sizes). This is yet another example of how we established methods that are "good enough" for the purpose of immediately testing hypotheses, but with future troubleshooting, these methods will undoubtedly become easier, more high-throughput, and more robust.

These and other experiments began to reveal that though *H. diminuta* could only regenerate from its neck, we could not identify a neck-specific stem cell population that explained this behavior. In fact, cells from both regeneration-competent and regeneration-incompetent regions display collective pluripotency and can rescue viability and regeneration in irradiated/stem cell-deficient tapeworms as long as these cells are transplanted into the neck (Rozario et al., 2019). It is becoming clear that microenvironmental signals in the neck play pivotal roles in regulating regeneration and/or stem cell behaviors. Clearly there is much to learn from these enigmatic creatures.

4.5 Even tapeworms need friends

However enthusiastic we may be, all would be for naught if we could not generate interest among our colleagues and the greater scientific community. Perhaps we are fortunate that both parasitologists and developmental biologists work under big tents. The study of parasitology involves cell and molecular biology, microbiology, immunology, physiology, chemistry, epidemiology, and ecology. Similarly, developmental biology is highly interdisciplinary. Through conferences, our work has brought us in contact with diverse researchers from both developmental biology and parasitology, which has improved our perspectives and scholarship while being enormously fun to boot! Early on, while we were just getting up and running, Phil suggested we write an R21 grant application, not just for the practical matter of receiving funding, but to gauge if there was enough interest in our system, ideas, and methodologies, to merit support from our peers on study sections. Thankfully, the support of the NIH-NIAID was and continues to be encouraging and generous. Yet funding alone will not determine the success of this model. The real test will be whether students and post-docs choose to push this model forward, accelerate the research, and make their own careers with *H. diminuta* as at least one of their models of choice. I hope very much that my new lab at the University of Georgia will serve as an incubator for the success of this model and for researchers who will move it to heights I have not yet imagined!

H. diminuta, though impressively large in size, is still a model in its infancy. On top of the potential for translational relevance, I am excited by the promise of what we stand to learn. The limited nature of *H. diminuta* regeneration could allow us to uncover mechanisms that both restrict and enhance regenerative ability. The potency of tapeworm stem cell subpopulations and their relationship to the microenvironment of the neck is still mysterious. Is there a pluripotent subpopulation or are multiple subpopulations able to regain/retain/maintain pluripotency as long as they receive the appropriate niche signals? What is the nature of the stem cell niche? What are the microenvironmental signals that regulate regeneration? How are segments initiated during regeneration and are they regulated by a segmentation clock? When are germ cells first formed and how are they coordinated with segmentation? These and many other questions are up for grabs. If you, dear reader, feel like wrangling tapeworms and unearthing their secrets holds a strangely irresistible (if somewhat queasy) appeal, the Rozario lab is looking for students and postdocs!

5. Concluding thoughts

This special issue exists because there is growing recognition that we need to complement the enormous advances we have gained from traditional model organisms with new emerging models that capture the breadth and diversity of the tree of life. Failure to do so would limit the scope of our knowledge and blind us to unique and extreme traits that already exist all around us (Goldstein & King, 2016). Realistically, we live in a world of limited resources so anyone attempting to establish a new/neglected model will be faced with this ultimate question—is it worth it? To us, we must choose our models because they are particularly well or uniquely suited to answer interesting scientific questions. With that criterion satisfied, the hard work—and it is hard work—will pave the way for an adventurous and rewarding ride!

Acknowledgments

We would like to acknowledge the many researchers not mentioned in this chapter whose excellent work could not be discussed in detail. We recognize that our success would not be possible without their contributions. None of this work would have been possible without the generous and flexible support of the Howard Hughes Medical Institute, of which PAN is an investigator. We are grateful to Melanie Issigonis for invaluable feedback in the writing of this manuscript. The figures were prepared by Matt Stefely at the Morgridge Institute for Research. Sincere thanks goes out to many colleagues and lab members who contributed significantly to the development of these models; specifically, the schistosome work was greatly aided by Fred Lewis, David Williams, Harini Iyer, Marla Tharp, Ryan King, Bo Wang, and Bram Lambrus, and the tapeworm work by Dick Davis, Jianbin Wang, Jennifer McDonald, and Edward Quinn.

References

Arai, H. P. (Ed.). (1980). *Biology of the tapeworm Hymenolepis diminuta*. New York: Academic Press, Inc.

Basch, P. F. (1991). *Schistosomes: Development, reproduction, and host relations*. New York: Oxford University Press.

Bely, A. E., & Nyberg, K. G. (2010). Evolution of animal regeneration: Re-emergence of a field. *Trends in Ecology & Evolution*, *25*, 161–170.

Boyle, J. P., Wu, X.-J., Shoemaker, C. B., & Yoshino, T. P. (2003). Using RNA interference to manipulate endogenous gene expression in *Schistosoma mansoni* sporocysts. *Molecular and Biochemical Parasitology*, *128*, 205–215.

Brehm, K., & Koziol, U. (2014). On the importance of targeting parasite stem cells in anti-echinococcosis drug development. *Parasite*, *21*, 72.

Brutto, O. H. D. (2012). Neurocysticercosis. *Continuum Lifelong Learning in Neurology*, *18*, 1392–1416.

Caira, J. N., & Jensen, K. (Eds.). (2017). *Planetary biodiversity inventory (2008–2017): Tapeworms from vertebrate bowels of the earth*. Lawrence, Kansas: Natural History Museum, University of Kansas.

Chandler, A. C. (1939). The effects of number and age of worms on development of primary and secondary infections in *Hymenolepis diminuta* in rats, and an investigation into the true nature of "Premunition" in tapeworm infections. *American Journal of Hygiene*, *29*, 105–114.

Cioli, D. (1976). Transfer of *Schistosoma mansoni* into the mesenteric veins of hamsters. *International Journal for Parasitology*, *6*, 349–354.

Cogswell, A. A., Collins, J. J., III, Newmark, P. A., & Williams, D. L. (2011). Whole mount in situ hybridization methodology for *Schistosoma mansoni*. *Molecular and Biochemical Parasitology*, *178*, 46–50.

Collins, J. N. R., & Collins, J. J., III. (2016). Tissue degeneration following loss of *Schistosoma mansoni cbp1* is associated with increased stem cell proliferation and parasite death in vivo. *PLoS Pathogens*, *12*, e1005963.

Collins, J. J., III, Hou, X., Romanova, E. V., Lambrus, B. G., Miller, C. M., Saberi, A., et al. (2010). Genome-wide analyses reveal a role for peptide hormones in planarian germline development. *PLoS Biology*, *8*, e1000509.

Collins, J. J., III, King, R. S., Cogswell, A., Williams, D. L., & Newmark, P. A. (2011). An atlas for *Schistosoma mansoni* organs and life-cycle stages using cell type-specific markers and confocal microscopy. *PLoS Neglected Tropical Diseases*, *5*, e1009.

Collins, J. J., III, Wang, B., Lambrus, B. G., Tharp, M. E., Iyer, H., & Newmark, P. A. (2013). Adult somatic stem cells in the human parasite *Schistosoma mansoni*. *Nature*, *494*, 476–479.

Collins, J. J., III, Wendt, G. R., Iyer, H., & Newmark, P. A. (2016). Stem cell progeny contribute to the schistosome host-parasite interface. *eLife*, *5*, 243.

Davis, R. E., Hardwick, C., Tavernier, P., Hodgson, S., & Singh, H. (1995). RNA trans-splicing in flatworms. Analysis of trans-spliced mRNAs and genes in the human parasite, *Schistosoma mansoni*. *Journal of Biological Chemistry*, *270*, 21813–21819.

Davis, R. E., & Hodgson, S. (1997). Gene linkage and steady state RNAs suggest trans-splicing may be associated with a polycistronic transcript in *Schistosoma mansoni*. *Molecular and Biochemical Parasitology*, *89*, 25–39.

Eckert, J., & Deplazes, P. (2004). Biological, epidemiological, and clinical aspects of echinococcosis, a zoonosis of increasing concern. *Clinical Microbiology Reviews*, *17*, 107–135.

Fedecka-Bruner, B. (1967). Etudes sur la regeneration des organes genitaux chez la planaire *Dugesia lugubris*. I. Regeneration des testicules apres destruction. *Bulletin Biologique de la France et de la Belgique*, *101*, 255–319.

Gao, J., Yang, N., Lewis, F. A., Yau, P., Collins, J. J., III, Sweedler, J. V., et al. (2019). A rotifer-derived paralytic compound prevents transmission of schistosomiasis to a mammalian host. *PLoS Biology*, *17*, e3000485.

Ghirardelli, E. (1965). Differentiation of the germ cells and regeneration of the gonads in planaria. In V. Kiortsis, & H. A. L. Trampusch (Eds.), *Regeneration in animal and related problems* (pp. 177–184). Amsterdam: North-Holland Publishing Co.

Goldstein, B., & King, N. (2016). The future of cell biology: Emerging model organisms. *Trends in Cell Biology*, *26*, 818–824.

Harris, A. R. C., Russell, R. J., & Charters, A. D. (1984). A review of schistosomiasis in immigrants in Western Australia, demonstrating the unusual longevity of *Schistosoma mansoni*. *Transactions of the Royal Society of Tropical Medicine and Hygiene*, *78*, 385–388.

Holroyd, N., & Sanchez-Flores, A. (2012). Producing parasitic helminth reference and draft genomes at the Wellcome Trust sanger institute. *Parasite Immunology*, *34*, 100–107.

Hornstein, L., Lederer, G., Schechter, J., Greenberg, Z., Boem, R., Bilguray, B., et al. (1990). Persistent *Schistosoma mansoni* infection in Yemeni immigrants to Israel. *Israel Journal of Medical Sciences*, *26*, 386–389.

Hyman, L. (1951). *The invertebrates (Vol. II): Platyhelminthes and Rhynchocoela*. New York: McGraw Hill.

Koziol, U., Rauschendorfer, T., Rodríguez, L. Z., Krohne, G., & Brehm, K. (2014). The unique stem cell system of the immortal larva of the human parasite *Echinococcus multilocularis*. *EvoDevo*, *5*, 10.

Krautz-Peterson, G., Bhardwaj, R., Faghiri, Z., Tararam, C. A., & Skelly, P. J. (2010). RNA interference in schistosomes: Machinery and methodology. *Parasitology*, *137*, 485–495.

Krautz-Peterson, G., Radwanska, M., Ndegwa, D., Shoemaker, C. B., & Skelly, P. J. (2007). Optimizing gene suppression in schistosomes using RNA interference. *Molecular and Biochemical Parasitology*, *153*, 194–202.

Lewis, F. (2001). Schistosomiasis. In *Current protocols in immunology* John Wiley & Sons, Inc. (chapter 19), Unit 19.1.1–19.1.28.

Lewis, F. A., Liang, Y., Raghavan, N., & Knight, M. (2008). The NIH-NIAID schistosomiasis resource center. *PLoS Neglected Tropical Diseases*, *2*, e267.

LoVerde, P. T. (2002). Sex and schistosomes: An interesting biological interplay with control implications. *Journal of Parasitology*, *88*, 3–13.

Lumsden, R. D. (1975). The tapeworm tegument: A model system for studies on membrane structure and function in host-parasite relationships. *Transactions of the American Microscopical Society*, *94*, 501–507.

Lumsden, R. D., & Specian, R. (1980). The morphology, histology, and fine structure of the adult stage of the Cyclophyllidean tapeworm *Hymenolepis diminuta*. In H. P. Arai (Ed.), *Biology of the tapeworm Hymenolepis diminuta* (pp. 157–280). New York: Academic Press, Inc.

McKerrow, J., & Salter, J. (2002). Invasion of skin by *Schistosoma* cercariae. *Trends in Parasitology*, *18*, 193–195.

Nowak, R. M., Jastrzębski, J. P., Kuśmirek, W., Sałamatin, R., Rydzanicz, M., Sobczyk-Kopcioł, A., et al. (2019). Hybrid de novo whole-genome assembly and annotation of the model tapeworm *Hymenolepis diminuta*. *Scientific Data*, *6*, 302.

Orrego, M. A., Verastegui, M. R., Vasquez, C. M., Koziol, U., Laclette, J. P., Garcia, H. H., et al. (2021). Identification and culture of proliferative cells in abnormal *Taenia solium* larvae: Role in the development of racemose neurocysticercosis. *PLoS Neglected Tropical Diseases*, *15*, e0009303.

Payet, B., Chaumentin, G., Boyer, M., Amaranto, P., Lemonon-Meric, C., & Lucht, F. (2009). Prolonged latent schistosomiasis diagnosed 38 years after infestation in a HIV patient. *Scandinavian Journal of Infectious Diseases*, *38*, 572–575.

Pearce, E. J., & MacDonald, A. S. (2002). The immunobiology of schistosomiasis. *Nature Reviews Immunology*, *2*, 499–511.

Rajkovic, A., Davis, R. E., Simonsen, J. N., & Rottman, F. M. (1990). A spliced leader is present on a subset of mRNAs from the human parasite *Schistosoma mansoni*. *Proceedings of the National Academy of Sciences of the United States of America*, *87*, 8879–8883.

Roberts, L. S. (1980). Development of *Hymenolepis diminuta* in its definitive host. In H. P. Arai (Ed.), *Biology of the tapeworm Hymenolepis diminuta* (pp. 357–423). New York: Academic Press, Inc.

Roberts, L. S., & Mong, F. N. (1973). Developmental physiology of Cestodes. XIII. Vitamin B_6 requirement of *Hymenolepis diminuta* during in vitro cultivation. *The Journal of Parasitology*, *59*, 101–104.

Rouhana, L., Weiss, J. A., Forsthoefel, D. J., Lee, H., King, R. S., Inoue, T., et al. (2013). RNA interference by feeding in vitro-synthesized double-stranded RNA to planarians: Methodology and dynamics. *Developmental Dynamics*, *242*, 718–730.

Rozario, T., & Newmark, P. A. (2015). A confocal microscopy-based atlas of tissue architecture in the tapeworm *Hymenolepis diminuta*. *Experimental Parasitology*, *158*, 31–41.

Rozario, T., Quinn, E. B., Wang, J., Davis, R. E., & Newmark, P. A. (2019). Region-specific regulation of stem cell-driven regeneration in tapeworms. *eLife*, *8*, e48958.

Sánchez Alvarado, A., Newmark, P. A., Robb, S. M., & Juste, R. (2002). The *Schmidtea mediterranea* database as a molecular resource for studying platyhelminthes, stem cells and regeneration. *Development*, *129*, 5659–5665.

Schiller, E. L. (1965). A simplified method for the in vitro cultivation of the rat tapeworm, *Hymenolepis diminuta*. *The Journal of Parasitology*, *51*, 516–518.

Severinghaus, A. E. (1928). Sex studies on *Schistosoma japonicum*. *Quarterly Journal of Microscopical Science*, *71*, 653–702.

Shiff, C. J., Cmelik, S. H. W., Ley, H. E., & Kriel, R. L. (1972). The influence of human skin lipids on the cercarial penetration responses of *Schistosoma haematobium* and *Schistosoma mansoni*. *The Journal of Parasitology*, *58*, 476.

Sikes, J. M., & Newmark, P. A. (2013). Restoration of anterior regeneration in a planarian with limited regenerative ability. *Nature*, *500*, 77–80.

Skelly, P. J., Da'dara, A., & Harn, D. A. (2003). Suppression of cathepsin B expression in *Schistosoma mansoni* by RNA interference. *International Journal for Parasitology*, *33*, 363–369.

Štefanić, S., Dvořák, J., Horn, M., Braschi, S., Sojka, D., Ruelas, D. S., et al. (2010). RNA interference in *Schistosoma mansoni* Schistosomula: Selectivity, sensitivity and operation for larger-scale screening. *PLoS Neglected Tropical Diseases*, *4*, e850.

Stirewalt, M., & Lewis, F. A. (1981). *Schistosoma mansoni*: Effect of rotifers on cercarial output, motility and infectivity. *International Journal for Parasitology*, *11*, 301–308.

Tsai, I. J., Zarowiecki, M., Holroyd, N., Garciarrubio, A., Sanchez-Flores, A., Brooks, K. L., et al. (2013). The genomes of four tapeworm species reveal adaptations to parasitism. *Nature*, *496*, 57–63.

Turton, J. A. (1972). The in vitro cultivation of *Hymenolepis diminuta*: The culture of 6-day-old worms removed from the rat. *Zeitschrift für Parasitenkunde*, *40*, 333–346.

Tyler, S., & Tyler, M. S. (1997). Origin of the epidermis in parasitic platyhelminths. *International Journal for Parasitology*, *27*, 715–738.

Wang, J., Chen, R., & Collins, J. J., III. (2019). Systematically improved in vitro culture conditions reveal new insights into the reproductive biology of the human parasite *Schistosoma mansoni*. *PLoS Biology*, *17*, e3000254.

Wang, J., & Collins, J. J., III. (2016). Identification of new markers for the *Schistosoma mansoni* vitelline lineage. *International Journal for Parasitology*, *46*, 405–410.

Wang, B., Collins, J. J., III, & Newmark, P. A. (2013). Functional genomic characterization of neoblast-like stem cells in larval *Schistosoma mansoni*. *eLife*, *2*, e00768.
Wang, J., Paz, C., Padalino, G., Coghlan, A., Lu, Z., Gradinaru, I., et al. (2020). Large-scale RNAi screening uncovers therapeutic targets in the parasite *Schistosoma mansoni*. *Science*, *369*, 1649–1653.
Wang, Y., Stary, J. M., Wilhelm, J. E., & Newmark, P. A. (2010). A functional genomic screen in planarians identifies novel regulators of germ cell development. *Genes & Development*, *24*, 2081–2092.
Wang, Y., Zayas, R. M., Guo, T., & Newmark, P. A. (2007). *Nanos* function is essential for development and regeneration of planarian germ cells. *Proceedings of the National Academy of Sciences of the United States of America*, *104*, 5901–5906.
Wendt, G. R., Collins, J. N. R., Pei, J., Pearson, M. S., Bennett, H. M., Loukas, A., et al. (2018). Flatworm-specific transcriptional regulators promote the specification of tegumental progenitors in *Schistosoma mansoni*. *eLife*, 7, e33221.
Wendt, G., Zhao, L., Chen, R., Liu, C., O'Donoghue, A. J., Caffrey, C. R., et al. (2020). A single-cell RNA-seq atlas of *Schistosoma mansoni* identifies a key regulator of blood feeding. *Science*, *369*, 1644–1649.
Whitfield, P. J. (1979). *The biology of parasitism: An introduction to the study of associating organisms*. Baltimore: University Park Press.
Wilson, E. B. (1937). *The cell in development and heredity* (3rd ed.). New York: Macmillan Co.
Zayas, R. M., Bold, T. D., & Newmark, P. A. (2005). Spliced-leader trans-splicing in freshwater planarians. *Molecular Biology and Evolution*, *22*, 2048–2054.
Zayas, R. M., Cebrià, F., Guo, T., Feng, J., & Newmark, P. A. (2010). The use of lectins as markers for differentiated secretory cells in planarians. *Developmental Dynamics*, *239*, 2888–2897.
Zayas, R. M., Hernández, A., Habermann, B., Wang, Y., Stary, J. M., & Newmark, P. A. (2005). The planarian *Schmidtea mediterranea* as a model for epigenetic germ cell specification: Analysis of ESTs from the hermaphroditic strain. *Proceedings of the National Academy of Sciences of the United States of America*, *102*, 18491–18496.

CHAPTER THIRTEEN

Slipper snail tales: How *Crepidula fornicata* and *Crepidula atrasolea* became model molluscs

Deirdre C. Lyons[a,*] and Jonathan Q. Henry[b]
[a]U.C. San Diego, Scripps Institution of Oceanography, La Jolla, CA, United States
[b]Marine Biological Laboratory, Marine Resources Center, Woods Hole, MA, United States
*Corresponding author: e-mail address: d1lyons@ucsd.edu

Contents

Abstract

Despite the great abundance and diversity of molluscs, only a few have attained "model research organism" status. One of those species is the slipper snail *Crepidula fornicata*. Its embryos were first used for classical lineage tracing studies in the late 19th century, and over a 100 years later they were "re-discovered" by our labs and used for modern fate mapping, gene perturbation, in vivo imaging, transcriptomics, and the first application of CRISPR/Cas9-mediated genome editing among the Spiralia/Lophotrochozoa. Simultaneously, other labs made extensive examinations of taxonomy, phylogeny, ecology, life-history, mode of development, larval feeding behavior, and responses to the environment in members of the family Calyptraeidae, which includes the genus

Current Topics in Developmental Biology, Volume 147
ISSN 0070-2153
https://doi.org/10.1016/bs.ctdb.2021.12.013

Crepidula. Recently, we developed tools, resources, and husbandry protocols for another, direct-developing species, *Crepidula atrasolea*. This species is an ideal "lab rat" among molluscs. Together these species will be valuable for probing the cellular and molecular mechanisms underlying molluscan biology and evolution.

1. Introduction

Molluscs are abundant, diverse, and commercially valuable to humans, yet none of the more than 90,000 extant species serves as a *singular* molluscan model for biomedical research. Instead, several model species have been studied, with each possessing unique qualities useful for research (Davison & Neiman, 2021; Lesoway & Henry, 2019b; Wanninger & Wollesen, 2018). In one prime example, marine snails in the genus *Crepidula* have been used to study a wide variety of questions ranging from molecular and cell biology to ecology (Henry, Collin, & Perry, 2010; Henry & Lyons, 2016; Lesoway & Henry, 2019b). For example, *C. fornicata* development was first studied in the late 1800s for classical lineage tracing (Conklin, 1897). Over 100 years later it was found to be an excellent species for cellular and molecular studies, especially because reagents for modern fate-mapping and gene perturbation could be delivered to their embryos, and the subsequent phenotypes readily studied during development (Henry, Collin, et al., 2010; Henry, Perry, & Martindale, 2010; Henry & Lyons, 2016; Henry, Lesoway, et al., 2017). *Crepidula* have also been the subject of a wide array of "eco-evo-devo" studies over the last 50+ years due to the diversity of their life history strategies and modes of development (Henry, Collin, et al., 2010). In this review we first introduce members of this group of snails and discuss how and why two species in particular were selected for more intensive study. We highlight recent insights into molluscan biology that came from studies with various *Crepidula* spp., and conclude by proposing future areas for research including biomineralization, sex determination, lineage tracing in adults, and transgenesis.

2. Natural history of the genus *Crepidula*

At least 53 *Crepidula* species have been recognized, although reassignments of previously classified *Crepidula* species to new genera like *Bostrycapulus* and *Crepipatella* have occurred since the first molecular phylogenetic analyses by Rachel Collin (Smithsonian Tropical Research Institute) (Collin, 2003a, 2005; Collin, Chaparro, Winkler, & Véliz, 2007).

The motivation of Rachel's work was to understand the evolution of mode of development in marine invertebrates using a group for which interspecific variation in development was already well-documented. Calyptraeid gastropods were selected, as K. Elaine Hoagland had revised the taxonomy of *Crepidula* in the 1970s (Hoagland, 1977) and had subsequently documented significant diversity in development across the group (Hoagland, 1978, 1984, 1986).

Crepidula snails are members of the subclass Caenogastropoda, belonging to the family Calyptraeidae, and are colloquially referred to as "slipper snails" because of the shoe-like shape of their shells (Fig. 1). Other common names include: slipper limpet, lady's slipper, baby's boats, boat shell, and quarterdeck shell. The shells of *Crepidula* are relatively simple and flattened, and possess an internal shelf or "deck" that separates the foot from the viscera

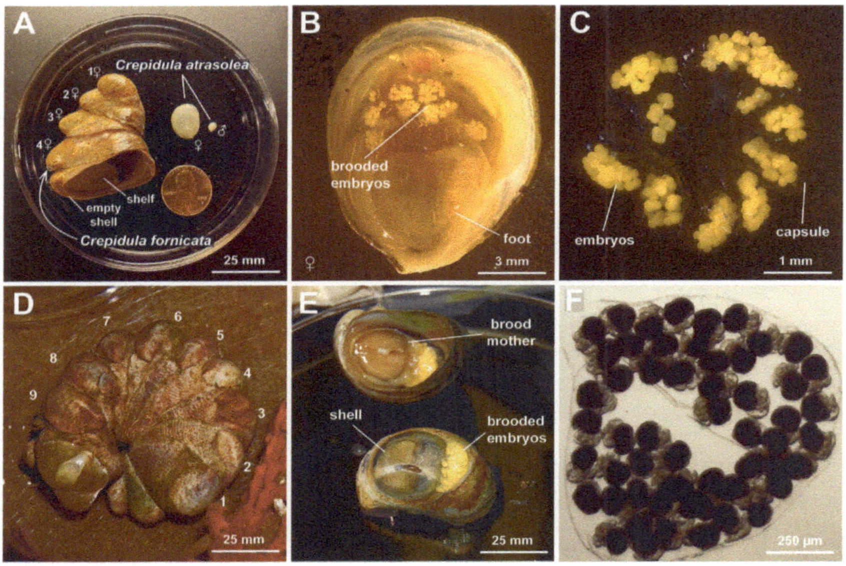

Fig. 1 Two species of *Crepidula* and their embryos. (A) Adults of *Crepidula fornicata* and *Crepidula atrasolea* reach different sizes at adulthood. For *C. fornicata*, groups of individuals stack on top of each other, whereas adults of *C. atrasolea* remain unstacked, but firmly attached to a substrate. Both species have males that will eventually become female over their lifetime. (B) Ventral view of a brooding adult female of *C. atrasolea*. (C) Clutch of brooded embryos of *C. atrasolea*. (D) Stack of *C. fornicata* adults numbered 1–9. (E) Brooded embryos of *C. fornicata* attached to the dorsal shell of another individual. Note: the mother who laid the brood is seen ventral side up in the background (at top of image). (F) Individual capsule containing embryos of *C. fornicata*.

(Fig. 1A, B, E; Henry, Collin, et al., 2010). In some species, individuals live together in curved stacks (Fig. 1A, D), while other species remain unstacked and firmly attached to a substrate by their foot (Fig. 1A and B). Taxonomy in this group can be challenging because of their relatively simple, plastic (influenced by environmental conditions) shells and the existence of cryptic species, though other characteristics like developmental mode and molecular signatures have been used to distinguish species (e.g., Collin, 2000).

Like other calyptraeids, adult slipper snails are sedentary filter feeders. They lift themselves slightly off the surface of the adjacent snail or other hard substrate beneath them, and use their gills to create currents that trap phytoplankton, which are transferred to the mouth and ingested. *Crepidula* are protandrous hermaphrodites, changing from male to female as the animal grows (Hoagland, 1977; Lesoway & Henry, 2019a). They live in temperate and tropical climates around the world in intertidal and shallow subtidal zones; perhaps because of their accessibility and wide distribution, various researchers work on a wide range of "local" species. Animals that settle and grow on the hulls of shipping vessels, or those that grow directly on oysters that are being transported for human consumption, have been introduced outside of their endemic ranges, making them excellent models for studying invasiveness (Moulin et al., 2007; Pechenik, Diederich, Browman, & Jelmert, 2017). Introduced populations can impact native species of *Crepidula* and damage local fisheries (e.g., oyster beds) by competing for habitat (Blanchard, 1997).

3. *Crepidula* are well-suited for studying the links between ecology, evolution, and development

Calyptraeid phylogeny is well-resolved and has been used to infer the direction of dispersal around the globe and the nature of speciation events (Collin, 2003a), as well as the evolution of modes of development (Collin, 2000, 2001, 2003a, 2003b, 2003c, 2004, 2019; Collin et al., 2007; Veliz, Winkler, Guisado, & Collin, 2012). The clade includes species with indirect development with planktonic feeding larvae, as well as direct developers, and also species that exhibit adelphophagy (cannibalism of one embryo by another) and poecilogony (intraspecific variation in developmental mode; McDonald, Collin, & Lesoway, 2014). This variation can be examined in light of the group's solid phylogeny. For example, the evolutionary loss of planktonic feeding larvae is generally thought to be irreversible, but mapping the modes of development onto the calyptraeid phylogeny suggested instead that there were at least three instances of re-gaining feeding larvae (Collin, 2004; Collin et al., 2007).

Within this framework, studies have also examined the (1) variation of feeding behaviors and structures (Collin, 2003; Collin & Moran, 2018; Collin, Shishido, Cornejo, & Lesoway, 2021), (2) the molecular differences that correlated with embryos that cannibalize sibling embryos vs those that are cannibalized (Lesoway, Abouheif, & Collin, 2016; Lesoway, Collin, & Abouheif, 2017), and (3) the relationship between life history variation and environmental factors (Carrillo-Baltodano & Collin, 2015; Collin & Salazar, 2010; Ly & Collin, 2021; Mérot & Collin, 2012). Other studies examined the way *Crepidula* species respond to stressors like low pH and nutritional reduction, to predict how marine species might respond to global climate change (Bogan, McMahon, Pechenik, & Pires, 2019; Maboloc & Chan, 2021; Maboloc & Chan, 2017; Reyes-Giler et al., 2021). Those studies measured the effects of such stressors on the rates of metamorphosis and larval growth in the F0 population, as well as measured the effects on the F1 generation (when the F0 have been exposed to stressful conditions) to identify multigenerational consequences of stress. The effects of microplastics on larval settlement and growth have also been examined (Lo & Chan, 2018).

4. Why did we choose *C. fornicata*?

We began using *C. fornicata* because of the advantages of working with their embryos, and because of the historical knowledge about their embryology that was gleaned over 125 years ago from the work of Edward Grant Conklin (1863–1952) at the Marine Biological Laboratory, Woods Hole, MA. Conklin was an American developmental biologist who studied cytological details of marine invertebrate embryos to understand developmental proximal causes–roughly akin to what we now call "mechanisms." Along with contemporaries like E.B. Wilson, F.R. Lillie, and T.H. Morgan, Conklin worked at a pivotal time in American biological research when natural history and morphological explanations of biology were unseated by approaches that used experimentation. As a graduate student with W. K. Brooks at Johns Hopkins University, Conklin traveled to the Marine Biological Laboratory to initially study siphonophores, but finding none there, he looked for other embryonic material and settled on *C. fornicata* and related species (Maienschein, 2015). Abundant on the east coast of

North America, *C. fornicata* (Linnaeus, 1758) lives in communal stacks that form characteristic arches (Fig. 1D), from which their species name is derived; *fornicata* comes from the Latin fornix, for arch. Conklin found that *C. fornicata* produced large broods (Fig. 1E and F) of synchronously-dividing embryos, making it easy to follow early cell divisions; nuclei and mitotic spindles were observable in fixed, stained, and sectioned specimens (Conklin, 1897; Steinert, 2016). The abundant material allowed Conklin to construct a developmental series and early embryonic cell lineage. Similar work in multiple mollusc and annelid species allowed Conklin, Wilson, Lillie and others the means to identify specific developmental cellular homologies and evolutionary relationships among animals with divergent body plans.

Conklin's meticulous studies of the early embryo led to his identification of a single cell that gave rise to mesoderm. That cell, which is born at the 25-cell stage, is now referred to as the 4d cell. Conklin recognized that it forms both endoderm and mesoderm, and that it is the first lineage in the embryo to exhibit bilateral symmetry. This finding was significant because Conklin's goal was to relate the early cell division of the egg to the formation of germ layers and axial properties of fully formed organs. Conklin made other important observations and contributions; though he nearly overlooked this, he reported that *C. fornicata* possesses a very small polar lobe, a transient cytoplasmic protrusion characteristic of some annelid and mollusc embryos, identifying another important similarity between the groups. His most enduring contribution may be his beautiful and detailed camera lucida drawings of developmental stages, and his improved spiralian nomenclature system that is still used in lineage studies today (Figs. 2 and 4).

5. How did evo-devo research with *C. fornicata* start up again?

After Conklin, no further lineage tracing studies used *C. fornicata* over the next century until Jonathan Henry's and Mark Martindale's work in the summer of 2004. They started working on *C. fornicata* rather accidentally–as Conklin had previously–when the animals they had planned to work with that summer in the Embryology course (taught at the Marine Biological Laboratory) were not gravid. Starting several years before that, Jon and Mark began a collaboration to apply modern lineage tracing techniques to less-studied invertebrate groups. Having first met in 1981 in graduate school at the University of Texas, Austin in the lab of Gary Freeman, they

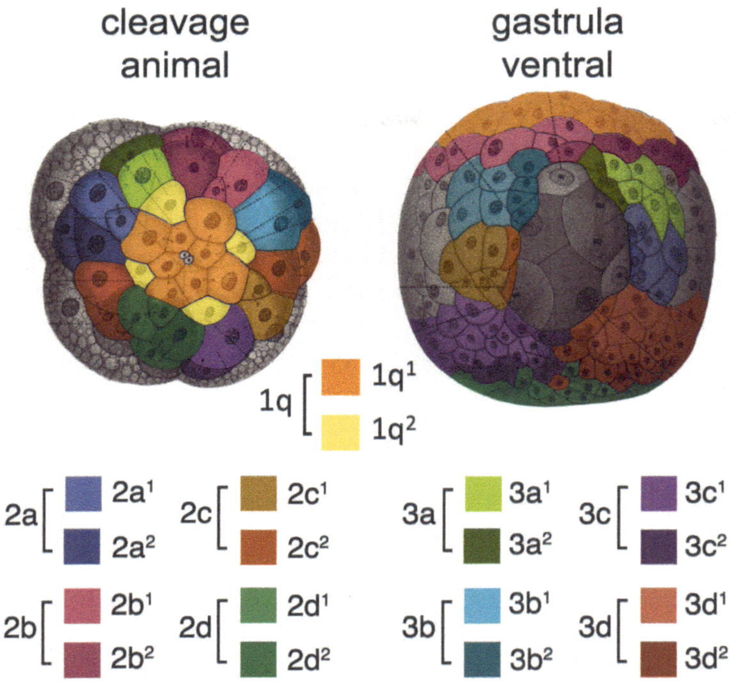

Fig. 2 *Crepidula fornicata* embryonic fate map. Conklin's (1897) original drawings, overlaid with our updated lineage-tracing results for the four founding quadrants, and the first three quartets of micromeres. *Data compiled from Hejnol, A., Martindale, M.Q., Henry, J.Q., 2007. High-resolution fate map of the snail Crepidula fornicata: The origins of ciliary bands, nervous system, and muscular elements.* Developmental Biology 305, *63–76. Henry, J.Q., Hejnol, A., Perry, K.J., Martindale, M.Q., 2007. Homology of ciliary bands in Spiralian Trochophores.* Integrative and Comparative Biology 47, *865–871. Lyons, D.C., Perry, K.J., Lesoway, M.P., Henry, J.Q., 2012. Cleavage pattern and fate map of the mesentoblast, 4d, in the gastropod Crepidula: A hallmark of spiralian development.* EvoDevo 3, *21. Lyons, D.C., Perry, K.J., Henry, J.Q., 2015. Spiralian gastrulation: Germ layer formation, morphogenesis, and fate of the blastopore in the slipper snail Crepidula fornicata.* EvoDevo 6, *24. Lyons, D.C., Perry, K.J., Henry, J.Q., 2017. Morphogenesis along the animal-vegetal axis: Fates of primary quartet micromere daughters in the gastropod Crepidula fornicata.* BMC Evolutionary Biology 17, *217.*

began work on experimental embryology and lineage tracing; this endeavor eventually turned into a collaboration that took-off during their respective postdoctoral years and continued after they received faculty positions. They promoted the use of a wide range of animals, including spiralians/lophotrochozoans (Henry, 2014), as important subjects for understanding the evolution of cell fate specification and axial patterning in metazoans (Henry & Martindale, 1998, 1999; Martindale & Henry, 1998).

Together they presented key findings about the regulative ability and fate map of nemerteans (Henry & Martindale, 1994, 1995, 1996; Martindale & Henry, 1995a), the cell lineage of acoels and polyclad turbellarians (Boyer, Henry, & Martindale, 1996; Henry, Martindale, & Boyer, 2000), the cell lineage of hemichordates (Henry, Tagawa, & Martindale, 2001), and the nature of inductive interactions, regulative development, regeneration, and the unique fate map of ctenophores (Martindale & Henry, 1995b, 1997), among others. Along with contemporaneous lineage tracing studies carried out on other spirally-cleaving organisms [including *Platynereis* (Ackermann, Dorresteijn, & Fischer, 2005), leeches (Stuart, Blair, & Weisblat, 1987; Weisblat & Shankland, 1985), and gastropod molluscs *Patella* and *Tritia (Ilyanassa)* (Damen & Dictus, 1994; Lambert, 2001; Render, 1991)] their work helped to revive an interest in using comparative lineage tracing to answer evo-devo questions in spiralians. Some of these fundamental questions remained the same as in Conklin's era, including: How are the organs and axial properties of larvae and adults related to the organization of the egg and early embryonic cells? How plastic (regulative) or fixed (mosaic) are cell lineages? How are cell lineages modified over the course of evolution to build disparate body-plans? On the other hand, the tools and intellectual framework had changed considerably; for example Conklin did not have antibiotics at his disposal and could not raise embryos outside of their capsules for very long, making experimentation difficult. Mark and Jon also had at their disposal high resolution microscopes, fluorescently tagged intra-cellular lineage tracers, and phylogenetic trees based on molecular characters, upon which one could map developmental character states and their transitions. As an undergraduate at Mount Holyoke College (MA), Dede Lyons read their papers, which sparked her interest in leveraging spiralians' homologous cell lineages to elucidate the mechanisms of developmental evolution (Lyons & Weisblat, 2009).

Like Conklin before them, Henry and Martindale found that *C. fornicata* embryos were excellent for lineage studies (Fig. 2). The large clutches of synchronously-developing eggs were easily freed from their capsules (Fig. 1E and F) with tweezers, and the embryos were "naked" inside, free of any additional maternal investments like fertilization envelopes. Thus, the embryos were amenable for microinjection and experimental manipulation. The relatively large zygotes (180 μm) proved easy to inject, as did the smaller daughter cells of the four basic embryonic quadrants A, B, C, D, and tiers of micromeres 1q through 4q (Figs. 1E and F, and 2). Jon and Mark, along with Andi Hejnol, leveraged these advantages to construct an initial modern cell lineage fate map for the major founding lineages in the embryo (Fig. 2; Hejnol, Martindale, & Henry, 2007; Henry, Hejnol, Perry, & Martindale, 2007). Jon and Mark also determined the developmental role of the small polar lobe that Conklin had first observed (Henry, Perry, & Martindale, 2006). *C. fornicata* produces two very small polar lobes during the first two rounds of cell division (Henry et al., 2006). They showed that the polar lobe usually fuses with the future D cell at the four-cell stage. Using a combination of lineage tracing, surgical removal of the polar lobe, blastomere isolations, and individual blastomere ablations, their experiments established that *C. fornicata* behaves like other "equal-cleaving" spiralian embryos and that the D-quadrant organizer, 4d (the cell that Conklin first identified as the mesentoblast), is conditionally specified from signals sent by the primary quartet micromeres. The "equal cleavage" nature of *Crepidula* development would prove to be an important factor for comparing developmental mechanisms with another well-studied caenogastropod, *Tritia (Ilyanassa) obsoleta*, which has larger polar lobes (Chan & Lambert, 2014; Lambert, Johnson, Hudson, & Chan, 2016; Rabinowitz, Chan, Kingsley, Duan, & Lambert, 2008). Jon and Mark continued working with *C. fornicata* embryos in subsequent summers, resulting in a detailed fate map, and in examinations of the developmental role of signaling pathways including B-catenin (Henry, Perry, & Martindale, 2010). Inspired by the work in *Tritia (Ilyanassa)* (Lambert & Nagy, 2001, 2003), Jon and his longtime collaborator, Kim Perry, also studied the role of MAPK signaling in organizer specification (Henry, Collin, et al., 2010; Henry & Lyons, 2016; Henry & Perry, 2008).

6. Growing the *Crepidula* evo-devo community

While Jon and Mark carried out their work at the MBL and taught in the Embryology course, they helped train a generation of students on embryological manipulation, intracellular injection, and comparative development. In earlier years, injected material would often be fixed and brought home to their labs at the University of Illinois or the University of Hawaii for further examination and documentation. This strategy was successful for some studies (the bulk of the images in their first *C. fornicata* lineage paper were taken by Andi Hejnol from fixed samples); however, more detailed lineage tracing experiments needed to be documented closer to the time of injection, or preferably on live samples, because the tracer signals tended to fade over time. When Dede took the Embryology course in the summer of 2009, Jon and Mark invited her to image live specimens as part of a study to determine the fate map of 4d sublineages. Jon invited Dede back in subsequent summers as a teaching assistant for the Embryology course to help complete the 4d lineage fate map (Lyons, Perry, Lesoway, Henry, 2012). This project grew into a long-term collaboration resulting in an series of detailed sub-lineage fate maps and live-imaging studies of morphogenesis (Henry & Lyons, 2016; Lyons & Henry, 2014; Lyons, Perry, Batzel, & Henry, 2020; Lyons, Perry, Henry, 2015, 2017; Lyons et al., 2015, 2017), and other studies examining the expression patterns of regulatory genes (Perry et al., 2015).

Through interactions with other students and visitors to the MBL and to the Embryology course, Jon encouraged and supported a number of other junior researchers to use *C. fornicata*, including Cristina Grande (Universidad Autónoma de Madrid, Spain) and her student Marta Truchado-Garcia (U.C. Berkeley, CA), Maryna Lesoway (U.C. San Diego, CA), Victoria Sleight (University of Aberdeen, Scotland), Nicole Webster (University of Saskatchewan, Canada), and others. This community continues to grow. Summaries of work through 2016 can be found in Henry and Lyons (2016). In the next sections we highlight several recent studies using *C. fornicata*.

7. Evolution of the spiralian embryonic organizer

Spiralians are an excellent group for examining the evolution of embryonic axial patterning (Lambert, 2008, 2010). The spiral cleavage program is conserved enough that individual cells can be compared between species, yet there is also enough variation in cleavage patterns and developmental mechanisms that one can use them to understand how axial patterning evolved (Hejnol, 2010; Henry, 2014; Lambert, 2010; Martín-Durán & Marlétaz, 2020). For example, experimental embryology studies revealed that the embryonic organizer is a different D-quadrant cell in *C. fornicata* (4d; Henry, Lyons, Perry, & Osborne, 2017; Henry, Lesoway, et al., 2017; Henry et al., 2006), mudsnail *Tritia (Illyanasa) obsoleta* (3D; Lambert, 2001; Lambert & Nagy, 2001), and the polychaete annelid *Capitella teleta* (2d; Amiel, Henry, & Seaver, 2013). The fact that organizer activity resides in cells of different birth order in these species provides a unique opportunity for comparative studies of developmental mechanisms.

As noted above, even though *C. fornicata* produces a small polar lobe, the specification of the 4d organizer is more similar to that of equal-cleaving species (e.g., *Lymnaea*, Martindale, 1986; Martindale, Doe, & Morrill, 1985) than to unequal-cleaving species (examples of the latter include *Tritia obsoleta* and *Capitella teleta*). Similar to the results found in experiments with equal-cleaving spiralian embryos (Martindale, 1986; Martindale et al., 1985; van den Biggelaar, 1977; van den Biggelaar & Guerrier, 1979), cell ablation experiments in *C. fornicata* revealed that the first-quartet micromeres (cells 1q and daughters 1q1) are necessary for organizer specification (Henry, Lyons, et al., 2017; Henry, Lesoway, et al., 2017). If these cells are deleted prior to the birth of the organizer (4d), the resulting embryos lack a dorsal-ventral axis, indicating misspecification of the D quadrant. Therefore, the D quadrant is specified by an inductive signal from these micromeres sometime between 5th (20-cell stage) and 6th (24-cell stage) cleavage (Henry, Lyons, et al., 2017, Henry, Lesoway, et al., 2017). Furthermore, there is evidence that the micromeres must themselves be primed, by even earlier signaling (Henry & Perry, 2008). This knowledge about the timing of D quadrant specification, and of organizer cell birth, makes *Crepidula* a valuable species to better understand and dissect these early developmental processes among spiralians, at single-cell resolution.

8. Axial patterning and the primary quartet micromeres

We carried out other recent studies focused on the morphogenesis and molecular specification of the primary quartet micromeres (1q or 1a-1d; Fig. 2). The primary quartet micromeres were the last major cell lineages of the embryo that we fate-mapped in detail (Lyons et al., 2017). The classical spiralian lineage tracing literature assumes that 1q2 (1a2-1d2) daughter cells of the primary quartet micromeres universally give rise to the prototroch (larval feeding organ). On the contrary, we found that the 1q2 cells in fact never contribute to the prototroch, and instead actually make a provisional epithelium that might allow the embryo to undergo morphogenesis—a sort of "bending" movement—during gastrulation stages. This work made several additional seminal contributions to the spiralian development field, including proposing a new model for how the presumptive mouth cells become displaced to the animal territory during development and how the anterior neural tissues rearrange during embryonic stages.

In another study, where we examined the role of BMP signaling in early *C. fornicata* development, we found that instead of having an obvious role in specification of the organizer itself, the BMP pathway mediates the proper "relay" of organizer signaling to specify cells within the 1q1 sublineages, which give rise to the eyes and the central nervous system (Lyons et al., 2020). The fact that we found BMP signaling to be dispensable for organizer specification and initial organizer signaling is significant: a series of recent papers examining the role of BMP in organizer function in *Tritia* (Lambert et al., 2016), *Capitella* (Lanza & Seaver, 2018; Webster, Corbet, Sur, & Meyer, 2021), and *Lottia* (Tan, Huan, & Liu, 2022) have found widely differing results, despite the highly conserved spiral cleavage program. It will be interesting to examine the mechanisms of BMP signaling in more depth in *C. fornicata*, as well as to identify the molecular nature of the organizer.

9. Unique sources of mesoderm

One interesting aspect of Spiralian development is that they have two different sources of mesoderm (Lyons & Henry, 2014). Some mesoderm is derived from an endodermal precursor as "endomesoderm." Endomesoderm is believed to represent the ancestral source of mesoderm within the Metazoa. In many spiralians, such as *C. fornicata* (but not all cases), endomesoderm arises from the fourth quartet daughter of the D quadrant, 4d (the "mesentoblast" cell, Lyons et al., 2012). This source gives rise to some endodermal tissue, such as the intestine, as well as many larval and adult mesodermal tissues, including: various muscles (e.g., the retractor muscle, foot muscles, etc.), the kidney complex, the heart, and primordial germ cells. A second source arises from ectodermal precursors as "ectomesoderm." The combination of ectomesodermal cells varies between different spiralian species, but typically they arise from ventral daughters of the 2nd and/or 3rd quartet micromeres (Fig. 2). In *Crepidula*, these cells arise specifically from the 3a2 and 3b2 cells. Ectomesodermal progeny undergo EMT as they leave the edge of the blastopore (Lyons et al., 2015). These cells contribute to the formation of various tissues, including the larval retractor muscle, the radial muscles in the larval velum, as well as scattered mesenchyme, and also pigment cells in the velum, in addition to a few unidentified progeny associated with the apical organ (J. Henry and K. Perry, Unpublished data). Little is known about the evolutionary origins of this alternate source of mesoderm and there are many unanswered questions. Do endo- and ectomesoderm share a common embryonic origin?

Did ectomesoderm arise by co-opting the same ancestral endomesodermal gene regulatory network (GRN) (Perry et al., 2015)? Alternatively, these two sources of mesoderm may be regulated by completely different GRNs. Furthermore, one can ask what genes regulate the process of EMT in the ectomesodermal progeny, and is this basic behavior regulated via a conserved EMT GRN (Osborne, Perry, Shankland, & Henry, 2018)? While there are some other examples of so-called ectomesenchyme that also undergo EMT, such as the neural crest in vertebrates, ectomesoderm appears to represent a spiralian novelty.

10. Shell development and biomineralization

Molluscs are well known for diversification of their shell structures (Kocot, Aguilera, McDougall, Jackson, & Degnan, 2016). The diversity is the result of organic extracellular matrix proteins (shell matrix proteins) directing the growth of mineral structures during larval and adult stages (Clark, 2020; Kocot et al., 2016). Shell matrix proteins (SMPs) have been identified by proteomics in many molluscs, but few have been studied in detail (Jackson & Degnan, 2016). *C. fornicata* is a good model for studying biomineralization at a mechanistic level. Intracellular fate-mapping showed that all 8 second-quartet daughter cells (2a1/2a2, 2b1/2b2, 2c1/2c2, 2d1/2d2; Fig. 2) contribute to the shell gland (Hejnol et al., 2007; Lyons et al., 2015). Screens for transcription factors and signaling pathway components led to the identification of at least 24 genes that are expressed in the *C. fornicata* shell gland, which makes the larval shell (Figs. 3 and 4; Perry et al., 2015; Osborne et al., 2018; Lyons et al., 2020; Truchado-Garcia, Caccavale, Grande, & D'Aniello, 2021). Using proteomics, we identified 185 SMPs occluded within the adult shell of *C. fornicata* (G. Batzel et al., in review). Differential gene expression analysis identified 20 SMP genes that are differentially expressed in the shell-producing mantle tissue (Figs. 3A and C, and 4A–C). Through in-situ hybridization we identified two genes, CfSMP1 and CfSMP2, that are expressed exclusively in the shell gland (Fig. 4D–F). These data reveal that the two most differentially expressed SMPs in the adult are also expressed in the shell gland. With this new knowledge about the terminal differentiation "nodes" (=SMPs), we can now carry out perturbation experiments of upstream "inputs" (=previously identified transcription factors) to build an experimentally-verified biomineralization GRN in a mollusc.

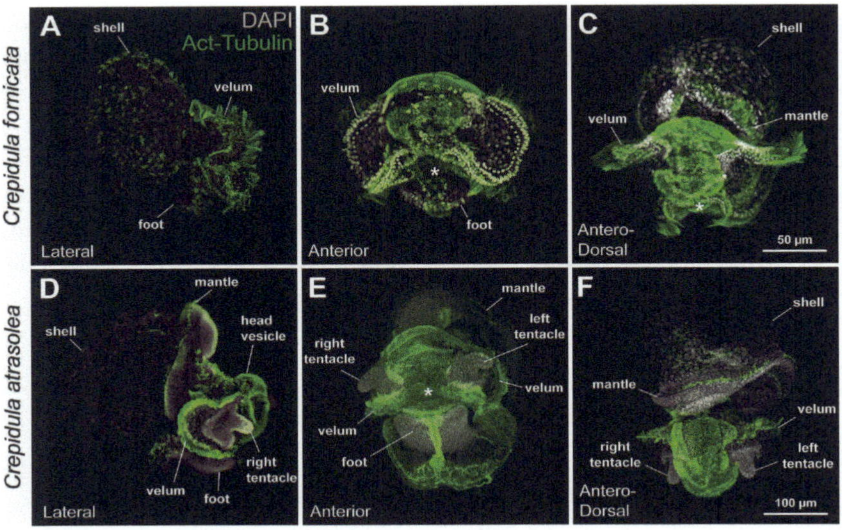

Fig. 3 Acetylated tubulin (Act-Tubulin) immunoreactivity in veliger larvae of *Crepidula fornicata* and *Crepidula atrasolea*. Green–Cilia, Acetylated Tubulin; Gray–Nuclei, DAPI. (A–C) Different views of a veliger larva of *C. fornicata*, with Act-tubulin immunoreactivity primarily found in cilia of the velum and the mantle. (D–F) Acetylated tubulin immunoreactivity in a veliger larva of *C. atrasolea*. Acetylated tubulin immunoreactivity is present in cilia of the velum and mantle edge. Note the differences in larval features between the two species, especially in regards to a reduced larval velum, presence of left and right tentacles, pronounced mantle cavity, and difference in overall veliger size in *C. atrasolea* compared to *C. fornicata*.

11. Sex determination

All calyptraeid gastropods are sequential hermaphrodites, meaning they transition from one sex to the other. As male-first (protandric) hermaphrodites, calyptraeids mature first to have an external penis and internal testis with sperm, before transitioning to having female external genitalia (female genital papilla) and producing oocytes. Sex determination in calyptraeids is environmentally determined, and the timing of sexual transitions is controlled by several factors including age, size, nutritional state, and social environment (reviewed in Lesoway & Henry, 2019a). Typically, males will delay their transition, when in the presence of females. In the absence of females, males change sex earlier and at smaller sizes (Lesoway & Henry, 2019a). Some species are less influenced by their social environment, which appears to correlate with the level of gregariousness of a particular species (Hoagland, 1978).

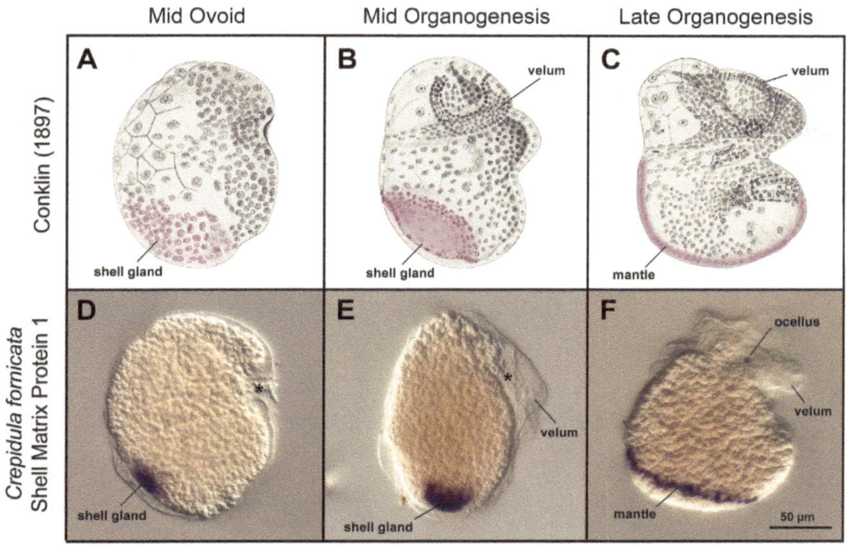

Fig. 4 Expression of *Crepidula fornicata* Shell Matrix Protein 1 (CfSMP1) during larval shell gland development. (A–C) Drawings of embryos (160–228 hpf) undergoing larval shell gland development. Purple shading indicates the cells that Conklin originally labeled as "shell gland" cells or "shell gland edge" cells in his original publication. (D–F) CfSMP1 expression in embryos of a similar stage as panels (A–C). CfSMP1 is found exclusively expressed in the shell gland (panels D–E) during shell gland development, and later (panel C) in the mantle edge in early veliger larvae. *Panels (A–C) modified from Conklin, E.G., 1897. The embryology of Crepidula: A contribution to the cell lineage and early development of some marine Gasteropods. Ginn.*

While the ecological and evolutionary consequences of sex determination in calyptraeids is of long-standing research interest (e.g., Collin, 1995, 2007), knowledge of the molecular underpinnings has lagged. Previous experiments implicated the left pedal ganglion in inducing male development (Le Gall, 1981; Le Gall & Streiff, 1975), including a partial characterization of a potential neuroendocrine peptide that triggers cell division in the gonad Le Gall, Feral, Lengronne, and Porchet (1987). Recent work has shown that targeting specific components of the retinoic acid pathway including retinoic acid receptors (RAR) and retinoid X receptors (RXR) induces penis development in juveniles, and blocking RXR reduces penis length (Lesoway & Henry, 2021). That study showed that both agonists and an antagonist of the RAR receptor increase penis length, suggesting that RAR may function in inhibitor release. Furthermore, exposure to tributyltin (TBT) resulted in high rates of penis induction in females of four different calyptraeid species. Interestingly, females of the small-bodied, solitary species *C. convexa* showed reduced rates of penis induction, correlating with reduced social control over sex determination. Interactions between levels of social control and specific pathways, including initiation of sexual transitions, remain areas for future exploration. More broadly, hermaphroditism, both sequential and simultaneous, has evolved repeatedly in gastropods and other molluscs (Lesoway & Henry, 2019a). Detailed knowledge of shared and divergent mechanisms will play an important role in elucidating evolutionary patterns of sexual development and sex determination in these and other animals.

12. *Crepidula atrasolea*: A complementary model system

While *C. fornicata* is a useful model system for study, another related *Crepidula* species has proved to be even more convenient. Working with *C. fornicata* at the Marine Biological Laboratory and at the University of Illinois, Jon and Kim Perry showed that they could tag endogenous Beta-catenin protein by CRISPR-mediated homology-directed repair (Perry & Henry, 2015). That study demonstrated for the first time the effectiveness of CRISPR-Cas9 mediated gene editing in a representative of the spiralian Lophotrochozoa, setting the stage for the development of transgenic lines carrying fluorescent reporters to study cell and developmental processes.

The F0s generated in the Perry and Henry (2015) study were not raised through metamorphosis to establish a stable line; the life cycle of *C. fornicata* is rather long (at least 1 year), and even if they wanted to raise animals in the lab, they lacked the necessary culturing infrastructure. Therefore, the present authors decided that it was necessary to develop an additional *Crepidula* species that could be easily reared in inland-aquaria throughout its entire life cycle, and one that would have a shorter generation time. Earlier, Rachel Collin had suggested that Jon and Mark use *C. atrasolea* (Figs. 1A–C and 3D–F), which she had recently described (Collin, 2000). This species is native to the warm waters of the southern Atlantic, and could be readily collected near the Smithsonian Marine Station (SMS) at Fort Pierce, Florida. It was first detected as a cryptic species that had previously been lumped with *Crepidula plana* by Elaine Hoagland, who examined the two with allozymes and described developmental differences (Hoagland, 1984, 1986), and who referred to it as *Crepidula* cf. *plana* without formally naming the species. Unlike *C. fornicata*, the larger embryos (280 μm) of *C. atrasolea* are direct-developing and hatch out as "crawl-away" juveniles (Fig. 3D–F). The adults do not form stacks, and grow to an adult size of about 1–2 cm (Fig. 1A–C). Importantly, their generation time is much shorter than that of *C. fornicata* at 4–6 months in captivity (Henry, Lesoway, & Perry, 2020; Henry, Lesoway, et al., 2017), comparable to the generation times of *Danio rerio* and *Xenopus tropicalis* (Showell & Conlon, 2009).

In the winter of 2015 Dede first collected *C. atrasolea* specimens from the waters around the Fort Pierce Marine station, and sent them to Jon's lab to rear. Jon made a trip to Fort Pierce in 2016 and supplemented the colony with additional specimens. All of the animals now in culture are derived from the animals collected during those two trips. Back in Illinois, Jon, Kim, and Maryna Lesoway, who was then a postdoc in Jon's lab, developed methods for raising the collected animals through their entire life cycle, described their development, made developmental transcriptomes, and ultimately developed methods for raising them in an automated, recirculating rack system (Henry et al., 2020; Henry, Lesoway, et al., 2017). With the implementation of automated aquatic systems (see below), it has become much easier to raise lots of animals through successive generations. We are now rearing animals in the 7th generation.

13. Development of an automated rack system to culture *Crepidula*

A major challenge is the amount of effort needed to rear marine animals in the laboratory, especially when one is rearing animals through successive generations and plans to maintain multiple inbred and transgenic lines. Recently, we succeeded in developing a fully automated marine aquatic rack system to rear *Crepidula atrasolea* through successive generations (Henry et al., 2020). This multi-tank system takes care of monitoring and adjusting all critical water quality parameters and also accomplishes automated feeding of commercially available liquid phytoplankton food (Fig. 5). Furthermore, the filtration system is designed to process waste and greatly minimize the need to carry out water changes. This aquatic system has decreased the overall amount of effort required to care for animals on a day to day basis and has allowed us to dramatically scale up production. It was instrumental in allowing us to maintain these animals throughout the Covid 19 pandemic with minimal hands-on effort. This system, and the

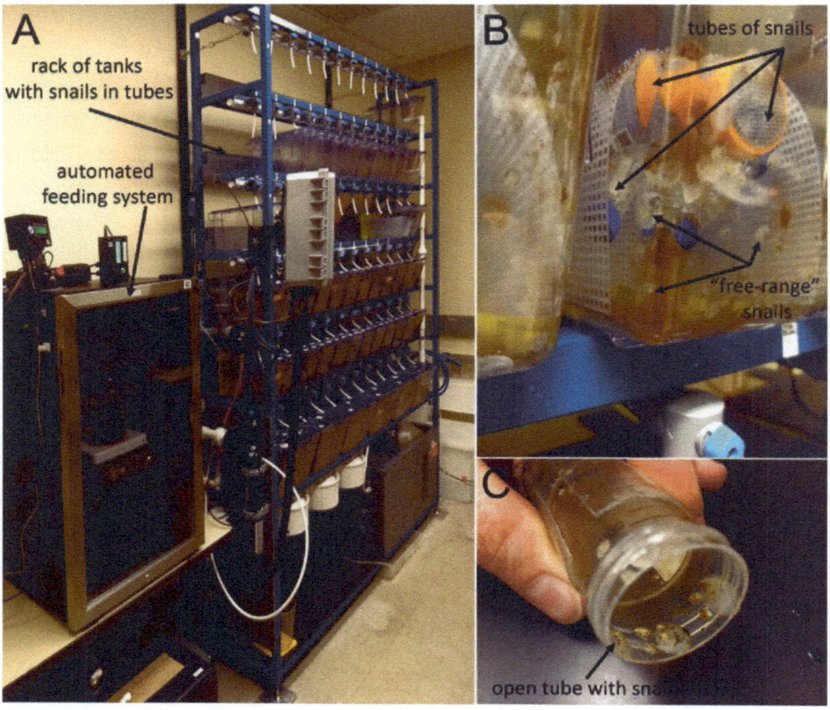

Fig. 5 Automated rack system to culture *Crepidula atrasolea.* (A) Recirculating rack with tanks, Iwaki controller, and automatic feeding apparatus. (B) Close up of individual tank, holding tubes of embryos, as well as individual snails. (C) Individual tube opened up to reveal a mating colony of snails. Details of the system can be found in Henry, Lesoway, et al. (2017).

protocols we have developed, will be of broad use to those working with different marine organisms. Jon Henry has recently moved to the Marine Biological Laboratory in Woods Hole, MA to develop additional systems and related tools to culture other aquatic species.

14. What's on the horizon for *Crepidula* research?

One of our highest priorities is sequencing and assembling the genomes for both species. While the *C. fornicata* genome is slated to be sequenced by the Darwin Tree of Life project (https://www.darwintreeoflife.org/), we are leading the effort to sequence the genome of *C. atrasolea.* Not only will these genome sequences form critical backbones for supporting other genomics studies like ATAC-seq, single cell

RNA-seq, and spatial transcriptomics, they will provide a valuable opportunity for comparative genomics. For example, we are eager to identify evolutionarily conserved regions within putative promoter sequences, and use them for making knock-in lines of *C. atrasolea*. With stable lines of transgenic animals expressing cell type-specific fluorescent proteins, or photoconvertible proteins like kaede, we could extend our lineage tracing studies to later stages including in the adult. To our knowledge, as of our writing, no lineage tracing studies linking embryonic cell lineages to adult organs have been published in molluscs. With the ability to undertake long-term lineage tracing, many interesting questions about adult molluscan biology could be addressed, including but not limited to: (1) how the shell gland tissue differentiates into the adult mantle tissue organ, (2) how the adult brain forms and grows, (3) how the immune system matures and responds to environmental changes, (4) how the primordial germ cells and gametes develop, and (5) how regeneration of adult tissues is regulated. The cellular and molecular underpinnings of these processes will be studied with gene perturbation studies, especially CRISPR/Cas9 genome editing, ideally with Cas9 being driven by inducible or tissue-specific promoters to perturb gene function in time and space. Access to a fast-growing, laboratory model like *C. atrasolea* enhances our ability to explore these late-stage developmental questions, which we hope will shed light on molluscan biology in general.

15. Join #TeamCrepidula

We encourage more people to use *Crepidula* species for their research and/or teaching. *C. fornicata* can be ordered from the Marine Biological Laboratory's Marine Resource Center. *Crepidula atrasolea* can be requested by emailing Dede Lyons. Our papers describe the methods for working with the adult and embryonic material, as well as accession numbers for transcriptomic data for *C. fornicata* and *C. atrasolea* (Henry, Collin, et al., 2010; Henry & Lyons, 2016; Henry, Lesoway, et al., 2017; Henry, Perry, Fukui, & Alvi, 2010; Truchado-Garcia, Harland, & Abrams, 2018). Additional information about rearing animals in closed aquaria, and collecting, handling, and raising embryos can be found in the supplement to the Henry, Collin, et al. (2010) paper and on Jon Henry's website (https://www.life.illinois.edu/henry/tools.html, see also https://github.com/HenryLabUIUC). Rachel Collin's lab has produced an excellent series of videos describing how to collect and work with marine invertebrates, including *Crepidula* spp. (https://www.youtube.com/user/CollinLabPanama/videos). We welcome anyone interested in joining the larger community working on molluscs and spiralians to follow the growing "SpiraliaBase" community (https://www.spiraliabase.org/; @SpiraliaBase), where many labs share their latest results and "hacks" for working with spiralian animals and their genomes.

Acknowledgments

We thank current lab members—especially Grant Batzel, Stephanie Neal, Hereroa Johnston, Maryna Lesoway, and Rebecca Lopez-Anido—for contributing ideas, images, and editing for this manuscript. We especially thank Kim Perry for her many years of work and collaboration on much of the research reviewed here. We thank Elaine Seaver for reminding J.Q. Henry and M.Q. Martindale that *C. fornicata* was readily available in the Marine Resource Center at the MBL. We acknowledge many Directors of the Marine Biological Laboratory's Embryology course (Drs. Joel Rothman, Richard Harland, Lee Niswander, Nipam Patel, Richard Behringer, Alejandro Sanchez-Alverado, Richard Schneider, David Sherwood, and Carole LaBonne) who deserve perennial thanks for their continued support in hosting us over so many summers. D.C. Lyons is indebted to Leslie Babonis and Seabird McKeon for helping to collect *C. atrasolea*, and to Dr. David R. McClay for his support. This work was supported by National Science Foundation grants IOS-1558061 and IOS-1827533 to J.Q. Henry (J.J. Henry). This work was also supported by National Science Foundation CAREER grant IOS-1943606 and National Institute of Health MIRA grant R35-GM133673 to D.C. Lyons. Lastly, we acknowledge the inspiration provided by Dr. Sebastian Beroe.

References

Ackermann, C., Dorresteijn, A., & Fischer, A. (2005). Clonal domains in postlarval Platynereis dumerilii (Annelida: Polychaeta). *Journal of Morphology*, *266*, 258–280.

Amiel, A. R., Henry, J. Q., & Seaver, E. C. (2013). An organizing activity is required for head patterning and cell fate specification in the polychaete annelid Capitella teleta: New insights into cell–cell signaling in Lophotrochozoa. *Developmental Biology*, *379*, 107–122.

Blanchard, M. (1997). Spread of the slipper limpet *Crepidula fornicata* (L. 1758) in Europe. Current state dans consequences. *Scientia Marina*, *61*, 109–118.

Bogan, S. N., McMahon, J. B., Pechenik, J. A., & Pires, A. (2019). Legacy of multiple stressors: Responses of gastropod larvae and juveniles to ocean acidification and nutrition. *The Biological Bulletin*, *236*, 159–173.

Boyer, B. C., Henry, J. Q., & Martindale, M. Q. (1996). Dual origins of mesoderm in a basal Spiralian: Cell lineage analyses in the Polyclad Turbellarian Hoploplana inquilina. *Developmental Biology*, *179*, 329–338.

Carrillo-Baltodano, A., & Collin, R. (2015). Crepidula slipper limpets Alter sex change in response to physical contact with conspecifics. *The Biological Bulletin*, *229*, 232–242.

Chan, X. Y., & Lambert, J. D. (2014). Development of blastomere clones in the Ilyanassa embryo: Transformation of the spiralian blastula into the larval body plan. *Development Genes and Evolution*, *224*, 159–174.

Clark, M. S. (2020). Molecular mechanisms of biomineralization in marine invertebrates. *The Journal of Experimental Biology*, *223*(Pt 11), jeb206961. https://doi.org/10.1242/jeb.206961.
Collin, R. (1995). Sex, size, and position: A test of models predicting size at sex change in the protandrous gastropod *Crepidula fornicata*. *The American Naturalist*, *146*, 815–831.
Collin, R. (2000). Phylogeny of the Crepidula plana (Gastropoda: Calyptraeidae) cryptic species complex in North America. *Canadian Journal of Zoology*, *78*, 1500–1514.
Collin, R. (2001). The effects of mode of development on phylogeography and population structure of North Atlantic Crepidula (Gastropoda: Calyptraeidae). *Molecular Ecology*, *10*, 2249–2262.
Collin, R. (2003). Worldwide patterns in mode of development in calyptraeid gastropods. *Marine Ecology Progress Series*, *247*, 103–122.
Collin, R. (2003a). Phylogenetic relationships among calyptraeid gastropods and their implications for the biogeography of marine speciation. *Systematic Biology*, *52*, 618–640.
Collin, R. (2003b). The utility of morphological characters in gastropod phylogenetics: An example from the Calyptraeidae. *Biological Journal of the Linnean Society. Linnean Society of London*, *78*, 541–593.
Collin, R. (2003c). Evolution of mode of development in Crepidula (Gastropoda: Calyptraeidae): Causes and consequences. *Marine Ecology Progress Series*.
Collin, R. (2004). Phylogenetic effects, the loss of complex characters, and the evolution of development in calyptraeid gastropods. *Evolution*, *58*, 1488–1502.
Collin, R. (2005). Crepidula badisparsa sp. nov.(Gastropoda: Calyptraeidae) from Bocas del Toro Province, Panama. *Caribbean Journal of Science*, *41*, 269–276.
Collin, R. (2007). Sex ratio, life-history invariants, and patterns of sex change in a family of protandrous gastropods. *Evolution*, *60*, 735–745.
Collin, R. (2019). Calyptraeidae from the Northeast Pacific (Gastropoda: Caenogastropoda). *Zoosymposia*, *13*, 107–130.
Collin, R., Chaparro, O. R., Winkler, F., & Véliz, D. (2007). Molecular phylogenetic and embryological evidence that feeding larvae have been reacquired in a marine gastropod. *The Biological Bulletin*, *212*, 83–92.
Collin, R., & Moran, A. (2018). Evolutionary transitions in mode of development. In T. J. Carrier, A. M. Reitzel, & A. Heyland (Eds.), *Evolutionary ecology of marine invertebrate larvae* (pp. 50–66). Oxford University Press. Chapter 4.
Collin, R., & Salazar, M. Z. (2010). Temperature-mediated plasticity and genetic differentiation in egg size and hatching size among populations of Crepidula (Gastropoda: Calyptraeidae). *Biological Journal of the Linnean Society. Linnean Society of London*, *99*, 489–499.
Collin, R., Shishido, C. M., Cornejo, A. J., & Lesoway, M. P. (2021). Ancestral form and function of larval feeding structures are retained during development of non-planktotrophic gastropods. *The International Journal of Developmental Biology*. https://doi.org/10.1387/ijdb.200154rc.
Conklin, E. G. (1897). *The embryology of Crepidula: A contribution to the cell lineage and early development of some marine Gasteropods*. Ginn.
Damen, P., & Dictus, W. J. (1994). Cell lineage of the prototroch of Patella vulgata (Gastropoda, Mollusca). *Developmental Biology*, *162*, 364–383.
Davison, A., & Neiman, M. (2021). Mobilizing molluscan models and genomes in biology. *Philosophical Transactions of the Royal Society of London. Series B, Biological Sciences*, *376*, 20200163.
Hejnol, A. (2010). A twist in time—The evolution of spiral cleavage in the light of animal phylogeny. *Integrative and Comparative Biology*, *50*, 695–706.
Hejnol, A., Martindale, M. Q., & Henry, J. Q. (2007). High-resolution fate map of the snail Crepidula fornicata: The origins of ciliary bands, nervous system, and muscular elements. *Developmental Biology*, *305*, 63–76.

Henry, J. Q. (2014). Spiralian model systems. *The International Journal of Developmental Biology*, *58*, 389–401.
Henry, J. J., Collin, R., & Perry, K. J. (2010). The slipper snail, Crepidula: An emerging lophotrochozoan model system. *The Biological Bulletin*, *218*, 211–229.
Henry, J. Q., Hejnol, A., Perry, K. J., & Martindale, M. Q. (2007). Homology of ciliary bands in Spiralian Trochophores. *Integrative and Comparative Biology*, *47*, 865–871.
Henry, J. Q., Lesoway, M. P., & Perry, K. J. (2020). An automated aquatic rack system for rearing marine invertebrates. *BMC Biology*, *18*, 46.
Henry, J. Q., Lesoway, M. P., Perry, K. J., Osborne, C. C., Shankland, M., & Lyons, D. C. (2017). Beyond the sea: Crepidula atrasolea as a spiralian model system. *The International Journal of Developmental Biology*, *61*, 479–493.
Henry, J. Q., & Lyons, D. C. (2016). Molluscan models: Crepidula fornicata. *Current Opinion in Genetics & Development*, *39*, 138–148.
Henry, J. Q., Lyons, D. C., Perry, K. J., & Osborne, C. C. (2017). Establishment and activity of the D quadrant organizer in the marine gastropod Crepidula fornicata. *Developmental Biology*, *431*, 282–296.
Henry, J. Q., & Martindale, M. Q. (1994). Establishment of the dorsoventral axis in nemertean embryos: Evolutionary considerations of spiralian development. *Developmental Genetics*, *15*, 64–78.
Henry, J. Q., & Martindale, M. Q. (1995). The experimental alteration of cell lineages in the nemertean Cerebratulus lacteus: Implications for the precocious establishment of embryonic axial properties. *The Biological Bulletin*, *189*, 192–193.
Henry, J. Q., & Martindale, M. Q. (1996). The establishment of embryonic axial properties in the nemertean, Cerebratulus lacteus. *Developmental Biology*, *180*, 713–721.
Henry, J. Q., & Martindale, M. Q. (1998). Evolution of cleavage programs in relationship to axial specification and body plan evolution. *The Biological Bulletin*, *195*, 363–366.
Henry, J. J., & Martindale, M. Q. (1999). Conservation and innovation in spiralian development. *Hydrobiologia*, *402*, 255–265.
Henry, J. Q., Martindale, M. Q., & Boyer, B. C. (2000). The unique developmental program of the acoel flatworm, Neochildia fusca. *Developmental Biology*, *220*, 285–295.
Henry, J. J., & Perry, K. J. (2008). MAPK activation and the specification of the D quadrant in the gastropod mollusc, Crepidula fornicata. *Developmental Biology*, *313*, 181–195.
Henry, J. J., Perry, K. J., Fukui, L., & Alvi, N. (2010). Differential localization of mRNAs during early development in the mollusc, Crepidula fornicata. *Integrative and Comparative Biology*, *50*, 720–733.
Henry, J. Q., Perry, K. J., & Martindale, M. Q. (2006). Cell specification and the role of the polar lobe in the gastropod mollusc Crepidula fornicata. *Developmental Biology*, *297*, 295–307.
Henry, J. Q., Perry, K. J., & Martindale, M. Q. (2010). β-Catenin and early development in the gastropod, Crepidula fornicata. *Integrative and Comparative Biology*, *50*, 707–719.
Henry, J. Q., Tagawa, K., & Martindale, M. Q. (2001). Deuterostome evolution: Early development in the enteropneust hemichordate, Ptychodera flava. *Evolution & Development*, *3*, 375–390.
Hoagland, K. E. (1977). *Systematic review of fossil and recent Crepidula and discussion of evolution of the Calyptraeidae, (1 of 3), Circa 1977, undated*. Academy of Natural Sciences of Drexel University.
Hoagland, K. E. (1978). Protandry and the evolution of environmentally-mediated sex change: A study of the Mollusca. *Malacologia*, *17*, 365–391.
Hoagland, K. E. (1984). Use of molecular genetics to distinguish species of the gastropod genus *Crepidula* (Prosobranchia: Calyptraeidae). *Malacologia*, *25*(2), 607–628.
Hoagland, K. E. (1986). Patterns of encapsulation and brooding in the Calyptraeidae (Prosobranchia: Mesogastropoda). *American Malacological Bulletin*, *4*, 173–183.

Jackson, D. J., & Degnan, B. M. (2016). The importance of evo-devo to an integrated understanding of molluscan biomineralisation. *Journal of Structural Biology*, *196*, 67–74.
Kocot, K. M., Aguilera, F., McDougall, C., Jackson, D. J., & Degnan, B. M. (2016). Sea shell diversity and rapidly evolving secretomes: Insights into the evolution of biomineralization. *Frontiers in Zoology*, *13*, 23.
Lambert, J. D. (2001). *Patterning a mollusc embryo: The roles of cell lineage and cell signaling in Ilyanassa obsoleta*. Ann Arbor, United States: The University of Arizona.
Lambert, J. D. (2008). Mesoderm in spiralians: The organizer and the 4d cell. *Journal of Experimental Zoology. Part B, Molecular and Developmental Evolution*, *310*, 15–23.
Lambert, J. D. (2010). Developmental patterns in spiralian embryos. *Current Biology*, *20*, R72–R77.
Lambert, J. D., Johnson, A. B., Hudson, C. N., & Chan, A. (2016). Dpp/BMP2-4 mediates signaling from the D-quadrant organizer in a Spiralian embryo. *Current Biology*, *26*, 2003–2010.
Lambert, J. D., & Nagy, L. M. (2001). MAPK signaling by the D quadrant embryonic organizer of the mollusc Ilyanassa obsoleta. *Development*, *128*, 45–56.
Lambert, J. D., & Nagy, L. M. (2003). The MAPK cascade in equally cleaving spiralian embryos. *Developmental Biology*, *263*, 231–241.
Lanza, A. R., & Seaver, E. C. (2018). An organizing role for the TGF-β signaling pathway in axes formation of the annelid Capitella teleta. *Developmental Biology*, *435*, 26–40.
Le Gall, S. (1981). Etude expérimentale du facteur morphogénétique contrôlant la différenciation du tractus génital mâle externe chez *Crepidula fornicata* L. (Mollusque hermaphrodite protandre). *General and Comparative Endocrinology*, *43*, 51–62.
Le Gall, S., Feral, C., Lengronne, C., & Porchet, M. (1987). Partial purification of the neuroendocrine mitogenic factor in the mollusk *Crepidula fornicata* L. *Comparative Biochemistry and Physiology. Part B, Biochemistry & Molecular Biology*, *86*, 393–396.
Le Gall, S., & Streiff, W. (1975). Protandric hermaphroditism in Prosobranch gastropods. In R. Reinboth (Ed.), *Intersexuality in the animal kingdom* (pp. 170–178). Berlin Heidelberg, Berlin, Heidelberg: Springer.
Lesoway, M. P., Abouheif, E., & Collin, R. (2016). Comparative transcriptomics of alternative developmental phenotypes in a marine gastropod. *Journal of Experimental Zoology. Part B, Molecular and Developmental Evolution*, *326*, 151–167.
Lesoway, M. P., Collin, R., & Abouheif, E. (2017). Early activation of MAPK and apoptosis in nutritive embryos of Calyptraeid gastropods. *Journal of Experimental Zoology. Part B, Molecular and Developmental Evolution*, *328*, 449–461.
Lesoway, M. P., & Henry, J. Q. (2019a). Sex determination, sexual development, and sex change in slipper snails. *Results and Problems in Cell Differentiation*, *68*, 155–182.
Lesoway, M. P., & Henry, J. Q. (2019b). Twisted shells, spiral cells, and asymmetries: Evo-devo lessons learned from gastropods. In L. Nuno de la Rosa, & G. Müller (Eds.), *Evolutionary developmental biology* (pp. 1–18). Cham: Springer. https://doi.org/10.1007/978-3-319-33038-9_179-1.
Lesoway, M. P., & Henry, J. Q. (2021). Retinoids promote penis development in sequentially hermaphroditic snails. *Developmental Biology*, *478*, 122–132.
Lo, H. K. A., & Chan, K. Y. K. (2018). Negative effects of microplastic exposure on growth and development of Crepidula onyx. *Environmental Pollution*, *233*, 588–595.
Ly, S., & Collin, R. (2021). Temperature-induced changes in hatching size of a tropical snail occur during oogenesis and can persist for several weeks. *The Biological Bulletin*, *240*, 16–22.
Lyons, D. C., & Henry, J. Q. (2014). Ins and outs of Spiralian gastrulation. *The International Journal of Developmental Biology*, *58*, 413–428.
Lyons, D. C., Perry, K. J., Batzel, G., & Henry, J. Q. (2020). BMP signaling plays a role in anterior-neural/head development, but not organizer activity, in the gastropod Crepidula fornicata. *Developmental Biology*, *463*, 135–157.

Lyons, D. C., Perry, K. J., & Henry, J. Q. (2015). Spiralian gastrulation: Germ layer formation, morphogenesis, and fate of the blastopore in the slipper snail Crepidula fornicata. *EvoDevo*, *6*, 24.
Lyons, D. C., Perry, K. J., & Henry, J. Q. (2017). Morphogenesis along the animal-vegetal axis: Fates of primary quartet micromere daughters in the gastropod Crepidula fornicata. *BMC Evolutionary Biology*, *17*, 217.
Lyons, D. C., Perry, K. J., Lesoway, M. P., & Henry, J. Q. (2012). Cleavage pattern and fate map of the mesentoblast, 4d, in the gastropod Crepidula: A hallmark of spiralian development. *EvoDevo*, *3*, 21.
Lyons, D. C., & Weisblat, D. A. (2009). D quadrant specification in the leech Helobdella: Actomyosin contractility controls the unequal cleavage of the CD blastomere. *Developmental Biology*, *334*, 46–58.
Maboloc, E. A., & Chan, K. Y. K. (2017). Resilience of the larval slipper limpet Crepidula onyx to direct and indirect-diet effects of ocean acidification. *Scientific Reports*, 7, 12062.
Maboloc, E. A., & Chan, K. Y. K. (2021). Parental whole life cycle exposure modulates progeny responses to ocean acidification in slipper limpets. *Global Change Biology*, *27*, 3272–3281. https://doi.org/10.1111/gcb.15647.
Maienschein, J. (2015). *Edwin Grant Conklin [WWW Document]*. MBL History Project digital exhibit https://history.archives.mbl.edu/browse/exhibits/edwin-grant-conklin.
Martindale, M. Q. (1986). The "organizing" role of the D quadrant in an equal-cleaving spiralian,*Lymnaea stagnalis* as studied by UV laser deletion of macromeres at intervals between third and fourth quartet formation. *International Journal of Invertebrate Reproduction and Development*, *9*, 229–242.
Martindale, M. Q., Doe, C. Q., & Morrill, J. B. (1985). The role of animal-vegetal interaction with respect to the determination of dorsoventral polarity in the equal-cleaving spiralian, *Lymnaea palustris*. *Wilhelm Roux' Archiv für Entwicklungsmechanik der Organismen*, *194*, 281–295.
Martindale, M. Q., & Henry, J. Q. (1995a). Modifications of cell fate specification in equal-cleaving nemertean embryos: Alternate patterns of spiralian development. *Development*, *121*, 3175–3185.
Martindale, M. Q., & Henry, J. Q. (1995b). Diagonal development: Establishment of the anal Axis in the ctenophore Mnemiopsis leidyi. *The Biological Bulletin*, *189*, 190–192.
Martindale, M. Q., & Henry, J. Q. (1997). Reassessing embryogenesis in the Ctenophora: The inductive role of e1 micromeres in organizing ctene row formation in the "mosaic"embryo, Mnemiopsis leidyi. *Development*, *124*, 1999–2006.
Martindale, M. Q., & Henry, J. Q. (1998). The development of radial and biradial symmetry: The evolution of bilaterality. *American Zoologist*, *38*, 672–684.
Martín-Durán, J. M., & Marlétaz, F. (2020). Unravelling spiral cleavage. *Development*, *147*, dev181081. https://doi.org/10.1242/dev.181081.
McDonald, K. A., Collin, R., & Lesoway, M. P. (2014). Poecilogony in the caenogastropod Calyptraea lichen (Mollusca: Gastropoda). *Invertebrate Biology*, *133*, 213–220.
Mérot, C., & Collin, R. (2012). Effects of stress on sex change in Crepidula cf. marginalis (Gastropoda: Calyptraeidae). *Journal of Experimental Marine Biology and Ecology*, *416–417*, 68–71.
Moulin, F. Y., Guizien, K., Thouzeau, G., Chapalain, G., Mülleners, K., & Bourg, C. (2007). Impact of an invasive species, Crepidula fornicata, on the hydrodynamics and transport properties of the benthic boundary layer. *Aquatic Living Resources*, *20*, 15–31.
Osborne, C. C., Perry, K. J., Shankland, M., & Henry, J. Q. (2018). Ectomesoderm and epithelial-mesenchymal transition-related genes in spiralian development. *Developmental Dynamics*, *247*, 1097–1120.
Pechenik, J. A., Diederich, C. M., Browman, H. I., & Jelmert, A. (2017). Fecundity of the invasive marine gastropod Crepidula fornicata near the current northern extreme of its range. *Invertebrate Biology*, *136*, 394–402.

Perry, K. J., & Henry, J. Q. (2015). CRISPR/Cas9-mediated genome modification in the mollusc, Crepidula fornicata: CRISPR/CAS9-mediated genome modification in Crepidula. *Genesis*, *53*, 237–244.
Perry, K. J., Lyons, D. C., Truchado-Garcia, M., Fischer, A. H. L., Helfrich, L. W., Johansson, K. B., et al. (2015). Deployment of regulatory genes during gastrulation and germ layer specification in a model spiralian mollusc Crepidula. *Developmental Dynamics*, *244*, 1215–1248.
Rabinowitz, J. S., Chan, X. Y., Kingsley, E. P., Duan, Y., & Lambert, J. D. (2008). Nanos is required in somatic blast cell lineages in the posterior of a mollusk embryo. *Current Biology*, *18*, 331–336.
Render, J. (1991). Fate maps of the first quartet micromeres in the gastropod Ilyanassa obsoleta. *Development*, *113*, 495–501.
Reyes-Giler, C. L., Benson, B. E., Levy, M., Chen, X., Pires, A., Pechenik, J. A., et al. (2021). The marine gastropod Crepidula fornicata remains resilient to ocean acidification across two life history stages. *Frontiers in Physiology*, *12*, 702864.
Showell, C., & Conlon, F. L. (2009). Natural mating and tadpole husbandry in the western clawed frog Xenopus tropicalis. *Cold Spring Harbor Protocols*, *2009*, prot5292. db. prot5292.
Steinert, B. (2016). Drawing embryos, seeing development. In *The Node*. https://thenode.biologists.com/drawing-embryos-seeing-development/discussion/.
Stuart, D. K., Blair, S. S., & Weisblat, D. A. (1987). Cell lineage, cell death, and the developmental origin of identified serotonin- and dopamine-containing neurons in the leech. *The Journal of Neuroscience*, 7, 1107–1122.
Tan, S., Huan, P., & Liu, B. (2022). Molluscan dorsal-ventral patterning relying on BMP2/4 and Chordin provides insights into spiralian development and bilaterian evolution. *Molecular Biology and Evolution*, *39*, msab322. https://doi.org/10.1093/molbev/msab322.
Truchado-Garcia, M., Caccavale, F., Grande, C., & D'Aniello, S. (2021). Expression pattern of nitric oxide synthase during development of the marine gastropod Mollusc, *Crepidula fornicata*. *Genes*, *12*, 314. https://doi.org/10.3390/genes12020314.
Truchado-Garcia, M., Harland, R. M., & Abrams, M. J. (2018). 3D-printable tools for developmental biology: Improving embryo injection and screening techniques through 3D-printing technology. *bioarchiv*. https://www.biorxiv.org/content/10.1101/376657v1.
van den Biggelaar, J. A. (1977). Development of dorsoventral polarity and mesentoblast determination in Patella vulgata. *Journal of Morphology*, *154*, 157–186.
van den Biggelaar, J. A., & Guerrier, P. (1979). Dorsoventral polarity and mesentoblast determination as concomitant results of cellular interactions in the mollusk Patella vulgata. *Developmental Biology*, *68*, 462–471.
Veliz, D., Winkler, F. M., Guisado, C., & Collin, R. (2012). A new species of Crepipatella (Gastropoda: Calyptraeidae) from northern Chile. *Molluscan Research*, *32*, 145–153.
Wanninger, A., & Wollesen, T. (2018). The evolution of molluscs. *Biological Reviews of the Cambridge Philosophical Society*, *94*, 102–115. https://doi.org/10.1111/brv.12439.
Webster, N. B., Corbet, M., Sur, A., & Meyer, N. P. (2021). Role of BMP signaling during early development of the annelid Capitella teleta. *Developmental Biology*, *478*, 183–204.
Weisblat, D. A., & Shankland, M. (1985). Cell lineage and segmentation in the leech. *Philosophical Transactions of the Royal Society of London. Series B, Biological Sciences*, *312*, 39–56.

CHAPTER FOURTEEN

Sifting through the mud: A tale of building the annelid *Capitella teleta* for EvoDevo studies

Elaine C. Seaver*
Whitney Laboratory for Marine Bioscience, University of Florida, St. Augustine, FL, United States
*Corresponding author: e-mail address: seaver@whitney.ufl.edu

Contents

Abstract

Over the last few decades, the annelid *Capitella teleta* has been used increasingly as a study system for investigations of development and regeneration. Its favorable properties include an ability to continuously maintain a laboratory culture, availability of a sequenced genome, a stereotypic cleavage program of early development, substantial regeneration abilities, and established experimental and functional genomics techniques. With this review I tell of my adventure of establishing the *Capitella teleta* as an emerging model and share examples of a few of the contributions our work has made to the fields of evo-devo and developmental biology. I highlight examples of conservation in developmental programs as well as surprising deviations from existing paradigms that highlight the importance of leveraging biological diversity to shift thinking in the field. The story for each study system is unique, and every animal has its own advantages and disadvantages as an experimental system. Just like most progress in science, it takes strategy, hard work and determination to develop tools and resources

Current Topics in Developmental Biology, Volume 147
ISSN 0070-2153
https://doi.org/10.1016/bs.ctdb.2021.12.018

for a less studied animal, but luck and serendipity also play a role. I include a few narratives to personalize the science, share details of the story that are not included in typical publications, and provide perspective for investigators who are interested in developing their own study organism.

1. History and context: Why *Capitella*?

For those who may want to work with a less studied animal to answer a question of biological interest or develop tools for a less studied animal, I share the story of why I chose to work with *Capitella teleta* (hereafter "*Capitella*"). The driver for me to work with annelids was a curiosity about the evolution of the segmented body plan. I had been exposed to insect neural development as a graduate student and during this time, I became interested in the segmented body plan and the different ways that insects generate new segments. As I broadened my perspective, I became excited about the question of the evolution of segmentation in bilaterians, and whether segmentation in chordates, annelids and arthropods was homologous (Davis & Patel, 1999; Seaver, 2003). At the time, new data about the cellular and molecular mechanisms underlying the segmentation process was emerging from chordates and insects, but far less was known about the development of annelid segments. What was known were descriptive studies in a handful of species as well as a detailed cellular description of leech segmentation (Shankland, 1991). My introduction to annelids was with leech embryos, and as a postdoctoral scholar, I was curious about the relative contribution of cell-cell signaling events versus cell autonomous specification in segment boundary formation. I performed an experimental embryology study that demonstrated cellular autonomy in patterning of segments in *Helobdella austinensis*. This lineage-based mechanism of segment formation is very different from the cell-cell signaling system utilized by insect embryos for segment boundary formation (Seaver & Shankland, 2000, 2001). However, the cleavage pattern in leech embryos exhibits substantial modifications relative to the more ancestral cleavage program for annelids, and leeches have an invariant number of segments and lack a capacity for adult regeneration (Dohle, 1999). I was eager to expand upon my work on leeches by investigating an annelid with a more "typical" spiral cleavage program that also continues to generate segments throughout its lifespan and can regenerate its segments.

Two additional factors influenced my choice to study segmentation in annelids. The first was the publication of the first animal phylogenies generated from DNA sequence data, which came out a few months after I began working with annelids (Aguinaldo et al., 1997). It was an exciting time because animal relationships drastically changed from previous associations based upon morphology. As is now well accepted, bilaterian animals become organized into three large superclades, the Deuterostomia, Ecdysosoa and Lophotrochozoa. Each of these superclades contained both segmented and unsegmented taxa, making the evolution of segmentation an even more compelling problem. These new animal relationships really solidified confidence in my decision, and I saw an opportunity. The second factor was the emergence of evo-devo as a nascent field in the United States, and there was new excitement for working with nontraditional animals. This swell of enthusiasm drew me in, and I began paddling to ride the wave.

Once I decided to study annelids, I then had to choose a specific annelid species. At the time, annelid relationships were contentious and therefore, it was not possible to choose a species that retained the greatest number of ancestral character traits with confidence. As an experimental scientist, I prioritized practical attributes relevant for developmental studies. My initial requirements were that the animal had to be readily culturable in the lab, produce enough eggs from a single female to perform an experiment with controls, be reproductive at least half of the year, and have a zygote of a reasonable diameter. I wanted to work with live material and thought that an animal with a limited reproductive season would present challenges for making regular progress.

My initial encounters with annelid candidates were at the Marine Biological Laboratory (MBL) in Woods Hole, MA. I took the Embryology course the first summer of my postdoc and stayed in Woods Hole for a few additional weeks to explore the nereid that lived in Eel Pond (Costello, Davidson, Eggers, Fox, & Henley, 1957). I enjoyed many late nights hunting for animals, but I was not convinced that I would be able to do the experiments I wanted to with this system. Fortuitously, I was introduced to Dr. Sue Hill from Michigan State University, a regular summer researcher at the MBL. Sue worked on regeneration in *Capitella* and agreed to teach me the culturing methods developed years earlier by Dr. Judy Grassle from Rutgers University. I spent the following summer as a "fellow" at the MBL working with Sue, and learned that it was possible to maintain animals in artificial sea water and that adults would reproduce in the lab throughout the year. That summer I performed simple experiments examining growth patterns during segment formation (Seaver, Thamm, & Hill, 2005), and by the end of the fellowship, I decided to go for it.

In the first few years of working with *Capitella*, I also considered other species and even performed some initial studies. However, the combination of attributes in *Capitella* was compelling. Thus, my decision to work with *Capitella* was influenced by careful choice as well as serendipity and a lucky encounter. I never looked back!

When I began working with *Capitella*, it was already known as a bioindicator of environmentally disturbed habitats. In 1969, an oil spill occurred in West Falmouth Harbor, MA, and it became one of the best studied oil spills due to its proximity to the MBL and WHOI (Reddy et al., 2002). The benthic marsh community was characterized in response to the spill, and Capitellid annelids were found to be the predominant benthic group that initially recolonized the impacted area (Grassle & Grassle, 1974). Capitellids quickly became well known as a bioindicator for environmentally disturbed habitats and as opportunistic species around the world. These studies also identified *Capitella teleta* (an unnamed species of the genus at the time) as a member of a sibling species complex (Grassle & Frederick Grassle, 1976). *Capitella*'s ability to withstand environmental insults provide unique opportunities to examine the connection between the development of *Capitella* and its ecology.

2. Phylogenetic position, body plan and life cycle of the annelid *Capitella teleta*

Capitella teleta (Blake, Grassle, & Eckelbarger, 2009), formerly known as *Capitella* sp. I, is a small marine, segmented worm that inhabits shallow marine estuarine environments. *Capitella* is a capitellid annelid and a member of the Spiralia, a morphologically diverse taxon of protostomes and one of the three bilaterian superclades (Fig. 1A). Phylogenomic studies place most annelids within either of two clades, the Sedentaria or the Errantia, terms that reflect the relative motility of the adults (Fig. 1B) (Struck et al., 2011; Weigert et al., 2014; Weigert & Bleidorn, 2016). *Capitella* is part of the Sedentaria along with other annelids that have been utilized for development and regeneration studies such as the leech *Helobdella robusta, Tubifex tubifex*, *Pristina leidyi, Hydroides elegans* and *Spirobranchus* (Fig. 1B).

The body plan of *Capitella* reflects its identity as a segmented worm and its burrowing lifestyle (Fig. 2A, B). As is typical for annelids, the body of *Capitella* is comprised of an anterior pre-segmental head region called the prostomium, numerous repeated units or segments in the trunk, and a terminal posterior region, the pygidium. Its trunk is composed of distinct thoracic and abdominal regions, with nine thoracic and 40 or more abdominal

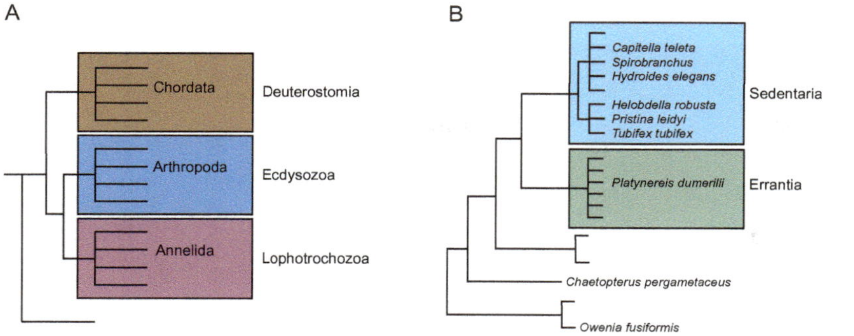

Fig. 1 Bilaterian and annelid phylogenies. (A) Schematized phylogenetic tree showing the three large clades that comprise the Bilateria (in boxes). The annelid *Capitella teleta* is a member of the Lophotrochozoa. Within the Deuterostomia (brown box), Ecdysozoa (blue box) and Lophotrochozoa (purple box) clades, there is a single segmented taxon (Chordata, Arthropoda, and Annelida, respectively) along with many unsegmented taxa. (B) Annelid relationships based on current phylogenomic studies (Struck et al., 2011; Weigert et al., 2014). Most annelids are assigned to either the Errantia (green box) or the Sedentaria (blue box). Named species are those that have been used for development or regeneration studies. *Capitella* is a member of the Sedentaria.

segments. *Capitella* is typical of burrowing worms in that the segments have highly reduced lateral appendages, called parapodia. Immediately anterior to the pygidium is a "posterior growth zone," from which new segments are continuously added throughout its lifespan. The newly formed segments immediately anterior of the posterior growth zone are smaller in width than more mature anterior segments. *Capitella* also has chitinous bristles called chaetae, a hallmark annelid character (Fig. 2C). Chaetae are used in burrowing, swimming or anchoring the body within its tube (Merz & Woodin, 2006). Chaetal morphology varies widely across species and is often used in species identification (Fauchald, 1977; Rouse & Pleijel, 2001). In *Capitella*, capillary chaetae are present in thoracic segments, while the abdominal segments bear hooded hook chaetae (Schweigkofler, Bartolomaeus, & von Salvini-Plawen, 1998). Like in other annelids, many structures are repeated along the body length of *Capitella*, including chaetae, appendages, nephridia, body wall muscle, and ganglia of the centralized nervous system. A dorsally positioned, muscular and ciliated pharynx permits *Capitella* to feed by ingesting sediment as it burrows through the substrate (Boyle & Seaver, 2009). Its digestive system is highly regionalized along its length, consisting of ectodermally derived foregut and hindgut, and a midgut of endodermal origin (Boyle & Seaver, 2008; Meyer, Boyle, Martindale, & Seaver, 2010).

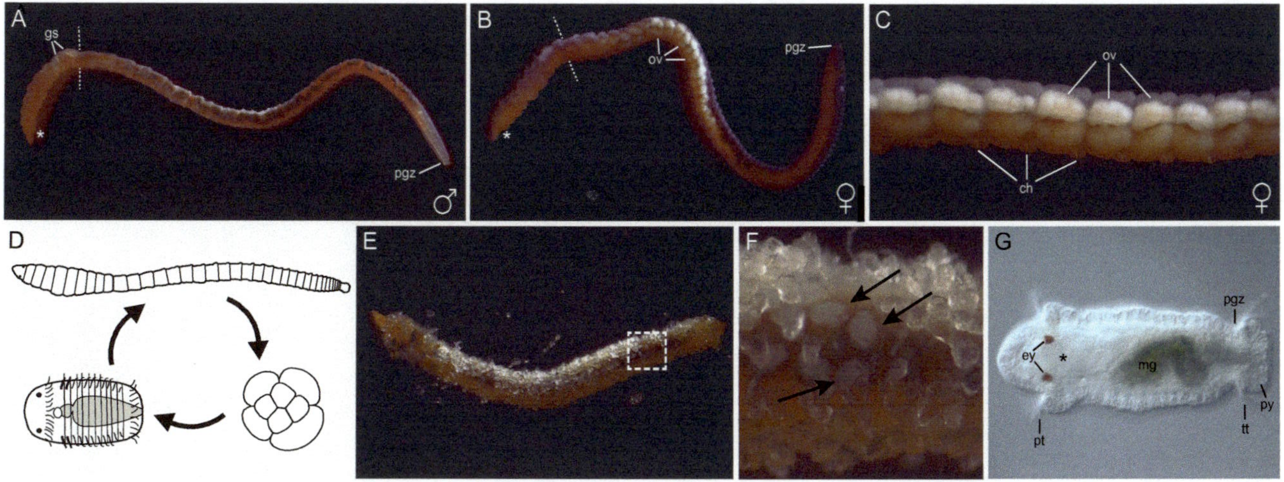

Fig. 2 Body plan and life cycle of *Capitella teleta*. (A, B) Segmented body of adult worms. (A) Male showing genital spines on the dorsal side of segments eight and nine. (B) Female has segmentally repeated paired ovaries (ov). (C) Enlarged view of numerous segments of an adult female showing paired ovaries (ov) in abdominal segments and segmentally repeated chaetae (ch) (ventral view). Length of adults is 2–5 cm. (D) Schematic of the biphasic life cycle. Top image is an adult worm. Bottom right shows an early-stage embryo (8 cell stage) and bottom left depicts a late-stage larva. (E) Brood tube containing developing larvae and female. (F) Enlarged view of brood tube (region depicted is dotted square in E) showing larvae contained within (arrows). Note that brood tube is transparent. (G) Late-stage larva in ventral view. Visible are the eye spots (ey), prototroch (pt), telotroch (tt), and olive midgut (mg). The length of the larva is approximately 500 μm. Anterior is to the left for all panels. Dotted lines in (A) and (B) indicate boundary between thoracic and abdominal segments. Asterisk indicates position of the mouth in panels (A), (B) and (G). Ch, chaetae; gs, genital spines; ov,ovaries; pgz, posterior growth zone; py, pygidium.

Capitella reproduces sexually and there are distinct males, females and hermaphrodites (Fig. 2A–D). Individuals can mate multiple times. Gametogenesis and the organization and morphology of reproductive organs have been well characterized (Blake et al., 2009; Eckelbarger & Grassle, 1983, 1987). Embryonic and early larval development occurs within a specialized tubular structure produced by the female, called the brood tube (Fig. 2E, F). The female lays zygotes within the brood tube and provides parental care by remaining inside the brood tube with the developing embryos until they emerge, approximately 9 days after fertilization. The tube itself is transparent and open at both ends but is typically covered with sand and mud debris from the local habitat. Larvae are competent to undergo metamorphosis upon emergence from the brood tube, but can spend several days in the water column, allowing for dispersal. Swimming larvae are positively phototactic (Butman, Grossie, & Buskey, 1988; Yamaguchi & Seaver, 2013). Notable features of late-stage larvae include two anterior pigmented eye spots, anterior (prototroch) and posterior (telotroch) ciliary bands used for swimming in the water column, a segmented trunk consisting of 13 identifiable segments, a ventral mouth, a ciliary band along the ventral midline (neurotroch), posterior growth zone, and a terminal pygidium in the posterior (Fig. 2G). *Capitella* larvae are lecithotrophic and do not feed. Rapid metamorphosis is induced by marsh sediments and includes loss of ciliary bands, body elongation and a change in behavior from swimming in the water column to burrowing in the sediment (Cohen & Pechenik, 1999; Dubilier, 1988). Following metamorphosis and initiation of feeding, juveniles undergo substantial growth, when the majority of segments present in reproductive adults are generated.

3. Favorable properties of *Capitella* for development and regeneration studies

A number of fundamental biological questions can be investigated with the annelid *Capitella teleta*. It is one of the systems with the most available molecular tools within the Spiralia (Martin-Duran & Marletaz, 2020), and has a number of favorable attributes for development and regeneration studies (Table 1). Its phylogenetic position means that findings from *Capitella* provide important perspective for understanding the evolution of developmental programs in the Bilateria that cannot be gleaned from only considering data from well established animal models of development.

Table 1 Favorable properties and available tools for *Capitella teleta*.

Feature	References
• Year round availability of embryonic material	Grassle and Frederick Grassle (1976), Seaver et al. (2005), Yamaguchi et al. (2016)
• Public genome sequence (330 Mbp)	Simakov et al. (2013)
• Development and regeneration staging systems	Seaver et al. (2005), de Jong and Seaver (2016)
• Microinjections	Meyer et al. (2010), Klann and Seaver (2019); Seaver (2016)
• Lineage tracing	Meyer and Seaver (2009, 2010), Meyer et al. (2010)
• Expression of exogenous RNA	Seaver (2016), Lanza and Seaver (2020a, 2020b)
• Gene knockdown (morpholino)	Klann and Seaver (2019), Lanza and Seaver (2020a, 2020b)
• CRISPR mutagenesis	Neal et al. (2019)
• Laser cell deletions	Pernet et al. (2012), Amiel et al. (2013), Yamaguchi and Seaver (2013), Yamaguchi et al. (2016)
• Blastomere dissociations	Carrillo-Baltodano and Meyer (2017)
• Live cell imaging	Meyer and Seaver (2010)
• Immunohistochemistry	Meyer and Seaver (2009), Yamaguchi and Seaver (2013), Meyer et al. (2015)
• WMISH and FISH	Seaver and Kaneshige (2006), Fröbius et al. (2008), Klann and Seaver (2019), Lanza and Seaver (2020a, 2020b)
• Single embryo RNAseq	Lanza and Seaver (2020a, 2020b)
• Single-cell RNAseq	Sur and Meyer (2021)

Culturing conditions for *Capitella* in the laboratory are straightforward, and breeding colonies can be maintained virtually anywhere. *Capitella* is reproductive year-round in the lab, facilitating experimental studies that require live material. Our adjustments to initial husbandry practices have optimized conditions for obtaining zygotes and early-stage embryos on a regular basis. Because all stages develop normally when raised in a dish of filtered sea water, embryos and larvae are readily accessible for observation and manipulation. Each female typically produces a brood of 100–300 embryos, and embryos from a single brood develop synchronously. Early development occurs through a stereotypical spiral cleavage program, and individual blastomeres can be uniquely identified and followed, both within and across species. Juvenile and adult animals can regenerate in the posterior direction following amputation at multiple locations along the anterior-posterior axis (de Jong & Seaver, 2016; Giani, Yamaguchi, Boyle, & Seaver, 2011). Juveniles are semi-transparent, providing the potential to perform live imaging on amputated worms. The life cycle is approximately 2 months at 16°C. When raised at room temperature, animals become reproductive in 6 weeks, a time frame permissive for establishing transgenic lines. One of the strengths of *Capitella* as a system is that it is possible to compare embryogenesis, adult growth and regeneration within a single animal.

We and others have generated a number of resources, techniques and descriptive information to facilitate mechanistic studies of development and regeneration in *Capitella* (Table 1). There is a published staging system for embryonic and larval development (Seaver et al., 2005), and for posterior regeneration (de Jong & Seaver, 2016). Zygotes and individual blastomeres can be microinjected with either lipophilic or aqueous dyes, RNA, oligonucleotides or DNA. A comprehensive fate map documents the embryonic origin of almost all larval tissues (Meyer et al., 2010; Meyer & Seaver, 2010). Furthermore, there are established cell manipulation techniques for early-stage embryos. These include targeted deletion of individual blastomeres and isolation of blastomeres and subsets of blastomeres from the rest of the embryo (Amiel, Henry, & Seaver, 2013; Carrillo-Baltodano & Meyer, 2017; Pernet, Amiel, & Seaver, 2012). Development of multiple tissues has been described by immunohistochemistry or whole mount in situ hybridization, including for the nervous system, digestive system, eyes, gonads, germline, and body wall muscle, and there are available molecular and immunohistological markers for multiple cell types (see Table 1 for references). Molecular studies in *Capitella* have benefited from the availability of a sequenced, slowly evolving genome (Simakov et al., 2013). Functional genomic techniques are established and include morpholino knockdown and CRISPR/Cas9 mutagenesis (Klann & Seaver, 2019; Lanza & Seaver, 2020a; Neal, De Jong, & Seaver, 2019). Exogenous mRNAs are expressed in the early stage embryo following microinjection into zygotes (Lanza & Seaver, 2020a; Seaver, 2016). Larval cell types have been mapped using single cell transcriptome sequencing (Sur & Meyer, 2021), and early-stage embryos have been analyzed using single embryo RNAseq (Lanza & Seaver, 2020a). Thus, *Capitella teleta* is experimentally tractable and has a combination of features and available resources that provide unique opportunities to investigate basic biological questions in development and regeneration. Future studies will benefit from additional tool advances such as transgenics, functional genomics for adult stages, and improvements to the genome sequence with current sequencing technologies.

4. Building the tools: Challenges and successes

Although studies in *Capitella* can now take advantage of a considerable number of techniques and resources, each one had to be individually established, often with considerable investment in effort, time and resources. The development of each technique/resource had its own combination of challenges along the way. Two particularly significant examples are the development of embryo microinjection and generation of a sequenced genome.

4.1 Microinjection

One of the biggest and more memorable technical challenges we faced with respect to building tools was developing zygotic and blastomere microinjections. An ability to microinject zygotes has substantial implications for the success of functional studies that depend upon reliable delivery of DNA and RNA into the embryo. When we started working on *Capitella*, there were no reports of successful microinjection for any polychaete, and I had been told by colleagues that attempts with other species had been unsuccessful. Our first challenge was to work out conditions to reliably obtain early-stage embryos. Although basic culturing conditions had been established, the priority had not previously been on reproductive output. Furthermore, *in vitro* fertilization had not been successful (and has yet to be). We eventually found that isolating females from males for a few days prior to mating dramatically increased the chances of obtaining zygotes. More recently, the addition of a light/dark regime has improved reliability and narrowed the time window required for obtaining zygotes.

Every embryo has its own unique properties that can confer specific challenges for successful microinjection. The *Capitella* zygote is approximately 200 μM in diameter (Grassle & Frederick Grassle, 1976), larger than many other polychaetes (60–120 μm). Therefore, size was not a limiting factor. However, both the outer egg membrane and the lack of turgor pressure of the *Capitella* embryo presented a challenge. It is not possible to penetrate the zygote with sharp glass needles without first softening the outer egg membrane. We tried a few different methods and settled on a modification of the demembranization protocol that had been used for the annelid *Chaetopterus* (Henry, 1986). The low turgor pressure of *Capitella* embryos in combination with the tough outer membrane required application of substantial force to penetrate the embryo, resulting in deformation of the embryo and sometimes exit of the needle on the backside of the embryo. Another issue was that it was difficult to remove the needle from the embryo without causing damage. We were further discouraged when the required length of time of exposure to the de-membranization solution resulted in abnormal development. It seemed like there were no conditions that would allow us to microinject and get healthy development. The breakthrough came after we switched to a different type of glass for making our micropipette needles. Up to that point we had been pulling needles using either borosilicate or aluminosilicate glass. We switched to quartz glass because it is harder than either borosilicate or aluminosilicate, and the harder glass allows for generation of a very fine needle tip with sufficient stiffness to pierce the outer egg membrane. Because quartz has a higher melting temperature than silicate glass, an instrument with a CO_2 laser is required to melt quartz pipette stock. At the time, there were no laser micropipette pullers on the entire island of Oahu, where my lab was located at the University of Hawai'i. Therefore, we had to purchase the puller without a guarantee that it would work. Fortunately, it made all the difference! Following this switch in glass, we were soon able to perform blastomere microinjections using diI as a lineage tracer, and generated a comprehensive fate map of larval tissues (Meyer et al., 2010). Generating a fate map had been a longstanding goal of mine, and it was exciting and very satisfying to finally achieve it. As an indirect consequence, quartz micropipettes have been successfully used for microinjection of other challenging embryos, including the annelid *Chaetopterus* and two acoel species, *Hofstenia miamia* and *Isodiametra pulchra*. What seemed like an impossible hurdle became possible largely as the result of switching to quartz glass for our microinjection needles. Furthermore, development of microinjection procedures generated opportunities to pursue functional genomic techniques that rely on successful delivery of exogenous materials into early-stage embryos.

4.2 Genomic resources

Another very important resource for *Capitella* research was the generation of a sequenced genome. We were fortunate that *Capitella teleta* was among the first lophotrochozoans to have its genome sequenced. Others included the parasites *Schistoshoma japonicum* and *mansoni*, the planarian *Schmidtea mediterranea*, the owl limpet *Lottia gigantea* and the leech *Helobdella robusta* (Berriman et al., 2009; Robb, Ross, & Alvarado, 2008; Simakov et al., 2013; Zhou et al., 2009). This is a story of being in the right place at the right time and having just enough data to convey the potential of our study system. In 2003, sequencing costs were prohibitive for an individual group (or groups) to sequence a genome, and the sequencing technology commonly used at that time was Sanger sequencing. The Joint Genome Institute (Department of Energy, Walnut Creek, CA) was interested in investigating animal diversity, and I was invited to a workshop at JGI to choose a lophotrochozoan species for genome sequencing. The criteria considered were phylogenetic position, genome size, presence of a community that could benefit from the project, and potential for experimental work in evo-devo. I had only begun my position as an assistant professor the previous year and although I had not yet published data on *Capitella*, I had established a stable animal culture and developed some promise of *Capitella* as an experimental system. At this workshop, I conveyed this potential by showing images of the larval nervous system using immunohistochemistry, gene expression patterns for a handful of genes by whole mount in situ hybridization (cloned out sans availability of a genome or transcriptome!), and early-stage embryos that had been microinjected with fluorescent lineage tracers. In the big picture, this was not super impressive, and even a few other annelids had more of a track record. Certainly, there were numerous publications of cellular and molecular development of the leech *Helobdella*. Another top candidate annelid was *Platynereis*, and several European labs were studying its development and had published gene expression studies. Since it was known that European labs were in the process of sequencing the *Platynereis* genome and there were no breeding colonies in North America, *Platynereis* was not chosen. Ultimately my unpublished data helped buoy *Capitella* into being one of three species that JGI sequenced for the project. These included one mollusk and two annelids, specifically *Lottia gigantea*, *Helobdella robusta* and *Capitella teleta* (https://mycocosm.jgi.doe.gov/Capca1/Capca1.home.html). Following public release of the sequence data, there was an unfortunate delay of 5–6 years before publication of the associated genome paper (Simakov et al., 2013). Ironically, the *Capitella* genome was sequenced prior to the formal description of *Capitella teleta*. We worked on it for nine years as *Capitella* sp. I before publication of the formal description (Blake et al., 2009). For us the JGI lophotrochozoan sequencing project fulfilled its promise: the *Capitella* genome has been widely used for genomic studies (phylogenomic analyses, gene family evolution, etc.) and evo devo work, and it has paved the way for us to develop morpholino knockdown and genome editing approaches and stimulated a growing community of *Capitella* researchers.

5. Attributes of the genome: Stable and slowly evolving

One of the biggest impacts of *Capitella* has been the widespread use of its genome sequence. The *Capitella* genome has been commonly used as a representative lophotrochozoan in metazoan phylogenomics, phylogenetic analyses within annelids, and in evolutionary analyses of specific gene families. This is due not only to the fact that it was one of the earliest available lophotrochozoan genomes, but importantly because it is a slowly evolving genome with a relatively stable gene content and genomic organization. For example, in an analysis of 21 metazoan genomes, the *Capitella* genome has retained the largest number of distinct gene clusters (Ryan et al., 2013). Most of the inferred ancestral bilaterian gene families are still present; thus, the gene content of *Capitella* reflects well the inventory of genes present in the last bilaterian ancestor (Simakov et al., 2013). Additionally, the *Capitella* genome exhibits substantial macrosynteny with the mollusk *Lottia* as well as with chordate genomes. Inclusion of lophotrochozoan genomic information has provided an important perspective on evolutionary trajectories within bilaterians and reconstruction of ancestral bilaterian genome characters. For example, genomes like *Capitella* have provided a tool for understanding patterns of gene loss across animal clades, challenging the idea that there is a correlation between gene complexity and morphological complexity (Albalat & Cañestro, 2016; Guijarro-Clarke, Holland, & Paps, 2020). In addition, numerous published analyses of gene families with key roles in development have utilized sequences from the *Capitella* genome. Other bioinformatic analyses have generated hypotheses for functional development studies (e.g., Cho, Vallès, Giani, Seaver, & Weisblat, 2010; Fröbius, Matus, & Seaver, 2008; Kenny et al., 2014). The *Capitella* genome has also been valuable in identifying novel clade-specific genes, such as is the case for the spiralian-specific genes *lophotrochin* and *trochin* (Wu et al., 2020). As additional annelid genomes are sequenced, the *Capitella* genome has served as a valuable comparison (Martín-Durán et al., 2021; Zakas, Harry, & Scholl, 2021). For example, it has contributed to building an understanding of a novel route to genome compaction in the case of the smallest annelid genome. This is impressive community impact for a genome sequenced to only 8 × coverage using Sanger sequencing!

6. Impacts on the field of spiralian development

Studies of *Capitella* early development have provided some important insights, both within the spiralian field and more broadly relevant to animal development and regeneration. Here I highlight a few examples of the more surprising findings as well as a story that illustrates why it is important to include perspective from multiple species to build an understanding of basic concepts in biology. When we began working on early development of *Capitella*, there was a notable lack of molecular developmental biology studies in non-leech annelids, and publications were dominated by studies in gastropod mollusks, with a few examples from nemerteans and a polyclad flatworm. By expanding sampling to include more annelids, we reasoned that we would either find compelling differences between annelids and mollusks or generate additional evidence of the strong conservation of the spiralian developmental program. We uncovered examples of both variation and strong conservation.

During early embryogenesis, *Capitella* undergoes a stereotypical cleavage program called spiral cleavage in which the relative position, size, orientation of the spindle plane and timing of birth of cells during early cleavages allows for identification of individual blastomeres (Hejnol, 2010; Henry & Martindale, 1999). This program of development can be found in annelids (including sipunculids and echiurans), mollusks, nemerteans, and some platyhelminthes, all members of a superclade of bilaterian animals called the Spiralia (Hejnol, 2010; Henry & Martindale, 1999). There is extraordinary diversity of adult body plans in spiralians, the spiral cleavage program is over 500 myr, yet in some cases the early cleavage program is so conserved that it is virtually impossible to distinguish to which taxon the animal belongs from examination of the early-stage embryo. The paradox between the conservation of the early cleavage program and the diversity of adult body plans has captured the interest of researchers since the 19th century (Wilson, 1898). Identification of individual cells and comparisons of homologous cells across species is facilitated by a standard nomenclature that was established and refined by early workers in the field (Conklin, 1897). In addition to the conserved cleavage patterns, there are similar descendant fates of embryonic cells across spiralians, and descendants occupy similar locations within the body. Indeed, the cleavage program was used by early workers to argue that this diverse assemblage of animals were descended from a common ancestor (Wilson, 1898).

Capitella exhibits general characteristics of spiralian development, both in its cleavage pattern and fate map. *Capitella* development was first described by Hugo Eisig at the Statione Zoologica in Naples (Eisig, 1898). The first cell division is unequal and produces a larger CD cell and a smaller AB cell. At the four-cell stage, one of the cells, known as the D cell, is visibly larger (Fig. 3A). During these first few divisions, the embryo generates a small transient cytoplasmic protrusion called a polar lobe. Polar lobes are present in early-stage embryos of many spiralians. Polar lobe contents are shunted into the CD and D blastomeres, resulting in unequal cleavages. The spiral cleavage pattern becomes readily apparent as the embryo passes from the four to the eight-cell stage, when the cleavage spindle is oriented at a 45 degree angle with respect to the position of the animal-vegetal axis. These divisions are asymmetric and produce a smaller set or quartet of daughter cells called 1a, 1b, 1c, and 1d "micromeres" towards the animal pole of the embryo, while their larger sisters called 1A, 1B, 1C and 1D macromeres are born towards the vegetal pole (Fig. 3B). The letter associated with each blastomere name refers to its quadrant of origin, and the associated number refers to the birth order of the cell. Micromeres are named with the letter corresponding to the quadrant of their birth in lower case letters and macromeres with uppercase letters. The vegetal macromeres produce three to four quartets of smaller daughter micromere cells through a series of subsequent asymmetric cell divisions.

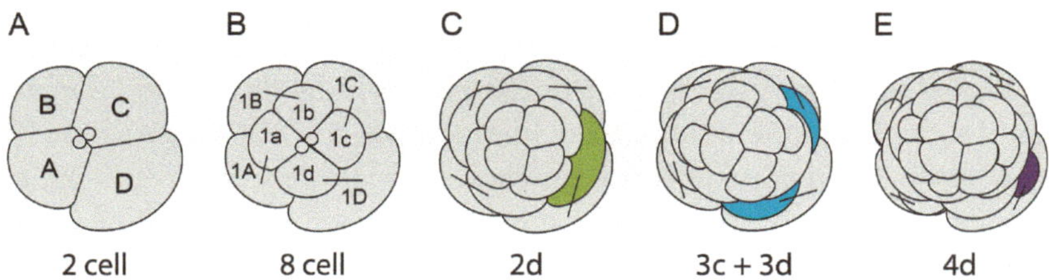

Fig. 3 Schematics of *Capitella* embryos at successive cleavage stages. Cells are named in panels (A) and (B) according to standard nomenclature for spiralian embryos. (A) Four-cell stage embryo. *Capitella* exhibits unequal spiral cleavage and D is larger than the other cells. (B) Eight-cell stage embryo showing birth of the first tier of smaller micromere cells resulting from a clockwise cleavage orientation. (C) Counterclockwise cleavages generate the second tier of micromeres, including cell 2d (green). 2d generates the trunk ectoderm and has embryonic organizing activity. (D) Trunk mesoderm is generated by the two cells 3c and 3d (blue). (E) The cell 4d (purple) is one of the 4th tier of micromeres and generates a few muscle fibers, the anus and germline. All panels are animal views. Straight lines indicate a mother-daughter relationship between cells, and the orientation of the line indicates the orthogonal spindle orientation. Identities of cells with color shading are indicated beneath the schematic.

In between each one of the macromere divisions, the mitotic spindles shift 90 degrees relative to the previous angle, oscillating between clockwise and counterclockwise orientations.

Our first big surprise with respect to *Capitella* early development came from the *Capitella* intracellular fate map study. We analyzed descendant clones with cellular resolution from the four cell stage through formation of the third quartet of micromeres. Many of the early blastomere fates in *Capitella* exhibit conservation with other spiralians, including several spiralian hallmark features (Meyer et al., 2010). For example, the first set of micromeres, the 1a, 1b, 1c, and 1d blastomeres, generate descendant clones that form quadrants of the head, and the larval eyes originate from 1a and 1c (Fig. 3B). In addition, virtually the entire trunk ectoderm and nervous system is generated by a single cell at the 16 cell-stage, called 2d or the primary somatoblast, and this is the case for a number of annelids that have been studied (Fig. 3C) (Ackermann, Dorresteijn, & Fischer, 2005; Anderson, 1959; Goto, Kitamura, Arai, & Shimizu, 1999; Henry & Martindale, 1999; Meyer & Seaver, 2010; Weisblat, Kim, & Stent, 1984; Wilson, 1892). The most surprising finding was the embryonic origin of the "endomesodermal" lineage, which in most spiralians gives rise to the majority of the trunk mesoderm, hindgut and germline and originates from the cell 4d. This is one of the most conserved features of the spiralian fate map. In *Capitella*, 4d generates the presumptive germline but only contributes to the anus and generates a few muscle fibers (Fig. 3E). Most of the trunk mesoderm instead arises from two other micromere cells, 3c and 3d (Fig. 3D). This represents an incomplete fate switch of 4d, since 4d generates some of the tissues generated by this cell in other animals. Interestingly, the historical cell lineage study of *Capitella* performed by Eisig (1898) also reported trunk mesoderm to originate from 3c and 3d, but these findings were viewed with suspicion by contemporaries. When I first presented these data at a national conference, I remember hearing an audible gasp from the audience!

The next set of surprises emerged from experimental manipulations designed to investigate the question of whether or not *Capitella* has an embryonic organizer. Patterning of the body axes represent some of the earliest events in embryogenesis, and in many animals this requires a localized signal that influences cell fates and global patterning in the rest of the embryo that is required for normal development. A striking characteristic of spiralian development is that organizing activity is typically localized to a single cell. Typical experimental demonstration of organizing activity in spiralians is by deletion of individual cells with resulting defects that are greater than would be expected from loss of the tissue that normally arises from the deleted cell, including absence of body axes. Prior to our study, this signal was known to emanate from a single D quadrant cell, localized to either the blastomere 3D or 4d (Cather & Verdonk, 1979; Clement, 1962; Henry, Perry, & Martindale, 2006; Martindale, 1986). Through laser deletion studies of individual blastomeres, we identified an organizing activity critical for bilateral symmetry and dorsal-ventral organization of the head, as well as patterning of the mesoderm, foregut and nervous system (Amiel et al., 2013). Similar to other spiralians, the organizing activity of *Capitella* is localized to the D quadrant; however, the cellular identity is different. We found that micromere 2d is a key cell responsible for organizing activity in *Capitella* (Fig. 3C). Furthermore, laser deletion experiments reveal that requirement of the organizing signal ends prior to the completion of the birth of the 2d daughters. Neither daughter cell is required for organizing activity (Amiel et al., 2013). Micromere 2d is born one to two cell divisions prior to 3D or 4d, and therefore our findings also identified a signaling event earlier than what had previously been reported for other spiralians. Such a change in timing of organizing activity reflects heterochrony. The implications of such a heterochronic shift are unclear, although precocious specification of cell fate facilitates direct development and an ability to respond to unstable environmental conditions. I proposed that precocious segregation of determinants into the 2d cell provides a mechanistic explanation for an evolutionary shift of organizing activity to an earlier time interval in *Capitella* relative to other spiralians (Seaver, 2017).

In subsequent studies, we identified a molecule that mediates the organizing signal to the rest of the embryo. In many animals, BMP signaling mediates the organizing signal that establishes the dorsal-ventral (D-V) axis, including in a few spiralians that have been examined (reviewed in De Robertis & Sasai, 1996; Ferguson, 1996). However, exposure of *Capitella* embryos to either exogenous BMP or chemical inhibitors of BMP signaling during the time of organizing signal results in embryos with a detectable D-V axis along with defects in nervous system development (Webster, Corbet, Sur, & Meyer, 2021). In contrast, inhibition of TGFß signaling by either chemical inhibitors or morpholino knockdown results in larvae that lack a D-V axis, demonstrating that TGFß is the primary signal that mediates organizing activity from 2d (Lanza & Seaver, 2018; Lanza & Seaver, 2020a). One might argue that *Capitella* development exhibits features that are not representative of other annelids. Our initial efforts to address this issue suggests that other annelids such as the early branching annelid *Chaetopterus pergamentaceus* (Fig. 1B) also utilize TGFß signaling for body axes patterning (Lanza & Seaver, 2020b). Furthermore, there is similar timing of organizing activity in *Chaetopterus* and *Capitella* embryos. Although results from two annelids is not definitive, it provides evidence that the *Capitella* results are shared, and hints of clade specific differences between annelids and mollusks.

In summary, we uncovered features of early embryogenesis in *Capitella* that dramatically differ from other spiralians, including a unique cellular origin of the trunk mesoderm, differences in the cellular identity and timing of the organizing signal, a restricted role of 4d, as well as utilization of a distinct signaling pathway to confer the D-V axis. Furthermore, the timing of organizing activity and birth of the cells that give rise to trunk mesoderm in *Capitella* are both shifted one cell division earlier and distributed to adjacent lineages (e.g., C quadrant) compared with that observed in other spiralians, a situation that is made permissive by early specification of the CD blastomere, typically found in unequal cleaving forms. In *Capitella*, D quadrant specification occurs by the eight cell stage (Yamaguchi, Dannenberg, Amiel, & Seaver, 2016). A precocious shift in embryogenesis can be associated with a direct developing life cycle (Hanken, Jennings, & Olsson, 1997; Martindale & Henry, 1995; Raff, 1996), and the retention of most larval features through metamorphosis into the juvenile phase in *Capitella* is consistent with this idea. A future challenge will be to link these and other variations in spiralian early development to the diversity of body plans within and across taxa to gain further insight into the relationship between genotype and phenotype.

7. Insights from neurogenesis studies

We were curious to build an understanding of the cellular and molecular underpinnings of neurogenesis in annelids and determine how these processes compare with neurogenesis in other animal clades. The centralized nervous system of *Capitella* is comprised of a bi-lobed brain connected to the ventral nerve cord by circumoral nerves, an organization typical for annelids (Meyer, Carrillo-Baltodano, Moore, & Seaver, 2015). The arrangement of nerves and ganglia in the ventral nerve cord is one that reflects the segmented body plan and is strikingly similar to the developing ventral nerve cord of many insects (Schmidt-Rhaesa, 2007). In *Capitella*, the adult nervous system begins to form during embryonic and larval stages, and the nervous system rudiment is readily accessible to observation and manipulation

(Meyer & Seaver, 2009). Neural progenitors of the brain are first detectable at the beginning of gastrulation. The *Capitella* life history provides an advantage for studies of development of the adult nervous system since the larval phase is short. In many other spiralians with long-lived larvae, development of the adult nervous system is not initiated until late larval stages or after metamorphosis, making it challenging to study neurogenesis in these animals.

An emerging theme from our neurogenesis studies in *Capitella* and those from others is that there are differences between the development of the brain and the ventral nerve cord. These differences are both molecular and cellular in nature and are manifest at distinct neurogenic stages. They include variation in the specification of neural fate, in cell behavior of neural progenitors, and in the genes associated with neurogenesis. We determined the embryonic origins of the nervous system during our fate mapping study; the brain and ventral nerve cord originate from different blastomeres in the early stage embryo (Meyer et al., 2010; Meyer & Seaver, 2010). The vast majority of the *Capitella* brain originates from two first quartet cells called $1c^1$ and $1d^1$ while the ventral nerve cord originates from a single cell, 2d. Further, head and trunk neural tissue appear to be specified differently (Carrillo-Baltodano & Meyer, 2017). This was demonstrated by dissociation of early stage blastomeres and raising them in isolation or in specific combinations of cells. These experiments provide evidence that head neural identity is autonomously specified whereas trunk neural tissue requires inductive signals from the 2Q macromeres (Carrillo-Baltodano & Meyer, 2017). Following segregation to a neural fate, neural progenitors in vertebrates and insects internalize by invagination or ingression, respectively, and then divide to produce a group of daughter cells that adopt neural or glial fates (Doe, Fuerstenberg, & Peng, 1998; Doe & Goodman, 1985; Hartenstein & Campos-Ortega, 1984; Jacobson & Rao, 2005). In contrast, we observed that neural progenitors of the *Capitella* brain divide only on the surface of the ectoderm, internalize as individual cells by ingression, and are postmitotic following internalization (Meyer & Seaver, 2009). The situation is slightly different in the ventral nerve cord, in which neural progenitors divide as surface cells, and some continue to divide after internalization (Sur, Renfro, Bergmann, & Meyer, 2020). Finally, there are differences in the relative onset of expression of neurogenic genes between the brain and ventral nerve cord. Specifically, the initiation of expression of *ash1* and *neuroD* homologs is later than the onset of *neurogenin* expression in the ventral nerve cord, but expression of these three genes is initiated at approximately the same time during brain development (Sur et al., 2020). From these data, it has been proposed that distinct developmental modules influence development of the brain and the ventral nerve cord.

Studies from diverse members of the Spiralia can provide an important perspective to understand the evolution of the bilaterian nervous system. There are both similarities and notable differences in neurogenesis between *Capitella* and vertebrates and insects. A conserved and general early cell fate decision during neurogenesis is the segregation between a neural and ectodermal fate. In insects and vertebrates, this initial specification relies on BMP signaling, which limits neural fate to either the ventral (insects) or dorsal side (vertebrates) of the embryo (De Robertis & Sasai, 1996; Wilson & Edlund, 2001). In contrast, both brain and ventral nerve cord form following exposure to exogenous BMP, demonstrating unexpectedly, that BMP does not inhibit neural fate specification in *Capitella* (Webster et al., 2021). Additionally, induction of the ventral nerve cord appears to occur independently of dorsal-ventral axis formation (Carrillo-Baltodano, 2019), unlike in vertebrates and insects, in which these events are closely linked (De Robertis & Kuroda, 2004; Holley et al., 1995; Wilson & Edlund, 2001). In other spiralians that have been studied, BMP signaling also does not appear to block neural specification (Denes et al., 2007; Kuo & Weisblat, 2011; Lambert, Johnson, Hudson, & Chan, 2016; Lyons, Perry, Batzel, & Henry, 2020).

As just highlighted, there are differences in the cellular and molecular characteristics of neurogenesis between *Capitella* and vertebrates and insects. At the same time, many of the genes involved in specification of the nervous system in vertebrates and insects are expressed in the developing neuroectoderm or central nervous system in *Capitella*, providing evidence of deep conservation of a common neurogenic toolkit. The progressive development of ganglia along the anterior-posterior axis in the trunk facilitates determination of the relative timing of onset of gene expression in the ventral nerve cord. Our analysis showed that homologs of the neurogenic genes *SoxB1*, *neurogenin*, *achaete scute*, *neuroD*, *musashi*, *elav1*, and *prospero* are expressed at distinct phases of neurogenesis in the brain and ventral nerve cord (Sur, Magie, Seaver, & Meyer, 2017). These genes show both similarities as well as differences of inferred function when compared with other bilaterian clades. One notable difference is represented by the expression of *neurogenin* in proliferating neural precursor cells in *Capitella* and another annelid, *Platynereis dumerilii* (Kerner, Simionato, Le Gouar, & Vervoort, 2009). In vertebrates, *neurogenin* homologs are instead expressed in cells that are exiting the cell cycle (Bertrand, Castro, & Guillemot, 2002; Lacomme, Liaubet, Pituello, & Bel-Vialar, 2012). Together, these data support the presence of a neurogenic GRN with deep conservation in the Bilateria that has been somewhat rearranged in different clades. Findings from *Capitella* and other annelids emphasize the importance of including data from the Spiralia to provide perspective in understanding the evolution of the bilaterian nervous system.

8. Regeneration of the germline

Among the segmented animal clades, annelids stand out in their ability to regenerate numerous organ systems following transverse amputation of the body (Özpolat & Bely, 2016). Furthermore, annelids display a wide range of regenerative abilities, including both anterior and posterior regenerative ability, providing opportunities to track evolutionary changes in regeneration potential (Bely, 2006). *Capitella* has the capability to replace complex structures and lost segments in the posterior direction as an adult ("tail regeneration") (de Jong & Seaver, 2016; Giani et al., 2011; Seaver, 2016; Seaver & de Jong, 2021). Regeneration potential changes during the life cycle of *Capitella*, as early-stage embryos generally do not regulate following deletion or separation of blastomeres (Amiel et al., 2013; Carrillo-Baltodano & Meyer, 2017).

One striking exception to the lack of regulation of missing structures in the embryo is an ability of *Capitella* to replace its germline, which we showed by targeted deletions of single blastomeres in early-stage embryos followed by analysis of reproductive status in adults (Dannenberg & Seaver, 2018). Male and female reproductive structures as well as gametogenesis have been well characterized in *Capitella* (Eckelbarger & Grassle, 1983, 1987). The embryological origin of the germline in spiralians is generally accepted to arise from the blastomere 4d (Lyons, Perry, Lesoway, & Henry, 2012; Rebscher, 2014). The germline in *Capitella* appears to also arise from 4d as shown from our intracellular fate mapping study (Meyer et al., 2010). Most larvae resulting from deletions of the germline precursor in early stage *Capitella* embryos lack expression of the stem cell markers *vasa* and *nanos* (Dill & Seaver, 2008) in the position of multipotent progenitor cells, as expected (Dannenberg & Seaver, 2018). However, a surprising result was that most adult animals raised from embryos with their germline precursors deleted have gametes and morphologically normal reproductive structures

(Dannenberg & Seaver, 2018). Furthermore, we observed fertile adults as assessed by production of morphologically normal larvae following mating. Because germ line precursors are absent in larvae following deletion of the germline precursor, the appearance of functional gametes in adults appears to be the result of somatic cell reprogramming. Regeneration of the germline has been reported in only a few bilaterians (Wessel, Morita, & Oulhen, 2020), and within spiralians, germline regeneration has been reported in platyhelminth flatworms (Pfister et al., 2008; Sato et al., 2006; Wang, Zayas, Guo, & Newmark, 2007). The presence of a stereotypic cleavage program, short larval phase, an ability to readily score for presence of reproductive structures and gametes, perform matings, and utilize the functional tools available for *Capitella* provide an opportunity to investigate the molecular basis of cell fate specification and cellular reprogramming throughout the life cycle, including germ line specification.

9. Unfinished business: Evolution of the segmented body plan

Although I have worked on a variety of projects with *Capitella*, my initial interest in annelids was driven by curiosity about the evolution of the segmented body plan within bilaterians. When I began working with annelids, there was particular excitement about this topic, and a debate as to whether the segments of vertebrates, annelids and arthropods were homologous (Davis & Patel, 1999; Seaver, 2003). An experimental embryological study I performed in leech embryos showed that the segment founder cells of trunk segments developed autonomously and did not require signaling from neighboring cells for formation of normal segment boundaries. This result is dramatically different from the signaling among adjacent cells by *hedgehog*, *wingless*, and *engrailed* required to form segment boundaries in insects. Since leech embryos have a distinct cleavage pattern relative to the spiral cleavage program present in other annelids, I was eager to add to this information by expanding these studies to polychaetes.

Our initial strategy was to clone annelid homologs of "segmentation" genes as defined by their roles in segment formation in *Drosophila* and other arthropods. At the time, the available approach to support or refute an argument of homology utilized comparative gene expression of candidate genes. The logic was that if genes are expressed in banded patterns in the ectoderm or mesoderm ("stripes") at presumptive segment borders, prior to overt segment formation, the genes could be inferred to have a role in the segmentation process. Then, if two animals share genes involved in the segmentation process, this could be interpreted as evidence of homology. My first publication in this area reported expression of the segment polarity gene *engrailed* in the early branching polychaete *Chaetopterus* (Seaver, Paulson, Irvine, & Martindale, 2001). *Ch-en* expression lacked segmental "stripes" in the ectoderm, but rather exhibited expression in subsets of neurons in the ganglia of the central nervous system as well as in mesoderm. We then cloned homologs of *engrailed* and other genes in the insect segmentation cascade from *Capitella*, and *engrailed* from the polychaete *Hydroides elegans* (Seaver & Kaneshige, 2006). To make a comparison with vertebrate somitogenesis, we also analyzed expression of homologs of *Notch*, *delta* and *hes* genes during segment formation in *Capitella* (Thamm & Seaver, 2008). Since this was the pre-genomic era, all of this work was done by degenerate and RACE PCR. Although we observed some interesting patterns and learned a lot about *Capitella* development from this work, none of these genes were expressed in patterns consistent with a role in segment formation. The strongest candidate was *engrailed* in *Capitella* since it showed a repeated ectodermal expression pattern around the time of segment formation. However, since expression was in a limited region and did not extend around the circumference of the body, it was not obviously involved in the process of segment boundary formation.

Around the same time, a group in Europe was examining some of these same genes in the polychaete *Platynereis dumerilii*. They showed banded ectodermal expression of *engrailed* and *wingless* associated with the chaetal sacs in the larva of *Platynereis* and interpreted their expression patterns as support for the homology of segments between annelids and arthropods (Prud'homme et al., 2003). A later *Platynereis* study examined *hedgehog* (*hh*) signaling and found that exposure to chemical inhibitors at early stages did not interfere with initial establishment of segment boundaries. Since exposure at later stages interfered with expression of segmental markers, *hh* was interpreted to have a role in segment maintenance (Dray et al., 2010). Our difference in opinion about the evolution of segmentation in the Bilateria became a public debate (Arendt, 2004)! It is plausible that *Capitella* and *Platynereis* utilize different mechanisms of segment formation, especially considering that they are positioned on distinct branches of the annelid tree (Fig. 1B). It is also possible that the function of *engrailed* in *Hydroides*, *Chaetopterus* and *Capitella* is different from that in *Platynereis*. However, I think the question of whether or not annelid and arthropod segmentation is homologous remains an unresolved issue. Interestingly, multiple *NK* genes show ectodermal striped patterns in *Platynereis* and there is at least one *NK* homolog in *Capitella* with an ectodermal striped pattern that appears around the time the first segments form (Saudemont et al., 2008); ECS, unpublished data). These are good segmentation gene candidates for functional studies. I predict that—at least for *Capitella*—the knockdown of homologs of arthropod segmentation genes will not interfere with segment formation. Fortunately, functional genomic tools now available for both *Capitella* and *Platynereis,* which allows one to test whether segment formation is perturbed following knockdown or knockout of target genes. Additionally, a forward approach could be used to identify genes involved in the segmentation process. The question that drew me to work with annelids years ago still fascinates me, and I look forward to answers generated with these new resources and approaches.

10. Reflections and recommendations

There is growing appreciation that data from a wide range of animals make an important contribution to our understanding of basic biological phenomena. Importantly, there are many questions best investigated by studying animals that exhibit biological properties relevant to a question of interest. It is also advantageous to consider characteristics such as cell size, optical transparency, life history properties, stereotyped cell lineages, ability to culture at ambient temperatures, and available genomic resources when choosing a new animal system. The perspective gained from studies of diverse groups of animals allows for interpretation of data within a phylogenetic context, establishes the direction of evolutionary change, and informs the generality of molecular mechanisms learned from detailed studies in a small number of animals. There are numerous examples in which consideration of data from a broad range of animals has changed the way we think about development and regeneration. Utilizing a comparative intellectual framework gives our results power to be bigger than comparisons among a handful of established genetic model systems.

There are several things to consider when deciding whether to begin working with a less studied animal. I have noticed differences in approach and thinking between those that work with established models and those that work with less studied animals. Some researchers struggle in making a transition to animal systems with fewer tools. When working with a less studied animal, it helps to have an independent and "pioneer spirit," and a willingness to problem solve. It is important to enjoy technique and protocol development. One must also be comfortable with a slower pace of research, a small research community, and fewer tools. However, it is fun to work on a new system. One thing that really drew me in is that there is plenty of opportunity for discovery ("elbow room") compared with some situations working on established animal models that have lots of people working on the same animal and maybe even on the same problem.

In choosing an animal for investigations of a particular question, it is critical to try to identify the best animal to answer the question. Ultimately the specific choice needs to balance a consideration of the animal's natural attributes ("nature's experiments") with practical considerations that either facilitate or impede performing the desired experiments. It is worthwhile to consider a few options early in the process, prioritizing development of animal husbandry methods. I recommend performing pilot experiments to establish feasibility of at least one technique early on that can be used to make real progress on answering the driving question. For the field of development and regeneration, functional genomics or live cell imaging are obvious technique choices. Do not try to do everything at once. I learned the hard way that it is not wise to try to develop tools for multiple animals at the same time. I initially started my lab working with three annelid species. However, with the resources I had as a junior professor, I quickly learned how unrealistic this approach was, and pivoted to a focus primarily on *Capitella*. Developing tools for one animal is challenging enough.

After carefully considering your passion for a particular biological problem and the characteristics of the animal system, go for it! No one system is perfect. There will be both unforeseen challenges and opportunities along the way. Fortunately, existing technologies can rapidly move a little-known study system to one with available genomic and functional tools; the potential for these quick advances is a big improvement from a decade or two ago. Working with collaborators that have distinct skill sets can help move the system forward, and there is plenty of talent looking to apply their skills to a new animal system. However, technologies will change, and it will be challenging to anticipate future conceptual or technical developments in the field. Therefore, it is key to pick a system that is not based on a single attribute or current technology. Your animal system of choice needs to have "staying power." Make an effort to keep your mind and eyes open, as there are lots of interesting and important things to discover if you pay attention. Be open to the unexpected. I strongly recommend diving in to work with a less studied animal that can address a question about which you are passionate. It is a fun adventure!

Acknowledgments

I acknowledge the many folks who have helped me along the way, particularly the efforts of all Seaver lab members past and present. I am grateful to Dr. Sue Hill for teaching me how to culture *Capitella* and to Dr. Judy Grassle for her generosity in supplying animals to me and the community for decades. I gratefully acknowledge the Evelyn and Melvin Spiegel fellowship for summer support at the MBL, which gave me the opportunity to begin working with *Capitella.* I thank my writing partner for fruitful discussions during the writing of this review. This is not a comprehensive review and my apologies for all the interesting *Capitella* studies that were not included here. My inspiration to study early spiralian development comes from the work of Mark Q. Martindale and Jon Q. Henry. I thank current lab members for helpful comments on this manuscript. I am grateful for support of our work on *Capitella* development from the National Science Foundation, which allowed us to try new things, follow our interests, and supported development of functional tools.

References

Ackermann, C., Dorresteijn, A., & Fischer, A. (2005). Clonal domains in postlarval *Platynereis dumerilii* (Annelida: Polychaeta). *Journal of Morphology, 266*, 258–280. https://doi.org/10.1002/jmor.10375.

Aguinaldo, A. M. A., Turbeville, J. M., Linford, L. S., Rivera, M. C., Garey, J. R., Raff, R. A., et al. (1997). Evidence for a clade of nematodes, arthropods and other moulting animals. *Nature, 387*, 489–493. https://doi.org/10.1038/387489a0.

Albalat, R., & Cañestro, C. (2016). Evolution by gene loss. *Nature Reviews. Genetics, 17*(7), 379–391. https://doi.org/10.1038/nrg.2016.39.

Amiel, A. R., Henry, J. Q., & Seaver, E. C. (2013). An organizing activity is required for head patterning and cell fate specification in the polychaete annelid *Capitella teleta*: New insights into cell-cell signaling in Lophotrochozoa. *Developmental Biology, 379*, 107–122. https://doi.org/10.1016/j.ydbio.2013.04.011.

Anderson, D. T. (1959). The embryology of the polychaete *Scoloplos armiger. Journal of Cell Science, s3-100*, 89–166. https://doi.org/10.1242/jcs.s3-100.49.89.

Arendt, D. (2004). Spiralians in the limelight. In *Genome biology* (pp. 1–3). BioMed Central. https://doi.org/10.1186/gb-2003-5-1-303.

Bely, A. E. (2006). Distribution of segment regeneration ability in the Annelida. *Integrative and Comparative Biology, 46*, 508–518. https://doi.org/10.1093/icb/icj051.

Berriman, M., Haas, B. J., Loverde, P. T., Wilson, R. A., Dillon, G. P., Cerqueira, G. C., et al. (2009). The genome of the blood fluke *Schistosoma mansoni. Nature, 460*, 352–358. https://doi.org/10.1038/nature08160.

Bertrand, N., Castro, D. S., & Guillemot, F. (2002). Proneural genes and the specification of neural cell types. *Nature Reviews. Neuroscience, 3*, 517–530. https://doi.org/10.1038/nrn874.

Blake, J. A., Grassle, J. P., & Eckelbarger, K. J. (2009). *Capitella teleta*, a new species designation for the opportunistic and experimental *Capitella* sp. I, with a review of the literature for confirmed records. *Zoosymposia, 53*, 25–53.

Boyle, M. J., & Seaver, E. C. (2008). Developmental expression of *foxA* and *gata* genes during gut formation in the polychaete annelid, *Capitella* sp. I. *Evolution and Development, 10*, 89–105. https://doi.org/10.1111/j.1525-142X.2007.00216.x.

Boyle, M., & Seaver, E. (2009). Evidence of a dorsal pharynx in the marine polychaete *Capitella teleta* (Polychaeta: Capitellidae). *Zoosymposia, 2*, 317–328.

Butman, C. A., Grossie, J. P., & Buskey, E. J. (1988). Horizontal swimming and gravitational sinking of *Capitella* sp. I (annelida: Polychaeta) larvae: Implications for settlement. *Ophelia, 29*, 43–57. https://doi.org/10.1080/00785326.1988.10430818.

Carrillo-Baltodano, A. M. (2019). *Understanding the evolution of animal body plans by looking at annelid neural development*. Clark University.

Carrillo-Baltodano, A. M., & Meyer, N. P. (2017). Decoupling brain from nerve cord development in the annelid *Capitella teleta*: Insights into the evolution of nervous systems. *Developmental Biology, 431*, 134–144. https://doi.org/10.1016/j.ydbio.2017.09.022.

Cather, J. N., & Verdonk, N. H. (1979). Development of *Dentalium* following removal of D-quadrant blastomeres at successive cleavage stages. *Wilhelm Roux's Archives of Developmental Biology, 187*, 355–366. https://doi.org/10.1007/BF00848469.

Cho, S. J., Vallès, Y., Giani, V. C., Seaver, E. C., & Weisblat, D. A. (2010). Evolutionary dynamics of the *wnt* gene family: A lophotrochozoan perspective. *Molecular Biology and Evolution, 27*, 1645–1658. https://doi.org/10.1093/molbev/msq052.

Clement, A. C. (1962). Development of *Ilyanassa* following removal of the D macromere at successive cleavage stages. *The Journal of Experimental Zoology, 149*, 193–215. https://doi.org/10.1002/jez.1401490304.

Cohen, R. A., & Pechenik, J. A. (1999). Relationship between sediment organic content, metamorphosis, and postlarval performance in the deposit-feeding polychaete *Capitella* sp. I. *Journal of Experimental Marine Biology and Ecology, 240*, 1–18. https://doi.org/10.1016/S0022-0981(99)00047-7.

Conklin, E. G. (1897). The embryology of *Crepidula*, a contribution to the cell lineage and early development of some marine gasteropods. *Journal of Morphology, XIII*, 1–210.

Costello, D. P., Davidson, M. E., Eggers, A., Fox, M. H., & Henley, C. (1957). *Methods for obtaining and handling marine eggs and embryos* (1st ed.). Lancaster, PA: Lancaster Press.

Dannenberg, L. C., & Seaver, E. C. (2018). Regeneration of the germline in the annelid *Capitella teleta*. *Developmental Biology, 440*. https://doi.org/10.1016/j.ydbio.2018.05.004.

Davis, G. K., & Patel, N. H. (1999). The origin and evolution of segmentation. *Trends in Genetics*. https://doi.org/10.1016/S0168-9525(99)01875-2.

de Jong, D. M., & Seaver, E. C. (2016). A stable thoracic Hox code and epimorphosis characterize posterior regeneration in *Capitella teleta*. *PLoS One, 11*. https://doi.org/10.1371/journal.pone.0149724.

De Robertis, E. M., & Kuroda, H. (2004). Dorsal-ventral patterning and neural induction in *Xenopus* embryos. *Annual Review of Cell and Developmental Biology*. https://doi.org/10.1146/annurev.cellbio.20.011403.154124.

De Robertis, E. M., & Sasai, Y. (1996). A common plan for dorsoventral patterning in Bilateria. *Nature, 380*, 37–40. https://doi.org/10.1038/380037a0.

Denes, A. S., Jékely, G., Steinmetz, P. R. H., Raible, F., Snyman, H., Prud'homme, B., et al. (2007). Molecular architecture of annelid nerve cord supports common origin of nervous system centralization in Bilateria. *Cell, 129*, 277–288. https://doi.org/10.1016/j.cell.2007.02.040.

Dill, K. K., & Seaver, E. C. (2008). *Vasa* and *nanos* are coexpressed in somatic and germ line tissue from early embryonic cleavage stages through adulthood in the polychaete *Capitella* sp. I. *Development Genes and Evolution, 218*. https://doi.org/10.1007/s00427-008-0236-x.

Doe, C. Q., Fuerstenberg, S., & Peng, C.-Y. (1998). Neural stem cells: From fly to vertebrates. *Journal of Neurobiology, 36*(2), 111–127. https://doi.org/10.1002/(SICI)1097-4695(199808)36:2.

Doe, C. Q., & Goodman, C. S. (1985). Early events in insect neurogenesis. I. Development and segmental differences in the pattern of neuronal precursor cells. *Developmental Biology, 111*, 193–205. https://doi.org/10.1016/0012-1606(85)90445-2.

Dohle, W. (1999). The ancestral cleavage pattern of the clitellates and its phylogenetic deviations. In *Reproductive strategies and developmental patterns in annelids* (pp. 267–283). Springe Netherlands. https://doi.org/10.1007/978-94-017-2887-4_16.

Dray, N., Tessmar-Raible, K., Le Gouar, M., Vibert, L., Christodoulou, F., Schipany, K., et al. (2010). Hedgehog signaling regulates segment formation in the annelid *Platynereis*. *Science (80-.), 329*, 339–342. https://doi.org/10.1126/science.1188913.

Dubilier, N. (1988). H_2S—A settlement cue or a toxic substance for *Capitella* sp. I larvae? *The Biological Bulletin*, *174*, 30–38. https://doi.org/10.2307/1541756.

Eckelbarger, K. J., & Grassle, J. P. (1983). Ultrastructural differences in the eggs and ovarian follicle cells of *Capitella* (Polychaeta) sibling species. *The Biological Bulletin*, *165*, 379–393. https://doi.org/10.2307/1541203.

Eckelbarger, K. J., & Grassle, J. P. (1987). Spermatogenesis, sperm storage and comparative sperm morphology in nine species of *Capitella*, *Capitomastus* and *Capitellides* (Polychaeta: Capitellidae). *Marine Biology*, *95*, 415–429. https://doi.org/10.1007/BF00409572.

Eisig, H. (1898). Zur Entwicklungsgeschichte der Capitelliden. *Mitteilungen aus der Zoologischen Station zu Neapel*, *13*, 1–292.

Fauchald, K. (1977). *The polychaete worms definitions and keys to the orders, families and genera.*

Ferguson, E. L. (1996). Conservation of dorsal ventral patterning in arthropods and chordates. *Current Opinion in Genetics & Development*, *6*, 424–431. https://doi.org/10.1016/S0959-437X(96)80063-3.

Fröbius, A. C., Matus, D. Q., & Seaver, E. C. (2008). Genomic organization and expression demonstrate spatial and temporal Hox gene colinearity in the lophotrochozoan *Capitella* sp. I. *PLoS One*, *3*. https://doi.org/10.1371/journal.pone.0004004.

Giani, V. C., Yamaguchi, E., Boyle, M. J., & Seaver, E. C. (2011). Somatic and germline expression of *piwi* during development and regeneration in the marine polychaete annelid *Capitella teleta*. *EvoDevo*, *2*, 10. https://doi.org/10.1186/2041-9139-2-10.

Goto, A., Kitamura, K., Arai, A., & Shimizu, T. (1999). Cell fate analysis of teloblasts in the *Tubifex* embryo by intracellular injection of HRP. *Development, Growth & Differentiation*, *41*, 703–713. https://doi.org/10.1046/j.1440-169X.1999.00469.x.

Grassle, J. P., & Frederick Grassle, J. (1976). Sibling species in the marine pollution indicator *Capitella* (polychaeta). *Science*, *192*, 567–569. https://doi.org/10.1126/science.1257794.

Grassle, J. F., & Grassle, J. P. (1974). Opportunistic life histories and genetic systems in marine benthic polychaetes. *Journal of Marine Research*, *32*, 253–284.

Guijarro-Clarke, C., Holland, P. W. H., & Paps, J. (2020). Widespread patterns of gene loss in the evolution of the animal kingdom. *Nature Ecology & Evolution*, *4*, 519–523. https://doi.org/10.1038/s41559-020-1129-2.

Hanken, J., Jennings, D. H., & Olsson, L. (1997). Mechanistic basis of life-history evolution in anuran amphibians: Direct development. *American Zoologist*, *37*, 160–171. https://doi.org/10.1093/icb/37.2.160.

Hartenstein, V., & Campos-Ortega, J. A. (1984). Early neurogenesis in wild-type *Drosophila melanogaster*. *Roux's Archives of Developmental Biology*, *193*(5), 308–325.

Hejnol, A. (2010). A twist in time-the evolution of spiral cleavage in the light of animal phylogeny. *Integrative and Comparative Biology*, *50*, 695–706. https://doi.org/10.1093/icb/icq103.

Henry, J. J. (1986). The role of unequal cleavage and the polar lobe in the segregation of developmental potential during first cleavage in the embryo of *Chætopterus variopedatus*. *Roux's Archives of Developmental Biology*, *195*, 103–116. https://doi.org/10.1007/BF00456106.

Henry, J. J., & Martindale, M. Q. (1999). Conservation and innovation in spiralian development. *Hydrobiologia*, *402*, 255–265. https://doi.org/10.1023/A:1003756912738.

Henry, J. Q., Perry, K. J., & Martindale, M. Q. (2006). Cell specification and the role of the polar lobe in the gastropod mollusc *Crepidula fornicata*. *Developmental Biology*, *297*, 295–307. https://doi.org/10.1016/j.ydbio.2006.04.441.

Holley, S. A., Jackson, P. D., Sasai, Y., Lu, B., de Robertis, E. M., Hoffmann, F. M., et al. (1995). A conserved system for dorsal-ventral patterning in insects and vertebrates involving sog and chordin. *Nature*, *376*, 249–253. https://doi.org/10.1038/376249a0.

Jacobson, M., & Rao, M. S. (2005). *Developmental neurobiology*. New York: Kluwer Academic/Plenum.

Kenny, N. J., Namigai, E. K. O., Dearden, P. K., Hui, J. H. L., Grande, C., & Shimeld, S. M. (2014). The lophotrochozoan TGF-α signalling cassette—Diversification and conservation in a key signalling pathway. *The International Journal of Developmental Biology, 58*, 533–549. https://doi.org/10.1387/ijdb.140080nk.

Kerner, P., Simionato, E., Le Gouar, M., & Vervoort, M. (2009). Orthologs of key vertebrate neural genes are expressed during neurogenesis in the annelid *Platynereis dumerilii*. *Evolution & Development, 11*, 513–524. https://doi.org/10.1111/j.1525-142X.2009.00359.x.

Klann, M., & Seaver, E. C. (2019). Functional role of pax6 during eye and nervous system development in the annelid *Capitella teleta*. *Developmental Biology, 456*. https://doi.org/10.1016/j.ydbio.2019.08.011.

Kuo, D. H., & Weisblat, D. A. (2011). A new molecular logic for BMP-mediated dorsoventral patterning in the leech *Helobdella*. *Current Biology, 21*, 1282–1288. https://doi.org/10.1016/j.cub.2011.06.024.

Lacomme, M., Liaubet, L., Pituello, F., & Bel-Vialar, S. (2012). NEUROG2 drives cell cycle exit of neuronal precursors by specifically repressing a subset of cyclins acting at the G1 and S phases of the cell cycle. *Molecular and Cellular Biology, 32*, 2596–2607. https://doi.org/10.1128/mcb.06745-11.

Lambert, J. D., Johnson, A. B., Hudson, C. N., & Chan, A. (2016). Dpp/BMP2-4 mediates signaling from the D-quadrant organizer in a spiralian embryo. *Current Biology, 26*, 2003–2010. https://doi.org/10.1016/j.cub.2016.05.059.

Lanza, A. R., & Seaver, E. C. (2018). An organizing role for the TGF-β signaling pathway in axes formation of the annelid *Capitella teleta*. *Developmental Biology, 435*, 26–40. https://doi.org/10.1016/j.ydbio.2018.01.004.

Lanza, A. R., & Seaver, E. C. (2020a). Functional evidence that activin/nodal signaling is required for establishing the dorsal-ventral axis in the annelid *Capitella teleta*. *Development, 147*. https://doi.org/10.1242/dev.189373.

Lanza, A. R., & Seaver, E. C. (2020b). Activin/nodal signaling mediates dorsal-ventral axis formation before third quartet formation in embryos of the annelid *Chaetopterus pergamentaceus*. *EvoDevo, 11*, 17. https://doi.org/10.1186/s13227-020-00161-y.

Lyons, D. C., Perry, K. J., Batzel, G., & Henry, J. Q. (2020). BMP signaling plays a role in anterior-neural/head development, but not organizer activity, in the gastropod *Crepidula fornicata*. *Developmental Biology, 463*, 135–157. https://doi.org/10.1016/j.ydbio.2020.04.008.

Lyons, D. C., Perry, K. J., Lesoway, M. P., & Henry, J. Q. (2012). Cleavage pattern and fate map of the mesentoblast, 4d, in the gastropod *Crepidula*: A hallmark of spiralian development. *EvoDevo, 3*, 1. https://doi.org/10.1186/2041-9139-3-21.

Martindale, M. Q. (1986). The "organizing" role of the D quadrant in an equal-cleaving spiralian, *Lymnaea stagnalis* as studied by UV laser deletion of macromeres at intervals between third and fourth quartet formation. *International Journal of Invertebrate Reproduction and Development, 9*, 229–242. https://doi.org/10.1080/01688170.1986.10510198.

Martindale, M. Q., & Henry, J. Q. (1995). Modifications of cell fate specification in equal-cleaving nemertean embryos: Alternate patterns of spiralian development. *Development, 121*, 3175–3185.

Martin-Duran, J. M., & Marletaz, F. (2020). Unravelling spiral cleavage. *Development, 147*. https://doi.org/10.1242/dev.181081.

Martín-Durán, J. M., Vellutini, B. C., Marlétaz, F., Cetrangolo, V., Cvetesic, N., Thiel, D., et al. (2021). Conservative route to genome compaction in a miniature annelid. *Nature Ecology & Evolution, 5*, 231–242. https://doi.org/10.1038/s41559-020-01327-6.

Merz, R. A., & Woodin, S. A. (2006). Polychaete chaetae: Function, fossils, and phylogeny. *Integrative and Comparative Biology, 46*, 481–496. https://doi.org/10.1093/icb/icj057.

Meyer, N. P., Boyle, M. J., Martindale, M. Q., & Seaver, E. C. (2010). A comprehensive fate map by intracellular injection of identified blastomeres in the marine polychaete *Capitella teleta*. *EvoDevo*, *1*. https://doi.org/10.1186/2041-9139-1-8.

Meyer, N. P., Carrillo-Baltodano, A., Moore, R. E., & Seaver, E. C. (2015). Nervous system development in lecithotrophic larval and juvenile stages of the annelid *Capitella teleta*. *Frontiers in Zoology*, *12*. https://doi.org/10.1186/s12983-015-0108-y.

Meyer, N. P., & Seaver, E. C. (2009). Neurogenesis in an annelid: Characterization of brain neural precursors in the polychaete *Capitella* sp. I. *Developmental Biology*, *335*. https://doi.org/10.1016/j.ydbio.2009.06.017.

Meyer, N. P., & Seaver, E. C. (2010). Cell lineage and fate map of the primary somatoblast of the polychaete annelid *Capitella teleta*. *Integrative and Comparative Biology*, *50*. https://doi.org/10.1093/icb/icq120.

Neal, S., De Jong, D. M., & Seaver, E. C. (2019). CRISPR/CAS9 mutagenesis of a single *r-opsin* gene blocks phototaxis in a marine larva. *Proceedings of the Royal Society B: Biological Sciences*, *286*. https://doi.org/10.1098/rspb.2018.2491.

Özpolat, B. D., & Bely, A. E. (2016). Developmental and molecular biology of annelid regeneration: A comparative review of recent studies. *Current Opinion in Genetics & Development*, *40*, 144–153. https://doi.org/10.1016/j.gde.2016.07.010.

Pernet, B., Amiel, A., & Seaver, E. C. (2012). Effects of maternal investment on larvae and juveniles of the annelid *Capitella teleta* determined by experimental reduction of embryo energy content. *Invertebrate Biology*, *131*. https://doi.org/10.1111/j.1744-7410.2012.00263.x.

Pfister, D., De Mulder, K., Hartenstein, V., Kuales, G., Borgonie, G., Marx, F., et al. (2008). Flatworm stem cells and the germ line: Developmental and evolutionary implications of macvasa expression in *Macrostomum lignano*. *Developmental Biology*, *319*, 146–159. https://doi.org/10.1016/j.ydbio.2008.02.045.

Prud'homme, B., De Rosa, R., Arendt, D., Julien, J. F., Pajaziti, R., Dorresteijn, A. W. C., et al. (2003). Arthropod-like expression patterns of engrailed and wingless in the annelid *Platynereis dumerilii* suggest a role in segment formation. *Current Biology*, *13*, 1876–1881. https://doi.org/10.1016/j.cub.2003.10.006.

Raff, R. A. (1996). *The shape of life: Genes, development and the evolution of animal form*. Chicago: University of Chicago Press.

Rebscher, N. (2014). Establishing the germline in spiralian embryos. *The International Journal of Developmental Biology*, *58*, 403–411. https://doi.org/10.1387/ijdb.140125nr.

Reddy, C. M., Eglinton, T. I., Hounshell, A., White, H. K., Xu, L., Gaines, R. B., et al. (2002). The West Falmouth oil spill after thirty years: The persistence of petroleum hydrocarbons in marsh sediments. *Environmental Science & Technology*, *36*, 4754–4760. https://doi.org/10.1021/es020656n.

Robb, S. M. C., Ross, E., & Alvarado, A. S. (2008). SmedGD: The *Schmidtea mediterranea* genome database. *Nucleic Acids Research*, *36*, 599–606. https://doi.org/10.1093/nar/gkm684.

Rouse, G. W., & Pleijel, F. (2001). *Polychaetes*. New York: Oxford University Press.

Ryan, J. F., Pang, K., Schnitzler, C. E., Nguyen, A. D., Moreland, R. T., Simmons, D. K., et al. (2013). The genome of the ctenophore *Mnemiopsis leidyi* and its implications for cell type evolution. *Science*, *342*. https://doi.org/10.1126/science.1242592.

Sato, K., Shibata, N., Orii, H., Amikura, R., Sakurai, T., Agata, K., et al. (2006). Identification and origin of the germline stem cells as revealed by the expression of *nanos*-related gene in planarians. *Development, Growth & Differentiation*, *48*, 615–628. https://doi.org/10.1111/j.1440-169X.2006.00897.x.

Saudemont, A., Dray, N., Hudry, B., Le Gouar, M., Vervoort, M., & Balavoine, G. (2008). Complementary striped expression patterns of NK homeobox genes during segment formation in the annelid *Platynereis*. *Developmental Biology*, *317*, 430–443. https://doi.org/10.1016/j.ydbio.2008.02.013.

Schmidt-Rhaesa, A. (2007). *The evolution of organ systems*. New York: Oxford University Press.
Schweigkofler, M., Bartolomaeus, T., & von Salvini-Plawen, L. (1998). Ultrastructure and formation of hooded hooks in *Capitella capitata* (Annelida, Capitellida). *Zoomorphology*, *118*, 117–128.
Seaver, E. C., & de Jong, D. M. (2021). Regeneration in the segmented annelid *Capitella teleta*. *Genes*, *12*(11), 1769. https://doi.org/10.3390/genes12111769.
Seaver, E. C. (2003). Segmentation: Mono- or polyphyletic? *The International Journal of Developmental Biology*, *47*, 583–595.
Seaver, E. C. (2016). Annelid models I: *Capitella teleta*. *Current Opinion in Genetics & Development*, *39*. https://doi.org/10.1016/j.gde.2016.05.025.
Seaver, E. C. (2017). Annelids shed light on the evolution of spiralian development. *Canadian Journal of Zoology*, *95*, 705–712. https://doi.org/10.1139/cjz-2016-0261.
Seaver, E. C., & Kaneshige, L. M. (2006). Expression of "segmentation" genes during larval and juvenile development in the polychaetes *Capitella* sp. I and *H. elegans*. *Developmental Biology*, *289*. https://doi.org/10.1016/j.ydbio.2005.10.025.
Seaver, E. C., Paulson, D. A., Irvine, S. Q., & Martindale, M. Q. (2001). The spatial and temporal expression of *Ch-en*, the *engrailed* gene in the polychaete *Chaetopterus*, does not support a role in body axis segmentation. *Developmental Biology*, *236*. https://doi.org/10.1006/dbio.2001.0309.
Seaver, E. C., & Shankland, M. (2000). Leech segmental repeats develop normally in the absence of signals from either anterior or posterior segments. *Developmental Biology*, *224*. https://doi.org/10.1006/dbio.2000.9781.
Seaver, E. C., & Shankland, M. (2001). Establishment of segment polarity in the ectoderm of the leech *Helobdella*. *Development*, *128*.
Seaver, E. C., Thamm, K., & Hill, S. D. (2005). Growth patterns during segmentation in the two polychaete annelids, *Capitella* sp. I and *Hydroides elegans*: Comparisons at distinct life history stages. *Evolution & Development*, *7*. https://doi.org/10.1111/j.1525-142X.2005.05037.x.
Shankland, M. (1991). Leech segmentation: Cell lineage and the formation of complex body patterns. *Developmental Biology*. https://doi.org/10.1016/0012-1606(91)90416-Z.
Simakov, O., Marletaz, F., Cho, S. J., Edsinger-Gonzales, E., Havlak, P., Hellsten, U., et al. (2013). Insights into bilaterian evolution from three spiralian genomes. *Nature*, *493*, 526–531. https://doi.org/10.1038/nature11696.
Struck, T. H., Paul, C., Hill, N., Hartmann, S., Hösel, C., Kube, M., et al. (2011). Phylogenomic analyses unravel annelid evolution. *Nature*, *471*, 95–98. https://doi.org/10.1038/nature09864.
Sur, A., Magie, C. R., Seaver, E. C., & Meyer, N. P. (2017). Spatiotemporal regulation of nervous system development in the annelid *Capitella teleta*. *EvoDevo*, *8*. https://doi.org/10.1186/s13227-017-0076-8.
Sur, A., & Meyer, N. P. (2021). Resolving transcriptional states and predicting lineages in the annelid *Capitella teleta* using single-cell RNAseq. *Frontiers in Ecology and Evolution*, *8*, 618007. https://doi.org/10.3389/fevo.2020.618007.
Sur, A., Renfro, A., Bergmann, P. J., & Meyer, N. P. (2020). Investigating cellular and molecular mechanisms of neurogenesis in *Capitella teleta* sheds light on the ancestor of Annelida. *BMC Evolutionary Biology*, *20*, 1–29. https://doi.org/10.1186/s12862-020-01636-1.
Thamm, K., & Seaver, E. C. (2008). Notch signaling during larval and juvenile development in the polychaete annelid *Capitella* sp. I. *Developmental Biology*, *320*. https://doi.org/10.1016/j.ydbio.2008.04.015.
Wang, Y., Zayas, R. M., Guo, T., & Newmark, P. A. (2007). Nanos function is essential for development and regeneration of planarian germ cells. *Proceedings of the National Academy of Sciences of the United States of America*, *104*, 5901–5906. https://doi.org/10.1073/pnas.0609708104.

Webster, N. B., Corbet, M., Sur, A., & Meyer, N. P. (2021). Role of BMP signaling during early development of the annelid *Capitella teleta*. *bioRxiv*. 2020.11.15.383695. https://doi.org/10.1101/2020.11.15.383695.

Weigert, A., & Bleidorn, C. (2016). Current status of annelid phylogeny. *Organisms, Diversity and Evolution*, *16*, 345–362. https://doi.org/10.1007/s13127-016-0265-7.

Weigert, A., Helm, C., Meyer, M., Nickel, B., Arendt, D., Hausdorf, B., et al. (2014). Illuminating the base of the annelid tree using transcriptomics. *Molecular Biology and Evolution*, *31*, 1391–1401. https://doi.org/10.1093/molbev/msu080.

Weisblat, D. A., Kim, S. Y., & Stent, G. S. (1984). Embryonic origins of cells in the leech *Helobdella triserialis*. *Developmental Biology*, *104*, 65–85. https://doi.org/10.1016/0012-1606(84)90037-X.

Wessel, G. M., Morita, S., & Oulhen, N. (2020). Somatic cell conversion to a germ cell lineage: A violation or a revelation? *Journal of Experimental Zoology. Part B, Molecular and Developmental Evolution*. https://doi.org/10.1002/jez.b.22952.

Wilson, E. B. (1892). The cell lineage of *Nereis*. A contribution to the cytogeny of the annelid body. *Journal of Morphology*, *6*, 361–480.

Wilson, E. B. (1898). Considerations on cell-lineage and ancestral reminiscence based on a re-examination of some points in the early development of annelids and polyclades. *Annals of the New York Academy of Sciences*, 1–27.

Wilson, S. I., & Edlund, T. (2001). Neural induction: Toward a unifying mechanism. *Nature Neuroscience*, *4*, 1161–1168. https://doi.org/10.1038/nn747.

Wu, L., Hiebert, L. S., Klann, M., Passamaneck, Y., Bastin, B. R., Schneider, S. Q., et al. (2020). Genes with spiralian-specific protein motifs are expressed in spiralian ciliary bands. *Nature Communications*, *11*, 1–11. https://doi.org/10.1038/s41467-020-17780-7.

Yamaguchi, E., Dannenberg, L. C., Amiel, A. R., & Seaver, E. C. (2016). Regulative capacity for eye formation by first quartet micromeres of the polychaete *Capitella teleta*. *Developmental Biology*, *410*, 119–130. https://doi.org/10.1016/j.ydbio.2015.12.009.

Yamaguchi, E., & Seaver, E. C. (2013). The importance of larval eyes in the polychaete *Capitella teleta*: Effects of larval eye deletion on formation of the adult eye. *Invertebrate Biology*, *132*. https://doi.org/10.1111/ivb.12034.

Zakas, C., Harry, N. D., Scholl, E. H., & Rockman, M. V. (2021). The genome of the poecilogonous annelid *Streblospio benedicti*. *bioRxiv*. 2021.04.15.440069. https://doi.org/10.1101/2021.04.15.440069.

Zhou, Y., Zheng, H., Chen, Y., Zhang, L., Wang, K., Guo, J., et al. (2009). The *Schistosoma japonicum* genome reveals features of host-parasite interplay. *Nature*, *460*, 345–351. https://doi.org/10.1038/nature08140.

CHAPTER FIFTEEN

Glossiphoniid leeches as a touchstone for studies of development in clitellate annelids

David A. Weisblat*
Department of Molecular & Cell Biology, University of California, Berkeley, CA, United States
*Corresponding author: e-mail address: weisblat@berkeley.edu

Contents

Abstract

My goals in this chapter are to share my enthusiasm for studying the biology of leeches, to place this work in context by presenting my rationale for studying non-traditional biological models in general, and to sample just three of the questions that intrigue me in leech biology, namely segmentation, genome evolution and neuronal fate specification. I first became excited about the idea of using leeches as a subject of investigation as an undergraduate in 1970 and have been engaged in this work since I arrived at Berkeley as a postdoc in 1976, intending to study leech neurobiology. Both my research interests and the rationale for the work have expanded greatly since then. What follows is a fragmentary personal and historical account–the interested reader may find more comprehensive treatments elsewhere (Kuo et al., 2020; Shankland & Savage, 1997; Shain, 2009; Weisblat & Huang, 2001; Weisblat & Kuo, 2009, 2014; Weisblat & Winchell, 2020).

Current Topics in Developmental Biology, Volume 147
ISSN 0070-2153
https://doi.org/10.1016/bs.ctdb.2021.12.019

1. Deep background

My approach to science, and life in general is shaped by an underlying philosophy that I would describe as a good-natured and occasionally optimistic nihilism, rooted in the firm knowledge that life on earth and even earth itself are ultimately doomed. As the sun ineluctably burns through its supply of hydrogen, and progresses to the red giant phase, in about 5 billion years, it is expected to expand to the point at which it will engulf the earth (https://skyandtelescope.org/astronomy-news/the-eventual-fate-of-our-solar-system/). In fact, however, the gradually increasing solar luminosity will render the planet uninhabitable long before then, well within 1 billion years (https://faculty.wcas.northwestern.edu/~infocom/The%20Website/evolution.html), which lends greater urgency to the sense of looming catastrophe.

In light of this inevitable 1–2 punch, what's the point in worrying about global warming or anything else? Here's where the optimism comes in. Like a pinball player who keeps at it, knowing they are destined to lose eventually, I see humanity's ultimate purpose as struggling valiantly against these long odds–if we can avoid the immediate threats of anthropogenic climate change, mass extinction, and the possible collapse of civilization, and if we can continue to progress scientifically and technologically–staving off the asteroids as we go–maybe, just maybe, we can reach the point of being able to fashion a way to escape from earth, and ultimately the solar system, with the intellectual and material resources needed to colonize new worlds…hopping from place to place until the Universe itself grows cold, or maybe even finding a wormhole to a more hospitable one.

It goes without saying that the scope of knowledge required for such an ambitious undertaking is immense. In addition to the many technological advances in other fields, it certainly includes a deep understanding of global biological diversity and all the incidental discoveries we might make along the way to such understanding. We should shoot for knowing not only the genomes, development, physiology, behavior and ecological relationships of all living things, but also their evolutionary histories and potential trajectories.

2. Origins of biodiversity

The diversification of species reflects the accumulation of phenotypic differences between reproductively isolated populations over time–differences in everything from the absolute and relative sizes and shapes of different body parts down to cellular and sub-cellular structures and processes within organ systems or organs that affect physiology and behavior.

These phenotypic changes affect how organisms interact with other organisms and their physical environment (through natural selection and individual fitness), and they themselves reflect changes in developmental processes by which phenotypic features arise.

Changes in developmental processes usually reflect changes in the genome, with cascading effects on all the processes mentioned above. But genomes, development, physiology and morphology can change in many different ways, and the effects of a single change in any one structure or process may have effects ranging from immeasurably minor to profound. For now at least, the linkages by which changes at one level effect changes at other levels are largely unpredictable. Moreover, while we scientists seek to manipulate complex systems systematically, by changing one variable while holding all others constant, nature suffers no such limitations.

To begin to appreciate this immense complexity, consider the fact that, using the molecular genetic machinery of life here on earth, a single moderately sized protein of 100 amino acids, with 20 possible amino acids at each position, could exist in any of 20^{100} different versions, far outstripping the estimated total of 10^{80} atoms in the universe from which they might be made. Now factor in the number of proteins encoded in a typical animal genome (typically 20,000–30,000), the numerous possibilities for alternative splicing and post-translational modifications, and the multifactorial genetic elements (cis and trans acting activators and repressors, and epigenetic modifications) that regulate the temporal, spatial and quantitative expression of each gene), and the functional consequences that changing any of these features may have, at levels ranging from individual proteins and macromolecular assemblies up to cellular and physiological processes.

Thus, we may view biological evolution as populations of genetically clustered and reproductively linked entities (species) progressing along a narrow trajectory through an (essentially) infinitely dimensioned matrix of evolutionary and developmental possibilities that I call Evo-Devo Space, subject only to the constraint that some of the individuals are reproductively successful at each point along the way. At each point in the matrix, there are so many possible "next steps" that it is impossible to even begin sampling all the possibilities, and pretty much impossible for a species to return to a previous state. By this argument, it would seem that even the incredible diversity of life that has or does exist on earth is an incomplete sampling of what is possible.

Practitioners in the field of evolutionary developmental biology (Evo-Devo) are charged with exploring Evo-Devo Space by expanding our knowledge beyond the intensively studied models. In what follows, I hope to explain how studies of leeches generally, and in particular of the glossiphoniid species *Helobdella austinensis* (Fig. 1), fit in to this enterprise.

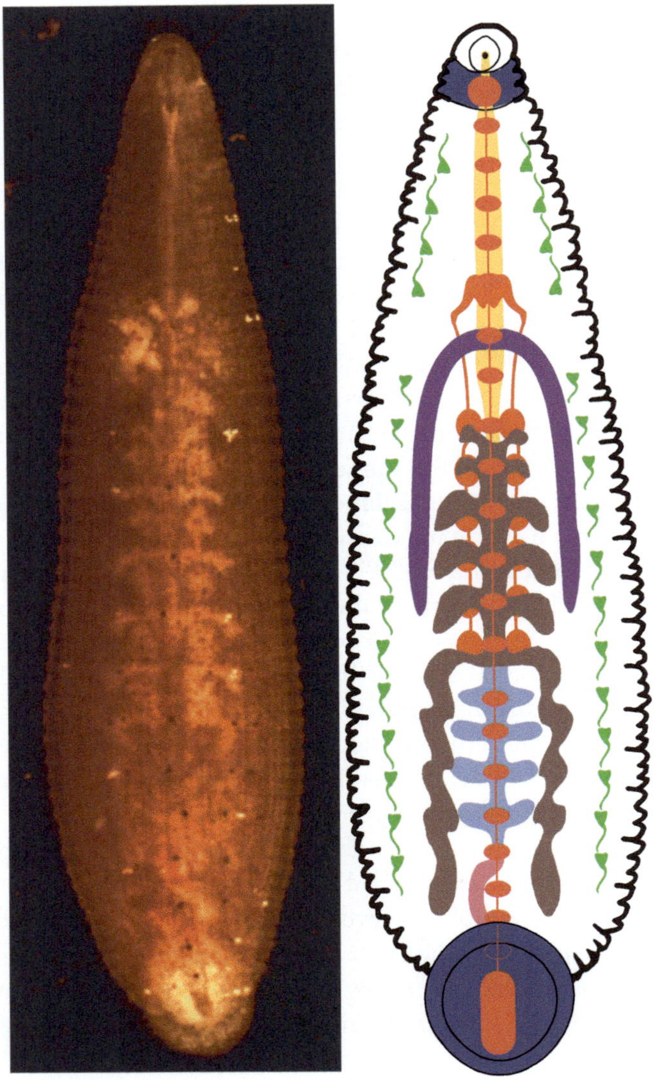

Fig. 1 Segmentation of the leech body plan. Photographic (*left*) and schematic (*right*) representations of a *Helobdella austinensis* adult (ventral views with anterior up). The rostral and caudal tagmata contain four and seven fused segments (*dark blue*), respectively, with the latter forming the rear sucker. The midbody tagma consists of 21 separate segments. Segmental muscles are not shown. Ectodermal segmentation is indicated by: 3 epidermal annuli in each segment (small *black bulges*, flanked by deeper constrictions of the body wall to mark segment boundaries); bilaterally symmetric ganglia of the ventral nerve cord, linked by connective nerves (small red ovals and *red lines*, respectively; due to segment fusion, the rostral and caudal tagmata possess larger

3. Ontogeny and phylogeny of leech developmental and neuro biology: 19th Century roots

Various cultures have used species of bloodsucking leeches as medical devices for centuries (Payton, 1981). In modern medicine, they are sometimes used for minimally invasive treatment of venous congestion after reconstructive surgery (Hackenberger & Janis, 2019). But the use of leeches in basic scientific investigations came in the late 19th century from two independent branches.

First, the pioneering American zoologist Charles O. Whitman studied embryonic cell lineages in *Clepsine marginata* (now *Hemiclepsis marginata*), for his PhD thesis at the University of Leipzig (1878, 1887). As described by Maienschein (1978), this work was carried on in the context of Ernst Haeckel's biogenetic "law" that the development of each species is a shortened and accelerated recapitulation of the adult forms of that species' ancestors, from unicellular and colonial organisms onwards through progressively more refined forms. A corollary of this law was that the early cell divisions are purely proliferative and that differentiation among cells somehow begins at gastrulation. Observations by Whitman and his contemporaries contradict this notion. They observed clearly unequal divisions in various embryos, starting with first cleavage in leech.

Whitman's work provided a remarkably accurate and detailed description of development in glossiphoniid leech species that was the starting point for my own postdoctoral work at Berkeley in the 1970s and 1980s. As part of his thesis, Whitman apparently guided the fabrication of a wonderful set of wax models representing leech embryos at various stages of development.

---

ganglia (large *red ovals*) than midbody segments; the nerve cord in midbody segment 21 (*dashed oval* and *line*) is obscured by the posterior sucker; segmental components of the peripheral nervous system are not shown. Mesodermal segmentation is indicated by the male genital organs and multiple paired testisacs (*orange*), which develop in midbody segments 5 and 8–13, respectively; the female reproductive system (*purple*), originating in midbody segment 6, with ovaries extending posteriorly; the excretory system of paired metanephridia (*green*) in midbody segments 2–5 and 8–18 (these two sets have reversed orientation relative to one another). Endodermal segmentation is indicated by ceca of the midgut, which form distinct segmental evaginations in the crop (taupe) and the intestine (*light blue*). Non-segmented gut structures include the proboscis and esophagus (*yellow*) and the rectum (*pink*). *Adapted from Weisblat, D.A. & Winchell, C.J. (2020). Segmentation in leech. In Chipman, A. ed.* Cellular processes in segmentation, *CRC Press. Chapter 7, pp. 153–181.*

At the time, this was a fairly common practice–commercial studios made and sold sets of wax models as hands-on teaching aids. A set of the models made based on Whitman's drawings were discovered and documented in 2017 (https://blogs.ucl.ac.uk/museums/2017/05/12/specimen-of-the-week-291-leech-embryo-models/).

Whitman came from a fascinating family. A biographical memoir presented to the National Academy of Sciences by Edward Morse (1912) documents a streak of fanatical obstinacy on his father's side of the family, including "one John Whitman (who) left his family at Rye, England, in 1572, crossed over to Ostend (Belgium) 'and there demeaning himself fanatically and offensively to Roman Catholics of that place was apprehended and consigned to the flames in which he perished without repining and in seeming exultation'.". Whitman's father carried on the trait, displaying a "fanatical adherence to the idea that the world was coming to an end at three o'clock P. M. (Eastern Standard Time?) on a certain date in 1843". Born in 1842, Charles Otis Whitman sought to enlist in the Union Army, but was prevented from doing so by his pacifist family. Despite his promising start on leech embryology, he became better known for later work with pigeons, as a founder of the field of ethology, for contributing to the establishment of modern zoology in Japan, as the first director of the Woods Hole Marine Biological Laboratory at Woods Hole, and for having contributed the last surviving passenger pigeons to the Chicago Zoo.

A second strand of research connecting with modern studies of leeches came from a Swedish physician and anatomist Gustaf Retzius (1891). In addition to other pursuits (much to his discredit, these included eugenics and advocating for the supremacy of the Nordic "race"), his wide-ranging neuroanatomical studies included accurate descriptions of individually identified neurons in the methylene blue-stained preparations of segmental ganglia of *Hirudo*.

4. 20th Century: From articulata to lophotrochozoa/spiralia

At the dawn of the golden age of invertebrate neurophysiology in the 1960s, Retzius' description of the leech nervous system came to the attention of scientists looking for experimentally tractable systems in which to study the electrical properties of neurons (Eckert, 1963; Hagiwara & Morita, 1962) and glia (Kuffler & Potter, 1964). In seminal publications,

John Nicholls (initially as a postdoc with Stephen Kuffler at Harvard Medical School) established the leech as a system for in depth investigations of how nervous systems function (e.g., Baylor & Nicholls, 1969; Nicholls & Baylor, 1968).

In addition to more than two decades of work from his own lab, which focused mainly on neural regeneration (Nicholls, 1987), Nicholls' remarkable record of teaching and mentoring has impacted the lives of countless people all over the world, most certainly and positively including my own. Not only did Nicholls' lectures get me excited about leeches as an undergraduate, he was also instrumental in persuading Gunther Stent (my future postdoctoral advisor) to take up the leech during Stent's (~1969) sabbatical at Harvard, while Nicholls was there with Kuffler, and in furthering my career in many ways later on.

The distinct neurobiological and embryological threads of leech research merged at Berkeley in the 1970s. After returning from his fateful sabbatical at Harvard Medical School, Gunther Stent recruited postdoc William B. Kristan Jr. from Donald Kennedy's lab at Stanford and established a highly productive group studying the neural circuits underlying such behavior as swimming and heartbeat in the medicinal leech *Hirudo* (Kristan & Stent, 1976; Stent et al., 1978; Thompson & Stent, 1976). This was the work that drew me to his lab as a postdoc, but by the time I arrived, Gunther's interests were expanding to questions of neurodevelopment, for which *Hirudo* was ill-suited due to the small size and difficulty in culturing of its embryo.

To get around this problem, Roy Sawyer, a pre-eminent natural historian and systematist with a passion for leeches (1986), had introduced several other species, including a giant species *Haementeria ghilianii* and a much smaller species *Helobdella triserialis*, both from the same family (Glossiphoniidae) as Whitman's *Clepsine*.

In his 19th century work, Whitman had observed a cytoplasmic reorganization in the leech zygote prior to first cleavage resulting in two domains of yolk-free cytoplasm (teloplasm), at the animal and vegetal poles, respectively. Then, during a series of stereotyped early divisions, teloplasm was distributed first to the D quadrant at the 4-cell stage, and thence to five bilateral pairs of stem cells (teloblasts), which he correctly concluded are the sources of segmental mesoderm and ectoderm for the leech (Fig. 2). Whitman's work revealed that each teloblast undergoes repeated and highly unequal cell divisions, generating a column of segmental founder cells (blast cells). On each side of the embryo, the blast cell columns coalesce into parallel arrays

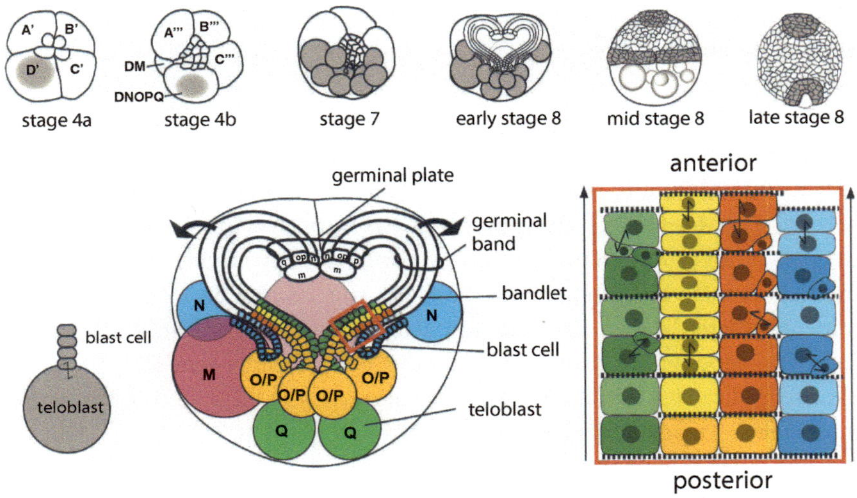

Fig. 2 Formation of the *Helobdella* posterior growth zone (PGZ). Top: Selected stages of *Helobdella* development. Yolk-deficient cytoplasm (teloplasm, gray shading in stage 4a through early stage 8) forms in the zygote and is segregated by unequal divisions to the D quadrant (stage 4) and thence to the five bilateral pairs of teloblasts (stages 7 and early 8). Teloblasts and their proximal blast cells constitute the PGZ. For clarity, the micromere-derived epithelium that covers the germinal bands and germinal plate (*gray shading* in stages mid and late 8) and the prospective dorsal territory behind them is not shown for early stage 8, but is depicted by small irregular contours for mid and late stage 8. *Bottom left*: Each teloblast undergoes repeated, highly unequal divisions, forming a column of segmental founder cells (blast cells) for that lineage. *Bottom center*: Detailed depiction of an early stage 8 embryo, showing that the five ipsilateral columns of teloblast-derived bandlets merge into a parallel array (germinal band) at the surface of the embryo in the vicinity of the animal pole, with the four ectodermal bandlets lying over the mesodermal bandlet (not visible here) on each side of the embryo. Arrows indicate the ventrovegetally directed gastrulation movements of the germinal bands as they coalesce into the germinal plate along the ventral midline. *Bottom right*: Schematic representation of stereotyped blast cell divisions and other events in the early germinal band (*red boxes*). Within the ectoderm, grandparental ventral (N) and dorsal (Q) lineages are depicted by alternating dark and light cells in the *blue* (n) and *green* (q) bandlets. Fate specification in the O/P equivalence group is represented by the transition from *orange/yellow* (equipotent o/p) lateral ectodermal precursors at the posterior end of the germinal band to distinct *orange* (O, ventrolateral) and *yellow* (P, dorsolateral) lineages more anteriorly. Horizontal *dotted lines* indicate that individual segmental complements in the O and P lineages are not yet in register with those in the N and Q lineages. Arrows indicate the rostral sliding of the n and q bandlets relative to the m, o and p bandlets in the most recently formed portion of the germinal bands. *Adapted from Weisblat, D.A. & Winchell, C.J. (2020). Segmentation in leech. In Chipman, A. ed.* Cellular processes in segmentation, *CRC Press. Chapter 7, pp. 153–181.*

(left and right germinal bands), which eventually zipper together in anteroposterior progression along the ventral midline of the embryo, forming a bilaterally symmetric germinal plate from which the structures of the segmental mesoderm and ectoderm arise. As cells proliferate within the germinal plate, it expands laterally. Eventually the lateral edges of the germinal plate meet and coalesce along the dorsal midline of the embryo, surrounding the developing gut and completing the tubular body plan of the adult leech. The midgut epithelium arises by mesodermally induced cellularization of nuclei at the surface of a syncytial yolk cell (SYC) (Nardelli-Haefliger & Shankland, 1993; Wedeen & Shankland, 1997). The SYC arises through a rather complex series of stepwise, cell-cell fusions during early development, that involves the A, B and C (non-teloblastogenic) quadrant macromeres, the 10 teloblasts after they have made all their blast cells, and sets of "supernumerary" blast cells, which are made just before the teloblasts stop dividing, and which do not contribute to segmental mesoderm or ectoderm (Desjeux & Price, 1999; Liu, Isaksen, Smith, & Weisblat, 1998; Zackson, 1982). At least the first step of this process (fusion of macromeres A''' and B''') is regulated inductively by signals from the D quadrant (Isaksen, Liu, & Weisblat, 1999) reminiscent of D quadrant signaling in the early development of equally cleaving mollusks.

Whitman speculated that different teloblasts inherited different tissue type-specific factors, so that, for example one pair of teloblasts would serve as the exclusive progenitors of the segmental nervous system. Whitman lacked the tools to test this hypothesis—my big success as a postdoc was to show that horseradish peroxidase could be used as a developmental cell lineage tracer, microinjected into blastomeres of the early *Helobdella* embryo, inherited by their progeny during subsequent development and then visualized histochemically to reveal the clone of the injected cell with single cell resolution. As quaint as it may seem, this really was a significant contribution for developmental biology in the 1970s (Weisblat, Sawyer, & Stent, 1978; Weisblat, Zackson, Blair, & Young, 1980)! The result of these experiments was somewhat contrary to Whitman's prediction–*each* of the five teloblasts generates a mixture of cell types, including neurons (Kramer & Weisblat, 1985; Weisblat, Harper, Stent, & Sawyer, 1980; Weisblat, Kim, & Stent, 1984). As has been reviewed extensively (e.g., Kuo et al., 2020; Shankland & Savage, 1997; Weisblat & Huang, 2001; Weisblat & Kuo, 2014), it turned out that each kind of teloblast generates specific sets of segmentally repeating progeny (designated as M, N, O, P and Q kinship groups), but each kinship group includes a mix of cell types (e.g., neurons and epidermal cells from the four distinct ectodermal lineages; muscles, gonads and nephridia, but a few neurons as well from the mesodermal lineage). Lineage tracing also resolved (by rendering moot) the question of how many segments are contained in the anteriormost region (the prostomium) of the leech body plan. Since the prostomium has distinct non-teloblastic origins, we could ignore it and settle on the intriguing value of 32 segments (4 fused segmental primordia anteriorly, 21 discrete midbody segments and 7 fused segmental primordia in the tail).

The excitement of investigating a brand new experimental system with a new set of tools was irresistible. I happily transitioned from neurobiology to neurodevelopment, and thence to pure development and evolutionary biology of leeches. To justify studying developmental biology in a system that was neither medically relevant nor genetically tractable, we emphasized the importance of comparative development to address the fundamental question of how the dramatically different body plans of modern animals are achieved. Even if we "solved" development down to the finest detail in *Drosophila* for example, we would still not know how leeches or any other kinds of animals develop, nor would we know how all the modern species evolved from a common ancestor, or what that ancestor was like (Weisblat, 1998).

To get at these latter questions, i.e., to understand the developmental changes underlying morphological changes that occur during evolution is really hard, especially for the great preponderance of species without a good fossil record. The big problem is that we are trapped in the present, without the ability to study the development of the evolutionary ancestor of any group, or any of the intermediates leading from that ancestor to its modern descendants. The only thing we can do is compare the development of the various modern species and interpret the developmental similarities and differences with respect to the tree by which they arose. Similarities represent features likely to have been present in the ancestor, while differences represent changes along one or both lineages.

For the first 20 years or so of this work, a significant effort was made to establish links between segmentation in *Helobdella* (which develops via holoblastic cleavages) and *Drosophila* (which develops via a syncytial blastoderm), first at the cellular and later (as more tools became available) at the molecular level. Perhaps the leech embryo counts out precisely 32 segments by means of five rounds of syncytial divisions in the teloblasts, ($2^5 = 32$), followed by stepwise parceling out of the nuclei into segmental founder cells, homologous to the syncytial blastoderm of *Drosophila*? Are the leech homologs of the well-known *Drosophila* segmentation genes (especially pair rule and segment polarity genes) used to create segments in leech embryos?

These questions were rooted in the Articulata Hypothesis, an evolutionary scenario which runs as follows. Since segmentation is such a sophisticated and complex process, it must have been very difficult to have evolved in the random, mutation-driven walk through Evo-Devo space in the first place. And since segmented body plans have been so successful, this feature would not have been lost once it did evolve. Thus, while segmentation may have arisen independently in the lineage leading to vertebrates among deuterostomes, it's likely that annelids (including leeches) and arthropods (including insects) as segmented protostomes, evolved from a segmented common ancestor.

Alternative evolutionary scenarios existed–for example, the Spiralian Hypothesis posits that mollusks, annelids and other taxa share a common ancestor that excludes arthropods based on shared patterns of early embryonic divisions called spiral cleavage. How do we decide which tree among the various possibilities is valid?

Or maybe this is not even the right question to be asking. Perhaps there is a more fundamental problem with this approach? You may have already realized that there is an inescapable circular logic inherent in using (developmentally derived) morphological characteristics to build a tree, and then using that tree to interpret developmental comparisons aimed at understanding how morphological characteristics evolve and diverge (Fig. 3, and see below for specific examples).

For me, this realization dawned gradually, beginning with confusion at the convoluted attempts of "grown-up" phylogeneticists to incorporate molecular sequence information along with other traits (morphological, embryological, behavioral, ecological, etc.) into traditional tree-building exercises. To this beginner, it seemed obvious that there was no way of knowing how to weight all these different kinds of information. Eventually, I came to appreciate that molecular phylogenies provide an approach to building phylogenetic trees without reference to developmentally derived traits at all (indeed, without even knowing the identity of the species under comparison), and that mental hygiene requires that we address *separately* the two questions: (1) what is the structure of the phylogenetic tree? and (2) what does this phylogenetic tree tell us about how development has changed during evolution? It's like how seeing your first digital watch allows you to realize that you've been using an analog watch all your life.

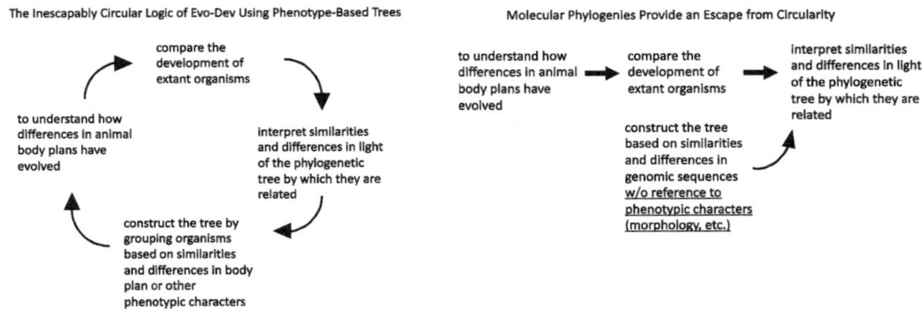

Fig. 3 The intellectual underpinnings of modern evolutionary developmental biology (Evo-Devo) lie in molecular methods which allow us to construct evolutionary hypotheses (phylogenetic trees) independent of developmentally-derived traits.

What I've described here is my personal intellectual journey to the Evo-Devo rationale, starting with minimal awareness of the field. Others have certainly reached the same conclusions independently and earlier than I did. In my view, this area exploded with the paper by Aguinaldo et al. (1997), whose molecular sequence comparisons led to two big conclusions. First and foremost was that the unsegmented, pseudocoelomate nematodes were *not* a deeply branching phylum as had long been assumed; rather, this taxon grouped with arthropods, tardigrades, onychophorans, nematomorphs, kinorhynchs and priapulids, a collection united by the fact that they all undergo molting of the external cuticle, and not their condition with respect to segmentation or their coelom.

A second, corollary result was the refutation of a close relationship between annelids and arthropods. This result, extensively confirmed since then, signified the death of the Articulata Hypothesis that had guided much of the previous work from my lab. Instead, some version of the Spiralian Hypothesis is supported, uniting annelids, mollusks and numerous other taxa into a second large grouping of protostome taxa called Lophotrochozoa and/or Spiralia. This grouping also includes Platyhelminthes, which were previously believed to represent a much deeper branch in the evolution of bilaterally symmetric animals. As will no doubt be discussed elsewhere in this volume, the separation of flatworms into Platyhelminthes (within Spiralia) and Acoelomorpha (as a possible sister group to other protostomes and deuterostomes) is yet another major result from molecular phylogenies (Hejnol et al., 2009; Mwinyi et al., 2010; Philippe, Brinkmann, Copley, et al., 2011). These and other dramatic changes in our view of animal phylogeny based on sequence comparisons rather than morphological comparisons reveal that body plans, and hence the developmental processes by which they arise, exhibit far greater plasticity during evolution than we had previously assumed.

Molecular phylogenetics has invigorated evolutionary developmental biology by allowing us to break out of the circular logic inherent in the traditional approaches, and by separating the bioinformatic challenges of tree-building from the embryological questions of how development works and how it changes over evolutionary time. It has also allowed us to reformulate the list of interesting questions based on our new understanding. For example, rather than striving to homologize segmentation processes between annelids and arthropods, we need to ask whether or not they share a segmented ancestor in the first place. If so, how is it that such an adaptive trait has been lost so many times, and if not, what is the range of mechanisms by which segmentation is achieved (Weisblat & Winchell, 2020; Zattara & Weisblat, 2020)?

At the phylum level, the Annelida were traditionally regarded as comprising three classes, polychaetes, oligochaetes and leeches, the inference being that these were all monophyletic taxa. In contrast, molecular phylogenies reveal that leeches arose from within oligochaetes, together forming a monophyletic group (Clitellata) that arose from within the polychaetes, corresponding to the annelid invasion of freshwater and terrestrial habitats. Other taxa that were previously regarded as distinct phyla also appear to have evolved from within the polychaetes, for example Echiura and Sipuncula, which thus appear to have secondarily lost segmentation (Struck et al., 2007). Understanding these evolutionary relationships allows us to properly define specific questions in the evolution of developmental changes that have occurred in these groups (Shain, 2009).

5. Segmentation

As mentioned above, our initial investigations of leech segmentation were framed and interpreted within the Articulata Hypothesis, i.e., that annelids and arthropods evolved from a segmented common ancestor. The production of segments from a teloblastic posterior growth zone (PGZ, defined here as the teloblasts and their pre-mitotic progeny) in clitellate annelids and certain crustaceans was initially seen as evidence of a shared segmented ancestor (Anderson, 1973). But this superficial similarity can equally well be interpreted as arising from convergence, starting from a more likely bilaterian feature–axial elongation from the posterior end–and the possibility that stem cell divisions are an ancient and/or common feature of that process.

When we began using injected cell lineage tracers to study the five bilateral pairs of stem cells (M, N, O, P and Q teloblasts) that serve as segmental precursors in *Helobdella*, we anticipated that within each lineage the sequentially produced progeny (m, n, o, p and q blast cells, respectively) would generate identical clones, giving rise invariably to a particular set of cells in each segment–thus, each segment would comprise a "polyclone," the sum of the five blast cell clones on each side. This prediction was wrong in several interesting ways.

First, while the individual clones are stereotyped, the individual clones in several lineages (M, O and P) interdigitate with their anterior and posterior neighbors along the anteroposterior axis (and with clones in different lineages along the mediolateral axis; Fig. 4). Thus, the segments in clitellate annelids arise from the formation of iterated, stereotyped clones in each lineage, and not because of boundaries to cell mingling imposed on fields of initially interchangeable cells (as in vertebrates and some arthropods). Thus, we now distinguish between lineage-driven segmentation (exemplified by clitellate annelids such as leeches) and boundary-driven segmentation (as in fruit fly and mouse, for example) (Weisblat & Winchell, 2020; Fig. 5), much as we began distinguishing digital watches from analog watches only after the former had been invented.

In a sense then (and as Gunther Stent pointed out early on), the conceptual problem of leech segmentation is "solved" by the production of the longitudinally iterated segmental founder cells (blast cells)—all that remains is the (dauntingly open-ended) task of elucidating the mechanisms of all the subsequent cellular events that lead to the stereotyped interdigitating clones (equal and unequal cell divisions, cell migrations and differentiation processes within each lineage), and how these vary among clitellate species.

A second surprise was that in the N (and Q) teloblast lineages, sequentially produced blast cells exhibit two distinct fates in precise alternation, beginning with the timing, orientation and symmetry of their first divisions, and are thus designated as nf and ns (or qf and qs) blast cells (Zackson, 1984). Thus, we call the N and Q teloblasts "grandparental stem cells" because at each division, the stem cell gives rise to the same type of cell as its lineal grandparent, in contrast to canonical stem cells (e.g., the M, O and P teloblasts), which give rise to same type of progeny as their parental stem cell at each division (Stent & Weisblat, 1985). So far as I know, apart from the

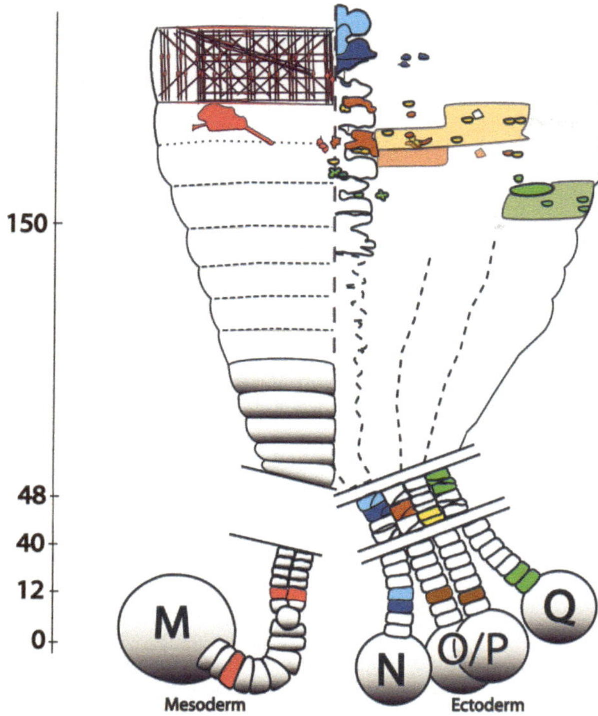

Fig. 4 Stereotyped spatial distribution of blast cell clones in parental (M, O, P) and grandparental (N, Q) stem cell lineages. Schematic representation of the *Helobdella* PGZ (lower portion of drawing) and germinal plate (upper portion), depicting the anteroposterior progression of segmentation in terms of individual blast cell clones; scale at left indicates the approximate age of the blast cell clones in hours for *Helobdella*. For simplicity, only the mesodermal (M) lineage is shown in the *left side* of the figure and only the four ectodermal (N, O, P, Q) lineages are shown in the *right side* of the figure; the ventral midline is indicated by the *dotted line* at the center of the drawing, while the edges correspond to the dorsal midline. In the mesodermal (M teloblast) lineage (*pink*), individual blast cells initiate their stereotyped pattern of divisions prior to entering the germinal band. Each cell makes one segment's worth of definitive progeny, including body wall muscles, nephridia, and a few ganglionic neurons, but the final clone of cells arising from a single m blast cell is distributed over three segments. The O and P lineages arise from initially equipotent O/P teloblasts and blast cells, which assume distinct O (orange) and P (*yellow*) fates within the germinal bands. Each individual o and p blast cell clone makes one segment's worth of definitive progeny for ventrolateral and dorsolateral ectoderm, respectively, including specific sets of ganglionic neurons, epidermis, and peripheral neurons, but the final clones of cells are distributed over two morphological segments. The N (*blue*) and Q (*green*) lineages also contribute specific sets of ganglionic neurons, epidermis, and peripheral neurons to ventral and dorsal ectoderm, respectively, but in each of these lineages, two blast cell clones are required to make a single segment's worth of definitive progeny (from Weisblat & Winchell, 2020).

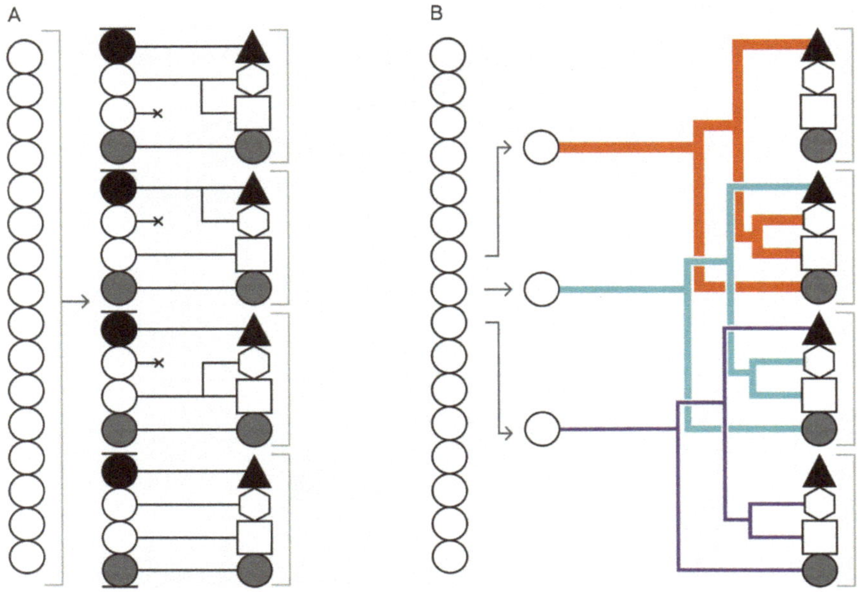

Fig. 5 Boundary-driven and lineage-driven segmentation. In boundary-driven segmentation (*left*), as exemplified in arthropods and vertebrates, segmental boundaries are imposed on initially unspecified regions of cells (syncytial blastoderm or pre-somitic mesoderm). Segmentally iterated structures arise within the segmental boundaries despite variable patterns of cell division, in part due to restrictions on the ability of cells to cross the boundaries. In lineage-driven segmentation (*right*), as exemplified in leeches and other clitellate annelids, stereotyped patterns of cell division among longitudinally arrayed precursor cells lead to segmentally repeating structures even if their clones interdigitate across morphological segment boundaries (from Weisblat & Winchell, 2020).

original observations in *C. elegans* mutants (Chalfie, Horvitz, & Sulston, 1981), such grandparental stem cell lineages have not been described outside the clitellate annelids, but do appear to be conserved throughout this clade (Storey, 1989a; Nakamoto, Arai, & Shimizu, 2000), i.e., for about 400 MY (Erséus et al., 2020).

A priori, the differences between nf and ns blast cells must either be present as the cells are born (suggesting the involvement of some sort of flip-flop mechanism in the parent teloblast) or are imprinted on initially equipotent n blast cells. It's frustrating to have made so little progress on this question, but we have at least generated some molecular cues, in the form of differentially expressed *cdc42x*, *sfrp1/2/5/c*, and *wnt16a* genes (Zhang, Kuo, & Weisblat, 2009; Sung-Jin Cho and Shinja Yoo, personal communications).

The question of how f and s blast cells assume distinct fates within the N and Q lineages leads to the related question of how blast cell clones come to assume segment-specific or region-specific fates along the A-P axis generally. This latter issue has been explored in a body of work initiated and led by Marty Shankland and colleagues, starting while Shankland was a postdoc with Gunther Stent at Berkeley, and then continuing at Harvard Medical School and the University of Texas in Austin. The outlines of this story are as follows.

First, Gunther Stent's grad student Saul Zackson made the seminal observation that the fixed number of 32 teloblast-derived segments in the leech body plan *cannot* be explained by the production of precise numbers of blast cells by the teloblasts; rather, each teloblast generates a variable number of "supernumerary blast cells" beyond the 32 or 64 cells required to make the full complement of segmental derivatives (for the M, O and P or N and Q lineages, respectively) (Zackson, 1982). It was originally assumed that these extra cells degenerate and die, but we now believe they fuse with the syncytial yolk cell that ultimately cellularizes to generate the definitive endoderm, dependent on signaling from visceral mesoderm (Desjeux & Price, 1999; Nardelli-Haefliger & Shankland, 1993; Wedeen & Shankland, 1997).

Shankland recognized the two blast cell fates (contributing to definitive segments versus degeneration/death) as an opportunity to investigate the perennial "nature-vs-nurture" issue that frames developmental fate decisions at levels ranging from gene expression all the way to personality development and career choices–to what extent is any given fate decision controlled by intrinsic versus extrinsic factors? To address this issue, he took advantage of the enhanced photosensitivity of cells labeled with fluoresceinated dextran lineage tracer to selectively ablate cells within labeled ectodermal bandlets near the point at which they coalesce to form the germinal band. Blast cells in the bandlet on the teloblast-proximal side of the lesion were delayed in entering the germinal bands and thus would often occupy a position posterior to where they would normally lie in the embryo (Fig. 6). Shankland named this phenomenon "blast cell slippage" and found that in this case, the blast cells assumed the fate appropriate to their new position–cells that would have normally contributed clones comprising numerous cells to segmental ectoderm, having slipped posterior to the germinal band into supernumerary blast cell territory, now failed to divide further and merged with the yolk cells, indicating that the fate of the shifted cells was determined by their new environment. In this case, nurture wins (Shankland, 1984).

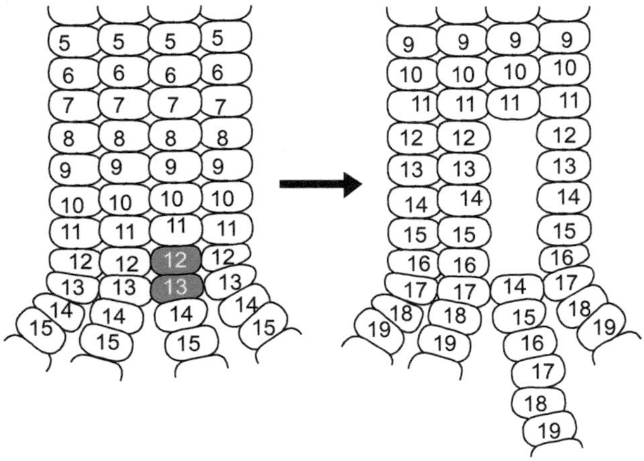

Fig. 6 Slippage experiments demonstrate that ectodermal blast cells acquire segment-specific identities well before their first mitoses. Diagrammatic representation of blast cells in the posterior (most recently formed) section of a germinal band. Numbers indicate the birth order of blast cells in each bandlet. (See Fig. 2 for details regarding the blast cell division patterns and the differences between the inner and outer bandlets in the germinal band.) When blast cells are killed as they enter the germinal band (*shading* at *left*), cells posterior to the lesion are often delayed in entering the germinal band. As a result, the cells posterior to the lesion become aligned with cells fated to contribute to more posterior segments (*right*), but still express markers appropriate to their originally fated position.

More exciting results came later, when Shankland and colleagues used the same approach to show that blast cells in each ectodermal lineage are committed to specific segmental identities autonomously of other teloblast lineages and in some cases well before the expression of HOX genes that are often associated with the specification of segmental or regional identities along the A-P axis (Kourakis et al., 1997). In brief, this approach entails using a precise descriptive approach to characterize rather subtle but clear distinctions between the contributions of blast cell clones in a given teloblast lineage to different segments (e.g., Berezovskii & Shankland, 1996; Blair, Martindale, & Shankland, 1990; Shankland, Martindale, Nardelli-Haefliger, Baxter, & Price, 1991) and then using the blast cell slippage technique to shift blast cells and their descendant clones to more posterior locations. Almost invariably the slipped blast cells gave rise to the progeny appropriate to their *normal* location. This work provides strong evidence that ectodermal blast cells assume segment-specific identities before they even enter the germinal bands and undergo their first divisions (Martindale & Shankland, 1990) and independent of signals from either laterally adjacent bandlets (Nardelli-Haefliger, Bruce, & Shankland, 1994) or serially adjacent segments (Seaver & Shankland, 2000).

Working with a different glossiphoniid leech species, Lidia Gleizer, then a graduate student in Gunther Stent's lab, showed that the same principle holds true for blast cells in the mesodermal (M teloblast) lineages, but with two important differences. One was that when one m blast cell is killed, the flanking cells in the bandlet usually come together to close the gap, so that posterior cells slip forward rather than backward within the germinal plate. The other is that the resultant mis-positioned m blast cell clones tend to return to their "correct" registration, presumably through interaction with the contralateral mesodermal bandlet (Gleizer & Stent, 1993). Curiously, Dan Shain and Françoise Huang found something operationally similar in the n bandlet–when ns is ablated the flanking cells slip posteriorly but when nf is ablated the flanking cells tend to close the gap (Shain, Stuart, Huang, & Weisblat, 2000). We interpreted this as reflecting an early difference between the two n blast cell phenotypes, so that the ns-to-ns interaction (after an nf ablation) leads to different cell behaviors than the nf-to-nf interaction (after an ns ablation).

The coexistence of parental and grandparental stem cells in the leech posterior growth has interesting consequences. In particular, since all the teloblasts produce blast cell progeny at about the same rate (and of about the same size), and start producing blast cells at about the same time, this means that the M, O and P lineages produce *segment equivalents* at twice the rate of the N and Q lineages, and that the blast cells in the N and Q lineages that are going to contribute a given segment are initially out of register in both time and space with respect to the con-segmental blast cells in the M, O and P lineages.

The spatial mismatch is handled by differential expansion of the m, o and p blast cell clones with respect to the pairs of n and q blast cell clones—and since the n and q blast cells are out of register with their con-segmental m, o and p blast cells, they are sliding anteriorly past these other cells as they enter into the parallel arrays of cells (germinal bands) on each side of the embryo (Weisblat & Shankland, 1985).

The temporal mismatch between the N/Q lineages and the M/O/P lineages is also intriguing. For the most anterior segments, the temporal mismatch is only an hour or so. But we know that the teloblast division rates remain unchanged throughout development up until the onset of supernumerary blast cell production (Wordeman, 1983; Shinja Yoo and Sophia Bylsma, personal communication) and the effects of the mismatch are cumulative, so for more posterior segments, the N and Q lineage founder cells are born a day or more after the con-segmental M, O and P founder cells. Also, so far as we know (which is not far at all), individual blast cell clones divide according to birthdate of the individual blast cells, rather than synchronizing the clones in each segment (Lans, Wedeen, & Weisblat, 1993). To the extent that this observation holds true, it has implications for such things as intercellular signaling across lineages to orchestrate cell division, differentiation and migration in the proliferating clones and the formation of connections among neurons arising from different lineages within the segmental ganglia of the nervous system.

The mis-match between the N/Q and M/O/P lineage poses even more severe issues for other clitellate annelids. It appears that that oligochaete *embryos* generate segments in the same manner as leeches, with five bilateral teloblast lineages that are clearly homologous in their patterning to those in leeches (Nakamoto et al., 2000; Storey, 1989a). But whereas leeches exhibit determinate growth, making a fixed number of segments during embryogenesis and no more, many oligochaetes exhibit post-embryonic growth and segmentation, adding segments from a posterior growth zone throughout their lives (Storey, 1989b; Zattara & Bely, 2013). This is presumably the ancestral annelid condition (Anderson, 1973). Do the N and Q stem cells of indeterminately growing oligochaetes divide more rapidly in post-embryonic growth? Or do the embryos of indeterminately growing species use a different set of stem cells for post-embryonic growth?

To me, these are some of the most interesting and answerable questions in the evolutionary development of the clitellate annelids, but technical challenges have so far stood in the way of answering them. First, determining the source of post-embryonic segments directly requires that injected lineage tracers survive much longer than the commonly used enzymes or fluorescent dextrans (Gimlich & Braun, 1985) or even genetically encoded tracers such as mRNAs or plasmids (Zhang & Weisblat, 2005; Gline, Kuo, Stolfi, & Weisblat, 2009)—for leech at least, we have now solved that problem using Minos-mediated transgenesis to achieve stable transgene insertions (Chris Winchell, personal communication). Another problem is that it's more challenging to work with the embryos of any of the oligochaetes that have been used to far. *Tubifex* is the most obvious candidate, thanks to the extensive body of work achieved by Prof. Takashi Shimizu at Hokkaido University (sadly for the field, now retired) and colleagues (e.g., Shimizu, 1982 and the many subsequent publications available through PubMed), but that embryo requires manual de-vitellinization prior to microinjection, and a fresh, custom-forged needle for each injection (Takashi Shimizu, personal communication)–even removing the embryos from the cocoon taxes the skills of those trained on *Helobdella* embryos.

Whether this challenge will be solved by the emergence of a suitably skilled practitioner, a new oligochaete model, or new technical work-arounds remains to be seen. In any case, surmounting these problems is well worth the effort, because it will also allow definitive experiments to reveal the cellular source(s) of regeneration and asexual reproduction in this biologically diverse group of animals (Bely chapter, this volume).

A third major result from early studies of cell lineage in leech segmentation was the "O-P equivalence group", which contradicts the assumption of the invariability of blast cell fates (Weisblat & Blair, 1984). This line of investigation emerged from observations of variability in the positions of the O and P teloblasts, precursors of ventrolateral and dorsolateral ectoderm, respectively. Whitman named the ectodermal O, P and Q teloblasts in alphabetical order based on their proximity to the N teloblast (the postulated neural precursor in the *Clepsine* embryo), and this convention was transferred to *Helobdella*. But lineage tracer experiments revealed that sometimes the nominal O teloblast (i.e., closest to the N teloblast) generated the P pattern of dorsolateral ectoderm and vice versa. A more careful analysis revealed that it was the positions of the nominal O and P teloblast-derived blast cells *within the germinal bands* that was strictly correlated with their ultimate fates–whichever cells lay next to the n blast cells generated O lineage progeny (ventrolateral ectoderm), while those that lay next to the q blast cells generated P lineage progeny (dorsolateral ectoderm).

A priori, this result could be interpreted two ways. If the O and P teloblasts were committed to their respective fates as we had assumed, but sometimes switched positions (during the mitosis of the precursor OP blastomere), then perhaps their pre-committed blast cell progeny were just finding their correct positions within the germinal bands. Alternatively, it could be that the nominal o and p blast cells were developmentally equipotent, with their final fates being determined by their positions within the germinal bands. Seth Blair and I showed that the o blast cells could assume the P fate when the nominal p blast cells were missing (Weisblat & Blair, 1984); Saul Zackson (1984) showed that the O and P fates correlated with differences in the timing and orientation of the blast cell divisions in the respective clones; Marty Shankland and I found that cells in the o blast cell clones lose their developmental plasticity in a stepwise manner during the stereotyped divisions in each clone (Shankland & Weisblat, 1984; Shankland, 1987a, 1987b); Robert Ho showed that the integrity of an embryonic epithelium covering the blast cells in the germinal bands is required for normal O and P fate assignment (Ho & Weisblat, 1987)–and there things stopped for more than a decade, except for Ian Quigley's discovery that expression of the putative transcription factor *six1/2a* promotes P cell fates (Quigley, Schmerer, & Shankland, 2010). Further ablation experiments aimed at determining the cellular source of the fate-determining signal were unsuccessful (Zackson, 1984). In retrospect, we realized that this failure was due to the mistaken assumption that the left and right germinal bands developed independently of one another (so that when teloblast ablations were carried out on one side of the embryo the germinal band on the contralateral side could be used as an internal control). This assumption overlooked the transient interactions between bandlets from opposite sides at the point where they enter their respective germinal bands.

The issue was finally clarified a dozen years later by Françoise Huang, a gifted and fiercely tenacious experimentalist (Huang & Weisblat, 1996). Through an exhaustive series of experiments, some of which entailed 10 microinjections *per experimental embryo* (combinations of ablations and lineage tracer injections), she showed that even a transient, short-range contact with q blast cells from *either* side of the embryo at the point where the bandlets enter the germinal bands, was necessary and sufficient to convert lateral ectodermal cells from a default O fate (ventrolateral ectoderm) to the P fate (dorsolateral ectoderm; Fig. 7). Another dozen or so years later came a second gifted post-doc, Dian-Han Kuo, who had the insight to see the Q-derived signal as an example of dorsoventral (D/V) patterning, which by homology with other embryos suggested the involvement of BMP signaling. Making full use of the emerging palette of molecular and genomic resources, Dian-Han showed the q blast cell-derived, P fate-inducing, signal is indeed a BMP, but with surprising deviations from the canonical models D/V patterning (Kuo & Weisblat, 2011; Kuo, Shankland, & Weisblat, 2012)–the key patterning gene is a *bmp5-8* homolog (rather than a *bmp2/4* homolog), the BMP antagonist is encoded by a *gremlin* homolog (rather than *sog* or *chordin*) and, as predicted by Françoise' work, the signal appears to be contact-mediated rather than diffusing. This latter observations raises a whole new set of questions about evolutionary and/or developmental changes in how BMP pro-peptides are processed (D.-H. Kuo, personal communication).

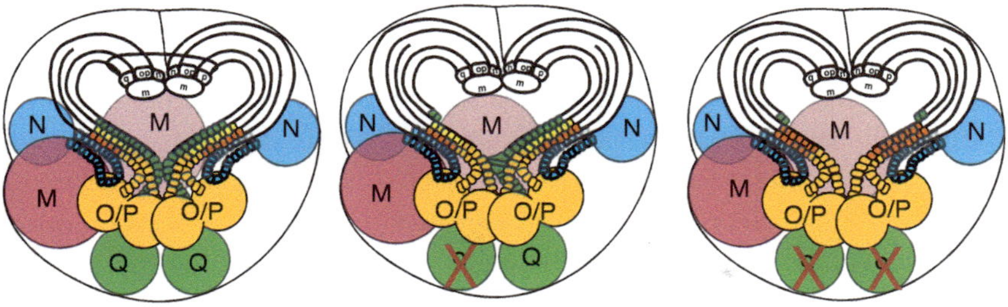

Fig. 7 A short range, Q lineage-derived signal specifies dorsolateral ectoderm in the O—P equivalence group. *Left*: schematic image of an early stage 8 embryo, showing the normal arrangement and fates of teloblasts and ectodermal blast cells within the germinal bands. On each side of the embryo, the N lineage (*blue*) and Q lineage (*green*) contribute ventral and dorsal ectoderm, respectively. The two O/P lineages are initially equipotent (*light orange*), but within the germinal band they differentiate into distinct ventrolateral (O, *dark orange*) and dorsolateral (P, *yellow*) lineages, driven by a short range BMP5–8 signal emanating from cells in the Q lineage. (See Figs. 2 and 4 for further details.) *Center*: following the ablation of a single Q teloblast (X), blast cells from the remaining q bandlet contact both of the presumptive p bandlets and the normal O and P fates are obtained on both sides. *Right*: following the ablation of both Q teloblasts, however, all four of the O/P-derived bandlets in the region of the dual ablation follow the ventrolateral O fate.

6. Molecular studies of "segmentation genes"

Molecular phylogenies render the Articulata Hypothesis highly un-parsimonious. Instead, molecular trees strongly favor the Spiralian Hypothesis. Rejecting the Articulata Hypothesis would also be consistent with the substantial body of work studying the expression of the leech homologs of the well-known *Drosophila* segmentation genes, such as *nanos*, *hedgehog*, *hunchback*, *engrailed*, *hairy-enhancer-of-split*, *wnt* and *even-skipped* (Cho, Vallès, Giani, Seaver, & Weisblat, 2010; Kang et al., 2003; Kang, Pilon, & Weisblat, 2002; Pilon & Weisblat, 1997; Savage & Shankland, 1996; Seaver & Shankland, 2001; Song, Huang, Chang, & Weisblat, 2002; Song, Huang, Gonsalves, & Weisblat, 2004). Numerous experiments almost invariably failed to support commonality at the molecular level between segmentation in annelids and arthropods. For example, immunohistochemical studies of the expression of an *engrailed* homolog revealed segmentally iterated rows of labeled cells that we assumed to be homologous with those observed for stripes of expression of *engrailed* and other segment polarity genes in *Drosophila* embryos, consistent with the Articulata

Hypothesis (Lans et al., 1993; Ramirez, Wedeen, Stuart, Lans, & Weisblat, 1995; Wedeen & Weisblat, 1991)—in retrospect, this conclusion exemplifies the wry, cautionary dictum that I attribute to John Gerhart "I'll see it when I believe it". Later, it was shown that: (1) this expression is not initiated until *after* morphologically distinct segmental structures are already in place (Shain et al., 2000); (2) the *hedgehog* homolog is not expressed in similar rows, or anywhere adjacent to the *engrailed*-positive cells (Kang et al., 2003); and (3) ablating *engrailed*-positive cells has no effect on segment polarity (Seaver & Shankland, 2001).

In another approach, leech homologs of *Drosophila* pair-rule genes (*hairy-enhancer-of-split* and *even-skipped*) were examined, by Mi Hye Song, who remarkably completed all the specific aims of a full NIH grant as a graduate student (Song et al., 2002, 2004)! Neither of these genes showed a pattern of expression corresponding to every other segment as we had hoped to see. Instead, both genes are expressed uniformly among cells of the PGZ. Intriguingly, expression of *even-skipped* homologs in the PGZ is observed in both polychaete and oligochaete annelids (de Rosa, Prud'homme, & Balavoine, 2005; Seaver, Yamaguchi, Richards, & Meyer, 2012).

These molecular investigations of segmentation were carried out before the advent of genome editing techniques, so functional studies were limited. Investigating the developmental roles of the canonical segmentation genes using CRISPR/Cas9, mRNA plasmid and transposon-mediated mis-expression, and transcriptional profiling should certainly prove fruitful.

7. Genome evolution

The first data for molecular phylogenies came from protein sequences, laboriously obtained in the 1960s by painstaking biochemical methods applicable only to highly abundant proteins (for a historical perspective, see Alvarez-Ponce, 2021). Fast forwarding through the invention and proliferation of DNA sequencing technologies and PCR (1970s to present), we are now able to assemble comprehensive gene inventories from cell- and tissue-specific transcriptomes and whole genomes.

Whole genome comparisons further supported the assignment of most protostome taxa into the two super-phyla Ecdysozoa and Lophotrochozoa, and uncovered a surprising result, namely the loss of macrosynteny in the leech genome relative to other animals (Simakov et al., 2013). Here, macrosynteny is defined as the conservation of gene linkages within chromosomes or large scale genomic scaffolds, independent of gene order within a chromosome or sub-chromosomal scaffold (Lv, Havlak, & Putnam, 2011). Putnam et al. (2007) discovered that shared macrosynteny can be detected across species as diverse as cnidarians and bilaterians, i.e., that genes on a genomic scaffold in one species are more likely to share a scaffold in another species. This is thought to reflect the notion that, while genomic inversions are relatively common, so that genes can effectively diffuse back and forth along a chromosome or chromosome arm over evolutionary timescales, translocations or other changes that move genes from one chromosome to another are much rarer.

To the extent that overall genome architecture plays a role in regulating development, relaxing constraints on genomic organization may permit embryos to explore Evo-Devo Space more ambitiously, thereby unveiling new variations of developmental mechanisms and processes. Coincidentally or not, another group in which extensive genome rearrangements are seen is the cephalopod mollusks, whose body plans differ dramatically from those of other animals (Albertin et al., 2015; Albertin & Simakov, 2020).

Available evidence suggests that the extensive genome rearrangements such as those observed for *Helobdella* are seen throughout the Clitellata. Perhaps these rearrangements were enabled by the changes in reproductive life histories associated with simultaneous hermaphroditism, and/or by ecological considerations associated with the transition from marine to freshwater and terrestrial habitats? Are these accelerated rearrangements continuing unabated, or do they reflect a transient condition at the base of at the Clitellata? Comparing genome architecture among various oligochaetes and leeches may provide answers to these questions.

The global genomic rearrangements in the lineage leading to leech extend even to the architecture of the famously conserved HOX cluster. Compared to the inferred ancestral lophotrochozoan cluster of 11 HOX genes in a single cluster (Simakov et al., 2013), the leech has 19 HOX genes, including duplications of some and deletions of others; moreover, its largest HOX cluster has only four HOX paralogs, and those are not immediately contiguous in the ancestral Lophotrochozoan HOX cluster (Simakov et al., 2013). To understand the possible consequences of this aspect of genome reorganization, we are using CRISPR/Cas9 mutagenesis to examine the functions and regulatory interactions among specific Hox genes, focusing on those central and posterior group genes that appear to be involved in the development of the highly specialized structures of the posterior gut and posterior sucker (Chris Winchell et al., personal communication).

8. Back to the future: Neural fate specification

Let us return to the question of how neuronal phenotypes arise within the leech segmental ganglia (which was the original impetus for my studies in leech development) and sketch out the strategy for how this issue is now being approached. Recall Whitman's idea, that N teloblasts inherit cytoplasmic factors that commit them to serving as neural precursors. The assumption was that the N teloblasts were the exclusive neural precursors. Even if Whitman's hypothesis had proved true, it failed to address the question of how diverse neuronal phenotypes were generated. This failure is understandable since awareness of the phenotypic diversity of neurons was yet to come in the 1870s.

We now know that *each* teloblast lineage ultimately generates blast cells that give rise to a *mixture* of cell types, including specific subsets of (individually identified) ganglionic neurons and glia, plus non-neuronal cells. Moreover, as outlined above, the situation is further complicated at multiple levels by the distinctions between mesodermal (M) and ectodermal (N, O/P and Q) precursors, between parental (M, O/P) and grandparental (N, Q) teloblasts, between f and s blast cells in the latter category, between and O and P fates for the O/P teloblasts, and among the region-specific or segment-specific fates of the blast cells within each lineage. A final level of complication arises from the assignment of segment-specific or region-specific identities to sequentially produced blast cells within each teloblast lineage.

Our current understanding of cell differentiation mechanisms during development provides a general notion of how different cell types and sub-types arise during development by an emergent process–akin to what was historically termed epigenesis, as distinct from pre-formationism (Van Speybroeck, De Waele, & Van de Vijver, 2002). For the terminally differentiated cell types that are the end product of normal development, goals are to define the differences in gene expression that account for the phenotypic differences among cells (Heath-Heckman et al., 2021), to understand how their transcriptional profiles are maintained (presumably by a positive feedback network of terminal transcription factors in post-mitotic cells and (most difficult) how the various cell types are established by dynamic patterns of gene expression in the mitotic lineages by which they arise. This has been fleshed out most completely perhaps for *C. elegans*. (Hobert & Kratsios, 2019; Taylor et al., 2021).

It seems reasonable to assume that each M, N, O/P and Q teloblast inherits a particular combination of transcription factors and other cellular components through the combined influences of unequal cell divisions and/or intercellular signaling processes and/or epigenetic chromatin modifications, that sets it on the path to generating its appropriate mix of blast cell types in the sequence. Similarly, for each blast cell, executing the program of mostly unequal divisions leading to the final clone of diverse progeny depends on additional fate-determining steps at each division along the way. Presumably this program results from a similar mix of cell intrinsic and cell extrinsic mechanisms.

The advent of high throughput transcriptional profiling (RNAseq) provides a promising approach to understanding the progression of cell states leading from the zygote through the various teloblast precursors to the four classes of teloblasts (M, N, O/P and Q). The stereotyped division patterns, relatively large size, and experimental accessibility of the individually identifiable blastomeres in these early stages of development should all suggest that RNAseq will provide a high resolution account for this aspect of neurogenesis in *Helobdella*; indeed, such studies are already underway (Dan Shain, personal communication). Moreover, for functional analysis, antisense approaches (siRNA and antisense morpholino oligomers), misexpression approaches (RNA and DNA injections; transient and germline transgenesis mediated by transposons) and genome editing (CRISPR-mediated G0 and germline mutagenesis) have all been used on this embryo (Gline et al., 2009; Kuo & Weisblat, 2011; Schmerer, Null, & Shankland, 2013; Todd, Kristan, & French, 2010; Zhang & Weisblat, 2005; Winchell et al. personal communication).

Applying these approaches to the many parallel pathways leading from the four classes of teloblasts to the seven major classes of blast cells (m, nf, ns, o, p, qf, qs) and thence to the roughly 200 phenotypes of differentiated neurons is technically challenging due to the large numbers and small sizes of the cells involved, but here too, the leech offers experimental advantages (e.g., combining RNASeq with tracer injections to mark cells in specific lineages) that should allow us to address these questions with a level of precision that is not obtained with most other embryos. It might be worthwhile to recognize that the common practice of assigning cells to particular developmental trajectories based solely on similarities of their transcriptional profiles at successive timepoints is potentially subject to the same type of logical errors that we encountered when constructing phylogenetic trees based on morphological similarities. How sure are we that cells cannot undergo large scale shifts in transcriptional profiles? This assumption can be rigorously tested in the *Helobdella* embryo, by combining RNASeq and tracer injections in the context of our detailed knowledge of the cell lineages.

Current work is aimed at exploiting *Helobdella* and the larger *Hirudo* species as complementary models in which to integrate physiological and molecular approaches to behavioral studies (Kuo et al., 2020). In brief, we can identify candidate genes by RNASeq of *Hirudo* neurons, validate their expression by in situ hybridization and identify orthologous genes in orthologous neurons in *Helobdella* (Heath-Heckman et al., 2021). For suitable candidates, we can then investigate the consequences on neuronal development and behavior of knocking out specific genes by CRISPR-mediated mutagenesis (Elizabeth Heath-Heckman and Rachel Kim, personal communication).

9. Eco-evo-devo

In addition to the universal questions of general interest in Evo-Devo, such as how segments (or embryonic axes or germ layers) form in this deep branch of bilaterians, the emergence of leeches (Hirudinea) from oligochaetes within Clitellata presents numerous avenues for fruitful research into specific processes of evolutionary history. Evolutionary changes in diet provide one example. Most oligochaetes are detritovores, typically consuming materials that are abundant but of relatively low nutritional value; in at least some cases they rely heavily on gut symbionts to provide essential amino acids (Larsen et al., 2016). A priori, these considerations suggest that they eat more or less constantly. A continuously growing oligochaete will also be adding gut tissue continually as well–this complicates the issue of functional specialization in the gut because of its continually shifting regional identities.

Leeches, by contrast, are predators or sanguivorous ectoparasites–their diets are of high nutritional value but much harder to come by. And as note above, they've lost the capacity for indeterminate growth, which would seem to simplify the problems of specializing different sections of the gut to a greater extent than in oligochaetes. Correlated with these features, the leech midgut is highly specialized–the anterior midgut (crop, typically six segments' worth) exhibits prominent segmentally iterated lateral extensions (caecae) that allows sanguivores like *Hirudo* to consume several times their own weight at a single feeding. And the posterior midgut/intestine (four segments' worth) has similar but smaller extensions, presumably specialized for digestion and absorption. How has this specialization of the midgut been achieved? We are currently investigating the involvement of HOX genes in development and differentiation of the midgut and hindgut, starting with the hypothesis that the gene duplications, deletions, and rearrangements of HOX genes relative to the inferred ancestral HOX cluster may have provided a substrate for evolutionary specialization of the leech gut (Chris Winchell et al. in preparation).

Finally, having established the utility of *Helobdella* as a model for studying development), we must also acknowledge that a full understanding of evolutionary development requires that we explore how developmentally driven changes in body plan and cellular phenotype affect behavior and physiology of the individual, and thus how it interacts with the environment and how this affects its reproductive success compared to others of its species. Such interactions constitute the selective forces that determine which of the underlying, stochastically generated genomic changes are propagated to serve as the substrate for further evolutionary tinkering (Jacob, 1977).

One obvious area of interest here is the simultaneous hermaphroditism that appears to have emerged at the base of the clitellate annelids. This feature has persisted throughout most of the clade despite what must be an increased physiological cost of developing and maintaining both female and male reproductive organs. Does the shared lineage of female and male germline precursors in hermaphrodites facilitate genome evolution and sympatric speciation (Cho, Valles, & Weisblat, 2014)? How are resources allocated between female and male functions? And how might this allocation differ for self-fertile species when raised as isolated individuals compared to interbreeding cohorts (Iyer, Rogers, Levine, Winchell, & Weisblat, 2019)?

The ability to characterize *Helobdella* reproduction and behavior in lab culture (Iyer et al., 2019; Kim et al., 2019) is a valuable complement to field studies (Davies, Wrona, & Linton, 1979; Wrona, Davies, Linton, & Wilkialis, 1981) and makes this a tractable group for ecological investigation. How are prey preferences established for specialist and generalist species? What are the consequences of dietary differences for growth and reproduction? How do the dynamics of leech and snail populations impact other components of freshwater ecosystems?

10. Conclusion

At this point in my exposition, I imagine my audience stealing progressively less furtive glances at their watches (cell phones), and edging toward the exit. Thus, I endeavor a succinct summary.

The central point of this essay is that leeches in general, and species in the glossiphoniid genus *Helobdella* in particular, are valuable models in which to explore a broad spectrum of questions concerning the evolution of developmental mechanisms and animal body plans. They are particularly interesting because they represent a vast constellation of animals known collectively as the Spiralia or Lophotrochozoa, that has been evolving independently of the more commonly and intensively studied chordate and ecdysozoan models for a non-trivial fraction (~4%) of the life of the universe. Thus, within the confines of planet Earth, this clade is about as close as we can get to studying alien life forms. Given the tremendous diversity of this group–consider the extremes of size and intelligence in cephalopod mollusks, or the potent regenerative powers of planaria–it should be clear that leeches are in no way typical or representative lophotrochozoans, any more than *Drosophila* or *Caenorhabditis* capture ecdysozoan diversity. What *Helobdella* does provide is an experimentally tractable model with which to elucidate the particular mixture of ancestral and derived processes that characterize its development, and which by comparison to other lophotrochozoan models, can help to distinguish the former from the latter (Weisblat & Kuo, 2009). If you have any questions, or are interested in studying leeches yourself, I'd be happy to help you get started.

Acknowledgments

I thank Chris Winchell for comments and countless valuable conversations. Work in my lab over the years has been supported by grants from the Basil O'Connor Fund of the March of Dimes Foundation the Human Frontiers Science Program, the National Aeronautics and Space Agency, the National Institutes of Health, and the National Science Foundation.

References

Aguinaldo, A. M., Turbeville, J. M., Linford, L. S., Rivera, M. C., Garey, J. R., Raff, R. A., et al. (1997). Evidence for a clade of nematodes, arthropods and other moulting animals. *Nature, 387*, 489–493.

Albertin, C. B., & Simakov, O. (2020). Cephalopod biology: At the intersection between genomic and organismal novelties. *Annual Review of Animal Biosciences, 8*, 71–90.

Albertin, C. B., Simakov, O., Mitros, T., Wang, Z. Y., Pungor, J. R., Edsinger-Gonzales, E., et al. (2015). The octopus genome and the evolution of cephalopod neural and morphological novelties. *Nature, 524*, 220–224.

Alvarez-Ponce, D. (2021). Richard Dickerson, molecular clocks, and rates of protein evolution. *Journal of Molecular Evolution, 89*, 122–126.

Anderson, D. T. (1973). *Embryology and phylogeny in annelids and arthropods*. Pergamon Press.

Baylor, D. A., & Nicholls, J. G. (1969). Chemical and electrical synaptic connexions between cutaneous mechanoreceptor neurones in the central nervous system of the leech. *The Journal of Physiology, 203*, 591–609.

Berezovskii, V. K., & Shankland, M. (1996). Segmental diversification of an identified leech neuron correlates with the segmental domain in which it expresses Lox2, a member of the Hox gene family. *Journal of Neurobiology*, *29*, 319–329.
Blair, S. S., Martindale, M. Q., & Shankland, M. (1990). Interactions between adjacent ganglia bring about the bilaterally alternating differentiation of RAS and CAS neurons in the leech nerve cord. *The Journal of Neuroscience*, *10*, 3183–3193.
Chalfie, M., Horvitz, H. R., & Sulston, J. E. (1981). Mutations that lead to reiterations in the cell lineages of *C. elegans*. *Cell*, *24*, 59–69.
Cho, S. J., Vallès, Y., Giani, V. C., Jr., Seaver, E. C., & Weisblat, D. A. (2010). Evolutionary dynamics of the wnt gene family: A lophotrochozoan perspective. *Molecular Biology and Evolution*, *27*, 1645–1658.
Cho, S. J., Valles, Y., & Weisblat, D. A. (2014). Differential expression of conserved germ line markers, and delayed segregation of male and female primordial germ cells in a hermaphrodite, the leech *Helobdella*. *Molecular Biology and Evolution*, *31*, 341–354.
Davies, R. W., Wrona, F. J., & Linton, L. (1979). A serological study of prey selection by *Helobdella stagnalis* (Hirudinoidea). *J Animal Ecol.*, *48*, 181–194.
de Rosa, R., Prud'homme, B., & Balavoine, G. (2005). Caudal and even-skipped in the annelid *Platynereis dumerilii* and the ancestry of posterior growth. *Evolution & Development*, 7, 574–587.
Desjeux, I., & Price, D. J. (1999). The production and elimination of supernumerary blast cells in the leech embryo. *Development Genes and Evolution*, *209*, 284–293.
Eckert, R. (1963). Electrical interaction of paired ganglion cells in the leech. *The Journal of General Physiology*, *46*, 573–587.
Erséus, C., Williams, B. S., Horn, K. M., Halanych, K. M., Santos, S. R., James, S. W., et al. (2020). Phylogenomic analyses reveal a Paleozoic radiation and support a freshwater origin for clitellate annelids. *Zool. Scripta*, *49*, 614–640.
Gimlich, R. L., & Braun, J. (1985). Improved fluorescent compounds for tracing cell lineage. *Developmental Biology*, *109*, 509–514.
Gleizer, L., & Stent, G. S. (1993). Developmental origin of segmental identity in the leech mesoderm. *Development*, *117*, 177–189.
Gline, S. E., Kuo, D. H., Stolfi, A., & Weisblat, D. A. (2009). High resolution cell lineage tracing reveals developmental variability in leech. *Dev. Dynamics*, *239*, 3139–3152.
Hackenberger, P. N., & Janis, J. E. (2019). A comprehensive review of medicinal leeches in plastic and reconstructive surgery. *Plastic and Reconstructive Surgery - Global Open.*, 7, e2555.
Hagiwara, S., & Morita, H. (1962). Electrotonic transmission between two nerve cells in leech ganglion. *Journal of Neurophysiology*, *25*, 721–731.
Heath-Heckman, E., Yoo, S., Cj, W., Pellegrino, M., Angstadt, J., Lammardo, V. B., et al. (2021). Transcriptional profiling of identified neurons in leech. *BMC Genomics*, *22*, 215.
Hejnol, A., Obst, M., Stamatakis, A., Ott, M., Rouse, G. W., Edgecombe, G. D., et al. (2009). Assessing the root of bilaterian animals with scalable phylogenomic methods. *Proceedings of the Biological Sciences*, *276*, 4261–4270.
Ho, R. K., & Weisblat, D. A. (1987). A provisional epithelium in leech embryo: Cellular origins and influence on a developmental equivalence group. *Developmental Biology*, *120*, 520–534.
Hobert, O., & Kratsios, P. (2019). Neuronal identity control by terminal selectors in worms, flies, and chordates. *Current Opinion in Neurobiology*, *56*, 97–105.
Huang, F. Z., & Weisblat, D. A. (1996). Cell fate determination in an annelid equivalence group. *Development*, *122*, 1839–1847.
Isaksen, D. E., Liu, N. J., & Weisblat, D. A. (1999). Inductive regulation of cell fusion in leech. *Development*, *126*, 3381–3390.

Iyer, R. G., Rogers, D. V., Levine, M. D., Winchell, C. J., & Weisblat, D. A. (2019). Reproductive differences among species, and between individuals and cohorts, in the leech genus *Helobdella* (Lophotrochozoa; Annelida; Clitellata; Hirudinida; Glossiphoniidae), with implications for reproductive resource allocation in hermaphrodites. *PLoS One*, *14*, e0214581.

Jacob, F. (1977). Evolution and tinkering. *Science*, *196*, 1161–1166.

Kang, D., Huang, F., Li, D., Shankland, M., Gaffield, W., & Weisblat, D. A. (2003). A hedgehog homolog regulates gut formation in leech (*Helobdella*). *Development*, *130*, 1645–1657.

Kang, D., Pilon, M., & Weisblat, D. A. (2002). Maternal and zygotic expression of a *nanos*-class gene in the leech *Helobdella robusta*: Primordial germ cells arise from segmental mesoderm. *Developmental Biology*, *245*, 28–41.

Kim, R. C., Le, D., Ma, K., Heath-Heckman, E. A. C., Whitehorn, N., Kristan, W. B., Jr., et al. (2019). Behavioral analysis of substrate texture preference in a leech, *Helobdella austinensis*. *J. Comp. Physiol. A*, *205*, 191–202.

Kourakis, M. J., Master, V. A., Lokhorst, D. K., Nardelli-Haefliger, D., Wedeen, C. J., Martindale, M. Q., et al. (1997). Conserved anterior boundaries of Hox gene expression in the central nervous system of the leech *Helobdella*. *Developmental Biology*, *190*, 284–300.

Kramer, A. P., & Weisblat, D. A. (1985). Developmental neural kinship groups in the leech. *The Journal of Neuroscience*, *5*, 388–407.

Kristan, W. B., Jr., & Stent, G. S. (1976). Peripheral feedback in the leech swimming rhythm. *Cold Spring Harbor Symposia on Quantitative Biology*, *40*, 663–674.

Kuffler, S. W., & Potter, D. D. (1964). Glia in the leech central nervous system: Physiological properties and neuron-glia relationship. *Journal of Neurophysiology*, *27*, 290–320.

Kuo, D. H., Heath-Heckman, E. A. C., Fernandez de Miguel, F., Szczupak, L., Todd, K., Weisblat, D. A., et al. (2020). A tale of two leeches: Toward the understanding of the evolution and development of behavioral neural circuits. *Evolution & Development*, *22*, 471–493.

Kuo, D. H., Shankland, M., & Weisblat, D. A. (2012). Regional differences in BMP-dependence of dorsoventral patterning in the leech *Helobdella*. *Developmental Biology*, *368*, 86–94.

Kuo, D. H., & Weisblat, D. A. (2011). A new molecular logic for BMP-mediated dorsoventral patterning, in the leech *Helobdella*. *Current Biology*, *21*, 1282–1288.

Lans, D., Wedeen, C. J., & Weisblat, D. A. (1993). Cell lineage analysis of the expression of an engrailed homolog in leech embryos. *Development*, *117*, 857–871.

Larsen, T., Ventura, M., Maraldo, K., Triadó-Margarit, X., Casamayor, E. O., Wang, Y. V., et al. (2016). The dominant detritus-feeding invertebrate in Arctic peat soils derives its essential amino acids from gut symbionts. *The Journal of Animal Ecology*, *85*, 1275–1285.

Liu, N. L., Isaksen, D. E., Smith, C. M., & Weisblat, D. A. (1998). Movements and stepwise fusion of endodermal precursor cells in leech. *Development Genes and Evolution*, *208*, 117–127.

Lv, J., Havlak, P., & Putnam, N. H. (2011). Constraints on genes shape long-term conservation of macro-synteny in metazoan genomes. *BMC Bioinformatics*, *12*(Suppl 9), S11.

Maienschein, J. (1978). Cell lineage, ancestral reminiscence, and the biogenetic law. *Journal of the History of Biology*, *11*, 129–158.

Martindale, M. Q., & Shankland, M. (1990). Intrinsic segmental identity of segmental founder cells of the leech embryo. *Nature*, *347*, 672–674. https://doi.org/10.1038/347672a0. PMID: 2215696.

Morse, E. (1912). *Biographical memoir of Charles Otis Whitman* (pp. 1842–1910). http://www.nasonline.org/publications/biographical-memoirs/memoir-pdfs/whitman-c-o.pdf.

Mwinyi, A., Bailly, X., Bourlat, S. J., Jondelius, U., Littlewood, D. T., & Podsiadlowski, L. (2010). The phylogenetic position of Acoela as revealed by the complete mitochondrial genome of *Symsagittifera roscoffensis*. *BMC Evolutionary Biology*, *10*, 309.

Nakamoto, A., Arai, A., & Shimizu, T. (2000). Cell lineage analysis of pattern formation in the *Tubifex* embryo. II. Segmentation in the ectoderm. *The International Journal of Developmental Biology*, *44*, 797–805.

Nardelli-Haefliger, D., Bruce, A. E., & Shankland, M. (1994). An axial domain of HOM/Hox gene expression is formed by morphogenetic alignment of independently specified cell lineages in the leech *Helobdella*. *Development*, *120*, 1839–1849.

Nardelli-Haefliger, D., & Shankland, M. (1993). Lox10, a member of the NK-2 homeobox gene class, is expressed in a segmental pattern in the endoderm and in the cephalic nervous system of the leech *Helobdella*. *Development*, *118*, 877–892.

Nicholls, J. G. (1987). In I. I. Vol (Ed.), *The search for connections. Magnus lecture series* Sinauer Associates Inc.

Nicholls, J. G., & Baylor, D. A. (1968). Specific modalities and receptive fields of sensory neurons in CNS of the leech. *Journal of Neurophysiology*, *31*, 740–756.

Payton, B. (1981). History of medicinal leeching and early medical references. In K. J. Muller, & S. G. S. NichollsJG (Eds.), *Neurobiology of the leech* (pp. 27–34). Cold Spring Harbor Press.

Philippe, H., Brinkmann, H., Copley, R. R., et al. (2011). Acoelomorph flatworms are deuterostomes related to Xenoturbella. *Nature*, *470*, 255–258.

Pilon, M., & Weisblat, D. A. (1997). A nanos homolog in leech. *Development.*, *124*, 1771–1780.

Putnam, N. H., Srivastava, M., Hellsten, U., Dirks, B., Chapman, J., Salamov, A., et al. (2007). Sea anemone genome reveals ancestral eumetazoan gene repertoire and genomic organization. *Science*, *317*, 86–94.

Quigley, I. K., Schmerer, M. W., & Shankland, M. (2010). A member of the six gene family promotes the specification of P cell fates in the O/P equivalence group of the leech *Helobdella*. *Developmental Biology*, *344*, 319–330.

Ramirez, F. A., Wedeen, C. J., Stuart, D. K., Lans, D. A., & Weisblat, D. A. (1995). Identification of a neurogenic sublineage required for CNS segmentation in an annelid. *Development*, *121*, 2091–2097.

Retzius, G. (1891). Zur Kenntnis des centralen Nervensystems der Würmer. *Biol. Unters. (NF)*, *2*, 1–28.

Savage, R. M., & Shankland, M. (1996). Identification and characterization of a hunchback orthologue, Lzf2, and its expression during leech embryogenesis. *Developmental Biology*, *175*, 205–217.

Sawyer, R. T. (1986). *Leech biology and behavior.* Oxford University Press.

Schmerer, M. W., Null, R. W., & Shankland, M. (2013). Developmental transition to bilaterally symmetric cell divisions is regulated by Pax-mediated transcription in embryos of the leech *Helobdella austinensis*. *Developmental Biology*, *382*, 149–159.

Seaver, E. C., & Shankland, M. (2000). Leech segmental repeats develop normally in the absence of signals from either anterior or posterior segments. *Developmental Biology*, *224*, 339–353.

Seaver, E. C., & Shankland, M. (2001). Establishment of segment polarity in the ectoderm of the leech *Helobdella*. *Development*, *128*, 1629–1641.

Seaver, E. C., Yamaguchi, E., Richards, G. S., & Meyer, N. P. (2012). Expression of the pair-rule gene homologs runt, Pax3/7, even-skipped-1 and even-skipped-2 during larval and juvenile development of the polychaete annelid *Capitella teleta* does not support a role in segmentation. *EvoDevo*, *3*, 8.

Shain, D. H. (Ed.). (2009). *Annelids in modern biology* Wiley-Blackwell.

Shain, D. H., Stuart, D. K., Huang, F. Z., & Weisblat, D. A. (2000). Segmentation of the central nervous system in leech. *Development, 127*, 735–744.

Shankland, M. (1984). Positional determination of supernumerary blast cell death in the leech embryo. *Nature, 307*, 541–543.

Shankland, M. (1987a). Differentiation of the O and p cell lines in the embryo of the leech. I. Sequential commitment of blast cell sublineages. *Developmental Biology, 123*, 85–96.

Shankland, M. (1987b). Differentiation of the O and P cell lines in the embryo of the leech. II. Genealogical relationship of descendant pattern elements in alternative developmental pathways. *Developmental Biology, 123*, 97–107.

Shankland, M., Martindale, M. Q., Nardelli-Haefliger, D., Baxter, E., & Price, D. J. (1991). Origin of segmental identity in the development of the leech nervous system. *Development. Supplement, 2*, 29–38.

Shankland, M., & Savage, R. M. (1997). Annelids, the segmented worms. In S. F. Gilbert, & A. M. Raunio (Eds.), *Embryology: Constructing the organism* (pp. 219–236). Sinauer Associates, Inc. chapter 12.

Shankland, M., & Weisblat, D. A. (1984). Stepwise commitment of blast cell fates during the positional specification of the O and P cell lines in the leech embryo. *Developmental Biology, 106*, 326–342.

Shimizu, T. (1982). Development in the Freshwater Oligochaete *Tubifex*. In F. W. Harrison, & R. R. Cowden (Eds.), *Developmental Biology of Freshwater Invertebrates* (pp. 283–316). AR Liss, Inc.

Simakov, O., Marletaz, F., Cho, S. J., Edsinger-Gonzales, E., Havlak, P., Hellsten, U., et al. (2013). Insights into bilaterian evolution from three spiralian genomes. *Nature, 493*, 526–531.

Song, M. H., Huang, F. Z., Chang, G. Y., & Weisblat, D. A. (2002). Expression and function of an even- skipped homolog in the leech *Helobdella robusta*. *Development, 129*, 3681–3692.

Song, M. H., Huang, F. Z., Gonsalves, F. C., & Weisblat, D. A. (2004). Cell cycle-dependent expression of a hairy and enhancer of split (hes) homolog during cleavage and segmentation in leech embryos. *Developmental Biology, 269*, 183–195.

Stent, G. S., Kristan, W. B., Jr., Friesen, W. O., Ort, C. A., Poon, M., & Calabrese, R. L. (1978). Neuronal generation of the leech swimming movement. *Science, 200*, 1348–1357.

Stent, G. S., & Weisblat, D. A. (1985). Cell lineage in the development of invertebrate nervous systems. *Annual Review of Neuroscience (Palo Alto, CA), 8*, 45–70.

Storey, K. G. (1989a). Cell lineage and pattern formation in the earthworm embryo. *Development, 107*, 519–531.

Storey, K. G. (1989b). The effects of ectoteloblast ablation in the earthworm embryo. *Development, 107*, 533–545.

Struck, T. H., Schult, N., Kusen, T., Hickman, E., Bleidorn, C., McHugh, D., et al. (2007). Annelid phylogeny and the status of *Sipuncula* and *Echiura*. *BMC Evolutionary Biology, 7*, 57.

Taylor, S. R., Santpere, G., Weinreb, A., Barrett, A., Reilly, M. B., Xu, C., et al. (2021). Molecular topography of an entire nervous system. *Cell*. https://doi.org/10.1016/j.cell.2021.06.023. S0092-8674(21)00758-3. Epub ahead of print 34237253.

Thompson, W. J., & Stent, G. S. (1976). Neuronal control of heartbeat in the medicinal leech. *Journal of Comparative Physiology, 111*, 281–307.

Todd, K. L., Kristan, W. B., Jr., & French, K. A. (2010). Gap junction expression is required for normal chemical synapse formation. *The Journal of Neuroscience, 30*, 15277–15285.

Van Speybroeck, L., De Waele, D., & Van de Vijver, G. (2002). Theories in early embryology: Close connections between epigenesis, preformationism, and self-organization. *Annals of the New York Academy of Sciences, 981*, 7–49.

Wedeen, C. J., & Shankland, M. (1997). Mesoderm is required for the formation of a segmented endodermal cell layer in the leech *Helobdella*. *Developmental Biology*, *191*, 202–214.

Wedeen, C. J., & Weisblat, D. A. (1991). Segmental expresssion of an *engrailed*-class gene during early development and neurogenesis in an annelid. *Development*, *113*, 805–814.

Weisblat, D. A. (1998). Embryonic development as a quasihistorical process. *The International Journal of Developmental Biology*, *42*, 475–478.

Weisblat, D. A., & Blair, S. S. (1984). Developmental indeterminacy in embryos of the leech *Helobdella triserialis*. *Developmental Biology*, *101*, 326–335.

Weisblat, D. A., Harper, G., Stent, G. S., & Sawyer, R. T. (1980). Embryonic cell lineages in the nervous system of the glossiphoniid leech *Helobdella triserialis*. *Developmental Biology*, *76*, 58–78.

Weisblat, D. A., & Huang, F. Z. (2001). An overview of glossiphoniid leech development. *Canadian Journal of Zoology*, *79*, 218–232.

Weisblat, D. A., Kim, S. Y., & Stent, G. S. (1984). Embryonic origins of cells in the leech *Helobdella triserialis*. *Developmental Biology*, *104*, 65–85.

Weisblat, D. A., & Kuo, D. H. (2009). *Helobdella* (leech): A model for developmental studies. In *Vol. 1. Emerging model organisms* (pp. 245–274). N.Y.: Cold Spring Harbor Laboratory Press, Cold Spring Harbor.

Weisblat, D. A., & Kuo, D. H. (2014). Developmental biology of the leech *Helobdella*. *The International Journal of Developmental Biology*, *58*, 429–443.

Weisblat, D. A., Sawyer, R. T., & Stent, G. S. (1978). Cell lineage analysis by intracellular injection of a tracer enzyme. *Science*, *202*, 1295–1298.

Weisblat, D. A., & Shankland, M. (1985). Cell lineage and segmentation in the leech. *Philosophical Transactions of the Royal Society of London. Series B, Biological Sciences*, *312*, 39–56.

Weisblat, D. A., & Winchell, C. J. (2020). Segmentation in leech. In A. Chipman (Ed.), *Cellular processes in segmentation* (pp. 153–181). CRC Press. Chapter 7.

Weisblat, D. A., Zackson, S. L., Blair, S. S., & Young, J. D. (1980). Cell lineage analysis by intracellular injection of fluorescent tracers. *Science*, *209*, 1538–1541.

Whitman, C. O. (1878). The embryology of *Clepsine*. *The Quarterly Journal of Microscopical Science*, *18*, 215–315.

Whitman, C. O. (1887). A contribution to the history of the germ layers in *Clepsine*. *Journal of Morphology*, *1*, 105–182.

Wordeman, L. (1983). Kinetics of primary blast cell production in the embryo of the leech *Helobdella triserialis*. Honors thesis In *Department of Molecular Biology*. Berkeley, CA: University of California.

Wrona, F. J., Davies, R. W., Linton, L., & Wilkialis, J. (1981). Competition and coexistance between *Glossiphonia complanata* and *Helobdella stagnalis* (Glossiphoniidae: Hirudinoidea). *Oecologia*, *48*, 133–137.

Zackson, S. L. (1982). Cell clones and segmentation in leech development. *Cell*, *31*, 761–770.

Zackson, S. L. (1984). Cell lineage, cell-cell interaction, and segment formation in the ectoderm of a glossiphoniid leech embryo. *Developmental Biology*, *104*, 143–160.

Zattara, E. E., & Bely, A. E. (2013). Investment choices in post-embryonic development: Quantifying interactions among growth, regeneration, and asexual reproduction in the annelid *Pristina leidyi*. *Journal of Experimental Zoology. Part B, Molecular and Developmental Evolution*, *320*, 471–488.

Zattara, E. E., & Weisblat, D. A. (2020). Cellular and molecular mechanisms of segmentation in Annelida: An open question. In A. Chipman (Ed.), *Cellular processes in segmentation* (pp. 71–97). CRC Press. Chapter 4.

Zhang, S. O., Kuo, D. H., & Weisblat, D. A. (2009). Grandparental stem cells in leech segmentation: Differences in CDC42 expression are correlated with an alternating pattern of blast cell fates. *Developmental Biology*, *336*, 112–121.
Zhang, S. O., & Weisblat, D. A. (2005). Applications of mRNA injections for analyzing cell lineage and asymmetric cell divisions during segmentation in the leech *Helobdella robusta*. *Development*, *132*, 2103–2113.

CHAPTER SIXTEEN

Journey beyond the embryo: The beauty of *Pristina* and naidine annelids for studying regeneration and agametic reproduction

Alexandra E. Bely*
Department of Biology, University of Maryland, College Park, MD, United States
*Corresponding author: e-mail address: abely@umd.edu

Contents

Abstract

Regeneration, asexual agametic reproduction, and other forms of postembryonic development are ecologically important and widely variable across animals, yet our understanding of this important aspect of animal diversity remains limited. A significant limitation has been the dearth of adequate study systems for exploring the mechanisms and evolution of these processes. Here I describe key parts of our journey in developing naid annelids as a study system for investigating the evolution and development of regeneration and fission. Naids are small freshwater annelids that provide numerous advantages for studying postembryonic development: they are small and reproduce readily by fission, they include species with diverse regenerative abilities, and many species are easy to culture. Among the naids, *Pristina leidyi* is a particularly useful study

Current Topics in Developmental Biology, Volume 147
ISSN 0070-2153
https://doi.org/10.1016/bs.ctdb.2021.12.020

species, being largely transparent and quite robust to a variety of experimental manipulations. Building on a sparse but long history of past research on these animals, we have developed this system by establishing methods and generating resources for working with them. Naids are yielding novel insights into the evolution of regeneration and fission, providing one of many examples of the value of developing new study species to enable the exploration of fundamental and understudied questions in biology. Establishing new study systems comes with challenges but is exciting and rewarding, and I provide perspectives from my own experiences with the hope of encouraging the further expansion of study systems in biology.

1. Introduction

Every species on this planet has stories to tell. The intensive study of a few organisms (the "model-system approach") is powerful, allowing extremely deep dives into some of these stories, but understanding the diversity of life forms and processes requires exploring well beyond this small handful of species (Duffy et al., 2021; Jenner & Wills, 2007). The intensively studied models do not—indeed, cannot—tell all: not only do they not represent existing diversity, but many questions simply cannot be addressed with them. This is why developing new study species is so exciting: it opens the way for addressing new questions.

I have felt fortunate to travel the exciting road of developing naid annelids, a group of small freshwater annelids, as a new system for studying postembryonic development, with a particular focus on the evolution of regeneration and asexual reproduction. Why did I find myself establishing a new study system? The fields of animal developmental biology and evo-devo have predominantly focused—understandably—on the study of embryos. But development does not end when embryogenesis ends. Postembryonic life stages are far from developmentally static: juveniles and adults can undergo widespread tissue homeostasis, growth, and even degrowth, and in many lineages postembryonic life stages can also build extensive parts of the body through regeneration and asexual agametic reproduction (e.g., fission, budding). These postembryonic developmental capabilities are widely variable across species and, where they are present, are often critically important to the ecologies of these animals, indicating that the evolution of postembryonic development has been a key aspect of the evolution of animals. And yet our understanding of postembryonic developmental processes and how they evolve has lagged far behind our understanding of embryonic development, in large part because of a dearth of adequate models and tools to study them.

A significant impediment to broader understanding of postembryonic development is that studying such processes is often technically challenging. Many animal groups have limited postembryonic developmental abilities, but even in species where such capabilities are present, studies are often seriously hampered by large body size, opaque tissues, and thickened outer body layers of juveniles and adults. Even animals that have beautifully transparent and experimentally accessible embryos often develop into awkwardly large and opaque adults, covered with thickened membranes, furs, scales, or shell, and with a strong will to do what postembryonic stages do—including move and feed—making it challenging to describe and manipulate postembryonic developmental processes occurring within them.

My interest in postembryonic development led me to naids. Naids are small, aquatic annelids that present numerous advantages for studying regeneration, agametic reproduction, and postembryonic development more generally. Naids reproduce by fission; most species can regenerate extensive parts of their bodies; they have small and largely transparent bodies with most organ systems being only a few cell layers thick; diverse species are easily cultured; and several species are generally tolerant of experimental conditions. Among the naids, *Pristina leidyi* (Smith) has been a particularly useful lab workhorse. Propelled by broad questions about how postembryonic development unfolds and evolves, my lab and I have worked with these animals, developing approaches and tools for their study, and gained interesting insights. It is particularly rewarding to now see additional researchers beginning to study these animals.

2. Beginning the journey

When I first started working on naids, I didn't realize I was establishing a new study system. I was simply a beginning graduate student, keen to explore questions that were exciting to me, eager to develop a new project that would be my own, and very fortunate to be in a graduate school environment that encouraged research independence and expected me to consider my questions first and foremost—to let my questions guide me to an appropriate study system.

I have loved biological diversity for as long as I can remember, with a particular fondness for soft-bodied invertebrates and their remarkable diversity of forms and life histories. I became particularly intrigued by the prevalence of asexual reproduction in freshwaters and entered graduate school thinking I might study the genetic underpinnings of parthenogenesis, perhaps in bdelloid rotifers. But I then became fascinated with asexual reproduction that is agametic—processes such as fission and budding that involve adult animals making new individuals without going through a gamete or embryo phase. As a beginning graduate student, I spent many months reading and thinking widely about invertebrate reproduction and development, huddled in a small, sparsely-lit shared office, feet literally on my desk for hours on end, reading books and articles. I scribbled questions and notes into a little notebook, in search of big, thesis-worthy questions to focus on. How do regeneration and agametic reproduction converge on the same endpoint as embryogenesis, but from such starkly different starting points? And how do these postembryonic developmental trajectories arise and evolve? I was hooked.

What followed was a search for an animal group in which I could study these questions. I was intrigued by the evolution of regeneration and agametic reproduction—processes that occurred not in tiny embryos but in adults—and this guided my search. First and foremost, I wanted to find an animal group that had both *extensive regenerative abilities and easily elicited agametic reproduction*. There were many possibilities to choose from. But if I was going to focus on these adult-level developmental processes, I saw clear benefits to studying an animal that was *small*—I essentially wanted to work on an adult that was the size of an embryo. If I wanted to have the clearest view of adult-level morphogenesis, I wanted to work on an animal that was *transparent*. I wanted to be able to look directly into my study animal to understand what was happening, rather than have to infer it. I wanted to work on a bilaterian—a species that had a clear head and tail, so that findings would be broadly relatable. I was a graduate student starting an unfunded project, so I wanted to work on an animal that was *inexpensive to maintain*. I also wanted to maximize my time for experiments, not animal maintenance, and so wanted an animal that was *simple to culture*. And then there was the *"fascination" factor*. I hoped to work on animals that would fuel my curiosity and sense of wonder—and that of others—through their inherent properties.

In the context of these deliberate choices, a dose of chance ultimately led me to naids. Across the hall from my lab was an evolutionary ecology lab that studied aquatic invertebrates, and a graduate student in that lab, with whom I shared a fascination with invertebrate life history evolution, suggested I consider a naid annelid he was working on for a study of senescence—*Paranais litoralis*. It reproduced asexually in a remarkable way—by building a new head and tail right in the middle of its body! And it could regenerate. Not only was it capable of remarkable developmental feats, it was also small, fairly transparent, and easy to culture. I was skeptical at first, unsure that something as boring as a "worm" should actually become the focus of my thesis. But I began culturing them and quickly appreciated their many qualities. I explored this group of annelids further by ordering fissioning annelids from several biological supply companies to expand my options, and quickly realized that multiple naid species were commonly used material in biology classrooms, signaling their ease of culture and lab tractability. I also ventured out into the field and found a remarkable diversity of naids easily accessible in local fresh and brackish waters. Whether I sampled sediments or vegetation; collected from ponds, streams, or mudflats; or sampled in freshwater or brackish water, I consistently found naids, often multiple species and in large abundance. And although all had wormy bodies, they were actually surprisingly diverse, and in many dimensions—they varied in morphology, behavior, diet, life history, and manifestations of fission (and, as I later learned, they also varied in regenerative abilities). As someone driven to study evolutionary diversity, I realized that this unassuming group of small worms could actually be an outstanding study system for comparative biology.

3. Rediscovering naids: A brief overview of a long history of naid research

Although as a graduate student I thought I had discovered a great new study system, I was far from the first to appreciate the qualities of these small annelids for studying postembryonic development. Digging into the literature, I found a modest but long history of research on naids and their regeneration and fission, dating back to the mid-1700s.

The earliest works were based almost exclusively on microscope observations and histology, yet they laid a strong foundation for understanding the dynamic processes involved in fission and regeneration. Naturalists of the 1700s such as Trembley, Réaumur, and Bonnet first described naids and documented their fission and regenerative abilities, including through striking illustrations (Rota, 2022a, 2022b). Semper (1876, cited in Krecker, 1923) first described cells in *Chaetogaster* that were later hypothesized to be important for regeneration and fission; Hepke (1897) appears to have been the first to infer cell migration during naid regeneration; and Bourne (1891), Brode (1898), and Galloway (1899) provided important descriptions of naid morphology, fission and regeneration, including in *Dero* and *Pristina*. Harper (1904) generated the earliest detailed description of the widespread tissue remodeling (morphallaxis) that occurs during fission and regeneration, working with *Stylaria*. Hyman (1916) worked with *Dero* to reveal metabolic gradients using cyanide and to describe in detail the axial patterns of regenerative potential. Other important descriptions of the processes of fission and regeneration were made, including by Hempelmann (1923) and Van Cleave (1937) for *Pristina*, by Aiyer (1924) and Hyman (1938) for *Allonais* (then *Nais*), by Chu and Pai (1944) and Chu (1945, 1946) for *Stylaria*, by Herlant-Meewis (1946) for *Nais*, by Meewis (1934) for *Chaetogaster*, and by Dehorne (1916) for a number of naid species. Many of these descriptive works, as well as later ones, aimed to understand both morphogenetic events involved in regeneration and fission (e.g., which original tissues gave rise to new tissues), as well as the role of annelid neoblasts (migratory mesodermal cells first described by Randolph (1891, 1892) in *Lumbriculus*) that were suspected of being regenerative stem cells and playing a key role in regeneration.

The application of X-ray irradiation to biological systems fueled an active period of experimental research in the first part of the 1900s, allowing research to move beyond primarily descriptive studies of naid post-embryonic development. Van Cleave (1934), O'Brien (1942, 1946), and Foulkes (1953) used X-ray irradiation to better understand regeneration and/or fission in *Pristina*, *Dero*, and *Allonais*, using both partial and total X-ray irradiation approaches. Much of this work focused on the role of neoblasts in annelid regeneration (comparable work was also being done in planarians during this period). These studies provided evidence that neoblasts migrate to both anterior and posterior wounds and a common conclusion was that these neoblasts directly contribute to naid regeneration.

By the middle of the 1900s, a quite extensive literature on annelid regeneration, and to a lesser degree on fission, had accumulated. A number of foundational reviews emerged in which naids featured prominently, among them Stephenson (1930), Hyman (1940), Berrill (1952), and Herlant-Meewis (1964). Important taxonomic works also emerged that covered or focused on the naids, including Stephenson (1930) and Sperber (1948). Research on naids then largely tapered off, with few studies on their fission and regeneration in the second half of the 1900s. I have appreciated that in this early phase of research on naids, women made significant contributions: Lucienne Dehorne (Paris), Christina Sperber (Uppsala), Libbie Hyman (Chicago), and Henriette Herlant-Meewis (Brussels) all produced seminal works on these animals. Naids had sparked their curiosity just as they had mine, and I wish I knew more about the lives and experiences of these women scientists working in such different times.

More recently, a few research groups have focused on regeneration and fission of naids. Bilello and Potswald (1974) described and quantified naid neoblasts, inferring their migration to wound sites, in *Ophidonais*. Drewes and Fourtner (1991) showed in *Dero* that fission and regeneration involve neurophysiological remodeling and reorganization of associated behavioral reflexes. Kharin, Zagainova, and Kostyuchenko (2006) studied patterns in the location of fission zones along the body axis and the origin of fission zone tissues. Naid ecology and life history parameters were also investigated, including by Learner, Lochhead, and Hughes (1978), Lochhead and Learner (1983), Parish (1981), Schierwater and Hauenschild (1990), Armendáriz (2000), Armendáriz and César (2001), and Ladle and Todd (2006). Working with *Stylaria*, Kaliszewicz (2003) and Kaliszewicz, Johst, Grimm, and Uchmański (2005) focused further on the importance of regeneration and fission to naid life-histories, documenting sublethal predation by aquatic insects and demonstrating that several aspects of fission are modulated by the presence of predator cues.

Although I would have so enjoyed live conversations with the community of early naid researchers, it has been wonderful to learn from these colleagues through their written works. And I have savored the handful of live conversations about fission and regeneration with other naid afficionados outside of my lab group—I can count these conversations on one hand. I was fortunate that the international aquatic oligochaete meetings, held every 3 years somewhere around the globe, occurred in the United States while I was a late-stage graduate student, and I made the 17 h drive from Stony Brook, New York to Presque Isle, Maine to attend. It was a small group—a few dozen people —but it was a thrill to connect with others who knew and loved the same group of animals. Although most of the people there worked on different annelid subgroups and did not focus on regeneration or fission, I relished a rare opportunity for live conversation about naids with Christer Erséus. I still fondly remember discussing with him the exciting possibility that naids could stem from multiple origins of fission—he had had that insight years before this was ultimately shown with molecular data (Envall, Källersjö, & Erséus, 2006). At that same meeting I was honored to meet aquatic oligochaete giant Ralph Brinkhurst. And I also met Charlie Drewes, an invertebrate neurophysiologist who was a wonderful promoter of the use of naids and other aquatic invertebrates for simple yet elegant teaching and research activities. Charlie had boundless positive energy and was a dedicated educator; he quickly and kindly became an informal mentor for me. His tragically early death was felt deeply by many.

The literature on naids was sparse, but long. Naids had served as excellent subjects for studying regeneration and agametic reproduction for more than a century and a half. But at the time my interest was first piqued, modern approaches had not yet been applied to this study system. I became excited to continue building on this foundation.

4. The beauty of *Pristina* and naidines as research study systems

Naids have many advantages for studying the evolution and underlying processes of regeneration and fission, and postembryonic development more broadly. Key among these are their small size, transparency, and segmented body organization; their ease of culture and experimental tractability; their robust fission and regenerative abilities; and the diversity of their postembryonic developmental repertoires. Among the naids, the species *P. leidyi* has been a particularly useful study species for laboratory work.

Naids represent not one but two closely related groups of small fissioning annelids (Fig. 1). The name "naid" (referring to the Greek water nymphs, hence the common name "water nymph worms") is an informal one we use to refer to two subfamilies: Pristininae (comprised of the genus *Pristina*) and Naidinae (comprised of about two dozen genera, including common ones

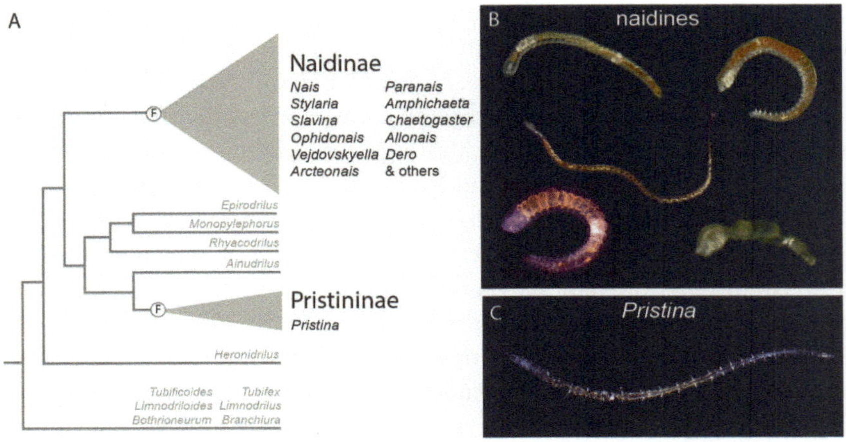

Fig. 1 Diversity of naid annelids. (A) Naids comprise two groups, the closely related subfamilies Naidinae and Pristininae. Each of these is inferred to have evolved fission separately. Origins of fission are marked with "F." Phylogenetic relationships are based on Envall et al. (2006). (B) Naidinae, as illustrated by specimens of five genera. Top row, left to right: *Arcteonais, Nais.* Middle row: *Dero*. Bottom row, left to right: *Paranais, Chaetogaster*. (C) Pristininae, as illustrated by *Pristina leidyi*. Anterior is left in all images. Images of *Dero* and *P. leidyi* are by E. Zattara; all others are by the author.

such as *Nais*, *Stylaria*, *Chaetogaster*, *Dero*, and *Paranais*). These two subfamilies, within the large clitellate family Naididae, were previously thought to represent one group, but molecular phylogenetic data demonstrated that these are actually two closely related but distinct groups, apparently stemming from two distinct origins of fission (Envall et al., 2006). Because species in these two groups share many features (e.g., small body size, similar ecologies, reproduction by fission), we continue to informally refer to them collectively as "naids" for convenience. Naids are common worldwide and easily accessible, living within sediments or on submerged surfaces in freshwater and estuarine habitats. Most species feed by ingesting sediments or grazing on algae and biofilms, but several lineages have become parasitic and one lineage has become an ambush predator (Brinkhurst & Jamieson, 1971; Stephenson, 1930).

Naids have flexible, segmented bodies and most species are largely transparent, allowing direct viewing of internal anatomy, organismal functioning, and even developmental processes (Fig. 1). Transparency is much more than experimentally convenient: time and time again, I have found that this transparency engages minds, fuels questions, and encourages an integrative understanding of the organism. The size of naids (most are about the size of an eyelash) makes them large enough to be seen with the naked eye and be handled with relative ease, yet small enough that whole animals fit easily on a cover-slipped slide, many hundreds fit in an Eppendorf tube, and thousands can be grown in a small section of your benchtop. Most of the length of the body is composed of a series of body segments possessing iteratively organized components of the major organ systems. These segments provide useful, stable landmarks in experimental studies. The anterior and posterior tips of the body are asegmental (prostomium and peristomium anteriorly, pygidium posteriorly). Naids have a thin outer cuticle, thin tissues, and significant coelomic cavities, which further facilitates the observation of internal body structures and penetration of reagents. They have a convenient level of complexity: they have all major organ systems of complex bilaterians (including a well-developed central nervous system with brain and ventral nerve cord ganglia, regionalized gut, closed circulatory system, multiple muscle layers, excretory organs, and sensory and locomotory structures) but each organ is only a few cell layers thick.

The ease and inexpensiveness of culturing naids is another significant advantage. To grow *P. leidyi*, all that is needed is springwater (made up as 1% seawater using artificial sea salts), spirulina powder, and a culture dish; some brown paper towel as substrate is helpful for them to crawl on but not even necessary (Bely & Wray, 2001; Zattara & Bely, 2011). *Pristina leidyi* grow optimally with water changes and feedings every few days to every few weeks, but cultures can tolerate weeks or even months of inattention. Under favorable culture conditions individuals will fission regularly, roughly doubling every week; it is therefore simple to generate hundreds or thousands of individuals for experiments. Because they reproduce clonally, it is possible to establish isogenic lines derived from a single individual, allowing studies to be carried out on genetically homogeneous material. Many other naid species can be maintained under similarly simple culture conditions, by feeding with spirulina, wheat or other grains, algae pellets, mud, or—in the case of the carnivorous *Chaetogaster*—cut pieces of other worms (Bely & Sikes, 2010). Several naid species are available from biological supply companies, including the strain of *P. leidyi* we have primarily worked on (originally from Carolina Biological Supply Company, where it is thought to have been cultured for over 60 years, Özpolat & Bely, 2015). A number of genera can reliably be found in freshwater or brackish water bodies around the globe and collecting them involves little more than bringing some substrate or submerged vegetation back to the lab to sort under a dissecting scope. Since these animals reproduce clonally, even one individual can be sufficient to establish a lab culture.

Naids, and *P. leidyi* in particular, are convenient subjects for experimental work. They are quite permeable to drugs and reagents, facilitating whole-mount protocols and enabling bath application of drugs. Their freshwater lifestyle means that reagent incubations of live worms can occur at low salinity, facilitating drug solvency. They can be temporarily anesthetized using a nicotine solution or permanently paralyzed using tetrodotoxin. *P. leidyi* is particularly amenable to and tolerant of experimental conditions: they can be maintained and even cultured in very small volumes (tens of microliters) and are robust to mechanical injury: whether subjected to a clean cut, widespread abrasion, or systemic damage, they often recover well, with high survival. Naids are aquatic and yet are largely mucous-free, facilitating many procedures on both live and fixed material. Over the years, we have established a number of protocols and generated resources for naids with a focus on *P. leidyi*, including efficient DNA/RNA extraction from small amounts of tissue, whole mount in situ hybridization, immunohistochemistry, BrdU and EdU incorporation, temporary and permanent immobilization, whole-mount long-duration live imaging, single worm culture, single worm drug incubation, microinjection, and microrespirometry (Bely & Wray, 2001; Özpolat, Sloane, Zattara, & Bely, 2016; Zattara & Bely, 2011;

Zattara, Turlington, & Bely, 2016; A.E. Bely et al. unpublished data). A fission/regeneration transcriptome was sequenced a number of years ago for *P. leidyi* (Nyberg, Conte, Kostyun, Forde, & Bely, 2012), and additional molecular resources are being developed by our lab as well as other labs who are beginning to work on these animals.

What drew me in to working on naids was their wonderfully dynamic adult bodies and extensive postembryonic capabilities. They are the "Hydras" of the Bilateria! Their postembryonic developmental repertoire is remarkably broad: many species can regenerate heads, regenerate tails, build heads and tails through fission, remodel tissue (morphallaxis), undergo tissue renewal, grow both in size and by adding new segments, and even shrink, with all of these occurring on the time scale of days. All parts of their body and all of their organ systems can be formed through several modes of development, making it possible to investigate how different developmental trajectories can build the same body part.

Under favorable conditions, naids continuously grow and fission. They undergo indeterminate growth, adding new segments, with all of their associated organ systems, from a proliferative subterminal posterior growth zone (segment addition zone). Many body regions also appear to be continuously renewed through homeostatic processes, as evidenced by widespread cell proliferation (Zattara & Bely, 2013). One of the most developmentally intriguing aspects of naid biology is their reproduction by fission. Most species fission by paratomy, forming worm chains that ultimately split apart (Fig. 2). During paratomy, new head and tail tissue is intercalated in the middle of the body, forming a new head for the posterior portion of the original worm and a new tail (including a new posterior growth zone) for the anterior portion of the original worm, thus forming transiently linked chains of individuals (Berrill, 1952; Zattara & Bely, 2011). Remarkably, these intercalated head and tail regions form in the middle of a functioning body: new tissue develops around the existing gut and organ systems such that the functional gut, blood vessels, musculature, and ventral nerve cord traverse the developing fission zone, maintaining functionality and integration between the anterior and posterior halves of the original worm until late in the fission process. The fission process also involves remodeling (morphallaxis) of tissue of the original animal, especially evident in the gut and gonads (Özpolat et al., 2016; Zattara & Bely, 2011).

The ability to repair physical damage is generally high in naids, with many species able to regenerate extensive anterior and posterior regions of the body following transverse amputation (Bely, 2006; Bely & Sikes, 2010)

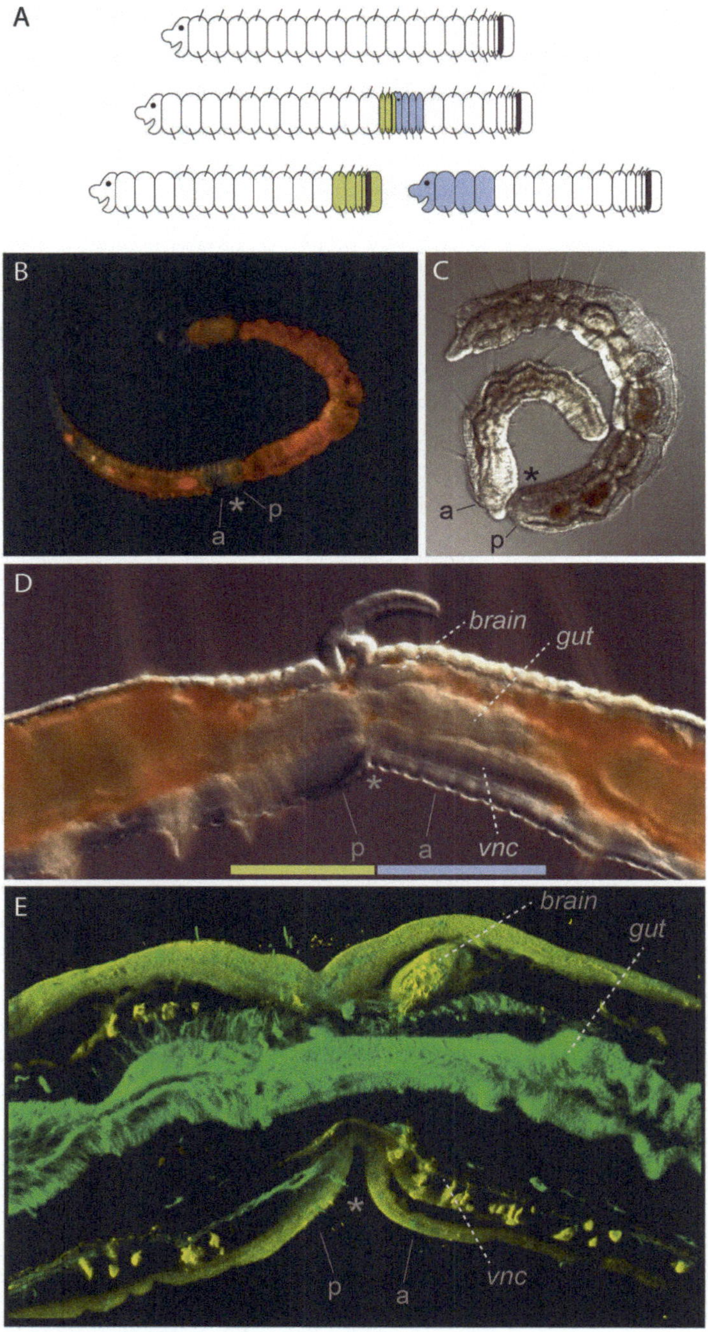

Fig. 2 See figure legend on opposite page.

(Fig. 3). All body regions can be replaced following injury, including asegmental termini, the brain, segments, all organ systems including gonads, and the posterior growth zone. Regeneration of new head and tail ends occurs through the formation of a terminal blastema, a mass of undifferentiated cells that forms at the wound site and from which the new structures develop, involving extensive local cell proliferation (Özpolat & Bely, 2016; Zattara & Bely, 2011). As with fission, regeneration can also involve significant remodeling of existing body regions to reestablish the adult phenotype, which is especially evident in the gut.

Although the phenomenal postembryonic capabilities of naids are what attracted me to these animals, what makes this group such a rich system to explore is the considerable variation across species. Although they all share a similar body organization, and all grow indeterminately by adding segments posteriorly, naid species vary substantially in mode of fission and regenerative abilities, as well as other features. With respect to fission, naids include species that undergo simple fission (breaking and then regenerating), "slow" paratomic fission (forming worm chains with only one fission zone developing at a time), and "fast" paratomic fission (forming worm chains with multiple fission zones developing at a time) (Dehorne, 1916), with the location of fission zones along the body axis being environmentally sensitive, physiologically responsive, and variable between species. Furthermore, fission, and specifically paratomic fission, has apparently evolved independently in the Pristininae and the Naidinae, making naids an excellent system to explore the convergent origins of a new developmental trajectory

Fig. 2 Paratomic fission in naids. (A) Diagram of paratomic fission. A new head (blue) and a new tail (green) are intercalated in the middle of the body, forming a chain of linked worms which ultimately separate. (B) Fissioning *Paranais litoralis*. Note fission zone (new tissue, which is clearer) in the middle of the specimen. (C) Fissioning *Pristina osborni*. Note the strong constriction, the plane of fission, in the middle of the body. (D) Late-stage fission zone of *Pristina leidyi*. Note the clear tissue of the fission zone, the new brain behind the fission plane, and the new anterior proboscis extending dorsally behind the fission plane. Blue and green bars respectively indicate extent of new anterior (head) and posterior (tail) tissue forming within the fission zone. (E) Late-stage fission zone of *Pa. litoralis* showing the continuity of the ciliated gut (tubulin-stained, green), the continuity of the longitudinal fibers of the ventral nerve cord (serotonin-stained, yellow), and the new brain just posterior to the fission plane. Panels are lateral views with anterior to the left (or pointing left). Asterisk—plane of fission (future plane of separation); a—new anterior tissue (head) forming in the fission zone; p—new posterior tissue (tail) forming in the fission zone; vnc—ventral nerve cord. Image in panel (D) is by E. Zattara; all others are by the author.

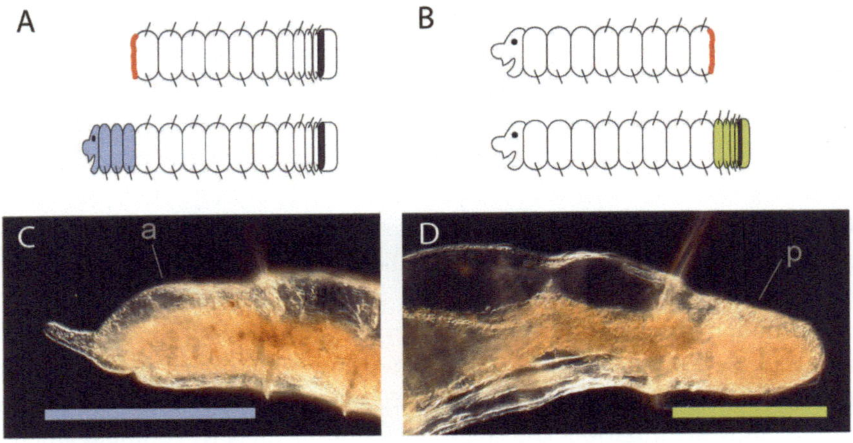

Fig. 3 Regeneration in naids. (A) Diagram of anterior regeneration. Following anterior tissue loss (red), a new head (blue) is formed. (B) Diagram of posterior regeneration. Following posterior tissue loss (red), a new tail (green) is formed. (C) New anterior region (head) regenerating in *P. leidyi*. (D) New posterior region (tail) regenerating in *P. leidyi*. Blue and green bars indicate extent of regenerated anterior and posterior tissue, respectively. Panels are lateral views with anterior to the left. a—new anterior tissue; p—new posterior tissue.

(Envall et al., 2006; Zattara, 2012). Regeneration abilities also vary dramatically across naids: some can regenerate a complete body from just a couple of excised mid-body segments, while others cannot reliably regenerate following even very minimal removals of tissue anteriorly or posteriorly (Bely, 2006; Bely & Sikes, 2010; Herlant-Meewis, 1964). Developmental comparisons across deep phylogenetic distances are hard to interpret rigorously; groups such as the naids—where postembryonic developmental variation abounds across relatively small evolutionary distances—are therefore particularly valuable systems in which to explore the evolution of postembryonic development.

Although naids have many qualities that make them outstanding for studies of postembryonic development, they do not have accessible embryos. Most naids are known to periodically reproduce sexually in the wild (Parish, 1981), but in the lab they reproduce asexually, requiring specific and largely unknown cues to reproduce sexually. Only a few studies have explored and described sexual maturation and embryo formation in the lab (Lochhead & Learner, 1984; Schierwater & Hauenschild, 1990), and although occasionally lab naid cultures will produce one or a few sexual individuals (often after periods of culture neglect), despite considerable efforts we have not been successful at eliciting sexual reproduction in any consistent way. Regardless, the embryos of naids are encased in cocoons and have a provisional epithelium covering much of the embryo, such that even if sexual reproduction could be reliably induced, naids are unlikely to be a good group in which to perform embryonic studies. The lack of accessible embryos is unfortunate: it precludes comparisons between embryonic and postembryonic development and has hampered certain lines of inquiry and the use of techniques that would depend on intracellular injections. Their virtues as an experimental developmental system lie firmly in the postembryonic realm, and there they shine.

5. Reckoning with adults: The challenges of mobility and sensitivity

Our understanding of postembryonic developmental processes has lagged far behind those of embryonic development—and it's no mystery why. By comparison with adults, embryos are docile life stages! Embryos may be finicky, vulnerable, and difficult to manipulate because of their size, but they also (mostly) stay where they are and do not adjust everything they are doing based on their context—what they sense in their environment, what else is happening in their bodies, and when and what they last ate. When I first started working with naid postembryonic development, I quickly learned that I would have to reckon with two pervasive challenges: mobility and high sensitivity (context dependence).

5.1 Taming transparent worms for time-lapse imaging

The favorable qualities of naids came together in a powerful way when we successfully established long-duration time-lapse imaging for these animals. It had been a long-standing goal of ours to record movies of regeneration and fission in naids. Doing so would enable us to directly observe the individual cell behaviors of postembryonic development, allowing us to test ideas that had been in the literature for close to a century, such as that regeneration elicits migration of "neoblast" cells (putative stem cells) toward the wound site. Naids are small and transparent, which are important advantages, but immobilization was a core challenge. How could we immobilize these adult animals long enough to record the building of new tissues and individual cell behaviors? Nicotine, our routine anesthetic, was effective only on the scale of minutes; ivermectin paralyzed animals for hours at a time but led to high mortality; and neither chilling them nor embedding them in agarose were sufficiently effective—their ethos is to move. I had tried to find a way to immobilize them as a graduate student and again as a postdoc, with no real success. My first graduate student James Sikes had—with much effort—managed to collect a series of images of an ivermectin-treated worm that revealed clear migration of cells toward a wound site; we were ecstatic, but the success rate was very low as ivermectin often caused mortality. We finally had a major breakthrough as a result of work on a different project. Ed Zattara, then a graduate student in my lab, was carrying out a large drug screen to identify possible disruptors of regeneration. Most drugs either killed worms or had no appreciable effect, but he noticed that one of the drugs that caused no effect on regeneration had a secondary effect—paralysis. Tetrodotoxin (TTX), a sodium channel blocker, immobilized worms without killing them. Realizing the significance of this finding, he optimized a mounting protocol involving low melting point agarose and halocarbon oil to allow the animals to survive on a microscope slide. Remarkably, the worms could survive days—immobilized and coverslipped—their general tolerance to experimental conditions being another key ingredient for our success.

The movies of naid postembryonic development were both beautiful and illuminating. Naids allowed themselves to be viewed over time, revealing their inner dynamic workings (without even the need for molecular tricks to follow and track individual cells). These movies revealed the incredibly dynamic nature of the wound response, allowing us to identify multiple migratory cell types, characterize morphogenetic movements, assess cell migration speeds and directions, and document cell division by migratory cells and other cell behaviors—but chief among these insights was the long-awaited direct confirmation of the migration of annelid neoblasts (Zattara et al., 2016). We called them "sliders" to be conservative, but neoblast-like cells did indeed migrate along the ventral nerve cord in response to amputation. Our colleagues from a century ago, who first suggested this from fixed material, would have been thrilled. But there were surprises too. These neoblast-like cells moved along multiple migration routes, not only along the ventral nerve cord; they moved at different speeds, suggesting possible heterogeneity of this cell population; and, most unexpectedly, some of these cells migrated *away* from wounds rather than toward them—hardly the expectation if their role was to help initiate and contribute to the regenerated tissues. Time-lapse imaging was also used by then postdoc Duygu Özpolat, to investigate cell movements during fission (Özpolat & Bely, 2015). Here again, she found the ventral nerve cord to be a significant cell migration route, and here again she found cells, in this case putative germ line cells, moving bidirectionally along the nerve cord. Based heavily on these invaluable movies, we proposed a model by which germ line is transmitted across asexual generations through migration of individual germ line cells along the nerve cord. These movies have raised numerous questions and much work remains to be done, but with these movies, naids have certainly proved their worth as valuable study systems for understanding key facets of postembryonic development.

5.2 Complex modulation of postembryonic development

Another particularly interesting—and challenging—aspect of studying postembryonic development in naids has been that these processes are strongly context dependent. Naids are not unique in this regard. Postembryonic life stages of most animals exist in a highly variable world and have complex physiological adaptations to modulate what their bodies do depending on their internal and external environments. This sensitivity adds a significant level of complexity to the study of postembryonic development. One of the advantages of naids is that, by virtue of their small bodies with relatively few tissue layers, this sensitivity can be assessed across the entire body of an animal, allowing broad understanding of these complex effects.

One of the most significant differences between embryonic and postembryonic stages of animals is that postembryonic stages typically feed. As such, postembryonic stages are typically energetically "open" systems, and adapted to be so, by contrast with embryos that are typically energetically "closed" systems, and adapted to be so. This fundamental difference has great implications for developmental studies of postembryonic processes such as body homeostasis, growth, wound-healing, regeneration, and fission. I came up against the challenges of studying the developmental biology of an "open" system almost immediately when I began working with naids. As a graduate student, my in situ hybridization signals were lousy until I realized that the nutritional state of *P. leidyi* had a major effect on the transcript levels, and thus in situ signal, of developmental genes. This was not a subtle effect. Without sufficient feeding, expression of my study genes was essentially undetectable. Since those early days, my lab and I have regularly come up against the challenges of the sensitivity of the phenomena we are studying. We routinely have to consider animal condition, nutritional state, injury history, and environmental conditions as key factors in our protocols and experiments. Depending on our goals, we devise ways to either directly study or factor out the high context-dependence of postembryonic development.

Research on naids has demonstrated complex modulation of postembryonic developmental processes by nutritional state as well as other conditions such as injury state and age. Importantly, different postembryonic processes (e.g., growth, fission, regeneration) are often impacted differentially by the same conditions. For example, at the whole organismal level, nutritional state strongly affects investment into growth, fission, and posterior regeneration but has no detectable effect on investment in anterior regeneration, possibly because anterior regeneration is adapted for occurring in an energetically "closed" system, unlike the others (Özpolat et al., 2016; Zattara & Bely, 2013; unpublished data). A loss in resource acquisition potential, either from decapitation or from fasting, can even lead to complete resorption of early fission zones (Zattara et al., 2016), even ones that have already made anlage of major organs, such as the brain. By contrast, under a broad range of conditions we have not recorded any evidence of resorption of regeneration blastemas. Feeding level impacts not only relative investment into postembryonic processes but can also affect body and organ size (including sexual organs), cell proliferation, and transcript levels (Özpolat & Bely, 2015; Özpolat et al., 2016; Zattara & Bely, 2011; Zattara et al., 2016; unpublished data). Naids have also demonstrated that injury can have rapid, body-wide effects. For example, injury causes a near complete shutdown of homeostatic cell proliferation across the body within just minutes after injury; in contrast, fission zone initiation and progression happens in a context of extensive homeostatic cell proliferation across the body (Zattara & Bely, 2011; Zattara et al., 2016). Naids develop gonads even as they fission, and the presence and size of gonads are also highly context dependent. Not only is gonad size strongly dependent on feeding, but there is a remarkably strong and unexpected "maternal" effect, with gonad size of the offspring of young parents (anterior worms) being much greater than those of the offspring of older parents (Özpolat et al., 2016). Collectively, these findings reveal complex and often very rapid physiological modulation of postembryonic development by both internal and external factors. Although this modulation can be challenging to work with, these studies are illuminating the nature of environmental and physiological responsiveness of adult-level processes.

6. Evolution of fission and regeneration: Contributions from naids

6.1 Evolution of fission: A novel developmental trajectory

Naids have been particularly useful for understanding the evolution of agametic reproduction. Nearly all animal species can reproduce gametically (sexual reproduction, parthenogenesis) but agametic alternatives to reproduction, such as fission and budding, also abound across animal phylogeny, occurring in roughly half of animal phyla (Brusca, Moore, & Shuster, 2016; Vorontsova & Liosner, 1960). Agametic reproduction is extremely important ecologically (e.g., it is intrinsic to colonial animals such as corals) and can strongly affect population and evolutionary dynamics, by virtue of generating clones. How does agametic reproduction evolve? How does a new, nonembryonic developmental trajectory for building an animal arise? Most standard animal models are not amenable for answering these questions because they either do not reproduce agametically, or if they do, so do their close relatives (but see, e.g., Tan et al., 2012).

Naids have been providing insight into these fundamental questions relating to the evolution of agametic reproduction. Ancestral state reconstruction analyses have shown that for naids, as well as for annelids more generally, regeneration abilities are ancestral and fission evolves in lineages with regenerative abilities, strongly supporting the "regeneration first" hypothesis for the origin of fission, at least in this phylum (Zattara & Bely, 2016). Evolutionary and developmental analyses also indicate that fission has evolved repeatedly across annelids, with *Pristina* and the naidines representing two closely related but distinct origins of fission (Envall et al., 2006; Zattara, 2012). At the developmental level, research on *P. leidyi* has revealed extensive similarities between the developmental processes of fission and regeneration, supporting the hypothesis that fission evolves through co-option of regenerative abilities (Bely & Wray, 2001; Zattara & Bely, 2011). Although these two processes make the same endpoint phenotype, they are not identical: detailed characterizations in *P. leidyi* have identified differences between regeneration and fission in gene expression, general morphogenesis, gut ciliation, muscle dynamics, neural dynamics, and inferred distribution and behavior of germline cells (Bely & Wray, 2001; Özpolat et al., 2016; Zattara & Bely, 2011), pointing to some key differences underlying the divergence of regeneration and fission. Furthermore, regeneration ability can be lost even as fission is retained, further indicating the distinctness of these two processes (Bely, 1999). Gene expression and time-lapse imaging studies in *P. leidyi* also suggest that the evolution of paratomic fission may have involved the evolution of a novel mechanism for germ line transmission, involving migration of presumptive germ line cells toward fission zones to transmit germ line across asexual generations (Özpolat & Bely, 2015). Unexpectedly, this work also indicates that fission does not preclude some level of sexual development, as gonads are routinely formed in fissioning worms, even though they do not progress to sexual maturation (Özpolat & Bely, 2015). The origin of fission is expected to lead to more complex resource allocation strategies, as animals can now invest resources into a larger number of possible postembryonic processes (e.g., growth, regeneration, and fission). Studies in *P. leidyi* have indeed demonstrated complex crosstalk between different postembryonic modes, with relative investment to fission and regeneration shown to be highly context dependent (e.g., sensitive to the position of the cut and the stage of fission) (Zattara & Bely, 2013). Outcomes range from fission acceleration and regeneration stalling (high fission:regeneration investment) to resorption of fission zones and progression of regeneration (low fission:regeneration investment). In sum, data from naids support a "duplication and divergence" model for the evolution of a novel developmental trajectory for making an animal body, providing broader insight into the developmental basis of life history evolution of animals.

6.2 Evolution of regeneration: Unexpected losses

Naids have also provided important insights into the evolution of regeneration. Regeneration ability is highly variable across animals, but our understanding of the evolutionary and developmental processes that lead to this variation remains limited (Bely & Nyberg, 2010; Gazave & Röttinger, 2021). Intensive study of regeneration mechanisms in a handful of animal models is revealing cellular and molecular processes underlying regeneration, but focal species are mostly deeply diverged from one another, limiting the inferences that can be made from comparisons between them. Comparisons across close relatives are needed to reveal the pattern and process of regeneration evolution (e.g., Seifert et al., 2012; Sikes & Newmark, 2013). How variable are regeneration abilities among close relatives? Is regeneration variation due to gains or losses? Are changes in regenerative ability closely tied to changes in other related postembryonic processes such as growth or fission? Research on naids has been important in beginning to answer these questions.

Because animals that can reproduce agametically typically have very high regenerative abilities (Berrill, 1952; Vorontsova & Liosner, 1960), the finding that a naid, *Paranais litoralis*, is incapable of regenerating anteriorly—even when it can fission—was highly unexpected (Bely, 1999). These animals routinely form heads in fission zones but cannot make those same structures if the head is experimentally removed. This finding has highlighted the distinctness of regenerative and fission processes: even though fission and regeneration are evolutionarily and developmentally closely related (Berrill, 1952; Kostyuchenko, Kozin, & Kupriashova, 2017; Vorontsova & Liosner, 1960), the two can be evolutionarily decoupled. Further comparative regeneration analyses have indicated that head regeneration ability has been lost many times, not only within naids but across annelids (Bely & Sikes, 2010; Zattara & Bely, 2016). Additional work in naids has shown that regenerative processes are halted at different points in different nonanteriorly regenerating lineages. In addition, in at least one lineage, head amputation specifically within young developing tissue (i.e., within heads developing within fission zones) can elicit some limited regeneration (Bely & Sikes, 2010). This finding that young tissue is more permissive to regeneration is consistent with findings in some other animals showing an age-related loss of regenerative ability and suggests an interesting hypothesis for regeneration loss—that regeneration loss may involve progressive restriction in regenerative ability across life stages. Nonanteriorly regenerating naids nevertheless wound heal successfully, pointing to the distinctness of wound healing and regenerative mechanisms. The finding that anteriorly amputated naids can still release offspring by fission has led to a new hypothesis for regeneration loss: the evolution of fission may dramatically decrease the adaptive value of anterior regeneration (because decapitated fissioning animals can still reproduce), leading to an essentially evolutionarily neutral loss of regenerative abilities (Bely, 2010; Bely & Sikes, 2010). Naids have also generated information about the developmental processes involved in annelid regeneration, providing a valuable point of comparison with other annelids as well as more distantly related animal groups (Gazave & Röttinger, 2021; Özpolat & Bely, 2016; Seaver, 2022).

7. Lessons learned while establishing a new study system

Investigating species on which few labs work is rewarding but also has real challenges. Reflecting on my own experiences, two noteworthy challenges relate to funding and community.

After an exciting—and funded—initial buildup of research momentum working with naids, and just as I thought we had established a strong foundation on which to build, I found myself unable to obtain additional funding. For years, I wrote grant after grant after grant, without success. My grants were well received, but functional tools commonly used in model systems had become a near de facto requirement for funding. In my lab we had certainly tried to develop these tools but had not had quick success. The challenge was that without significant funding, I could not invest in developing functional tools. But without functional tools, I could not obtain significant funding. Without others working on my study system, I also could not easily benefit from technique development in other labs. I found myself in a difficult funding Catch-22. A similiar story was surely playing out in many other non-model system labs at the time. The NSF is now working to address this broad problem through a relatively new program (the NSF EDGE program) that directly funds technique development, providing a way out of the Catch-22. The reality is that technique development takes considerable personnel time and money. If funding agencies do not directly fund such efforts, yet require these techniques for funding success, then the system will strongly favor labs already holding ample resources, labs that can afford to divert significant resources to unfunded efforts. As a scientific community, we need to remain aware of the ongoing need to directly fund technique development in order to realize the vast potential of lesser studied systems.

A second challenge I have experienced working on a new study system relates to community. My initial focus on naids and postembryonic development was motivated by a love of scientific exploration. I am drawn to large knowledge gaps—to lesser studied questions and organisms. But because research communities in biology are primarily focused around shared study organisms or shared questions, I have found myself at the edges of existing communities. As a graduate student, I found papers on asexual reproduction and regeneration in naids—but most of these dated many decades back. As I quipped at my graduate defense, most of my closest colleagues were long dead! On the conceptual front, because I worked on the evolution and development of asexual reproduction and regeneration, I felt most connected to the young and flourishing evo-devo community. But the key question motivating the core of evo-devo has been: what is the developmental basis of morphological novelty? And the key question I was focused on was in many ways orthogonal: how can distinct developmental processes evolve that converge on the same endpoint morphology? I was working on adults rather than embryos, and on the evo-devo of trajectories rather than morphologies. Although there are challenges to working at the edges of research communities, I have come to appreciate that the edges of established communities are exciting and fertile ground.

Despite some real challenges, I have found working on a new system incredibly rewarding. By focusing on broad open questions and a new (or, rather, rediscovered) study system, my lab group and I have been pushed to think broadly about our biological questions. We have had the pleasure of working across levels of biological organization—considering cellular, molecular, physiological, organismal, ecological, and evolutionary facets of our questions. And we have experienced the joys of working in an uncrowded field. On such an open scientific landscape, our research has a quality of broad exploration that is deeply satisfying.

8. Final remarks on establishing a new study system

For those considering starting a new system, I share here some advice based on my own experiences and learned perspectives. First, focus on broad and important questions. Focusing on processes that are unusual but not unique (in my case, regeneration and agametic reproduction) is a sweet spot, drawing interest while having relevance well outside of the study system. Second, choose an organism (or group of organisms) that has not one but several major advantages. Research methodologies and expectations are continually evolving, and having more than one advantage in your pocket will reduce your vulnerabilities and give your study system staying power. Third, be creative in defining your community. Your most fruitful and helpful connections are likely to be scattered across different parts of the scientific landscape, much more so than for researchers working on heavily studied groups. View yourself not as an outsider but as a nucleus for new connections. Fourth, embrace collective leadership. Working on a new system means that there will be many fruitful avenues to explore in your research. Giving people who work with you the freedom to pursue research threads that are most exciting to them provides numerous advantages. And finally, allow yourself to be drawn forward by questions and organisms that are inherently fascinating to you. Doing so will sustain you through the challenges you will inevitably encounter, and science is most joyous and productive when we are fascinated and hungry for answers.

Acknowledgments

I am very grateful for the wonderful people in my lab group who have worked with me across the years and who have contributed in numerous ways to developing naids as a study system. Among these, I would like to express particular appreciation to Ed Zattara, Duygu Özpolat, and James Sikes, outstanding scientists and colleagues who have shared my excitement about naids and helped lay the foundations for working on this study system. I would also like to acknowledge Corey Rennolds, Joseph Mack, and James Nowotny who are each expanding the scope of work on naids in new and exciting ways. I am grateful for insights, suggestions, support, and encouragement from colleagues over the years, especially Elaine Seaver, Charlie

Drewes, David Weisblat, Mark Martindale, Chris Lowe, Ehab Abouheif, and Greg Wray. Sincere thanks to Bob Goldstein and Mansi Srivastava for their vision and efforts to assemble contributions on this important topic.

References

Aiyer, K. S. P. (1924). Reproduction in the aquatic oligochaete, *Nais paraguayensis*. *Annals and Magazine of Natural History, Series 9, 14*, 615.

Armendáriz, L. C. (2000). Population dynamics of *Stylaria lacustris* (Linnaeus, 1767) (Oligochaeta, Naididae) in Los Talas, Argentina. *Hydrobiologia, 438*, 217–226.

Armendáriz, L. C., & César, I. I. (2001). The distribution and ecology of littoral oligochaeta and aphanoneura (Annelida) of the natural and historical reserve of Isla Martín García, Río de la Plata River, Argentina. *Hydrobiologia, 463*, 207–216.

Bely, A. E. (1999). Decoupling of fission and regenerative capabilities in an asexual oligochaete. *Hydrobiologia, 406*, 243–251.

Bely, A. E. (2006). Distribution of segment regeneration ability in the Annelida. *Integrative and Comparative Biology, 46*, 508–518.

Bely, A. E. (2010). Evolutionary loss of animal regeneration: Pattern and process. *Integrative and Comparative Biology, 50*, 515–527.

Bely, A., & Nyberg, K. (2010). Evolution of animal regeneration: Re-emergence of a field. *Trends in Ecology & Evolution, 25*, 161–170.

Bely, A. E., & Sikes, J. M. (2010). Latent regeneration abilities persist following recent evolutionary loss in asexual annelids. *Proceedings of the National Academy of Sciences of the United States of America, 107*(4), 1464–1469.

Bely, A. E., & Wray, G. A. (2001). Evolution of regeneration and fission in annelids: Insights from *engrailed*- and *orthodenticle*-class gene expression. *Development, 128*, 2781–2791.

Berrill, N. J. (1952). Regeneration and budding in worms. *Biological Reviews, 27*, 401–438.

Bilello, A. A., & Potswald, H. E. (1974). A cytological and quantitative study of neoblasts in the naid *Ophidonais serpentina* (Oligochaeta). *Wilhelm Roux' Archiv fur Entwicklungsmechanik der Organismen, 174*, 234–249.

Bourne, A. G. (1891). Notes on the naidiform Oligochaeta; containing a description of new species of the genera *Pristina* and *Pterostylarides*, and remarks upon cephalization and gemmation as generic and specific characters in the group. *The Quarterly Journal of Microscopical Science, s2–32*, 335–356.

Brinkhurst, R. O., & Jamieson, G. B. M. (1971). Aquatic Oligochaeta of the world. In *Internationale Revue der gesamten Hydrobiologie und Hydrographie*. Edinburgh: Oliver & Boyd.

Brode, H. W. (1898). *Morphology of Dero vaga*.

Brusca, R. C., Moore, W., & Shuster, S. M. (2016). *Invertebrates* (3rd ed.). Sunderland, MA: Sinauer Associates.

Chu, J. (1945). A genealogical analysis of fission in *Stylaria fossularis*. *Proceedings. Zoological Society of London, 115*, 194–206.

Chu, J. (1946). Further studies on the relation between natural fission and regeneration in *Stylaria fossularis*. *Proceedings. Zoological Society of London, 116*, 229–239.

Chu, J., & Pai, S. (1944). The relations between natural fission and regeneration in *Stylaria fossularis* (Annelida). *Physiological Zoology, 17*, 159–166.

Dehorne, L. (1916). Les naidimorphes et leur reproduction asexuée. *Archives de Zoologie Experimentale et Générale*, 25–157.

Drewes, C. D., & Fourtner, C. R. (1991). Reorganization of escape reflexes during asexual fission in an aquatic oligochaete, *Dero digitata*. *Journal of Experimental Biology, 260*, 170–180.

Duffy, M. A., García-Robledo, C., Gordon, S. P., Grant, N. A., Green, D. A., Kamath, A., et al. (2021). Model systems in ecology, evolution, and behavior: A call for diversity in our model systems and discipline. *American Naturalist, 198*, 53–68.

Envall, I., Källersjö, M., & Erséus, C. (2006). Molecular evidence for the non-monophyletic status of Naidinae (Annelida, Clitellata, Tubificidae). *Molecular Phylogenetics and Evolution, 40*, 570–584.

Foulkes, R. H. (1953). Regeneration of the anterior end of *Aulophorus furcatus* (Naididae) with special reference to effect of X-rays. *Biological Bulletin, 105*, 80–86.

Galloway, T. W. (1899). Observations on non-sexual reproduction in *Dero vaga*. *Bulletin of the Museum of Comparative Zoölogy, 35*, 1–140.

Gazave, E., & Röttinger, E. (2021). 7th Euro Evo Devo meeting: Report on the "evolution of regeneration in Metazoa" symposium. *Journal of Experimental Zoology Part B: Molecular and Developmental Evolution, 336*, 89–93.

Harper, E. H. (1904). Notes on regulation in *Stylaria lacustris*. *Biological Bulletin, 6*, 173–190.

Hempelmann, F. (1923). Kausal-analytische Untersuchungen uber das Auftreten vergrosserter Borsten und die Lage der Teilungszone bei Pristina. *Archiv fur mikroskopische Anatomie, 98*, 379.

Hepke, P. (1897). Über histo- und organogenetische Vorgange bei den Regenerationsprocessen der Naiden. *Zeitschrift für Wissenschaftliche Zoologie, 63*, 263–291.

Herlant-Meewis, H. (1946). Reconstitution du germen chez *Lumbricillus lineatus* (Enchytraeides). *Archives de biologie, Paris, 57*, 197–306.

Herlant-Meewis, H. (1964). Regeneration in annelids. In M. Abercrombie, & J. Brachet (Eds.), *Vol. 4. Advances in morphogenesis* (pp. 155–215). New York: Academic Press.

Hyman, L. H. (1916). An analysis of the process of regeneration in certain microdrilous oligochaetes. *Journal of Experimental Zoology, 20*, 99–163.

Hyman, L. H. (1938). The fragmentation of *Nais paraguayensis*. *Physiological Zoology, 11*, 126–143.

Hyman, L. H. (1940). Aspects of regeneration in annelids. *American Naturalist*, 513–527.

Jenner, R. A., & Wills, M. A. (2007). The choice of model organisms in evo-devo. *Nature Reviews. Genetics, 8*, 311–319.

Kaliszewicz, A. (2003). Sublethal predation on *Stylaria lacustris*: A study of regenerative capabilities. *Hydrobiologia, 501*, 83–92.

Kaliszewicz, A., Johst, K., Grimm, V., & Uchmański, J. (2005). Predation effects on the evolution of life-history traits in a clonal oligochaete. *American Naturalist, 166*, 409–417.

Kharin, A. V., Zagainova, I. V., & Kostyuchenko, R. P. (2006). Formation of the paratomic fission zone in freshwater oligochaetes. *Russian Journal of Developmental Biology, 37*, 354–365.

Kostyuchenko, R. P., Kozin, V. V., & Kupriashova, E. E. (2017). Regeneration and asexual reproduction in annelids: Cells, genes, and evolution. *Biology Bulletin of the Russian Academy of Sciences, 43*, 185–194.

Krecker, F. H. (1923). Origin and activities of the neoblasts in the regeneration of microdrilous annelida. *Journal of Experimental Zoology, 37*, 27–46.

Ladle, R. J., & Todd, P. A. (2006). Sex or sanctuary: How do asexual worms survive the winter? *Hydrobiologia, 559*, 395–399.

Learner, M. A., Lochhead, G., & Hughes, B. D. (1978). A review of the biology of British Naididae (Oligochaeta) with emphasis on the lotic environment. *Freshwater Biology, 8*, 357–375.

Lochhead, G., & Learner, M. A. (1983). The effect of temperature on asexual population growth of three species of Naididae (Oligochaeta). *Hydrobiologia, 98*, 107–112.

Lochhead, G., & Learner, A. (1984). The cocoon and hatchling of *Nais variabilis* (Naididae: Oligochaeta). *Freshwater Biology, 14*, 189–193.

Meewis, H. (1934). Eléments régénérateurs dans le bourgeonnement pygidial et dans la zone de scissiparité chez *Chaetogaster diaphanus*. *Annales de la Societe Royale Zoologique de Belgique*, *65*, 9–39.

Nyberg, K. G., Conte, M. A., Kostyun, J. L., Forde, A., & Bely, A. E. (2012). Transcriptome characterization via 454 pyrosequencing of the annelid *Pristina leidyi*, an emerging model for studying the evolution of regeneration. *BMC Genomics*, *13*, 287.

O'Brien, J. P. (1942). Studies on the effects of X-rays on regeneration in the fragmenting oligochaete, *Nais paraguayensis*. *Growth*, *6*, 203–229.

O'Brien, J. P. (1946). Studies on the cellular basis of regeneration in *Nais paraguayensis*, and the effects of X-rays thereon. *Growth*, *10*, 25–44.

Özpolat, B. D., & Bely, A. E. (2015). Gonad establishment during asexual reproduction in the annelid *Pristina leidyi*. *Developmental Biology*, *405*(1), 123–136.

Özpolat, B. D., & Bely, A. E. (2016). Developmental and molecular biology of annelid regeneration: A comparative review of recent studies. *Current Opinion in Genetics and Development*, *40*, 144–153.

Özpolat, B. D., Sloane, E. S., Zattara, E. E., & Bely, A. E. (2016). Plasticity and regeneration of gonads in the annelid *Pristina leidyi*. *EvoDevo*, 7, 22.

Parish, J. (1981). Reproductive ecology of Naididae (Oligochaeta). *Hydrobiologia*, *83*, 115–123.

Randolph, H. (1891). The regeneration of the tail in *Lumbriculus*. *Zoologischer Anzeiger*, *14*, 154–156.

Randolph, H. (1892). The regeneration of the tail in *Lumbriculus*. *Journal of Morphology*, 7, 317–344.

Rota, E. (2022a). How the discovery of oligochaete regeneration during the Enlightenment was pivotal to the advancement of annelid research. *Zoological Journal of the Linnean Society*. In press.

Rota, E. (2022b). Exploring annelids in the age of Enlightenment and beyond: Classification and bioluminescence. *Zoological Journal of the Linnean Society*. In press.

Schierwater, B., & Hauenschild, C. (1990). A photoperiod determined life-cycle in an oligochaete worm. *Biological Bulletin*, *178*, 111–117.

Seaver, E. C. (2022). Sifting through the mud: A tale of building the annelid *Capitella teleta* for EvoDevo studies. *Current Topics in Developmental Biology*. In this issue.

Seifert, A. W., Kiama, S. G., Seifert, M. G., Goheen, J. R., Palmer, T. M., & Maden, M. (2012). Skin shedding and tissue regeneration in African spiny mice (*Acomys*). *Nature*, *489*, 561–565.

Semper, C. (1876). Die Verwandschaftsbeziehungen der gegliederten Tiere. *Arbeiten aus dem Zoologisch-Zootomischen Institut in Würzburg*, *3*.

Sikes, J. M., & Newmark, P. A. (2013). Restoration of anterior regeneration in a planarian with limited regenerative ability. *Nature*, *500*, 77–80.

Sperber, C. (1948). A taxonomical study of the Naididae. *Zoologiska Bidrag från Uppsala*, *28*, 1–296.

Stephenson, J. (1930). *The Oligochaeta*. Oxford: Clarendon Press.

Tan, T. C. J., Rahman, R., Jaber-Hijazi, F., Felix, D. A., Chen, C., Louis, E. J., et al. (2012). Telomere maintenance and telomerase activity are differentially regulated in asexual and sexual worms. *Proceedings of the National Academy of Sciences of the United States of America*, *109*, 4209–4214.

Van Cleave, C. D. (1934). The effects of X-radiation on the reconstitution of Stenostomum tenuicauda and some other worms. *Biological Bulletin*, *67*, 304–314.

Van Cleave, C. D. (1937). A study of the process of fission in the naid *Pristina longiseta*. *Physiological Zoology*, *10*, 299–314.

Vorontsova, M. A., & Liosner, L. D. (1960). *Asexual propagation and regeneration*. London: Pergamon Press.

Zattara, E. E. (2012). *Regeneration, fission and the evolution of developmental novelty in naid annelids*. PhD dissertation College Park: Biology Department, University of Maryland.
Zattara, E. E., & Bely, A. E. (2011). Evolution of a novel developmental trajectory: Fission is distinct from regeneration in the annelid *Pristina leidyi*. *Evolution & Development*, *13*, 80–95.
Zattara, E. E., & Bely, A. E. (2013). Investment choices in post-embryonic development: Quantifying interactions among growth, regeneration, and asexual reproduction in the annelid *Pristina leidyi*. *Journal of Experimental Zoology Part B: Molecular and Developmental Evolution*, *320*, 471–488.
Zattara, E. E., & Bely, A. E. (2016). Phylogenetic distribution of regeneration and asexual reproduction in Annelida: Regeneration is ancestral and fission evolves in regenerative clades. *Invertebrate Biology*, *135*, 400–414.
Zattara, E. E., Turlington, K. W., & Bely, A. E. (2016). Long-term time-lapse live imaging reveals extensive cell migration during annelid regeneration. *BMC Developmental Biology*, *16*, 6.

CHAPTER SEVENTEEN

Streblospio benedicti: A genetic model for understanding the evolution of development and life-history

Christina Zakas*
North Carolina State University, Raleigh, NC, United States
*Corresponding author: e-mail address: czakas@ncsu.edu

Contents

Current Topics in Developmental Biology, Volume 147
ISSN 0070-2153
https://doi.org/10.1016/bs.ctdb.2021.12.021

Abstract

Investigating developmental evolution usually requires comparing differences across related species to infer how phenotypic change results from embryological modifications. However, when comparing organisms from different environments, ecologies, and evolutionary histories there can be many confounding factors to finding a genetic basis for developmental differences. In the marine annelid *Streblospio benedicti,* there are two distinct types of offspring with independent developmental pathways that converge on the same adult phenotype. To my knowledge, *S. benedicti* is the only known species that has heritable (additive) genetic variation in developmental traits that results in alternative life-history strategies. Females produce either hundreds of small, swimming and feeding larvae, or dozens of large, nonfeeding larvae. The larvae differ in their morphology, ecology, and dispersal potential. This developmental dimorphism makes *S. benedicti* a unique and useful model for understanding how genetic changes result in developmental modifications that ultimately lead to overall life-history differences. Because the offspring phenotypes of *S. benedicti* are heritable, we can use forward genetics within a single evolutionary lineage to disentangle how development evolves, and which genes and regulatory mechanisms are involved.

1. Choosing the best system to answer the big questions

Streblospio benedicti (Webster, 1879) is apparently, and previously unbeknownst to me, colloquially named the "bar-gilled mud worm." This underwhelming and drab moniker does not capture a worm that appears more like a miniaturized East-Asian dragon. But due to the fascinating variation and morphological innovation in the Annelida, "dragon worm" and even "mud dragon" have already been taken. While it misses out on fantastical branding, from an evolutionary and developmental perspective there is something truly unusual about *S. benedicti.*

S. benedicti is common in the benthic estuaries of North America and it is easily identifiable based on the paired feeding palps and branchia on its head, which are characteristic of the spionid annelids. I started working on *S. benedicti* specifically because it was one of the few worms I, as a novice annelid zoologist, could definitively find and identify from pictures during my graduate work. (The ratio of head adornments to body length firmly places *S. benedicti* in the small dragon group.)

There are many attractive aspects to working on evolutionary biology in *S. benedicti*. As a lab animal, they are small, ~1 cm (Fig. 1), and easy to maintain. They can survive harsh conditions such as a bucket in the trunk of a car for many hours, and have lived in basements, porches, and kitchen counters for weeks at a time. They require relatively little maintenance compared to other invertebrates. I recommend picking a model that can live through a lab

Fig. 1 Brooding planktotrophic adult female of *S. benedicti*. Scale bar is 1 mm.

shutdown (as they did for me during hurricane Sandy in 2012, and the COVID-19 pandemic), or that simply lets you take a vacation. Their generation time is about 6 weeks, which is reasonable if you work on mice. Overall, it's a stellar lab model.

As a common US East Coast annelid, there have been many studies documenting *S. benedicti's* morphology and reproduction. But the work of Lisa Levin and her colleagues at North Carolina State University, which was published through the 1980s and 1990s, was what hooked many of us who have researched it since. There are a handful of Streblophiles who have worked to better our understanding of developmental evolution in this system and I detail aspects of their work here.

For me, I mostly came to this system by luck and an interest in marine larval life-history. I did not set out to make *S. benedicti* a genetic model for EvoDevo when I started working with it in 2007 as a graduate student, but as my dissertation advisor will attest, I was stubbornly attached. (I know John P. Wares meant this as a complement.) I came to work on quantitative genetics and genomics in this system by true convergence through independent origins. Matthew Rockman was studying *S. benedicti* genetic diversity when we met at a special symposium on *Poecilogony* at the Society of Integrative and Comparative Biology in 2011 (Knott & McHugh, 2012), just after I defended my PhD on *S. benedicti* population genetics. We constructed the first annelid genetic map in this system and a chromosome-scale genome assembly. There are now multiple transcriptome sets and population sequences available. We've also developed techniques for husbandry and maintenance of these worms. *S. benedicti* has become an important genetic and genomic model for researching development, embryology and life-history evolution.

2. Poecilogony: A collection of exceptions

Understanding how life-history strategies evolve is of paramount importance to understanding many basic biological questions. How do differences in fitness-related traits arise when selection is strong and development is driven by a conserved toolkit? What aspects of development and regulation can be "tinkered" with to create novel change and what must be constrained to produce viable offspring? To answer these questions, we must study models that possess intrinsic modularity and flexibility in their life-history strategy. In marine invertebrates, this flexibility in development is called *poecilogony*.

S. benedicti is probably best known for being a true example of *poecilogony*. The term was coined by Giard (1905) to describe marine organisms that had two different larval types within a single species. The existence of larval dimorphisms raised many intriguing possibilities for embryologists and larval biologists, however most species that were described as poecilogonous were subsequently revealed by molecular barcoding to be pairs of cryptic species, each with a single larval type (Chia, Gibson, & Qian, 1996; Hoagland & Robertson, 1988). Poecilogony, which is now used to describe different larval modes within a single species, has been shown in only a few Lophotrochozoan taxa: Six are spionid annelids (reviewed in: Oyarzun & Brante, 2015). In gastropods, six are sacoglossan sea slugs (reviewed in: Krug, 2007, 2009), with a caenogastropod mollusk (McDonald, Collin, & Lesoway, 2014) and one a pulmonate (Wiggering, Neiber, Gebauer, & Glaubrecht, 2020). Of these animals, most have plasticity in development, where a single individual can alter the type of offspring based on environmental cues like seasonality, or even by producing multiple larval types within a clutch. These examples of poecilogony occur on the individual level. Over decades of research, only *S. benedicti* maintains a difference in larval development that is heritable, where the larval differences are due to additive genetic variation and females cannot change their offspring type (Levin, Zhu, & Creed, 1991; Zakas, Deutscher, Kay, & Rockman, 2018).

2.1 Marine life-history strategies

To truly understand the exceptional life-history strategy of poecilogony, it is necessary to consider more typical marine invertebrate life-histories. Marine invertebrates are an ideal focal group for studying variation in life-history strategy, as key traits like egg size have far more variation within a taxon than terrestrial animals (Strathmann, 1990). Because of this (and because they are fascinating), marine invertebrate life-history strategies have been studied extensively. Many marine invertebrate lineages contain species that have evolved *indirect development,* which is development via a larval stage that is morphologically and ecologically different from the adult form. A notable trend among these indirect developing animals is that despite all the taxonomic, ecological, and selective pressures that differ between them, two predominant life-history strategies dominate. These strategies are essentially opposite ends of the life-history spectrum for indirect developers and include a suite of trade-offs between egg-size, fecundity, development time, feeding mode, defenses and more. Essentially, the life-histories of indirect developing marine invertebrates can be distilled into two categories: those that feed in the plankton—*planktotrophy*—and those that do not (yolk-feeding)—*lecithotrophy*. Notably, almost all marine invertebrates that produce larvae fall into these two categories with very few exceptions (Allen & Pernet, 2007). *S. benedicti* is one such exception. Being poecilogonous, the species produces both planktotrophic and lecithotrophic larvae.

While the terms planktotrophy and lecithotrophy are assigned based on feeding-mode of larvae, for *S. benedicti* we often use the terms to encapsulate the full larval life-history syndrome. Planktotrophic larvae have small eggs, take longer to develop, are obligately feeding, and have variety of morphological and developmental differences compared to the lecithotrophic larvae (Table 1). Nuances certainly exist within a mode. For example, the large-egged, quickly developing mode is often referred to as "lecithotrophic," when in fact some populations of this morph are actually *facultative planktotrophs,* meaning they can feed although do not require food to complete development (Pernet & McArthur, 2006). I will refer to them as lecithotrophic here. However, despite some population level variation, the larval type of a female is a fixed trait, not plastic, so it does not change with maternal food availability, temperature, or larval feeding (Levin & Creed, 1986). The work of Levin et al. (1991) demonstrates that these larval characteristics are heritable, with planktotrophic larvae growing into females that pass on planktotrophic alleles.

2.2 Larval evolution

Evolutionary shifts in development from planktotrophy to lecithotrophy have occurred across many taxa. It is seemingly canonical to assume that in marine lineages with both larval types, feeding larvae are plesiomorphic

Table 1 Life-history traits and trade-offs for the two larval types of *S. benedicti* as summarized in Levin and Bridges (1994).

	Planktotrophs	**Lecithotrophs**
Life-history traits		
Egg size (diameter μm)	60–70	100–200
Larvae/brood	68–463	15–70
Larvae size (length μm)	250–300	550–650
Larval development		
Planktonic period	2.5–7 weeks (~1 week)	<8 days (~1–2 days)
Larval feeding	Obligate planktotrophy	Facultative planktotrophy or lecithotrophy
Larval morphology		
Chaetae	Present (~20/pouch)	Absent
Pronounced anal cirri	Present	Absent

Parentheses denote my lab observations.

and the evolutionary loss of elaborate feeding structures is parsimonious (Smith, Zigler, & Raff, 2007; Strathmann, 1985; Wray, 1996), yet the reverse has been possible in some clades (reviewed in Collin & Moran, 2018). How these rapid and dramatic shifts occur at the morphological and genomic level is an open question. However, an organism's life-history strategy informs our expectations of its evolutionary trajectory. For example, we see that lecithotrophic species tend to have smaller population sizes, less gene flow (Palumbi, 1995; Shanks, 2009), and more frequent extinctions (Valentine & Jablonski, 1986) than their planktotrophic relatives. We can also predict which environments will favor one strategy over the other. Thorson's rule (1950) for example, is a classic observation that larval type is correlated with latitude and sea temperature: high latitudes and low temperatures favor lecithotrophy while the tropical waters have more planktotrophic species. We expect that a combination of factors such as planktonic patchiness, food availability, predation rates, location in a range, and others will create an adaptive landscape that favors one strategy over another. The most interesting facet of these trade-offs is that while any combination of conditions can exist in a geographic location or over time, these two opposing life-history strategies repeatedly emerge. Despite the multitude of phenotypic changes that must occur to switch a larval type from one strategy to another (again, as a syndrome and not just feeding-ability), and the presumably strong selection against intermediate strategies, these developmental switches have occurred rapidly and repeatedly.

3. Embryology, larval morphology, and developmental differences

To illustrate the complexities of evolving a new life-history strategy and developmental program, I will first focus on differences between the two types in the egg and embryo, then discuss the major morphological changes to the larval body plan between the planktotrophic and lecithotrophic morphs.

3.1 Gametogenesis and reproduction

S. benedicti reaches sexual maturity in about a month following settlement in the benthos (Grassle & Grassle, 1974). Males and females produce gametes throughout their body, with the first gametogenic segment starting around 9–11 setigers (chaetae-bearing segments; Levin & Creed, 1986). When gravid, the difference between the sexes is obvious as gametes can be seen through the clear body wall (Fig. 1). Fertilization happens internally, and embryos are passed to a dorsal brood pouch. The brood pouch is a thin, transparent epithelial layer external to the main body of the female (on setigers 17–45: Rice, 1991), where the embryos are brooded until release as swimming larvae (Table 1, Dean, 1965).

In the lab, reproduction happens year-round, although it is not necessarily continuous. Mated females will produce 1–5 broods in their lifetime in the lab. The number of broods, on average, is not different between the two developmental types (Levin, Caswell, DePatra, & Creed, 1987) suggesting most of the trade-offs in fecundity and survivorship over a female's lifetime are due to differences in larval type.

3.2 Egg size evolution and oogenesis

The egg is the *pièce de resistance* for a myriad of evolutionary and developmental questions; it is central to life-history theory as well as developmental and reproductive biology. Wray and Raff (1991) identified the production of large, yolk-filled eggs as a key innovation for the transition from planktotrophy to lecithotrophy (also see: Hart, 1995; Levitan, 2000; Moran & McAlister, 2009; Strathmann, 1977). In many marine invertebrates, parental investment, particularly offspring or egg size, can vary wildly even among closely related taxa (Marshall, Allen, & Crean, 2008). Egg size is a key life-history trait in *S. benedicti* as the different larval types originate from eggs of different sizes. So, from the very beginning of development we expect that embryological and morphological differences will be initiated by alternate maternal provisioning.

The eggs of lecithotrophs are ~8× bigger by volume than the planktotrophs (Table 1). In *S. benedicti* females egg size is highly heritable and insensitive to environmental perturbations (Levin et al., 1991). Presumably the oocytes are differentially packaged in terms of lipids, proteins and RNAs; although this is still not known. Studies of *S. benedicti* have revealed oogenesis occurs rapidly, with vitellogenesis taking place entirely within the ovary (Eckelbarger, 1980). The size of yolk granules and relative proportion of heterosynthetically-derived yolk may differ between planktotrophic and lecithotrophic forms of *S. benedicti* (Eckelbarger, 1986; Rice & Levin, 1998). I am repeatedly surprised that we know so little about oogenesis in *S. benedicti* given its utility in experimentally addressing fundamental questions about egg evolution. I am especially interested in how maternal RNA provisioning differs between the oocytes of each type, because this comparison will allow us to better understand how variation in maternal packaging shapes downstream developmental differences.

3.3 Spermatogenesis

Males produce spermatophore packets that they deposit on the benthic mud by sticking their heads out of their mud tubes, holding spermatophores up in their branchia, and letting them drift away (Supplementary Video 1 in the online version at https://doi.org/10.1016/bs.ctdb.2021.12.021). These spermatophores are made from paired nephridia simultaneously discharging spermatophores, which fuse and form a paired structure resembling a V with sperm tuffs at the tip (Fig. 2; Supplementary Video 2 in the online version at https://doi.org/10.1016/bs.ctdb.2021.12.021, described in Rice & Levin, 1998). On occasion I've seen a female pick one up with their palps and branchia, same as they do with mud particles when feeding, and retreat into her tube, but fertilization has not been described. Females can store sperm for weeks to months as evidenced by isolated-mated females in the lab. Rice (1981) noted that the sperm contain cytoplasmic platelets that may store energy, and even postulated that sperm storage may enhance the introduction success of *S. benedicti* from a single mated female.

Fig. 2 Spermatophores on the surface of mud have a V shape and translucent sperm tips. Scale bar is 100 μm.

There is potential for female selection on spermatophores. For example, Mahon and Dauer (2005) described chemosensing in *S. benedicti* but only as it relates to organic content. Schulze, Rice, Simon, and Karl (2000) reports that there may be some mate-potential sensing as well: males stopped producing spermatophores in the presence of a distantly related female (based on population of origin). We have found that females presented with no male and spermatophores only will rarely mate but will readily do so when in the same container as the male producing spermatophores, which was also reported by Rice (1991). Given that females can store sperm for long periods, there is also potential for sperm-competition during fertilization.

3.4 Early embryogenesis

The taxonomic restriction of poecilogony to the spionids and gastropods suggests there is something spiralian embryo-specific about this ability to maintain alternative development types within a species.

Spiral cleavage is characterized by the conserved arrangement of blastomeres during early embryogenesis (Henry & Martindale, 1999). Animals with spiral cleavage diversified more than 500 million years ago, and comprise one of the largest clades of metazoan diversity (Dunn et al., 2008; Halanych, Dahlgren, & McHugh, 2002).

Spiralian development has been well described, but we know little about the embryonic development of *S. benedicti* despite the compelling potential for intraspecific ontogenetic comparisons. *S. benedicti* is an unequal spiral cleaver for both egg types, meaning there is autonomous specification of cell fates by maternal factors in the egg. Cells "A,B,C, and D" are the signifiers of the four cell-stage Spiralian embryo. The maternal factors are asymmetrically distributed in a large polar lobe, which is a cytoplasmic extension of the vegetal pole, to bring ooplasmic determinants to the D quadrant during the first and second cleavage (Fig. 3). The maternal determinants contained in the polar lobes act to specify the D quadrant's unique developmental role (Guerrier, Van Den Biggelaar, Van Dongen, & Verdonk, 1978; Henry, Perry, & Martindale, 2006), determining many key fates such as the dorsoventral axis (Henry & Martindale, 1994), the embryonic organizer (Lambert & Nagy, 2001), the germline (Özpolat, Handberg-Thorsager, Vervoort, & Balavoine, 2017), the gut (from the 2d lineage), and chaetal sacs (also from the 2d lineage; Dorresteijn, 2005).

3.5 Comparative embryology

In *S. benedicti* we see some notable differences in embryogenesis between the two types (McCain, 2008). But do these differences in embryogenesis lead to different larval types? In the Spiralian embryo, comparison across species with alternate larval types has highlighted three evolutionary modifications in the embryo that may lead to lecithotrophic larvae: (1) increase in 1d (or 1c) micromere size compared to the macromere (Jones, Stankowich, & Pernet, 2016); (2) creating a 2d cell larger than the 2D cell (Guralnick & Lindberg, 2001); and (3) early division of D relative to ABC

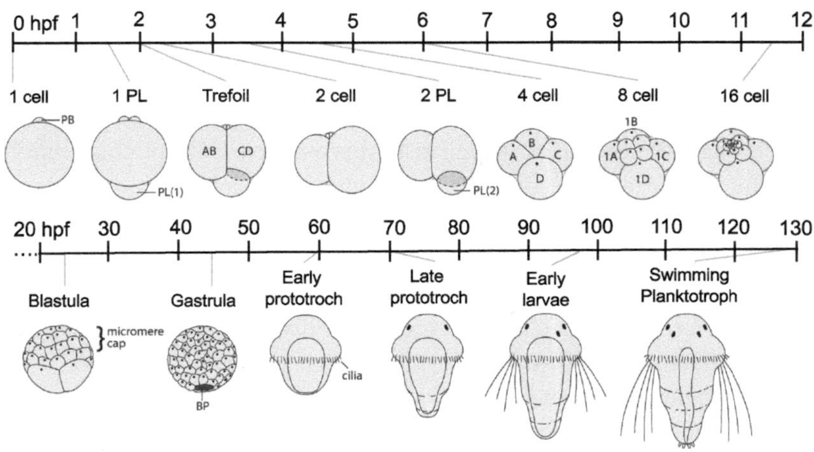

Fig. 3 Planktotrophic timeline of *S. benedicti* development at 20 °C. BP, blastopore; PB, polar bodies; PL, polar lobe.

(Freeman & Lundelius, 1992). These modifications may be adaptive based on larval strategy: lecithotrophs possibly benefit from increased allocation of vegetal yolk to the macromeres giving rise to the future gut, and/or planktotrophs may require rapid division of future ectodermal cells for swimming and feeding (Emlet, 1991).

If we compare planktotrophic to lecithotrophic embryos within *S. benedicti* we see that these three conditions are met. McCain (2008) found that the relative proportion of the egg cytoplasm segregated into the polar lobe (containing the maternal factors) differs between morphs. Also, the large, 1D macromere divided differently in planktotrophs and lecithotrophs with the former having a larger 2d:2D cell ratio than the latter, where the 2d and 2D cells are equal in size. It is not just the relative size that is different but also the rate of cleavage; there is precocious division of the 1D quadrant in the planktotroph. Such intraspecific comparisons can indicate which embryo modifications are predicted to have downstream ontogenetic effects.

The timing and cell size difference during early embryogenesis between the morphs is suggestive of regulatory change, although, whether cell fate-specification happens differently between the two morphs is still unknown. But significant modifications in gene regulatory networks can occur rapidly among closely related lineages, leading to profound changes in development (Peter & Davidson, 2011). In other poecilogonous spionids there is epigenetic modification (methylation differences) between the types (Kesäniemi, Heikkinen, & Knott, 2016). I think it is likely that changes in gene expression through maternal factors profoundly influence development and *S. benedicti* is a tractable model to explore the developmental impact of maternal allocation and autonomous specification.

3.6 Gut formation

As marine invertebrate life-histories are categorized by larval feeding ability, it is pertinent to focus on gut formation. As the larger-egged offspring do not rely on exogenous nutrients for development, feeding pressure is relaxed and so too is selection on functional early guts (Wray, 1996). Therefore, the key evolutionary modification of increased egg size may lead to delayed development of a functional gut in annelids (Boyle & Seaver, 2010; Pernet & McHugh, 2010). We see this in *S. benedicti,* where feeding ability and onset differ between the two morphs (Dauer, Mahon, & Sardá, 2003; Pernet & McArthur, 2006). Gibson, MacDonald, and Dufton (2010) shows there is a heterochronic shift in timing of through-gut formation, where the yolky, large-egged lecithotrophs develop a complete gut later relative to the planktotrophy (5 segment stage vs 3 segment; Pernet & McHugh, 2010). As an evolutionary comparison, *Streblospio shrubsolii* (a European relative) has a 200–300 μm egg diameter, obligate lecithotrophy, and develops a functional gut later at the 16-segment stage. These functional morphological changes often proceed to evolutionary diversification across related taxa (Lyons, Kaltenbach, & McClay, 2012; Perry et al., 2015; Pernet & McHugh, 2010).

From an embryological perspective, there are physical mechanisms that can create such a delay in organogenesis. The larval and juvenile midgut is derived from the embryonic endoderm, which is constructed from the larger, yolky, macromeres during Spiralian development (Hejnol, Martindale, & Henry, 2007; Meyer & Seaver, 2010). The equivalent cell progenitors in the lecithotroph take longer to divide than the planktotroph and add a possible mechanistic explanation to this heterochronic shift. The larger eggs lead to a disproportionate increase in allocation of cytoplasm to the embryonic macromeres, which eventually form the endodermal midgut in annelids (Boyle & Seaver, 2010; McCain, 2008; Pernet, Amiel, & Seaver, 2012). One possibility is that with the increased size and slower cleavage the formation of a functional gut is delayed, which may be adaptive (Freeman & Lundelius, 1992; Henry & Martindale, 1999; Lillie, 1899; Strathmann, Staver, & Hoffman, 2002). Gut formation here is a classic example of a developmental constraint leading to a heterochronic shift.

3.7 Chaetae

Chaetal diversity is fundamental to our understanding of annelid systematics. Larval chaetae are long chitinous bristles originating from paired chaetal sacs and are one of the most striking differences between the two larval morphs (Fig. 4). Planktotrophs have many long, serrated, chitinous, chaetae (~15 per chaetal sac) while they are completely absent from lecithotrophs (Fig. 4). However, larval chaetal sacs are present in both morphs and segmental chaetal sacs are first observed in three-segment larvae of both morphs (Gibson et al., 2010). Within the chaetal sacs, there are clusters of elongate follicles composed of lateral cells and a single chaetoblast that produces each chaeta. Chaetal sac development in lecithotrophic larvae appeared to go through an early period of suspended development. The chaetal sacs are strictly larval tissues and do not contribute to adult features, and swimming chaetae are lost at metamorphosis.

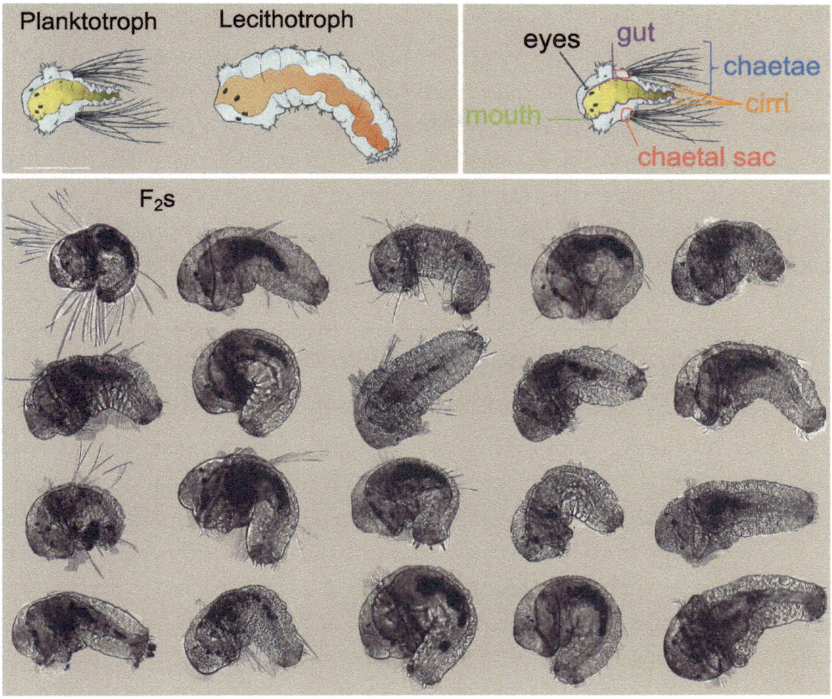

Fig. 4 *S. benedicti* larval types. Cartoons show the larval phenotypes of *S. benedicti* at the larval swimming stage when they are released from the maternal brood pouch. There is morphological variation in F_2s, which are second generation intercrosses of the parental types. Larval images taken with Zeiss Axio Imager M2 with a DIC 20 × objective. Scale bar is 200 μm.

Chaetae may function primarily as defense against pelagic predation, which has been demonstrated in annelids (Pennington & Chia, 1984). Larvae exhibit a startle response when agitated (Bezares-Calderón et al., 2018). Larvae flare their chaetae, changing their shape and size in the water, which may actively deter predators or simply change the gape-size necessary to consume them (Bashevkin & Morgan, 2020). I have also seen them flare chaetae to change directions when swimming (they are positively phototactic, personal observation) and to aid in particle capture. It is also possible that chaetae are costly to produce when not necessary. The loss of chaetae in the lecithotrophic morphs could be due to drift under relaxed selection, but the stark contrast between larval types seems to suggest otherwise. While it has not been specifically tested in *S. benedicti*, it seems likely chaetae size and number is under selection.

3.8 Cirri

Another morphological difference that has been noted between the two larval morphs is the presence or absence of pronounced anal cirri, which are posterior protrusions that only swimming planktotrophic larvae possess (Fig. 4). Within the cirri, planktotrophs have bacillary cells with rodlike rhabdites, and these bacillary cells are not present in the anal cirri of lecithotrophic larvae. The anal cirri and bacillary cells are lost in 11–13-segment planktotrophic larvae, again making this a larval-stage specific distinction (Gibson et al., 2010).

4. Natural distribution and ecological factors

4.1 Ecosystem engineers and environmental indicators

S. benedicti is a benthic deposit feeder in estuarian habitats that constructs soft tubes in the settlement by secreting mucus. It is also a primary colonizer in disturbed or stressed habitats making them useful indicators of environmental change. *S. benedicti* has been used as an indicator of stressed marsh environments including low oxygen, petroleum hydrocarbons, high organic matter, sewage content, or oil spills (Bridges, Levin, Cabrera, & Plaia, 1994; García-Arberas & Rallo, 2004; Levin, 1986; Llansó, 1991). However, the extent of these environmental effects on growth and reproduction are unknown and understanding their effect in *S. benedicti* could aid in creating better assays to measure environmental health.

4.2 Geographic distribution

S. benedicti is one of the most common annelids in the US marshes. Experimental field studies have shown that densities exceeding 400,000 per m^2 can be attained within azoic sediment boxes in less than 10 days (McCall, 1977). *S. benedicti* has a wide geographic range spanning from Nova Scotia to Florida on the US East Coast and into the Gulf of Mexico and Caribbean (Foster, 1971; Levin, 1984; Schulze et al., 2000; Zakas & Wares, 2012).

Notably, both reproductive types are found on the US East Coast in a patchy distribution that cannot be well explained by geographic differences or latitude. Within a patch (a distance of a few feet in which I collect mud), the larval type is usually homogenous, although one known exception is the Baruch Marine Lab, SC where both types coexist, and have since at least 2003 (Bruno Pernet, personal communication). Other populations in close proximity, ∼1–2km apart, can have almost opposite predominance of development types. It is unclear if these populations are ephemeral. The repeated sampling I've done sporadically since 2007 when I began collecting *S. benedicti* suggests population larval mode has stayed the same, although there could be seasonal fluctuations and turnover.

There are populations through the US West Coast as well (Mahon, Mahon, Dauer, & Halanych, 2009; Zakas & Wares, 2012), although these West Coast populations are likely a recent (early 1900s) introduction (Carlton, 1979; Fonseca-Genevois & Cazaux, 1987; Zakas & Wares, 2012). Only lecithotrophic individuals have been found on the US West Coast. This could be a result of selection in the new environment, or a consequence of human mediated transport. The most likely introduction is through anthropogenic trans-continental shipping or ballast water. Lecithotrophs, not needing to feed, may be more likely to survive that transport.

S. benedicti has also been introduced in a number of ports worldwide (Brazil: Lana, Santos, Oliveira, Garraffoni, & Radashevsky, 2006; Atlantic coast of Europe: Sardá & Martin, 1993; South America: Foster, 1971; Hawaii: Bruno Pernet, personal communication) *S. benedicti japonica*, which is considered a subspecies (Imajima, 1990), has been described from the Yatsu tidelands, Japan. The development mode in these locations has not been reported. With further documentation, it will be interesting to determine the role that larval mode plays in introduction and invasion dynamics in these locations.

While environment can certainly influence development mode through selection, it is also interesting to consider the role of developmental mode on geographic distribution. As I mentioned, the marine larval literature identifies broad environmental variables that may favor one strategy over the other. It is even possible that fluctuating environments may select for the maintenance of this polymorphism as a form of bet-hedging. (Although, that may suggest more animals in similar environments should be poecilogonous.) A novel finding about population demographics using *S. benedicti* as a model is that asymmetric dispersal between populations is sufficient to maintain a persistent and evolutionarily stable larval dimorphism (Zakas & Hall, 2012). This outcome suggests that factors such as range limits and offshore current advection can play a part in maintaining the geographical distribution of larval modes we see. Certainly, more investigation into the developmental mode in new locations is warranted.

4.3 Speciation

At first pass, it may seem the patchy larval-type distribution in *S. benedicti* is simply a case of spatial population divergence with ongoing hybridization, but this is not the case. By any species concept to date, *S. benedicti* constitutes a single species. Reproduction between types happens readily with no adverse effects on offspring noted in the lab. Presumably, natural selection on the F_1 larvae may be large, but there is no obvious pre- or postzygotic reproductive barriers. (I refer to the cross offspring as F_1s since the definition of *hybrid* is not fitting as it implies crosses from different species, which these are not.) Phylogenetically, individuals group closest together based on population of origin and geography rather than development mode (Schulze et al., 2000; Zakas & Wares, 2012), and individuals of either larval type can even share identical mitochondrial COI haplotypes (Mahon et al., 2009; Schulze et al., 2000). Population demographic modeling indicates that genomic differentiation is best explained by population and geographic differences rather than larval mode (Zakas & Rockman, 2015; Zakas, Schult, McHugh, Jones, & Wares, 2012). Our subsequent genetic work has demonstrated that we do not expect areas of extensive rearrangement in the *S. benedicti* genome between the two types (Zakas et al., 2018). Instead, regions that contribute to developmental differences may be islands in an otherwise homogeneous genomic background (Zakas & Rockman, 2021). This is to say, the divergence in larval mode in *S. benedicti* is more akin to *ecotype* differentiation than speciation, but even that definition is not apt as the larval differences are not necessarily divided along ecological or geographical boundaries. It remains possible that the dimorphism we see in *S. benedicti* is a case of incipient species happening in real time. However, reproductive isolation has not occurred.

5. Assortative mating and intermediate larvae

Whether there is assortative mating happening in *S. benedicti* is an outstanding question, in both senses of the word. (I recall this being a large point of contention during my graduate qualifications. But over a decade later, we still do not know if there are mating preferences in these worms.) Mating preference is an essential issue, because nonrandom mating can explain the maintenance of divergent larval modes, even with geneflow between populations.

Due to the patchy population structure seen in natural habitats, it is possible that alternative morphs seldom overlap. However, we now know from my dissertation work that there is gene flow between planktotrophic and lecithotrophic populations (Zakas & Wares, 2012). This suggests that there are two major questions about assortative mating to be answered: (1) is there evidence for nonrandom mating based on development mode? And (2) is it due to mate preference, or just population demographics? These two non-exclusive possibilities raise important questions about sexual selection and population genetic assumptions.

Intermediate offspring between the discrete morphs have been described in natural populations. A small percentage of individuals collected from such populations are reported to be F_1s between the two forms because they produced larvae with intermediate development traits (Levin, 1984; Levin & Huggett, 1990). But I caution against designating F_1 status to a field collected individual without knowing the distribution of traits within that site because there can be substantial variation within a type across populations.

As adults the two types can and will readily mate. Crosses happen in both directions, alternating maternal and paternal roles. This is, in my opinion, the most compelling reason to work on *S. benedicti:* viable crosses in both directions allow *S. benedicti* to be used as a genetic model for life-history and developmental evolution. We can experimentally determine the functional mechanism of transitions between developmental modes using these reciprocal F_1 crosses. It is the only model I am aware of where such comparisons are possible, opening avenues for answering major questions in developmental genetics.

In the lab, our F_1 crosses have been between individuals from different oceans (Long Beach, CA and Bayonne, NJ). F_1s are viable with no noticeable fitness disadvantage in the lab. F_1s can possess intermediate phenotypes, depending on the trait in question. But because egg size and thus larval size is maternally determined, the F_1s end up looking mostly like their mothers with some minor variations in chaetae number, length, and anal cirri (maternal background effects on these traits are described in Zakas & Rockman, 2014; other egg size effects: Bridges & Heppell, 1996). This is because egg size and provisioning is maternally determined and cannot be genetically separated until the following (F_2) generation. The F_1 females will produce intermediate egg sizes, with individual variation, that are discretely intermediate to the planktotrophic and lecithotrophic egg sizes (described fully in Zakas et al., 2018). In our F_1 lab crosses, we see F_2 larval traits segregate widely, generating most combinations of possible morphologies (Fig. 4). Interestingly, while there is a wide distribution in the larval phenotypes, the F_2s almost never fully recapitulate the parental phenotypic values, meaning the F_2s rarely get the right combination of traits to look like a typical planktotroph or lecithotroph (Zakas & Rockman, 2021).

In my field experience I have rarely, if ever, collected a female whose larvae were not easily classifiable as stereotypically planktotrophic or lecithotrophic, and I have not encountered the intermediate variations that I can generate in the lab. It is possible I have collected F_1s with phenotypes resembling pure parentals, because these are extremely difficult to distinguish without systematic measurements, but nonetheless I do not find these intermediate phenotype combinations in natural populations. This raises questions about the relative roles of mate preference, frequency-dependent selection and ecological selection against these intermediate larval types that are yet to be answered.

6. Population genetics, selection, and frequency-dependence

Random mating is a key assumption in many standard population genetic analyses, but we know this assumption is being violated in most populations of *S. benedicti*. Because of the homogenous population structure, individuals will mate with the same type by default. However, where the two types do overlap there could be mate choice preferences as well. In these areas of overlap, if random mating occurs it will impede local adaptation, and the extent this happens depends on the genetic architecture of the traits involved. For *S. benedicti*, larval phenotypes are modular with an independent genetic basis for each larval trait (Levin et al., 1991; Zakas et al., 2018). This means that at the population level random mating between the morphs will result in a variety of offspring phenotype combinations, many of which are maladaptive. In this case, we might predict selection for pre- or postzygotic reproductive barriers to reduce the production of unfit intermediates, but we do not see evidence for such mating barriers in the lab.

6.1 Frequency dependent selection

There can be nonrandom mating that does not result in selection for mating barriers. If there is frequency-dependent selection, where the fitness of an allele depends on its frequency in the population, maladaptive intermediate offspring would be rare and the selective pressure to form maintain barriers would be low (reviewed in Nosil, 2012). Given the modularity of larval alleles, the most adaptive combination of traits for a larva is dependent on ecological selective pressures, but also the parental genotypes present in the population. Consider an example: if an environment that favors lecithotrophic alleles is predominantly occupied by adults carrying planktotrophic alleles, the expectation is that a rare individual carrying the favored lecithotrophic alleles would have high fitness and the lecithotrophic alleles would sweep to fixation. But this is not necessarily the case. Random mating means the offspring from such a cross would possess maladaptive trait combinations: the favored lecithotrophic alleles would be expressed in a planktotrophic maternal background (or vice versa). A small larva with no chaetae (a possible outcome of this cross: Zakas & Rockman, 2014), would not have a fitness advantage in this environment. This singular example is meant to illustrate the more complex population-level dynamics of selection on life-history traits. Knowing the genetic architecture of developmental traits better informs our understanding of a population's ability to adapt to new or changing environments.

6.2 Maternal genetic effects

The modularity of larval traits, with some traits such as egg size and lipid content being determined by the maternal genome, and others such as chaetae length being determined by the zygotic genome, leads to higher-order genetic interactions (Zakas et al., 2018; Zakas & Rockman, 2021). The maternal background effects the expression of a trait, and its selective value, which is a dynamic that plays out across multiple generations. I find this epigenetic interaction to be a fascinating component of life-history evolution. An individual's phenotype depends on the genotype of its parents, not simply in terms of inheritance of alleles, but also due to parental genetic effects. We have seen some F_1 offspring, for example, where genotypes that can produce a large number of chaetae in a planktotrophic maternal background, will produce none in a lecithotrophic maternal background (Zakas & Rockman, 2014). While less intuitive, we have also found paternal background effects on larval traits like offspring size (Zakas & Rockman, 2021). The dependence of offspring phenotype on three interacting genomes critically changes our expectations about a population's response to selection (Wade, 1998; Wolf, 2000). We have only begun to explore these dynamics in *S. benedicti*, but it is an ideal system to genetically dissect the components of maternal effects and epigenetic interactions.

I expect that this field of maternal effect genetics is the next step for *S. benedicti*. It's of tremendous interest to evolution and medicine to determine how the role of maternal provisioning during embryogenesis contributes to developmental plasticity. I also expect that changes in maternal provisioning during oogenesis may be a factor in maintaining poecilogonous development. The role of parental genetic effects on evolutionary developmental shifts is unknown, and we have only begun to explore what is possible in this system.

7. Caveats and tools for emerging model systems

I would be remiss not to mention some of the extensive limitations in working on an emerging model system. *S. benedicti* is useful as a genetic model, but much harder to use for development and embryological studies. Some of the biological issues will always persist; egg numbers are limited (Table 1), fertilization is internal, and broods cannot be timed. But other impediments such as tool and method availability are being removed. For example, we have a full-length reference genome assembled from Illumina sequencing, Hi-C, Chicago and long reads (Zakas, Harry, Scholl, & Rockman, 2022). It is annotated from long-read RNAs of multiple stages and tissues. We have also made progress developing methods for embryo injections and gene functional validations (thanks to Duygu Özpolat and the MBL). I hope that establishing these resources will inspire more researchers to use *S. benedicti* in their work so that these avenues will expand.

8. New possibilities and next steps

It is an exciting time for embracing emerging models in genetics and development as technological advances have certainly enabled leaps in progress for many new systems. Interestingly, in the spiralian community a consensus model has not emerged, but rather a set of diverse study organisms has persisted. The extensive research in other Lophotrochozoan systems has contributed to making *S. benedicti* a compelling model and in turn, *S. benedicti* can provide a unique insight into many of the evolutionary and developmental questions in these systems. As with all models, there are questions where *S. benedicti* is useful (hopefully I have imparted my enthusiasm), and others where it is far less ideal. But always, choosing the best model depends on the question, and there are many questions for which *S. benedicti* is exemplary.

Acknowledgments

Thank you to all the people who have let me keep buckets of worms in their homes or labs; Thanks to all the folks who think weird worms are cool and helped by collecting, sharing, or listening. Thanks to J. Deutscher for Figs. 3 and 4 drawings; S.M. Cole and N.D. Harry for images; B. Pernet for feedback on the manuscript; M.V. Rockman for feedback and video. MBL Whitman Fellowship and NCSU startup for tool development support; M. Srivastava and B. Goldstein for editing.

References

Allen, J. D., & Pernet, B. (2007). Intermediate modes of larval development: Bridging the gap between planktotrophy and lecithotrophy. *Evolution & Development, 9*, 643–653.

Bashevkin, S. M., & Morgan, S. G. (2020). Predation and competition. In K. Anger, S. Harzsch, & M. Thiel (Eds.), *Vol. 7. Developmental biology and larval ecology* (p. 360). New York: Oxford University Press.

Bezares-Calderón, L. A., Berger, J., Jasek, S., Verasztó, C., Mendes, S., Gühmann, M., et al. (2018). Neural circuitry of a polycystin-mediated hydrodynamic startle response for predator avoidance. *eLife, 7*, e36262.

Boyle, M. J., & Seaver, E. C. (2010). Expression of FoxA and GATA transcription factors correlates with regionalized gut development in two lophotrochozoan marine worms: *Chaetopterus* (Annelida) and *Themiste lageniformis* (Sipuncula). *EvoDevo, 1*, 1–18.

Bridges, T. S., & Heppell, S. (1996). Fitness consequences of maternal effects in *Streblospio benedicti* (Annelida: Polychaeta). *American Zoologist, 36*, 132–146.

Bridges, T. S., Levin, L. A., Cabrera, D., & Plaia, G. (1994). Effects of sediment amended with sewage, algae, or hydrocarbons on growth and reproduction in two opportunistic polychaetes. *Journal of Experimental Marine Biology and Ecology, 177*, 99–119.

Carlton, J. T. (1979). *History, biogeography, and ecology of the introduced marine and estuarine invertebrates of the Pacific Coast of North America.* PhD dissertation Davis: University of California.

Chia, F. S., Gibson, G., & Qian, P. Y. (1996). Poecilogony as a reproductive strategy of marine invertebrates. *Oceanologica Acta, 19*, 203–208.

Collin, R., & Moran, A. (2018). Evolutionary transitions in mode of development. In T. J. Carrier, A. M. Reitzel, & A. Heyland (Eds.), *Evolutionary ecology of marine invertebrate larvae.* Oxford University Press.

Dauer, D. M., Mahon, H. K., & Sardá, R. (2003). Functional morphology and feeding behavior of *Streblospio benedicti* and *S. shrubsolii* (Polychaeta: Spionidae). In *Advances in polychaete research* (pp. 207–213). Springer.

Dean, D. (1965). On the reproduction and larval development of *Streblospio benedicti* Webster. *The Biological Bulletin, 128*, 67–76.

Dorresteijn, A. (2005). Cell lineage and gene expression in the development of polychaetes. In *Morphology, molecules, evolution and phylogeny in polychaeta and related taxa* (pp. 1–22). Springer.

Dunn, C. W., Hejnol, A., Matus, D. Q., Pang, K., Browne, W. E., Smith, S. A., et al. (2008). Broad phylogenomic sampling improves resolution of the animal tree of life. *Nature, 452*, 745–749.

Eckelbarger, K. J. (1980). An ultrastructural study of oogenesis in *Streblospio benedicti* (spionidae), with remarks on diversity of vitellogenic mechanisms in polychaeta. *Zoomorphologie, 94*, 241–263.

Eckelbarger, K. J. (1986). Vitellogenic mechanisms and the allocation of energy to offspring in polychaetes. *Bulletin of Marine Science, 39*, 426–443.

Emlet, R. B. (1991). Functional constraints on the evolution of larval forms of marine invertebrates: Experimental and comparative evidence. *American Zoologist, 31*, 707–725.
Fonseca-Genevois, V., & Cazaux, C. (1987). *Streblospio benedicti* Webster, 1879 (Annèlide Polychéte) dans l'estuaire de la Loire: biologie et écologie. *Cahiers de Biologie Marine, 28*, 231–261.
Foster, N. M. (1971). Spionidae (Polychaeta) of the Gulf of Mexico and the Caribbean Sea. *Studies on the Fauna of Curaçao and Other Caribbean Islands, 36*, 1–183.
Freeman, G., & Lundelius, J. W. (1992). Evolutionary implications of the mode of D quadrant specification in coelomates with spiral cleavage. *Journal of Evolutionary Biology, 5*, 205–247.
García-Arberas, L., & Rallo, A. (2004). Population dynamics and production of *Streblospio benedicti* (Polychaeta) in a non-polluted estuary in the Basque coast (Gulf of Biscay). *Scientia Marina, 68*, 193–203.
Giard, A. (1905). La poecilogonie. In *Compte-rendu des séances du sixième Congrès international de zoologie, tenu à Berne du 14 au 16 août 1904* (pp. 617–646).
Gibson, G., MacDonald, K., & Dufton, M. (2010). Morphogenesis and phenotypic divergence in two developmental morphs of *Streblospio benedicti* (Annelida, Spionidae). *Invertebrate Biology, 129*, 328–343.
Grassle, J. F., & Grassle, J. P. (1974). Opportunistic life histories and genetic systems in marine benthic polychaetes. *Journal of Marine Research, 32*, 253–284.
Guerrier, P., Van Den Biggelaar, J. A. M., Van Dongen, C. A. M., & Verdonk, N. H. (1978). Significance of the polar lobe for the determination of dorsoventral polarity in *Dentalium vulgare* (da Costa). *Developmental Biology, 63*, 233–242.
Guralnick, R. P., & Lindberg, D. R. (2001). Reconnecting cell and animal lineages: What do cell lineages tell us about the evolution and development of Spiralia? *Evolution, 55*, 1501–1519.
Halanych, K. M., Dahlgren, T. G., & McHugh, D. (2002). Unsegmented annelids? Possible origins of four lophotrochozoan worm taxa. *Integrative and Comparative Biology, 42*, 678–684.
Hart, M. W. (1995). What are the costs of small egg size for a marine invertebrate with feeding planktonic larvae? *The American Naturalist, 146*, 415–426.
Hejnol, A., Martindale, M. Q., & Henry, J. Q. (2007). High-resolution fate map of the snail *Crepidula fornicata*: The origins of ciliary bands, nervous system, and muscular elements. *Developmental Biology, 305*, 63–76.
Henry, J. Q., & Martindale, M. Q. (1994). Establishment of the dorsoventral axis in nemertean embryos: Evolutionary considerations of spiralian development. *Developmental Genetics, 15*, 64–78.
Henry, J. J., & Martindale, M. Q. (1999). Conservation and innovation in spiralian development. *Hydrobiologia, 402*, 255–265.
Henry, J. Q., Perry, K. J., & Martindale, M. Q. (2006). Cell specification and the role of the polar lobe in the gastropod mollusc *Crepidula fornicata*. *Developmental Biology, 297*, 295–307.
Hoagland, K. E., & Robertson, R. (1988). An assessment of poecilogony in marine invertebrates: Phenomenon or fantasy? *The Biological Bulletin, 174*, 109–125.
Imajima, M. (1990). Spionidae (Annelida, Polychaeta) from Japan V. The genera *Streblospio* and *Dispio*. *Bulletin of the National Science Museum, 16*, 155–163.
Jones, C., Stankowich, T., & Pernet, B. (2016). Allocation of cytoplasm to macromeres in embryos of annelids and molluscs is positively correlated with egg size. *Evolution & Development, 18*, 156–170.
Kesäniemi, J. E., Heikkinen, L., & Knott, K. E. (2016). DNA methylation and potential for epigenetic regulation in *Pygospio elegans*. *PLoS One, 11*, e0151863.

Knott, K. E., & McHugh, D. (2012). Introduction to symposium: Poecilogony—A window on larval evolutionary transitions in marine invertebrates. *Integrative and Comparative Biology*, *52*, 120–127.
Krug, P. J. (2007). Poecilogony and larval ecology in the gastropod genus *Alderia*. *American Malacological Bulletin*, *23*, 99–111.
Krug, P. J. (2009). Not my "type": Larval dispersal dimorphisms and bet-hedging in opisthobranch life histories. *The Biological Bulletin*, *216*, 355–372.
Lambert, J. D., & Nagy, L. M. (2001). MAPK signaling by the D quadrant embryonic organizer of the mollusc *Ilyanassa obsoleta*. *Development*, *128*, 45–56.
Lana, P., Santos, C., Oliveira, V., Garraffoni, A., & Radashevsky, V. (2006). Checklist of polychaete species from Paraná state (Southern Brazil). *Check List*, *2*, 30.
Levin, L. A. (1984). Multiple patterns of development in *Streblospio benedicti* Webster (Spionidae) from three coasts of North America. *The Biological Bulletin*, *166*, 494–508. https://doi.org/10.2307/1541157.
Levin, L. A. (1986). Effects of enrichment on reproduction in the opportunistic polychaete *Streblospio benedicti* (Webster): A mesocosm study. *The Biological Bulletin*, *171*, 143–160.
Levin, L. A., & Bridges, T. S. (1994). Control and consequences of alternative developmental modes in a poecilogonous polychaete. *American Zoologist*, *34*, 323–332.
Levin, L. A., Caswell, H., DePatra, K. D., & Creed, E. L. (1987). Demographic consequences of larval development mode: Planktotrophy vs. lecithotrophy in *Streblospio benedicti*. *Ecology*, *68*, 1877–1886.
Levin, L. A., & Creed, E. L. (1986). Effect of temperature and food availability on reproductive responses of *Streblospio benedicti* (Polychaeta: Spionidae) with planktotrophic or lecithotrophic development. *Marine Biology*, *92*, 103–113.
Levin, L. A., & Huggett, D. V. (1990). Implications of alternative reproductive modes for seasonality and demography in an estuarine polychaete. *Ecology*, *71*, 2191–2208.
Levin, L. A., Zhu, J., & Creed, E. (1991). The genetic basis of life-history characters in a polychaete exhibiting planktotrophy and lecithotrophy. *Evolution*, *45*, 380–397.
Levitan, D. R. (2000). Optimal egg size in marine invertebrates: Theory and phylogenetic analysis of the critical relationship between egg size and development time in echinoids. *The American Naturalist*, *156*, 175–192.
Lillie, F. R. (1899). Adaptation in cleavage. In *Biological lectures from the Marine Biological Laboratory, summers of 1897–98*. Boston: Ginn.
Llansó, R. J. (1991). Tolerance of low dissolved oxygen and hydrogen sulfide by the polychaete *Streblospio benedicti* (Webster). *Journal of Experimental Marine Biology and Ecology*, *153*, 165–178.
Lyons, D. C., Kaltenbach, S. L., & McClay, D. R. (2012). Morphogenesis in sea urchin embryos: Linking cellular events to gene regulatory network states. *Wiley Interdisciplinary Reviews: Developmental Biology*, *1*, 231–252.
Mahon, H. K., & Dauer, D. M. (2005). Organic coatings and ontogenetic particle selection in *Streblospio benedicti* Webster (Spionidae: Polychaeta). *Journal of Experimental Marine Biology and Ecology*, *323*, 84–92.
Mahon, A. R., Mahon, H. K., Dauer, D. M., & Halanych, K. M. (2009). Discrete genetic boundaries of three *Streblospio* (Spionidae, Annelida) species and the status of *S. shrubsolii*. *Marine Biology Research*, *5*, 172–178.
Marshall, D. J., Allen, R. M., & Crean, A. J. (2008). The ecological and evolutionary importance of maternal effects in the sea. In *Vol. 46. Oceanography and marine biology: An annual review* (pp. 203–250). CRC Press.
McCain, E. R. (2008). Poecilogony as a tool for understanding speciation: Early development of *Streblospio benedicti* and *Streblospio gynobranchiata* (Polychaeta: Spionidae). *Invertebrate Reproduction and Development*, *51*, 91–101.

McCall, P. L. (1977). *Community patterns and adaptive strategies of the infaunal benthos of Long Island Sound*.
McDonald, K. A., Collin, R., & Lesoway, M. P. (2014). Poecilogony in the caenogastropod *Calyptraea lichen* (Mollusca: Gastropoda). *Invertebrate Biology*, *133*, 213–220.
Meyer, N. P., & Seaver, E. C. (2010). Cell lineage and fate map of the primary somatoblast of the polychaete annelid *Capitella teleta*. *Integrative and Comparative Biology*, *50*, 756–767.
Moran, A. L., & McAlister, J. S. (2009). Egg size as a life history character of marine invertebrates: Is it all it's cracked up to be? *The Biological Bulletin*, *216*, 226–242.
Nosil, P. (2012). *Ecological speciation*. Oxford University Press.
Oyarzun, F. X., & Brante, A. (2015). A new case of poecilogony from South America and the implications of nurse eggs, capsule structure, and maternal brooding behavior on the development of different larval types. *The Biological Bulletin*, *228*, 85–97.
Özpolat, B. D., Handberg-Thorsager, M., Vervoort, M., & Balavoine, G. (2017). Cell lineage and cell cycling analyses of the 4d micromere using live imaging in the marine annelid *Platynereis dumerilii*. *eLife*, *6*, e30463.
Palumbi, S. R. (1995). Using genetics as an indirect estimator of larval dispersal. In *Ecology of marine invertebrate larvae*. CRC Press.
Pennington, J. T., & Chia, F.-S. (1984). Morphological and behavioral defenses of trochophore larvae of Sabellaria cementarium (Polychaeta) against four planktonic predators. *The Biological Bulletin*, *167*, 168–175.
Pernet, B., Amiel, A., & Seaver, E. C. (2012). Effects of maternal investment on larvae and juveniles of the annelid *Capitella teleta* determined by experimental reduction of embryo energy content. *Invertebrate Biology*, *131*, 82–95.
Pernet, B., & McArthur, L. (2006). Feeding by larvae of two different developmental modes in *Streblospio benedicti* (Polychaeta: Spionidae). *Marine Biology*, *149*, 803–811.
Pernet, B., & McHugh, D. (2010). Evolutionary changes in the timing of gut morphogenesis in larvae of the marine annelid *Streblospio benedicti*. *Evolution & Development*, *12*, 618–627.
Perry, K. J., Lyons, D. C., Truchado-Garcia, M., Fischer, A. H., Helfrich, L. W., Johansson, K. B., et al. (2015). Deployment of regulatory genes during gastrulation and germ layer specification in a model spiralian mollusc *Crepidula*. *Developmental Dynamics*, *244*, 1215–1248.
Peter, I. S., & Davidson, E. H. (2011). A gene regulatory network controlling the embryonic specification of endoderm. *Nature*, *474*, 635–639.
Rice, S. A. (1981). Spermatogenesis and sperm ultrastructure in three species of *Polydora* and in *Streblospio benedicti* (Polychaeta: Spionidae). *Zoomorphology*, *97*, 1–16.
Rice, S. A. (1991). Reproductive isolation in the *Polydora ligni* complex and the *Streblospio benedicti* complex (Polychaeta: Spionidae). *Bulletin of Marine Science*, *48*, 432–447.
Rice, S. A., & Levin, L. A. (1998). Streblospio gynobranchiata, a new spionid polychaete species (Annelida: Polychaeta) from Florida and the Gulf of Mexico with an analysis of phylogenetic relationships within the genus *Streblospio*. *Proceedings of Biological Society of Washington*, *111*, 694–707.
Sardá, R., & Martin, D. (1993). Populations of *Streblospio* (Polychaeta: Spionidae) in temperate zones: Demography and production. *Journal of the Marine Biological Association of the United Kingdom*, *73*, 769–784.
Schulze, S. R., Rice, S. A., Simon, J. L., & Karl, S. A. (2000). Evolution of poecilogony and the biogeography of North American populations of the polychaete *Streblospio*. *Evolution*, *54*, 1247–1259.
Shanks, A. L. (2009). Pelagic larval duration and dispersal distance revisited. *The Biological Bulletin*, *216*, 373–385.
Smith, M. S., Zigler, K. S., & Raff, R. A. (2007). Evolution of direct-developing larvae: Selection vs loss. *BioEssays*, *29*, 566–571.

Strathmann, R. R. (1977). Egg size, larval development, and juvenile size in benthic marine invertebrates. *The American Naturalist*, *111*, 373–376.
Strathmann, R. R. (1985). Feeding and nonfeeding larval development and life-history evolution in marine invertebrates. *Annual Review of Ecology and Systematics*, *16*, 339–361.
Strathmann, R. R. (1990). Why life histories evolve differently in the sea. *American Zoologist*, *30*, 197–207.
Strathmann, R. R., Staver, J. M., & Hoffman, J. R. (2002). Risk and the evolution of cell-cycle durations of embryos. *Evolution*, *56*, 708–720.
Thorson, G. (1950). Reproductive and larval ecology of marine bottom invertebrates. *Biological Reviews*, *25*, 1–45.
Valentine, J. W., & Jablonski, D. (1986). Mass extinctions: Sensitivity of marine larval types. *Proceedings of the National Academy of Sciences of the United States of America*, *83*, 6912–6914.
Wade, M. J. (1998). The evolutionary genetics of maternal effects. In *Vol. 21. Maternal effects as adaptation* (pp. 827–839).
Wiggering, B., Neiber, M. T., Gebauer, K., & Glaubrecht, M. (2020). One species, two developmental modes: A case of geographic poecilogony in marine gastropods. *BMC Evolutionary Biology*, *20*, 1–10.
Wolf, J. B. (2000). Gene interactions from maternal effects. *Evolution*, *54*, 1882–1898.
Wray, G. A. (1996). Parallel evolution of nonfeeding larvae in echinoids. *Systematic Biology*, *45*, 308–322.
Wray, G. A., & Raff, R. A. (1991). The evolution of developmental strategy in marine invertebrates. *Trends in Ecology & Evolution*, *6*, 45–50.
Zakas, C., Deutscher, J. M., Kay, A. D., & Rockman, M. V. (2018). Decoupled maternal and zygotic genetic effects shape the evolution of development. *eLife*, *7*, e37143.
Zakas, C., & Hall, D. W. (2012). Asymmetric dispersal can maintain larval polymorphism: A model motivated by *Streblospio benedicti*. *Integrative and Comparative Biology*, *52*, 197–212.
Zakas, C., Harry, N. D., Scholl, E. H., & Rockman, M. V. (2022). The genome of the poecilogonous annelid *Streblospio benedicti*. *Genome Biology and Evolution*, *14*(2).
Zakas, C., & Rockman, M. V. (2014). Dimorphic development in *Streblospio benedicti*: Genetic analysis of morphological differences between larval types. *The International Journal of Developmental Biology*, *58*, 593–599.
Zakas, C., & Rockman, M. V. (2015). Gene-based polymorphisms reveal limited genomic divergence in a species with a heritable life-history dimorphism. *Evolution & Development*, *17*, 240–247.
Zakas, C., & Rockman, M. V. (2021). Baby makes three: Maternal, paternal, and zygotic genetic effects shape larval phenotypic evolution. *Evolution*, *75*(7), 1607–1608.
Zakas, C., Schult, N., McHugh, D., Jones, K. L., & Wares, J. P. (2012). Transcriptome analysis and SNP development can resolve population differentiation of *Streblospio benedicti*, a developmentally dimorphic marine annelid. *PLoS One*, *7*, e31613.
Zakas, C., & Wares, J. P. (2012). Consequences of a poecilogonous life history for genetic structure in coastal populations of the polychaete *Streblospio benedicti*. *Molecular Ecology*, *21*, 5447–5460.

CHAPTER EIGHTEEN

The arm of the starfish: The far-reaching applications of *Patiria miniata* as a model system in evolutionary, developmental, and regenerative biology

Anne Meyer and Veronica Hinman*
Department of Biological Sciences, Carnegie Mellon University, Pittsburgh, PA, United States
*Corresponding author: e-mail address: veronica@cmu.edu

Contents

Abstract

Many species of echinoderms have long been considered model research organisms in biology. Historically, much of this research has focused on the embryology of sea urchins and the use of their extensive gene regulatory networks as a tool to understand

Current Topics in Developmental Biology, Volume 147
ISSN 0070-2153
https://doi.org/10.1016/bs.ctdb.2022.01.006

how the genome controls cell state specification and patterning. The establishment of *Patiria miniata*, the bat sea star, as a research organism has allowed us to expand on the concepts explored with sea urchins, viewing these genetic networks through a comparative lens, gaining great insight into the evolutionary mechanisms that shape developmental diversity. Extensive molecular tools have been developed in *P. miniata*, designed to explore gene expression dynamics and build gene regulatory networks. Echinoderms also have a robust set of bioinformatic and computational resources, centered around echinobase.org, an extensive database containing multiomic, developmental, and experimental resources for researchers. In addition to comparative evolutionary development, *P. miniata* is a promising system in its own right for studying whole body regeneration, metamorphosis and body plan development, as well as marine disease.

Sea stars are a well-known, and charismatic, group of animals that sit within the phylum Echinodermata. Due to their characteristic endoskeleton, this phylum has a remarkably extensive fossil record that dates back to the base of the Cambrian Period and provides a robust temporal perspective for evolutionary innovations within lineages. Echinodermata originated approximately 530–524 million years ago (Smith & Paul, 1988), and underwent a rapid radiation into five major classes: crinoids (feather stars), ophiuroids (brittle stars), asteroids (sea stars), holothurians (sea cucumbers), and echinoids (sea urchins) (Telford et al., 2014). There is a striking amount of morphological variation between these groups, not only in the better-known adult forms, but also in their larval stages, and life cycles. While all modern forms have an adult form with radial (and often pentaradial) symmetry, many larval forms are bilateral and share similarities in axis patterning with their deuterostome relatives (Angerer, Yaguchi, Angerer, & Burke, 2011; Lapraz, Haillot, & Lepage, 2015; Müller, 1854).

The morphological diversity among echinoderm species, in conjunction with the well documented historical record, was an initial motivation to bring these sea stars into the lab as a comparative system for studying evolutionary change and innovation. As we discuss below, sea stars also have some fascinating biological properties that provide an exceptional opportunity to study some poorly understood mechanisms of cell specification, patterning, proportion control and morphogenesis.

1. Historical use of sea stars in biological research

It's important to realize that, while sea stars have never been considered mainstream research organisms, they have been used to study many aspects of biology for over a hundred years. The easy accessibility of gametes, the relatively large size of the oocyte and embryo, as well as their optical transparency have made these animals appealing to many scientists asking a diversity of questions.

In the 19th century, a variety of research was done using sea stars, reflecting the intellectual trends of the time. Paleontologists synthesized the extensive echinoderm fossil record (Clark & Twitchell, 1915; Gregory, 1899) and laid the groundwork for the establishment of an exceptionally detailed and extensive echinoderm fossil record which has been an asset in our modern understanding of their evolutionary history. Zoologists sought to characterize and describe the diversity and distribution of these taxa (Gray, 1866) and build an organismal biology basis for future work (Andrews, 1897; Prouho, 1891; Romanes, 1883). Embryologists also began probing sea star development, characterizing the early stages of cleavage (Agassiz, 1865; Goto, 1895; Khaner, 1993).

It was in the 20th century that sea star experimental research rose in prominence. Much of this can be credited to Robert Chambers, an influential developmental and cell biologist at the Marine Biological Laboratory, who developed an extensive set of tools to observe and manipulate ova in a variety of marine organisms, including sea stars (Chambers, 1922; Goldring, 1979). At this time, Chambers and others, like noted cell biologist and neurophysiologist R.S. Lillie, also conducted extensive research into the molecular mechanism of fertilization (Chambers, 1923; Lillie, 1912, 1926). Élie Metchnikoff was awarded the Nobel Prize in 1908 for his discovery of phagocytic cells, stemming, at least in part, from observations of the blastocoelar cells in sea star larvae (Metchnikoff, 1893). Vertebrate developmental biologists were also drawn to the simplicity of the sea star, e.g., noted zoologist and geneticist HH Newman utilized starfish to understand the mechanisms governing twinning events (Newman, 1922).

The molecular biology boom of the 1960s and 70s enabled researchers to explore the role of individual elements in a cell like never before. The large eggs of sea stars made them a popular research system at this time, and work done on sea stars has made critical contributions to our understanding of molecular cell biology. For example, extensive work was done to characterize the induction of meiosis and egg maturation (Shen & Steinhardt, 1976; Stevens, 1970), as well as the mechanisms and dynamics of the post-fertilization cortical reaction (Holland, 1980), having major implications for our understanding of vesicle dynamics.

Of particular note, is the advances in our understanding of the cell cycle and meiotic divisions derived from work with sea stars. The Kanatani group was one of the first to identify a meiosis-inducing hormone (Kanatani, Shirai, Nakanishi, & Kurokawa, 1969). Later work in the late 1990s by Kishimoto group would establish the concept tof M phase promoting factors, hugely influencing the direction of cell cycle research (Kishimoto, 1999).

1.1 Comparative regulatory networks and the birth of evo-devo

The unification of phylogenetics, molecular genetics, and evolution, during the Modern Synthesis of the mid-20th century greatly influenced the ways in which scientists explored evolution. There was a resurgence in using comparative methods to understand the mechanism of evolution that gave rise to new phenotypes. The Second Modern Synthesis of the later 20th century approaches these concepts with an appreciation of developmental processes. It is here that evolutionary development emerges as a field, focusing on how embryonic development evolves and how it has come to shape modern biological diversity (Müller, 2007).

One theme that quickly emerged is the importance of understanding how the genome controls the developmental process and encodes the information necessary to establish distinct transcriptomic states in different cells. Pioneering work by the Davidson group at Caltech developed the concept of hierarchical networks of regulatory interactions between genes (i.e., a Gene Regulatory Network, GRN) that can be used to describe development and the differential deployment of genes needed to establish phenotypes (Davidson et al., 2002a; Erwin & Davidson, 2009).

Davidson and colleagues used the sea urchin as a research organism to establish the most comprehensive developmental GRN at the time (Davidson et al., 2002b). There is a long, parallel history of the use of many species of sea urchin in development, cell and molecular biology (Ernst, 1997). This provided an extensive research literature that started with embryological phenomenology, spanned through to molecular studies of gene function, and culminated in the establishment of the developmental gene regulatory network (Davidson et al., 2002a; Peter, Faure, & Davidson, 2012; Rafiq, Cheers, & Ettensohn, 2012; Saudemont et al., 2010; Saunders & McClay, 2014).

Conceptualizing development through the lens of the GRN had a profound impact on the ways people thought about development, but also provided an extraordinary opportunity to think about the evolution of development through comparing the GRNs of different species.

This initiated the search for a comparative system for the sea urchin GRN, which quickly settled on the sea star, *P. miniata*, commonly known as bat star, which was abundant off the Coast of California, and the subsequent development of a series of protocols to facilitate GRN research in the bat star.

2. Sea stars in the lab

The bat star, *P. miniata* (Brandt, 1835), is one of the most extensively used sea stars for experimental studies. *P. miniata* has a typical biphasic life-style, consisting of a bilaterally symmetrical bipinnaria larvae and a pentaradial benthic adult form (Fig. 1A) (McEdward & Miner, 2001). The adults are easily collected in the wild along the western coast of North America from Alaska to the Gulf of California and readily maintained in aquaria in either natural or artificial seawater aquaria (Fig. 2A). These organisms can be purchased commercially from multiple collectors. Adults require feeding about once a week, and can be fed shrimp, shellfish or other seafood available at any grocery store. The adults have long lifespans and the same individuals may be maintained and used for gamete procurement for many years in the lab if housed in optimal conditions.

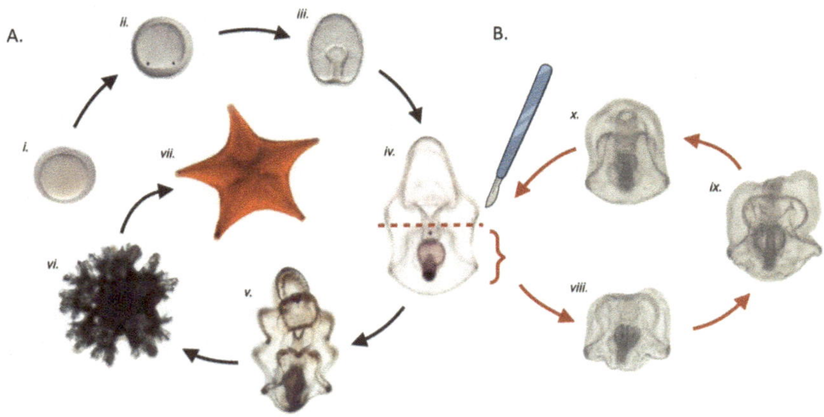

Fig. 1 The life cycle of *P.miniata* and larval regeneration series. The development of *P. minitata* (A). Stages represented are: zygote (i), blastula (ii), gastrula (iii), juvenile larvae (iv), mature larvae (v), metamorphosing larvae (vi), and pentaradial adult (vii). The regeneration of the posterior half of a young P.miniata larvae (B). To induce regeneration, a young larva (pictured in A.iv) is bisected along the transverse plane. Stages pictured are 1 day post-bisection (viii), 4 days post-bisection (ix) and 8 days post-dissection (x).

Fig. 2 *P. miniata* in the lab. *P. miniata* can be kept in large (shown here 100 gal) saltwater tanks (A). A female has been induced to spawn by injection of 1-Methyladenine and within one to two hours many thousands of mature oocytes will be spawned by the five oviducts. (B). De-jellied fertilized single cell zygotes have been rowed into straight lines on a 5 mL petri dish of seawater ready for microinjection. This shows approximately 1000 rowed zygotes which can be injected in 10 to 20 min (C). DNA constructs containing enhancers and reporter proteins are routinely used to probe regulatory DNA function. Zygotes injected with DNA constructs will express them mosaically, due to clonal inheritance. Shown here is gastrula stage expressing GFP in clonal population of cells driven by Sox4 enhancer which marks early neurons (D).

Cultures of embryos can be set up any time of the year, but are most successful and voluminous during the North American spring, the period during which they naturally spawn in the wild. Cheatle et al. have described laboratory culture methods for this species (Cheatle Jarvela & Hinman, 2014). Fig. 3B shows an individual that has been induced to spawn by injecting 1-methyladenine (purchased from Spectrum Chemical) into the body cavity to a final concentration of around 10 μM. *P. miniata* cultures can be very large, consisting of up to hundreds of thousands of embryos, making this species ideal for high throughput experiments. Cultures can be maintained in natural or artificial sea water and stored at 10–15 °C to mimic natural conditions, although they will also do well at room temperature as long as these conditions are stable. Once they develop into feeding larvae (after 3 days), they can be fed algae (*Rhodomonas lens*), which can be

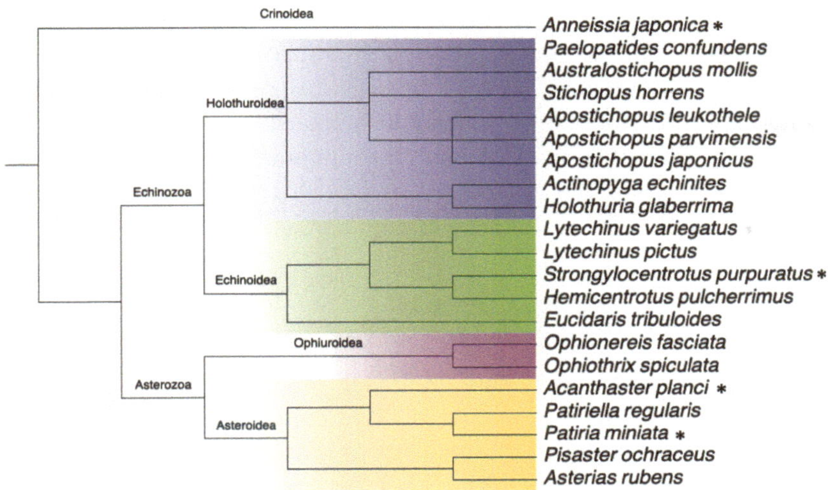

Fig. 3 Phylogenetic tree of echinoderm bioinformatic resources. A tree showing the phylogenetic distribution of species echinoderms with nuclear genome assemblies available on NCBI. Echinozoa are composed of Holothuroidea (sea cucumbers) and Echinoidea (sea urchins). The Asterozoans are made up of Ophiuroidea (brittle stars) and Asteroidea (sea stars). Asterix indicate species with genomes available through Echinobase.

administered directly to the dish (Lowe & Wray, 2000). Under the proper conditions, *P. miniata* can also be cultured through metamorphosis (Cameron & Holland, 1983).

3. Developing the molecular toolkit

3.1 Spatio-temporal examination of gene expression

The first step in comparing GRNs between selected species is to compare the spatio-temporal expression of orthologous genes. One particularly useful and widely used tool is RNA whole mount in situ hybridization (WMISH). Multiple WMISH protocols, developed for other marine larvae worked well in *P. minita* as they have no impenetrable membranes and little yolk (Holland, Holland and Holland, 1996), but a proteinase K-free, high stringency protocol was settled upon in early 2000s for ease of use and reliability (Hinman, Nguyen, & Davidson, 2003; Perillo, Paganos, Spurrell, Arnone, & Wessel, 2021). WMISH has been instrumental in understanding how the expression of regulatory genes establish distinct regional patterns during embryonic development. Further work has also been done to adapt these situ protocols to determine the expression of more than one gene at once using double fluorescent in situ hybridization (Yankura, Martik, Jennings, & Hinman, 2010). The ability to visualize the expression domains of multiple gene transcripts simultaneously has allowed for a more precise description of regulatory domains throughout embryos (Fresques & Wessel, 2018; Materna, Swartz, & Smith, 2013; Yankura, Koechlein, Cryan, Cheatle, & Hinman, 2013; Yankura et al., 2010).

3.2 Quantifying gene expression

The ability to accurately, robustly, and efficiently quantify levels of gene expression is another invaluable tool for GRN studies. This allows for the assessment of differential gene expression in control vs experimental states needed for determining the regulatory relationships between genes. Traditional techniques like bulk RNA-seq are widely used, as well as other, smaller-scale and economic methods, like nanostring assays (Cheatle Jarvela, Yankura, & Hinman, 2016; Gildor, Cary, Lalzar, & Hinman, 2019). The ability to generate large synchronized cultures allows for ample RNA preparations. There are extensive archives of probes available that can be used in real time quantitative-PCR as well as nanostring sequencing available at Echinobase.org.

3.3 Perturbing gene expression

A key requirement for functional and GRN studies is to be able to specifically perturb gene expression. There is an array of small molecule inhibitors to well-known signaling pathways that can be used and have the advantage of allowing for simple bathing by addition to the sea water culture. These can be administered at any time, through mixing with the sea water culture medium. This method has been heavily utilized in understanding how WNT and Delta signaling establishes different embryonic territories in the sea star (McCauley, Akyar, Saad, & Hinman, 2015).

Most perturbations, however, rely on microinjecting nucleic acids and their derivatives into the oocytes. Fortunately *P. miniata* oocytes, and even early embryos, are very easy to microinject and many thousands of embryos can be injected in a sitting. While often a bottleneck in establishing new research organisms, *P. miniata* proved very easy to inject, with the largest hurdle being to find ways to anchor the buoyant embryos in place. We have published protocols elsewhere (Cheatle Jarvela & Hinman, 2014) on the microinjection, which briefly, require removing the jelly coat with pH

4.0 sea water and rowing the oocytes onto plastic dishes using a hand drawn rowing pipette that is angled to gently push the fertilized zygote onto a plastic dish where they will stick naturally (Fig. 2C). A concentrated solution of the reagent is injected into an egg using a small diameter needle controlled by a micromanipulator where we estimate 25–80 pL can be delivered reliably into the oocyte. A thorough walkthrough of this protocol is available in Cheatle Jarvela and Hinman (2014). Individuals injected with DNA constructs tend to be mosaics, with the injected constructs being incorporated randomly into a cell lineage (Fig. 2D) (Hough-Evans, Britten, & Davidson, 1988).

Morpholino modified oligomers (MOs) are a popular and effective tool for inhibiting the translation of specific genes by base-pairing with the target RNAs (Summerton & Weller, 1997). These molecules are introduced into *P. miniata* eggs following fertilization and take effect when the target gene is expressed. MOs have been valuable in deducing the regulatory targets of transcription factors and understanding how they impact downstream expression patterns. While there has been some question in some systems of the specificity of these reagents, we have found antisense morpholino oligos to be highly effective and specific as long as careful controls are in place (Fresques & Wessel, 2018; Hinman & Davidson, 2007; Hinman, Nguyen, Cameron, & Davidson, 2003). These include using at least two independent oligos and appropriate negative controls following the guidance of (Materna, 2017).

Overexpression can also be used to study gene function, but with the caveat that overexpressed RNAs are likely to have off-target effects (Krieg & Melton, 1984). Overexpression can be most useful when induced in two main fashions, either through the introduction of *in vitro* RNA or a DNA construct driving exogenous gene expression into the *P. miniata* egg. The direct injection of these molecules can also be valuable for inferring homology functional between orthologs or the impact of different alleles. For example, the inability of *P. miniata* Alx1 to rescue a knockdown of its ortholog in *S. purpuratus,* helped reconstruct the evolutionary history of this transcription factor, key to producing the *S. purpuratus*-specific larval skeleton (Khor & Ettensohn, 2017). Alternatively introducing dominant negative RNAs can be used to perturb function (e.g., shown in Sherwood and McClay (2001) for Notch) and making fusion RNAs to add dominant repressors such as engrailed domains can be used to transform a transcription factor from an activator to a repressor (Oliveri, Carrick, & Davidson, 2002).

The rise of CRISPR gene editing has enabled more targeted perturbations of gene expression. In echinoderms, CRISPR protocols have been established for use in sea urchins (Lin, Oulhen, Wessel, & Su, 2019). Further development of this technology will enable researchers to examine how both non-coding and coding elements impact gene expression in living organisms.

3.4 BAC recombinant libraries as a resource for studying GRNS

P. miniata, as typical for many species of echinoderms, readily express reporter genes that have been injected into the oocyte as DNA clones. The injected DNA is either incorporated into the genome or resides as extra chromosomal DNA that is inherited through cell divisions (McMahon, Novak, Britten, & Davidson, 1984). As the DNA may not be incorporated until after several cell divisions have passed, expression of the reporter may be mosaic (Hough-Evans et al., 1988). Multiple fluorescent proteins have also been used in order to localize gene expression (Cheatle Jarvela et al., 2014; Cheatle Jarvela et al., 2016) including at a subcellular level (Hamdoun & Foltz, 2019). Of particular note is the use of the photoconvertible protein kaede. Its ability to irreversibly change fluorescent color of existing proteins offers an invaluable resource for the detection of new gene expression activity (Poon, Fries, Wessel, & Yajima, 2019).

The relative ease and utility of DNA constructs has been extended by the use of BAC (bacterial artificial chromosomes) clones. BAC vectors hold large inserts of up to 350 Kbp (Monaco & Larin, 1994; Shizuya et al., 1992) and therefore libraries can be made where the locus of a coding gene can be flanked by 100 Kbp of DNA or more. This is likely, in most cases, to therefore contain all of the non-coding regulatory sequence needed to drive expression of the gene of interest (McMahon et al., 1984). There are some well-established protocols that use in vitro homologous recombination to exchange the endogenous coding sequence with a variety of reporters (Thomason et al., 2007). Two BAC libraries for *P. miniata* have been made and arrayed, (1) 50 kb fragments with 6 × genome coverage and (2) 150 kb fragments with 33 × genome coverage. Through screening these libraries (either computationally or experimentally), genes and regions of interest can be identified and readily recombined into reporter constructs for introduction into the organism using a standardized and detailed protocol (Buckley, Dong, Cameron, & Rast, 2018). These libraries, as well as nine already recombineered reporter vectors, are freely available for researchers and can be found and requested through Echinobase.org.

3.5 Elucidating regulatory interactions

The extensive BAC library provides an invaluable resource for studying the cis regulatory control of specific genes as homolgous recombination can also be used to delete or rearrange putative non-coding regulatory sequences (Buckley et al., 2018). These experiments can then identify enhancer regions and examine their control of spatiotemporal expression analyzed.

Several next-generation sequencing methods have also been adapted for use in *P. miniata*. ChIP-seq has provided valuable insight into transcription factor binding activity. This technique has been used to identify regulatory interactions and associate these with adjacent genes to build a regulatory network (Cary, Cheatle Jarvela, Francolini, & Hinman, 2017). In addition, protocols for ATAC seq have been adapted to allow for the examination of chromatin structural dynamics (Magri et al., 2021) and open chromatin associated with active regulatory DNA. These epigenetic datasets are currently available through Echinobase.org and are integrated in the website's genome browser feature (Cary, Cameron, & Hinman, 2018). Through these next-generation tools, a better understanding of the regulatory interactions between genes can be established.

4. Developing the bioinformatic toolkit

4.1 Sequencing the *P. miniata* genome

The production of a chromosome length genome has significantly improved the ability to identify regulatory elements and probe more deeply into the role of genome organization. Initial assemblies were done using short reads from 454 and illumina sequencing through the Baylor College of Medicine and Caltech (Kudtarkar & Cameron, 2017). In July of 2012, a 770.2 Mb assembly of this genome was uploaded to NCBI.

In 2020, the newest version of the genome was produced. In this version, 150 × coverage PacBio long read sequencing plus HiC for scaffolding were used to assemble and refine the previous assembly and resulted in a marked increase in contig and scaffold size. This 608 Mb genome comprises 30 scaffolds, with a scaffold N50 value of 23 million bases and more than 90% of the reads residing on 23 scaffolds that likely represent the chromosomes. Extensive RNA-Seq reads allowed were used as part of the NCBI annotation pipeline to develop gene models. These RNA-seq reads, as well as other 'omics resources are integrated with the genome browser feature at echinobase.org (Cary et al., 2018).

4.2 A brief history of Echinobase

In March 2005, the first echinoderm genome, that of *Strongylocentrotus purpuratus,* was sequenced. Two years later, it was brought online, as SpBase, an online genomics resource for the growing research community (Cameron, Samanta, Yuan, He, & Davidson, 2009). The importance of generating this community resource for synthesizing genomic and other data cannot be overstated. It brought together different datasets under one platform, unified naming conventions and served as a resource to build community and share knowledge. As this community grew and new research directions expanded, interest turned to sequencing genomes of other echinoderms. At the time of writing, 21 different species have nuclear genome assemblies available on NCBI, representing a diverse sampling of echinoderm families (Fig. 4A). In 2017, SpBase metamorphosed into EchinoBase, adding 'omics resources for other echinoderm species, like the Green Sea Urchin (*Lytechinus variegatus*), the Warty Sea Cucumber (*Parastichopus parvimensis*), the Spiny Brittle Star (*Ophiothrix spiculata*), and the Slate Pencil Urchin (*Eucidaris tribuloides*) ((Kudtarkar & Cameron, 2017). Three years later, in

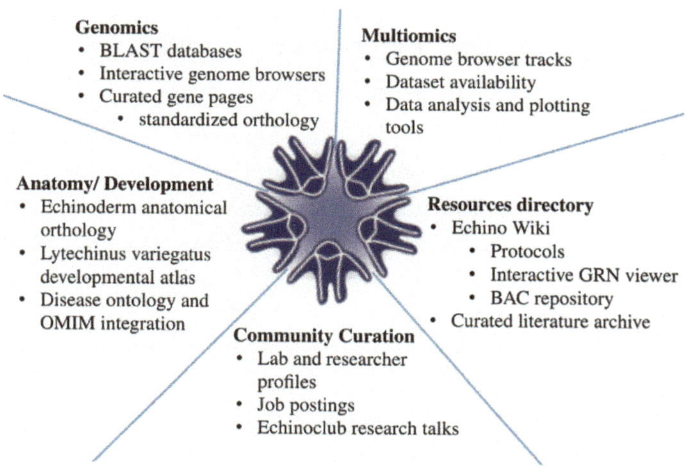

Fig. 4 Summary of resources available on echinobase.org. Echinobase.org is an open-access echinoderm research database that seeks to provide centralized access to relevant resources. It provides robust genomic resources, in addition to publicly-available multi-omics datasets and analysis tools. Echinobase also houses a repository of experimental resources and standardized anatomical resources. FInally, this website also emphasizes the importance of connecting an international research community and has features relevant to networking and information sharing.

2020, a new version of Echinobase (Echinobase.org) was launched, aiming to provide a comprehensive and accessible resource for the international echinoderm research community.

4.3 A data-sharing infrastructure for echinoderm biology

Today, Echinobase has evolved into a powerful, open-access data-sharing platform for echinoderm genomics (summarized in Fig. 4B). This website provides a comprehensive, centralized, database of genomic, anatomical, and evolutionary data. Echinobase joined forces with Xenbase to develop a clone of the Xenbase underlying database organization and programing, and to populate this with echinoderm genome data (Cary et al., 2018).

Echinobase's genomic repertoire centers on four different echinoderms: *S. purpuratus* (purple sea urchin), *A. planci* (crown of thorns starfish), *A. japonica* (sea lily), and *P. miniata* (bat sea star). These diverse taxa have been sampled throughout the phylum to allow a robust examination of echinoderm genomics. Integrated within echinobase, is an interactive genome browser, powered by JBROWSE (Buels et al., 2016) and BLAST databases. Furthermore, an array of optional data tracks can be overlaid on the genome, in addition to user-uploaded tracks. These include time-series RNA seq data sets and ATAC-seq peaks, as well as predicted transcription factor motifs.

One of the greatest strengths of Echinobase are the extensively curated gene pages. For a given gene, users are provided with instant access to sequence information, spatiotemporal expression details, associated research papers, species orthologs, functional ontologies, and links to relevant external resources. The explicite nomenclature guidelines ensure that orthologous genes are readily identifiable and named in a way that facilitates confident phyla-level cross species comparisons (Foley et al., 2021).

In addition to genomic and genetic assets, Echinobase offers a plethora of resources for developmental biology. The Echinoderm Anatomical Ontology (ECAO) provides a standardized vocabulary for describing echinoderm anatomy and processes embryonic development. These terms are organized in a hierarchical manner that contextualizes how a given structure developed over time and space. Echinobase also links out to the interactive tools for the visualization of developmental gene regulatory networks housed at ISB by William Longabaugh. This functionality is achieved through the use of BioTapestry, an open-source tool used to model and visualize gene regulatory networks (Longabaugh, 2012).

Echinobase also acts as a repository for experimental tools. The EchinoWiki pages house a variety of established experimental protocols and recommended external resources. It also contains an archive of qPCR primers, antibodies, morpholino, and BAC sequences and is currently curating guide RNAs. EchinoWiki also has educational resources, such as Echinobase user tutorials and outreach activity ideas.

Finally, Echinobase contains features meant to strengthen the connections between the echinoderm community. Through its networking features, users can explore the user profiles of researchers, browse available jobs, and explore relevant research associations. To further forge connections within the community, EchinoClub was established in 2021, where researchers come together once a month to share and discuss their research.

5. Frontiers of sea star research

5.1 Cell and fertilization biology

P. miniata have historically been used to study gametes and fertilization dynamics (King, 1913; Newman, 1921) and remain a popular model today. Their large, transparent eggs have been an asset in describing the composition and dynamics of egg cells before and during fertilization events (Zazueta-Novoa, Onorato, Reyes, Oulhen, & Wessel, 2016). A comparative approach has also been taken, utilizing multiple species to identify conserved aspects of gamete identity and behavior (Oulhen, Xu, & Wessel, 2014). In addition, new technology also allows for increasingly sophisticated studies on membrane biophysics, giving key insights into cellular dynamics and mechanics (Wigbers et al., 2021).

5.2 Insights into regeneration

Perhaps one of the most commonly known biological phenomena of sea stars is the ability of the adults to regenerate lost arms. Indeed, echinoderms generally have exceptional capacities to regenerate and are important research organisms for studying regeneration (Carnevali, 2006; Vickery, Vickery, Amsler, & McClintock, 2001). The ability of larvae to undergo whole body regeneration (Fig. 1B) on a time-scale of days has established them as a promising research organism for studying the process and mechanisms of regeneration and healing (Vickery et al., 2001). This enabled the exploration of the mechanisms controlling and enacting this regeneration process. Importantly, in the early 1900s, Theoror Mortensen also showed that the larval echinoderms are also able to regenerate and are even able to go through metamorphosis (Mortensen, 1921). Comparative transcriptomics has shown that regeneration-associated genes in Plathelminths and Cnidarians function similarly in *P. miniata* (Cary, Wolff, Zueva, Pattinato, & Hinman, 2019), shedding light on how regenerative capacity fits into the metazoan phylogeny.

Thus the emerging new direction of research using *P. miniata* as a model system to study regeneration promises key insights into the mechanisms and evolutionary history of deuterostome regeneration. This experimentally-convenient system can enable questions about the processes of regeneration and wound healing on an organismal and cellular level and molecular level. Comparisons across species of varying regenerative capabilities can help elucidate the causes and consequences of different regenerative capacities across metazoans (Cary et al., 2019).

5.3 Metamorphosis

Echinoderms are an ideal system for studying the evolution of life histories and body plan development. This phylum contains both direct and indirect developers, facilitating comparative studies to understand the molecular basis of metamorphosis and the evolutionary forces surrounding life cycles (Byrne, 2006). In addition, the transition from larval bilateral symmetry to adult pentaradial symmetry (Fig. 1A) positions *P. miniata* as a fascinating system to study how body plans evolve and are genomically controlled.

5.4 Wasting disease

Recently, sea stars have become of great interest in the study of aquatic disease. Wasting disease is a condition where adult sea stars develop lesions which expand as their bodies begin to decay and their appendages detach. It has proven to be an interesting case study for aquatic disease research, as researchers debate whether this condition is caused by a virus or by environmental factors (Aquino et al., 2020; Hewson et al., 2014). Regardless of the cause, the mass deaths of sea stars on the pacific coast of North America has allowed researchers to investigate the ecological consequences of keystone species dying off (Menge et al., 2016; Moritsch, 2021).

6. Conclusion

Echinoderms have been long appreciated for their ability to answer questions across many different fields of biology. Consequently, substantial literature exists describing the development and evolution of different families in this phyla. Through development of extensive molecular and genomic toolkits, ease of embryonic manipulation, optical transparency, and an emerging knowledge of developmental process, *P. miniata* has emerged as an exciting and pliable model system. From comparative evolution development to the molecular basis regeneration, *P. miniata* offers a unique system to probe critical questions in biology.

Acknowledgements

Thank you to Olga Zueva, Allison Meyer, and Johnathan Lee Andrades for their feedback in drafts of the manuscript. We acknowledge funding from the NSF (award number 1557431) and NIH (grant number 1P41HD071837).

References

Agassiz, A. (1865). *Embryology of the starfish*. Boston, MA.

Andrews, G. F. (1897). Some spinning activities of protoplasm in starfish, and sea-urchin eggs. *Journal of Morphology, 12*, 367–389. https://doi.org/10.1002/jmor.1050120204.

Angerer, L. M., Yaguchi, S., Angerer, R. C., & Burke, R. D. (2011). The evolution of nervous system patterning: Insights from sea urchin development. *Development, 138*, 3613–3623. https://doi.org/10.1242/dev.058172.

Aquino, C. A., Besemer, R. M., DeRito, C. M., Kocian, J., Porter, I. R., Raimondi, P. T., et al. (2020). Evidence that microorganisms at the animal-water Interface Drive Sea star wasting disease. *Frontiers in Microbiology, 11*, 610009. https://doi.org/10.3389/fmicb.2020.610009.

Brandt, J. F. (1835). *Prodromus descriptionis animalium ab H. Mertensio observatorum : Fascic. I. Polypos, Acalephas Discophoras et Siphonophoras, nec non Echinodermata continens*. Acad. imp. sci.

Buckley, K. M., Dong, P., Cameron, R. A., & Rast, J. P. (2018). Bacterial artificial chromosomes as recombinant reporter constructs to investigate gene expression and regulation in echinoderms. *Briefings in Functional Genomics, 17*, 362–371. https://doi.org/10.1093/bfgp/elx031.

Buels, R., Yao, E., Diesh, C. M., Hayes, R. D., Munoz-Torres, M., Helt, G., et al. (2016). JBrowse: A dynamic web platform for genome visualization and analysis. *Genome Biology, 17*, 66. https://doi.org/10.1186/s13059-016-0924-1.

Byrne, M. (2006). Life history diversity and evolution in the Asterinidae. *Integrative and Comparative Biology, 46*, 243–254. https://doi.org/10.1093/icb/icj033.

Cameron, R. A., & Holland, N. D. (1983). Electron microscopy of extracellular materials during the development of a sea star, Patiria miniata (Echinodermata: Asteroidea). *Cell and Tissue Research, 234*, 193–200. https://doi.org/10.1007/BF00217412.

Cameron, R. A., Samanta, M., Yuan, A., He, D., & Davidson, E. (2009). SpBase: The sea urchin genome database and web site. *Nucleic Acids Research, 37*, D750–D754. https://doi.org/10.1093/nar/gkn887.

Carnevali, M. D. C. (2006). Regeneration in echinoderms: Repair, regrowth, cloning. *Invertebrate Survival Journal, 3*(1), 64–76.

Cary, G. A., Cameron, R. A., & Hinman, V. F. (2018). Echinobase: Tools for echinoderm genome analyses. *Methods in Molecular Biology, 1757*, 349–369. https://doi.org/10.1007/978-1-4939-7737-6_12.

Cary, G. A., Cheatle Jarvela, A. M., Francolini, R. D., & Hinman, V. F. (2017). Genome-wide use of high- and low-affinity Tbrain transcription factor binding sites

during echinoderm development. *Proceedings of the National Academy of Sciences of the United States of America, 114*, 5854–5861. https://doi.org/10.1073/pnas.1610611114.
Cary, G. A., Wolff, A., Zueva, O., Pattinato, J., & Hinman, V. F. (2019). Analysis of sea star larval regeneration reveals conserved processes of whole-body regeneration across the metazoa. *BMC Biology, 17*, 16. https://doi.org/10.1186/s12915-019-0633-9.
Chambers, R. (1922). A micro injection study on the permeability of the starfish egg. *The Journal of General Physiology, 5*, 189–193. https://doi.org/10.1085/jgp.5.2.189.
Chambers, R. (1923). The mechanism of the entrance of sperm into the starfish egg. *The Journal of General Physiology, 5*, 821–829. https://doi.org/10.1085/jgp.5.6.821.
Cheatle Jarvela, A. M., Brubaker, L., Vedenko, A., Gupta, A., Armitage, B. A., Bulyk, M. L., et al. (2014). Modular evolution of DNA-binding preference of a Tbrain transcription factor provides a mechanism for modifying gene regulatory networks. *Molecular Biology and Evolution, 31*, 2672–2688. https://doi.org/10.1093/molbev/msu213.
Cheatle Jarvela, A. M., & Hinman, V. (2014). A method for microinjection of Patiria miniata zygotes. *Journal of Visualized Experiments*. e51913. doi:https://doi.org/10.3791/51913.
Cheatle Jarvela, A. M., Yankura, K. A., & Hinman, V. F. (2016). A gene regulatory network for apical organ neurogenesis and its spatial control in sea star embryos. *Development, 143*, 4214–4223. https://doi.org/10.1242/dev.134999.
Clark, W. B., & Twitchell, M. W. (1915). *The mesozoic and cenozoic echinodermata of the United States*. Monograph Govt. print. off. https://doi.org/10.3133/m54.
Davidson, E. H., Rast, J. P., Oliveri, P., Ransick, A., Calestani, C., Yuh, C.-H., et al. (2002a). A genomic regulatory network for development. *Science, 295*, 1669–1678. https://doi.org/10.1126/science.1069883.
Davidson, E. H., Rast, J. P., Oliveri, P., Ransick, A., Calestani, C., Yuh, C.-H., et al. (2002b). A provisional regulatory gene network for specification of endomesoderm in the sea urchin embryo. *Developmental Biology, 246*, 162–190. https://doi.org/10.1006/dbio.2002.0635.
Ernst, S. G. (1997). A century of sea urchin development. *American Zoologist, 37*, 250–259. https://doi.org/10.1093/icb/37.3.250.
Erwin, D. H., & Davidson, E. H. (2009). The evolution of hierarchical gene regulatory networks. *Nature Reviews. Genetics, 10*, 141–148. https://doi.org/10.1038/nrg2499.
Foley, S., Ku, C., Arshinoff, B., Lotay, V., Karimi, K., Vize, P. D., et al. (2021). Integration of 1:1 orthology maps and updated datasets into Echinobase. *Database: The Journal of Biological Databases and Curation, 2021*. https://doi.org/10.1093/database/baab030.
Fresques, T. M., & Wessel, G. M. (2018). Nodal induces sequential restriction of germ cell factors during primordial germ cell specification. *Development, 145*. https://doi.org/10.1242/dev.155663.
Gildor, T., Cary, G. A., Lalzar, M., Hinman, V. F., & Ben-Tabou de-Leon, S. (2019). Developmental transcriptomes of the sea star, Patiria miniata, illuminate how gene expression changes with evolutionary distance. *Scientific Reports, 9*, 16201. https://doi.org/10.1038/s41598-019-52577-9.
Goldring, I. P. (1979). Robert Chambers: 1881–1957. *American Zoologist, 19*, 1271–1273. https://doi.org/10.1093/icb/19.4.1271.
Goto, S. (1895). Preliminary notes on the embryology of the starfish (asterias pallida). *Proceedings of the American Academy of Arts and Sciences, 31*, 333. https://doi.org/10.2307/20020634.
Gray, J. E. (1866). *Synopsis of the species of starfish in the British museum*. J. Van Voorst.
Gregory, J. W. (1899). Ii.—On *lindstromaster* and the classification of the palæasterids. *Geological Magazine, 6*, 341–354. https://doi.org/10.1017/S0016756800142384.
Hamdoun, A., & Foltz, K. (2019). *Echinoderms Part B, Volume 151* (1st ed.). Elsevier.
Hewson, I., Button, J. B., Gudenkauf, B. M., Miner, B., Newton, A. L., Gaydos, J. K., et al. (2014). Densovirus associated with sea-star wasting disease and mass mortality. *Proceedings*

of the National Academy of Sciences of the United States of America, 111, 17278–17283. https://doi.org/10.1073/pnas.1416625111.

Hinman, V. F., & Davidson, E. H. (2007). Evolutionary plasticity of developmental gene regulatory network architecture. *Proceedings of the National Academy of Sciences of the United States of America, 104*, 19404–19409. https://doi.org/10.1073/pnas.0709994104.

Hinman, V. F., Nguyen, A. T., Cameron, R. A., & Davidson, E. H. (2003). Developmental gene regulatory network architecture across 500 million years of echinoderm evolution. *Proceedings of the National Academy of Sciences of the United States of America, 100*, 13356–13361. https://doi.org/10.1073/pnas.2235868100.

Hinman, V. F., Nguyen, A. T., & Davidson, E. H. (2003). Expression and function of a starfish Otx ortholog, AmOtx: A conserved role for Otx proteins in endoderm development that predates divergence of the eleutherozoa. *Mechanisms of Development, 120*, 1165–1176. https://doi.org/10.1016/j.mod.2003.08.002.

Holland, L. Z., Holland, P. W. H., & Holland, N. D. (1996). Revealing homologies between body parts of distantly related animals by in situ hybridization to developmental genes: Amphioxus versus vertebrates. In J. D. Ferraris, & S. R. Palumbi (Eds.), *Molecular zoology: Advances, strategies, and protocols* (pp. 267–282). New York: Wiley-Liss.

Holland, N. D. (1980). Electron microscopic study of the cortical reaction in eggs of the starfish (patria miniata). *Cell and Tissue Research, 205*, 67–76. https://doi.org/10.1007/BF00234443.

Hough-Evans, B. R., Britten, R. J., & Davidson, E. H. (1988). Mosaic incorporation and regulated expression of an exogenous gene in the sea urchin embryo. *Developmental Biology, 129*, 198–208. https://doi.org/10.1016/0012-1606(88)90174-1.

Kanatani, H., Shirai, H., Nakanishi, K., & Kurokawa, T. (1969). Isolation and indentification on meiosis inducing substance in starfish Asterias amurensis. *Nature, 221*, 273–274. https://doi.org/10.1038/221273a0.

Khaner, O. (1993). The potency of the first two cleavage cells in echinoderm development: The experiments of Driesch revisited. *Roux's Archives of Developmental Biology, 202*, 193–197. https://doi.org/10.1007/BF02427879.

Khor, J. M., & Ettensohn, C. A. (2017). Functional divergence of paralogous transcription factors supported the evolution of biomineralization in echinoderms. *eLife, 6*. https://doi.org/10.7554/eLife.32728.

King, W. O. R. (1913). *Artificial parthenogenesis and fertilization*. The University of Chicago Press.

Kishimoto, T. (1999). Activation of MPF at meiosis reinitiation in starfish oocytes. *Developmental Biology, 214*, 1–8. https://doi.org/10.1006/dbio.1999.9393.

Krieg, P. A., & Melton, D. A. (1984). Functional messenger RNAs are produced by SP6 in vitro transcription of cloned cDNAs. *Nucleic Acids Research, 12*, 7057–7070. https://doi.org/10.1093/nar/12.18.7057.

Kudtarkar, P., & Cameron, R. A. (2017). Echinobase: An expanding resource for echinoderm genomic information. *Database: The Journal of Biological Databases and Curation, 2017*. https://doi.org/10.1093/database/bax074.

Lapraz, F., Haillot, E., & Lepage, T. (2015). A deuterostome origin of the Spemann organiser suggested by nodal and ADMPs functions in echinoderms. *Nature Communications, 6*, 8927. https://doi.org/10.1038/ncomms9927.

Lillie, R. S. (1912). Certain means by which starfish eggs naturally resistant to fertilization may be rendered normal and the physiological conditions of this action. *The Biological Bulletin, 22*, 328–346. https://doi.org/10.2307/1535968.

Lillie, R. S. (1926). The activation of starfish eggs by acids. *The Journal of General Physiology, 8*, 339–367. https://doi.org/10.1085/jgp.8.4.339.

Lin, C.-Y., Oulhen, N., Wessel, G., & Su, Y.-H. (2019). CRISPR/Cas9-mediated genome editing in sea urchins. *Methods in Cell Biology, 151*, 305–321. https://doi.org/10.1016/bs.mcb.2018.10.004.

Longabaugh, W. J. R. (2012). BioTapestry: A tool to visualize the dynamic properties of gene regulatory networks. *Methods in Molecular Biology, 786*, 359–394. https://doi.org/10.1007/978-1-61779-292-2_21.

Lowe, C. J., & Wray, G. A. (2000). Rearing larvae of sea urchins and sea stars for developmental studies. *Methods in Molecular Biology, 135*, 9–15. https://doi.org/10.1385/1-59259-685-1:9.

Magri, M. S., Voronov, D., Ranđelović, J., Cuomo, C., Gómez-Skarmeta, J. L., & Arnone, M. I. (2021). ATAC-seq for assaying chromatin accessibility protocol using echinoderm embryos. *Methods in Molecular Biology, 2219*, 253–265. https://doi.org/10.1007/978-1-0716-0974-3_16.

Materna, S. C. (2017). Using morpholinos to probe gene networks in sea urchin. *Methods in Molecular Biology, 1565*, 87–104. https://doi.org/10.1007/978-1-4939-6817-6_8.

Materna, S. C., Swartz, S. Z., & Smith, J. (2013). Notch and nodal control forkhead factor expression in the specification of multipotent progenitors in sea urchin. *Development, 140*, 1796–1806. https://doi.org/10.1242/dev.091157.

McCauley, B. S., Akyar, E., Saad, H. R., & Hinman, V. F. (2015). Dose-dependent nuclear β-catenin response segregates endomesoderm along the sea star primary axis. *Development, 142*, 207–217. https://doi.org/10.1242/dev.113043.

McEdward, L. R., & Miner, B. G. (2001). Larval and life-cycle patterns in echinoderms. *Canadian Journal of Zoology, 79*, 1125–1170. https://doi.org/10.1139/z00-218.

McMahon, A. P., Novak, T. J., Britten, R. J., & Davidson, E. H. (1984). Inducible expression of a cloned heat shock fusion gene in sea urchin embryos. *Proceedings of the National Academy of Sciences of the United States of America, 81*, 7490–7494. https://doi.org/10.1073/pnas.81.23.7490.

Menge, B. A., Cerny-Chipman, E. B., Johnson, A., Sullivan, J., Gravem, S., & Chan, F. (2016). Sea star wasting disease in the keystone predator Pisaster ochraceus in Oregon: Insights into differential population impacts, recovery, predation rate, and temperature effects from long-term research. *PLoS One, 11*, e0153994. https://doi.org/10.1371/journal.pone.0153994.

Metchnikoff, E. (1893). *Lectures on the comparative pathology of inflammation delivered at the Pasteur institute in 1891*. New York: Dover Publications.

Monaco, A. P., & Larin, Z. (1994). YACs, BACs, PACs and MACs: Artificial chromosomes as research tools. *Trends in Biotechnology, 12*, 280–286. https://doi.org/10.1016/0167-7799(94)90140-6.

Moritsch, M. M. (2021). Expansion of intertidal mussel beds following disease-driven reduction of a keystone predator. *Marine Environmental Research, 169*, 105363. https://doi.org/10.1016/j.marenvres.2021.105363.

Mortensen, T. (1921). *Studies of the development and larval forms of echinoderms*. G. E. C. Gad.

Müller, J. (1854). I.— *On the structure of the* echinoderms. *Annals and Magazine of Natural History, 13*, 1–24. https://doi.org/10.1080/03745485709496297.

Müller, G. B. (2007). Evo-devo: Extending the evolutionary synthesis. *Nature Reviews. Genetics, 8*, 943–949. https://doi.org/10.1038/nrg2219.

Newman, H. H. (1921). On the development of the spontaneously parthenogenetic eggs of asterina (patiria) miniata. *The Biological Bulletin, 40*, 105–117. https://doi.org/10.2307/1536616.

Newman, H. H. (1922). Normal versus subnormal development in patiria miniata. A caution to laboratory embryologists. *The Biological Bulletin, 43*, 1–9. https://doi.org/10.2307/1536687.

Oliveri, P., Carrick, D. M., & Davidson, E. H. (2002). A regulatory gene network that directs micromere specification in the sea urchin embryo. *Developmental Biology, 246*, 209–228. https://doi.org/10.1006/dbio.2002.0627.

Oulhen, N., Xu, D., & Wessel, G. M. (2014). Conservation of sequence and function in fertilization of the cortical granule serine protease in echinoderms. *Biochemical and Biophysical Research Communications*, *450*, 1135–1141. https://doi.org/10.1016/j.bbrc.2014.05.098.

Perillo, M., Paganos, P., Spurrell, M., Arnone, M. I., & Wessel, G. M. (2021). Methodology for whole mount and fluorescent RNA in situ hybridization in echinoderms: Single, double, and beyond. *Methods in Molecular Biology*, *2219*, 195–216. https://doi.org/10.1007/978-1-0716-0974-3_12.

Peter, I. S., Faure, E., & Davidson, E. H. (2012). Predictive computation of genomic logic processing functions in embryonic development. *Proceedings of the National Academy of Sciences of the United States of America*, *109*, 16434–16442. https://doi.org/10.1073/pnas.1207852109.

Poon, J., Fries, A., Wessel, G. M., & Yajima, M. (2019). Evolutionary modification of AGS protein contributes to formation of micromeres in sea urchins. *Nature Communications*, *10*, 3779. https://doi.org/10.1038/s41467-019-11560-8.

Prouho, M. H. (1891). On the sense of smell in starfish. *Annals and Magazine of Natural History*, 7, 306–308. https://doi.org/10.1080/00222939109460616.

Rafiq, K., Cheers, M. S., & Ettensohn, C. A. (2012). The genomic regulatory control of skeletal morphogenesis in the sea urchin. *Development*, *139*, 579–590. https://doi.org/10.1242/dev.073049.

Romanes, G. J. (1883). Observations on the physiology of echinodermata. *Journal of the Linnean Society of London, Zoology*, *17*, 131–137. https://doi.org/10.1111/j.1096-3642.1883.tb02044.x.

Saudemont, A., Haillot, E., Mekpoh, F., Bessodes, N., Quirin, M., Lapraz, F., et al. (2010). Ancestral regulatory circuits governing ectoderm patterning downstream of nodal and BMP2/4 revealed by gene regulatory network analysis in an echinoderm. *PLoS Genetics*, *6*, e1001259. https://doi.org/10.1371/journal.pgen.1001259.

Saunders, L. R., & McClay, D. R. (2014). Sub-circuits of a gene regulatory network control a developmental epithelial-mesenchymal transition. *Development*, *141*, 1503–1513. https://doi.org/10.1242/dev.101436.

Shen, S., & Steinhardt, R. A. (1976). An electrophysiological study of the membrane properties of the immature and mature oocyte of the Batstar, Patiria miniata. *Developmental Biology*, *48*, 148–162. https://doi.org/10.1016/0012-1606(76)90053-1.

Sherwood, D. R., & McClay, D. R. (2001). LvNotch signaling plays a dual role in regulating the position of the ectoderm-endoderm boundary in the sea urchin embryo. *Development*, *128*, 2221–2232.

Shizuya, H., Birren, B., Kim, U. J., Mancino, V., Slepak, T., Tachiiri, Y., et al. (1992). Cloning and stable maintenance of 300-kilobase-pair fragments of human DNA in Escherichia coli using an F-factor-based vector. *Proceedings of the National Academy of Sciences of the United States of America*, *89*, 8794–8797. https://doi.org/10.1073/pnas.89.18.8794.

Smith, A. B., & Paul, C. R. C. (1988). Fossil evidence for the relationships of extant echinoderm classes and their times of divergence. In C. R. C. Paul, & A. Z. B. Smith (Eds.), *1. Echinoderm phylogeny and evolutionary biology* (pp. 85–97). Oxford: Clarendon Press.

Stevens, M. (1970). Procedures for induction of spawning and meiotic maturation of starfish oocytes by treatment with 1-methyladenine. *Experimental Cell Research*, *59*, 482–484. https://doi.org/10.1016/0014-4827(70)90658-0.

Summerton, J., & Weller, D. (1997). Morpholino antisense oligomers: Design, preparation, and properties. *Antisense & Nucleic Acid Drug Development*, 7, 187–195. https://doi.org/10.1089/oli.1.1997.7.187.

Telford, M. J., Lowe, C. J., Cameron, C. B., Ortega-Martinez, O., Aronowicz, J., Oliveri, P., et al. (2014). Phylogenomic analysis of echinoderm class relationships supports Asterozoa. *Proceedings of the Biological Sciences, 281*. https://doi.org/10.1098/rspb.2014.0479.

Thomason, L., Court, D. L., Bubunenko, M., Costantino, N., Wilson, H., Datta, S., et al. (2007). Recombineering: Genetic engineering in bacteria using homologous recombination. *Current Protocols in Molecular Biology*. https://doi.org/10.1002/0471142727.mb0116s78. Chapter 1, Unit 1.16.

Vickery, M. C., Vickery, M. S., Amsler, C. D., & McClintock, J. B. (2001). Regeneration in echinoderm larvae. *Microscopy Research and Technique, 55*, 464–473. https://doi.org/10.1002/jemt.1191.

Wigbers, M. C., Tan, T. H., Brauns, F., Liu, J., Swartz, S. Z., Frey, E., et al. (2021). A hierarchy of protein patterns robustly decodes cell shape information. *Nature Physics*. https://doi.org/10.1038/s41567-021-01164-9.

Yankura, K. A., Koechlein, C. S., Cryan, A. F., Cheatle, A., & Hinman, V. F. (2013). Gene regulatory network for neurogenesis in a sea star embryo connects broad neural specification and localized patterning. *Proceedings of the National Academy of Sciences of the United States of America, 110*, 8591–8596. https://doi.org/10.1073/pnas.1220903110.

Yankura, K. A., Martik, M. L., Jennings, C. K., & Hinman, V. F. (2010). Uncoupling of complex regulatory patterning during evolution of larval development in echinoderms. *BMC Biology, 8*, 143. https://doi.org/10.1186/1741-7007-8-143.

Zazueta-Novoa, V., Onorato, T. M., Reyes, G., Oulhen, N., & Wessel, G. M. (2016). Complexity of yolk proteins and their dynamics in the sea star Patiria miniata. *The Biological Bulletin, 230*, 209–219. https://doi.org/10.1086/BBLv230n3p209.

CHAPTER NINETEEN

Saccoglossus kowalevskii: Evo-devo insights from the mud

Jessica Gray[a], Jens H. Fritzenwanker[b], Doreen D. Cunningham[b], and Christopher J. Lowe[c,*]

[a]Department of Systems Biology, Harvard Medical School, Boston, MA, United States
[b]Department of Biology, Georgetown University, Washington, DC, United States
[c]Hopkins Marine Station, Department of Biology, Stanford University, Pacific Grove, CA, United States
*Corresponding author: e-mail address: clowe@stanford.edu

Contents

Abstract

Hemichordates have long been recognized as a critical group for addressing hypotheses of chordate origins. Historically this was due to anatomical traits that resembled those of chordates, most strikingly the dorsolateral gill slits. As molecular data and phylogenetic analyses were found to support a close phylogenetic relationship between hemichordates and chordates within the deuterostomes, interest was revived in hemichordates. In particular, *Saccoglossus kowalevskii* has been developed as a molecular model to represent hemichordate developmental biology. Herein, we highlight the considerations when choosing a particular species to study and the challenges we encountered when developing *S. kowalevskii*. We discuss our findings and how method and tool development enabled them, and how we envision expanding our repertoire of molecular tools in the future. Establishing a new model organism comes with many obstacles—from identifying a reliable season to collect animals, to developing modern molecular techniques. The *Saccoglossus* research community has benefited greatly from the collaborations and teamwork established over the years. As a result, *Saccoglossus* is well positioned to contribute to a new century of evolutionary developmental (evo-devo) research.

Current Topics in Developmental Biology, Volume 147
ISSN 0070-2153
https://doi.org/10.1016/bs.ctdb.2022.01.004

1. Historical focus on *Saccoglossus*

The representation of animal diversity in the field of comparative developmental and cellular biology has accelerated over the last decade, largely due to the advances in sequencing and functional genomic technologies that have significantly lowered the bar of entry, both financially and experimentally, for new model species. Nevertheless, selecting a species to represent the developmental biology of a clade of animals is an important and challenging decision (Bolker, 2012). There are both practical and biological or evolutionary considerations. Practical issues include animal procurement, ease of obtaining gametes, length of life cycle, and how easily the embryos can be manipulated. Evolutionary/biological ones: is the animal a good representative of the clade it is selected to represent, or does it possess a particular biological property that studies are targeting, not represented in established animal models. However, many of the major lineages of animals fall far short of the list of biological properties desirable for conventional model species—their adults are difficult to collect, their reproduction very seasonal, their life cycles very long, and many may be tied to work in restricted environments such as at marine stations. However, despite these practical drawbacks, the potential evolutionary insights into body plan evolution due to phylogenetic position, or particular insights due to the uniqueness of elements of their biology, can be so compelling that these drawbacks must be overlooked. Hemichordates are a good example of one such group. There is a range of biological challenges for species that have currently been developed for developmental studies, yet the evolutionary motivations due to their proposed morphological and phylogenetic ties with chordates, and another non-chordate representative of deuterostomes, make them a compelling group for comparative developmental studies.

Hemichordates are a phylum closely related to echinoderms (Bromham & Degnan, 1999; Cameron, Garey, & Swalla, 2000; Cannon et al., 2014; Furlong & Holland, 2002) and classically allied with chordates to form the deuterostomes (Bourlat et al., 2006; Dunn et al., 2008). This exclusively marine phylum is divided into two classes; the small, mostly colonial Pterobranchs, and the solitary enteropneusts (commonly known as acorn worms), which are marine worms ranging widely in size and habitat (Tassia et al., 2016). The study of the developmental and cellular biology of hemichordates has a long and rich history with seminal contributions to the work from some of the most recognizable names in developmental biology and genetics (Bateson, 1884, 1885, 1886; Kowalevsky, 1866; Morgan, 1894). Most developmental studies have focused on enteropneusts, largely due to the scarcity of pterobranchs, but also due to the much lower fecundity of the individual pterobranch zooids from the few species that have been described (Sato, Bishop, & Holland, 2008). The developmental biology of enteropneusts is split into two main categories defined by the major differences in life history; the majority of enteropneusts are indirect developers. A smaller group, the Harrimanids, is defined by direct development and a rapid progression to the juvenile without a larval body plan (Rottinger & Lowe, 2012; Tagawa, 2016). Working in either group has advantages and disadvantages. The advantage of working on a direct developer is that the development of the adult body plan is occurring during embryogenesis rather than during the terminal phase of larval development and metamorphosis, which may occur several months after hatching, making the mechanisms of adult body plan assembly accessible to embryonic manipulations (Lowe, 2021).

S. kowalevskii has been developed as a representative species of the Harrimanidae. It has a long history in the literature on developmental biology of hemichordates (Bateson, 1884, 1885; Colwin & Colwin, 1950, 1951, 1953). It is a species that is present along much of the eastern seaboard of the USA occupying an intertidal habitat making adult collection straightforward. William Bateson was the first person who comprehensively described its development during his work in the Chesapeake Bay in the 1880s in a series of manuscripts investigating the early developmental biology and the importance in studying the origin of chordates (Bateson, 1884, 1885, 1886). At the time it was known as *Balanoglossus kowalevskii*, and all his observations were on embryos collected from the field as he was unable to induce spawning and carry out controlled fertilizations. Bateson was so struck by morphological similarities between hemichordates and chordates that he elevated them to a basal class of chordates (Bateson, 1886). It was Bateson who highlighted the short dorsal rod projecting into the proboscis, now called the stomochord, as a potential homolog of the notochord, and when considered along with the collar cord and also the gill slits, came up with the name Hemichordata. However, despite his contributions, he later forcefully rejected the evolutionary implications of his work and the importance of comparative embryology for understanding chordate origins (Hall, 2005)—not such an auspicious start for *S. kowalevskii*. It was not until the 1950s that further detailed work on development was carried out by Arthur and Laura Colwin, at the Marine Biological Laboratory (MBL) at Woods Hole, in a series of elegant experimental papers. Their early work focused on experimental embryology to investigate the developmental capacity of blastomeres during early cleavage stages, and the early establishment of axes (Colwin & Colwin, 1950, 1951, 1953). Later work focused on fertilization and sperm structure (Colwin & Colwin, 1963a, 1963b). Their work in embryology laid strong foundations for later molecular work. It also created an important link between this species and the MBL, and not insignificantly established a protocol for spawning and fertilization (Colwin & Colwin, 1962).

The next phase of work on the development of *S. kowalevskii* began in 1998 when John Gerhart and Marc Kirschner initiated work on this species at the MBL. Chris Lowe joined a year later as their postdoc. The revolution in evo-devo was well underway and the goals of the project were to establish *S. kowalevskii* as a new molecular model to represent hemichordate developmental biology, and to explicitly address hypotheses of chordate origins. In addition, in most bilaterian body plan comparisons, deuterostomes were largely represented solely by chordates, as echinoderm adults have such a derived body architecture. The addition of hemichordates broadened the sampling of deuterostome clades. At the time, *Ptychodera flava*, an indirect developing species, had already emerged as a viable hemichordate model, but Gerhart and Kirschner considered that direct development offered some significant experimental advantages for understanding the development of the adult body plan, and afforded more direct comparisons with vertebrates. In indirect-developing species, embryogenesis gives rise to a larval body plan and formation of the adult body plan happens post-embryonically, after several months living in the plankton, making vertebrate comparisons more challenging (Rottinger & Lowe, 2012). There are other enteropneust populations available intertidally in the US, but many are very challenging to collect, or more remotely located. Chris Lowe had tried unsuccessfully to identify other species for developmental studies, including the indirect-developing species *Balanoglossus aurantiacus* from North Carolina, that took two people several hours to dig out a single intact adult worm! *S. kowalevskii* also requires digging in the intertidal "mud" to collect animals, but the populations on Cape Cod reside predominantly in soft sandy sediment in saltwater ponds near the shore (see Fig. 1 and Supplementary material in the online version at https://doi.org/10.1016/bs.ctdb.2022.01.004). The relative ease of collection of *S. kowalevskii*, with a large population within close proximity of the MBL, and with basic embryological work already characterized, made the choice ultimately quite obvious.

Fig. 1 *S. kowalevskii* field collection. (A) After digging sand from the top 1 to 1.5 ft into a sifter the sand is gently removed to check for animals. (B) *S. kowalevskii* leaves filtered sand in curled spirals (castings) on the seafloor which can be spotted on calm days. (C) *S. kowalevskii* can be found in sandy areas but is also found in decaying plant matter. (D) A male *S. kowalevskii* with bright orange testis. (E) A gravid female *S. kowalevskii* with gray eggs. (F) Group picture September 2009.

2. Reviving a model

While *S. kowalevskii* appeared to be an ideal model organism to explore questions of deuterostome evolution, an evo-devo research program required access to embryos. Despite the history of research described above, there was little concrete guidance for generating embryos on demand, as required for an intensive comparative biology approach. The first challenges were to find animals, induce spawning reliably, fertilize eggs, and raise embryos through juvenile stages. Only with that foundation could methods for determining developmental gene expression be established. Early efforts at induction and fertilization failed because they were attempted outside of spawning season - based on reports of summer collections by the Colwins who, like many academics of the time, were only able to carry out experiments between teaching responsibilities in the Spring and Fall. The site they successfully collected from during July and August remains undiscovered. Even after collecting hundreds of animals across multiple sites in the Woods Hole area and a literal call for help to the retired Colwins, all attempts at induction failed despite applying a wide range of methods and receiving assistance from multiple people. At the MBL, Rindy Jaffe assisted in successfully isolating active sperm from male worms for *in vitro* fertilization attempts but the eggs failed to fertilize. After one particularly frustrating summer, Marc Kirschner asked collectors at the MBL to send worms to his lab at Harvard. Mature worms were boxed up and placed on a bus north to Boston in the fall. Surprisingly, these animals spawned easily after a standard induction protocol of heating and cooling, and resulted in numerous successful fertilizations. These successes pointed to an explanation for failure and a reason to continue working with *Saccoglossus*—the early summer was not an ideal time for collecting gravid worms that are capable of producing mature eggs. Our current modern research program resulted from this combination of perseverance, observation, and luck. Once the primary gravid season on Cape Cod was identified as September, eggs and embryos became readily accessible. A second spawning season has since been identified in May—with the interesting twist that ambient water temperatures are much lower and all animal husbandry is best carried out at lower temperatures. The window for finding gravid animals in the field in May is also smaller- making the ability to maintain animals at the MBL integral to spring fieldwork. The early methods of *Saccoglossus* husbandry for molecular work were summarized in Lowe, Tagawa, Humphreys, Kirschner, and Gerhart (2004).

As with many emerging model organisms, there was little or no initial funding, making the first choices for resource development key. One of the most significant decisions early in the development of the project was to invest in an extensive expressed sequence tag (EST) collection (Freeman et al., 2008; Lowe et al., 2003). High quality cDNA libraries produced from a range of developmental stages were normalized, arrayed and sequenced. In the era of Illumina sequencing, this seems rather trivial, but at the time, Sanger sequencing was very expensive, and sequencing 200,000 EST clones was a large expense. This achievement was only possible through collaboration in the scientific community, and we greatly benefited from a collaboration with Eric Lander's lab in generating our first ESTs. Developing collaborations early in the process of model organism development relies heavily on active advocacy for your research program and the ability to communicate why it is important to be able to work with your organism.

These arrayed clones allowed the initial identification of homologs to vertebrate genes of interest, and provided the basis for initial gene expression assays and, later, manipulation. In particular, the development of *in situ* hybridization methods for visualizing gene expression enabled a comprehensive analysis of patterning genes in *Saccoglossus* (Lowe et al., 2004).

These expression patterns rapidly generated hypotheses about deuterostome-specific evolution of the anteroposterior and dorsoventral body axes as well as the central nervous system (Aronowicz & Lowe, 2006; Lowe et al., 2003, 2006).

An important next step in establishing *Saccoglossus* as a model that could provide mechanistic insights was the ability to test the functional effects of overexpression or knockdown of genes during development. The ability to perturb function can reveal differences or commonalities in gene function in development across species and uncover changes in genetic logic or regulatory networks linked to evolutionary changes. Here again, the biology of the *Saccoglossus* proved challenging. Eggs are particularly prone to irrecoverable damage upon injection-the most common approach to siRNA or morpholino delivery. Early efforts converged on a 'harpoon method', which was high throughput but low output (as measured by living embryos). Although this method worked, the high failure rate remained a frustrating point. In one critical test at the MBL, Marc Kirschner and Mark Terasaki each did injections for an afternoon. At the end of the day, both had the same number of living injected embryos, but Kirschner had injected roughly 20 times as many eggs. Since they felt that the harpoon method had been optimized as much as possible, the decision was made to develop a new injection protocol based on the platform of Jaffe and Terasaki (2004), working closely with Mark Terasaki. Through trial and error the group identified a 4 min window beginning approximately 3 min after fertilization that resulted in higher viability after injection. The relatively low throughput and time constraints of injection in this method created a bottleneck for experimental work despite the relatively large number of eggs per spawn. Hard work (and some brute force) by members of the community has allowed this to be dramatically scaled up. Injections now can be used as needed for siRNAs, morpholinos, RNA and transgenics (see Fig. 2). A copy of our current "field guide" with instructions on utilizing developed tools, including animal collection and husbandry, can be found in Supplementary material in the online version at https://doi.org/10.1016/bs.ctdb.2022.01.004 and a link to a live version which gets updated once a season can be found on the Lowe lab homepage (https://lowe.stanford.edu/).

One downside to this early injection approach is that many genes are used repeatedly during development. Thus, genes that are overexpressed or knocked down by injection may have catastrophic early phenotypes that obscure later stage patterning and signaling. Drug treatments—specific protein or pathway inhibitors and activators—enable experimental access to

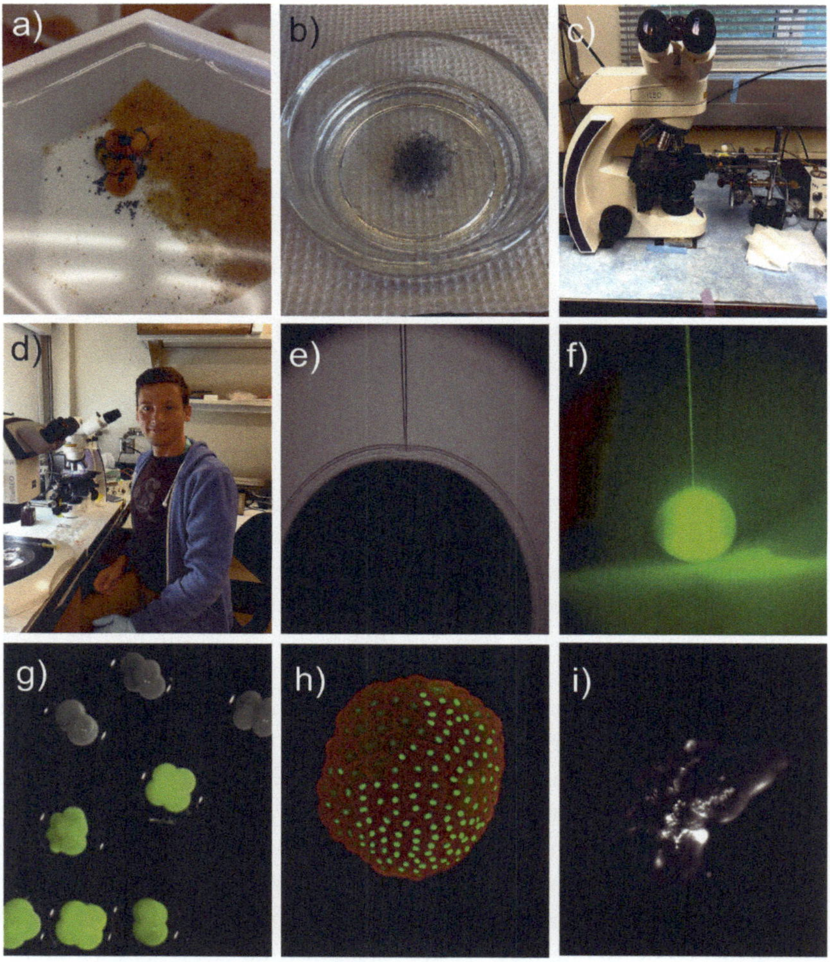

Fig. 2 *S. kowalevskii* injections. (A) A female *S. kowalevskii* is laying eggs after overnight heat induction. (B) A bowl of freshly collected unfertilized eggs to be used for fertilizations and injections. (C–D) Injection setup. (E) Individual fertilized egg and injection needle shown at 40 × magnification. (F) A fertilized egg being injected with siRNA and fluorescent tracer. (G) Screening of embryos for successful injection at early cleavage stages. (H) A blastula embryo injected with RNA for histone-GFP and membrane-mCherry. (I) A transgenic embryo expressing gfp driven by a *S. kowalevskii* promoter.

Saccoglossus developmental genes during discrete time windows outside of early development. Conveniently we found that *Saccoglossus* embryos respond to drugs suspended in the seawater, for example, BMP2/4 protein in sea water to dorsalize embryos. While this does not provide spatial restriction of genetic perturbation, it continues to be a valuable tool for unraveling genetic pathways at different stages of development as well as identifying critical sensitivity windows in development for specific pathways (Darras et al., 2018; Darras, Gerhart, Terasaki, Kirschner, & Lowe, 2011; Fritzenwanker, Uhlinger, Gerhart, Silva, & Lowe, 2019; Green, Norris, Terasaki, & Lowe, 2013.

3. Summary of early results and conclusions

Early tool development enabled rapid functional insight into outstanding questions of deuterostome evolution. Below we will briefly describe the important findings made possible as we developed research tools to establish *S. kowalevskii* as a model organism. A wider ranging review of the evolutionary and developmental insights from these data, including results from other hemichordates, was covered recently in a separate review (Lowe, 2021). The ability to visualize gene expression in the *Saccoglossus* embryo provided immediate insight into the anteroposterior and neural patterning of deuterostomes. An *in situ* expression analysis of conserved chordate neural patterning genes revealed a shared AP arrangement despite the largely diffuse character of the hemichordate nervous system (Lowe et al., 2003). The study further suggested the potential of a deuterostome common ancestor with a diffuse nervous system, or at the very least a modified role for these patterning factors in hemichordates, challenging the common view on deuterostome nervous system evolution at the time (Arendt & Nubler-Jung, 1994; De Robertis & Sasai, 1996; Geoffroy Saint-Hilaire, 1822; Reichert & Simeone, 2001). Unlike chordates these patterning genes are not restricted to a region of the ectoderm but are expressed circumferentially around the embryo's surface. Thus, despite an underlying conservation of pattern, these findings raised questions regarding the connection between patterning of the ectoderm and the nervous system (Hejnol & Lowe, 2015). This in turn has inspired more thorough research addressing nervous system patterning evolution in deuterostomes.

A second longstanding question in evolution concerned dorsoventral patterning and the dorsal or ventral positioning of the central nervous system in vertebrates and invertebrates respectively, as discussed by E. Geoffroy Saint-Hilaire in 1822 in the context of a DV axis inversion (Geoffroy Saint-Hilaire, 1822). In both groups Bmp and its antagonists such as Chordin are expressed on opposing dorsal or ventral sides, with the nervous system forming on the side opposite the side of Bmp expression

(De Robertis & Sasai, 1996). Since *Saccoglossus* lacks a clear centralized nervous system, it presents a novel opportunity to investigate not only dorsoventral patterning but also the mechanisms of nervous system organization.

An *expression* analysis of Bmp/Chordin-related genes confirmed conservation of an oppositional axis and suggested a likely ancestral role for these pathways in dorsoventral patterning. However, not all dorsoventral patterning genes shared conserved expression in the dorsoventral axis. In addition, while *Saccoglossus* bmp2/4 is expressed on the dorsal midline, similar to Drosophila and other invertebrates, and inverse to vertebrates, it lacks a clear centralized nervous system on either the dorsal or ventral side, and has cord-like condensations of nerves on both midlines. Expression analysis alone could not resolve a functional role for BMP signaling in either dorsoventral or neural patterning. The use of functional assays in this initial study (here overexpression of bmp2/4 protein and siRNA knockdowns) demonstrated a developmental role for BMPs signaling in dorsoventral patterning but not in neural specification (Lowe et al., 2006), thus revealing an evolutionary divergence when compared to other deuterostomes like vertebrates.

These early studies cemented the importance and usefulness of *Saccoglossus* for providing insight into deuterostome patterning and evolution (reviewed in Lowe, 2008). After these initial successes, research in *Saccoglossus* expanded dramatically by utilizing newly developed tools such as knockdowns using siRNA and morpholinos, over-expression, and also classical tools of embryological manipulations like tissue explants, single blastomere injections and labeling. Importantly, these new tools helped us to look more comprehensively at changes in developmental patterning and pathways.

Research into the nervous system has branched into understanding the neurogenic origins of the hemichordate nervous system (Cunningham & Casey, 2014), as well as the conservation of signaling centers responsible for patterning the vertebrate brain (Pani et al., 2012). A deeper understanding of body plan organization and germ layer induction has also broadened the comparative scope of hemichordate research to include germ layer induction and patterning (Darras et al., 2011; Green et al., 2013), AP body axis patterning (Darras et al., 2018) and posterior axis elongation (Fritzenwanker et al., 2019). While most of the focus of the comparative work has been on developmental patterning, many of the findings have had implications for testing classical hypotheses of morphological affinities with chordates such as whether the dorsal collar cord or stomochord are homologous to the chordate dorsal cord and stomochord respectively.

The molecular data has largely not supported these classical hypotheses (Lowe et al., 2003, 2006; Satoh et al., 2014), with one notable exception; the pharyngeal gill slits. Anatomical, developmental and genomic data all support gills as a deuterostome-specific trait (Gillis, Fritzenwanker, & Lowe, 2012; Simakov et al., 2015).

More recently the *Saccoglossus* community has worked to expand genetic tools following the publication of the genome (Simakov et al., 2015). Chip-Seq and ATAC-seq can be used to uncover promoter & enhancer regions using markers for open and closed chromatin, to broaden our understanding of gene regulation at key developmental stages. These regions can be tested by transgenesis in *Saccoglossus* (Minor et al., 2019) or tested across broad phylogenetic distances using cross-species enhancer swaps to look at differences/conservation of gene regulation and how that relates to the evolution of morphology. This is particularly useful given the morphological differences between the *Saccoglossus* and vertebrate nervous systems. In a striking example, *Saccoglossus* and mouse shh enhancers were validated to function reciprocally in patterning expression in the brain (Yao et al., 2016). Understanding changes in gene regulation at this level can begin to untangle the complex regulatory mechanisms underlying evolutionary change within the deuterostome lineage.

4. Expanding horizons

While past work has established *Saccoglossus* as a strong comparative developmental model within the deuterostomes, there are areas where one can imagine pushing the boundaries further. Most obviously, this includes establishing a standard protocol for CRISPR technology. In the absence of lab-based multi-generational husbandry, analysis is currently restricted to F0 animals (transient analysis), but these will still provide invaluable insight into development. Preliminary work has been done in this direction and will presumably be integrated into the *Saccoglossus* toolkit in the near term. The recently developed transgenic technology can also now be integrated into larger research questions (Minor et al., 2019). One unexplored option is to use existing BAC libraries to ensure accurate transgene expression. This approach would also be a valuable tool for discovering important regulatory elements for genes of interest and better refine comparative gene regulatory networks as has been shown in echinoderms (Buckley, Dong, Cameron, & Rast, 2018).

At the core of our research programs up to this point has been the ability to perturb gene function and interpret the effects on developmental pathways. Until recently these interpretations have been limited primarily to analysis of selected candidate or marker genes—by *in situ* or qPCR. We have now begun to apply RNA-Seq and, more recently, single-cell transcriptomics for downstream analyses. These approaches can provide unbiased insight into the broader networks and developmental trajectories underlying differences with vertebrates and even uncover previously unsuspected genes of importance. Analyses are underway using these approaches to address the fundamental question of the evolution of ectoderm from distributed neural cell lineages during development.

Now that we have a solid foundation for developmental studies we can also begin to look beyond the embryo. As in many species, experimental targeting of postembryonic growth has many challenges. There are a wide range of developmental processes that would become accessible experimentally with the application of targeted, specific disruption of gene function at late developmental stages, including regeneration, mesoderm patterning and late neural patterning. We can begin now to develop the approaches that will make these questions accessible as well.

Lastly, the genome continues to be improved, with the integration of newer sequencing technologies such as PacBio long reads and the utilization of chromosome conformation capture (Hi-C) to establish scaffolds and gain insight into large-scale genomic structure and organization. An improved genome and gene models will further seed progress and enable more focused comparisons of regulatory elements of interest. As these resources grow, however, so does a need for investment in bioinformatics infrastructure. Such an infrastructure will not only aid *Saccoglossus* researchers, but also make it easier to share information with the broader scientific community and increase intellectual accessibility of the organism.

5. Building a research community

Saccoglossus is a seasonal spawner and attempts to "domesticate" the animal have failed thus far (although it remains an area of interest). As a result, a pop-up community of researchers convenes at the MBL one or two times a year in sync with seasonal access to gravid worms. To establish an efficient work environment for doing research, we have found it most productive to work together across lab groups—forming a transient "acorn worm" metagroup. Labs and individuals work together to collect worms, and coordinate to induce spawning and collect eggs in the lab at MBL. The group also establishes an organizing schedule for when to collect animals relative to changing tides. While the work schedule remains demanding (somewhat by necessity for an organism that is accessible for experiments only a few weeks at a time), the addition of programmable incubators and improved husbandry techniques has eased the overall workload. The coordination of labor has been a crucial element and allows the *Saccoglossus* team to spend each day contributing to a successful collecting season while managing individual research goals. This teamwork approach has been one of the critical pieces that has allowed researchers to push forward on multiple fronts.

This community focus extends to larger scale method and resource building. Given the seasonal constraints of working with *Saccoglossus* we have also adopted an approach of building tools in parallel to pursuing more immediate experimental goals. A central focus from the beginning has been improving injection protocols to increase the number of successfully injected embryos. Similarly, significant progress has been made in husbandry and the reliable production of eggs. Once learning the basics of *Saccoglossus* protocols everyone is encouraged to optimize methods and share new techniques. As a result, the tools themselves continue to evolve (and improve) with each season. More recently this approach has extended to developing the next generation of methods such as transgenics, which will be valuable tools for the entire community going forward. While a new technique may be the goal of one person (or a lab), encouraging multiple people's contributions can increase and improve data for validating new techniques. This approach continually builds resources for the next set of experimental goals.

The MBL has also become an integral factor in the success of *Saccoglossus*. The *Saccoglossus* group now has a seasonally dedicated space at the MBL that incorporates space for sea tables with running sea water to maintain spawned adults for egg collection, group microscopy for screening of injected animals, and individual lab workstations for the intense multi-week period of experiments. This shared space further facilitates the sharing of information and coordination during the season.

In some ways, these field seasons have resulted in a return to more traditional rhythms of collegial research—with focused bursts of sample collection during spawning seasons; leaving the rest of the year to analyze data and plan future experiments. Additionally, the common experience of a field season encourages real time sharing of results to encourage discussion of hypotheses and plan new strategies.

6. Take aways

Biologists increasingly recognize that the standard model organisms are insufficient for answering particular evolutionary questions and for providing a broad, unbiased view of animal diversity. Choosing the "right" organism is a first step towards answering these questions. We have worked to develop *Saccoglossus* as a research model because it is uniquely positioned to provide insight into deuterostome evolution and chordate origins. As we learn more about development across the spectrum of animal diversity, we can begin to identify what makes individual species and lineages unique and to determine broad principles underlying developmental and evolutionary change.

Choosing the right organism often means working with an animal that does not have a history and community of research to provide support or information. More recently tools such as cheap accessible sequencing (genomic and RNA) and CRISPR have made it feasible to work in almost any animal. Still, a bottleneck for experimental work in novel embryos is establishing the "basics" like optimizing husbandry and embryo care. First steps should include establishing protocols for working with your organism in the most basic manner-if you can raise and breed the animal in the lab, all the better. It is equally essential in the modern molecular genetic era to identify genes, determine their pattern and timing of expression and create tools to modify them in order to reveal underlying developmental networks. The specific approaches for each of these stages of model development will vary from organism to organism and be determined in part by their own unique strengths and limitations.

The path of progress for *Saccoglossus* has not been without obstacles. Success has been as dependent on luck and persistence as conscious decisions. Up to this point the number of researchers working with the animal worldwide at any given time has been small relative to more "mature" model organisms but an agreement to work as a community has assisted greatly in rapidly establishing *Saccoglossus* as a model organism. Research is accelerating as we build off the foundations established over the last several years and expand collaborations with experts in other fields.

Perhaps not surprisingly discoveries in *Saccoglossus* have led to more questions amid newly discovered complexities. Homologies suggested by anatomy such as the stomochord relationship to the notochord have not held up with closer developmental and genetic scrutiny. Similarly, the search for signatures of vertebrate-specific cell types such as neural crest has thus far failed to identify an ancestral precursor. The relatively straightforward conservation of AP neural patterning gene order is complicated by fundamental differences in morphology. And the separation of ectodermal and nervous system patterning leaves unanswered the evolutionary relationships between conserved patterning genes and final cell type or tissue determination. As we uncover more information about *Saccoglossus* development we anticipate finding additional areas of divergence that can be explored to learn about chordate and deuterostome origins. A recently published paper has suggested that hemichordates may in fact not be members of the deuterostome clade, but instead places them apart from both protostomes and chordates (Kapli et al., 2021), at a phylotypically advantageous position to illuminate the cnidarian-bilaterian transition, while remaining ancestral to chordates, just more distant. This evolving deuterostome phylogeny may affect interpretations of data moving forward, but would not alter the usefulness of developing *Saccoglossus* as a model.

Saccoglossus is entering a new phase as a research organism. While the last 20 years have allowed us to address some of the long-standing questions of vertebrate evolution, we expect we will continue to discover unanticipated complexities and twists in the "path" of evolution as we learn more. In addition, we expect that *Saccoglossus* can contribute to new fields including post-embryonic development and regeneration, as well as novel comparisons of life history driven evolutionary consequences. The last two decades have put us in a strong position to fulfill the promise first hinted at by Bateson almost 150 years ago.

Acknowledgments

We would like to thank Marc Kirschner and John Gerhart for initiating the *Saccoglossus* research program and the *Saccoglossus* research community that has contributed to its further development: Jochanan Aronowicz, Sebastien Darras, Robert M. Freeman, Andrew Gillis, Paul Gonzalez, Stephen A. Green, Kenneth Halanych, Rindy Jaffe, Kristen Kwan, José Andrade López, Heather Marlow, Sabrina Kaul-Strehlow, Paul Minor, Tetsuto Miyashita, Rachael Norris, Ariel Pani, Leonid Peshkin, Ryan Range, Elena Silva, Francois Spitz, Thomas Stach, Michael Tassia, Mark Terasaki, Andrew Verardo, Marcin Wlizla; We are also indebted to our MBL research family: Louis Kerr, Herb Luther, Jim McIlvain (Zeiss), Adair Oesterle (Sutter Instruments), Martha Peterson, Rudi Rottenfusser (Zeiss), Ann Woolford, the leadership of MBL, and Pam and Wallace Stark; and finally the funding agencies that have supported members of the team: NASA, NSF, NIH, CZI.

References

Arendt, D., & Nubler-Jung, K. (1994). Inversion of dorsoventral axis? *Nature, 371*, 26.

Aronowicz, J., & Lowe, C. J. (2006). Hox gene expression in the hemichordate *Saccoglossus kowalevskii* and the evolution of deuterostome nervous systems. *Integrative and Comparative Biology, 46*, 890–901.

Bateson, W. (1884). Early stages in the development of Balanoglossus (sp. incert.). *Quarterly Journal of Microscopical Science, 24*, 208–236.

Bateson, W. (1885). Later stages in the development of *Balanoglossus Kowalevskii* with a suggestion as to the affinities of the Enteropneusta. *Quarterly Journal of Microscopical Science, 25*, 81–128.

Bateson, W. (1886). The ancestry of the chordata. *Quarterly Journal of Microscopical Science, 26*, 535–571.

Bolker, J. (2012). Model organisms: There's more to life than rats and flies. *Nature, 491*, 31–33.

Bourlat, S. J., Juliusdottir, T., Lowe, C. J., Freeman, R., Aronowicz, J., Kirschner, M., et al. (2006). Deuterostome phylogeny reveals monophyletic chordates and the new phylum Xenoturbellida. *Nature, 444*, 85–88.

Bromham, L. D., & Degnan, B. M. (1999). Hemichordates and deuterostome evolution: Robust molecular phylogenetic support for a hemichordate + echinoderm clade. *Evolution & Development, 1*, 166–171.

Buckley, K. M., Dong, P., Cameron, R. A., & Rast, J. P. (2018). Bacterial artificial chromosomes as recombinant reporter constructs to investigate gene expression and regulation in echinoderms. *Briefings in Functional Genomics, 17*, 362–371.

Cameron, C. B., Garey, J. R., & Swalla, B. J. (2000). Evolution of the chordate body plan: New insights from phylogenetic analyses of deuterostome phyla. *Proceedings of the National Academy of Sciences of the United States of America, 97*, 4469–4474.

Cannon, J. T., Kocot, K. M., Waits, D. S., Weese, D. A., Swalla, B. J., Santos, S. R., et al. (2014). Phylogenomic resolution of the hemichordate and echinoderm clade. *Current Biology: CB, 24*, 2827–2832.

Colwin, A. L., & Colwin, L. H. (1950). The developmental capacities of separated early blastomeres of an enteropneust, *Saccoglossus kowalevskii*. *Journal of Experimental Zoology, 115*, 263–295.

Colwin, A. L., & Colwin, L. H. (1951). Relationships between the egg and larva of *Saccoglossus kowalevskii* (enteropneusta): Axes and planes; general prospective significance of the early blastomeres. *Journal of Experimental Zoology, 117*, 111–137.

Colwin, A., & Colwin, L. (1953). The normal embryology of *Saccoglossus kowalevskii*. *Journal of Morphology, 92*, 401–453.

Colwin, L. H., & Colwin, A. L. (1962). Induction of spawning in Saccoglossus kowalevskii (Enteropneusta) at Woods Hole. *Biological Bulletin, 123*. 493–493.

Colwin, A. L., & Colwin, L. H. (1963a). Role of the gamete membranes in fertilization in Saccoglossus kowalevskii (Enteropneusta). I. The acrosomal region and its changes in early stages of fertilization. *Journal of Cell Biology, 19*, 477–500.

Colwin, L. H., & Colwin, A. L. (1963b). Role of the gamete membranes in fertilization in Saccoglossus kowalevskii (Enteropneusta). II. Zygote formation by gamete membrane fusio. *Journal of Cell Biology, 19*, 501–518.

Cunningham, D., & Casey, E. S. (2014). Spatiotemporal development of the embryonic nervous system of *Saccoglossus kowalevskii*. *Developmental Biology, 386*, 252–263.

Darras, S., Fritzenwanker, J. H., Uhlinger, K. R., Farrelly, E., Pani, A. M., Hurley, I. A., et al. (2018). Anteroposterior axis patterning by early canonical Wnt signaling during hemichordate development. *PLoS Biology, 16*, e2003698.

Darras, S., Gerhart, J., Terasaki, M., Kirschner, M., & Lowe, C. J. (2011). Beta-catenin specifies the endomesoderm and defines the posterior organizer of the hemichordate Saccoglossus kowalevskii. *Development, 138*, 959–970.

De Robertis, E. M., & Sasai, Y. (1996). A common plan for dorsoventral patterning in Bilateria. *Nature, 380*, 37–40.

Dunn, C. W., Hejnol, A., Matus, D. Q., Pang, K., Browne, W. E., Smith, S. A., et al. (2008). Broad phylogenomic sampling improves resolution of the animal tree of life. *Nature, 452*, 745–749.

Freeman, R. M., Wu, M., Cordonnier-Pratt, M.-M., Pratt, L. H., Gruber, C. E., Smith, M., et al. (2008). cDNA sequences for transcription factors and signaling proteins of the hemichordate *Saccoglossus kowalevskii*: Efficacy of the expressed sequence tag (EST) approach for evolutionary and developmental studies of a new organism. *The Biological Bulletin, 214*, 284–302.

Fritzenwanker, J. H., Uhlinger, K. R., Gerhart, J., Silva, E., & Lowe, C. J. (2019). Untangling posterior growth and segmentation by analyzing mechanisms of axis elongation in hemichordates. *Proceedings of the National Academy of Sciences of the United States of America, 116*, 8403–8408.

Furlong, R. F., & Holland, P. W. H. (2002). Bayesian phylogenetic analysis supports monophyly of ambulacraria and of cyclostomes. *Zoological Science (Tokyo), 19*, 593–599.

Geoffroy Saint-Hilaire, E. (1822). Considérations générales sur les vertebrés. *Mémoires du Muséum d'histoire Naturelle, 9*, 89–119.

Gillis, J. A., Fritzenwanker, J. H., & Lowe, C. J. (2012). A stem-deuterostome origin of the vertebrate pharyngeal transcriptional network. *Proceedings of the Royal Society B: Biological Sciences, 279*, 237–246.

Green, S. A., Norris, R. P., Terasaki, M., & Lowe, C. J. (2013). FGF signaling induces mesoderm in the hemichordate *Saccoglossus kowalevskii*. *Development, 140*, 1024–1033.

Hall, B. K. (2005). Betrayed by Balanoglossus: William Bateson's rejection of evolutionary embryology as the basis for understanding evolution. *Journal of Experimental Zoology, 304B*, 1–17.

Hejnol, A., & Lowe, C. J. (2015). Embracing the comparative approach: How robust phylogenies and broader developmental sampling impacts the understanding of nervous system evolution. *Philosophical Transactions of the Royal Society of London. Series B, Biological Sciences, 370*.

Jaffe, L. A., & Terasaki, M. (2004). Quantitative microinjection of oocytes, eggs, and embryos. *Methods in Cell Biology, 74*, 219–242.

Kapli, P., Natsidis, P., Leite, D. J., Fursman, M., Jeffrie, N., Rahman, I. A., et al. (2021). Lack of support for Deuterostomia prompts reinterpretation of the first Bilateria. *Science Advances, 7*.

Kowalevsky, A. (1866). Anatomie des Balanoglossus. *Mémoires de l'Académie Impériale des Sciences de Saint Pétersbourg, 7*, 16.

Lowe, C. J. (2008). Molecular genetic insights into deuterostome evolution from the direct-developing hemichordate *Saccoglossus kowalevskii*. *Philosophical Transactions of the Royal Society of London. Series B, Biological Sciences, 363*, 1569–1578.

Lowe, C. J. (2021). Molecular insights into deuterostome evolution from hemichordate developmental biology. *Current Topics in Developmental Biology, 141*, 75–117.

Lowe, C. J., Tagawa, K., Humphreys, T., Kirschner, M., & Gerhart, J. (2004). Hemichordate embryos: Procurement, culture and basic methods. *Methods in Cell Biology, 74*, 171–198.

Lowe, C. J., Terasaki, M., Wu, M., Freeman, R. M., Jr., Runft, L., Kwan, K., et al. (2006). Dorsoventral patterning in hemichordates: Insights into early chordate evolution. *PLoS Biology, 4*, e291.

Lowe, C. J., Wu, M., Salic, A., Evans, L., Lander, E., Stange-Thomann, N., et al. (2003). Anteroposterior patterning in hemichordates and the origins of the chordate nervous system. *Cell*, *113*, 853–865.

Minor, P. J., Clarke, D. N., López, J. M. A., Fritzenwanker, J. H., Gray, J., & Lowe, C. J. (2019). I-SceI meganuclease-mediated transgenesis in the acorn worm, *Saccoglossus kowalevskii*. *Developmental Biology*, *445*, 8–15.

Morgan, T. (1894). Development of Balanoglossus. *Journal of Morphology*, *9*, 1–86.

Pani, A. M., Mullarkey, E. E., Aronowicz, J., Assimacopoulos, S., Grove, E. A., & Lowe, C. J. (2012). Ancient deuterostome origins of vertebrate brain signalling centres. *Nature*, *483*, 289–294.

Reichert, H., & Simeone, A. (2001). Developmental genetic evidence for a monophyletic origin of the bilaterian brain. *Philosophical Transactions of the Royal Society of London. Series B, Biological sciences*, *356*, 1533–1544.

Rottinger, E., & Lowe, C. J. (2012). Evolutionary crossroads in developmental biology: Hemichordates. *Development*, *139*, 2463–2475.

Sato, A., Bishop, J. D., & Holland, P. W. (2008). Developmental biology of pterobranch hemichordates: History and perspectives. *Genesis*, *46*, 587–591.

Satoh, N., Tagawa, K., Lowe, C. J., Yu, J. K., Kawashima, T., Takahashi, H., et al. (2014). On a possible evolutionary link of the stomochord of hemichordates to pharyngeal organs of chordates. *Genesis*, *52*, 925–934.

Simakov, O., Kawashima, T., Marletaz, F., Jenkins, J., Koyanagi, R., Mitros, T., et al. (2015). Hemichordate genomes and deuterostome origins. *Nature*, *527*, 459–465.

Tagawa, K. (2016). Hemichordate models. *Current Opinion in Genetics & Development*, *39*, 71–78.

Tassia, M. G., Cannon, J. T., Konikoff, C. E., Shenkar, N., Halanych, K. M., & Swalla, B. J. (2016). The global diversity of hemichordata. *PLoS One*, *11*, e0162564.

Yao, Y., Minor, P. J., Zhao, Y.-T., Jeong, Y., Pani, A. M., King, A. N., et al. (2016). Cis-regulatory architecture of a brain signaling center predates the origin of chordates. *Nature Genetics*, 1–8.

CHAPTER TWENTY

The invertebrate chordate amphioxus gives clues to vertebrate origins

Linda Z. Holland* and Nicholas D. Holland
Marine Biology Research Division, Scripps Institution of Oceanography, University of California San Diego, La Jolla, CA, United States
*Corresponding author: e-mail address: lzholland@ucsd.edu

Contents

Abstract

Amphioxus (cepholochordates) have long been used to infer how the vertebrates evolved from their invertebrate ancestors. However, some of the body part homologies between amphioxus and vertebrates have been controversial. This is not surprising as the amphioxus and vertebrate lineages separated half a billion years ago–plenty of time for independent loss and independent gain of features. The development of new techniques in the late 20th and early 21st centuries including transmission electron microscopy and serial blockface scanning electron microscopy in combination with in situ hybridization and immunocytochemistry to reveal spatio-temporal patterns of gene expression and gene products have greatly strengthened inference of some homologies (like those between regions of the central nervous system), although others (like nephridia) still need further support. These major advances in establishing homologies between amphioxus and vertebrates, together with strong support from comparative genomics, have firmly established amphioxus as a stand-in or model for the ancestral vertebrate.

Current Topics in Developmental Biology, Volume 147
ISSN 0070-2153
https://doi.org/10.1016/bs.ctdb.2021.12.011

1. Introduction

The title of this volume *Emerging model systems in developmental biology* raises the question of just what is a model system? The term began appearing in biomedical publications in the 1950s. At first, it referred to non-human mammals and procedures applied to them for generating information relevant to human health. Subsequently, some non-mammalian animals were added to the list of models because of their advantages for elucidating cellular and molecular mechanisms underlying human disease. By now, "model system" has acquired quite a different meaning in the field of evodevo, where it is applied to extant animals that can be considered proxies for long-extinct ancestors of modern clades. For example, we use cephalochordates (amphioxus or lancelets) as a stand-in or model for the ancestral chordate to illustrate conditions at the beginning of vertebrate evolution. This approach is favored by the extremely slow rate of evolution in cephalochordates (Yue, Yu, Putnam, & Holland, 2014) and the general structural similarities of body plans in chordates. In spite of these advantages, however, it is important to keep two things always in mind: first that the cephalochordate and vertebrate lineages split half a billion years ago, which is plenty of time for independent loss or gain of characters in both lineages and second that, on a greatly compressed time-scale, amphioxus is related to us as our great uncle—not as our grandfather.

Pallas (1774), who initially described amphioxus, mistakenly classified it with sea slugs. Subsequently, however, Costa (1834) correctly recognized its affinities with fishes. During the rest of the 19th century, amphioxus was regarded by some researchers as a simple, but honest fish, but by others as a debased creature that had degenerated from a properly fishy ancestor. Kowalevsky (1867), who first described amphioxus development, found that its cleavage stages and embryos closely resembled those of developing sea urchins, but the subsequent larval stages had much in common with developing vertebrates. In his 1867 paper, Kowalevsky avoided phylogenetic speculations, but even before his embryological results were published, his friend Metschnikov (1866) began stressing the importance of amphioxus for understanding the invertebrate-to-vertebrate transition from a Darwinian point of view, an evolutionary scenario that Haeckel later raised virtually to the status of a cult (Hopwood, 2015; Stokes & Holland, 1998).

During the latter part of the 19th century, the phylum Chordata was usually considered to include vertebrates (where amphioxus resided as the simplest fish), tunicates, and often hemichordates as well (Bateson, 1885).

Ultimately, Herdman (1904) divided the chordates into three subphyla: Cephalochordata, Tunicata, and Vertebrata. Amphioxus has been considered to be an invertebrate chordate ever since. The present review discusses how the use of modern cephalochordates as models or substitutes for long-extinct ancestral chordates has provided insight into the evolution of vertebrates from their invertebrate ancestors.

2. The phylogenetic position of cephalochcordates

In the 20th century, most classifications based on morphology placed amphioxus as sister group of the vertebrates, and relegated tunicates, with their very simple larvae, to a basal position within the chordates. Phylogenetic analyses with 18S rDNA seemed at first to support this arrangement (Turbeville, Schulz, & Raff, 1994; Wada & Satoh, 1994). However, subsequent analyses with large sets of nuclear genes, which generally evolve relatively slowly, convincingly reversed the positions of tunicates and amphioxus, placing tunicates as the sister group of vertebrates, and situating amphioxus as the basal branch of the Chordata (Fig. 1) (Delsuc, Brinkmann, Chourrout, & Philippe, 2006). Additional studies revealed that, compared to amphioxus and vertebrates, tunicates have exceptionally high rates of evolution (Delsuc et al., 2018; Giribet, 2018; Tsagkogeorga, Turon, Galtier, Douzery, & Delsuc, 2010) along with a considerable loss of genes, which may help explain why the five major clades of tunicates (appendicularians, ascidians, salps, doliolids, and pyrosomes) are so very diverse. Not only are the five groups morphologically quite different (Fig. 1), they occupy benthic habitats ranging from the intertidal zone to the deep sea as well as midwater habitats over a wide bathymetric range. Tunicate life styles are also diverse. Ascidians are sessile while the other four groups are pelagic; moreover, although most tunicates are filter feeders, some doliolids and ascidians occupying the deep sea capture large prey (Monniot & Monniot, 1991; Robison, Raskoff, & Sherlock, 2005). Tunicates also have a variety of reproductive modes; appendicularians reproduce only sexually, but most of the others have a complex alternation of sexual and asexual generations. In ascidians, asexual reproduction my well be related to their pronounced ability to regenerate.

As noted above, vertebrates and amphioxus are evolving relatively slowly—amphioxus even more slowly than the vertebrates (Yue et al., 2014). As a likely consequence, amphioxus and vertebrates share far more characters with each other than either does with any tunicate. There are only three cephalochordate genera (*Branchiostoma, Asymmetron* and *Epigonichthys*).

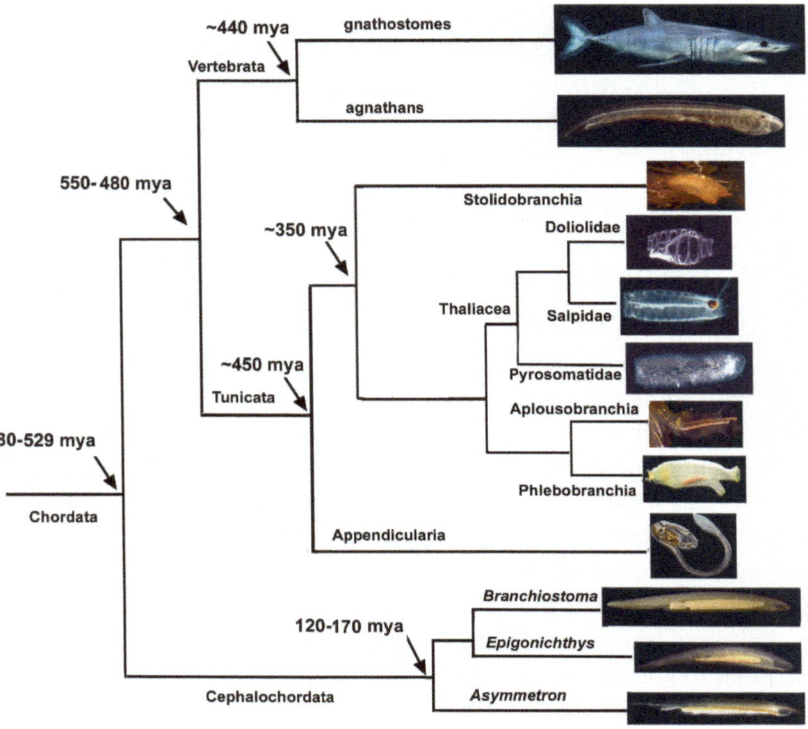

Fig. 1 Chordate phylogeny showing the three genera of cephalochordates (amphioxus), the five clades of tunicates (Stolidobranchia, Thaliacea, Aplousobranchia, Phlebobranchia, Appendicularia) and the two main groups of vertebrates–jawless fish (agnathans) and jawed vertebrates (gnathostomes), exemplified by the lamprey and shark, respectively. The cephalochordate and vertebrate plus tunicate split in the Cambrian while tunicates and vertebrates split in the early Ordovician.

All are morphologically similar, all reproduce exclusively sexually, and, with the exception of one population of *Asymmetron interferum* located at ~200 m near a whale-fall off Japan (Kon et al., 2007) all occupy similar near-shore habitats. The most obvious intergenic difference is that *Asymmetron* and *Epigonichthys* have gonads only on the right side of the animal, whereas *Branchiostoma* has a row of gonads on each side of the body. Most evodevo studies on amphioxus have concerned three of the approximately 30 species of *Branchiostoma* (*B. floridae, B. belcheri* and *B. lanceolatum*) and one of *Asymmetron (A. lucayanum)*.

Amphioxus has a much simpler body plan than vertebrates. It resembles a small, eyeless, limbless, jawless fish. The amphioxus central nervous system (CNS) is estimated to have only about 20,000 neurons (Nicol & Meinertzhagen, 1991), compared to hundreds of thousands in some fishes and up to billions in large mammals. In addition, the ~500-mb genomes of *Branchiostoma* species are small in comparison to those of vertebrates, which, with the exception of secondarily reduced genomes in puffer fish, typically range from 1.5 gb to >3 gb. The differences in genome sizes are due chiefly to non-protein-coding DNA. The number of genes in amphioxus species is between 20,000–24,000 (Putnam et al., 2008)—about the same number estimated for the human genome (Salzberg, 2018).

3. Rate of evolution and developmental mode

The rate of evolution and the developmental mode appear to be correlated. Slow-evolving bilaterians like amphioxus and vertebrates have indeterminate cleavage and regulative development, in which cell fates become fixed only gradually as development proceeds; this mode is likely ancestral for bilaterians (Valentine, 1997). In contrast, the fast-evolving bilaterians like tunicates and nematodes have fixed cell lineages and very early decisions of cell fate (mosaic or determinate development). Thus, for several vertebrates, mutation or inhibition of the function of early-expressed genes, compared to late-expressed ones, gives a considerably higher percentage of lethal phenotypes (Roux & Robinson-Rechavi, 2008; Uchida, Uesaka, Yamamoto, Takeda, & Irie, 2018). Similarly, early-expressed genes are evolving more slowly than later-expressed ones in the insect *Drosophila* (Artieri, Haerty, & Singh, 2009). Another study showed that there were relatively few duplicates and newly evolved genes among those expressed in early zebrafish development, although regulatory elements for transcription factors were most conserved in mid-development (Piesecka, Lichocki, Moretti, Beregmann, & Robinson-Rechavi, 2013). The same authors found that the lowest level of conservation of several parameters– gene sequences, gene age, expression, orthology and regulatory elements–was in late larvae and adults. In agreement with these results, Liu and Robinson-Rechavi (2018) compared zebrafish, mouse, *Caenorhabditis elegans* and *Drosophila,* and found that in general, phyletic age and duplicability were most conserved in early development, while sequence evolution was most conserved in mid-development; however, this study excluded genes expressed very early in development as they include a high percentage of maternal genes. Taken together, these studies suggest that a major constraint on the rate of evolution is the developmental mode—that is only after cell fates are determined are mutations more likely to be non-lethal and possibly selected for. Thus, when cell fates are determined very, very early as in tunicates or *C. elegans*, there is likely more opportunity for mutations in later-expressed genes to become fixed and mediate phenotypic change, than when cell fates are determined gradually over a longer period of development as in amphioxus and vertebrates.

The slow evolutionary rate makes amphioxus the best available substitute or model for the long-extinct ancestral chordate to address the question of how vertebrates evolved from their invertebrate ancestors. Extant amphioxus species have many similarities to Cambrian fossils proposed to have cephalochordate affinities like *Pikaia, Haikouella, Myllokunmingia, Yunnanozoa* and *Metaspriggina* (Conway Morris & Caron, 2014; Mallatt & Chen, 2003), although, unlike amphioxus, most of these fossils (*Pikaia* is an exception) clearly have paired eyes (Fig. 2). Most consequentially, the moderate degree

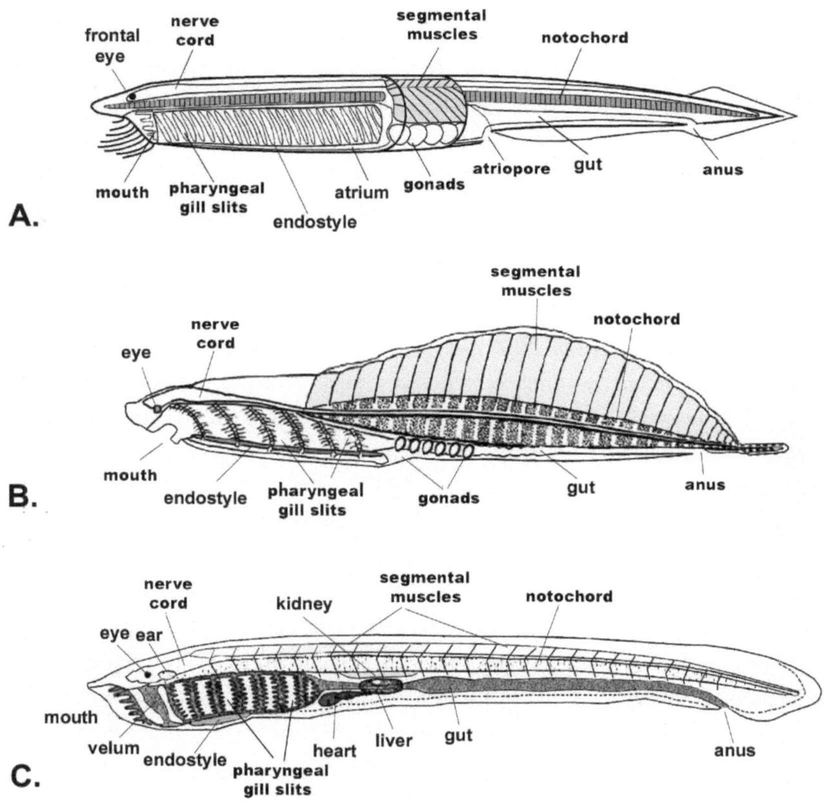

Fig. 2 Diagrams of (A) amphioxus (after Holland, Laudet, & Schubert, 2004), (B) *Haikouella*, a possible early craniate from the Cambrian, (C). lamprey larva (ammocoete). Anterior to the left. *Panel B: Adapted from fig. 1A in Mallatt, J., & Chen, J. Y. (2003). Fossil sister group of craniates: Predicted and found.* Journal of Morphology, 258, *1–31.*

of divergence of amphioxus and vertebrates, both in morphology and in molecular genetics, is favorable when considering homologies. Yue et al. (2016) have referred to the strategy of comparing moderate differences—those close enough to be recognized as similar, but different enough to avoid being boringly obvious—as the "Goldilocks principle."

4. Insights amphioxus has yielded about how vertebrates evolved

Here we will discuss the evolution of three structures—the central nervous system (CNS), neural crest, and kidney—to illustrate the problems of inferring homologies over large phylogenetic distances and how progress has been made as new techniques have been developed. The arguments concerning evolution of these structures are long-standing. They were initially addressed in the 19th and early 20th centuries with light microscopy. However, marked progress toward answering these and other long-standing questions about how the vertebrates evolved from an amphioxus-like ancestor awaited techniques developed in the last half of the 20th century including transmission electron microscopy (TEM)—especially 3-D reconstructions from volume electron microscopy, and developmental genetics, in which developmental gene expression patterns are used as phenotypic characters.

4.1 The techniques

No single technique can provide firm insights into questions of homology—that is whether two features descended from the same feature in a common ancestor. However, inferences of homology are strengthened when two or more methods point to the same answer. The approaches of TEM and developmental genetics each have their advantages and disadvantages.

4.1.1 Advances in microscopy

TEM provides very high resolution. However, it is very time-consuming to create 3-D images from serial TEM images, and without such an overview, the detailed relationships among structures can easily be missed. A 21st century variation on the technique is serial blockface scanning microscopy (SBSEM), which is less time-consuming and lends itself to construction of 3-D images. In SBSEM, specimens are fixed and embedded as for TEM but with contrast-enhancing material in the fixative (Deerinck, Bushong, Lev-Ram, Tsien, & Ellisman, 2010). The specimen chamber of the SEM includes a microtome (e.g., Gatan 3-view). A section, typically 50–250 nm thick, is cut from the top of the block to expose a new blockface, which is scanned to yield backscattered electrons that are recorded as images. The alternation of cutting and scanning is repeated to create a series of images, each looking superficially like a transmission electron micrograph. Resolution is usually not quite as good as that of conventional TEM, but a series of hundreds to thousands of images can be collected in several days to a few weeks. Computer programs can then generate 3-D reconstructions from the data set. This method has been applied to cephalochordates to reveal fine-structural details of cell populations occupying large tissue volumes of the larval kidney (Holland, 2018), adult peripheral nervous systyem (Holland & Somorjai, 2020a), and adult notochord (Holland & Somorjai, 2020b), but has not yet been applied to the amphioxus CNS.

4.1.2 Molecular genetics

Gene-expression patterns determined by in situ hybridization or antibody labeling contribute new morphological characters useful for inferring homologies. Here one simply uses the expression of a gene as a character of special quality and relative position, which are two of the chief criteria for homology (Remane, 1956). If one or more genes are expressed in corresponding organs, then proposed homologies are strengthened. One advantage of making comparisons within a single phylum—Chordata in our case—is that overall body plans are similar enough so that there is rarely any question about which of the component organs in the species compared correspond to one another. It is likely that this approach is effective because the genes transcribed in a particular body region are components of (and consequently proxies for) gene regulatory networks directing development and maintenance of the phenotype of the same region. Such suppositions are probably valid and will likely be confirmed someday—but only after a daunting amount of additional work on the structure and function of genetic/epigenetic networks and their interrelationship with phenotypic characters.

In situ hybridization and antibody labeling come with caveats. First, many genes are expressed in a number of domains, some or all of which may change over the course of development; therefore, expression of a single gene in structures that are morphologically dissimilar, cannot be taken as firm evidence of homology. Second, there can be technical problems such as false positives or low signals, which are especially problematic for genes coding for trans-membrane proteins typically with a low turnover. Third, there are problems caused by patterns of gene evolution. Expression patterns can diverge over time, gene networks can change with the cooption of new genes into the network or the loss of old ones, and non-homologous structures can converge with the recruitment into them of parts of gene networks typically expressed elsewhere in the organism. For example, some genes such as *Nodal, Lefty, Pitx* interact as part of a core network mediating asymmetry whether it is between left and right as in amphioxus (Ono, Koop, & Holland, 2018; Soukup et al., 2015) and vertebrates (Tian & Meng, 2006) or in chirality as in a snail (Grande & Patel, 2009). As another example, *Pax, Six, Eya,* and *Dach* interact in a variety of developing organs and tend to be involved in cell proliferation and/or cell migration (Kozmik et al., 2007).

4.2 Evolution of the CNS

The use of gene expression (augmented with a wiring diagram for cephalochordate neuroanatomy) for elucidating evolution of chordate nerve cords has worked so well for larval amphioxus that it is fair to consider this organ system in amphioxus as a "poster child" for evodevo. The amphioxus nerve cord, like that of vertebrates, is derived from the dorsal embryonic ectoderm that folds up to form a tube—the dorsal, hollow nerve cord. However, unlike the vertebrate nerve cord, that of amphioxus does not have a large brain. The adult amphioxus brain has only a slightly enlarged anterior end termed the cerebral vesicle with an estimated 20,000 neurons (Nicol & Meinertzhagen, 1991) (Fig. 3), compared to over 100,000 neurons in the brain of a larval zebrafish and many more in the brains of adults (Keller, Ahrens, & Freeman, 2015).

Opinions concerning homology of the amphioxus and vertebrate nerve cords have run the gamut from the two being entirely homologous to their having no evolutionary relationship. For example, Ayers (1890) and Balfour (1885), who considered amphioxus to be a primitive vertebrate, argued that the anterior part of the amphioxus brain was essentially equivalent to the vertebrate brain but lacked a telencephalon, while others argued that amphioxus lacked any equivalent of the vertebrate brain (Berrill, 1955; Garstang, 1960). The argument as to the relationship of amphioxus to vertebrates was still unresolved in the late 20th century. For example, (Berrill, 1987) argued that amphioxus was not ancestral to vertebrates and that its nerve cord lacked anything that might be considered a rudimentary brain. In contrast, (Gans & Northcutt, 1983) considered the ancestral vertebrate to be amphioxus-like with a dorsal nerve cord that had a hindbrain but lacked a forebrain and midbrain.

Gene expression patterns together with TEM have done much to settle this argument; these studies have indicated that the amphioxus CNS

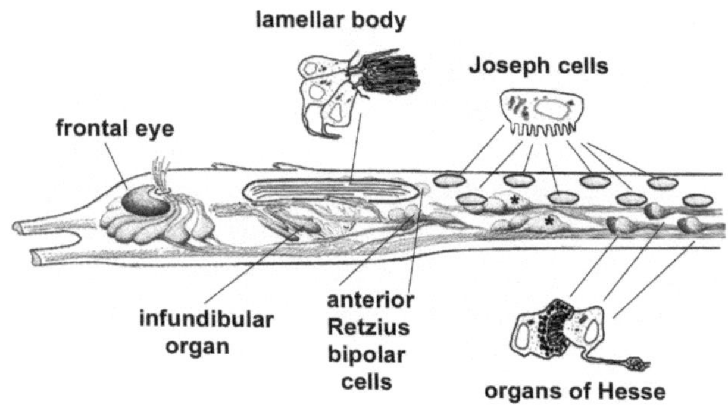

Fig. 3 Diagram of the anterior portion of the amphioxus CNS showing major features. Rostral nerves extend to the left. The four types of photoreceptors include two ciliary ones, the frontal eye, including a pigment cell and 12 ciliated photoreceptor cells, and the lamellar body as well as two types of microvillar photoreceptors—the Joseph cells and the organs of Hesse. The latter include pigment cells. Also shown are the infundibular organ and the Retzius bipolar cells, possibly homologous to the Rohon-Beard cells of vertebrates. Indicated with stars are two giant neurons in the primary motor center. *Adapted from fig. 1B in Lacalli, T. C. (1996). Landmarks and subdomains in the larval brain of Branchiostoma: Vertebrate homologs and invertebrate antecedants.* Israel Journal of Zoology, 42 *Supplement, 131–146. and fig. 2 in Lacalli, T. C. (1996). Frontal eye circuitry, rostral sensory pathways, and brain organization in amphioxus larvae: Evidence from 3D reconstructions.* Philosophical Transactions of the Royal Society B, 351, *243–263.*

includes likely homologs of the vertebrate diencephalic forebrain, a possible midbrain, a hindbrain and a spinal cord (Figs. 3 and 5). Genes involved in establishing the major anterior/posterior and medio/lateral divisions of the brain generally turn on at the gastrula through mid-neurula stages before there is overt differentiation of the neural tube. The divisions inferred from these gene expression patterns have complemented inferences from the microanatomy of larval brains. Taken together, these studies have indicated that the major divisions of the vertebrate brain were already established in the chordate ancestor.

The patterns of gene expression in the early amphioxus neural tube have recently been elaborated upon and reviewed (Albuixech-Crespo et al., 2017; Holland & Holland, 2021). In brief, nested expression of *Hox* genes in the developing CNS is comparable in amphioxus and vertebrates, indicating that amphioxus has a likely homolog of the vertebrate hindbrain and spinal cord (Holland, Holland, Williams, & Holland, 1992; Schubert, Holland, Laudet, & Holland, 2006). Similarly, abutting domains of anterior *Otx* expression and posterior *Gbx* expression at the posterior end of the amphioxus cerebral vesicle demarcates the equivalent of the midbrain/hindbrain boundary (MHB) (Castro, Rasmussen, Holland, Holland, & Holland, 2006) (Fig. 4). Other homologs of genes expressed at the vertebrate MHB are not expressed at the corresponding region in amphioxus larvae, suggesting that the likely MHB homolog in amphioxus does not have organizer ability. In addition, abutting *Fezf* and *Irx* expression domains in the cerebral vesicle have suggested a homolog of the zona limitans intrathalamica, but one lacking organizer properties as *Hedgehog*, which helps confer organizer properties to

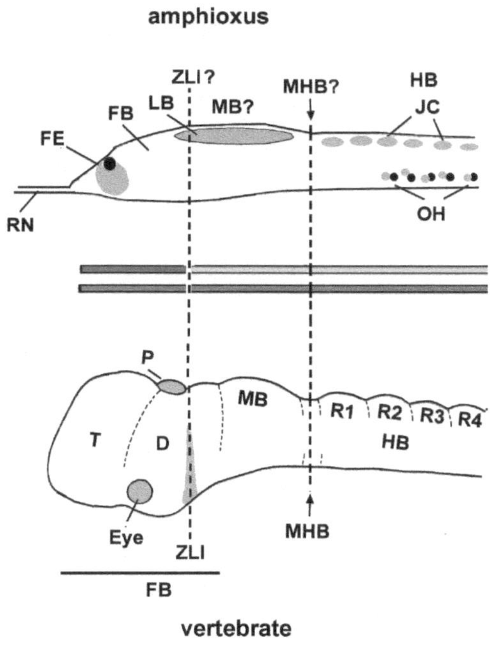

Fig. 4 Anterior/posterior patterning in the brains of amphioxus (*top*) and a generalized vertebrate (*bottom*). The Zona limitans intrathalamica (ZLI) is an organizing center located in the center of the vertebrate diencephalon at the boundary between anterior expression of *Fezf* and posterior expression of *Irx*. These genes are similarly expressed in the amphioxus brain. However, *Shh*, which is expressed at that boundary in vertebrates, is not similarly expressed in amphioxus. A second boundary is the midbrain/hindbrain boundary (MHB) marked by abutting domains of *Otx* and *Gbx*. However, genes that confer organizer properties on the vertebrate MHB such as *FGF* are not similarly expressed in amphioxus, suggesting that this region probably is not an organizer in amphioxus. Abbreviations; D, diencephalon; FB, forebrain; FE, frontal eye; HB, hindbrain; JC, Joseph cells; LB, lamellar body; MB, midbrain; OH, organs of Hesse; P, vertebrate pituitary; R1–4, rhombomeres (subdivisions of the hindbrain); RN, rostral nerve; T, telencephalon.

the vertebrate ZLI is not expressed at the boundary between the domains of *Fezf* and *Irx* in amphioxus (Irimia et al., 2010). Dorso-ventral patterning genes are also similarly expressed in the CNS of amphioxus and vertebrate (Albuixech-Crespo et al., 2017).

Microanatomy of differentiating amphioxus larvae and adults not only has confirmed the major CNS divisions inferred from gene expression, but has also revealed homologies of individual structures in the nervous systems of amphioxus and vertebrates. For a review of the morphology of the larval and adult nervous systems and probable homologies with the vertebrate CNS see (Wicht & Lacalli, 2005). These likely homologies have typically been supported by gene expression as the structures begin to differentiate. For example, Thurston Lacalli used serial TEM to make 3-D reconstructions of individual cells and their connections in the nerve cord of a late larva. Such a detailed study was possible because amphioxus larvae are very small (at metamorphosis they are only 4–5 mm long. First, Lacalli mapped the cell bodies and the axons of the medial photoreceptor located at the anterior tip of the CNS (Lacalli, 2013; Lacalli, Holland, & West, 1994). Although this ciliary photoreceptor is not image-forming, Lacalli argued that it is homologous to the paired eyes of vertebrates, which are ciliary photoreceptors located in the diencephalon, and that the posterior end of the amphioxus cerebral vesicle, where the axons of the photoreceptor neurons terminate, may be homologous to the vertebrate tectum, which receives input from the paired eyes (Lacalli, 1996). The idea of homology of the amphioxus frontal eye and the vertebrate paired eyes was strengthened by comparable patterns of gene expression. Not only do the photoreceptor cells of the frontal eye, like those of the vertebrate eyes, express *Pax6* before they are overtly differentiated (Glardon, Holland, Gehring, & Holland, 1998), they also express homologs of several other genes expressed in the developing vertebrate retina (Vopalensky et al., 2012).

Lacalli also did 3-D imaging of a second ciliary photoreceptor, the lamellar body, located dorsally near the posterior end of the cerebral vesicle (Fig. 3). This part of the cerebral vesicle also expresses *Pax6* early in development (Glardon et al., 1998). Structurally, the lamellar body closely resembles the larval pineal in the lamprey, which is located in the roof of the diencephalon (Lacalli et al., 1994; Pu & Dowling, 1981).

Another diencephalic structure in vertebrates is the subcommissural organ located dorsally near the pineal (Fig. 3). It secretes glycoproteins, especially SCO-spondin, into the cerebrospinal fluid, that form Reissner's fiber extending posteriorly down the lumen of the brain and spinal cord. Similar proteins are also produced ventrally in the larval floor plate and in the flexural organ at the anterior end of the floor plate (Meiniel et al., 2008; Richter et al., 2001; Rodríguez, del Brío León, Riera, Menendez, & Schoebitz, 1996). In the amphioxus cerebral vesicle, a fiber with the same immunohistochemistry as Reissner's fiber is secreted by cells located ventrally below the lamellar body at the anterior end of the floor plate (Olsson, 1955; Olsson, Yulis, & Rodriguez, 1994). Lacalli, following the terminology of Olsson (1962), called these cells "infundibular cells" (Fig. 3) (Lacalli & Kelly, 2002). Not surprisingly, like the vertebrate sub-commissural organ, these cells have recently been found to express the SCO-spondin gene (Bozzo et al., 2021). However, not only are the amphioxus infundibular cells immunocytochemically similar to the sub-commissural organ, they are also immunocytochemically like secretory cells in the vertebrate flexural organ (Olsson et al. (1994). Because SCO-spondin is made only by the ventrally-located infundibular cells in the amphioxus CNS but by both the dorsal sub-commissural organ and the ventral flexural organ as well as the floorplate itself in the vertebrate brain (Lopez-Avalos et al., 1997; Rodríguez et al., 1996), inference of homologies between the amphioxus infundibular cells and the vertebrate sub-commissural and/or flexural organs is complicated. Olsson (1958) suggested that the vertebrate sub-commissural and flexural organs and the amphioxus infundibular organ are all descended from a single large glandular area in the brain of an ancestral chordate. However, positional similarities between the vertebrate flexural organ and the amphioxus infundibulum suggest that the ancestral chordate brain may have had a single group of ventral cells secreting components of Reissner's fiber and leave open the question of the origins of the vertebrate sub-commissural organ.

Posterior/ventral to the lamellar body, Lacalli identified the "primary motor center" including two types of motor neurons—the DC or dorsal compartment ones that innervate superficial paraxial muscles and the VC or ventral compartment motorneurons that innervate the deep paraxial muscles. In early larvae, these manifest as two rows of cells, one on each side of the floor plate; the DC motor neurons are paired left and right, while the VC motor neurons are staggered (Lacalli & Kelly, 1999). In vertebrates, the branchial motor neurons, which innervate muscles derived from the branchial arches, are located in the hindbrain. A number of studies have shown that the amphioxus motor neurons express genes that are characteristically expressed in vertebrate motor neurons. These include VACHT and VGLUT, markers for cholinergic and glutamatergic neurons, respectively (Candiani, Moronti, Ramoino, Schubert, & Pestarino, 2012; Lacalli & Candiani, 2017) and several transcription factors including *OligA* and *Nkx6* (Ren et al., 2020).

However, the amphioxus CNS does not have homologs of all structures in the vertebrate CNS and vice versa—the vertebrate CNS lacks some structures present in the amphioxus CNS. For example, most authors consider that the amphioxus CNS lacks a telencephalon. In vertebrates, the telencephalon develops into the cerebrum. In lampreys and hagfish, the olfactory bulbs are at the anterior end of the cerebrum and the dorsal part of the cerebrum, the pallium, receives input from the eyes. Amphioxus lacks olfactory bulbs; it has been suggested that chemosensation may instead be due to scattered ectodermal sensory cells (Poncelet & Shimeld, 2020). A report on gene expression in the anterior CNS of adults by E. Benito-Gutierrez and colleagues, that is not yet formally published, argues that a telencephalon-like region forms in the amphioxus brain only after metamorphosis due to an expansion of late-differentiating neurons. While their data are consistent with amphioxus having a telencephalon-like region, vertebrate homologs of these amphioxus genes are expressed not only in the telencephalon, but also in other regions of the vertebrate brain. For example, in vertebrates, *Fox G1* is expressed in the telencephalon, and also in the optic stalks associated with the diencephalon, while *Lhx9* is also expressed in the diencephalon, mesencephalon and hindbrain, and *Lhx2* is also expressed in the diencephalon, pineal, mesencephalon, and hindbrain (Atkinson-Leadbeater, Bertolesi, Johnston, Hehr, & McFarlane, 2009; Chou & Tole, 2019; Smith et al., 2017). In addition, while some glutamatergic neurons are generated in the vertebrate telencephalon, others are generated in the upper rhombic lip in the hindbrain (Hibi & Shimizu, 2012). What is needed is a detailed neuroanatomical map of the adult amphioxus CNS which could show, for example, whether there are neurons in the anteriormost CNS of amphioxus that appear likely to receive input from the frontal eye similar to vertebrates in which cells in the lateral pallium receive input from the optic nerve Suryanarayana, Robertson, Wallén, and Grillner (2017). Techniques for creating such fine-scale 3-D images of individual neurons and their projections have been developed in recent years but are extremely time-consuming and expensive (reviewed in Landhuis, 2020).

In addition to the ciliary frontal eye and lamellar body photoreceptors, amphioxus has two types of microvillar photoreceptors that the vertebrate CNS lacks–the dorsal Joseph cells posterior to the lamellar body and the ventral Organs of Hesse, each associated with a pigment cell (Lacalli et al., 1994) (Fig. 3). Another amphioxus novelty is the way the CNS innervates the paraxial muscles. In vertebrates, the paraxial muscles are innervated by the ventral nerves (ventral roots) emanating from the CNS. In contrast, what had been thought to be the ventral roots of nerves emerging from the amphioxus central CNS were shown by TEM really to be projections of the paraxial muscles that synapse directly with the CNS (Flood, 1966, 1968). Another difference between amphioxus and vertebrates is that while their notochords are considered homologous, the amphioxus notochord consists of muscles, while in vertebrates, the notochord consists of vacuolated cells (Holland & Somorjai, 2020b; Stemple, 2005). These notochordal muscles in amphioxus, like the paraxial muscles, synapse directly with motor neurons in the CNS.

4.3 The evolution of neural crest (Fig. 5)

A major innovation of vertebrates is neural crest—pluripotent cells that migrate away from the dorsal neural tube and give rise to many different cell types including pigment cells, cells of the adrenal medulla, parts of cranial ganglia and facial cartilage. A key gene expressed in migratory neural crest in vertebrates in *distalless (dlx2a)*; knockdown of *dlx2a* in zebrafish shows that it is essential for development of neural crest derivatives including the sensory ganglia (Sperber, Saxena, Hatch, & Ekker, 2008). It has been suggested that ascidian tunicate larvae may have cells evolutionarily related to vertebrate neural crest since a few cells migrating away from the vicinity of the larval neural tube differentiate into pigment cells (Jeffery, Strickler, & Yamamoto, 2004), while ectopic expression of *twist,* which mediates epithelial/mesenchymal transition in vertebrate neural crest (Qin, Xu, He, Qin, & Xu, 2012), in the developing neural tube of a larval ascidian can induce cells to migrate away from the neural tube (Abitua & Levine, 2015). In ascidians, one *Distalless* gene is expressed in anterior ectoderm including neuroectoderm, although its function hasn't been determined (Harafuji, Keys, & Levine, 2002). Although amphioxus lacks cells like neural crest that migrate individually away from the neural tube, the ectoderm adjacent the neural plate has some neural crest characteristics (Fig. 4) (Holland et al., 1996). At the early neurula stage, this ectoderm, which expresses *distalless*, detaches from both sides of the neural plate and migrates as two sheets over the open neural plate to fuse in the dorsal midline (Fig. 4). However, these *distalless*-expressing cells in amphioxus remain ectodermal and neither migrate individually nor differentiate into other types of tissue as in vertebrates. This suggests that the common ancestor of amphioxus, tunicates and vertebrates may have had a migratory, *distalless*-expressing cell population near the neural plate. However, because there are no clear

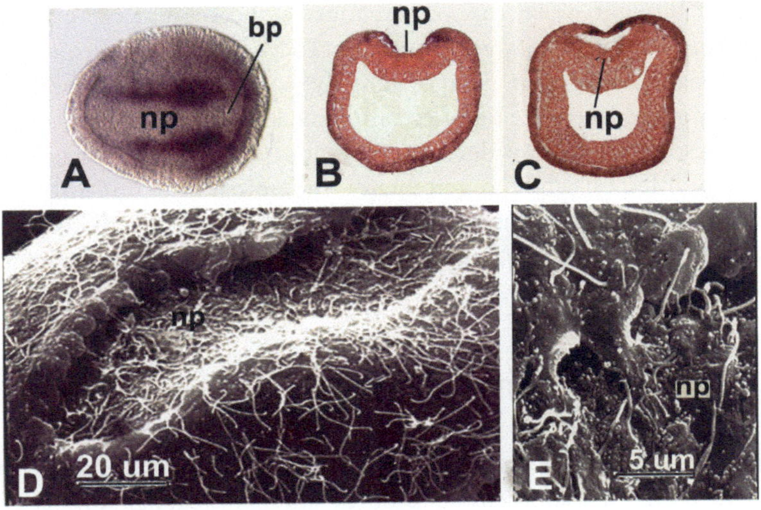

Fig. 5 Neurulation in amphioxus. (A–C) In situ hybridization for *Distalless*. (A) Dorsal view of a late gastrula/early neurula. Anterior to left. *Distalless* is expressed in the ectoderm that has detached from the neural plate and is "walking" over it. (B) Cross section through an embryo at the same stage as in panel (A). *Distalless* is strongly expressed in the ectoderm moving over the neural plate (np) and more weakly expressed in the remainder of the ectoderm. (C) Cross-section through a slightly later neurula than in panel (B). The *Distalless*-expressing ectoderm cells have finished walking over the neural plate (np), which is beginning to round up into a neural tube. (D) Scanning electron micrograph of the dorsal side of an early neurula as in A. All of the ectodermal cells are ciliated. (E) Higher magnification of the ectoderm walking over the neural plate showing the lamellipodia at the leading edge of the ectoderm. *Panels C–E: From Holland, N. D., Panganiban, G., Henyey, E. L., & Holland, L. Z. (1996). Sequence and developmental expression of AmphiDll, an amphioxus distal-less gene transcribed in the ectoderm, epidermis and nervous system: Insights into evolution of craniate forebrain and neural crest.* Development, 122, *2911–2920 with permission from Company of Biologists.*

intermediates between this ectodermal cell population in amphioxus, the migratory cells near the neural plate in ascidians and vertebrate neural crest, it is not possible to reconstruct the common ancestor of these three cell populations.

In sum, the dorsal, hollow nerve cord is one of the defining characters of the Phylum Chordata, and has been considered homologous among them for a long time. It is the regional homologies between the amphioxus and vertebrate nerve cords that have been in doubt. Together, the microanatomy and gene expression have indicated that the telencephalon is likely a vertebrate invention, and that while the midbrain/hindbrain boundary and zona limitans intrathalamica were probably present in the common ancestor of amphioxus and vertebrates, they both acquired organizer properties in the vertebrate lineage. In spite of the origins of vertebrate neural crest being somewhat ambiguous, this is a microanatomy and evo-devo success story.

4.4 Evolution of kidneys (Figs. 6 and 7)

Whereas microanatomy plus gene expression patterns have provided gratifying insights into the evolution of chordate neural tissues considered above in Sections 4.2–4.3, the approach has provided less definite, but still interesting data for other tissues. This can be illustrated with the example of chordate kidney evolution. Few organs in the animal kingdom have been more controversial than the cephalochordate kidney. Kidney function was initially attributed to a variety of amphioxus body parts, as summarized diagrammatically by Holland (2017). Finally, by the end of the 19th century, the amphioxus kidney (or more exactly, kidneys) had been correctly identified: these are a single, relatively large nephridium (Hatschek's nephridium) just anterior to the mouth and numerous smaller nephridia of Weiss-Boveri, one associated with each pharyngeal slit. The two sorts of kidneys first appear in the early larvae and persist for life. Both, when mature, are built along similar lines. Hatschek's nephridium is an elongate tube that is blind anteriorly but opens posteriorly via a nephridiopore into the pharynx just dorsal to the mouth, while each nephridium of Weiss-Boveri is a vesicle with a lobate outline that opens via a nephridiopore at the dorsal edge of a pharyngeal gill slit (Fig. 6).

Around the turn of the 20th century, Goodrich (1902) thought that, following its early development, each kidney had only single opening that was its nephridiopore, whereas Boveri (1904) claimed that the kidneys had additional openings draining nearby coelomic cavities. In the end, Goodrich proved to be correct. Unfortunately, he also thought incorrectly that amphioxus kidneys arise embryologically from ectoderm. He arrived at this conclusion because his light microscopic studies showed that many of the cells comprising the kidney had a very long apical process consisting of a flagellum surrounded by a palisade of equally long microvilli alternating with narrow slots (Fig. 6C). This arrangement resembles the solenocytes (literally, channel cells) that are the key component of the protonephridial kidneys of larvae and adults of a number of non-chordate invertebrates. This similarity convinced Goodrich that amphioxus kidneys are homologous to these

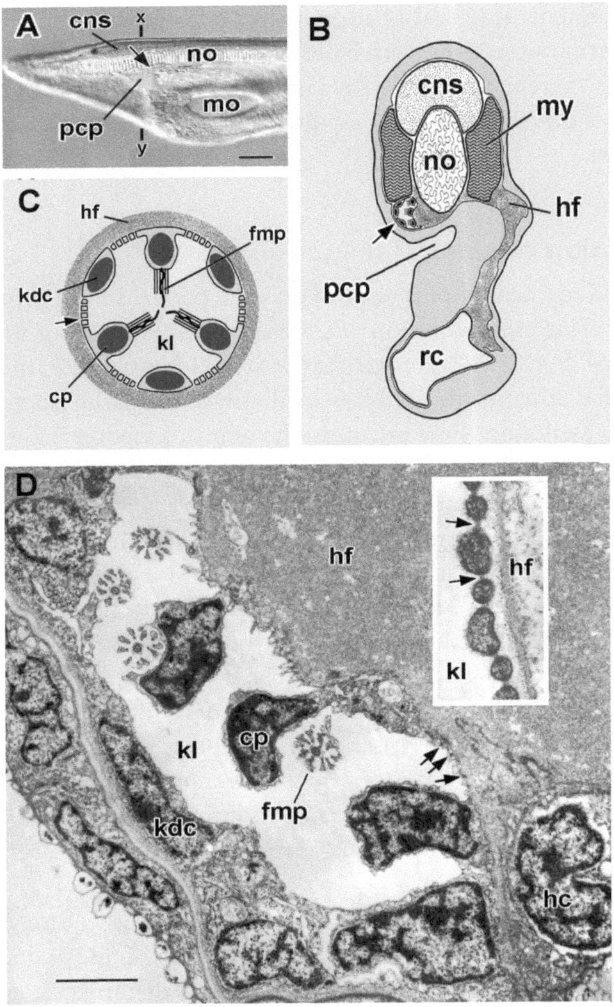

Fig. 6 Amphioxus nephridia. (A) Amphioxus larva (5 days old); side view of anterior end. Abbreviations: cns, central nervous system; no, notochord; mo, mouth; pcp, preoral ciliated pit. The arrow indicates the level of Hatschek's nephridium; x and y show the plane of section of the diagram in panel (B). Scale bar = 50 μm. (B) Larva in diagrammatic cross-section thorough level x–y in panel (A). The arrow indicates Hatschek's nephridium. Abbreviations: cns, central nervous system; no, notochord; my, myomere; hf, haemal fluid; rc, rostral coelom; pcp, preoral ciliated pit. (C) Schematic cross section of amphioxus kidney. Arrow indicates slit/diaphragms. (D) TEM of Hatsheck's nephridium in larval amphioxus. Arrows indicate a slit/diaphragm in basal region of a cyrtopodocyte. Abbreviations: hf: haemal fluid; hc: haemocyte; cp: cyrtopodocyte (with one of its slit/diaphragms arrowed); fmp: flagellar/microvillar process; kl, kidney lumen; fmp; kdc: kidney duct cell, kdc. Scale bar = 2 μm. (D) (inset) enlargement of basal region of cyrtopodocyte with slit/diaphragms (arrowed). *Panel D: Original TEM courtesy of T.C. Lacalli.*

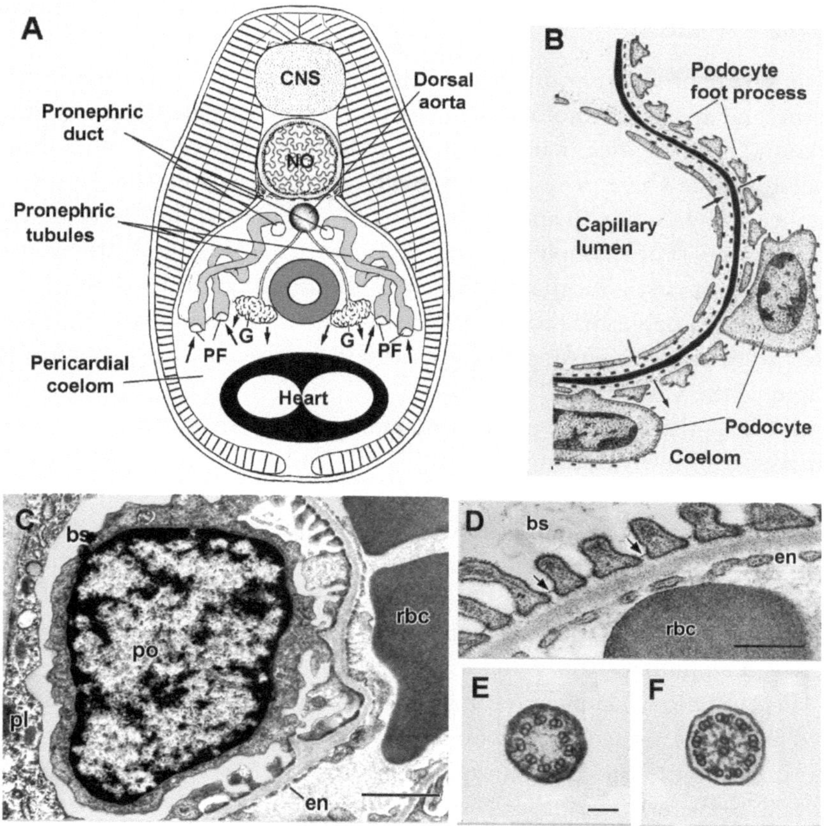

Fig. 7 Vertebrate nephridia. (A) Diagram of a cross-section through the pronephric kidney of an ammocoete larva of a lamprey, consisting of paired glomeruli (G) that filter waste from blood vessels extending from the dorsal aorta. Waste is expelled from the glomeruli into the pericardial coelom. Coelomic fluid is then taken up into the pronephric tubules by the ciliated pronephric funnels (PF). Arrows indicate the direction of flow of fluid from the kidney. The pronephric tubules empty into the cloaca. NO = notochord, CNS = central nervous system. (B) Cross-section of a glomerulus of a lamprey nephridium. The podocytes, like the crytopodocytes of amphioxus, have foot processes with gaps between them. The capillary is lined with an endothelium. Wastes are filtered from the blood through extracellular material termed the lamina densa into the coelom. (C) TEM of mammalian Bowman's capsule and adjacent glomerulus. Abbreviations: bs, Bowman's space; po, podocyte; pl, parietal layer, rbc red blood cell in glomerulus; en, endothelium. Scale bar = 2 μm. (D) Enlargement of basal region of podocyte in C with slit/diaphragms arrowed. Bs, bowman's space; en, endothelium; rbc, red blood corpuscle. Scale bar = 1 μm. (E) TEM of cross section of a primary cilium from the embryonic mouse neural tube. (D) Schematic cross section of amphioxus kidney. (F) TEM of cross section of a conventional cilium/flagellum. Scale bar in panel (E) (also applicable to F) = 0.1. *Panel B: Reprinted with permission from Ellis, L.C., & Youson, J. H. (1991). The anionic charge barrier in the renal corpuscle of the pronephros in the lamprey, Petromyzon marinus L.* The Anatomical Record, *231, 178–184. Panel C: Adapted from https://blogs.zeiss.com/microscopy/en/scanning-electron-microscopy-pathology/. Panel E: Reprinted with permission from Willaredt, M.A., Hasenpusch-Theil, K., Gardner, H.A.R., Kitanovic, I., Hirschfeld-Warneken, et al. (2008). A crucial role for primary cilia in cortical morphogenesis.* Journal of Neuroscience, 28, *12887–12900.*

invertebrate protonephridia, which arise during embryology from the ectoderm, and are not homologous to the podocyte-based nephridia of vertebrates. Thus, the idea that amphioxus kidneys are a legacy from a distant ancestor somewhere deep in the tree of life, was vigorously and repeatedly publicized by Goodrich and became widely accepted in the first half of the 20th century. For example, Smith (1953) claimed, "the vertebrate nephron could not by any conceivable stretch of the imagination be derived from Amphioxus's nephridial complex." In his view, this circumstance was sufficient "to reduce the Amphioxus theory to a jelly in all discussions of the origin of the vertebrates."

It is true that even the pronephric kidney of agnathan vertebrates differs somewhat from that of amphioxus (Fig. 7A and B). The functioning adult kidneys of teleost fish and amphibians differ even more, as the pronephric kidney is succeeded in development by a mesonephros, while in amniotes, a metanephros succeeds the mesonephros in development and becomes the definitive kidney. Even so, not long after the strong words of Smith (1953), transmission electron microscopy began to show a number of commonalities between amphioxus and vertebrate kidneys.

For amphioxus kidneys, Kümmel and Brandenberg (1961) and Welsch (1975) used transmission electron microscopy (TEM) to show that many of the cells project not only an apical flagellar/microvillar process but also extend basolateral cytoplasmic regions that are sheet-like and perforated with slit/diaphragms (Fig. 6). To this extent, kidney cells in amphioxus resemble podocytes in vertebrate kidneys (Fig. 7 B—D). In the light of this new information, the filtration cells in the amphioxus kidney were given the compound name of cyrtopodocyte ("cyrto" is Greek for fishing creel, an apt name suggested by the palisade of long microvilli alternating with slots). Amphioxus cyrtopodocytes in a larval Hatschek's nephridium are comparable in many ways to vertebrate podocytes comprising the visceral wall of Bowman's capsule (Fig. 6D; 7B—D). Although no functional data are available for the amphioxus excretory system, there can be little doubt that the slit diaphragms of the cyrtopodocytes filter materials passing from the haemal fluid into the kidney lumen just as the slit/diaphragms of vertebrate podocytes filter material from the blood into Bowman's space.

During the same period, TEM was used to show details of the pronephric kidneys in both ammocoete larvae and adults of the lamprey (McMillan, 2011; Youson & McMillan, 1970a; Youson & McMillan, 1970b; Youson & McMillan, 1971a; Youson & McMillan, 1971b; Youson & McMillan, 1971c). The gross anatomy of lamprey kidneys is quite different from that of amphioxus kidneys (Fig. 7A). Lamprey kidneys consist of a glomus or glomerulus that contains many blood vessels deriving from the dorsal aorta. Waste passes from the blood into the glomerulus and thence into the pericardial coelom. The coelomic fluid is then taken up by the pronephric tubules that send the waste posteriorly to the cloaca. However, the glomerular cells themselves, except for lacking the microvillar collar, quite closely resemble those of the pronephric kidney in amphioxus, suggesting a common origin (Fig. 7B).

Even in the light of the similarities in fine-structure data, a sticking point was that vertebrate kidneys are clearly of mesodermal origin while Goodrich's idea persisted that amphioxus kidneys were ectodermal—either entirely (Ruppert, 1996) or at least in part (Stach & Eisler, 1998). It was only recently that fine-structural studies of the appropriate developmental stages finally established that amphioxus kidneys are entirely mesodermal in origin and that the central cavity in each one is a coelom (Holland, 2018; Kaji, Reimer, Morov, Kuratani, & Yasui, 2016). This space within each amphioxus kidney is moderately voluminous—intermediate between the relatively large coelomic space (which doubles as the pericardial coelom) of the pronephric kidney of lampreys (Ellis & Youson, 1989) and the low-volume of each Bowman's space in kidneys of many other vertebrates (Scott & Quaggin, 2015).

However, in spite of the similarities between the basal regions of vertebrate podocytes and amphioxus cyrtopococytes and in spite of their both being mesodermal in origin, establishing a firm homology between the two cell types remains elusive. The problem is that vertebrate podocytes lack the apical flagellar/microvillar process that arises from the apex of each amphioxus cyrtopodcyte. Unfortunately for discussions of homology, nothing definite is yet known about the function of the apical process—its structure seems altogether too complicated for merely impelling fluid down the kidney lumen to the nepridiopore. The only recent development tending to blur the difference between amphioxus and vertebrate kidneys has been the discovery of a long-overlooked primary cilium arising from the podocyte apex in many vertebrates. A primary cilium is very short, immotile, and has a relatively simple internal structure (Fig. 7E) contrasting with the more complex organization of microtubules and related proteins in a conventional cilium or flagellum (Fig. 7F). It has been proposed that primary cilia are mechnoreceptors (likely flow meters in kidneys) and/or act as signaling hubs for growth factors (Nishimura, Kasahara, Shiromizu, Watanabe, & Inagaki, 2019). To date, a primary cilium has been demonstrated on podocytes of adult lampreys (Miyoshi, 1978), adult fishes (Ojeda, Icardo, & Domezain, 2003; Zuasti, Agulleiro, & Hernandez, 1983) and fetal mammals (Ichimura, Kurihara, & Sakai, 2010). At present, it is far from clear how, if at all, primary cilia are related phylogenetically to conventional cilia and flagella. Moreover, the primary cilium at the apex of the vertebrate podocyte is not surrounded by even a hint of a microvillar collar. Thus, the flagellar/microvillar apparatus of amphioxus kidneys continues to fit poorly into scenarios for the history of filtration cells during chordate kidney evolution.

In spite of these morphological differences between amphioxus and vertebrate kidneys, the expression of several homologous genes in the developing amphioxus kidney and in the vertebrate pronephric kidney is consistent with their being homologous. These include *Pax2/5/8*, *Notch* and *Lim1/5* (Heller & Brändli, 1999; Holland et al., 2001; Karavanov, Saint-Jeannet, Karavanova, Taira, & Dawid, 1996; Kozmik et al., 1999; Langeland, Holland, Chastain, & Holland, 2006; Osório, Megías, Pombal, & Rétaux, 2006). *Pax2/5/8* is upstream of *Lim1/5* in the core network of pronephric kidney specification (Boualia et al., 2013). However, given the divergent anatomy of amphioxus kidneys and the pronephric ones of the lamprey, the conserved expression of two or three genes, while consistent with kidney homology, is insufficient evidence of such homology. The problem is that, as noted above, some genes like *Pax, Six, Eya* often work together as part of a gene network and are, therefore, co-expressed in development of a variety of tissues and organs. For example, these genes are co-expressed in pharyngeal development in amphioxus and are also co-expressed in development of cells in the CNS (Kozmik et al., 2007).

The patterns discussed above underscore that homologies inferred from gene expression are more certain when the anatomy of the structures is also similar. However, it should be kept in mind that homologous structures can diverge both anatomically and in gene expression. A good example of the latter is the larval gut in starfish and sea urchins, which are morphologically similar and almost certainly both evolved from the gut in an ancestral echinoderm larva. However, while parts of the gene network making the larval gut are conserved between the two echinoderms, others have diverged (Hinman & Davidson, 2007). For chordate kidneys, because relatively little is known about their gene expression, and their morphology is fairly different in amphioxus compared to lamprey, convergence between them cannot be ruled out. Even so, the similarities they do share, such as the podocyte foot processes, suggest that they likely share a common origin in the ancestral chordate.

5. The future of amphioxus as a "model organism"

The studies cited above have clearly demonstrated that amphioxus is the best living proxy for the ancestral chordate. It is also a highly tractable experimental organism. Warm water species of *Branchiostoma,* including *B. floridae* and *B. belcheri* have been raised through several generations in the laboratory (Holland & Li, 2021). Ripe animals can be maintained in the laboratory at winter temperatures. Maturation of gametes is induced by a + 5-8 °C shift in temperature and subsequent spawning by darkness. In addition to its high fecundity, short life-cycle (egg to breeding adult in 2–3 months) and long life span (more than 3–4 years), *B. floridae* has proven amenable to mutagenesis (Holland & Li, 2021; Hu, Li, Wang, & Wang, 2017; Ren et al., 2020; Zhong, Herrera-Úbeda, Garcia-Fernàndez, Li, & Holland, 2020).

Because amphioxus is so interesting and tractable, numerous biological and genomic resources have been developed. Amphioxus genomes are on the order of 500–600 mb. Genome assemblies are publically available for 3 species of *Branchiostoma* (*B. floridae, B. lanceolatum,* and *B. belcheri*). Transcriptome sequences have been published for *Asymmetron luayanum, B. floridae, B. lanceolatum* and *B. belcheri*. In addition, gridded cDNA libraries have been constructed for *B. floridae* and *B. belcheri*, and clones have been end-sequenced (EST). cDNA clones of *B. floridae* are available on request from LZ Holland or J.K. Yu (jkyu@ gate.sinica.edu.tw). There are several genome databases including http://genome.bucm.edu.cn/lancelet/ which is primarily for *B. belcheri*, but also contains *B. floridae* and *B. lanceolatum* sequences. A database that is primarily for *B. lanceolatum*, but also contains sequences for *B. floridae, B. belcheri,* and *B. japonicum* is https://www.ebi.ac.uk/ena/browser/home. Finally, https://mycocosm.jgi.doe.gov/Brafl1/Brafl1.home.html contains both diploid and haploid genome sequences for *B. floridae.*

6. Future directions

There will always be some question as to exactly how much amphioxus has changed since its lineage split from the line leading to tunicates and vertebrates. The best answer is "probably not a whole lot." It is quite telling that *Asymmetron* and *Branchiostoma* are so very much alike even though they split at least 120 mya (Yue et al., 2014). The main differences are in size (*Branchiostoma* is larger), in *Asymmetron* having gonads only on the right side and *Branchiostoma* having them on both sides, and in *Asymmetron*, but not *Branchiostoma*, using moonlight to trigger gamete maturation and spawning (Holland, 2011; Holland & Holland, 2010). However, the two genera can hybridize and make viable larvae that are morphologically intermediate between their parents (Holland, Holland, & Heimberg, 2015). That is surprising in light of the low viability of hybrids of more closely related species such as those of *Drosophila melanogaster* and *Drosophila simulans*, which diverged only ~0.25 mya. With *D. simulans* mothers and *D. melanogaster* fathers, the F1females die as embryos while the F1 males are sterile (Barbash, 2010).

The best that can be done to decide whether a particular trait in amphioxus is an apomorphy or was present in the chordate ancestor is to obtain more data, not only on amphioxus species but also on developing agnathans and cartilaginous fish. Lamprey embryos can readily be obtained during the spring breeding season, which ranges from about 1 week for the brook lamprey *Lethenteron japonicum,* up to about a month for the sea lamprey, *Petromyzon marinus.* Embryos take about 4.5 days to reach the gastrula (Nikitina, Bronner-Fraser, & Sauka-Spengler, 2009). Ammocoete larvae can be fished from streams. However, raising ammocoetes through metamorphosis is impractical as lampreys spend 3–7 years as ammocoetes before the onset of metamorphosis. Even so, several labs in Japan, USA and Europe have published evodevo studies of lamprey embryos revealing many parallels with amphioxus embryos together with many vertebrate apomorphies such as neural crest, paired eyes, lack of paraxial muscles in the head. In contrast, hagfish embryos are very difficult to obtain. Only one laboratory in the world: that of Shigeru Kuratani at Riken in Japan has obtained hagfish embryos in recent years. His student Kinya Ota was the first to obtain hagfish embryos since they were last collected by Conel 90 years ago (reviewed in Holland, 2007). In Japan, the adults are collected in August, left in tanks in the lab or in cages in the ocean until spawning occurs about October/November. The embryos develop very, very slowly. Only in February/March is it clear that some embryos are developing. See https://natureecoevocommunity.nature.com/posts/31326-slimy-hox.

7. Conclusion

In sum, in the absence of a time machine (preferably an undersea model), reconstructing the ancestral chordate can never be an exact science, but when the organisms being compared are evolving relatively slowly, and two or more types of data point to the same answer, homologies can be inferred with a reasonable certainty.

In the last 30 years since amphioxus was first spawned in the laboratory, major questions of evolution such as how the vertebrate brain arose from the nerve cord of an ancestral chordate have been largely answered by a combination of microanatomy and gene expression. Other questions such as how kidneys evolved or whether the vertebrate head muscles evolved from the anterior paraxial muscles of the ancestral chordate remain under discussion. Inferring homologies between phylogenetically distant organisms is facilitated when both the anatomy and gene expression are similar–the more they differ, the lower the confidence in homology.

Technical advancements, especially the development of laboratory spawning and continuous breeding cultures of several species of amphioxus have led to an increase in the number of laboratories studying amphioxus and basal vertebrates such as the lamprey from fewer than five world-wide to about 30 today. It is noteworthy that most of these laboratories are in east Asia and Europe. It is likely that so few of these laboratories are in the USA because funding for developmental biology has focused on the six "model organisms"—the mouse, chick, zebrafish, the frog *Xenopus laevis,* the fly *Drosophila melanogaster* and the nematode *Caenorhabditis elegans*. About 15 years ago at a workshop for graduate students from the Americas sponsored by the National Science Foundation and the Society for Developmental Biology to introduce them to several "emerging model organisms," one student asked how long it took the workshop instructors from the USA to obtain funding for research on their organisms. They all replied, "About five years." That is 5 years of paying for the research out of personal funds or perhaps funds intended for another purpose. As a result, given a tight job market and the emphasis of universities in the USA on extramural funding, few students are willing to risk studying an "emerging model organism." This situation will need to change if evolutionary developmental biology in the USA is to prosper. The present volume on emerging model organisms is a start toward raising awareness that these organisms are very much worth studying.

Acknowledgments

This work was supported in part by a grant from the National Science Foundation of the U.S.A. (IOS 1952567). We thank Thurston Lacalli and Esteban Rodríguez for their helpful comments on the infundibulum and Reissner's fiber.

References

Abitua, P. B., & Levine, M. (2015). The pre-vertebrate origins of neurogenic placodes. *Nature, 524*, 462–465.

Albuixech-Crespo, B., López-Blanch, L., Burguera, D., Maeso, I., Sánchez-Arrones, L., Moreo-Bravo, J. A., et al. (2017). Molecular regionalization of the developing amphioxus neural tube challenges major partitions of the vertebrate brain. *PLoS Biology, 15*. article e2001573.

Artieri, C. G., Haerty, W., & Singh, R. S. (2009). Ontogeny and phylogeny: Molecular signatures of selection, constraint, and temporal pleiotropy in the development of *Drosophila*. *BMC Biology*, 7. article 42.

Atkinson-Leadbeater, K., Bertolesi, G. E., Johnston, J. A., Hehr, C. L., & McFarlane, S. (2009). FGF receptor dependent regulation of *lhx9* expression in the developing nervous system. *Developmental Dynamics, 238*, 367–375.

Ayers, H. (1890). Concerning vertebrate cephalogenesis. *Journal of Morphology, 4*, 221–245.

Balfour, F. M. (1885). *A treatise on comparative embryology* (Vols. 1 & 2). London: MacMillan. 591 pp.

Barbash, D. A. (2010). Ninety years of *Drosophila melanogaster* hybrids. *Genetics, 186*, 1–8.

Bateson, W. (1885). The later stages in the development of *Balanoglossus kowalevskii*, with a suggestion as to the affinities of the Enteropneusta. *Quarterly Journal of Microscopical Science, 25*(Supplement), 81–122.

Berrill, N. J. (1955). *The origin of vertebrates* (p. 257). Oxford: Clarendon Press.

Berrill, N. J. (1987). Early chordate evolution. Part 1. Amphioxus, the riddle of the sands. *International Journal of Invertebrate Reproduction and Development, 11*, 1–14.

Boualia, S. K., Gaitan, Y., Tremblay, M., Sharma, R., Cardin, J., et al. (2013). A core transcriptional network composed of *Pax2/8*, *Gata3* and *Lim1* regulates key players of pro/mesonephros morphogenesis. *Developmental Biology, 382*, 555–566.

Boveri, T. (1904). Bemerkungen über den Bau der Nierenkanalchen des Amphioxus. *Anatomischer Anzeiger, 25*, 599–604.

Bozzo, M., Lacalli, T. C., Obino, V., Caicci, F., Marcenaro, E., et al. (2021). Amphioxus neuroglia: Molecular characterization and evidence for early compartmentalization of the developing nerve cord. *Glia, 24*, 23982.

Candiani, S., Moronti, L., Ramoino, P., Schubert, M., & Pestarino, M. (2012). A neurochemical map of the developing amphioxus nervous system. *BMC Neuroscience, 13*, article 59.

Castro, L. F. C., Rasmussen, S. L. K., Holland, P. W. H., Holland, N. D., & Holland, L. Z. (2006). A *Gbx* homeobox gene in amphioxus: Insights into ancestry of the ANTP class and evolution of the midbrain/hindbrain boundary. *Developmental Biology, 295*, 40–51.

Chou, S.-J., & Tole, S. (2019). *Lhx2*, an evolutionarily conserved, multifunctional regulator of forebrain development. *Brain Research, 1705*, 1–14.

Conway Morris, S., & Caron, J. (2014). A primitive fish from the Cambrian of North America. *Nature, 512*, 419–422.

Costa, O. G. (1834). *Cenni Zoologici ossia Descrizione Sommaria delle Specie Nuove di Animali Discoperti in Diverse Contrade del Regno nell'Anno 1834* (p. 90). Napoli: Azzolino.

Deerinck, T. J., Bushong, E. A., Lev-Ram, V., Tsien, R. Y., & Ellisman, M. H. (2010). Enhancing serial block-face scanning electron microscopy to enable high resolution 3-D nanohistology of cells and tissues. *Microscopy and Microanalysis, 16*, 1138–1139.

Delsuc, F., Brinkmann, H., Chourrout, D., & Philippe, H. (2006). Tunicates and not cephalochordates are the closest living relatives of the vertebrates. *Nature, 439*, 965–968.

Delsuc, F., Philippe, H., Tsagkogeorga, G., Simion, P., Tilak, M. K., et al. (2018). A phylogenomic framework and timescale for comparative studies of tunicates. *BMC Biology, 16*. article 39.

Ellis, L. C., & Youson, J. H. (1989). Ultrastructure of the pronephric kidney in upstream migrand sea lamprey, *Petromyzon marinus* L. *American Journal of Anatomy*, *185*, 429–443.
Flood, P. R. (1966). A peculiar mode of muscular innervation in amphioxus. Light and electron microscopic studies of the so-called ventral roots. *Journal of Comparative Neurology*, *126*, 181–218.
Flood, P. R. (1968). Structure of the segmental trunk muscles in amphioxus, with notes on the course and "endings" of the so-called ventral root fibers. *Zeitschrift für Zellforschung und Mikroskopische Anatomie*, *84*, 389–416.
Gans, C., & Northcutt, R. G. (1983). Neural crest and the origin of vertebrates: A new head. *Science*, *220*, 268–274.
Garstang, W. (1960). In W. Yust (Ed.), *Amphioxus in Encyclopaedia Britannica* (14th ed., pp. 843–846). Chicago: Encyclopaedia Britannica.
Giribet, G. (2018). Phylogenomics resolves the evolutionary chronicle of our squirting closest relatives. *BMC Biology*, *16*, article 49.
Glardon, S., Holland, L. Z., Gehring, W. J., & Holland, N. D. (1998). Isolation and developmental expression of the amphioxus *Pax-6* gene (*AmphiPax-6*): Insights into eye and photoreceptor evolution. *Development*, *125*, 2701–2710.
Goodrich, E. S. (1902). On the structure of the excretory organs of amphioxus (part 1). *Quarterly Journal of Microscopical Science*, *45*, 493–501.
Grande, C., & Patel, N. H. (2009). Nodal signalling is involved in left-right asymmetry in snails. *Nature*, *457*, 1007–1011.
Harafuji, N., Keys, D. N., & Levine, M. (2002). Genome-wide identification of tissue-specific enhancers in the *Ciona* tadpole. *Proceedings of the National Academy of Sciences USA*, *99*, 6802–6805.
Heller, N., & Brändli, A. W. (1999). *Xenopus Pax-2/5/8* orthologues: Novel insights into *Pax* gene evolution and identification of *Pax-8* as the earliest marker for otic and pronephric cell lineages. *Developmental Genetics*, *24*, 208–219.
Herdman, W. A. (1904). Ascidians and amphioxus. In S. F. Harmer, & H. E. Shipley (Eds.), *Vol. 7. The Cambridge natural history* (pp. 35–138). London: Macmillan. 760 pp.
Hibi, M., & Shimizu, T. (2012). Development of the cerebellum and cerebellar neural circuits. *Developmental Neurobiology*, *72*, 282–301.
Hinman, V. F., & Davidson, E. H. (2007). Evolutionary plasticity of developmental gene regulatory network architecture. *Proceedings of the National Acadamy of Sciences USA*, *104*, 19404–19409.
Holland, L. Z., Abi Rached, L., Tamme, R., Holland, N. D., Inoko, H., et al. (2001). Characterization and developmental expression of the amphioxus homolog of *Notch* (*AmphiNotch*): Evolutionary conservation of multiple expression domains in amphioxus and vertebrates. *Developmental Biology*, *232*, 493–507.
Holland, L. Z., & Holland, N. D. (2021). Cephalochordates: A window into vertebrate origins. *Current Topics in Developmental Biology*, *141*, 119–147.
Holland, L. Z., Laudet, V., & Schubert, M. (2004). The chordate amphioxus: An emerging model organism for developmental biology. *Cellular and Molecular Life Science*, *61*, 2290–2308.
Holland, L. Z., & Li, G. (2021). Laboratory culture and mutagenesis of amphioxus (*Branchiostoma floriae*). *Methods in Molecular Biology*, *2219*, 1–29.
Holland, N. D. (2007). Hagfish embryos again—The end of a long drought. *BioEssays*, *29*, 833–836.
Holland, N. D. (2011). Spawning periodicity of the lancelet, *Asymmetron lucayanum* (Cephalochordata), in Bimini, Bahamas. *Italian Journal of Zoology*, *78*, 478–486.
Holland, N. D. (2017). The long and winding path to uderstanding kidney structure in amphioxus–a review. *International Journal of Developmental Biology*, *61*, 683–688.

Holland, N. D. (2018). Formation of the initial kidney and mouth opening in larval amphioxus studied with serial blockface scanning electron microscopy (SBSEM). *EvoDevo, 9*. article 16.

Holland, N. D., & Holland, L. Z. (2010). Laboratory spawning and development of the Bahama lancelet, *Asymmetron lucayanum* (Cephalochordata): Fertilization through feeding larvae. *Biological Bulletin, 219*, 132–141.

Holland, N. D., Holland, L. Z., & Heimberg, A. (2015). Hybrids between the Florida amphioxus (*Branchiostoma floridae*) and the Bahamas lancelet (*Asymmetron lucayanum*): Developmental morphology and chromosome counts. *Biological Bulletin, 228*, 13–24.

Holland, N. D., Panganiban, G., Henyey, E. L., & Holland, L. Z. (1996). Sequence and developmental expression of *AmphiDll*, an amphioxus *distal-less* gene transcribed in the ectoderm, epidermis and nervous system: Insights into evolution of craniate forebrain and neural crest. *Development, 122*, 2911–2920.

Holland, N. D., & Somorjai, I. M. L. (2020a). The sensory peripheral nervous system in the tail of a cephalochordate studied by serial blockface scanning electron microscopy. *Journal of Comparative Neurology, 528*, 2569–2582.

Holland, N. D., & Somorjai, I. M. L. (2020b). Serial blockface SEM suggests that stem cells may participate in adult notochord growth in an invertebrate chordate, the Bahamas lancelet. *EvoDevo, 11*, article 22.

Holland, P. W. H., Holland, L. Z., Williams, N. A., & Holland, N. D. (1992). An amphioxus homeobox gene: Sequence conservation, spatial expression during development and insights into vertebrate evolution. *Development, 116*, 653–661.

Hopwood, N. (2015). The cult of amphioxus in German Darwinism; or our gelatinous ancestors in Naples' blue and balmy bay. *History & Philosophy of Science, 36*, 371–393.

Hu, G., Li, G., Wang, H., & Wang, Y. (2017). Hedgehog participates in the establishment of left-right asymmetry during amphioxus development by controlling *Cerberus* expression. *Development, 144*, 4694–4703.

Ichimura, K., Kurihara, H., & Sakai, T. (2010). Primary cilia disappear in rat podocytes during glomerular development. *Cell & Tissue Reserch, 341*, 197–209.

Irimia, M., Pineiro, C., Maeso, I., Gomez-Skarmeta, J. L., Casares, F., & Garcia-Fernàndez, J. (2010). Conserved developmental expression of *Fezf* in chordates and *Drosophila* and the origin of the zona limitans intrathalamica (ZLI) brain organizer. *EvoDevo, 1*. article 7.

Jeffery, W. R., Strickler, A. G., & Yamamoto, Y. (2004). Migratory neural crest-like cells form body pigmentation in a urochordate embryo. *Nature, 43*, 696–699.

Kaji, T., Reimer, J. D., Morov, A. R., Kuratani, S., & Yasui, K. (2016). Amphioxus mouth after dorso-ventral inversion. *Zoological Letters, 2*. article 2.

Karavanov, A. A., Saint-Jeannet, J. P., Karavanova, I., Taira, M., & Dawid, I. B. (1996). The LIM homeodomain protein Lim-1 is widely expressed in neural, neural crest and mesoderm derivatives in vertebrate development. *The International Journal of Developmental Biology, 40*, 453–461.

Keller, P. J., Ahrens, M. B., & Freeman, J. (2015). Light-sheet imaging for systems neuroscience. *Nature Methods, 12*, 27–29.

Kon, T., Nohara, M., Yamanoue, Y., Fujiwara, Y., Nishida, M., & Nishikawa, T. (2007). Phylogenetic position of a whale-fall lancelet (Cephalochordata) inferred from whole mitochondrial genome sequences. *BMC Evolutionary Biology*, 7, 1–12.

Kowalevsky, A. (1867). *Entwickelungsgeschichte des Amphioxus lanceolatus. Mémoires de l'Académie impériale des sciences de St. Pétersbourg. Vol. Sér. VII), 11* (pp. 1–17). (+ pl.I-III).

Kozmik, Z., Holland, N. D., Kalousova, A., Paces, J., Schubert, M., & Holland, L. Z. (1999). Characterization of an amphioxus paired box gene, *AmphiPax2/5/8*: Developmental expression paterns in optic support cells, nephidium, thyroid-like structures and pharyngeal gill slits, but not in the midbrain-hindbrain boundary region. *Development, 126*, 1295–1304.

Kozmik, Z., Holland, N. D., Kreslova, J., Oliveri, D., Schubert, M., et al. (2007). *Pax-six-Eya-Dach* network during amphioxus development: Conservation in vitro but context specificity in vivo. *Developmental Biology, 306*, 143–159.

Kümmel, G., & Brandenberg, J. (1961). Die Reusengeisselzellen (Cyrtocyten). *Zeitschrift für Naturforschug, 16b*, 692–697.

Lacalli, T., & Candiani, S. (2017). Locomotory control in amphioxus larvae: New insights from neurotransmitter data. *EvoDevo, 8*. article 4.

Lacalli, T. C. (1996). Frontal eye circuitry, rostral sensory pathways, and brain organization in amphioxus larvae: Evidence from 3D reconstructions. *Philosophical Transactions of the Royal Society B, 351*, 243–263.

Lacalli, T. C. (2013). Looking into eye evolution: amphioxus points the way. *Pigment Cell and Melanoma Research, 26*, 162–164.

Lacalli, T. C., Holland, N. D., & West, J. E. (1994). Landmarks in the anterior central nervous system of amphioxus larvae. *Philosophical Transactions of the Royal Society London B, 344*, 165–185.

Lacalli, T. C., & Kelly, S. J. (1999). Somatic motoneurones in amphioxus larvae: Cell types, cell position and innervation patterns. *Acta Zoologica Stockholm, 80*, 113–124.

Lacalli, T. C., & Kelly, S. J. (2002). Floor plate, glia and other support cells in the anterior nerve cord of amphioxus larvae. *Acta Zoologica Stockholm, 83*, 87–98.

Landhuis, E. (2020). Probing fine-scale connections in the brain. *Nature, 586*, 631–633.

Langeland, J. A., Holland, L. Z., Chastain, R., & Holland, N. D. (2006). An amphioxus LIM-homeobox gene, *AmphiLim1/5*, expressed early in the invaginating organizer region and later in differentiating cells of the kidney and central nervous system. *International Journal of Biological Sciences, 2*, 110–116.

Liu, J., & Robinson-Rechavi, M. (2018). Developmental constraints on genome evolution in four bilaterian model species. *Genome Biology and Evolution, 10*, 2266–2277.

Lopez-Avalos, M. D., Cifuentes, M., Grondona, J. M., Miranda, E., Perez, J., & Fernandez-Llebrez, P. (1997). Rostral floor plate (flexural organ) secretes glycoproteins immunologically similar to subcommissural organ glycoproteins in dogfish (*Scyliorhinus canicula*) embryos. *Developmental Brain Research, 102*, 69–75.

Mallatt, J., & Chen, J. Y. (2003). Fossil sister group of craniates: Predicted and found. *Journal of Morphology, 258*, 1–31.

McMillan, D. B. (2011). Histology of the kidney. In A. P. Farrell (Ed.), *Encyclopedia of fish physiology. From genome to environment* (pp. 1395–1410). Academic Press.

Meiniel, O., Meiniel, R., Lalloue, F., Didier, R., Jauberteau, M.-O., et al. (2008). The lenghthening of a giant protein: When, how, and why? *Journal of Molecular Evolution, 66*, 1–10.

Metschnikov, E. (1866). A. Kovalevsky: le developpement de *l'Amphioxus lanceolatus*. Saint-Petersbourg, 1865, in-octavio. *Archives des Sciences Physiques et Naturelles, 27*, 193–195.

Miyoshi, M. (1978). Scanning electron microscopy of the renal corpuscle of the mesonephros in the lamprey, *Entosphenus japonicus* Martens. *Cell & Tissue Research, 187*, 105–113.

Monniot, C., & Monniot, F. (1991). Discovery of a new lineage among the deep-sea ascidians—a carnivorous Ascidiidae. In *312. Comptes Rendus de l'Académie des Sciences, Sér. III. Sciences de la vie—life sciences* (pp. 378–383).

Nicol, D., & Meinertzhagen, I. A. (1991). Cell counts and maps in the larval central nervous system of the ascidian *Ciona intestinalis* (L.). *Journal of Comparative Neurology, 309*, 415–429.

Nikitina, N., Bronner-Fraser, M., & Sauka-Spengler, T. (2009). *Culturing lamprey embryos*. Cold Spring Harbor Protocols. (1) pdb. protocol 5122.

Nishimura, Y., Kasahara, K., Shiromizu, T., Watanabe, M., & Inagaki, M. (2019). Primary cilia as signaling hubs in health and disease. *Advances. Science, 6*. article 1801138.

Ojeda, J. L., Icardo, J. M., & Domezain, A. (2003). Renal corpuscle of the sturgeon kidney: An ultrastructural, chemical dissection, and lectin-binding study. *Anatomical Record, 273A*, 563–573.

Olsson, R. (1955). Structure and development of Reissner's fibre in the caudal end of amphioxus and some lower vertebrates. *Acta Zoologica Stockholm, 36*, 167–198.

Olsson, R. (1958). *The subcommissural organ*. Doctoral dissertation Stockholm University. 16 pp.

Olsson, R. (1962). The infundibular cells of amphioxus and the question of fibre-forming secretions. *Arkiv för Zoologi, 15*, 347–355.

Olsson, R., Yulis, R., & Rodriguez, E. M. (1994). The infundibular organ of the lancelet (*Branchiostoma lanceolatum*, Acrania): An immunocytochemnical study. *Cell & Tissue Research, 277*, 107–114.

Ono, H., Koop, D., & Holland, L. Z. (2018). *Nodal* and *Hedgehog* synergize in gill slit formation during development of the cephalochordate *Branchiostoma floridae*. *Development, 145*. article 162586.

Osório, J., Megías, M., Pombal, M. A., & Rétaux, S. (2006). Dynamic expression of the LIM-homeodomain gene *Lhx15* through larval brain development of the sea lamprey (*Petromyzon marinus*). *Gene Expression Patterns, 6*, 873–878.

Pallas, P. S. (1774). *Spicilegia zoologica. Vol 1.* Berlin: G. A. Lange. Fascicle 10. 41 pp.

Piesecka, B., Lichocki, P., Moretti, S., Beregmann, S., & Robinson-Rechavi, M. (2013). The hourglass and the early conservation models—co-existing patterns of developmental constraints in vertebrates. *PLoS Genetics, 9*. article e1003476.

Poncelet, G., & Shimeld, S. M. (2020). The evolutionary origins of the vertebrate olfactory system. *Open Biology, 10*, 200330.

Pu, G. A., & Dowling, J. E. (1981). Anatomical and physiological characteristics of pineal photoreceptor cell in the larval lamprey, *Petromyzon marinus*. *Journal of Neurophysiology, 46*, 1018–1038.

Putnam, N. H., Butts, T., Ferrier, D. E. K., Furlong, R. F., Hellsten, U., et al. (2008). The amphioxus genome and the evolution of the chordate karyotype. *Nature, 453*, 1064–1071.

Qin, Q., Xu, Y., He, T., Qin, C., & Xu, J. (2012). Normal and disease-related biological functions of *Twist1* and underlying molecular mechanisms. *Cell Research, 22*, 90–106.

Remane, A. (1956). *Die Grundlagen des naturlichen Systems, der vergleichenden Anatomie und der Phylogenetik: theoretische Morphologie und Systematik*. Leipzig: Geest & Portig. 364 pp.

Ren, Q., Zhong, Y., Huang, X., Leung, B., Xing, C., et al. (2020). Step-wise evolution of neural patterning by *hedgehog* signalling in chordates. *Nature Ecology & Evolution, 4*, 1247–1255.

Richter, H. G., Munoz, R. I., Millan, C. S., Guinazu, M. F., Yulis, C. R., & Rodríguez, E. M. (2001). The floor plate cells from bovine express the mRNA encoding for SCO-spondin and its translation products. *Molecular Brain Research, 93*, 137–147.

Robison, B. H., Raskoff, K. A., & Sherlock, R. E. (2005). Adaptations for living deep: A new bathypelagic doliolid, from the eastern North Pacific. *Journal of the Marine Biological Association of the United Kingdom, 85*, 595–602.

Rodríguez, E. M., del Brío León, M. A., Riera, P., Menendez, J., & Schoebitz, K. (1996). The floor plate of the hindbrain is a highly specialized gland. Immunocytochemical and ultrastructural characteristics. *Brain Research. Developmental Brain Research, 97*(2), 153–168.

Roux, J., & Robinson-Rechavi, M. (2008). Developmental constraints on vertebrate genome evolution. *PLoS Genetics, 4*. article e1000311.

Ruppert, E. E. (1996). Morphology of Hatschek's nephridium in larval and juvenile stages of *Branchiostoma virginiae* (Cephalochordata). *Israel Journal of Zoology, 42*(Supplement), 161–182.

Salzberg, S. L. (2018). Open questions: How many genes do we have? *BMC Biology*, *16*. article 94.

Schubert, M., Holland, N. D., Laudet, V., & Holland, L. Z. (2006). A retinoic acid-*Hox* hierarchy controls both anterior/posterior patterning and neuronal specification in the developing central nervous system of the cephalochordate amphioxus. *Developmental Biology*, *296*, 190–202.

Scott, R. P., & Quaggin, S. E. (2015). The cell biology of renal filtration. *Journal of Cell Biology*, *209*, 199–210.

Smith, H. W. (1953). *From fish to philosopher*. Boston: Little, Brown, & Company. 264 pp.

Smith, R., Huang, Y.-T., Tian, T., Vojtasova, D., Mesalles-Naranjo, O., Pollard, S. M., et al. (2017). The transcriptio factor *Foxg1* promotes optic fissue closure in the mouse by suppressing *Wnt8b* in the nasal optic stalk. *Journal of Neuroscience*, *37*, 7975–7993.

Soukup, V., Yong, L. W., Lu, T. M., Huang, S. W., Kozmik, Z., & Yu, J.-K. (2015). The Nodal signaling pathway controls left-right asymmetric development in amphioxus. *EvoDevo*, *6*. article 5.

Sperber, S. M., Saxena, V., Hatch, G., & Ekker, M. (2008). Zebrafish *dlx2a* contributes to hindbrain neural crest survival, is necessary for differentiation of sensory ganglia and functions with *dlx1a* in maturation of the arch cartilage elements. *Developmental Biology*, *314*, 59–70.

Stach, T., & Eisler, K. (1998). The ontogeny of the nephridial system of the larval amphioxus (*Branchiostoma lanceolatum*). *Acta Zoologica Stockholm*, *79*, 113–118.

Stemple, D. L. (2005). Structure and function of the notochord: An essential organ for chordate development. *Development*, *132*, 2503–2512.

Stokes, M. D., & Holland, N. D. (1998). The lancelet. *American Zoologist*, *86*, 552–560.

Suryanarayana, S. M., Robertson, B., Wallén, P., & Grillner, S. (2017). The lamprey pallium provides a blueprint of the mammalian layered cortex. *Current Biology*, *27*, 3264–3277.

Tian, T., & Meng, A. M. (2006). *Nodal* signals pattern vertebrate embryos. *Cellular &Molecular Life Sciences*, *63*, 672–685.

Tsagkogeorga, G., Turon, X., Galtier, N. D., Douzery, E. J., & Delsuc, F. (2010). Accelerated evolutionary rate of housekeeping genes in tunicates. *Journal of Molecular Evolution*, *71*, 153–167.

Turbeville, J. M., Schulz, J. R., & Raff, R. A. (1994). Deuterostome phylogeny and the sister group of the chordates: Evidence from molecules and morphology. *Molecular Biology and Evolution*, *11*, 648–655.

Uchida, Y., Uesaka, M., Yamamoto, T., Takeda, H., & Irie, N. (2018). Embryonic lethality is not sufficient to explain hourglass-like conservation of vertebrate embryos. *EvoDevo*, *9*. article 7.

Valentine, J. W. (1997). Cleavage patterns and the topology of the metazoan tree of life. *Proceedings of the National Academy of Sciences USA*, *94*, 8001–8005.

Vopalensky, P., Pergner, J., Liegertova, M., Benito-Guitiérrez, E., Arendt, D., & Kozmik, Z. (2012). Molecular analysis of the amphioxus frontal eye unravels the evolutionary origin of the retina and pigment cells of the vertebrate eye. *Proceedings of the Natlional Academy of Sciences USA*, *109*, 15383–15388.

Wada, H., & Satoh, N. (1994). Details of the evolutionary history from invertebrates to vertebrates, as educed from the sequences of 18S rDNA. *Proceedings of the National Academy of Sciences USA*, *91*, 1801–1804.

Welsch, U. (1975). The fine structure of the pharynx, cyrtopodocytes and digestive caecum of amphioxus (*Branchiostoma lanceolatum*). *Symposia of the Zoological Society of London*, *36*, 17–41.

Wicht, H., & Lacalli, T. C. (2005). The nervous system of amphioxus: Structure, development, and evolutionary significance. *Candian Journal of Zoology*, *83*, 122–150.

Youson, J. H., & McMillan, D. B. (1970a). The opisthonephric kidney of the sea lamprey of the great lakes, *Petromyzon marinus* L. I. The renal corpuscle. *American Journal of Anatomy*, *127*, 207–231.

Youson, J. H., & McMillan, D. B. (1970b). The opisthonephric kidney of the sea lamprey of the great lakes, *Petromyzon marinus* L. II. Neck and proximal segments of the tubular nephron. *American Journal of Anatomy*, *127*, 233–257.

Youson, J. H., & McMillan, D. B. (1971a). The opisthonephric kidney of the sea lamprey of the great lakes, *Petromyzon marinus* L. III. Intermediate, distal, and collecting segments of the ammocoete. *American Journal of Anatomy*, *130*, 55–71.

Youson, J. H., & McMillan, D. B (1971b). The opisthonephric kidney of the sea lamprey of the great lakes, *Petromyzon marinus* L. IV. Intermediate, distal, and collecting segments of the adult. *American Journal of Anatomy*, *130*, 281–303.

Youson, J. H., & McMillan, D. B. (1971c). The opisthonephric kidney of the sea lamprey of the great lakes, *Petromyzon marinus* L. V. The archinephric duct. *American Journal of Anatomy*, *131*, 289–313.

Yue, J.-X., Kozmikova, I., Ono, H., Nossa, C. W., Kozmik, Z., et al. (2016). Conserved noncoding elements in the most distant genera of cephalochordates: The goldilocks principle. *Genome Biology & Evolution*, *8*, 2387–23405.

Yue, J.-X., Yu, J.-K., Putnam, N. H., & Holland, L. Z. (2014). The transcriptome of an amphioxus, *Asymmetron lucayanum*, from the Bahamas: A window into chordate evolution. *Genome Biology & Evolution*, *6*, 2681–2696.

Zhong, Y., Herrera-Úbeda, C., Garcia-Fernàndez, J., Li, G., & Holland, P. W. H. (2020). Mutation of amphioxus *Pdx* and *Cdx* demonstrates conserved roles for *ParaHox* genes in gut, anus and tail patterning. *BMC Biology*, *18*, 68.

Zuasti, A., Agulleiro, B., & Hernandez, F. (1983). Ultrastructure of the kidney of the marine teleost *Sparus auratus*: The renal corpuscle and the tubular nephron. *Cell & Tissue Research*, *228*(1), 99–106.

Further reading

Lacalli, T. C. (2004). Sensory systems in amphioxus: A window on the ancestral chordate condition. *Brain Behavior & Evolution*, *64*, 148–162.

CHAPTER TWENTY-ONE

Big insight from the little skate: *Leucoraja erinacea* as a developmental model system

J. Andrew Gillis[a,b,*], Scott Bennett[b], Katharine E. Criswell[a], Jenaid Rees[a], Victoria A. Sleight[c], Christine Hirschberger[a], Dan Calzarette[b], Sarah Kerr[d], and Jeremy Dasen[e]

[a]Department of Zoology, University of Cambridge, Cambridge, United Kingdom
[b]Marine Biological Laboratory, Woods Hole, MA, United States
[c]School of Biological Sciences, University of Aberdeen, Aberdeen, United Kingdom
[d]Wesleyan University, Middletown, CT, United States
[e]Department of Neuroscience and Physiology, NYU School of Medicine, Neuroscience Institute, NY, United States
*Corresponding author: e-mail address: jag93@cam.ac.uk

Contents

Current Topics in Developmental Biology, Volume 147
ISSN 0070-2153
https://doi.org/10.1016/bs.ctdb.2021.12.016

Abstract

The vast majority of extant vertebrate diversity lies within the bony and cartilaginous fish lineages of jawed vertebrates. There is a long history of elegant experimental investigation of development in bony vertebrate model systems (e.g., mouse, chick, frog and zebrafish). However, studies on the development of cartilaginous fishes (sharks, skates and rays) have, until recently, been largely descriptive, owing to the challenges of embryonic manipulation and culture in this group. This, in turn, has hindered understanding of the evolution of developmental mechanisms within cartilaginous fishes and, more broadly, within jawed vertebrates. The little skate (*Leucoraja erinacea*) is an oviparous cartilaginous fish and has emerged as a powerful and experimentally tractable developmental model system. Here, we discuss the collection, husbandry and management of little skate brood stock and eggs, and we present an overview of key stages of skate embryonic development. We also discuss methods for the manipulation and culture of skate embryos and illustrate the range of tools and approaches available for studying this system. Finally, we summarize a selection of recent studies on skate development that highlight the utility of this system for inferring ancestral anatomical and developmental conditions for jawed vertebrates, as well as unique aspects of cartilaginous fish biology.

black, leathery, tough,
devil's purse begets a skate.
flat, sharp, gazing up...

Anonymous

1. Introduction

Skates, with their remarkable body plan and unusual looking egg cases ("mermaid's" or "devil's purses"), have captured the imaginations of scientists, fishermen, natural historians and artists alike for centuries (Fig. 1). Skates, along with sharks, rays and holocephalans, belong to the cartilaginous fish (chondrichthyan) lineage of jawed vertebrates. Cartilaginous fishes are the sister group to the bony vertebrates (ostheichthyans), the latter including the four most intensively studied vertebrate developmental model systems (i.e., mouse, chick, frog and zebrafish) (Fig. 2). In the 19th and early 20th centuries, descriptive studies of cartilaginous fish anatomy and

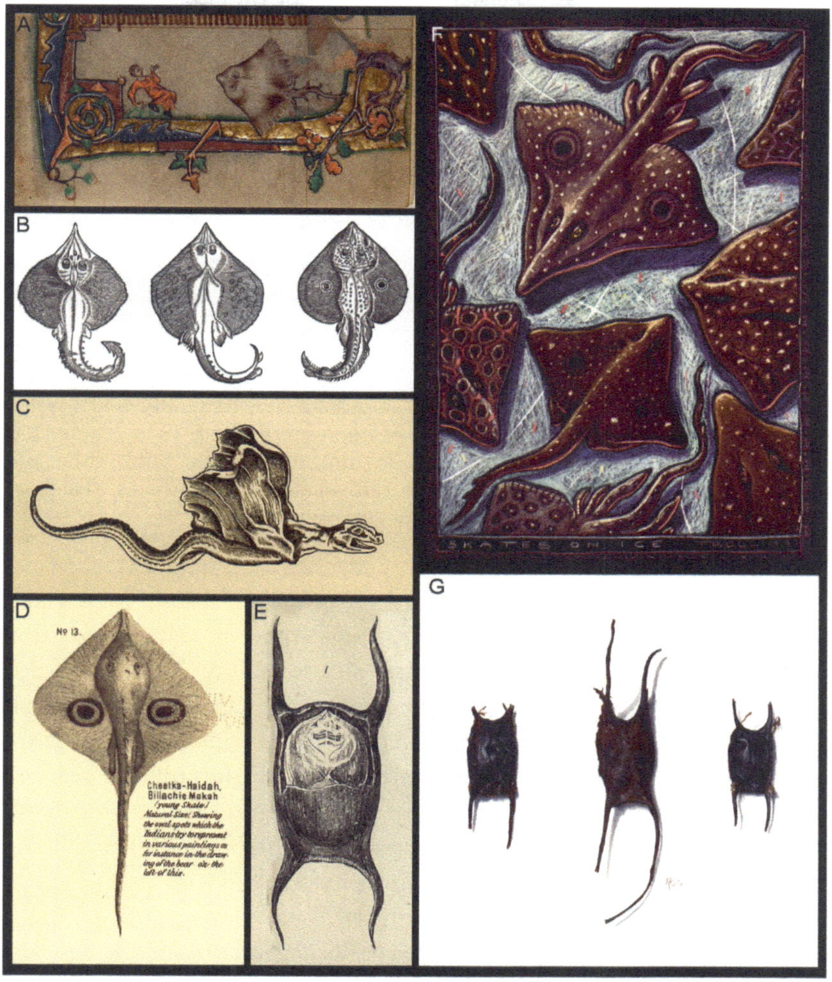

Fig. 1 Skates in art and literature from the 14th century to present. From top left to bottom right: (A) Skate attacks a man (fo.68r) from *The Macclesfield Psalter*, circa 1330. (B) Skate illustrations from Conrad Gessner's *Historium Animalium*, circa 1551–1558. (C) Illustration of a "Jenny Haniver." Touted as a seamonster by fisherman, such specimens were revealed in *Historium Animalium* to be the dried and deformed carcasses of skates. (D) Drawing of a skate from *The Haidah Indians of the Queen Charlottes Islands, British Columbia* (1874) by James Gilchrist Swan. The eyespots from skate wings are noted here as inspiration for the ovoid form that commonly appears in Northwest Coast Native American art. (E) Drawing of a skate egg and embryo from *The common objects of the sea shore; including hints for an aquarium* (1864) by John George Wood. (F) "Skates on ice" (1991) by Ray Troll (https://www.trollart.com). (G) "Skate egg cases" (2020) by Rebecca Gelernter/Near Bird Studios (https://www.nearbirdstudios.com). *Panel A: Reproduced with permission from The Fitzwilliam Museum, Cambridge.*

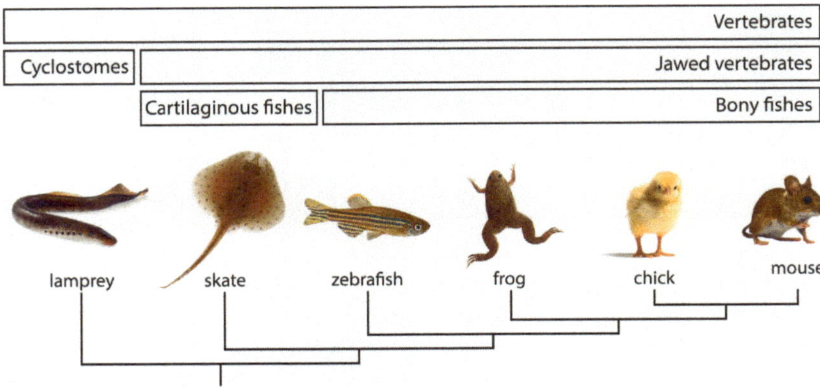

Fig. 2 Evolutionary relationships of selected vertebrate developmental model systems. Vertebrates include the jawless cyclostomes (represented here by a lamprey) and the jawed vertebrates. Jawed vertebrates, in turn, include the bony "fishes" (e.g. zebrafish, frog, chick and mouse) and the cartilaginous fishes (sharks, skates, rays and holocephalans—represented here by the little skate).

embryology abounded (e.g., Balfour, 1878; Dean, 1906; reviewed by Wourms, 1997), owing to their relatively large embryos, and their presumed retention of ancestral body plan features and suitability as a proxy for the ancestral jawed vertebrate. However, it is now appreciated that both cartilaginous and bony vertebrate lineages have evolved considerably since their divergence from a common ancestor approximately 450 million years ago. Accordingly, all cartilaginous and bony vertebrate taxa may be regarded as mosaics of derived and ancestral character states, and only comparative studies, with appropriate taxon sampling in a phylogenetic context, permit the inference of anatomical and developmental conditions present in the last common ancestor of jawed vertebrates (Gillis & Shubin, 2009).

Over the past several decades, two oviparous (egg-laying) cartilaginous fishes have emerged, foremost, as experimentally tractable developmental model systems: the lesser-spotted dogfish (*Scyliorhinus canicula*) and the little skate (*L. erinacea*) (Fig. 3). The dogfish has a lengthy history of embryological study in Europe and occurs in abundance in the Northeast Atlantic Ocean and Mediterranean Sea. The little skate, on the other hand, is commonly encountered in relatively shallow habitats of the Northwest Atlantic, from Nova Scotia, Canada to North Carolina, U.S.A. and has long been used for neuroanatomical, physiological—and, more recently, embryological—research in North America. The utility of the dogfish as an embryological model system has been reviewed previously (Coolen et al., 2008), so the

Fig. 3 An adult male specimen of the little skate (*Leucoraja erinacea*). ZMH 119935 (ex ISH 3690–1979). *Photograph by Dr. Simon Weigmann.*

focus of this article will instead be the little skate. The aims of this review article are three-fold. First, we provide an overview of the collection and husbandry of little skate brood stock, and the anatomy, management and staging of skate eggs. Second, we will present selected stages of skate embryonic development, key developmental events that are occurring at each of these stages, and the approximate developmental times needed to reach each stage. This is not intended as a detailed normal series for the little skate, but rather a broad overview of skate development that will help those new to the system in pinpointing the developmental stage(s) most relevant for their investigations. We will also provide a summary of methods for the observation, manipulation and culture of both early and later stages of skate embryonic development. Finally, we will review selected studies of little skate development, illustrating how this system has illuminated ancestral features of the jawed vertebrate body plan, and showcasing how unique aspects of skate biology have yielded important insight into mechanisms of vertebrate development, growth and regeneration.

2. Collection and husbandry of little skate brood stock

The collection methods described here will focus specifically on the methods employed by the Marine Biological Laboratory's (MBL) Marine Resources Department (MRD) and the R/V Gemma, MBL's 50 ft trawling vessel. Many of the observations reported here are anecdotal, but nevertheless illustrate established practices that serve to maintain healthy skate brood stock at the MBL, and to ensure reliable, year-round availability of skate eggs for developmental biology research.

The little skate is common in the northwest Atlantic Ocean. They are often found on sandy, gravel bottoms where they can camouflage easily. The popularity of little skate as a commercial species has grown in recent years, and while stocks in areas such as the Gulf of Maine are declining, they remain relatively abundant in the local waters surrounding Cape Cod, Martha's Vineyard, and Nantucket. The MRD collects skates in Vineyard Sound, the narrow body of water between Martha's Vineyard and the Elizabeth Islands (Fig. 4A). These relatively shallow, inshore waters

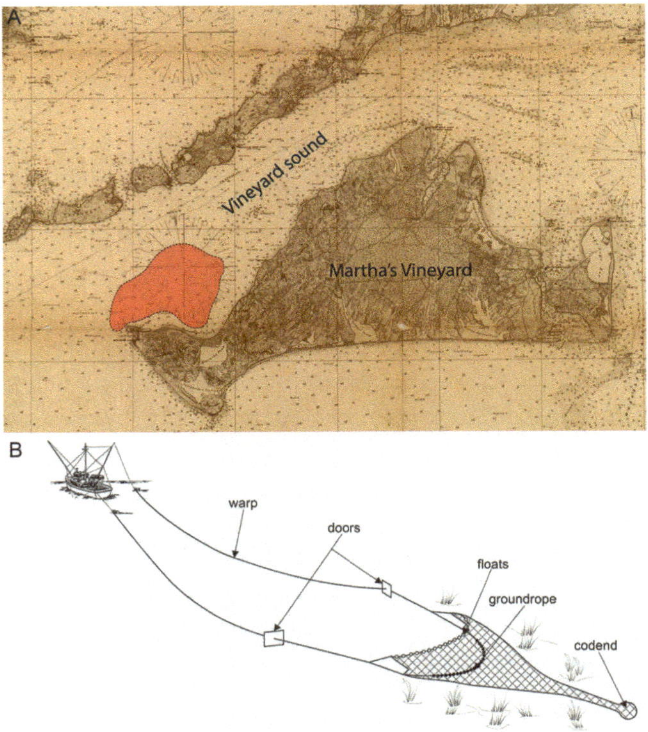

Fig. 4 Collection of little skate brood stock. (A) Adult skates are fished from Vineyard Sound, between Martha's Vineyard and the Elizabeth Islands (area indicated in red). (B) Schematic representation of the otter trawl method used to fish for adult skates. *Panel A: Map modified from "Muskeget Channel to Buzzards Bay and entrance to Vineyard Sound, Mass Coast chart no. 112", Norman B. Leventhal Map & Education Center, Boston Public Library. Panel B: From Kombiadou, K., & Krestinitis, Y.N. (2011) Simulating the fate of mechanically eroded masses in the Thermaikos Gulf.* Continental Shelf Research, 31, *817–831.*

are historically productive fishing grounds, and many commercially valuable species can be collected around under-sea mounts pursuing aggregations of seasonal prey.

Skates are collected by otter trawl (Fig. 4B), a common commercial method for catching ground fish. The net is approximately 100 ft long with a bifurcated end closed by purse strings. The top of the otter trawl is held open by floats, and the bottom is weighed down with chain. Two steel "doors," secured to either side of the net mouth, pull the net open, horizontally, as it is towed through the water. As the chain bottom runs along the sea floor it disrupts the top layer of sediment forcing skates and other demersal species into the mouth of the net. Skates are typically caught at depths of 40-80 ft. at a speed of around 3 knots. This relatively slow speed is essential so as to disrupt buried skates but not pull the net over them. Although skates can be caught year-round, the optimal seasons are Fall and Spring, when ambient water temperatures rarely exceed 14 °C. Anecdotally, we have found that skates caught during warmer months and exposed to water of ~20 °C for prolonged periods of time can exhibit neurological abnormalities that often result in morbidity and death shortly after collection.

As the net is pulled up it is emptied into a large sorting tank on the deck. Skates are moved to a live well from the general sorting tank using flat nets to avoid entanglement. An engine-driven water pump supplies the live well with circulating ambient sea water. Collection trips can run up to and often longer than three hours and use of the live well limits stress during transport and optimizes the long-term health of the skates. During offloading, skates are placed in seawater filled fish totes and wheeled directly into the holding facility located at the end of the dock.

At the MBL Marine Resources Center (MRC), skate brood stock tanks are maintained at 15 ± 0.5 °C, which optimizes the rate of egg production while mitigating the adverse effects of stress and developmental defects associated with increased water temperature (e.g., reduced hydroxyapatite density within the mineralized tesserae of the fin skeleton—Di Santo, 2019). The bottoms of each brood stock tank are covered with sand. This allows adults to bury themselves and gives an optimized environment for females to deposit embryos. The sand is cleaned daily and replaced periodically throughout the year.

Each brood stock tank holds between 15 and 20 adult females, and 1–2 adult males. Male skates possess specialized intromittent organs called claspers, which extend from the base of the pelvic fins. Fertilization in the little skate, as in other species of cartilaginous fishes, is achieved by copulation, with males inserting one of their claspers into the female cloacal aperture, thereby transferring sperm into the female reproductive tract (Luer & Gilbert, 1985; Wourms & Demski, 1993). However, skate copulation is only occasionally observed in brood stock tanks, and the vast majority of eggs are instead fertilized from sperm that is retained within the female reproductive tract from mating in the wild. Skate brood stock are fed a consistent diet of chopped capelin and squid three days a week, and a weekly Mizuri vitamin supplement. Since brood stock temperatures are maintained year-round, there is no seasonal reduction in feeding and consistent egg production is maintained independent of normal seasonal fluctuations. Adult females are kept as brood stock for several months, until they cease laying fertile eggs, at which point they are released and replaced with fresh stock.

3. Management of little skate egg stocks

Once acclimated to captivity, a healthy adult female skate will lay approximately two eggs every 1–2 weeks. Skate eggs take the form of tough, leathery cases, and are generally dark amber or brown in color, and each egg case typically contains a single embryo (Fig. 5A). Skate eggs possess four horns or "tendrils"—one extending from each corner of the roughly rectangular egg chamber—with a pair of straight and curly tendrils demarcating the "anterior" and "posterior" ends of the egg, respectively. Note that, in this instance, the terms "anterior" and "posterior" are used according to convention and with reference to the egg case, but do not relate in any way to the axis of the embryo inside the egg. During oviposition, eggs are extruded, anterior end first, through the cloaca, and are deposited on the bottom of the tank, covered with a stringy, fibrous material that may serve to anchor the egg to the substrate in the wild. However, this material may be readily removed by hand, to facilitate candling of the eggs for staging.

After collection from the bottom of brood stock tanks, skate eggs are maintained at 15 °C in floating baskets according to week of oviposition. Maintenance of skate eggs in highly porous baskets with ample water flow is of vital importance, as later stage skate embryos require water flow through the egg case for respiration (see below). Little skate embryos develop relatively slowly (~20 weeks from oviposition to hatching at 15 °C), and are highly asynchronous in their rates of development, so must be staged individually during selection for experimental work. Grouping skate eggs by week of oviposition allows for rapid and coarse screening of large numbers

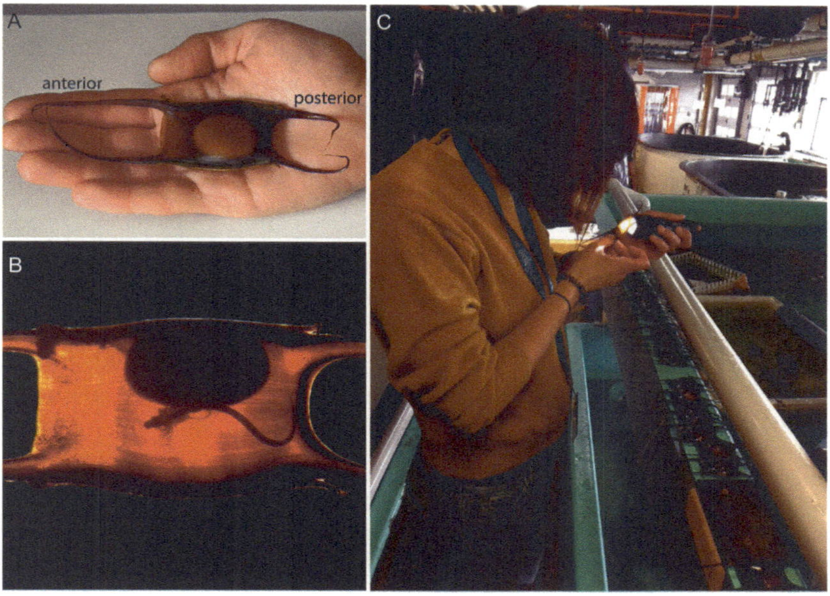

Fig. 5 Staging little skate eggs by candling. (A) Little skate egg cases have an "anterior" end with straight tendrils, and a "posterior" end with curly tendrils. (B, C) Shinning a bright flashlight up through the bottom surface of the egg ("candling") allows visualization of the embryo and yolk through the translucent top surface.

of eggs and identification of one or more baskets of interest, before more careful staging and selection of individual eggs for a particular experiment. In addition to having an anterior-posterior polarity, skate eggs also have a distinct top and bottom, with the top side indicated by the direction of curvature of the posterior tendrils. The top surface of the skate egg case is much more translucent than the bottom surface, and so skate eggs may be most easily staged by candling—i.e., shining a bright (waterproof) flashlight up through the bottom, which will illuminate the yolk and embryo and permit visual staging through the translucent top surface of the egg (Fig. 5B and C).

Skate eggs possess thin slits along the lateral edges of the tendrils, and these slits are initially sealed, ensuring that the developing embryo is housed inside a water-tight compartment. For the first several weeks of development, little skate eggs can therefore be completely removed from the water for staging, without disrupting the embryo. Skate embryos develop atop a large yolk sac, which will slowly rotate inside the egg chamber such that the embryo is always facing up, and so maintenance of skate eggs inside floating baskets often results in embryos that are not positioned optimally (relative to the top surface of the egg) for candling. Early eggs may be removed from their baskets and left resting out of the water (e.g., on a table, benchtop or tank edge) flat on their bottom surfaces for 5–10 min. This will reorient the yolk and embryo into an optimal position for candling and observation through the top surface of the egg (Fig. 5C).

After several weeks of development (see below), the slits along the edges of the tendrils open. At this point, the egg case continues to serve a protective function for the developing embryo and its yolk supply, but the embryo is developing in seawater, which flows freely through the egg chamber. At these later stages of development, seawater will drain out of the egg chamber through the slits if eggs are removed from the water, exposing the embryos to air. Brief exposure to air will not harm the embryo, but retention of air bubbles inside the egg chamber will cause the egg to float when returned to the water, and this could lead to desiccation of the embryo inside. Eggs with open tendril slits must therefore be "burped" when returned to the water, by holding the egg with the anterior or posterior end pointing up and gently squeezing the egg chamber to expel any air and refill the egg chamber with seawater.

4. Selected stages of skate embryonic development

Below, we present an overview of selected stages of little skate embryonic development, with notes about particular features of embryological and/or practical significance. There is not currently a normal staging series for the little skate, though a complete staging table has been published for the lesser spotted dogfish, *S. canicula* (Ballard, Mellinger, & Lechenault, 1993), and an abbreviated table for the winter skate, *Leucoraja ocellata* (Maxwell, Fröbisch, & Heppleston, 2008). Here, we use a combination of informal nomenclature, as well as references to roughly corresponding *S. canicula* or *L. ocellata* stages or stage ranges, to describe key stages of little skate development. We also give an indication of the developmental time needed to reach each of these stages (at 15 °C) in days post-oviposition (dpo), though as mentioned above, skate eggs do not develop synchronously, and there can be considerable variation in rates of development (even between sibling eggs). The timings provided are, therefore, necessarily approximate, and all embryos must always be individually staged more precisely by candling before use.

4.1 Blastodisc to neurula (~S4–17; ~0–10 dpo)

At the time of oviposition, the little skate embryo is a blastodic of several hundred cells on top of a large, undivided yolk mass (Fig. 6A). While gastrulation has not been studied in the little skate, it has been studied extensively in the shark, *S. canicula* (Coolen et al., 2007; Coolen, Menuet, & Mazan, 2009; Godard et al., 2014; Godard & Mazan, 2013; Sauka-Spengler, Baratte, Lepage, & Mazan, 2003; Sauka-Spengler, Plouhinec, & Mazan, 2004; Vandebroek, 1936), and it appears (based on external morphology) that the process of gastrulation is likely largely conserved between sharks and skates. Briefly, in *S. canicula*, the blastodic overlies a subgerminal cavity (likely equivalent to the blastocoel), and thickens along its posterior margin, forming a bi-layered overhang of columnar epithelium. An indentation subsequently appears at the center of this posterior thickening, and this indentation marks the axial mesoderm progenitors (i.e., the "notochord triangle"; Criswell, Coates, & Gillis, 2017b).

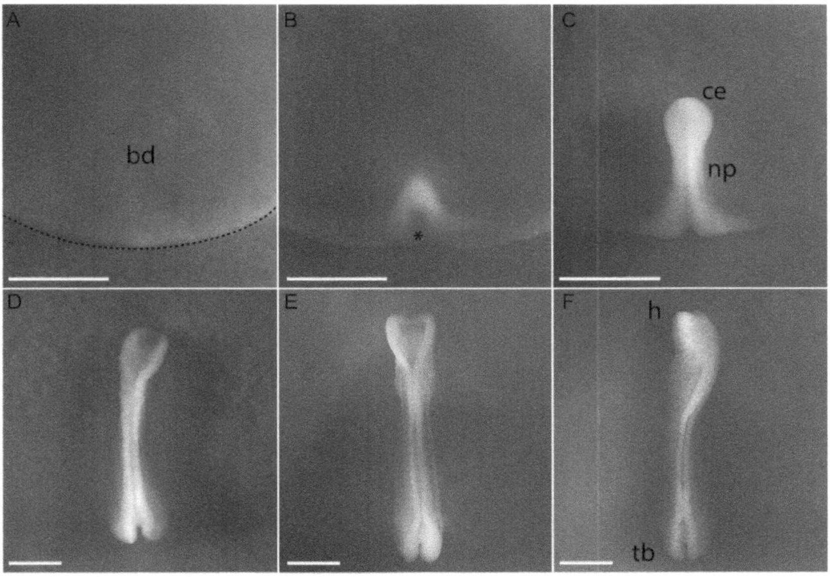

Fig. 6 Gastrulation and neurulation in embryos of the little skate. (A) When skate eggs are first laid, the embryo is a blastodisc (*bd*) on top of a large yolk sac. (B) Mesendoderm is internalized by involution over the posterior margin of the blastodisc, and onset of axial elongation is marked by the appearance of axial mesodermal progenitors in the form of a triangle (the "notochord triangle," indicated by *). (C) A cephalic enlargement (*ce*) appears at the anterior end of the neural plate (np). (D–F) As the neural tube closure proceeds, the embryo acquires a distinct head (h) and tailbud (tb). Scale bars: 1 mm.

Internalization of the mesendoderm occurs by involution, along the posterior margin of the blastodisc, on either side of the notochord triangle. Axial elongation follows, with the appearance of a cephalic enlargement at the anterior extreme of the neural plate. Neural folds from either side of the neural plate fuse at the dorsal midline, giving rise to a dorsal neural tube with a prominent central lumen, coincident with the onset of somitogenesis. By S17 of *S. canicula* development, neurulation is complete, with the embryo possessing distinct brain and spinal cord regions of the neural tube, and up to 17 pairs of somites.

In the early little skate embryo, a thickening along the posterior edge of the blastodisc (comparable to that described in *S. canicula*) is formed by ~5–6 dpo. A prominent notochord triangle is apparent by 8dpo (Fig. 6B), and lineage tracing has confirmed that this domain gives rise to the embryonic notochord. Formation of the neural plate, neural tube closure, somitogenesis and axial elongation (Fig. 6C–F) appear broadly as described in *S. canicula*, and neurulation is complete by ~10 dpo.

4.2 Line-on-yolk (~S18; ~11–12 dpo)

At this stage of little skate development, the embryo is apparent as a prominent line on the yolk (~3–4 mm in length) upon candling, with a subtle dorsal curvature, but is not yet moving (Fig. 7A). This stage is notable because neural tube closure is complete, and cranial neural crest cells are specified within the dorsal neural tube but have not yet undergone epithelial-to-mesenchymal transition (Martik et al., 2019; Sleight & Gillis, 2020). In the skate, as in other deuterostomes, pharyngeal arches are delineated by an iterative series of endodermal pouches, which bulge out from the foregut endoderm and contact the overlying surface ectoderm. In non-chordate deuterostomes and fishes, these pharyngeal endodermal pouches fuse with overlying ectoderm, giving rise to gill slits. At the line-on-yolk stage of skate development, two pharyngeal endodermal pouches have formed (delineating the future mandibular and hyoid arches), but these pouches have not yet contacted the overlying surface ectoderm. Finally, this stage of skate development is characterized by the presence of molecularly distinct domains of cranial paraxial and lateral plate mesoderm (Sleight & Gillis, 2020), and ~20–30 pairs of somites.

4.3 Wiggler (~S19–20; ~15–17 dpo)

This stage of skate development is so named because it is the stage at which the embryo begins to move. Movement is initially observed as a side-to-side

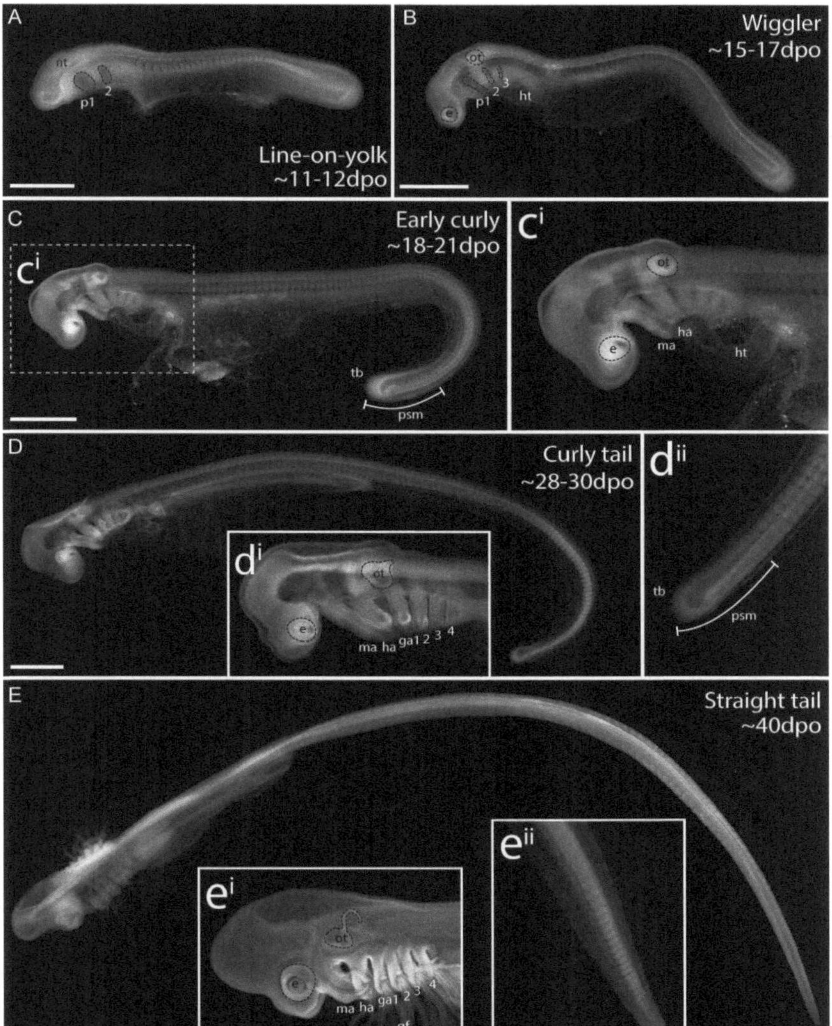

Fig. 7 Post-gastrulation embryonic development of the little skate. (A) At the line-on-yolk stage, neural tube (*nt*) closure is complete, and the embryo possesses two pharyngeal pouches (*p1,2*). (B) At wiggler stage, the embryo begins to move, and possesses a distinct eye (*e*), otic vesicle (*ot*), heart (*ht*) and three pharyngeal pouches (*p1,2,3*). (C) Early curly embryos are characterized by a curved tail with a tail bud (*tb*) following the presomitic mesoderm (*psm*), a broad attachment to the yolk on the ventral side, and distinct mandibular (*ma*) and hyoid (*ha*) arches. (D) Curly tail embryos still possess a curved tail with a rounded tail bud (*tb*) and some unsegmented presomitic mesoderm (*psm*), but their attachment to the yolk has narrowed to form a yolk stalk. At this stage, the mandibular, hyoid and first 2–3 gill arches (*ga1–3*) have fully formed, and gill buds (precursors of the external gill filaments) become apparent on the hyoid and gill arches. (E) By straight tail stage, somitogenesis has terminated, and the posterior extreme of the embryo has transformed from a rounded tail bud to a pointed tail tip. Embryos possess mandibular (*ma*), hyoid (*ha*) and gill arches 1–4 (*ga1–4*), and long external gill filaments (*gf*). Scale bars: 1 mm.

twitching of the head, which then elaborates into a regular, alternating bending of the body axis. The tail has elongated to the point that it now begins to exhibit a downward bending, at an angle of ~30–45° to the main body axis (Fig. 7B). At this stage, three pharyngeal endodermal pouches have formed (but have not yet fused with surface ectoderm). Cranial neural crest cell migration is occurring, with three distinct streams (mandibular, hyoid, and branchial) of neural crest cells originating from the hindbrain region and populating the developing pharyngeal arches. Neural crest cell migration has also begun from the rostral trunk neural tube, while the caudal trunk neural crest cells are specified (but not yet migrating) (Gillis, Alsema, & Criswell, 2017). Wigglers possess a beating heart, an otic vesicle and up to 60 pairs of somites.

4.4 Early curly (~S21–23; ~18–21 dpo)

At the early curly stage (Fig. 7C), posterior elongation and somitogenesis continue, with embryos possessing 80–90 pairs of somites. The tail takes on a pronounced curvature and terminated with a rounded tail bud that sits at the end of the presomitic mesoderm. At this stage, the embryo still sits directly on top of the yolk, with a broad attachment between the ventral surface of the embryo and the yolk sac epithelium. The mandibular and hyoid arches are fully formed, and gill arches 1–3 are delineated by pharyngeal pouches (though these pouches have not yet fused with surface ectoderm). Cranial neural crest cells have populated the pharyngeal region, and trunk neural crest cell migration is occurring.

4.5 Curly tail (~S24; ~28–30 dpo)

By this stage of development (Fig. 7D), the tail remains prominently curved with a rounded tail bud as posterior elongation and somitogenesis continue (embryos possess 100–120 pairs of somites), but the embryonic attachment to the yolk sac has constricted into a yolk stalk (rather than a broad attachment along the ventral side of the embryo). The mandibular, hyoid and first 2–3 gill arches have fully formed, and it is during this stage that dorsoventral patterning of the pharyngeal arches, via a conserved transcriptional network including endothelin signaling, bmp signaling and nested expression within the pharyngeal arches of a host transcription factors and co-factors, is underway (Gillis, Modrell, & Baker, 2013; Hirschberger, Sleight, Criswell, Clark, & Gillis, 2021). Other notable embryonic features at this stage include the appearance of distinct lateral line placodes (Gillis et al., 2012), and the external gill filaments from hyoid and gill arch epithelium (Gillis & Tidswell, 2017). It is at this stage that embryos may be carefully cultured *ex ovo* (see next section for details).

4.6 Straight tail (∼S25–27; ∼40 dpo)

This 2–3-week period of skate development (Fig. 7E) is marked by the termination of somitogenesis (with a final count of ∼130–140 somites), the straightening of the tail, and the transition from a bulbous tail bud to a pointed tail tip (Criswell & Gillis, 2020). By this period, all pharyngeal arches have formed, and the hyoid arch and gill arches 1–4 begin to undergo a lateral expansion (ultimately giving rise to the interbranchial septa that derive from these arches) (Gillis & Hall, 2016). This period of development also sees the appearance and outgrowth of pectoral and pelvic fins, which initiate as slight protrusions from the trunk and then expand into half-moon-shaped fin buds, as well as elongation of the cranial and trunk lateral line sensory ridges (Baker, Modrell, & Gillis, 2013; Gillis et al., 2012), and initial differentiation of the spiral valve intestine (Theodosiou & Oppong, 2019).

4.7 S29 (∼50 dpo)

This stage of skate development (Fig. 8A) is notable because it is the stage at which the egg case tendril slits open, and seawater begins to flow through the egg. Embryos at S29 possess both long, external gill filaments and internal gill lamellae. At this stage, the pectoral fin buds have changed from a half-moon to a more elongate, rounded rectangular shape, with the anterior edge of the

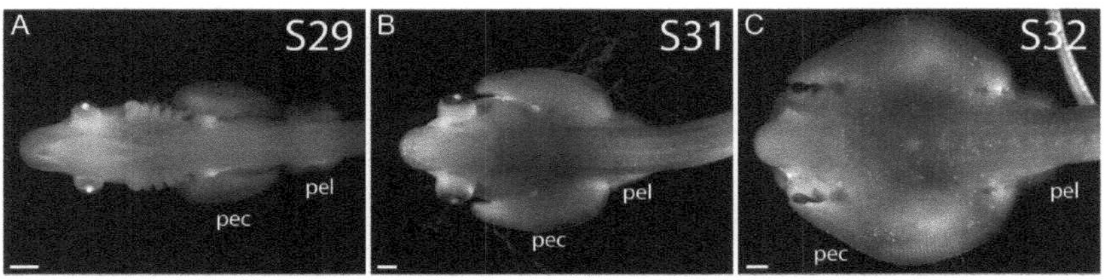

Fig. 8 Later stages of little skate development. (A) At stage (S)29, the pectoral (pec) and pelvic (pel) fins have taken on an elongate and slightly rectangular shape, with the pectoral fin extending as far forward as gill arch 4. In this embryo, external gill filaments have been dissected away to reveal fin morphology. (B) By S31, the pectoral fins are undergoing a pronounced anterior expansion, and by (C) S32, the pectoral fins have fused with the rostrum, forming the closed pectoral fin disk that is characteristic of adult morphology. Scale bars: 1 mm.

pectoral fin bud sitting at approximately the level of gill arch 4. At S29, the pectoral fin bud prominently expresses *Shh* in the posterior zone of polarizing activity (Dahn, Davis, Pappano, & Shubin, 2007), and *Fgf8* in the apical ectodermal ridge (Gillis, Dahn, & Shubin, 2009a), with first onset of these signaling center expression patterns occurring during the previous extended straight tail stage.

4.8 S30/31 (~60-65 dpo)

This stage is of considerable significance for skeletal development, as it includes preskeletal mesenchymal condensation and early differentiation of the cartilaginous endoskeleton of the head and trunk (Gillis, Dahn, & Shubin, 2009b; Marconi, Hancock-Ronemus, & Gillis, 2020). At this stage, the pectoral fins are expanding rostrally (Fig. 8B), with this expansion driven by a novel anterior apical ectodermal ridge-like signaling center (Nakamura et al., 2015; Barry & Crow, 2017—and see below). However, the pectoral fins have not yet fused into a pectoral fin disc at the tip of the rostrum. The axial skeleton takes the form of a continuous, unsegmented cartilaginous tube that surrounds the notochord and spinal cord (Criswell, Coates, & Gillis, 2017a) and rostrally, includes a differentiating synarcual (a compound element that derives from several axial skeletal segments, and that forms from the anterior vertebral column—Johanson, Martin, Fraser, & James, 2019). This stage also sees the onset of tooth and dermal denticle development (with these appearing initially as epithelial thickenings or placodes) (Miyake, Vaglia, Taylor, & Hall, 1999; Rasch et al., 2020), differentiation of lateral line sense organs (mechanosensory neuromasts and electrosensory ampullary organs) (Gillis et al., 2012), and the onset of clasper development in males (O'Shaughnessy, Dahn, & Cohn, 2015).

4.9 S32 (from ~80 dpo)

This stage is marked by the fusion of the pectoral fins at the tip of the rostrum, to form the pectoral fin disc that is characteristic of batoid fishes (Fig. 8C). By S32, most organ systems have differentiated, as have most elements of the endoskeleton, and the axial skeleton has subdivided into distinct vertebrae (Criswell et al., 2017a).

4.10 S33-hatching (from ~90 to 100 dpo)

During this extended period of development, the embryo grows to fill the inside of the egg chamber. Over this period, we see near-complete resorption of the yolk, appearance of pigmentation in the epidermis, eruption of teeth and dermal denticles (Gillis et al., 2017; Underwood et al., 2015), and development of superficial calcified plates ("tesserae") on the surface of the cartilaginous endoskeleton (Marconi et al., 2020).

Skates begin to hatch from their eggs after ~5 months of development at 15 °C. As the embryos near hatching, they position themselves inside the egg such that their rostrum and tail are pressed against the anterior and posterior ends of the egg chamber, respectively. Little skate hatching occurs at night, with hatchlings pushing against the posterior wall of the egg chamber with their tail, and emerging, head-first, through a seam in the egg case that is located between the two anterior (straight) tendrils (Video 1 in the online version at https://doi.org/10.1016/bs.ctdb.2021.12.016). At the time of hatching, little skates resemble miniature adults, and are completely self-sufficient. Hatchlings will continue to feed off of internalized remnants of their yolk supply for approximately two weeks after emerging from the egg, at which point they will begin to feed independently.

5. Observation, manipulation and culture of little skate embryos

Owing to its slow rate of embryonic development, lengthy time to sexual maturity, relatively large adult size and complex husbandry requirements, the little skate is not currently a genetically tractable model system (that is, in the sense of maintaining stable transgenic lines). However, its oviparity, with large embryos that are robust and tolerant of observation and manipulation, renders the little skate particularly well suited to a range of experimental embryological manipulations and approaches, comparable to those typically used in chicken embryos. As with many emerging model systems, the limiting factor for experimental work with skate embryos has been challenges associated with the successful culture of embryos following experimental manipulation. The embryos of oviparous chondrichthyans have long been viewed as particularly challenging in this regard, and consequently much of the early literature on chondrichthyan embryology provides beautiful, detailed descriptions of embryonic development, though with little evidence of underlying developmental mechanisms, cell movements or tissue interactions. For example, until relatively recently, the only cell lineage tracing experiments in a chondrichthyan system were those of Vandebroek (1936), who used vital dye labeling of early *S. canicula* embryos to track cell movements during gastrulation. These experiments were extremely insightful and did accurately demonstrate some cell movements around the blastodic and gastrulating *S. canicula* embryo. However, labeled embryos only survived for a few days, and this precluded the longer-term mapping of early embryonic tissues to later-forming differentiated organs and cell types.

When we began working with the little skate as an embryological model system, we spent a considerable amount of time testing and optimizing methods for manipulation and culture of both early and later stage skate embryos. It is now very feasible to manipulate any stage of skate embryo, and to culture manipulated embryos for as long as is needed. Much of this comes down to considerations of the embryonic environment inside the egg case and conducting manipulations (and subsequently culturing manipulated embryos) in ways that closely maintain or mimic this environment. Broadly speaking, approaches to manipulation of skate embryos differ depending on whether the manipulations are taking place *in ovo* (i.e., from blastodisc to early curly stage) or *ex ovo* (i.e., from curly tail stage onward), and we will outline both of these approaches below.

5.1 *In ovo* manipulation of early skate embryos

Skate embryos develop atop a large yolk sac, which initially occupies approximately half of the volume of the egg chamber and maintaining the integrity of this yolk is of utmost importance when carrying out embryonic manipulations. From blastodisc to early curly stages, the embryonic yolk is extremely soft and delicate, and if the embryo is removed from the egg case, the yolk will quickly break, leading to death of the attached embryo. At oviposition, the yolk and embryo inside the egg chamber are surrounded and supported by a clear gelatinous substance (a "jelly layer"), which fills the space between the yolk and the eggshell on all sides. Over the first few weeks of development, this jelly layer gradually thins, and by curly tail stage, the embryo and yolk may be carefully removed from the egg with the yolk remaining intact (see below). Prior to this stage, all manipulations must occur *in ovo*, with the embryo remaining inside its protective jelly layer.

We have found that *in ovo* manipulation of early skate embryos may be readily carried out by windowing the top surface of the egg case (Fig. 9A), and applying gentle pressure to the bottom surface, which raises the embryo up to the level of the window and thins the jelly layer such that the

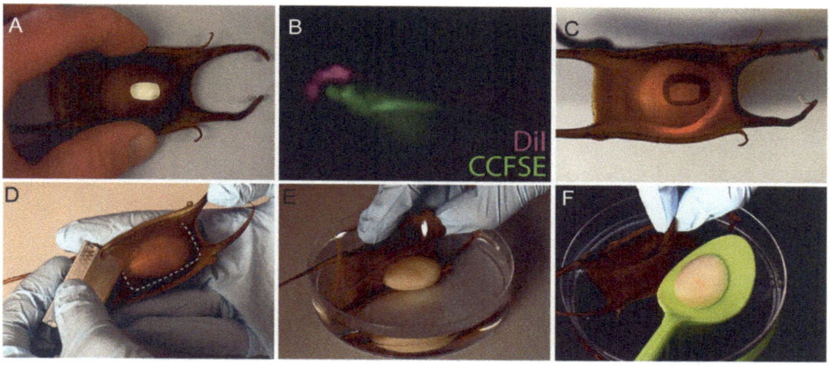

Fig. 9 Accessing skate embryos for *in ovo* and *ex ovo* manipulation. (A) To manipulate early skate embryos *in ovo*, the embryo may be accessed through a small window in the top surface of the egg case. (B) Application of gentle pressure to the bottom of the egg will raise the embryo to the window, but care must be taken not to disrupt the transparent jelly layer that sits between the embryo and the eggshell. Manipulations must be conducted through this jelly layer. In this example, the cranial neural tube has been injected with CM-DiI, and the pharyngeal cavity with CCFSE. (C) Egg cases may then be sealed using piece of donor eggshell fixed in place with a cyanoacrylate adhesive and returned to the seatable for growth to the desired stage. (D) For *ex ovo* manipulation of later stage skate embryos, egg cases may be opened along three sides with a razor blade, (E) the embryo and yolk may be gently poured into a dish of seawater. (F) Once a manipulation is complete, the embryo and yolk may be moved back into the egg case for growth using a plastic soup spoon.

embryo may be manipulated (using fine instruments or needles) through the jelly (Fig. 9B). Once an embryonic manipulation is complete, the egg may be sealed using a piece of donor eggshell (pre-cut and sized slightly larger than the window) and a cyanoacrylate adhesive (e.g., Krazy glue gel) (Fig. 9C). It is imperative that the seal around the donor eggshell be air- and watertight, and it is best to avoid air bubbles inside the re-sealed egg chamber as much as possible (as larger air bubbles can lead to desiccation of the jelly layer and/or embryo and can cause the egg to float upon return to the tank). Maintaining gentle pressure on the egg to keep the jelly layer up to the level of the window while re-sealing the egg will ensure a tight seal and will protect the integrity of the jelly layer and yolk. Once the window has been sealed and the adhesive has dried, eggs may be returned to a sea table and left to develop until the desired developmental stage is reached. For *in ovo* manipulations, care must be taken not to puncture the yolk or disrupt the jelly layer, but sterile technique or addition of antibiotics before resealing are not needed. We have used this approach extensively to label the endoderm (Gillis & Tidswell, 2017), mesoderm (Criswell et al., 2017b;

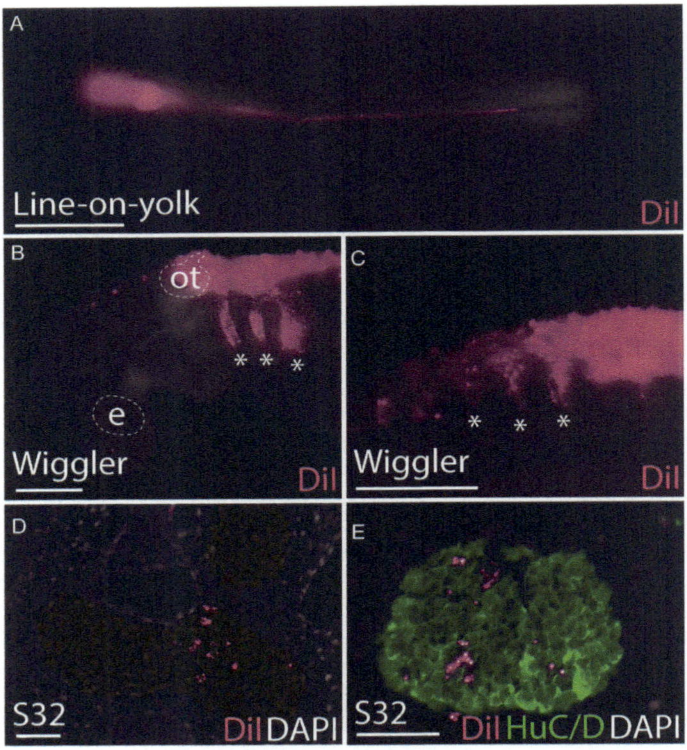

Fig. 10 CM-DiI labeling of the little skate neural tube *in ovo*. (A) The lumen of the neural tube in skate embryos may be microinjected with CM-DiI, to label cells of the neural tube as well as pre-migratory neural crest cells. (B) Five days post-injection, CM-DiI-labeled neural crest cells may be seen migrating from the cranial and (C) trunk neural tube. (D) Several weeks later (at S32), CM-DiI-labeled, neural crest-derived cells may be recovered within the cartilages of the gill arches, and (E) in dorsal root ganglia. Scale bars: a, 1 mm; b, 500 μm; c, 300 μm; d–e, 25 μm.

Gillis et al., 2017; Sleight & Gillis, 2020) and neural crest cells (Gillis et al., 2017; Sleight & Gillis, 2020) of the early skate embryo with lipophilic dyes, and to recover contributions of these germ layers to organs and cell types as late as embryonic stages 32/33 (e.g., Fig. 10A–E).

5.2 *Ex ovo* manipulation of later stage skate embryos

By curly tail stage, the yolk sac epithelium has strengthened to the point that the embryo may be removed from its egg case, and observations or manipulations may be performed *ex ovo*. The egg chamber may be opened by simply cutting the top surface along three sides with a razor blade, and the resulting flap lifted (Fig. 9D), so that the embryo and its yolk may be

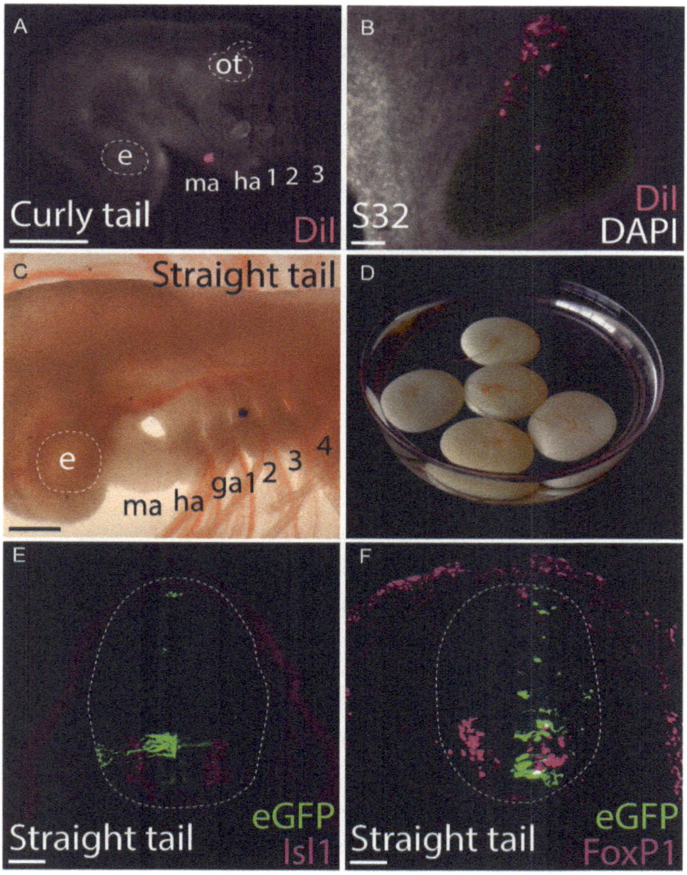

Fig. 11 *ex ovo* experimental manipulation of skate embryos. From curly tail stage onward, embryos may be removed from and manipulated outside of their egg cases. (A) Focal labeling of ventral mandibular arch mesenchyme with CM-DiI at the curly tail stage results in (B) CM-DiI-labeled chondrocytes within the lower jaw (Meckel's) cartilage at S32. (C) An example of surgical implantation of an Affi-Gel blue bead beneath the epithelium of the second gill arch. (D) For systemic treatment with precise concentrations of small molecules, skate embryos may be reared in petri dishes of drug-containing seawater. (E) Expression of enhanced green fluorescent protein (eGFP) driven by the cytomegalovirus (CMV) promotor in cells of the skate neural tube following lentiviral transfection or (F) plasmid electroporation at the straight tail stage. Images in (E) and (F) by Myungin Baek and Heekyung Jung, respectively. Scale bars: a,c, 500 μm; b,e,f, 25 μm.

gently poured into a petri dish of seawater (Fig. 9E). Embryos at these later stages of development are actively wiggling but may be temporarily immobilized by anesthesia with a 100 mg/L solution of MS-222 in seawater. Once anesthetized, embryos may be microinjected (Fig. 11A and B), or otherwise surgically manipulated (Fig. 11C), then recovered in fresh seawater before being returned to their egg case. Embryos at these stages may be moved between petri dishes or transferred from a petri dish back into their opened egg case, using a plastic soup spoon (taking care when transferring not to pinch or damage the yolk) (Fig. 9F). At these stages, it is not necessary to re-seal the egg case before returning it to a sea table or aquarium. Although the egg case tendril slits do not open until S29, embryos from curly stage onward may be maintained in untreated natural or artificial seawater and will continue to develop normally under these conditions. However, care must be taken to ensure that the embryo's tail does not become trapped in the egg case flap.

5.3 Molecular manipulation of skate embryos

Small molecules or other pharmacological agents may be delivered to early skate embryos *in ovo* (Dahn et al., 2007; Gillis et al., 2009a; Gillis & Hall, 2016), into the space between the jelly layer and the embryos—either by injecting with a high gauge needle and syringe through the egg case and jelly layer, or by injecting underneath the jelly layer though an egg window and re-sealing—with the caveat that only a single dose may be delivered this way (so if the drug being used is reversible, this will inevitably result in a transient effect). Systemic drug treatments may also be performed from the curly tail stage of development, either by injecting the drug into the egg chamber (prior to the point at which the egg chamber becomes open to surrounding seawater), or by removing embryos from their egg cases as described above and rearing them in petri dishes of drug-containing seawater (Fig. 11D). The latter approach allows for regular changes of drug and seawater, and therefore sustained treatment at more precise concentrations. For *ex ovo* drug treatments or controls, we typically maintain up to 6 embryos in a standard 100 cm Petri dish in 40 mL of seawater.

Finally, it has recently been demonstrated that misexpression experiments may be carried out in skate embryos using amniote expression vectors (e.g., green fluorescent protein expression driven by the human cytomegalovirus promoter/enhancer) and tissue-specific transfection by injection of lentiviruses (Fig. 11E) or by electroporation (Fig. 11F; Jung et al., 2018). With these approaches, and with the recent development of single plasmid systems for efficient conditional misexpression (Chu, Nguyen, Smith, Vavrušová, & Schneider, 2020) and CRISPR-Cas9 genome editing (Gandhi et al., 2021) in embryos of traditionally non-genetic model organisms, skates now hold even greater promise as a system in which both embryological and sophisticated molecular and functional genomic experiments are feasible.

6. Skates offer insight into ancestral and derived jawed vertebrate developmental mechanisms

As a member of the sister group to the bony vertebrates, skates are well situated to inform ancestral anatomical conditions and developmental mechanisms for jawed vertebrates. For example, ancestral jawed vertebrate skeletal or neuroanatomical conditions may be inferred based on conservation between skate and other, more widely studied bony vertebrate model systems, such as mouse, zebrafish, chick, and frog. But equally, and like all taxa, skates also exhibit derived anatomical and developmental conditions that present opportunity to investigate, for example, the cellular and molecular underpinnings of adaptation to a benthic lifestyle or extended post-embryonic growth and repair of a cartilaginous endoskeleton. Here, we will briefly discuss examples of how research on skate development has shed light on the nature of the last common ancestor of jawed vertebrates, on a striking example of anatomical novelty, and on unique aspects of skeletal biology that may have relevance to the treatment of human disease.

6.1 Skates inform ancestral mechanisms of vertebral development and segmentation

The vertebral column is a defining feature of the jawed vertebrate body plan, but the embryonic origin, evolution and homology of vertebral skeletal components across jawed vertebrate lineages remains unresolved. Teleost fishes (e.g., zebrafish and salmon) possess vertebral centra consisting of an inner bony layer that derives from the notochord sheath, an outer bony layer that is secreted by cells of paraxial mesodermal (i.e., somitic) origin, and neural/hemal arches that form by endochondral ossification of somite-derived progenitors (Fleming, Keynes, & Tannahill, 2004; Grotmol, Kryvi, Nordvik, & Totland, 2003; Nordvik, Kryvi, Totland, & Grotmol, 2005). Tetrapod vertebrae, on the other hand, are composed of a centrum that forms as a single, solid endochondral ossification and neural/hemal aches, all of which are entirely of somitic origin (Bagnall, Sanders, Higgins, & Leam, 1988; Piekarski & Olsson, 2014). We previously found that skate vertebrae are composed of centra with three distinct layers: an inner layer of hyaline cartilage, a middle areolar calcification, and an outer layer of hyaline cartilage that is continuous with the neural and hemal arches (Criswell et al., 2017a). However, in the absence of data on the embryonic origin of vertebral tissue layers in cartilaginous fishes, it is impossible to infer ancestral somitic vs notochordal contributions to the vertebral skeleton of jawed vertebrates.

To directly test the embryonic origin of the skate vertebral skeleton, we used the lipophilic dye CM-DiI to label notochordal and somite progenitors in gastrulation and curly tail skate embryos, respectively (Criswell et al., 2017b). We then tested for contributions of these tissues to the vertebral skeleton at S32. We found that all three layers of the vertebral centrum, and neural/hemal arches, are of somitic origin in skate. While the notochord persists in the center of the skate axial skeleton, we found no evidence of a notochordal contribution (either cellular, or by secretion of extracellular matrix) to skate vertebral centra. These findings indicate that the vertebral skeleton of jawed vertebrates was ancestrally of somitic origin, with the secondary acquisition of a notochordal contribution to the inner vertebral centrum in teleost fishes (Fig. 12A).

Another example of developmental divergence between the axial skeletons of tetrapods and teleost fishes relates to mechanisms of vertebral segmentation and the relationship between somites and vertebrae. In tetrapods, somites do not correspond 1:1 with vertebrae - rather, there is a frameshift between somites and vertebrae, such that one vertebra derives from the posterior and anterior halves of two adjacent somites (Bagnall, Higgins, & Sanders, 1988; Piekarski & Olsson, 2014; Ward, Evans, & Stern, 2017). This process is termed "resegmentation", and results in the preservation of original somite boundaries at the center of each vertebra. However, in zebrafish, cells from adjacent somites mix as they migrate, resulting in a lack of clear resegmentation and a loss of somite boundaries within the vertebral skeleton (Morin-Kensicki, Melancon, & Eisen, 2002). Skates exhibit a unique mode of vertebral development, in which somite-derive mesenchyme condenses around the neural tube and notochord, and differentiates into a continuous cartilaginous tube, which then subdivides into discrete vertebrae (Criswell et al., 2017a). By tracking the fate of neighboring somites with CM-DiI and SP-DiOC$_{18}$, we found in skate that somite boundaries are retained through differentiation of the continuous cartilaginous tube and vertebral segmentation and sit in the middle of individual vertebrae (Criswell & Gillis, 2020). This indicates that skates undergo tetrapod-like resegmentation in the transition from somites to vertebrae. Thus, despite a complex evolutionary history, with extensive

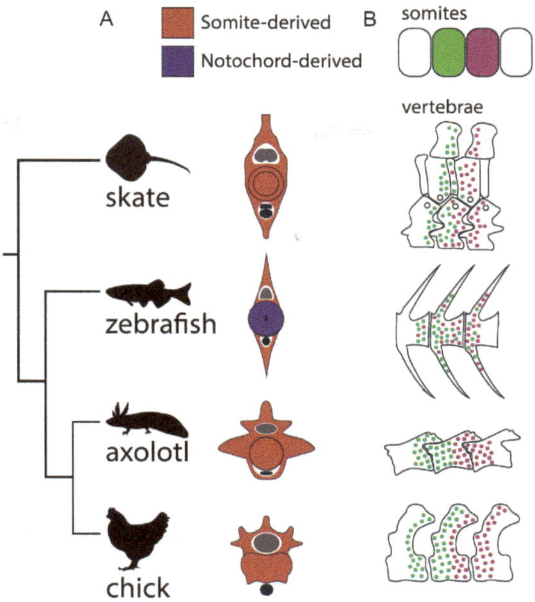

Fig. 12 Development and evolution of the jawed vertebrate axial skeleton. (A) Tetrapods possess vertebral bodies that derive entirely from somites, while zebrafish vertebral centra are of dual somitic and notochordal origin. Skate vertebral bodies derive entirely from somites, pointing to an ancestral somitic origin of vertebrae in jawed vertebrates, and secondary acquisition of a notochord-derived inner centrum in teleosts. (B) The little skate exhibits tetrapod-like axial skeletal resegmentation, with somite boundaries preserved in the middle of vertebral bodies. Resegmentation is therefore an ancestral feature of the jawed vertebrate axial skeleton.

convergent or parallel evolution of vertebral subcomponents (Arratia, Schultze, & Casciotta, 2001; Fleming, Kishida, Kimmel, & Keynes, 2015), the vertebral skeleton of jawed vertebrates appears to have an ancestral somitic origin and mode of segmentation that is deeply conserved and shared between cartilaginous fishes and tetrapods (Fig. 12B).

6.2 Neural substrates for bipedal locomotion are ancestral for jawed vertebrates

In addition to pectoral fin-driven swimming, skates are capable of generating walking-like behaviors on the seafloor using their pelvic fins (Koester & Spirito, 2003). Pelvic fin-driven locomotion displays features analogous to those of tetrapods, including left-right alternation and reciprocal contraction of pelvic flexor and extensor muscles. In tetrapods this sequence of muscle activation is controlled by specialized types of limb-level spinal motor neurons and interneurons, which were thought to have evolved near the time tetrapods transitioned to land (Dasen, De Camilli, Wang, Tucker, & Jessell, 2008).

Through comparative expression analyses of transcription factors and synaptic specificity determinants, it was shown that skates express the majority of genes known to be essential for the development of tetrapod limb circuits (Jung et al., 2018). One significant difference between skates and tetrapods is the relative distribution of fin/limb-innervating motor neurons, which in skates extend along the length of the body, whereas in tetrapods they occupy segments adjacent to forelimb and hindlimb. These differences appear to be due changes in the expression of Hox transcription factors. Due to a natural deletion of the *HoxC* cluster, skates lack the *Hoxc9* gene which in tetrapods is necessary to repress the specification of limb-innervating motor neurons in non-limb segments (Jung et al., 2010; King, Gillis, Carlisle, & Dahn, 2011; Turner et al., 2019). Despite these differences, these observations suggest that basic elements of the gene regulatory networks essential for limb control were established in the common ancestor of all vertebrates with paired appendages (Fig. 13).

6.3 The developmental basis of unique skate pectoral and pelvic fin morphologies

One of the most remarkable anatomical features of batoid cartilaginous fishes (skates and rays) is their dorsoventrally flattened, disk-like body shape. During embryonic development, this body form arises via anterior expansion of the pectoral fins, and subsequent fusion of these expanded pectoral fins to the head. This unusual body plan likely evolved as an adaptation to a benthic lifestyle, but the developmental mechanisms underlying this exceptional fin morphology (Fig. 14A) have long been unclear.

In bony vertebrates, fin and limb but outgrowth is maintained by an epithelial signaling center called the apical ectodermal ridge (AER). The AER is located at the leading edge of the outgrowing fin/limb bud, and expresses genes encoding members of the fibroblast growth factor (FGF) family of signaling molecules (Sun, Mariani, & Martin, 2002). Expression of *fgf8* is a conserved feature of the AER of developing skate fins (Gillis et al., 2009b) (Fig. 14B). However, Nakamura et al. (2015) recently demonstrated that skates possess a persistent AER-like signaling center at the anterior margin of the expanding pectoral fins that overlies a domain of enhanced

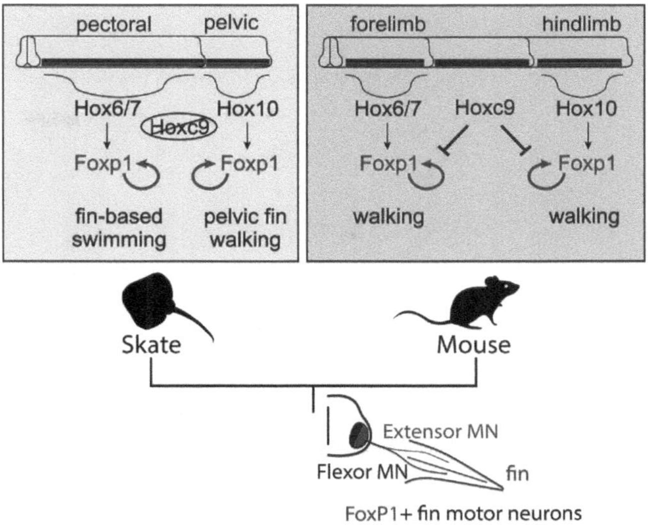

Fig. 13 Conserved motor neuron subtypes and ambulatory locomotion in skate and mammals. In tetrapods, limb motor neurons express a characteristic suite of *Hox* genes and *FoxP1*, with distinct neuronal subtypes innervating extensor and flexor muscle in the limb to facilitate walking. Homologous motor neuron subtypes and circuits are found in the spinal cord and fins of the little skate, which exhibit a unique mode of pelvic fin-based ambulatory locomotion (pelvic fin "walking"). These findings indicate that walking has evolved independently in skate and tetrapods, though via deeply conserved neural substrates that were present in the last common ancestor of jawed vertebrates. *Figure modified from Jung, H., Baek, M., D'Elia, K.P., Boisvert, C., Currie, P.D., & Tay, B.H., et al. (2018). The ancient origins of neural substrates for land walking.* Cell, 172, *667–682.*

mesenchymal cell proliferation during pectoral fin expansion. This persistent AER domain expresses *wnt3* (Fig. 14B), and pharmacological inhibition of Wnt signaling disrupts anterior expansion of the pectoral fins (Fig. 14C, C'). It therefore appears as though a specialized subdomain of the pectoral fin AER promotes expansion and fusion of skate pectoral fins, in this way contributing to the skate's unique body plan. Interestingly, a circular body form with disk-like pectoral fins that are fused and supported to the snout tip by fin radials has evolved convergently within batoids, in skates and stingrays (Aschliman et al., 2012), offering an opportunity to test whether an extreme body plan adaptation arose via homologous, convergent, or parallel developmental mechanisms (Hall, 2007).

Another distinguishing feature of the fins of cartilaginous fishes is the clasper that extends from the posterior base of the male pelvic fins

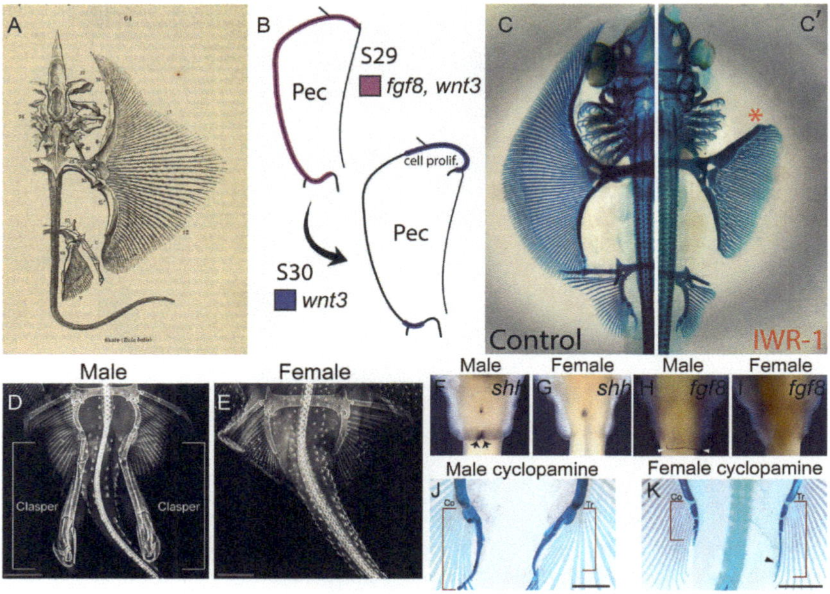

Fig. 14 Mechanisms of anterior pectoral fin expansion and clasper development in the little skate. (A) Skates possess a flattened, disk-like body form, which results from the anterior expansion and fusion of the pectoral fin to the side of the head. (B) Nakamura et al. (2015) found that a persistent, *wnt3*-expressing subdomain of the apical ectodermal ridge is associated with enhanced proliferation and disproportionate expansion of the anterior pectoral fin. (C–C′) Disruption of wnt signaling with the small molecule inhibitor IWR-1 inhibits anterior expansion of the pectoral fin (red asterisk). (D) Male skates possess claspers that project from their posterior pelvic fins, while (E) female skates do not. (F–I) Male clasper buds express *shh* and *fgf8* in patterns reminiscent of those from developing pectoral and pelvic appendages. (J) Inhibition of hedgehog signaling with cyclopamine results in truncation of the male clasper skeleton, and (K). Application exogenous Shh protein is sufficient to induce development of a clasper skeleton from the female pectoral fin. *A: Modified after Owen, R. (1866). Anatomy of Vertebrates I. Fishes and Reptiles. London, UK: Longmans, Green, and Co. Images (D–K) from O'Shaughnessy, K.L., Dahn, R.D., & Cohn, M.J. (2015) Molecular development of chondrichthyan claspers and the evolution of copulatory organs. Nature Communications, 6, 6698..*

(Fig. 14D and E). Claspers are copulatory organs that are used for sperm transfer during internal fertilization. These structures arose along the jawed vertebrate stem (based on their presence in some placoderms—Ahlberg, Trinajstic, Johanson, & Long, 2009; Trinajstic, Boisvert, Long, Maksimenko, & Johanson, 2014; Long et al., 2014), though among extant vertebrates they are only found in cartilaginous fishes. In some taxa, claspers are adorned with spurs or denticles that function to anchor the organ inside a female's cloaca during copulation. Once inside the cloaca, sperm is transferred from the male cloaca through a clasper tube and is expelled out of the distal end by secretions that are pumped from a muscular proximal clasper gland (reviewed by Wourms, 1977). The clasper is supported by a cartilaginous extension of the male pelvic fin endoskeleton (Daniel, 1922), leading to questions about how the molecular mechanisms patterning paired appendages may have been modified, redeployed or extended in males to direct skeletogenesis in this remarkable secondary sex characteristic.

O'Shaughnessy et al. (2015) discovered that the clasper skeleton of male skates develops under the control of an extended phase of the conserved gnathostome fin/limb developmental program. Claspers initiate as buds projecting from the caudal pelvic fins of male skates at S31, with these buds retaining prolonged expression of several fin patterning genes, including *shh* and *fgf8* (Fig. 14F–I), in a manner not seen in stage-matched female pelvic fins. They also show that Shh signaling is required for normal development of the clasper skeleton in males (Fig. 14J), and that application of exogenous Shh is sufficient to induce ectopic clasper-like skeletal elements in female pelvic fins (Fig. 14K). During limb development, Hand2 is a direct transcriptional regulator of *Shh* expression in the posterior zone of polarizing activity (Galli et al., 2010), and O'Shaughnessy and colleagues found that *hand2* is also expressed broadly within the male clasper bud. Interestingly, *hand2* and *shh* expression in male clasper buds are sensitive to perturbations in androgen receptor (AR) signaling, owing to the binding of AR to two newly discovered androgen response elements in the *hand2* locus. Together, these findings suggest that claspers evolved as elaborations of the male pelvic fin skeleton in cartilaginous fishes via acquisition of an androgen signaling regulatory input acting upstream of the conserved, *Hand2*-dependent jawed vertebrate fin and limb patterning program.

6.4 Adult chondrogenesis and spontaneous cartilage repair in the skate

Cartilage is a skeletal tissue that consists of a single cell type (the chondrocyte) embedded within an extensive extracellular matrix (reviewed by Gillis, 2019). In mammals, cartilage is predominantly an embryonic tissue, making up the anlage of the cranial, axial and appendicular endoskeleton. Most mammalian cartilage is replaced by bone through a process known as endochondral ossification, with cartilage persisting in just a few sites within the skeleton—for example, at the ends of long bones, as articular cartilage

(Decker, 2017). Mammalian articular cartilage is avascular, aneural and lacks resident progenitor cells, and therefore has a relatively poor capacity for spontaneous repair (Hunziker, 1999). Consequently, injuries to articular cartilage tend to fill with scar tissue, and can be lasting, often leading to chronic and debilitating conditions like osteoarthritis.

Cartilaginous fishes, on the other hand, possess an endoskeleton that is composed entirely of cartilage, and that remains cartilaginous throughout life (Fig. 15A and B). It is often assumed that the cartilaginous skeleton of sharks and skates reflects retention of a primitive condition (i.e., a primitive absence of bone). However, recent fossil evidence points instead to an origin of endochondral ossification along the jawed vertebrate stem, and therefore to a loss of bone in cartilaginous fishes (Brazeau et al., 2020). Like many fishes, skates also exhibit an indeterminate mode of growth, and this led us to wonder whether this group might have evolved a mechanism to grow and repair their cartilaginous skeleton (Marconi et al., 2020).

Using label retention experiments, we found that juvenile and adult skates possess a cartilage progenitor cell type that resides within the perichondrium (the fibrous sheath that enwraps the cartilaginous endoskeleton). These progenitor cells co-express genes encoding the pro-chondrogenic transcription factors Sox5, Sox6 and Sox9 Fig. 15C), and give rise to new chondrocytes throughout adulthood (Fig. 15D and E). To test whether the skate's sustained postembryonic chondrogenesis correlates with an ability to repair injury to the adult cartilage skeleton, we conducted a series of cartilage injury experiments, in which we removed a wedge of cartilage from the metapterygium using a biopsy punch, and then assessed extent of repair over the following year. We found that by 2 months post-operation (mpo), cartilage injury sites had filled with fibrous connective tissue (Fig. 15F and G). By 3mpo, this tissue began to differentiate into repair cartilage, which was composed of chondrocytes that expressed type II collagen comparable to adjacent non-injured chondrocytes (Fig. 15H–J). By 12 months post-operation, cartilage injuries were completely filled with repair cartilage. Interestingly, mechanical perturbation of the perichondrium induced a strong ectopic chondrogenic response from this tissue, indicating that, in skate, the perichondrium is a source of new cartilage during both normal growth and spontaneously repair cartilage injuries.

Current stem cell-based approaches for *in vitro* engineering of cartilage for transplantation into articular cartilage injuries hold promise but have limitations. Given that mammalian cartilage typically functions as a transient model for endochondral bone, many embryonic or induced pluripotent

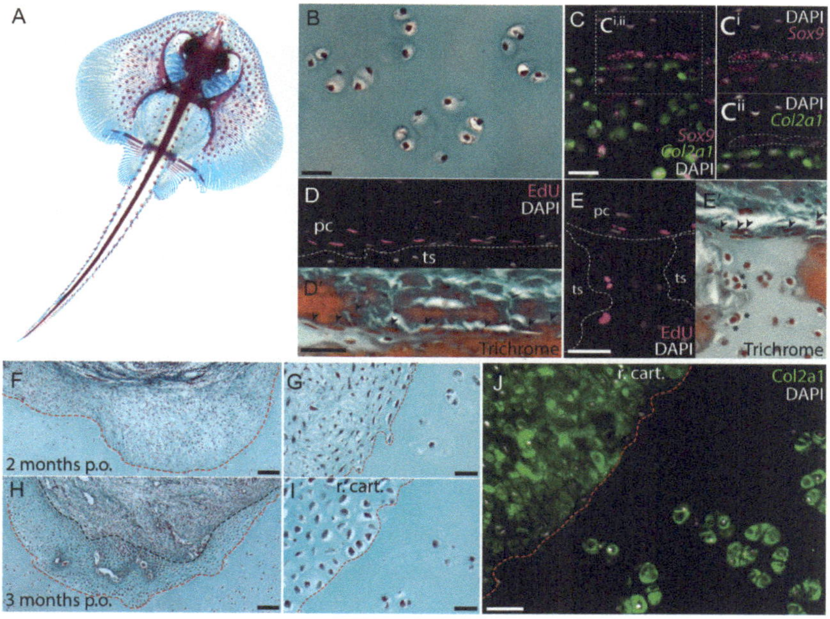

Fig. 15 Postembryonic cartilage growth and repair in the little skate. (A) A skeletal preparation of a little skate hatchling, with cartilage stained blue and mineralized tissues (e.g. the calcified surface of some endoskeletal elements) stained red. The vast majority of the skate endoskeleton is composed of (B) hyaline cartilage. (C) Hatchling and adult skates possess cartilage progenitor cells in the perichondrium that express genes encoding pro-chondrogenic transcription factors, such as Sox9. (D) These perichondral progenitor cells are label-retaining, and (E) give rise to chondrocyte progeny throughout adulthood. (D′ and E′) are the same sections as (D) and (E), respectively, but trichrome stained after EdU detection. Perichondral progenitor cells are indicated with arrowheads, and progenitor-derived chondrocytes are indicated with asterisks. (F, G) Adult skates are also capable of spontaneous repair of cartilage injury. Two months after a cartilage biopsy (months post-operation, or p.o.) the injury site is largely infilled with connective tissue, and by (H, I) 3 months p.o., this connective tissue begins to differentiate into hyaline cartilage. (J) Repair cartilage and chondrocytes express type II collagen, like uninjured chondrocytes. In (F–J), the red dashed line indicates the extent of the cartilage injury, and the black dashed line in (H) indicates the extent of repair cartilage. Scale bars: b,c, 25 μm; d,e, 50 μm; f, 100 μm; g, 30 μm; h, 100 μm; i, 30 μm; j, 50 μm.

stem cell-based strategies for cartilage repair are plagued by undesirable osteogenic differentiation. The skate offers a unique model for studying adult chondrogenesis and the postembryonic generation of cartilage as a terminal product (rather than as a transient step toward bone), during both normal growth and injury repair. Further research on the molecular and cellular basis of bone loss and cartilage growth in skates could therefore inspire novel *in vitro* and *in vivo* approaches to mammalian cartilage repair, and novel therapeutic strategies for articular cartilage injury and osteoarthritis.

7. Conclusion

"Science is a bit like the joke about the drunk who is looking under a lamppost for a key that he has lost on the other side of the street, because that's where the light is."

Noam Chomsky, letter to Robert F. Barsky, 14 June 1993

Reasons for studying unconventional or emerging model organisms in developmental biology are many (Goldstein & King, 2016). Often, organisms are selected for study because they occupy important phylogenetic positions, which may permit comparative analyses with other taxa and inference of ancestral conditions for a given clade. But perhaps more than anything else, emerging model organisms are critically important for studying the development of organs, tissues, cell types or physiological processes that are simply not found in more widely studied model systems—effectively offering natural forward genetics experiments, which allow us to uncover the basis in development and genetics of phenotypes that have arisen over an evolutionary timescale. Through the course of evolution, nature has arrived at elegant solutions to innumerable biological problems. By turning to natural experiments and looking where answers are most likely to lie (rather than where it is easiest to look), we stand the greatest chance of observing, appreciating and learning from such solutions.

From a relatively small body of published work (and with technical advances consisting of little more than a plastic soup spoon and some Krazy glue), the little skate has contributed to a greater understanding of the development and evolution of the jaw, gills, paired fins, vertebrae, teeth, dermal denticles, neural crest cells, gut, intromittent organs and the central and peripheral nervous systems of vertebrates. With the exception of cartilage (Marconi et al., 2020) and kidney (Elger et al., 2003), the regenerative abilities of the little skate remain largely unexplored, but undoubtedly hold great promise. By combining an ever-growing array of tools for genome editing and other assorted molecular manipulations (many of which are increasingly efficient and not host taxon-specific) and the established embryological approaches outlined here, we are optimistic that the little skate will continue to serve as an increasingly valuable model system for studies of cartilaginous fish and jawed vertebrate cell and developmental biology.

Acknowledgments

We thank Rebecca Gelernter, Ray Troll, the Fitzwilliam Museum, Dr. Simon Weigmann, Dr. Myungin Baek, Dr. Heekyung Jung and Prof. Martin Cohn for providing images. We also thank the staff of the Marine Resources Center and the Embryology community of the Marine Biological Laboratory—particularly Profs. Nipam Patel, Lee Niswander, Richard Behringer, Alejandro Sanchez Alvarado, Dave Sherwood, Rich Schneider and Jon Henry—for generously supporting efforts to develop the skate as a developmental model system.

References

Ahlberg, P., Trinajstic, K., Johanson, Z., & Long, J. (2009). Pelvic claspers confirm chondrichthyan-like internal fertilization in arthrodires. *Nature*, *460*, 888–889.

Arratia, G., Schultze, H. P., & Casciotta, J. (2001). Vertebral column and associated elements in dipnoans and comparison with other fishes: Development and homology. *Journal of Morphology*, *250*, 101–172.

Aschliman, N. C., Nishida, M., Miya, M., Inoue, J. G., Rosana, K. M., & Naylor, G. J. (2012). Body plan convergence in the evolution of skates and rays (Chondrichthyes: Batoidea). *Molecular Phylogenetics and Evolution*, *63*, 28–42.

Bagnall, K. M., Higgins, S. J., & Sanders, E. J. (1988). The contribution made by a single somite to the vertebral column: Experimental evidence in support of resegmentation using the chick-quail chimaera model. *Development*, *103*, 69–85.

Bagnall, K. M., Sanders, E. J., Higgins, S. J., & Leam, H. (1988). The effects of somite removal on vertebral formation in the chick. *Anatomy and Embryology (Berlin)*, *178*, 183–190.

Baker, C. V., Modrell, M. S., & Gillis, J. A. (2013). The evolution and development of vertebrate lateral line electroreceptors. *Journal of Experimental Biology*, *216*, 2515–2522.

Balfour, F. M. (1878). *A monograph on the development of elasmobranch fishes*. London: Macmillan and Co.

Ballard, W. W., Mellinger, J., & Lechenault, H. (1993). A series of normal stages for development of *Scyliorhinus canicula*, the lesser spotted dogfish *(Chondrichthyes: Scyliorhinidae)*. *Journal of Experimental Zoology*, *267*, 318–336.

Barry, S. N., & Crow, K. D. (2017). The role of *HoxA11*and *HoxA13* in the evolution of novel fin morphologies in a representative batoid (*Leucoraja erinacea*). *EvoDevo*, *8*, 24.

Brazeau, M. D., Giles, S., Dearden, R. P., Jerve, A., Ariunchimeg, Y., Zorig, E., et al. (2020). Endochondral bone in an Early Devonian 'placoderm' from Mongolia. *Nature Ecology and Evolution*, *4*, 1477–1484.

Chu, D., Nguyen, A., Smith, S. S., Vavrušová, Z., & Schneider, R. A. (2020). Stable integration of an optimized inducible promoter system enables spatiotemporal control of gene expression throughout avian development. *Biology Open*, *9*, bio055343.

Coolen, M., Menuet, A., Chassoux, D., Compagnucci, C., Henry, S., Lévèque, L., et al. (2008). The dogfish *Scyliorhinus canicula*: A reference in jawed vertebrates. *Cold Spring Harbor Protocols*, *2008*. https://doi.org/10.1101/pdb.emo111.

Coolen, M., Menuet, A., & Mazan, S. (2009). Towards a synthetic view of axis specification mechanisms in vertebrates: Insights from the dogfish. *Comptes Rendus Biologies*, *332*, 210–218.

Coolen, M., Sauka-Spengler, T., Nicolle, D., Le-Mentec, C., Lallemand, Y., Da Silva, C., et al. (2007). Evolution of axis specification mechanisms in jawed vertebrates: Insights from a chondrichthyan. *PLoS One*, *2*, e374.

Criswell, K. E., Coates, M. I., & Gillis, J. A. (2017a). Embryonic development of the axial column in the little skate, *Leucoraja erinacea*. *Journal of Morphology*, *278*, 300–320.

Criswell, K. E., Coates, M. I., & Gillis, J. A. (2017b). Embryonic origin of the gnathostome vertebral skeleton. *Proceedings of the Royal Society B: Biological Sciences*, *284*, 20172121.

Criswell, K. E., & Gillis, J. A. (2020). Resegmentation is an ancestral feature of the gnathostome vertebral skeleton. *eLife*, *9*, e51696.

Dahn, R. D., Davis, M. C., Pappano, W. N., & Shubin, N. H. (2007). Sonic hedgehog function in chondrichthyan fins and the evolution of appendage patterning. *Nature*, *445*, 311–314.

Daniel, J. F. (1922). *The elasmobranch fishes*. Berkeley, CA: University of California Press.

Dasen, J. S., De Camilli, A., Wang, B., Tucker, P. W., & Jessell, T. M. (2008). Hox repertoires for motor neuron diversity and connectivity gated by a single accessory factor, FoxP1. *Cell*, *134*, 304–316.

Dean, B. (1906). Chimaeroid fishes and their development. *Carnegie Institution of Washington Publications*, *32*, 1–172.

Decker, R. S. (2017). Articular cartilage and joint development from embryogenesis to adulthood. *Seminars in Cell and Developmental Biology*, *62*, 50–56.

Di Santo, V. (2019). Ocean acidification and warming affect skeletal mineralization in a marine fish. *Proceedings of the Royal Society B: Biological Science*, *286*, 20182187.

Elger, M., Hentschel, H., Litteral, J., Wellner, M., Kirsch, T., Luft, F. C., et al. (2003). Nephrogenesis is induced by partial nephrectomy in the elasmobranch *Leucoraja erinacea*. *Journal of the American Society of Nephrology*, *14*, 1506–1518.

Fleming, A., Keynes, R., & Tannahill, D. (2004). A central role for the notochord in vertebral patterning. *Development*, *131*, 873–880.

Fleming, A., Kishida, M. G., Kimmel, C. B., & Keynes, R. J. (2015). Building the backbone: The development and evolution of vertebral patterning. *Development*, *142*, 1733–1744.

Galli, A., Robay, D., Osterwalder, M., Bao, X., Bénazet, J. D., Tariq, M., et al. (2010). Distinct roles of Hand2 in initiating polarity and posterior Shh expression during the onset of mouse limb bud development. *PLoS Genetics*, *6*, e1000901.

Gandhi, S., Li, Y., Tang, W., Christensen, J. B., Urrutia, H. A., Vieceli, F. M., et al. (2021). A single-plasmid approach for genome editing coupled with long-term lineage analysis in chick embryos. *Development*, *148*, dev193565.

Gillis, A. (2019). The development and evolution of cartilage. *Reference Module in Life Sciences*. Elsevier.

Gillis, J. A., Alsema, E. C., & Criswell, K. E. (2017). Trunk neural crest origin of dermal denticles in a cartilaginous fish. *Proceedings of the National Academy of Sciences of the United States of America*, *114*, 13200–13205.

Gillis, J. A., Dahn, R. D., & Shubin, N. H. (2009a). Shared developmental mechanisms pattern the vertebrate gill arch and paired fin skeletons. *Proceedings of the National Academy of Sciences of the United States of America*, *106*, 5720–5724.

Gillis, J. A., Dahn, R. D., & Shubin, N. H. (2009b). Chondrogenesis and homology of the visceral skeleton in the little skate, *Leucoraja erinacea* (Chondrichthyes: Batoidea). *Journal of Morphology*, *270*, 628–643.

Gillis, J. A., & Hall, B. K. (2016). A shared role for sonic hedgehog signalling in patterning chondrichthyan gill arch appendages and tetrapod limbs. *Development*, *143*, 1313–1317.

Gillis, J. A., Modrell, M. S., & Baker, C. V. (2013). Developmental evidence for serial homology of the vertebrate jaw and gill arch skeleton. *Nature Communications*, *4*, 1436.

Gillis, J. A., Modrell, M. S., Northcutt, R. G., Catania, K. C., Luer, C. A., & Baker, C. V. (2012). Electrosensory ampullary organs are derived from lateral line placodes in cartilaginous fishes. *Development*, *139*, 3142–3146.

Gillis, J. A., & Shubin, N. H. (2009). The evolution of gnathostome development: Insight from chondrichthyan embryology. *Genesis*, *47*, 825–841.

Gillis, J. A., & Tidswell, O. R. (2017). The origin of vertebrate gills. *Current Biology*, *27*, 729–732.

Godard, B. G., Coolen, M., Le Panse, S., Gombault, A., Ferreiro-Galve, S., Laguerre, L., et al. (2014). Mechanisms of endoderm formation in a cartilaginous fish reveal ancestral and homoplastic traits in jawed vertebrates. *Biology Open*, *3*, 1098–1107.

Godard, B. G., & Mazan, S. (2013). Early patterning in a chondrichthyan model, the small spotted dogfish: Towards the gnathostome ancestral state. *Journal of Anatomy*, *222*, 56–66.

Goldstein, B., & King, N. (2016). The future of cell biology: Emerging model organisms. *Trends in Cell Biology*, *26*, 818–824.

Grotmol, S., Kryvi, H., Nordvik, K., & Totland, G. K. (2003). Notochord segmentation may lay down the pathway for the development of the vertebral bodies in the Atlantic salmon. *Anatomy and Embryology (Berlin)*, *207*, 263–272.

Hall, B. K. (2007). Homoplasy and homology: Dichotomy or continuum? *Journal of Human Evolution*, *52*, 473–479.

Hirschberger, C., Sleight, V. A., Criswell, K. E., Clark, S. J., & Gillis, J. A. (2021). Conserved and unique transcriptional features of pharyngeal arches in the skate (*Leucoraja erinacea*) and evolution of the jaw. *Molecular Biology and Evolution*. https://doi.org/10.1093/molbev/msab123.

Hunziker, E. B. (1999). Articular cartilage repair: Are the intrinsic biological constraints undermining this process insuperable? *Osteoarthritis and Cartilage*, *7*, 15–28.

Johanson, Z., Martin, K., Fraser, G., & James, K. (2019). The synarcual of the little skate, *Leucoraja erinacea*: Novel development among the vertebrates. *Frontiers in Ecology and Evolution*, *7*, 12.

Jung, H., Baek, M., D'Elia, K. P., Boisvert, C., Currie, P. D., Tay, B. H., et al. (2018). The ancient origins of neural substrates for land walking. *Cell*, *172*, 667–682.

Jung, H., Lacombe, J., Mazzoni, E. O., Liem, K. F., Jr., Grinstein, J., Mahony, S., et al. (2010). Global control of motor neuron topography mediated by the repressive actions of a single hox gene. *Neuron*, *67*, 781–796.

King, B. L., Gillis, J. A., Carlisle, H. R., & Dahn, R. D. (2011). A natural deletion of the HoxC cluster in elasmobranch fishes. *Science*, *334*, 1517.

Koester, D. M., & Spirito, C. P. (2003). Punting: An unusual mode of locomotion in the little skate, *Leucoraja erinacea* (Chondrichthyes: Rajidae). *Copeia*, 553–561.

Long, J. A., Mark-Kurik, E., Johanson, Z., Lee, M. S. Y., Young, G. C., Min, Z., et al. (2014). Copulation in antiarch placoderms and the origin of gnathostome internal fertilization. *Nature*, *517*, 196–199.

Luer, C. A., & Gilbert, P. W. (1985). Mating behavior, egg deposition, incubation period, and hatching in the clearnose skate, *Raja eglanteria*. *Environmental Biology of Fishes*, *13*, 161–171.

Marconi, A., Hancock-Ronemus, A., & Gillis, J. A. (2020). Adult chondrogenesis and spontaneous cartilage repair in the skate, *Leucoraja erinacea*. *Elife*, *9*, e53414.

Martik, M. L., Gandhi, S., Uy, B. R., Gillis, J. A., Green, S. A., Simoes-Costa, M., et al. (2019). Evolution of the new head by gradual acquisition of neural crest regulatory circuits. *Nature*, *574*, 675–678.

Maxwell, E. E., Fröbisch, N. B., & Heppleston, A. C. (2008). Variability and conservation in late chondrichthyan development: Ontogeny of the winter skate (*Leucoraja ocellata*). *Anatomical Record*, *291*, 1079–1087.

Miyake, T., Vaglia, J. L., Taylor, L. H., & Hall, B. K. (1999). Development of dermal denticles in skates (Chondrichthyes, Batoidea): Patterning and cellular differentiation. *Journal of Morphology*, *241*, 61–81.

Morin-Kensicki, E. M., Melancon, E., & Eisen, J. S. (2002). Segmental relationship between somites and vertebral column in zebrafish. *Development*, *129*, 3851–3860.

Nakamura, T., Klomp, J., Pieretti, J., Schneider, I., Gehrke, A. R., & Shubin, N. H. (2015). Molecular mechanisms underlying the exceptional adaptations of batoid fins. *Proceedings of the National Academy of Sciences of the United States of America*, *112*, 15940–15945.

Nordvik, K., Kryvi, H., Totland, G. K., & Grotmol, S. (2005). The salmon vertebral body develops through mineralization of two preformed tissues that are encompassed by two layers of bone. *Journal of Anatomy*, *206*, 103–114.

O'Shaughnessy, K. L., Dahn, R. D., & Cohn, M. J. (2015). Molecular development of chondrichthyan claspers and the evolution of copulatory organs. *Nature Communications*, *6*, 6698.

Piekarski, N., & Olsson, L. (2014). Resegmentation in the Mexican axolotl, *Ambystoma mexicanum*. *Journal of Morphology*, *275*, 141–152.

Rasch, L. J., Cooper, R. L., Underwood, C., Dillard, W. A., Thiery, A. P., & Fraser, G. J. (2020). Development and regeneration of the crushing dentition in skates (Rajidae). *Developmental Biology*, *466*, 59–72.

Sauka-Spengler, T., Baratte, B., Lepage, M., & Mazan, S. (2003). Characterization of Brachyury genes in the dogfish *S. canicula* and the lamprey *L. fluviatilis*. Insights into gastrulation in a chondrichthyan. *Developmental Biology*, *263*, 296–307.

Sauka-Spengler, T., Plouhinec, J.-L., & Mazan, S. (2004). Gastrulation in a chondrichthyan, the dogfish *Scyliorhinus canicula*. In C. D. Stern (Ed.), *Gastrulation from cells to embryo* (pp. 151–156). New York: Cold Spring Harbor Laboratory Press.

Sleight, V. A., & Gillis, J. A. (2020). Embryonic origin and serial homology of gill arches and paired fins in the skate, *Leucoraja erinacea*. *eLife*, *9*, e60635.

Sun, X., Mariani, F. V., & Martin, G. R. (2002). Functions of FGF signalling from the apical ectodermal ridge in limb development. *Nature*, *418*, 501–508.

Theodosiou, N. A., & Oppong, E. (2019). 3D morphological analysis of spiral intestine morphogenesis in the little skate, *Leucoraja erinacea*. *Developmental Dynamics*, *248*, 688–701.

Trinajstic, K., Boisvert, C., Long, J., Maksimenko, A., & Johanson, Z. (2014). Pelvic and reproductive structures in placoderms (stem gnathostomes). *Biological Reviews of the Cambridge Philosophical Society*, *90*, 467–501.

Turner, N., Mikalauskaite, D., Barone, K., Flaherty, K., Senevirathne, G., Adachi, N., et al. (2019). The evolutionary origins and diversity of the neuromuscular system of paired appendages in batoids. *Proceedings of the Royal Society B: Biological Sciences*, *286*, 20191571.

Underwood, C. J., Johanson, Z., Welten, M., Metscher, B., Rasch, L. J., Fraser, G. J., et al. (2015). Development and evolution of dentition pattern and tooth order in the skates and rays (batoidea; chondrichthyes). *PLoS One*, *10*, e0122553.

Vandebroek, G. (1936). Les mouvements morphogénétiques au cours de la gastrulation chez *Scyllium canicula* Cuv. H. Vaillant-Carmanne. *Archives of Developmental Biology*, *47*, 499–582.

Ward, L., Evans, S. E., & Stern, C. D. (2017). A resegmentation-shift model for vertebral patterning. *Journal of Anatomy*, *230*, 290–296.

Wourms, J. P. (1977). Reproduction and development in chondrichthyan fishes. *American Zoologist*, *17*, 379–410.

Wourms, J. P. (1997). The rise of fish embryology in the nineteenth century. *American Zoologist*, *37*, 269–310.

Wourms, J. P., & Demski, L. S. (1993). The reproduction and development of sharks, skates, rays and ratfishes: Introduction, history, overview, and future prospects. *Environmental Biology of Fishes*, *38*, 7–21.

CHAPTER TWENTY-TWO

The Axolotl's journey to the modern molecular era

Karen Echeverri[a], Jifeng Fei[b], and Elly M. Tanaka[c,*]
[a]Bell Center for Regenerative Biology and Tissue Engineering, Marine Biological Laboratory, Woods Hole, MA, United States
[b]Guangdong Provincial People's Hospital, Guangdong Academy of Medical Sciences, Guangzhou, China
[c]Institute of Molecular Pathology (IMP), Vienna Biocenter (VBC), Vienna, Austria
*Corresponding author: e-mail address: elly.tanaka@imp.ac.at

Contents

Abstract

The salamander *Ambystoma mexicanum*, commonly called "the axolotl" has a long, illustrious history as a model organism, perhaps with one of the longest track records as a laboratory-bred vertebrate, yet it also holds a prominent place among the emerging model organisms. Or rather it is by now an "emerged" model organism, boasting a full cohort molecular genetic tools that allows an expanding community of researchers in the field to explore the remarkable traits of this animal including regeneration, at cellular and molecular precision—which had been a dream for researchers over the years. This chapter describes the journey to this status, that could be helpful for those developing their respective animal or plant models.

Current Topics in Developmental Biology, Volume 147
ISSN 0070-2153
https://doi.org/10.1016/bs.ctdb.2021.12.010

1. Introduction—Reemerging the classic axolotl model into the molecular era

(Elly Tanaka)

During the explosive growth of molecular biology in the late 20th century emerging techniques were rapidly applied to invertebrates such as drosophila, and vertebrates such as mouse, frog and zebrafish. These activities engendered the field of molecular embryology but the axolotl remained somewhat recalcitrant. Those features that had made the animal attractive for classical work such as large yolky eggs and slow cell cycle hindered direct application of molecular protocols to axolotl embryos. Additionally, regeneration of *postembryonic* body structures such as the limb and the tail are the axolotl's seductive value but molecular methods were difficult to apply to adult tissues in any animal due to the complex cohorts of tissues and presence of substantial extracellular matrix. Incisive progress in the form of molecular genetics of regeneration appeared practically insurmountable. Regeneration is characterized by the changes in the adult tissue after amputation including influx of immune cells, migration of different cells layers that ultimately yield growth of new tissues. On one hand the source cells for regeneration had to be disentangled from noncontributing cells, and genetic manipulation of the regenerating cells had to be achieved, which basically required the use of genetic fate mapping and transduction of genetic material into those cells to express and downregulate molecular factors.

At the turn of this century as in situ hybridization, DNA transfection, and viral systems were further developed for mammalian systems, several brave groups among a dwindling salamander community carried the flag and reported impressive and important work characterizing gene expression during regeneration, using biolistic DNA transfection of blastema cells to study the role of retinoic acid receptors in limb segment positional identity, viral transduction to study the contribution of muscle cells to the blastema, and the role of embryonic morphogens in regeneration such as SHH and WNT (Gardiner, Blumberg, Komine, & Bryant, 1995; Kawakami et al., 2006; Kumar, Velloso, Imokawa, & Brockes, 2000; Pecorino, Lo, & Brockes, 1994; Roy, Gardiner, & Bryant, 2000). These methods were usually either cumbersome, inefficient, hard to reproduce, suffered from silencing *in vivo*, or damaged the transduced cells, limiting their broad use.

By the dawn of the 21st century everexpanding methodologies to transduce somatic cells and viral methods became available, followed by the next

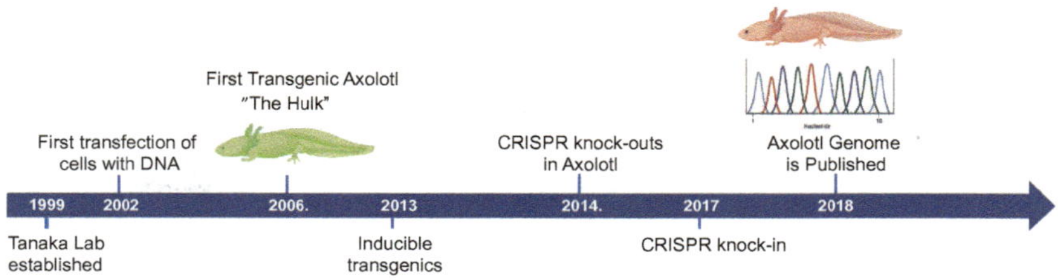

Fig. 1 Timeline of technical developments pioneered by the Tanaka lab that have driven the advancement of the axoltol as a genetically tractable research organism.

generation sequencing, and genome editing revolutions. Our account describes how we and others developed these methods for axolotls coupled with the goal to answer fundamental questions in regeneration biology which has allowed each of us and others to establish robust research programs focusing on the molecular mechanisms of axolotl regeneration. Fig. 1 provides an overview timeline as each of these methods and their associated scientific findings were published.

Stepping back, it may be possible to extract a few insights for those with similar ambitions. The first might be "never say never." This is exemplified by Karen Echeverri's account of developing electroporation to express gene products in regenerating tissues. At the beginning of 2000s we had heard several people in the field conclude that salamander cells might not be transfectable, after they had tried the standardly available methods such as calcium phosphate precipitation and lipofection. With the broadening implementation of electroporation in chick embryos, and descriptions of ionotophoresis, Karen's clever and dedicated effort to get this working in the axolotl tail was successful. Electroporating single spinal cord stem cells to follow their fate over time was not simple. It involved many attempts at pulling needles, engineering tilting stages to see the spinal cord lumen, and calling Jeffrey Corwin to learn that MS222 anesthetic was more toxic to the animals than benzocaine, which was critical for animal health during timelapse imaging. Having been successful, she could then apply the method not only to the spinal cord, but to testing the molecular basis of positional identity during limb regeneration (Echeverri & Tanaka, 2005). We also then applied electroporation successfully to cultured cells. In my mind, this experience addresses the oft heard anxiety of whether it is wise to potentially send PhD or postdoc careers to the graveyard developing methods. If the primary goal of the project is to answer a key question in the field by developing a method, then it is our strong belief that it will be possible to develop that method or another proxy method to address that question and even partial steps toward that goal will be publishable and appreciated if the question is important enough. Answering a question and developing a method to do so requires full commitment to analyze all the failures and to keep pushing with the attitude that something will work. This activity is not a side project.

Perhaps another insight concerns the role of emerging technologies and colleagues. As described below, transgenesis had been previously tried in salamanders, and had not been stable, but the key suggestion from wizard molecular biologist Francis Stewart that only certain promoters avoided silencing, allowed us to succeed. Similarly, we and others had consistently pursued gene sequence information for molecular studies, first in the form of expressed sequence tags (ESTs), then transcriptomes with the strong support of Bianca Haberman (Habermann et al., 2004; Putta et al., 2004). Another key turning point came when we were fortunate enough to be colleagues with Eugene Myers whose mathematical fascination to assemble genomes with newly emerging long-read technologies coupled with his generous intent to help young researchers in regeneration brought forth the planaria and axolotl genome assemblies that had been impossible to that point (Grohme et al., 2018; Nowoshilow et al., 2018). The presence of an exquisite molecular biologist, Ji-Feng Fei, in the lab was critical at this juncture, as he generated incredibly high-quality genomic DNA that showed the best output from PacBio machines that anyone had seen to that point. In parallel he was developing all the genome editing methodologies (gene mutation, gene knock-in) with remarkable care, testing many conditions, many different variants, following F0 animals into F1s and F2s with rigorous, quantitative genotyping and phenotyping that, not only filled up our axolotl colony, but lead to the breakthroughs in the field and a set of manuscripts that could withstand a very demanding review process (Fei et al., 2014; Nowoshilow et al., 2018). Many might have wondered why Ji-Feng was spending so much effort on following things through, but these quantitative observations provide a rock-solid basis for robust genome editing in our field. These anecdotes stress the importance of deep understanding and quantitative, rigorous treatment of these technology "attempts." They also highlight the importance of connecting to colleagues in emerging technologies whether locally or by broadly by attending meetings focusing on those topics. Staying narrowly focused on the traditional scope of a given model organism may restrict one's perspective on what is really possible.

We sincerely hope that our highly personal chapter describing the development of transient transfection, transgenesis, viral transduction, genome assembly and genome editing for the axolotl provide an amusing but also informative and inspiring account for how it is possible to address fundamental questions in regeneration biology at the molecular level in an animal previously considered as lagging in the molecular era.

2. Live cell imaging and functional studies by transient transfection

(Karen Echeverri)

2.1 The historical starting point

A longstanding question in the field of regeneration has been, where do the cells come from to regenerate the new tissue? In the early days of the axolotl regeneration field tools to work in these animals were sparse. The first insights into the how cells reacted to injury and where the information to tell cells what must be regenerated was gained from classical transplantation experiments. Some of the first indications that blastema cells contained positional information came from studies where larval limb blastema cells were dissociated, reaggregated and then transplanted onto fin tissue, surprisingly a limb then grew from that position, suggesting that the cells contained all the information they needed to regenerate (Stocum, 1968). Other experiments using similar transplantation methods further showed positional information is set up very early during blastema formation, if a distal blastema is transplanted to a proximal stump then these cells regenerate only distal elements (Iten & Bryant, 1975; Stocum & Melton, 1977). These early grafting experiments gave important insights into regeneration and similar grafting techniques were later used to address from where the cells actually come to form the blastema. Elegant transplantation of triploid skin cells to diploid animals allowed the tracing and identification of cells population contributing to the limb blastema (Muneoka, Fox, & Bryant, 1986; Muneoka, Wise, Fox, & Bryant, 1984). These studies showed that 43% of the blastema appeared to be derived from derma skin cells, while skeletal accounted for only about 2% of the total cells in the blastema (Muneoka et al., 1986). These types of experiments gave researchers incredibly important insights into the complex regeneration process. However traditional transplantation studies give limited understanding into the cellular and molecular events that are occurring to drive the regenerative process. After reading many of the classical transplantation papers as a starting graduate student, the biggest question in my mind was where do cells come from and what do they give rise to during regeneration? I was fortunate to hear a fascinating talk from Ted Salmon at EMBL showing beautiful images of cell division in newt lung epithelial cells (Skibbens, Skeen, & Salmon, 1993; Waterman-Storer, Salmon, & Salmon, 2000; Waters, Skibbens, & Salmon, 1996) which inspired me to think about how we could possibly label cells *in vivo* in a live animals and image them during the course of regeneration. Thankfully Elly was completely onboard with this idea and in reading classical regeneration experiments carried out in a range of different salamander species it became clear to us that if I wanted to visualize cell behavior *in vivo* the white axolotl would be the most suited species for this. Beautiful TEM work from Elizabeth Hay suggested that terminally dedifferentiated muscle fibers in the salamander limb in fact dedifferentiated in response to limb amputation. From high quality serial images of the regenerating limb blastema she inferred that the muscle fibers pinched off of single cells, which then entered the blastema (Hay, 1959). The idea that terminally differentiated cells would dedifferentiate was a highly controversial idea. However given that muscle fibers are so easily visible and accessible in the axolotl tail this seemed like a logical place to start and an exciting question to revisit. And so began the adventures of getting the first white axolotls from the Stock Center in Bloomington Indiana safely to Germany and working out *in vivo* cell labeling and imaging.

2.2 The path to fluorescently labeling the first cells *in vivo* in the axolotl

The next challenge was how we would label cells in a differentiated animal, many of our colleagues were labeling cells in a developing embryo but differentiated cells remained a challenge. Elly contacted a former colleague at University College London from her days as a Postdoc there, Jon Clark. Jon had worked with labeling cells in a range of organisms including some of the very first *in vivo* labeling of neurons in axolotls with Nigel Holder (Clarke, 1999; Clarke, Alexander, & Holder, 1988; Clarke & Tickle, 1999; Holder, Clarke, Kamalati, & Lane, 1990; McGonnell, Clarke, & Tickle, 1998; Tawk, Bianco, & Clarke, 2009). Thankfully he didn't think it was a crazy idea and was willing to host us for a week in his lab. Many late nights later under Jon's patient guidance Elly and I became experts at injecting cells, perfecting the technique of gently peeling back the axolotl skin to give better access to the muscle cells and less anguish over blocked needles and broken tips. There was great excitement when we got our first muscle cells labeled and great disappointment hours later when the cells disappeared!! We discovered our conditions were too harsh and the cells gradually died, but at least we had a baseline to work from. We packed our bags and figured out the logistics of bringing axolotls back to Germany again, with some successes under our belt and a resolve from Jon to continue to help us. Early experiments with the first iterations of the lipophilic dye were ambiguous as the dye could be easily based taken up and passed to close by cells. Eventually we worked out conditions where with microinjection of rhodamine dextran we could successfully label muscle cells with no adverse effects, the labeled cells remained for days. This technique allowed us for the first time to visualize the reaction of muscle fibers to injury and live image them after amputation. This technique gave insights into muscle regeneration; however the dye was rapidly diluted and the cells could not be traced for the duration of tail regeneration (Echeverri, Clarke, & Tanaka, 2001).

2.3 Plasmid transfection: The next frontier

In Jon's basement windowless lab at UCL he first introduced us to the labeling cells using iontophoresis (Fraser, 1996). This was a simple technique of using electrical charge to aid molecules to enter cells. This could be cheaply set up in a lab using a 6 V battery and crocodile clips, it worked to label the muscle cells but they were very sensitive to the current and ultimately simple injection of dyes worked best. However we really wanted to be able to label cells for longer and image them over longer time-courses. In meantime many labs were moving beyond iontophoresis and developing electroporation as a more powerful technique for transfecting cells.

Electroporation is a technique that works by applying a short train of electrical pulses which form temporary pores in the cell membrane allowing charged molecules like DNA to enter the cell, while noncharged molecules may also enter by passive diffusion once the pores are formed (Neumann, Scaefer-Ridder, Wang, & Hofschneider, 1982; Neumann, Toensing, Kakorin, Budde, & Frey, 1998; Potter, 1988). Electroporation is used extensively by chick embryologists and more recently for *in-ovo* electroporation in mouse (Itasaki, Bel-Vialar, & Krumlauf, 1999; Swartz, Eberhart, Mastick, & Krull, 2001).

In many animals models bulk transfection of cells is achieved by microinjecting plasmid DNA into the area of interest and the applying an external electrical pulse using tweezer electrodes (Harada, Omi, & Nakamura, 2017; McLennan & Kulesa, 2019; Saito & Nakaatsuji, 2001; Tawk, Tuil, Torrente, Vriz, & Paulin, 2002; Voiculescu & Stern, 2017). Advances on this standard technique have shown that a resolution of single cell electroporation can be achieved using a microelectrode technique that can achieve labeling of small groups or cells or even down to a single cell in the brain of Xenopus (Haas, Sin, Javaherian, Li, & Cline, 2001).

Luckily for us the people developing electroporation were very open and willing to help, so under guidance from Kurt Haas in Holly Cline's lab at the time, I found myself immersed in the world of grass stimulators and electroporators, measuring voltages, tips sizes and amplitudes! What seemed like an easy straight forward technique in other systems especially the chick, proved to be more challenging in axolotls. Why axolotl cells were so much harder to transfect remains a bit of a mystery, we speculate it may be because of more mucus on skin or the cells are a surrounded by a different composition of ECM. The axolotl spinal cord is in fact a very small structure and the cells proven difficult to see, one major breakthrough for us was developing a tilted stage which allowed us to finally visualize the cells in the spinal cord and to insert an electrode that acted both as the microinjection pipette and the electrode. However through perseverance and endless testing of voltages, pulse durations, the first cells in the spinal cord were eventually successfully labeled and could be imaged over long periods of time without too much reduction in the signal due to the stability and perdurance of the GFP protein (Echeverri & Tanaka, 2003). The adaptation of this technique to the axolotl allowed cells within the spinal cord to be labeled using plasmid DNA and imaged for the first time *in vivo* during the time-course of regeneration could be accomplished (Echeverri & Tanaka, 2002, 2003). The cell cycle in axolotl cells is relatively long in comparison to many other species, ranging from 24 to 72 h depending on the cell type and it is a state of development or regeneration, which means transfected cells are visible for much longer than in cells with rapid short cell cycle length (Rost et al., 2016).

Over time the electroporation technique itself has been further refined allowing higher efficacy of transfection to be achieved (Rodrigo Albors & Tanaka, 2015) and it has additionally shown to be an effective method for knocking down genes using morpholinos (Schnapp & Tanaka, 2005). Electroporation in axolotls has been a technique that has ultimately supported the elucidation of some key molecular insights into limb regeneration. We rapidly established that the same conditions used to label axolotl spinal cord cells could also be used to introduce plasmids and other molecules into a range of cells types, additionally other labs showed that the technique could be applied to other types of salamanders (Berg et al., 2010; Kumar, Godwin, Gates, Garza-Garcia, & Brockes, 2007).

The use of this technology in the context of limb regeneration allowed cells in different positions of the blastema to be labeled and confirmed previous transplantation data showing that the proximodistal axis is established very early in the blastema (Echeverri & Tanaka, 2005). Work from Mercader et al. elegantly illustrated how this technique could be used to overexpress proximally expressed Meis genes in the distal limb blastema leading to a proximalization of the cells and to introduce a morpholino to knock-down Meis expression, clearing illustrating that Meis genes are a target of retinoic acid (Mercader, Tanaka, & Torres, 2005). This technique has played a significant role in elucidating many of the key molecular pathways that are essential for initiating a regenerative response and patterning the regenerate. Recent work on the role of the immune system in limb regeneration has used electroporation to overexpress IL-8 in the blastema, resulting in increased recruitment of myeloid cells (Tsai, Baselga-Garriga, & Melton, 2019). The most recent example of the functionality of this system has shown how it can be used to label and examine the role of filopodia in blastemal cells, optogenic constructs could be electroporated into limb blastema cells and activated by light to change filopodia length (Zhang et al., 2021).

While the majority of uses to-date of electroporation has been to introduce plasmids to label cells or overexpress certain genes, it also has been used to knock-out genes. One caveat of studying genes in axolotl regeneration is that they can be necessary for regeneration, so in some cases knocking them out in development is not a viable option. In the early days of the development of the electroporation technology it was shown to be efficient for knocking down genes using morpholinos. Morpholinos are a very useful tool to reduce levels of proteins, however in some cases a knock-out of the gene can be more powerful. Electroporation has been shown to be an efficient technique for also introducing CRISPR guides complexed with the CAS9 protein into cells in the axolotl to specifically knock-out a gene, this has been shown to work well in the context of spinal cord regeneration (Fei et al., 2016, 2014).

The ability to label cells and knock-out or overexpress genes in the axolotl has paved the way for the axolotl to enter the stage as a genetically tractable model system. As interest waned in the axolotl due to lack of tools in comparison to systems like the zebrafish, the ability to label cells and modulate gene expression led to increased interest in the system again. However challenges remain, the axolotl can be tricky to image for long periods of time. The animal must be kept in an aqueous solution, the tissues are delicate but yet must be held down to prevent floating and going out of focus.

Imaging cells closer to the lens like skin and muscle is relatively easy while cells located deep in the blastema are more challenging due to light scattering. Advances with microscopy techniques allowing gentler deeper and faster imaging over long time periods will be very important for taking *in vivo* imaging in the axolotl to the next level.

3. The trick to transgenics and its transformative influence on studying regenerative cells

(Elly Tanaka and Ji-Feng Fei)

One of the fundamental questions in the field of regenerative biology is where do the blastema cells come from? In the other words, does the blastema comprise a ubiquitous pluripotent stem cell population or tissue specific progenitors? In order to address this issue, an ideal approach is to genetically track the origin and destiny of the blastema cells. For this purpose, creation of transgenic axolotls was very critical. Transgenesis means introducing extra, generally exogenous piece of DNA into genome and creation of the stable germline transmission of integrated sequences. It allows for manipulating activities of functional genes or labeling of cells to study their roles and behaviors during tissue regeneration.

An endeavor of the Tanaka lab from the earliest days was to make transgenic axolotls to address these questions. Elly was convinced that we should adapt the REMI sperm injection method from frogs given the similar physiology of the animals but this was not very practical as axolotl sperm have giant genomes packaged into long, thin sperm that are prone to damage. Luckily, Werner Straube, a starting PhD student having learned zebrafish transgenesis through plasmid injection wanted to give it a try in axolotl and succeeded right away to produce CMV promoter-driven GFP-expressing F0 transgenics. However, as previously reported for β-gal plasmid-injected transgenic salamanders, the signal faded over time, which was interpreted as the plasmid remaining episomal rather than integrated. By then Werner, who had another project, had graciously handed over the reins to Lidia Sobkow. During her first thesis advisory committee, Francis Stewart told her that the CMV promoter is often silenced in mouse transgenics, and we should try the CAGGs promoter. Furthermore, he told us about yet unpublished work implementing the SceI meganuclease to promote integration and reduce tandem copies in medaka (Thermes et al., 2002). With this lucky advice, and many trials of finding good ways to dejelly axolotl eggs for injection, Lidia succeeded in making stable "hulk" transgenic axolotls that expressed GFP in virtually all tissues (albeit to iffering levels) and passed the transgene through the germline (Sobkow, Epperlein, Herklotz, Straube, & Tanaka, 2006). Interestingly as this work was taking place Ueda and colleagues demonstrated sperm injection-mediated transgenesis in the Japanese newt (Ueda, Kondoh, & Mizuno, 2005). Taking advantage of our axolotl *CAGGS:eGFP* transgenic axolotl and tissue transplantation technique (Kragl et al., 2009; Nacu, Knapp, Tanaka, & Epperlein, 2009), Kragl and colleagues have achieved the labeling of major limb tissues. Further tracing of their cell fate has revealed that each tissue tends to give rise to corresponding tissue relevant to their origin during regeneration (Kragl et al., 2009). This study has solved a long-lasting question in the field what is the potency of a limb blastema cell and reached a conclusion that the blastema is composed of lineage-restricted progenitors, instead of pluripotent stem cells.

Although combining constitutive transgenesis and tissue transplantation is very powerful using such an approach can never exclude the potential problem of grafting undesired cells during transplantation. Generation of transgenics using a tissue specific promoter could achieve genetic labeling of particular cell/tissue type. At that point Shahryar Khattak joined as a postdoc who was deeply determined to develop Cre-ER^{T2}/loxP-mediated genetic fate mapping in axolotl, initially with the goal of fate mapping muscle vs satellite cell contribution to the limb blastema. Shahryar, with the assistance of Tobias Richter and then Maritta Schuez injecting millions of eggs, showed that well-characterized tissue-specific promoters from other animals such as mouse, drove specific expression in axolotl. He made many important tissue-specific GFP-expressing lines that are used by the field to this day (Khattak, Schuez, et al., 2013). He also systematically tried not only Cre, but also FRT and other recombinase systems (codon optimized) for axolotl, and concluded that Cre worked most efficiently and could induce recombination of a floxed GFP—Cherry loxP reporter transgenic that he created. Again, Shahryar's success was characterized by thoroughness and care. He insisted on the transgenics always expressing a fluorescent reporter so that he could examine the animals by sections, cell by cell to confirm if the transgenic expressed the gene specifically in the correct location. This also allowed him to quantify the mosaicism of the F0 transgenic. In the search for a way to express gene x (such as Cre) together with a fluorescent protein, Shahryar, introduced the use of T2A and P2A peptides very early after their publication in our transgenics, which has been a crucial aspect of our strategy.

3.1 Inducible Cre-transgenics were not so easy…

Frustration came in trying to put together the muscle-specific and satellite-specific promoters with the *Cre-ER*^{T2} gene. After many trials, many promoters, many transgenics, we realized that somehow the *ER*^{T2} gene often induced silencing of the transgene. Although Shahryar explored many solutions such as insulator elements, it was difficult to overcome this problem with the available muscle promoters. Shahryar even developed a working virus-transduction system for this purpose (see Section 4) but the packaging limit made implementation difficult. Furthermore, the *Pax7* locus and other satellite cell-specific gene loci were poorly characterized, making it impossible to make the two tissue specific *Cre-ER*^{T2} transgenics that we had envisioned at that time. Nonetheless, we were able to achieve our scientific goal, by using the *CAGGS:Cre-ER*^{T2} transgenic in a tricky way. We exploited the ability to graft blastemas from one animal to another and the fact that muscle fibers represent the product of fusion of muscle progenitors. To label limb muscle fibers, we grafted a blastema from the *CAGGS:Cre-ER*^{T2} onto the amputated stump of a *CAGGs:loxP-GFP-Cherry* transgenic (and vice versa) which showed muscle-specific expression, and allowed us to assess the contribution of muscle dedifferentiation to axolotl limb regeneration (Sandoval-Guzman et al., 2014).

3.2 Harvesting the power of transgenic technologies

Since the development of the first transgenic axolotl *CAGGS: eGFP*, this line was soon used to elucidate cell lineages of several major organs, including the teeth (Soukup, Epperlein, Horacek, & Cerny, 2008), hematopoiesis system (Lopez et al., 2014), bone tissues (McCusker, Diaz-Castillo, Sosnik, Phan, & Gardiner, 2016; Piekarski, Gross, & Hanken, 2014) during early development and regeneration in axolotls. Visualizing live cell activity has extra advantages over the static image from the fixed sections. Transplanting of lateral plate mesoderm tissue from double transgenic axolotls harboring a brainbow transgene and an inducible *CAGGS: Cre-ER*^{T2} alleles, it could be achieved the sparse, specific brainbow-labeling of the connective tissue (Currie et al., 2016; Gargioli & Slack, 2004). Connective tissue has been indicated as one of the major cell sources contributing to blastema formation (Currie et al., 2016; Dunis & Namenwirth, 1977; Kragl et al., 2009) and may carry positional information during limb regeneration (Nacu et al., 2013). Through live-imaging of the brainbow (multiple-color fluorescence) labeled connective tissue cells, Currie and colleagues has identified this cell population gives rise to only the connective tissue lineage during digit tip blastema formation and regeneration (Currie et al., 2016). It further strengthened the concept that tissue-specific progenitor cells are the major sources during axolotl tissue regeneration.

In the meanwhile, many other transgenic axolotls have been reported in the past years on solving other issues in regeneration of axolotls. Maden lab has generated a retinoic (RA) acid reporter using a characterized RA-response enhancer, which allow the monitoring of the endogenous RA activity during regeneration (Monaghan & Maden, 2012). It can be expanded to visualize the activity of other signaling molecules when particular enhancers are applied. Furthermore, although Cre-LoxP based conditional gene expression system is well established in axolotls, however, one of the drawbacks, particularly to overexpress functional genes, is that turning on the gene expression is irreversible. To overcome this obstacle, Whited and colleagues has established a more flexible *E. coli* Lac operon based inducible gene expression system in axolotls, which allows turning on and off of the target gene expression under the control of IPTG (Whited, Lehoczky, & Tabin, 2012).

In order to address the roles and mechanisms of how connective tissue participates in regeneration, Prayag Murawala was able to generate an inducible *Prrx1:Cre-ER*^{T2}, and a *Col1A2:Cre-ER*^{T2} transgenic axolotl, in which a well characterized mouse *Prrx1* and *Col1A2* enhancer was employed to achieve the specific expression of gene of interest (GOI) in connective tissue (Gerber et al., 2018). When crossed with a *CAGGS: floxed eGFP-Cherry* reporter line followed by a pulse treatment of Tamoxifen, the Cherry-labeled PRRX1-postive connective tissue cells gives rise to only the connective tissue lineage during limb regeneration, which provides further evidence of the lineage specific origin of blastema (Gerber et al., 2018). Interestingly, analysis of the gene expression profile of these fluorescent-labeled connective tissue cells using single cell RNA sequencing technique has revealed that these cells are reprogrammed back to an embryonic like status during regeneration (Gerber et al., 2018). This study has discovered a new mechanism—the programming feature of progenitor cells prior to/during blastema formation and regeneration, that could potentially present a general mechanism of tissue regeneration.

Although applying transgenic axolotls have shown its broad application potential for answering diverse questions during axolotl tissue regeneration, and many exciting and encouraging progress have been achieved, this technique still faces many challenges. Axolotl transgenesis is established on the basis of I-Sce1 or mTol2 transposes system (Khattak et al., 2014; Khattak,

Schuez, et al., 2013; Sobkow et al., 2006). In either case, the integration of the transgene-cassette into axolotl genome is random. This random integration may lead to the uneven expression of transgene in the targeted tissue, or even silencing of the transgene. Therefore, it is very necessary to characterize a safe-harbor, as the *Rosa26* locus in mice (Soriano, 1999), to hold proper transgene expression in axolotls. Nonetheless, it is clear and can be easily predicted that more transgenics will be produced to facilitate the understanding of the molecular and cellular mechanisms of tissue regeneration in axolotls.

4. Viral transduction systems to study limb regeneration

(Elly Tanaka)

Compared to other aquatic vertebrates, the axolotl boasts multiple viral transduction systems—perhaps related to the fact that axolotl is primarily used to study the postembryonic phenomenon of regeneration. Stephan Roy during his postdoc in the Bryant laboratory described the ability of the vaccinia virus to overexpress sonic hedgehog powerfully enough to generate numerous ectopic digits, providing a standard assay in the field for assessing effectiveness of viral expression in the blastema (Roy et al., 2000). He pointed out however that as vaccinia kills the infected cells, the method is appropriate mainly for testing extracellular factors.

Moving into the 21 century, two integrating retroviruses were developed for use in regeneration. Jessica Whited during her postdoc in the Tabin lab demonstrated the effectiveness of pseudotyped MMLV retrovirus for stably expressing genes in the blastema and cultured axolotl cells (Whited et al., 2013). VSV-G pseudotyped viruses were used to infect dividing blastema cells and assess the proportion of different tissue-specific progenitors in the blastema at different times of regeneration. As an integrating virus at low copy number, expression of SHH in the blastema yielded modest changes in metacarpal morphogenesis. Moving further she cleverly used transgenics to express the TVA ecotropic receptor on endothelial cells driven by the PECAM promoter, so that they could be infected with an ASLV-A glycoprotein pseudotyped MMLV virus. In parallel, Shahryar Khattak demonstrated that the axolotl blastema and cultured cells are efficiently transduced by the human foamy virus retrovirus compared to lentivirus which showed loss of expression in the limb (Khattak, Sandoval-Guzman, et al., 2013). Muscle-specific expression could be achieved by driving GFP expression with a muscle-specific actin promoter. Both viral systems which infect dividing cells have the advantage of integrated expression, but the disadvantage of strict viral genome size limits for efficient packaging, which are required for *in vivo* infection.

In the search for a viral system with larger packaging limits and stronger expression for driving gene expression to obtain strong phenotypes, I decided to try pseudotyped baculovirus, which, while not integrating, could accommodate inserts up to 35 kb, and as a multicopy episomal system should drive strong expression, which, conveniently makes the system S1 instead of S2 genetic safety category in many situations. Knowing from the foamy and MMLV virus work that very high titers are required for *in vivo* transduction, my first attempts involved concentrating commercial BacMam preparations on a centricon and injecting blastemas which immediately showed strong GFP expression that persisted through regeneration and showed expression in cartilage. Based on this observation, Catarina Oliveira started her PhD thesis on developing baculovirus for axolotl regeneration studies. With this as her primary goal, she exhaustively tested viral production methods, pseudotyping methods and virus concentration eventually converging to a protocol where she pseudotyped baculovirus with a VSV-GED mutant which showed less toxicity, and a concentration through sucrose gradient to purify virus (Oliveira et al., 2018). This work allowed her not only to demonstrate efficient transduction *in vivo* and *in vitro*, a tropism of the virus to connective tissue and satellite cells, as well as strong digit duplication after SHH expression in the blastema, it allowed her to make important contributions to two functional studies in limb regeneration (Nacu, Gromberg, Oliveira, Drechsel, & Tanaka, 2016; Wagner et al., 2017). Firstly, she used her viruses encoding SHH and FGF8 to collaborate with Eugen Nacu and Elena Gromberg to demonstrate that the dependency of limb regeneration on anterior and posterior tissue represents the implementation of a SHH-FGF8 positive feedback loop by inducing limbs from anterior and posterior regions of the limb by expressing these factors respectively (Nacu et al., 2016). Secondly, she used the baculovirus to show the role of BMP-signaling in myogenic cell cycle-reentry during newt limb regeneration by expression of BMP receptor mutants (Wagner et al., 2017).

5. Overcoming a major obstacle—Assembly of the giant axolotl genome

Despite the pace of scientific and technological progress studying regeneration and developing the axolotl model system researchers in the growing axolotl field still faced a problem to be generally being perceived to be working on a viable model—and that was the genome and the ability to manipulate the genome. The axolotl has a similar gene complement to other vertebrates, and therefore, as next generation technologies including 454 sequencing and Illumina sequencing became available, the community did manage to implement reasonable transcriptome. However the unassembled genome always loomed, and this was a daunting task considering the 32 gB genome. Sequencing of selected clones in bacterial artificial chromosomes had shown the presence of large intergenic and intronic sequences full of repeats, which meant that short read technologies would not suffice to obtain a high quality assembly (Smith et al., 2009). Indeed exploration of short read genome sequencing by the Voss lab even coupled to purification of specific chromosomes to ameliorate the genome size problem demonstrated the difficulty of assembling the genome (Keinath et al., 2015).

Luckily Eugene Myers who had ended up as a Director at the Max Planck Institute in Dresden decided to put his mathematical skills to the problem of genome assembly long, but error prone reads emerging from Pacific Biosciences technology. Eugene applied to the Max Planck Society to obtain the machines and consumables to sequence this giant genome along with the planaria and was fortunately successful. Working with a newly founded junior group headed by Siegfried Schloissnig, they worked on new assembly algorithms that got around unusual patterns of sequencing artifacts generated by the technology, without severing the long reads as other programs tended to do. One cannot underestimate the luck we had being a direct colleague of a leading mathematician in the genome assembly field. In 2016 we received the call that a first preliminary assembly of the axolotl data was finished. We were so astonished with the result that we decided to annotate the assembly. Then the headaches began. As the axolotl genome is 10 times the size of the human genome, none of the software used to analyze genomes was written in enough bit depth to process the data. Furthermore, as the introns were so long, ab initio gene prediction using standard software did not function. We had to find many work arounds such as splitting the genome into subparts to run software such as repeat annotation, and we had to settle with annotating genes based on mapping RNASeq data to the genome. But after many late night and weekend exchanges between Sergej Nowoshilow, Michael Hiller and I, we could annotate the genome including the genes, transposable elements (with George King). This work describing the genome assembly combined with Ji-Feng Fei's TALEN and CRISPR generated extensive mutant analysis of the axolotl *Pax7* locus showing that it substituted for a missing *Pax3* generated tremendous excitement as provided the two missing elements for axolotl as a molecular genetic model organism to enable the community to work on the genetic basis of axolotl traits (Nowoshilow et al., 2018).

We were in luck again when we wanted to take the initial assembly and scaffold it to chromosome level using HiC contact data. HiC scaffolding is based on the principle when the genome is cross-linked, if one looks at any one place in the genome, the frequency of contacts to that place falls off as an exponential with distance. This is observed in HiC heat maps at the strong signal at the diagonal with falling off density from the diagonal (Burton et al., 2013; Kaplan & Dekker, 2013; Lieberman-Aiden et al., 2009). Using this method, the many contigs of a typical genome assembly can be ordered and oriented up to a contiguous chromosome scale assembly. This scaffolding effort was greatly assisted by the chromosomal scale meiotic map scaffold that had been in the mean-time produced by Jeramiah Smith and Randal Voss's groups (Smith et al., 2019). Unfortunately the meiotic data could not orient the many of the DNA pieces along the assembly, and the resolution of the method could not allow inclusion of all contigs in the scaffolding. Akane Kawaguchi, a postdoc interested in epigenetics of regeneration, as a first step in her own aims, realized she needed this chromosome scale assembly. She adapted the protocols to perform HiC preps on axolotl cells, importantly including figuring how to synchronize AL1 cells in mitosis, since the mitotic HiC data increased the number of informative contacts for assembly tremendously. These developments took well over 1 year. In the meantime Randal Voss and Jeramiah Smith also generated HiC data. During this time, talking to HiC experts including Erez Lieberman-Aiden, Leonid Mirney, Job Dekker and Stephan Mundlos as they passed by as seminar speakers, as well as talking to local experts, Gordana Wutz, Roman Stocits and Anton Golodorodko gave us many important tips on how to go on and also how to analyze our data. At the beginning of our HiC journey, the concepts and data seemed like learning a foreign language but we persevered and became fully ensconced in this endeavor. We certainly benefited from having moved from Dresden to the Vienna Biocenter which is a hot bed of research groups examining large scale chromosome structure. In addition, we luckily could recruit computer scientist Siegfried Schloissnig to Vienna, who put his mind to the HiC scaffolding problem. Again, the available softwares could not be used because they were not written in a way to be fast enough to scaffold this giant genome, and the calculation of contact frequencies is highly disturbed in axolotl due to the presence of the many repetitive regions. Therefore expert computer programming to accelerate the processing so that many different ways of trying to calculate the contact frequencies to gain accurate scaffolding could be tried and in the end a satisfying chromosomal scale assembly (Schloissnig et al., 2021). But finally we are proud that the community can celebrate the existence of a high quality, chromosome level genome assembly that is enviable in quality for any model organism.

Transcriptome and then genome data can be overwhelming in volume so databases to easily search and access the data are essential. As we accumulated the data, we were lucky that Sergej Nowoshilow, with much input from the biologists such as Prayag Murawala, constructed a highly convenient portal called https://www.axolotl-omics.org that allows all researchers to query the genome with repeat annotation and associated gene models both in the form of BLAST searching, through a UCSC implemented genome browser, and the HiGlass browser for HiC data (Nowoshilow & Tanaka, 2020). This portal nicely complements the salamander, Sal-site (https://ambystoma.uky.edu/quick-links/sal-site) constructed by Randal Voss and colleagues (Baddar, Woodcock, Khatri, Kump, & Voss, 2015). Reflecting that axolotl is fully in the genomic age, researchers recently formalized gene nomenclature in this organism (Nowoshilow, Fei, Voss, Tanaka, & Murawala, 2021).

6. The remarkable, enabling era of genome editing

(Ji-Feng Fei)

Withthe availability of gene sequence information and the tremendous progress in genome editing, making axolotl mutants, and using the genome sequence to precisely knock-in gene sequences into the genome became a reality. Editing of defined genomic locus is an important approach for creating genetic modified animals. In the past, unlike the broadly used animal models e.g. mice, it is impossible to make genetic knockout or knock-in axolotls due to the lack of proper genome editing technology. Moreover, RNA interference also does not work in axolotls. Morpholino is the only technology could be applied to axolotls to inhibit the expression of the targeted genes (Schnapp & Tanaka, 2005). Therefore, establishing an efficient genetic loss of function analysis in axolotls is in great demand in order to deeply investigate the roles of given genes during tissue regeneration. Luckily, the rapid development of genome engineering techniques provides such an opportunity to axolotls. Soon after the demonstrations of precision genome modification of the Zinc-Finger Nucleases (ZFNs) on desired genomic loci in mammalian cells and animal models (Doyon et al., 2008; Santiago et al., 2008), we started the attempt of establishing ZFN technique in axolotls, following a ZFN screening protocol published from Joung's lab (Wright et al., 2006). We teamed up with Andrew Oates lab who is using zebrafish as a model to study the rhythmic control of embryonic body segment formation. After running over half a year between two labs on the fourth floor of MPI-CBG, we achieved the first precise modification at axolotl *Oct4* genomic locus using obtained ZFN pair. However, one of the major problems we were facing at that time is the early embryonic lethality of the ZFN injected embryos, that is likely due to the unspecific toxicity of ZFN and made it very difficult to generate global gene knockout axolotls. While we were still optimizing the ZFN injection condition, a new alternative technique Transcription Activator-Like Effector Nucleases (TALENs) was developed and rapidly applied in varied species for generation of genetic modified organisms, such as mice and zebrafish (Bedell et al., 2012; Sander et al., 2011; Sung et al., 2013). After carrying out the first pilot experiment by introducing TALENs against either the endogenous *Tyrosinase* gene or the *eGFP* transgene into freshly laid eggs, we observed typical loss of function phenotypes corresponding to the relevant genes, that was truly a very exciting moment. We found that TALEN is generally more efficient and less toxic than ZFN. Importantly, the modified genomic alleles in F0 TALEN founders can be transmitted to F1 progeny when crossed with DD axolotls.

However, the progress of technology never stops. In the year of 2013, two laboratories reported successful genome engineering in eukaryotic cells using a new system Clustered Regulatory Interspaced Short Palindromic Repeat (CRISPR)/Cas9 (Cong et al., 2013; Mali et al., 2013), right after we have established TALEN approach in the lab. It took us a while to decide whether we should compare TALEN and CRISPR/Cas9 system. Amazingly, the first attempt applying CRISPR/Cas9 in axolotls to knockout *Tyrosinase* gene, surprisingly yield an extremely high efficiency (Fei et al., 2014; Flowers, Timberlake, Mclean, Monaghan, & Crews, 2014). Nearly all injected F0 axolotls showed massive, even complete loss of pigment, and developed properly, which were never observed from ZFN and TALEN injections. Considering it is relatively easy to synthesize gRNA, CRISPR/Cas9 system is definitely a better choice, in most of the cases, for the generation of gene knockout axolotls. Having demonstrated effective disruption of an easily scorable pigment phenotype, we moved on to mutate a gene, *Sox2*, involved in spinal cord stem cell function for which we had raised and antibody so we could monitor loss of expression. While *Sox2* is embryonic lethal in mouse, to our great surprise, F0 axolotl *Sox2* crispants with penetrant modification survived to larval stages. Careful examination of the spinal cord showed loss of SOX2 expression and lack of spinal cord regeneration. Further analysis showed that while in embryos, *Sox2* and its paralog *Sox3* are coexpressed in the embryo, during spinal cord regeneration *Sox3* is not present and the process seems to be solely dependent on *Sox2*. Our observations disclosed the critical role of *Sox2* gene on neural stem cell rapid amplification during spinal cord regeneration (Fei et al., 2014). The availability of TALEN and CRISPR/Cas9-mediated gene knockout provides an opportunity for deep dissection of the gene function during development and regeneration in axolotls.

6.1 Tissue-specific expression via knock-in methods

One of the major challenges is how to achieve tissue specific gene expression in transgenics. When using classical transgenic method, it generally requires a well characterized tissue specific promoter to drive the expression of GOI. However, such well-defined promoter/enhancer is normally not available. In early days, while Khattak and colleagues were trying to generate inducible *Sox2:Cre-ER*^{T2} and *Pax7:Cre-ER*^{T2} transgenic axolotls for cell fate mapping of SOX2+ neural stem cells and PAX7+ satellite cells during regeneration, it is failed to obtain any transgenics with proper GOI expression using available promoters of these two genes (Khattak, Schuez, et al., 2013). It may be due to the fact that the expression of *Sox* and *Pax7* family genes are well known controlled by complex regulatory sequences. And it is very hard to isolate the complete promoter/enhancer sequence of such genes for transgenesis. Taking the advantages of CRISPR/Cas9 technology that could introduce a double strand DNA break (DSB) at the desired axolotl genomic loci, at the presence of exogenous DNA, this donor DNA can integrate into the DSB site (Albadri, Del Bene, & Revenu, 2017). Under this principle, either the *Cherry* reporter gene or inducible *Cre-ER*^{T2} cassettes were inserted into axolotl *Sox2* or *Pax7* genomic loci (Fei et al., 2017). In all of the knock-in lines generated, GOI expression can be faithfully detected in the expected tissues/cells. Satellite cell fate mapping using inducible *Pax7: Cre-ER*^{T2} line has shown that these cells are the major source contributing to axolotl limb regeneration (Fei et al., 2018, 2017). Extended application of the CRISPR/Cas9 mediated knock-in was also used in a recently work from Whited lab, in which they inserted the *Cre* recombinase into *Eyes absence 2* locus to trace the limb blastema cell during development

(Sousounis et al., 2020). Overall, although the CRISPR/Cas9 mediated knock-in in axolotl in a newly developed approach, it has been demonstrated to be a valuable method in studying the mechanism of tissue regeneration in axolotls.

So far, the majority achievements of using CRISPR/Cas9 system has been to generate global gene knockouts (Fei et al., 2018, 2014; Flowers, Sanor, & Crews, 2017; Flowers et al., 2014; Sanor, Flowers, & Crews, 2020). It is still challenging to establish tissue specific gene knockout in axolotls but has been achieved using delivery of CAS9-guide RNA complexes into targeted tissue by electroporation. In transgenesis, generation of a loss of function floxed allele, then combining with a tissue specific Cre line is the most conventional way to achieve tissue specific knockout of targeted gene. Until now, no floxed allele on endogenous gene has been reported in axolotls. Technically, it is not easy to insert two loxP sites separately into axolotl genome to embrace a coding exon and how to do this still needs to be explored. An alternative approach is to introduce CAS9-guide RNA complexes into targeted tissue by electroporation or viral transduction to knockout given genes. It has been demonstrated that delivery of CAS9-guide RNA complexes targeting *Sox2* by electroporation led to efficient gene knock out in spinal cord ependymal cells, phenocopying the *Sox2* mutant phenotype (Fei et al., 2016; Lou, Wang, Long, Liu, & Fei, 2019). So far this approach has been less efficient in the blastema, and we are actively working on alternative methods. Overall, more efforts are still needed to further optimize the CRISPR/Cas9 mediated genomic modification for diverse purposes in axolotls.

7. Parallel developments in other salamanders

The axolotl has been and continues to be a highly popular vertebrate due to the ease of breeding and rearing the animal in laboratory conditions and now with powerful molecular genetic tools. But it is important to keep in mind that regeneration has traditionally been studied in several salamander species. In parallel to developments in the axolotl, transient transfection in the eye, as well as sperm injection-mediated transgenesis were developed in the Japanese newt Pyrrhogaster (Casco-Robles, Yamada, Miura, & Chiba, 2010; Hayashi et al., 2001). However, the long generation time of this animal has made genetic analysis more difficult. Recently, *Pleurodeles waltl* has reemerged as a newt species for molecular regeneration studies, with gene sequence information, robust TALEN and CRISPR-mediated gene mutation (Elewa et al., 2017; Hayashi & Takeuchi, 2015; Matsunami et al., 2019). The availability of these two animal models will allow further investigation of common and divergent aspects of development and regeneration in this family.

8. Concluding remarks

Regeneration in the axolotl remains a fascinating problem to this day with the advent of understanding molecular mechanism still at its infancy. The enabling developments that we and others undertook were a labor of love for the model and the scientific problem which ameliorated worries about publications which always came in the end. We hope that this account helps others to take on this gratifying mission for their respective models and biologies.

Acknowledgments

We would like to thank all the amazing colleagues from the lab and the community who worked together on and supported over the years this wonderful model organism. K.E. is supported by NIH NICHD (R01HD092451) and the Owens Family Foundation. Work in the JFF laboratory is supported by grants from National Key R&D Program of China 2019YFE0106700, the Natural Science Foundation of China 31970782, Key-Area Research and Development Program of Guangdong Province 2018B030332001, 2019B030335001.

E.T. is an ERC Advanced Investigator (RegGeneMems 742046).

References

Albadri, S., Del Bene, F., & Revenu, C. (2017). Genome editing using CRISPR/Cas9-based knock-in approaches in zebrafish. *Methods*, *121-122*, 77–85.

Baddar, N. W., Woodcock, M. R., Khatri, S., Kump, D. K., & Voss, S. R. (2015). Sal-Site: Research resources for the Mexican axolotl. *Methods in Molecular Biology*, *1290*, 321–336.

Bedell, V. M., Wang, Y., Campbell, J. M., Poshusta, T. L., Starker, C. G., Krug, R. G., 2nd, et al. (2012). In vivo genome editing using a high-efficiency TALEN system. *Nature*, *491*, 114–118.

Berg, D. A., Kirkham, M., Beljajeva, A., Knapp, D., Habermann, B., Ryge, J., et al. (2010). Efficient regeneration by activation of neurogenesis in homeostatically quiescent regions of the adult vertebrate brain. *Development*, *137*, 4127–4134.

Burton, J. N., Adey, A., Patwardhan, R. P., Qiu, R., Kitzman, J. O., & Shendure, J. (2013). Chromosome-scale scaffolding of de novo genome assemblies based on chromatin interactions. *Nature Biotechnology*, *31*, 1119–1125.

Casco-Robles, M. M., Yamada, S., Miura, T., & Chiba, C. (2010). Simple and efficient transgenesis with I-SceI meganuclease in the newt, Cynops pyrrhogaster. *Developmental Dynamics*, *239*, 3275–3284.

Clarke, J. D. (1999). Using fluorescent dyes for fate mapping, lineage analysis, and axon tracing in the chick embryo. *Methods in Molecular Biology*, *97*, 319–328.

Clarke, J. D., Alexander, R., & Holder, N. (1988). Regeneration of descending axons in the spinal cord of the axolotl. *Neuroscience Letters*, *89*, 1–6.

Clarke, J. D., & Tickle, C. (1999). Fate maps old and new. *Nature Cell Biology*, *1*, E103–E109.

Cong, L., Ran, F. A., Cox, D., Lin, S., Barretto, R., Habib, N., et al. (2013). Multiplex genome engineering using CRISPR/Cas systems. *Science*, *339*(6121), 819–823. https://doi.org/10.1126/science.1231143.

Currie, J. D., Kawaguchi, A., Traspas, R. M., Schuez, M., Chara, O., & Tanaka, E. M. (2016). Live imaging of axolotl digit regeneration reveals spatiotemporal choreography of diverse connective tissue progenitor pools. *Developmental Cell*, *39*, 411–423.

Doyon, Y., McCammon, J. M., Miller, J. C., Faraji, F., Ngo, C., Katibah, G. E., et al. (2008). Heritable targeted gene disruption in zebrafish using designed zinc-finger nucleases. *Nature Biotechnology*, *26*, 702–708.

Dunis, D. A., & Namenwirth, M. (1977). The role of grafted skin in the regeneration of x-irradiated axolotl limbs. *Developmental Biology*, *56*, 97–109.

Echeverri, K., Clarke, J. D., & Tanaka, E. M. (2001). In vivo imaging indicates muscle fiber dedifferentiation is a major contributor to the regenerating tail blastema. *Developmental Biology*, *236*, 151–164.

Echeverri, K., & Tanaka, E. M. (2002). Ectoderm to mesoderm lineage switching during axolotl tail regeneration. *Science*, *298*, 1993–1996.

Echeverri, K., & Tanaka, E. M. (2003). Electroporation as a tool to study in vivo spinal cord regeneration. *Developmental Dynamics*, *226*, 418–425.

Echeverri, K., & Tanaka, E. M. (2005). Proximodistal patterning during limb regeneration. *Developmental Biology*, *279*, 391–401.

Elewa, A., Wang, H., Talavera-Lopez, C., Joven, A., Brito, G., Kumar, A., et al. (2017). Reading and editing the Pleurodeles waltl genome reveals novel features of tetrapod regeneration. *Nature Communications*, *8*, 2286.

Fei, J. F., Knapp, D., Schuez, M., Murawala, P., Zou, Y., Singh, S. P., et al. (2016). Tissue- and time-directed electroporation of CAS9 protein–gRNA complexes in vivo yields efficient multigene knockout for studying gene function in regeneration. *NPJ Regenerative Medicine*, *1*, 16002.

Fei, J. F., Lou, W. P., Knapp, D., Murawala, P., Gerber, T., Taniguchi, Y., et al. (2018). Application and optimization of CRISPR-Cas9-mediated genome engineering in axolotl (Ambystoma mexicanum). *Nature Protocols*, *13*, 2908–2943.

Fei, J. F., Schuez, M., Knapp, D., Taniguchi, Y., Drechsel, D. N., & Tanaka, E. M. (2017). Efficient gene knockin in axolotl and its use to test the role of satellite cells in limb regeneration. *Proceedings of the National Academy of Sciences of the United States of America*, *114*, 12501–12506.

Fei, J. F., Schuez, M., Tazaki, A., Taniguchi, Y., Roensch, K., & Tanaka, E. M. (2014). CRISPR-mediated genomic deletion of Sox2 in the axolotl shows a requirement in spinal cord neural stem cell amplification during tail regeneration. *Stem Cell Reports*, *3*, 444–459.

Flowers, G. P., Sanor, L. D., & Crews, C. M. (2017). Lineage tracing of genome-edited alleles reveals high fidelity axolotl limb regeneration. *eLife*, *6*.

Flowers, G. P., Timberlake, A. T., Mclean, K. C., Monaghan, J. R., & Crews, C. M. (2014). Highly efficient targeted mutagenesis in axolotl using Cas9 RNA-guided nuclease. *Development*, *141*, 2165–2171.

Fraser, S. E. (1996). Iontophoretic dye labeling of embryonic cells. *Methods in Cell Biology*, *51*, 147–160.

Gardiner, D. M., Blumberg, B., Komine, Y., & Bryant, S. V. (1995). Regulation of HoxA expression in developing and regenerating axolotl limbs. *Development*, *121*, 1731–1741.

Gargioli, C., & Slack, J. M. (2004). Cell lineage tracing during Xenopus tail regeneration. *Development*, *131*, 2669–2679.

Gerber, T., Murawala, P., Knapp, D., Masselink, W., Schuez, M., Hermann, S., et al. (2018). Single-cell analysis uncovers convergence of cell identities during axolotl limb regeneration. *Science*, *362*.

Grohme, M. A., Schloissnig, S., Rozanski, A., Pippel, M., Young, G. R., Winkler, S., et al. (2018). The genome of Schmidtea mediterranea and the evolution of core cellular mechanisms. *Nature*, *554*, 56–61.

Haas, K., Sin, W. C., Javaherian, A., Li, Z., & Cline, H. T. (2001). Single-cell electroporation for gene transfer in vivo. *Neuron*, *29*, 583–591.

Habermann, B., Bebin, A. G., Herklotz, S., Volkmer, M., Eckelt, K., Pehlke, K., et al. (2004). An Ambystoma mexicanum EST sequencing project: Analysis of 17,352 expressed sequence tags from embryonic and regenerating blastema cDNA libraries. *Genome Biology*, *5*, R67.

Harada, H., Omi, M., & Nakamura, H. (2017). In ovo electroporation methods in chick embryos. *Methods in Molecular Biology*, *1650*, 167–176.

Hay, E. D. (1959). Electron microscopic observations of muscle dedifferentiation in regenerating Amblystoma limbs. *Developmental Biology*, *1*, 555–585.

Hayashi, T., & Takeuchi, T. (2015). Gene manipulation for regenerative studies using the Iberian ribbed newt, Pleurodeles waltl. *Methods in Molecular Biology*, *1290*, 297–305.

Hayashi, T., Yamagishi, A., Kuroiwa, A., Mizuno, N., Kondoh, H., & Okamoto, M. (2001). Highly efficient transfection system for functional gene analysis in adult amphibian lens regeneration. *Development, Growth & Differentiation*, *43*, 361–370.

Holder, N., Clarke, J. D., Kamalati, T., & Lane, E. B. (1990). Heterogeneity in spinal radial glia demonstrated by intermediate filament expression and HRP labelling. *Journal of Neurocytology*, *19*, 915–928.

Itasaki, N., Bel-Vialar, S., & Krumlauf, R. (1999). Shocking developments in chick embryology: Electroporation and in ovo gene expression. *Nature Cell Biology*, *1*, E203–E207.

Iten, L. E., & Bryant, S. V. (1975). The interaction between the blastema and stump in the establishment of the anterior–posterior and proximal–distal organization of the limb regenerate. *Developmental Biology*, *44*, 119–147.

Kaplan, N., & Dekker, J. (2013). High-throughput genome scaffolding from in vivo DNA interaction frequency. *Nature Biotechnology*, *31*, 1143–1147.

Kawakami, Y., Rodriguez Esteban, C., Raya, M., Kawakami, H., Marti, M., Dubova, I., et al. (2006). Wnt/beta-catenin signaling regulates vertebrate limb regeneration. *Genes & Development*, *20*, 3232–3237.

Keinath, M. C., Timoshevskiy, V. A., Timoshevskaya, N. Y., Tsonis, P. A., Voss, S. R., & Smith, J. J. (2015). Initial characterization of the large genome of the salamander Ambystoma mexicanum using shotgun and laser capture chromosome sequencing. *Scientific Reports*, *5*, 16413.

Khattak, S., Murawala, P., Andreas, H., Kappert, V., Schuez, M., Sandoval-Guzman, T., et al. (2014). Optimized axolotl (Ambystoma mexicanum) husbandry, breeding, metamorphosis, transgenesis and tamoxifen-mediated recombination. *Nature Protocols*, *9*, 529–540.

Khattak, S., Sandoval-Guzman, T., Stanke, N., Protze, S., Tanaka, E. M., & Lindemann, D. (2013). Foamy virus for efficient gene transfer in regeneration studies. *BMC Developmental Biology*, *13*, 17.

Khattak, S., Schuez, M., Richter, T., Knapp, D., Haigo, S. L., Sandoval-Guzman, T., et al. (2013). Germline transgenic methods for tracking cells and testing gene function during regeneration in the axolotl. *Stem Cell Reports*, *1*, 90–103.

Kragl, M., Knapp, D., Nacu, E., Khattak, S., Maden, M., Epperlein, H. H., et al. (2009). Cells keep a memory of their tissue origin during axolotl limb regeneration. *Nature*, *460*, 60–65.

Kumar, A., Godwin, J. W., Gates, P. B., Garza-Garcia, A. A., & Brockes, J. P. (2007). Molecular basis for the nerve dependence of limb regeneration in an adult vertebrate. *Science*, *318*, 772–777.

Kumar, A., Velloso, C. P., Imokawa, Y., & Brockes, J. P. (2000). Plasticity of retrovirus-labelled myotubes in the newt limb regeneration blastema. *Developmental Biology*, *218*, 125–136.

Lieberman-Aiden, E., van Berkum, N. L., Williams, L., Imakaev, M., Ragoczy, T., Telling, A., et al. (2009). Comprehensive mapping of long-range interactions reveals folding principles of the human genome. *Science*, *326*, 289–293.

Lopez, D., Lin, L., Monaghan, J. R., Cogle, C. R., Bova, F. J., Maden, M., et al. (2014). Mapping hematopoiesis in a fully regenerative vertebrate: The axolotl. *Blood*, *124*, 1232–1241.

Lou, W. P., Wang, L., Long, C., Liu, L., & Fei, J. F. (2019). Direct gene knock-out of axolotl spinal cord neural stem cells via electroporation of CAS9 protein-gRNA complexes. *Journal of Visualized Experiments*, (149). https://doi.org/10.3791/59850.

Mali, P., Yang, L., Esvelt, K. M., Aach, J., Guell, M., DiCarlo, J. E., et al. (2013). RNA-guided human genome engineering via Cas9. *Science*, *339*(6121), 823–826. https://doi.org/10.1126/science.1232033.

Matsunami, M., Suzuki, M., Haramoto, Y., Fukui, A., Inoue, T., Yamaguchi, K., et al. (2019). A comprehensive reference transcriptome resource for the Iberian ribbed newt Pleurodeles waltl, an emerging model for developmental and regeneration biology. *DNA Research*, *26*, 217–229.

McCusker, C. D., Diaz-Castillo, C., Sosnik, J., Phan, A. Q., & Gardiner, D. M. (2016). Cartilage and bone cells do not participate in skeletal regeneration in Ambystoma mexicanum limbs. *Developmental Biology*, *416*, 26–33.

McGonnell, I. M., Clarke, J. D., & Tickle, C. (1998). Fate map of the developing chick face: Analysis of expansion of facial primordia and establishment of the primary palate. *Developmental Dynamics*, *212*, 102–118.

McLennan, R., & Kulesa, P. M. (2019). In ovo electroporation of plasmid DNA and morpholinos into specific tissues during early embryogenesis. *Methods in Molecular Biology*, *1976*, 71–82.

Mercader, N., Tanaka, E. M., & Torres, M. (2005). Proximodistal identity during vertebrate limb regeneration is regulated by Meis homeodomain proteins. *Development*, *132*, 4131–4142.

Monaghan, J. R., & Maden, M. (2012). Visualization of retinoic acid signaling in transgenic axolotls during limb development and regeneration. *Developmental Biology*, *368*, 63–75.

Muneoka, K., Fox, W. F., & Bryant, S. V. (1986). Cellular contribution from dermis and cartilage to the regenerating limb blastema in axolotls. *Developmental Biology*, *116*, 256–260.

Muneoka, K., Wise, L. D., Fox, W. F., & Bryant, S. V. (1984). Improved techniques for use of the triploid cell marker in the axolotl, Ambystoma mexicanum. *Developmental Biology*, *105*, 240–245.

Nacu, E., Glausch, M., Le, H. Q., Damanik, F. F., Schuez, M., Knapp, D., et al. (2013). Connective tissue cells, but not muscle cells, are involved in establishing the proximo-distal outcome of limb regeneration in the axolotl. *Development*, *140*, 513–518.

Nacu, E., Gromberg, E., Oliveira, C. R., Drechsel, D., & Tanaka, E. M. (2016). FGF8 and SHH substitute for anterior-posterior tissue interactions to induce limb regeneration. *Nature*, *533*, 407–410.

Nacu, E., Knapp, D., Tanaka, E. M., & Epperlein, H. H. (2009). Axolotl (*Ambystoma mexicanum*) embryonic transplantation methods. *Cold Spring Harbor Protocols*, *2009*. pdb.prot5265.

Neumann, E., Scaefer-Ridder, M., Wang, Y., & Hofschneider, P. H. (1982). Gene transfer into mouse lyoma cells by electroporation in high electric fields. *The EMBO Journal*, *1*, 841–845.

Neumann, E., Toensing, K., Kakorin, S., Budde, P., & Frey, J. (1998). Mechanism of electroporative dye uptake by mouse B cells. *Biophysical Journal*, *74*.

Nowoshilow, S., Fei, J. F., Voss, R. S., Tanaka, E. M., & Murawala, P. (2021). Gene and transgenics nomenclature for the laboratory axolotl—Ambystoma mexicanum. *Developmental Dynamics*. https://doi.org/10.1002/dvdy.351.

Nowoshilow, S., Schloissnig, S., Fei, J. F., Dahl, A., Pang, A. W. C., Pippel, M., et al. (2018). The axolotl genome and the evolution of key tissue formation regulators. *Nature*, *554*, 50–55.

Nowoshilow, S., & Tanaka, E. M. (2020). Introducing www.axolotl-omics.org—An integrated -omics data portal for the axolotl research community. *Experimental Cell Research*, *394*, 112143.

Oliveira, C. R., Lemaitre, R., Murawala, P., Tazaki, A., Drechsel, D. N., & Tanaka, E. M. (2018). Pseudotyped baculovirus is an effective gene expression tool for studying molecular function during axolotl limb regeneration. *Developmental Biology*, *433*, 262–275.

Pecorino, L. T., Lo, D. C., & Brockes, J. P. (1994). Isoform-specific induction of a retinoid-responsive antigen after biolistic transfection of chimaeric retinoic acid/thyroid hormone receptors into a regenerating limb. *Development*, *120*, 325–333.

Piekarski, N., Gross, J. B., & Hanken, J. (2014). Evolutionary innovation and conservation in the embryonic derivation of the vertebrate skull. *Nature Communications*, *5*, 5661.

Potter, H. (1988). Electroporation in biology: Methods, applications, and instrumentation. *Analytical Biochemistry*, *174*, 361–373.

Putta, S., Smith, J. J., Walker, J. A., Rondet, M., Weisrock, D. W., Monaghan, J., et al. (2004). From biomedicine to natural history research: EST resources for ambystomatid salamanders. *BMC Genomics*, *5*, 54.

Rodrigo Albors, A., & Tanaka, E. M. (2015). High-efficiency electroporation of the spinal cord in larval axolotl. *Methods in Molecular Biology*, *1290*, 115–125.

Rost, F., Rodrigo Albors, A., Mazurov, V., Brusch, L., Deutsch, A., Tanaka, E. M., et al. (2016). Accelerated cell divisions drive the outgrowth of the regenerating spinal cord in axolotls. *eLife*, *5*.

Roy, S., Gardiner, D. M., & Bryant, S. V. (2000). Vaccinia as a tool for functional analysis in regenerating limbs: Ectopic expression of Shh. *Developmental Biology*, *218*, 199–205.

Saito, T., & Nakaatsuji, N. (2001). Efficient gene transfer into the embryonic mouse brain using in vivo electroporation. *Developmental Biology*, *240*, 237–246.

Sander, J. D., Cade, L., Khayter, C., Reyon, D., Peterson, R. T., Joung, J. K., et al. (2011). Targeted gene disruption in somatic zebrafish cells using engineered TALENs. *Nature Biotechnology*, *29*, 697–698.

Sandoval-Guzman, T., Wang, H., Khattak, S., Schuez, M., Roensch, K., Nacu, E., et al. (2014). Fundamental differences in dedifferentiation and stem cell recruitment during skeletal muscle regeneration in two salamander species. *Cell Stem Cell*, *14*, 174–187.

Sanor, L. D., Flowers, G. P., & Crews, C. M. (2020). Multiplex CRISPR/Cas screen in regenerating haploid limbs of chimeric axolotls. *eLife*, *9*.

Santiago, Y., Chan, E., Liu, P. Q., Orlando, S., Zhang, L., Urnov, F. D., et al. (2008). Targeted gene knockout in mammalian cells by using engineered zinc-finger nucleases. *Proceedings of the National Academy of Sciences of the United States of America*, *105*, 5809–5814.

Schloissnig, S., Kawaguchi, A., Nowoshilow, S., Falcon, F., Otsuki, L., Tardivo, P., et al. (2021). The giant axolotl genome uncovers the evolution, scaling, and transcriptional control of complex gene loci. *Proceedings of the National Academy of Sciences of the United States of America*, *118*.

Schnapp, E., & Tanaka, E. M. (2005). Quantitative evaluation of morpholino-mediated protein knockdown of GFP, MSX1, and PAX7 during tail regeneration in Ambystoma mexicanum. *Developmental Dynamics*, *232*, 162–170.

Skibbens, R. V., Skeen, V. P., & Salmon, E. D. (1993). Directional instability of kinetochore motility during chromosome congression and segregation in mitotic newt lung cells: A push-pull mechanism. *The Journal of Cell Biology*, *122*, 859–875.

Smith, J. J., Putta, S., Zhu, W., Pao, G. M., Verma, I. M., Hunter, T., et al. (2009). Genic regions of a large salamander genome contain long introns and novel genes. *BMC Genomics*, *10*, 19.

Smith, J. J., Timoshevskaya, N., Timoshevskiy, V. A., Keinath, M. C., Hardy, D., & Voss, S. R. (2019). A chromosome-scale assembly of the axolotl genome. *Genome Research*, *29*, 317–324.

Sobkow, L., Epperlein, H. H., Herklotz, S., Straube, W. L., & Tanaka, E. M. (2006). A germline GFP transgenic axolotl and its use to track cell fate: Dual origin of the fin mesenchyme during development and the fate of blood cells during regeneration. *Developmental Biology*, *290*, 386–397.

Soriano, P. (1999). Generalized lacZ expression with the ROSA26 Cre reporter strain. *Nature Genetics*, *21*, 70–71.

Soukup, V., Epperlein, H. H., Horacek, I., & Cerny, R. (2008). Dual epithelial origin of vertebrate oral teeth. *Nature*, *455*, 795–798.

Sousounis, K., Bryant, D. M., Martinez Fernandez, J., Eddy, S. S., Tsai, S. L., Gundberg, G. C., et al. (2020). Eya2 promotes cell cycle progression by regulating DNA damage response during vertebrate limb regeneration. *eLife*, *9*.

Stocum, D. L. (1968). The urodele limb regeneration blastema: A self-organizing system. I. Morphogenesis and differentiation of autografted whole and fractional blastemas. *Developmental Biology*, *18*, 457–480.

Stocum, D. L., & Melton, D. A. (1977). Self-organizational capacity of distally transplanted limb regeneration blastemas in larval salamanders. *The Journal of Experimental Zoology*, *201*, 451–461.

Sung, Y. H., Baek, I. J., Kim, D. H., Jeon, J., Lee, J., Lee, K., et al. (2013). Knockout mice created by TALEN-mediated gene targeting. *Nature Biotechnology*, *31*, 23–24.

Swartz, M., Eberhart, J., Mastick, G. S., & Krull, C. E. (2001). Sparking new frontiers: Using in vivo electroporation for genetic manipulation. *Developmental Biology*, *233*, 13–21.

Tawk, M., Bianco, I. H., & Clarke, J. D. (2009). Focal electroporation in zebrafish embryos and larvae. *Methods in Molecular Biology*, *546*, 145–151.

Tawk, M., Tuil, D., Torrente, Y., Vriz, S., & Paulin, D. (2002). High-efficiency gene transfer into adult fish: A new tool to study fin regeneration. *Genesis*, *32*, 27–31.

Thermes, V., Grabher, C., Ristoratore, F., Bourrat, F., Choulika, A., Wittbrodt, J., et al. (2002). I-SceI meganuclease mediates highly efficient transgenesis in fish. *Mechanisms of Development*, *118*, 91–98.

Tsai, S. L., Baselga-Garriga, C., & Melton, D. A. (2019). Blastemal progenitors modulate immune signaling during early limb regeneration. *Development*, *146*.

Ueda, Y., Kondoh, H., & Mizuno, N. (2005). Generation of transgenic newt Cynops pyrrhogaster for regeneration study. *Genesis*, *41*, 87–98.

Voiculescu, O., & Stern, C. D. (2017). Manipulating gene expression in the chick embryo. *Methods in Molecular Biology*, *1565*, 105–114.

Wagner, I., Wang, H., Weissert, P. M., Straube, W. L., Shevchenko, A., Gentzel, M., et al. (2017). Serum proteases potentiate BMP-induced cell cycle re-entry of dedifferentiating muscle cells during newt limb regeneration. *Developmental Cell*, *40*(608–617), e606.

Waterman-Storer, C. M., Salmon, W. C., & Salmon, E. D. (2000). Feedback interactions between cell-cell adherens junctions and cytoskeletal dynamics in newt lung epithelial cells. *Molecular Biology of the Cell*, *11*, 2471–2483.

Waters, J. C., Skibbens, R. V., & Salmon, E. D. (1996). Oscillating mitotic newt lung cell kinetochores are, on average, under tension and rarely push. *Journal of Cell Science*, *109*(Pt. 12), 2823–2831.

Whited, J. L., Lehoczky, J. A., & Tabin, C. J. (2012). Inducible genetic system for the axolotl. *Proceedings of the National Academy of Sciences of the United States of America*, *109*, 13662–13667.

Whited, J. L., Tsai, S. L., Beier, K. T., White, J. N., Piekarski, N., Hanken, J., et al. (2013). Pseudotyped retroviruses for infecting axolotl in vivo and in vitro. *Development*, *140*, 1137–1146.

Wright, D. A., Thibodeau-Beganny, S., Sander, J. D., Winfrey, R. J., Hirsh, A. S., Eichtinger, M., et al. (2006). Standardized reagents and protocols for engineering zinc finger nucleases by modular assembly. *Nature Protocols*, *1*, 1637–1652.

Zhang, Z., Denans, N., Liu, Y., Zhulyn, O., Rosenblatt, H. D., Wernig, M., et al. (2021). Optogenetic manipulation of cellular communication using engineered myosin motors. *Nature Cell Biology*, *23*, 198–208.

CHAPTER TWENTY-THREE

A remarkable rodent: Regeneration and reproduction in spiny mice (*Acomys*)

Ashley W. Seifert[a,b,*] and Peter Temple-Smith[c]
[a]Department of Biology, University of Kentucky, Lexington, KY, United States
[b]Department of Veterinary Anatomy and Physiology, University of Nairobi, Nairobi, Kenya
[c]Department of Obstetrics & Gynecology, Monash University, Clayton, VIC, Australia
*Corresponding author: e-mail address: awseifert@uky.edu

Contents

Abstract

Although certain organisms are chosen and employed to better understand a specific problem in biology (so-called model organisms), sometimes an animal model reveals its' biomedical importance by happenstance. In many ways, the advent of spiny mice (*Acomys*) as an emerging model to study regeneration and menstruation stands as a case study in scientific pseudoserendipity (Diaz de Chumaceiro, 1995). As we recount

ISSN 0070-2153
https://doi.org/10.1016/bs.ctdb.2021.12.017

in this chapter, the discovery of these phenotypes, while not entirely accidental, was nonetheless unexpected. In addition to recounting how we uncovered these unusual mammalian traits, we outline recent work by our groups and others that has begun to outline the cellular and genetic mechanisms underlying bonafide mammalian tissue regeneration and a human-like mode of reproduction in spiny mice.

1. Introduction

Every kind of organism has a special place in the epistemological sun.

E.O. Wilson

Relatively recent diversification among murid species provides a strong context for comparative studies seeking to understand the genomic basis for speciation as well as the cellular and genetic basis for lineage and clade-specific traits. Despite these advantages, other than *Mus*, *Rattus* and a handful of other species, murid rodents remain woefully understudied from the standpoint of developmental and reproductive biology.

Commonly referred to as spiny mice, *Acomys* are an enigmatic genus of rodents that are taxonomically grouped with *Lophuromys*, *Deomys* and *Uranomys* in the subfamily Deomyinae (Alhajeri, Hunt, & Steppan, 2015). Together with the Gerbillinae, Murinae and the single species clade Lophiomyinae, these four subfamilies represent the most diverse group of mammals on earth (Steppan & Schenk, 2017). This is all the more impressive given the relative youth of the murid clade ($\sim$18 mya) and that radiations in these three multispecies subfamilies occurred within the last 10 million years (Fig. 1) (Steppan & Schenk, 2017).

Among murids, *Acomys* stand alone in possessing a suite of unique biological traits that are uncommon even among rodents (e.g., precocial young, highly social group dynamics, flexible sleep patterns, etc.) and several reviews have covered these topics in detail (Brunjes, 1990; D'Udine & Alleva, 1988; Haughton, Gawriluk, & Seifert, 2015; Shafrir, Ziv, & Kalman, 2006). Adding to this list of unique murid traits found in *Acomys*, work by our respective groups uncovered two additional phenotypes: complex tissue regeneration and a human-like pattern of reproduction (Bellofiore et al., 2017; Bellofiore, Rana, Dickinson, Temple-Smith, & Evans, 2018; Gawriluk et al., 2016; Seifert et al., 2012). As our respective discoveries have gained attention among biologists and biomedical researchers, a growing community of researchers has begun to embrace *Acomys* as a useful model to study these and other traits.

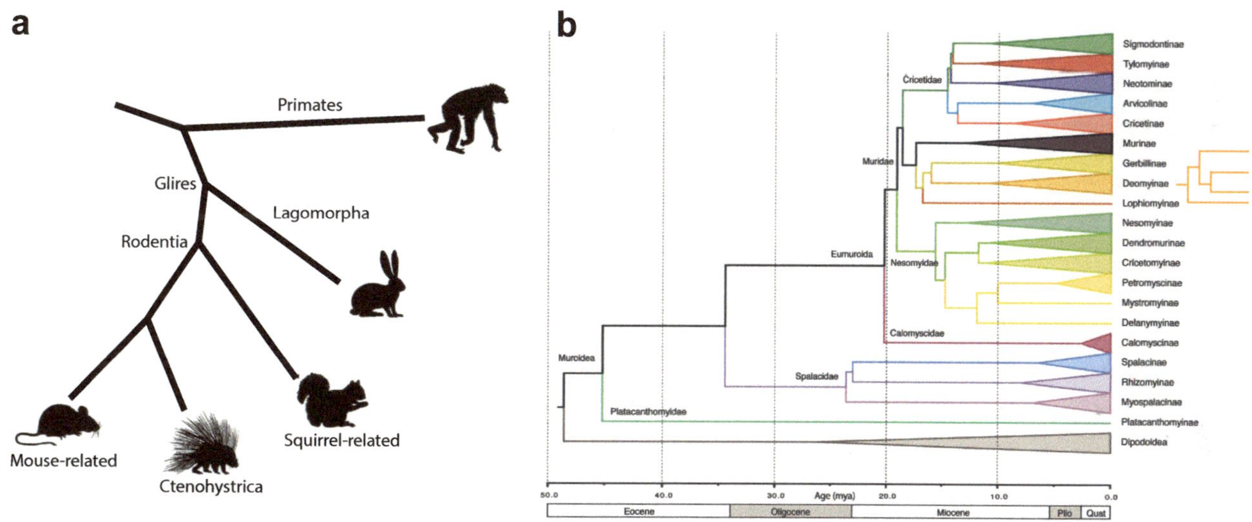

Fig. 1 Phylogenetic summary of murid subfamilies and taxonomic relationships of *Acomys*. (A) General relationship between primates, lagomorphs and rodents where all three groups shared a common ancestor approximately 90 million year ago (timetree.org). (B) Maximum-likelihood chronogram depicting taxonomic relationships among the major muroid subfamilies. *Acomys* are grouped with *Lophuromys*, *Deomys* and *Uranomys* in the subfamily Deomyinae. The Deomyinae, Gerbillinae, Murinae and the single species clade Lophiomyinae, comprise the Muridae, the largest family of rodents and mammals. General taxonomic relationships in (A) based on a detailed phylogenetic reconstruction of rodents inferred from a molecular supermatrix based on 11 mitochondrial and nuclear genes (Fabre, Hautier, Dimitrov, & Douzery, 2012). *Panel (B) adapted from Steppan, S. J. and Schenk, J. J. (2017). Muroid rodent phylogenetics: 900-species tree reveals increasing diversification rates. PLoS One 12, e0183070. under Creative Commons Attribution License (CC BY 4.0).*

August Krogh opined that one or several animals exist to conveniently study any scientific problem. He also lamented these animals would remain unknown to the persons most needing them and he encouraged scientists to "*...apply to the zoologists to find them and to lay our hands on them*" (Krogh, 1929). Writing years later on the importance of returning to taxon-centered research, E.O. Wilson wrote that the so-called inverse Krogh rule applied with equal force and that "*...for every organism there exists a problem for the solution of which it is ideally suited and others for which it is useless*" (Wilson, 1989). As adherents to approaches that some might consider antagonistic, we believe both strategies expand the horizon for scientific discovery and in doing so, unshackles those toiling away as they practice the progression of natural science.

2. Regnerative biology

Discovery is to see what everybody else has seen and think what nobody else has thought.

***Albert Szent-Györgyi, from his book,* Bioenergetics**

These little guys jump right out of their skin!

Truman Young

2.1 A chance encounter catalyzes a curiosity driven scientific adventure

Whether at meetings or in casual conversation, scientists and nonscientists frequently ask to know the story of how I (Ashley) came to study regeneration in *Acomys*. Depending on the context, I usually recount an abridged version of how I traveled to Kenya and ultimately discovered their regenerative abilities. Devoid of decorative hyperbole, the complete story presents either an impassioned defense for pursuing curiosity driven science or a cautionary tale of the career risks associated with this approach. Whichever lesson one derives, much like the observation of female menstruation in these animals, both remind us that sometimes discoveries lie hidden in plain sight.

The spiny mouse regeneration story was born in the Biology department at the University of Florida. Nearing the end of my doctoral studies, I had chosen to remain in Gainesville to pursue postdoctoral research. Although staying at one's doctoral institution is generally frowned upon when pursuing a career in academia, two factors were prominent in my decision to stay in Gainesville, and both are pertinent to the story which follows. First, as I mulled over options for choosing a postdoctoral lab, my partner was in the midst of graduate school and my heart was intent on remaining with her until she completed her dissertation research. Second, although there were many biological problems intriguing enough to consider studying as a postdoc, none were as seductive to me as trying to understand interspecific variation in regenerative ability (i.e., depending on the species, injury elicits either fibrotic repair or regenerative healing in response to tissue damage). Fortunately, Malcolm Maden, a prominent regeneration biologist, had recently moved his lab from King's College, London to the University of Florida. As a member of my thesis committee, we were already acquainted and sitting together in his office one afternoon, I made an impassioned pitch to join his lab to pursue regeneration research. While he was delighted to have me join his group, we needed to come up with a way to support my salary. Luckily, fate intervened in the form of an opening on a training grant and the die was cast. Heart and head aligned, I was thrilled to have my future postdoc plans settled.

With my dissertation defense looming, I decided that a well-deserved break would be in order before I launched into my postdoc research. What better way to celebrate defending my thesis than traveling to Kenya to help my partner Megan collect field data for her own dissertation. As an ecologist studying how plants defend themselves from herbivores, she was set to start her field season at Mpala Research Centre in Laikipia. Naturally, Megan was excited to have an extra pair of hands in the field and I was more than happy to lend them. Of course, there was the small matter of paying for my trip that needed to be settled before I went anywhere. Little did I know, a chance encounter was about to trigger a chain of events that would put me on a plane to Kenya and set the course for my postdoctoral research.

Around the time I had settled my postdoctoral plans, Truman Young, a community ecologist from UC Davis who had spent his career working at Mpala, had been invited to our department to give a seminar. Truman was Megan's academic grandfather and the evening after his talk we struck up conversation at the post-seminar social. After introducing myself, I shared my post-graduation plans to spend several months at Mpala with Megan. Not knowing that I was a developmental biologist, he asked me what I planned to work on during my postdoc. After a bit of back and forth, I started describing how species can heal injuries quite differently and that for my postdoc I planned to find out why. At some point, the conversation pivoted to the concept that regeneration might be an adaptive trait and I explained that this idea had a rich history which had formed the basis for a contentious debate between Thomas Hunt Morgan and August Weisman (Esposito, 2013). Like Morgan, I believed (and still do) that the distribution of animal regeneration could not be explained by purely adaptive means.

Before I could continue, he stroked his mustache and blurted out something about "these mice in Kenya" and then suggested that these rodents might be relevant to my larger research question. "When you handle them, these little guys jump right out of their skin," he said. Dumfounded, I asked him what he meant. He recounted a decades-old conversation with Steve Takata, a graduate student working at Mpala who had trapped spiny mice and who had observed what he described as a skin shedding behavior. Truman recounted how supposedly routine handling would result in massive skin tears and in the most extreme cases, near complete degloving. The working hypothesis was that it could act as predator defense mechanism, that is, if the mouse survived its injuries. "When you go over to Kenya this summer why don't you trap some? Maybe they regenerate their skin?"

I thought about the conversation for the rest of the night. Why would a mammal evolve skin that could tear so easily and, in the process, readily expose themselves to infection? Losing a tail sheath or an entire tail was one thing, and many small mammals were known to exhibit tail loss (Shargal, Rath-Wolfson, Kronfeld, & Dayan, 1999), but losing large pieces of skin was an entirely different matter. My mind was filled with questions. How could these rodents maintain water balance? How did they fight infection? How did they heal their wounds? *Did* they heal their wounds? While the skeptic in me thought the behavior implausible, the dreamer in me found the potential to uncover a biological explanation seductive.

If there was credence to this behavior, then someone must have published an account somewhere. Over the next several weeks, I poured through primary research papers and also dipped into the gray literature world of technical reports and theses. My efforts turned up nothing beyond the occasional account of tail loss. I reached out to Megan's advisor, Todd Palmer, who had overlapped with Steve Takata during his own dissertation field work. He didn't remember conversations about skin shedding, so he sent out an email inquiry to other researchers working at Mpala. Later that week he forwarded me a picture taken by Jake Goheen, a mammalogist working at Mpala, that showed a spiny mouse missing a large piece of back skin. It appeared that Truman's memory was spot on.

Soon enough I had outlined what seemed like a straightforward plan: go to Kenya, catch spiny mice, investigate skin tearing and if possible, characterize wound healing. With the plan hatched, all I needed to do was find a way to pay for it! With no data to speak of except for a single picture, I found whatever fellowships I could and applied for money (Explorers Club, Lewis and Clark Foundation, National Geographic). Given the fellowship timelines, at best they would reimburse the project, although regrettably none panned out. A few weeks later I pitched my project idea to Todd. I reckoned that one way or another, characterizing what was happening with spiny mice in the field would make for an interesting manuscript. In exchange for authorship, I asked for some logistical support in the field and two thousand dollars to cover my plane ticket and to offset my accommodations. To my surprise, he agreed to this arrangement and with help from Megan, I set about the task of lining up research permits to trap rodents in Kenya.

Having previously conducted research in Uganda, I acutely understood the problems associated with working in East Africa. Starting a project from scratch in Africa involves all kinds of logistical hurdles in addition to the normal challenges of simply carrying out the research. While Todd and Mpala Research Centre could provide some of what I needed in the field, trapping mice that no one actively worked with would prove to be problematic. As I slowly worked to line up the necessary Kenyan permits from afar, two more introductions would facilitate key components necessary for the proposed experimental work.

The first introduction concerned the wound healing experiments I hoped to complete. Luckily, my advisors introduced me to the developmental biologist and wound healing expert Paul Martin. My graduate advisor Martin (Marty) Cohn and Malcolm were very well acquainted with Paul, and with no expertise in wound healing between the three of us, we arranged a call so I could discuss my ideas with Paul. Several months before heading into the field, Paul and I connected by phone in Marty's office. After explaining the basis for my expedition, Paul offered insight into the most appropriate paradigm for a pilot wound healing experiment and was happy to answer my endless barrage of questions. I wanted to be as prepared as possible in case I was able to trap enough animals. Before the conversation ended, I asked him what he thought about the potential for skin regeneration in spiny mice given their apparent propensity for skin autotomy. He responded by proposing a bet that the likely outcome of their skin injuries would be normal fibrotic repair (not regeneration), although he conceded they might heal faster. Whatever the outcome, he wished me well on my expedition and told me to look at their inflammatory response if things got interesting.

The second introduction concerned collaborative support from within Kenya. Despite the viewpoint that science knows no borders, the fact remains that foreign scientists are guests. Permits aside, evolving national restrictions on visiting scientists are a recent development stemming from a historical legacy of exploitation and economic abuse, real or perceived. While I hoped to establish connections with Kenyan scientists once I arrived, another chance encounter provided an introduction that established a close collaboration, and friendship that I enjoy to this day. Ketan Patel, a British muscle biologist, had recently visited Marty while on holiday in the States. Ketan and Marty had been at the University of Reading together before Marty joined the Biology Department at Florida and Ketan and Malcolm were friends from the developmental biology community in England. One day in the lab, I mentioned to Ketan my pending trip to Kenya. As it turned out, one of his former students had graduated from the veterinary medicine program at the University of Nairobi and still had contacts there. After several emails, I was introduced to Stephen Kiama, then Chairman of the Department of Veterinary Anatomy and Physiology at the University of Nairobi. Professor Kiama had a longstanding interest in monocytes dating back to his own dissertation research in Bern and he found the nascent spiny mouse project particularly intriguing. With my departure weeks away, we arranged to meet in Nairobi when I landed to discuss project logistics. With seemingly all my ducks in a row, I successfully defended my thesis and packed my bags for the field.

2.2 A trapping horror resolves any doubt

When I arrived in Nairobi, I promptly made my way to the University of Nairobi's Chiromo Campus. Professor Kiama and I clicked right away. We leisurely exchanged personal histories, chatted about our research interests and even managed to wade into a bit of national politics. The Department of Veterinary Anatomy and Physiology had all the hallmarks of a classical Zoology department and walking the corridors I felt perfectly at home. Down the hall, a large theater ringed with skeletons stood idle for a laboratory in Comparative Anatomy. At the rear of the building stood an old animal house separated from the main building by a courtyard. Inside one of the rooms was a small colony of naked mole rats. Another room concealed a group of pancake tortoises (*Malacochersus tornieri*) in large sandboxes.

Later that morning Kiama introduced me to one of his graduate students, John Kimani, who showed me around the department. Besides the animals themselves, the inhalant anesthetic halothane was the most critical item I needed to procure if I was to perform any wound healing experiments.

Luckily, John helped me acquire a bottle of halothane from a pharmacy in the Central Business District; a half-day affair that introduced me to Nairobi's notorious traffic. Everything, it seemed, was falling neatly into place.

The next day, we were off to Laikipia for 2 months. Despite the aforementioned string of coincidences, introductions, and a bit of good luck, I was fully prepared for any number of small disasters which might occur in the field. What I didn't predict was an abdication of support from an American researcher who had promised me access to his Sherman traps and tips on trapping *Acomys*. When I arrived at Mpala, he needed the traps for himself and honestly had no idea what type of habitat *Acomys* preferred. I was beside myself. Disappointed as I was, there was no use wallowing in my misfortune. Instead, I spent days in the field with Megan measuring *Acacia* branches and tending to a seemingly endless array of spine punctures through my skin. As days turned into weeks, I immersed myself in her work while I soaked in one of the most amazing landscapes on Earth. On off days, I built wooden cages to house animals I didn't have and waited, albeit a bit impatiently, for the possibility of getting Sherman traps. Perhaps I *was* nuts for assuming a project I had spent months arranging with absolutely no experience in small mammal trapping or wound healing would simply work out because I wanted it too. Then, after 3 weeks of badgering and pleading, I finally acquired 50 Sherman traps.

At first, trapping *Acomys* seemed impossible. The savannah was expansive, and with 50 traps I wasn't catching much of anything. The spiny mice that researchers did catch in small mammal surveys were few and far between and comprised a very small percentage of the endemic species diversity. Late one afternoon after lamenting my inability to trap *Acomys*, I was reminded that Steve Takata had worked on kopjes (Fig. 2). Kopjes are very old rock

Fig. 2 *A. kempi* and *A. percivali* inhabit rock outcroppings in Kenya. (A) Rock outcroppings called kopjes dot the savannah landscape in Laikipia county, Kenya. (B) A single kopje emerging from the red soil, acacia dominated savanna. (C) *A. kempi* and *A. percivali*. *A. kempi* is notable for the rufous color of hairs covering its dosum. *A. percivali* retains a gray coat throughout adulthood.

formations that dot the landscape like islands (Fig. 2). That night I set 50 traps around, and on, a kopje named Jessel's Rock at the border of Peter Jessel's Farm and Mpala. When I returned the next morning, all the traps that we had set next to the kopje were empty. However, of the 25 traps we set *on* the kopje, I could already see that several trap doors were closed. Among these, several contained a small, slate gray rodent with a white belly and spiny hairs covering its entire body. This was *Acomys percivali*, Percival's spiny mouse (Fig. 2) (Kingdon, 1984). As I removed that first spiny mouse from the trap, a horror ensued. Partly because I had almost no experience trapping rodents and partly because I was nervous, I awkwardly grabbed the spiny mouse in the middle of its back. The animal used its legs to push off my hand and a piece of skin was torn from its back. As the spiny mouse fell to the ground and took off running into the rocks, I stared down at a little piece of skin in my hand. I couldn't believe it. The spiny mouse had escaped by tearing off its own skin.

That night I set my traps on the kopje again. Returning the next morning, I once again found that many of my traps were occupied. This time, I simply kept the spiny mice I captured inside the traps until releasing them into a large wooden enclosure I had found back at the field station. In addition to *A. percivali,* I had also trapped *Acomys kempi*, a rufous colored species that was slightly larger by comparison, but undoubtably spiny. This was Kemp's spiny mouse (Fig. 2). Over the next several weeks, I decided to set my traps exclusively on kopjes. In this manner, I figured out that *Acomys* preferred rocky habitat and that they were not the least bit trap shy. In a matter of weeks, I had already captured as many animals as I could hold in the compartmentalized wooden boxes I had constructed. With time remaining, I could now focus on the wound healing experiments I had hoped to conduct.

Observing and working with the animals at Mpala, I soon determined that their skin was uniformly weak. While I would later quantify the skin's tensile strength (Seifert, Kiama, et al., 2012), for now it was obvious that the skin was abnormally weak. I also followed animals with variable sized wounds and noticed that the skin appeared to heal on a typical time course (Clark, 1996). Moreover, being loose-skinned mammals, contraction from the wound margins accounted for most of the wound closure I observed just as Paul had predicted. Given the limited duration of my trip, I hadn't been able to complete anything more than a cursory assessment of the healing process in small, irregular wounds. Time was running out so I collected what samples I could to bring back to the States. I had, however, transported

50 animals back to the animal house in the Department of Veterinary Anatomy and Physiology in the hopes I would return. In the meantime, I figured that John could attempt to breed them to gather information on their reproduction.

Back in Malcolm's lab at Florida, a study describing fragile skin in geckos became a template for my histological investigations of the unwounded skin (Bauer, Russell, & ShadwickK, 1989). Geckos have a fracture plane in their dermis that allows them to lose and rapidly regenerate the upper dermis in response to autotomy. However, there was absolutely no evidence for this in spiny mouse skin. In retrospect, it was not surprising because the natural injuries I observed appeared to be full-thickness defects. Unfortunately, my wound healing experiments had gone poorly, and I returned with only a handful of tissue samples from relatively early time points. Based on wound healing studies in mice, this would still provide important material to look at tissue inflammation, but with the material I had it was impossible to systematically address the long-term outcome of tissue loss and healing (Mori, Shaw, & Martin, 2008). After completing what I considered a proof-of-principle study, tissue regeneration was far from my mind. For the uninitiated, wound healing is messy business, and from all the papers I read, the outcome in mammals was always the same. In fact, as I tunneled deeper into the wound healing and regeneration literature, it became apparent that with the exception of fetal wound healing models, no one had described an adult skin healing model where the outcome was regenerative healing. This revelation shifted my attention to investigating skin healing in salamanders (Seifert, Monaghan, Voss, & Maden, 2012), although I was still very much committed to tying up the spiny mouse project into something publishable.

2.3 Pseudoserendipity and the discovery of skin regeneration in wild *Acomys*

If Truman was the catalyst for the spiny mouse project, Malcolm was the fuel. Rather than discourage my curiosity, Malcolm had actively encouraged it. During my first summer, he joined me for several weeks in Kenya to see the animals for himself and to help me with the wound healing experiments. Working side-by-side together had validated my belief that young scientists should be encouraged to study what captivates them regardless of the associated career risks. When no organizations would fund my project proposals to return the following summer, Malcolm provided research funds to help support the work. I had mapped out a plan to complete two components of the project which I thought would help me wrap up the spiny mouse story. First, I would quantify skin weakness using a tensometer at the University of Nairobi. Second, I would complete a rigorous assessment of wound healing that included deep time points to assess the full extent of skin repair. Although the results would likely end up being far more descriptive than explanatory, a paper would bring this interesting behavior to the broader scientific community. The following summer I spent 2 months in the field and 1 month in Nairobi. While new adventures and disasters were plentiful, I managed to find a mechanical engineer in Nairobi who could help me record tensile strength from skin samples. Equally important, I was able to collect tissue samples that would allow me to completely characterize the wound repair process. With a full parcel of tissue samples in hand, I headed back to Florida for bench work.

Although I had observed remarkable healing in the spiny mice, my general skepticism to a healing outcome other than fibrotic repair obscured the notion of regeneration. For the wound healing experiments, I settled on a full-thickness, excisional model (1.5 cm diameter wounds). Gross observation of healing skin wounds in Kenya appeared to confirm my general expectation that contraction would dramatically shrink the wound bed which would later fill with scar tissue to complete repair in the center.

When I finally settled into a dissecting scope to prepare my tissue samples, I worked through the tubes in a temporal sequence. Uninjured skin followed by early healing time points and finally my 4-week samples. As I carefully emptied each parafilm wrapped tube into a petri dish of saline, my early time points were typical with a barren wound bed surrounded by clipped hairs at the wound margins. When I began to look at samples collected 4 weeks after injury, I couldn't believe my eyes. All throughout the wound bed I could see the pigmented tips of large spiny hairs erupting through the epidermis alongside much thinner guard hairs (Fig. 3). Cautiously, I leaned back in my chair and chuckled. The wound bed I was looking at didn't resemble the wound induced hair follicle neogenesis observed by some investigators when very large wounds were made in mice (termed WIHN). In those instances, new hairs emerged in small bunches from the center of the wound and were separated from the margins by a ring of scar tissue (Wang et al., 2015). Cutting parallel to the hairs, I could clearly see two hair types extending from the margins across the entire area (Fig. 3). What had apparently occurred in these skin wounds was regenerative healing instead of fibrotic repair! Although the discovery was not purely accidental, it was definitely an unexpected result. To embrace it was another

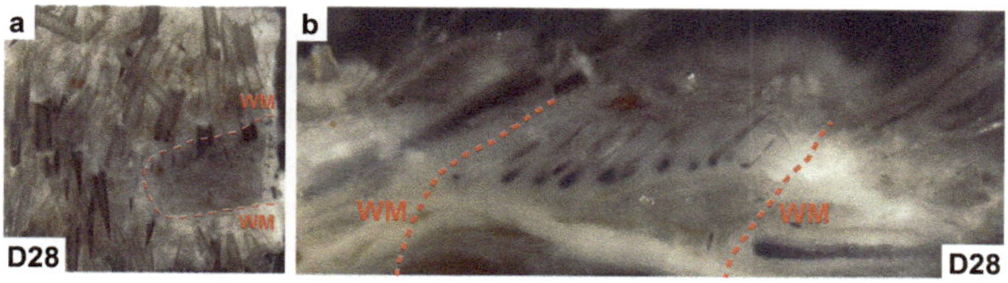

Fig. 3 *De novo* hair follicle regeneration in *Acomys*. (A) A piece of wounded back skin recovered 28 days after the original 1.5 cm diameter, full thickness skin injury. Large, spiny hairs in unwounded skin are visible outside the wound margins (WM) and the tips of newly generated hairs can be seen emerging from within the wound bed. (B) Longitudinal cut through wound bed center showing regenerated spiny (awl) and guard hairs across the entire bed. Dotted lines depict the wound margins (WM).

matter altogether. As the gravity of the discovery slowly sunk in, it was replaced with a mixture of excitement and doubt. Now I would need to convince myself that the spiny mice were indeed doing something remarkable. I would spend the rest of the year trying to do just that.

As I set about working up my dataset, I read every published report I could find pertaining to regeneration in mammals. This had the intended consequence of making me an expert on a topic at the fringes of regenerative biology where, with few exceptions, the animal models of choice were *Hydra*, planarians, zebrafish, salamanders, and newts. But there *were* exceptions relevant to my work, and three in particular would inform the project. The first was the mammalian digit tip model originally described by Richard Borgens (Borgens, 1982). Amputation of the digit tip (bone and soft tissue) in young adult mice induced a healing response that was sensitive to the level of amputation: an amputation distal to the nail bed induced a regenerative response whereas an amputation proximal to the nail bed stimulated fibrotic repair (Borgens, 1982; Fernando et al., 2011; Neufeld & Zhao, 1995). Amputations through more proximal elements never regenerated (Neufeld, 1980). One thing that struck me as odd though was that this model of regeneration was conserved across all those mammals in which it had been examined, including humans (Illingworth, 1974; Muneoka, Allan, Yang, Lee, & Han, 2008; Neufeld, 1980; Singer, Weckesser, Géraudie, Maier, & Singer, 1987). The second was the WIHN model of hair follicle neogenesis (rev. in Wang et al., 2015). The defining feature of this wound healing model are *de novo* generated hair follicles in the center of the wound separated from the wound margins by a ring of scar tissue. In addition to the lack of new hair follicles at the margins, regeneration was age-dependent with very large wounds necessary in older animals to generate new hair follicles (Ito et al., 2007). Moreover, it was also apparent that the new hair follicles in this model were unpigmented and lacked arrector pili muscles (Billingham & Russell, 1956), a fact which suggested that this phenomenon represented incomplete regeneration. The third model was a complete hole punch through the ear pinna. As originally described in the rabbit ear, this model demonstrated the progressive regeneration of cartilage, hair and full-thickness skin (Vorontsova, 1960). Subsequent investigations in rabbits showed that regeneration was sex-dependent, with males more likely to close the open holes than females (Joseph & Dyson, 1966; Williams-Boyce & Daniel, 1980). Cross-species comparisons suggested that hole closure alone, without appropriate histological evaluation of the new tissue, could confound a conclusion that regeneration occurred (Williams-Boyce & Daniel, 1986). Together, these models convinced me I needed to include multiple healing models, provide side-by-side comparison with identical injuries in lab mice, include both sexes in my analysis, and where possible, use older animals.

With these ideas in mind, I laid out a plan and headed back to Kenya for a third summer. I could partially reject the notion that *Acomys* exhibited WIHN because the samples I already collected showed new follicles across the entire wound bed (Fig. 3). Moreover, I found new arrector pili muscles and sebaceous glands associated with the regenerated spiny hairs (awl hairs) and guard hairs. What I needed to add was a comparison of similar sized wounds in mice. Thus, I added the observation that when I made identical large wounds in very young Swiss Webster mice, the result was a scar with no new hair follicles in the wound bed. While completing these wound healing experiments, I focused my remaining attention on adding an ear punch assay to the dataset I had in hand. Work on the Murphy's Roth Large (MRL/+) strain of mice had suggested that complex tissue regeneration in mammals was simply a matter of genetics as these and closely related inbred strains could rapidly close small, 2 mm ear holes (Clark, Clark, & Heber-Katz, 1998; Kench et al., 1999). Having recently come across a paper demonstrating that middle-aged C57BL/6 mice could close 2 mm ear holes just fine (Reines, Cheng, & Matzinger, 2009), I became suspicious that something other than regeneration might be able to explain the closure of small holes. Based on this intuition, I chose to use a 4 mm hole to test regenerative ability in *Acomys* and Swiss Webster mice. Using males and females to test for a sex bias, I punched the ears and waited. During the first 2 weeks the ear holes looked similar between the species and I resigned myself to accept the possibility that this assay would fail to elicit a regenerative response from *Acomys*. What a difference a week can make. Between the second and third week a profound disparity emerged between the two species. Notably, new tissue began filling in the *Acomys* ear holes. Approximately 40 days after the initial punch, the spiny mouse ear holes had filled in with new tissue whereas in the Swiss mice, the holes remained open and unchanged lest for a bit of new scar tissue that had formed in the margins. Whatever excitement I experienced seeing new hair follicles emerge in skin wounds paled in comparison to what I witnessed for the first time that summer. I now understood that what I had uncovered was something truly remarkable.

After many months of careful preparation, I drafted a manuscript containing the data I had collected. The paper presented the skin tearing behavior, quantification of skin weakness, comparative wound healing in small and large wounds, and finally, the ear pinna regeneration experiments. As peer-review goes, it was a slog, although in retrospect the reviews were quite positive. They were rightfully skeptical and wanted to see a deeper molecular characterization of regeneration while also providing more "mechanistic data." Luckily, the editor was keen on the manuscript and understood the limitations of working with animals in the field (i.e., I couldn't FACS sort ear pinna cells or isolate and expand fibroblast lines for cross-species transplantation, etc.). And so again, with a modicum of luck, it was accepted, and I had the distinct pleasure of introducing the scientific community to a fascinating new mammalian regeneration model (Seifert, Kiama, et al., 2012).

2.4 The model matures: Understanding complex tissue regeneration in a mammal

If the animal model was going to mature to the point where I could do more controlled experiments, I needed to address tractability. While working with animals in Kenya would allow me to collect more extensive comparative data between *Acomys* and nonregenerative sympatric species, I needed to establish a laboratory colony in the States if I wanted to begin unpacking the cellular and molecular basis for enhanced regenerative ability in mammals. Again, luck would facilitate progress on this front. After poking around online, I discovered that spiny mice could sometimes be acquired in the pet trade. As it turned out, Ben Cole, a former technician in a neighboring lab managed a reptile shop in Gainesville called Pangea. He also helped care for a very large snake collection owned by Dr. Elliot Jacobson, a retired veterinary pathologist, who, as it turned out, just happened to have a modest colony of pet *Acomys*. When I mentioned to Ben that I was looking for spiny mice he brought me with him to Elliot's house one evening. Elliot had inherited his animals in the early 1990s from the Phoenix Zoo and he was more than happy to give me some breeders, especially after I explained why I was trying to start a colony. In fact, Elliot eventually gave me all his animals, some 40 in total, and Malcolm and I began breeding them in our homes as pets. Ultimately, we would donate these pets to our respective Universities as research animals and they would eventually become the founders for my colony at the University of Kentucky and Malcolm's colony at the University of Florida.

When I started my own lab in Kentucky, spiny mice allowed me to focus on trying to understand the mechanistic basis for interspecific variation in regenerative ability. In attempting to derive solutions to this problem, spiny mice had suddenly provided a system in which to develop new hypotheses while also testing existing ones. Although the skin phenotype had initially propelled the project, it was the ear pinna injury model that excited me most. As a composite tissue, the ear pinna contains skin, adipose tissue, skeletal muscle, nerves and elastic cartilage (Gawriluk et al., 2016). This provides a means to examine the regenerative capacity for these tissues as an assemblage, similar to appendage regeneration in other vertebrate models, but also for each of these tissue types individually. Moreover, external ear pinnae are a mammalian synapomorphy (Manley, 2012), making them a perfect structure to assess regenerative ability across a wide range of species. The pinna is easy to access and because of the elastic cartilage core, ear holes exhibit little to no contraction during healing. Lastly, there are really only two basic mechanisms to repair the ear injury: regenerative healing or fibrotic repair. Almost no other regeneration model described thus far affords this type of comparison between closely related species other than the lack of tail regeneration in certain lizards (Dunoyer, Seifert, & Van Cleve, 2020) and the loss of head regenerative ability in certain planarian and annelid species (Bely & Sikes, 2010; Needham, 1952).

Naturally, Malcolm had also become fascinated with the *Acomys* model, and with his own animals in Florida he was keen to continue using them to explore regeneration. While I felt it was within my rights to simply take the entire system with me to establish my own lab, who was I to selfishly claim the system as my own. If Malcolm wanted to continue working with *Acomys*, then I thought we just needed to establish boundaries to avoid direct competition. I laid out my reasons for wanting to use the ear pinna and asked if he could leave my students and I to focus on this model moving forward.

In exchange, I would abstain from working on the skin for the foreseeable future. Anything else was fair game for either lab. With a gentleman's agreement in place a deal was struck.

As a student of the historical and contemporary regeneration literature, I relished the idea of developing a new system, and potentially a community, that might make an impact in our field. With a burgeoning *Acomys* colony in Kentucky, I established two parallel spiny projects. The first sought to explore the immune response to injury across healing types and species. Specifically, I wanted to test the hypothesis that loss of regenerative capacity in mammals resulted from a bias in the injury response toward adaptive immunity and a strong inflammatory response (Godwin & Brockes, 2006; King, Neff, & Mescher, 2012; Mescher & Neff, 2005). The second centered on activation of post-mitotic (differentiated) cells at the injury site and their ability to re-enter the cell cycle, proliferate and coordinate regeneration of the missing tissues. Although dedicated stem cell pools maintain mammalian tissues undergoing lifelong physiological turnover (e.g., intestine, blood, epidermis, hair follicles, etc.) (Levi & Morrison, 2008), there was no solid evidence that cells in differentiated mammalian tissues could undergo the type of dedifferentiation and proliferation commonly observed during epimorphic regeneration in salamanders, newts and zebrafish (Hay & Fischman, 1961; Lo, Allen, & Brockes, 1993; Oberpriller & Oberpriller, 1974; Poss, Wilson, & Keating, 2002; Sandoval-Guzman et al., 2014). Instead, mammalian cells were generally regarded as refractory to the triggers that stimulated cell proliferation in regenerating species (Pajcini, Corbel, Sage, Pomerantz, & Blau, 2010; Tanaka, Gann, Gates, & Brockes, 1997). With these two lines of inquiry as guideposts, my new students and I set ourselves to work.

First, using our 4mm ear punch assay, we established the regenerative equivalency of our captive *A. cahirinus* with the wild-caught *A. percivali* and *A. kempi* (Gawriluk et al., 2016; Seifert, Kiama, et al., 2012) and our characterization was independently reproduced by Gustavo Tiscornia's group at the University of Algarve (Matias et al., 2016). These studies also demonstrated that wild murids and laboratory mouse strains, including MRL/MpJ mice, could not close 4mm ear punches, let alone regenerate functional tissue. In fact, our study uncovered that 2mm holes were small enough to be filled with scar tissue which underscored the importance of coupling ear closure rates with cellular analysis when assessing regenerative *versus* fibrotic repair. An important feature of epimorphic regeneration observed in other vertebrate and invertebrate systems is that under normal environmental conditions, and among healthy animals, regenerative healing always occurs in response to injury. Likewise, in nonregenerating species, regeneration never occurs. Here again, results in *Acomys* reflected other regeneration models and also revealed that males and females regenerated equally well (Gawriluk et al., 2016; Matias et al., 2016; Seifert, Kiama, et al., 2012).

Cellular and genetic comparison of tissue healing in *Acomys* and *Mus* established key aspects of the regenerative microenvironment that were consistent with epimorphic regeneration. First, whereas *Mus* exhibited few cycling cells and almost none that were mitotically active, *Acomys* activated a robust proliferative population that persisted during new tissue formation (Gawriluk et al., 2016; Seifert, Kiama, et al., 2012). Second, gene and protein data demonstrated that extracellular matrix production during fibrotic repair was dominated by type I collagen with lesser contributions from other wound-induced ECM molecules (e.g., *Tenascin-C*, *Mmps*, *Col3a1*, etc.). In contrast, the wound environment observed during regeneration in spiny mice was rich in Fibronectin and matrix remodeling enzymes (*Mmp9*, *Mmp13*) while also maintaining high levels of Tenascin-C. This demonstrated a regenerative ECM composition that was rich in molecules supporting proliferation and migration and including those that serve to resist collagen aggregation (Calve, Odelberg, & Simon, 2010; Calve & Simon, 2012; Monaghan et al., 2012; Satoh, Hirata, & Makanae, 2012; Satoh, Hirata, & Satou, 2011; Seifert, Monaghan, et al., 2012). When vertebrate appendages regenerate, newly generated epidermis covering the injury needs to expand as new tissue is generated beneath. This neoepidermis covering the injury is often referred to as the wound epidermis and similar to wound healing, it is required for regeneration to proceed. Here again, our RNA-seq and immunohistochemical data showed that a transitional epidermis formed early during healing in *Mus* and *Acomys*. Whereby the epidermis quickly differentiated into a stratified, squamous epithelium during fibrotic repair, it remained in a transitional state, devoid of the lamina densa (i.e., the lower portion of the basement membrane) until the ear hole closed in *Acomys*. Lastly, a key requirement for complex tissue regeneration is the presence of an intact nerve supply that facilitates blastema formation and supports cell proliferation (rev. in Sinigaglia & Averof, 2019; Stocum, 2011). During new tissue formation in the ear pinna, we observed the invasion of peripheral axons from the proximal margin of the ear, a finding later corroborated by Malcolm Maden (Maden, 2018) and we ultimately demonstrated that new axons had innervated the regenerated tissue

(Gawriluk et al., 2016). Thus, the collective observation of localized cell proliferation, a growth promoting ECM, formation and retention of a transitional epidermis and the regeneration of peripheral nerves, cemented *Acomys* as a bonafide vertebrate model of complex tissue regeneration.

Building on this work, a comparative analysis of regeneration and fibrotic repair using *Acomys* and other mammals has provided a deeper understanding of the salient features that contribute to the trajectory of healing in different mammals. Results from several labs examining *Acomys* and non-regenerative murids have detailed phenotypic differences within inflammatory and adaptive immune cell populations that have helped refine existing ideas about how these cells may influence regenerative healing (Brant, Yoon, Polvadore, Barbazuk, & Maden, 2016; Cyr et al., 2019; Gawriluk et al., 2020; Simkin, Gawriluk, Gensel, & Seifert, 2017). While studies in amphibians and fetal mammals have fed the idea that strong inflammatory and adaptive immune responses are antagonistic to regeneration (Godwin & Brockes, 2006; King et al., 2012; Mescher & Neff, 2005), data from *Acomys* suggests a more nuanced interpretation of the existing data. For example, a detailed examination of the inflammatory microenvironment over a 20 day time course in *Acomys* and *Mus* showed that while regeneration was associated with significantly reduced levels of two pro-inflammatory cytokines (IL-6 and CXCL1) compared to fibrotic repair, both types of healing elicited a common set of pro-inflammatory cytokines and chemokines (Gawriluk et al., 2020). A time course analysis of innate immune cells determined that neutrophils were significantly more abundant during fibrotic repair immediately after injury, but that macrophages arrived in equal numbers and persisted regardless of healing type (Simkin et al., 2017). Although macrophages are required for regeneration in salamanders and zebrafish (Godwin, Pinto, & Rosenthal, 2013; Petrie, Strand, Tsung-Yang, Rabinowitz, & Moon, 2014), a finding we extended to *Acomys* (Simkin et al., 2017), what remains less clear is how they specifically regulate regeneration. Recent work in our lab suggests that interspecific differences in macrophage subtypes can partly explain how macrophages direct local cell populations to differentially respond to injury (unpublished data). Comparative *in vitro* and *in vivo* experiments examining circulating and resident macrophages reveal that *Acomys* macrophages uniquely secrete several pro-angiogenic and antifibrotic factors that contribute to accelerated revascularization of the injury site and production of pro-regenerative matrix remodeling enzymes (unpublished data). Similarly, a phenotypic analysis of circulating and bone-marrow derived neutrophils from wild and captive *Acomys* and *Mus* revealed that *Acomys* neutrophils are significantly more phagocytic and that humoral killing capacity is significantly stronger in spiny mice (Cyr et al., 2019). Together, these data point to subtle alterations in the innate immune response that maintain effective pathogen clearance and inflammatory resolution, while simultaneously promoting a regenerative microenvironment. It is also worth noting that our cursory analysis of T cell behavior during regenerative healing supports that *Acomys* exhibit a strong adaptive immune response to injury, possibly even stronger than the T-cell infiltration observed in laboratory and wild-caught murids (Gawriluk et al., 2020). Further characterization of immune cell phenotypes and manipulating specific macrophage populations in the context of regenerative healing and fibrotic repair is the natural evolution of where this work is headed.

Although phenotype switching among immune cells, and the secreted factors they supply, regulate the intensity and time course of inflammation (rev. in Eming, Hammerschmidt, Krieg, & Roers, 2009; Martin & Leibovich, 2005), inflammation remains a key feature of the general injury response. Fibrotic repair and regenerative healing seemingly represent two extremes of the final healing process, and yet the degree to which the innate immune response and inflammation promotes one healing type over the other remains unresolved. The ability to study how an identical injury results in two different healing types suggests three general models that can explain how the early responses to tissue damage might differentially affect healing outcomes. First, if the endpoints of fibrotic repair and regeneration lie along a smooth continuum, where local cells have the potential to heal by either means, then identical injuries could either trigger a response that is unique enough for each outcome that the two paths immediately diverge toward fibrosis or regeneration or instead, injury could trigger a response that is generally similar and instead differs primarily in its magnitude (Aloysius, Saxena and Seifert, 2021). Our current results revealing differences in immune cell phenotype points to the former and suggests that altering macrophage phenotypes during fibrotic repair could, in principle, shift the healing outcome. A third possibility is that fibrotic repair and regeneration represent two totally different responses that occur in a tissue-specific or species-specific manner and are relatively independent of the local innate immune response. Thus, upon the resolution of inflammation each healing outcome proceeds according to cell autonomous mechanisms regulated by local responding cell types (e.g., endothelial, keratinocyte, fibroblast, etc.) (Phan et al., 2020).

Support for distinct healing trajectories following inflammatory resolution comes from the observation that during epimorphic regeneration local cells must proliferate to generate new tissue. Unlike regeneration in many invertebrate systems where pluripotent stem cells are present in adult animals, regeneration in vertebrates relies on lineage-restricted cells in differentiated tissues to act as progenitor cells for new tissue formation (rev. in Kragl et al., 2009; Monaghan & Maden, 2013). Hereto, *Acomys* is beginning to offer insight into how cell intrinsic properties among species may confer differences in regenerative ability. In an attempt to explore the proliferative capacity of connective tissue fibroblasts across regenerating (spiny mice and rabbits) and nonregenerating (mice and rats) mammals, we have uncovered a remarkable capacity for cells from regenerating species to resist oxidative stress (Saxena, Vekaria, Sullivan, & Seifert, 2019). Whereas cells from mice and rats activated a senescence program in response to oxidative stress, exogenous hydrogen peroxide was detoxified by a remarkably efficient glutathione redox reaction. Moreover, this resilience to oxidative stress extended to their mitochondria as well. Metabolic analysis of ear pinna fibroblasts suggested that differentiated cells from spiny mice and rabbits rely on aerobic glycolysis to meet their energy demands, a property usually observed in stem cells (Saxena et al., 2019). Our previous work also documented widespread activation of p21 and p27 throughout healing ear tissue in *Mus* suggesting that injury stimulates cellular senescence during fibrotic repair (Gawriluk et al., 2016). In contrast, we rarely observed upregulation of these tumor suppressor proteins in regenerating tissue, although we detected them in nearby differentiated tissue. Ongoing work in our lab has extended these findings *in vivo* and provides a mechanism that would seem to explain how injury stimulates cell cycle progression in cells that are poised to proliferate (unpublished data). Together with our immunological work, a picture is emerging for how cell autonomous and noncell autonomous factors have evolved in the *Acomys* lineage to facilitate complex tissue regeneration in response to injury.

2.5 Potential for regeneration in other tissue and organ systems

After leaving the University of Florida, Malcolm moved forward with a strategy to assess regenerative ability in the skin and other organ systems. Analyzing skin healing in *A. cahirinus*, his lab documented parity with our original work in *A. percivali* and *A. kempi* (Brant, Lopez, Baker, Barbazuk, & Maden, 2015; Brant et al., 2016). Building on this work, Cheng Ming Choung's lab confirmed the regenerative ability of *A. cahirinus* skin and showed that all hair types are regenerated and that they are pigmented in contrast with the WIHN mode of healing in C57BL/6 mice which only regenerate unpigmented zig-zag hairs (Jiang, Harn, Ou, Lei, & Chuong, 2019). Recent work from this same group showed that tissue stiffness across the wound bed appears to help regulate the natural regenerative properties of spiny mouse skin (Hans et al., 2021). These studies confirm the uniqueness of *Acomys* skin regeneration and in light of mouse studies which have pinpointed specific fibroblast phenotypes that confer enhanced regenerative ability (Driskell et al., 2013; Jiang et al., 2018; Rognoni et al., 2016) provide a platform for understanding the behavior of dermal fibroblasts across these systems. From the standpoint of utilizing this type of information to stimulate regeneration, multiple groups have recently demonstrated how enriching specific fibroblast subsets in skin wounds can be used to promote regeneration in adult mouse wounds (Mascharak et al., 2021; Phan et al., 2020). It will be important to determine if similar fibroblast subsets are enriched in *Acomys* through comparative scRNA-seq experiments or if they deploy a unique strategy to facilitate endogenous skin regeneration.

In addition to the skin, recent work in *Acomys* has also demonstrated a propensity of other organs to resist fibrotic damage. In response to injury, the kidney, heart and spinal cord appear to activate cytoprotective mechanisms which allow these animals to resist damaging fibrosis that would otherwise compromise organ function in a mouse or rat (Koopmans et al., 2021; Okamura et al., 2018; Peng et al., 2021; Streeter et al., 2020). The heart provides an intriguing example. Recent work from Malcolm's group suggested that in response to a myocardial infarct, the *Acomys* heart can recover a good deal of heart function as measured by echocardiography (Qi et al., 2021). More recently, a collaborative effort in Kentucky in parallel with a project at Hubrecht University led by Kerstin Bartscherer's group documented a remarkable resistance to heart injury. In response to a myocardial infarct, either by a permanent left anterior descending artery (LAD) ligation (Peng et al., 2021) or an ischemia-reperfusion infarct (Koopmans et al., 2021), our respective groups observed recovery from identical injuries which proved lethal to inbred and outbred mouse strains. In both studies, proliferation of endothelial cells and rapid angiogenesis were observed which contributed to a well-vascularized infarct area that allowed the heart to recover ventricular outflow. Our group found that a large fraction of *Acomys* cardiomyocytes exhibited a T-type calcium channel that is normally found in embryonic cardiomyocytes and the Koopman study was able to demonstrate an expansion of the local cardiomyocyte pool. Obviously, more work is needed to determine the cellular mechanisms at play in this in other organs, but with an increasing battery of validated reagents for *Acomys* and a fully annotated genome, the horizon for these studies appears bright.

3. Reproduction and developmental biology

In the fields of observation chance favors only the prepared mind.
Louis Pasteur, from a lecture delivered at Universite de Lille, France (07.12.1854).

3.1 Prologue

While I (Peter) have been fortunate during my research career to have had a wealth of scientific experiences in the field of biology, I entered the field of spiny mouse reproduction rather late and by chance. I have always been interested in the intimate and fascinating process of reproductive biology, especially among the hairy, milk-producing mammals. Although my graduate project was initially to be on the reproductive biology of green turtles, that opportunity, also by chance, was withdrawn due to the unexpected disappearance of my designated PhD supervisor which happened 3 weeks before I was due to move from my alma mater, the University of Tasmania, to the Australian National University to start my PhD. I was then offered, as last-minute alternatives, two new projects from the Department of Zoology: a physiological study of a beautiful semi-aquatic reptile, the eastern water dragon (*Intellagama leseurii*), or a field study of that enigmatic, semi-aquatic icon of the Australian fauna, the platypus, *Ornithorhynchus anatinus*. While I was sure that the eastern water dragon would make for an interesting study species, I immediately knew that the platypus would be my choice, and I rapidly penned a letter to accept what would become the first PhD field study on the platypus. Four years of field work produced my PhD (Temple-Smith, 1973) which led to a 3-year Ford Foundation Post-doctoral Fellowship in the United States at Cornell University researching sperm maturation and fertilization in rabbits, hamsters, Australian brush-tailed possums (that I had brought with me from Australia) and Virginian opossums with Professor J. Michael Bedford, a Cambridge graduate in veterinary medicine who became an important mentor not only for my postdoctoral fellowship but the rest of my academic career. On my return to take up a position at Monash University I was able to extend my research interests into mammalian reproduction with projects on species from the three primary mammalian lineages: the monotremes (Temple-Smith & Grant, 2001), marsupials (Temple-Smith, 2003) and eutherians (Verma, Holland, Temple-Smith, & Verma, 2012; Wang, Catt, Pangestu, & Temple-Smith, 2009), including human reproduction and IVF (Temple-Smith, Southwick, Yates, Trounson, & De Kretser, 1985).

Although I had been doing laboratory research on other rodent models, my exposure to spiny mouse biology happened through a fortuitous collaboration with a young and enthusiastic early career researcher, Hayley Dickinson, who had completed her PhD on renal function in the spiny mouse (*A. cahirinus*) (Dickinson, Moritz, Wintour, Walker, & Kett, 2007) and was expanding into pregnancy and developmental studies using these animals. At the time of our encounter, I had no idea about, nor direct interest in, the spiny mouse as a research model. However, this was about to change very rapidly during our joint supervision of a PhD student and the completely unexpected findings from her research that spiny mice menstruated (Bellofiore et al., 2017; Bellofiore & Evans, 2019).

Until very recently, there was scant interest in the spiny mouse as a model for studies of reproduction or developmental biology. Regarded as just another estrous-cycling murid rodent requiring special security features for captive colonies because of its status as a pest species in Australia, the costs of maintaining spiny mice did not justify the expenditure when compared with the advantages of other laboratory murids such as *Mus musculus*. And *Mus* has the advantage of a full battery of molecular reagents, transgenesis, biomarkers, genetic strains and disease models, where comparatively few exist for the spiny mouse model. Until recent discoveries showed the potential of the spiny mouse to provide certain more effective human-disease models, there has been little incentive to entice researchers working with mammals to consider adopting spiny mice into their studies.

The spiny mouse colony at Monash University was the brainchild of David Walker, a perceptive, enthusiastic and well-respected physiologist and neurobiologist with a passion for the comparative approach to study problems in biomedical research. Having understood the opportunities offered by the spiny mouse as a research model for developmental biology research, he started the colony with five founder pairs that he managed to import from the University of Amsterdam in 2002. The Amsterdam colony had been started from four pairs of founders acquired from the University of Geneva suggesting that the current Monash colony has extremely low genetic diversity from years of bottlenecking and inbreeding. As I have recently found in my short personal journey of discovery with the spiny mouse, establishing a spiny mouse colony in Australia was a portentous affair! David Walker's prescience in starting this colony changed the

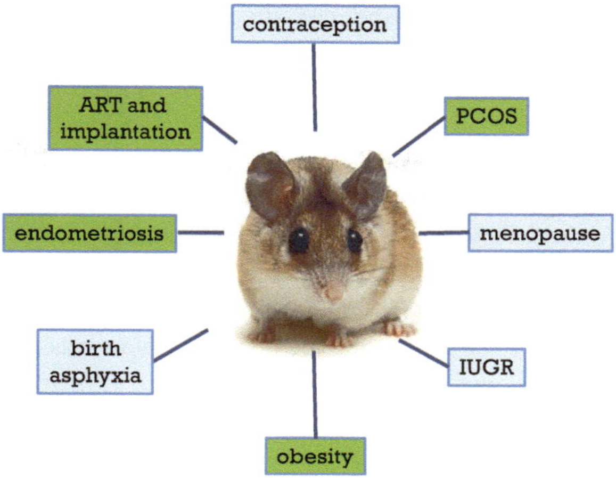

Fig. 4 Current (blue) and potential (green) spiny mouse models for women's health research. Assisted reproductive technologies (ART), intrauterine growth retardation (IUGR), polycystic ovary syndrome (PCOS).

direction of animal modeling in neurobiology and development while also contributing a new model for reproductive research, especially for several aspects of women's health (Fig. 4).

3.2 Menstruation and premenstrual tension: Pseudoserendipitous discoveries in female spiny mouse reproductive biology

The perception of spiny mice as an inferior model for women's health studies compared to laboratory mice and rats changed abruptly with the completely unexpected discovery that *A. cahirinus* females from our captive colony naturally menstruated in a very human-like fashion (Bellofiore, Cousins, Temple-Smith, Dickinson, & Evans, 2018). This observation set spiny mice apart, not only from *Mus* and *Rattus*, but from all other rodents thus far studied.

The story behind this startling discovery began with a suggestion to a bright young PhD student, Nadia Bellofiore, that she map out in more detail the estrous cycle of female spiny mice in the Monash University colony. Although a paper on the spiny mouse estrous cycle had been published in 1981 (Peitz, 1981), we thought characterization of the estrous cycle could be improved using more recently developed techniques. Although we didn't know it at the time, our seemingly innocuous suggestion to re-examine the spiny mouse estrous cycle would produce a totally unpredicted outcome for our understanding of female reproduction in this nonmodel rodent. Importantly, the discovery would elevate this relatively unknown species into more global scientific prominence as a new, rapidly emerging animal model for studying many aspects of human reproduction, especially as they relate to women's health (Fig. 4), in a profoundly new way.

Why was the demonstration of a menstrual cycle in the spiny mouse such an unexpected discovery? In part, because the spiny mouse remains the only rodent in which a menstrual cycle has been rigorously documented out of 2277 described species from 481 genera and 33 families (Carleton & Musser, 2005; Honeycutt, 2009). Although the cycle duration was short compared to women, (average cycle length, *Acomys* = 9 days; human = 28 days), we also discovered that endometrial changes characterizing the human menstrual cycle were mirrored in the spiny mouse endometrium during menstruation (Fig. 5). This discovery arose from naturally skeptical researchers in our group following clues that many would have dismissed based on an accepted adherence to dogma. The dogma in this case was that all rodents have estrous cycles; an idea so entrenched in the field that the occurrence of a menstruating rodent was considered unlikely, if not impossible.

The pathway to this finding followed the classical steps involved in evidence-based research. One day, Nadia approached her supervisors (Hayley Dickenson and me), about blood contamination in the vaginal smears she had been taking from a group of spiny mice in her project. Not surprisingly, Nadia was worried that there was a mechanical problem with her technique. We suggested that she practice again her lavage technique using laboratory mice, which she did and was confident, from the lack of blood in her practice *Mus* samples, that her technical skill in vaginal lavage was now competent. When she returned to vaginal lavage in the female spiny mice, she reported again that she was finding blood in the samples. This was the light-bulb moment for us. Maybe the bleeding wasn't an artifact at all! Had she instead, unlikely as it might seem, collected evidence that female *A. cahirinus* might have a menstrual cycle rather than the standard rodent estrous cycle? If so, we needed to collect additional experimental evidence. We then discovered that it was not uncommon for the animal house staff managing our colony to find traces of blood on the floor of the cages. However, when this had happened in the past, they had simply reported the blood to their superiors (management and researchers) who gave permission to euthanize these females over concerns that aggression was causing internal injuries. While this decision may have been appropriate under these circumstances, on at least some occasions, it is possible that the animal technicians may have been inadvertently finding menstrual blood from overt bleeders!

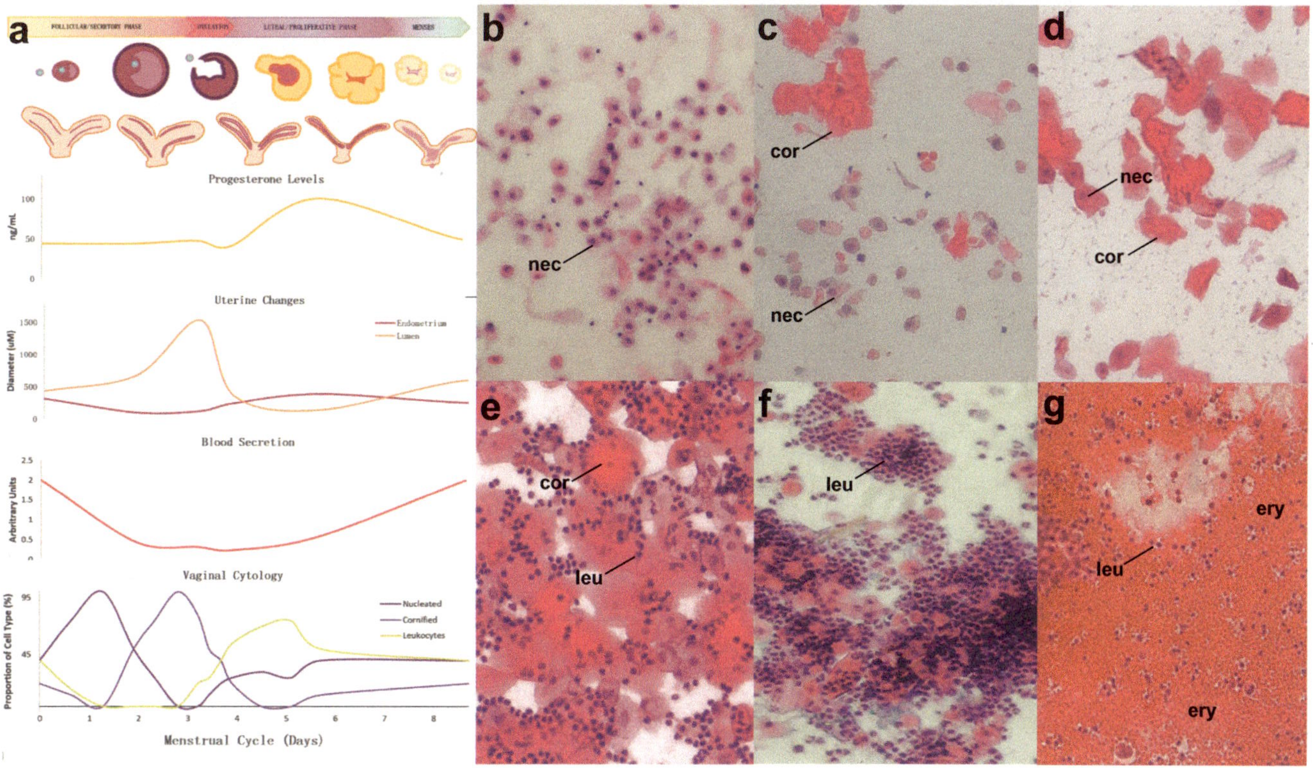

Fig. 5 See figure legend on next page.

Since captive colonies of spiny mice have existed in other parts of the world, it is interesting to contemplate why this discovery had not been made sooner. As the proverb suggests, discovery in science *does* favor the prepared. Given the infrequency with which menstruation has been observed in mammals, (menstruation is restricted to ~2% of all mammal species) perhaps it seemed preposterous to recognize it in a rodent. An alternative explanation is that our colony, due to its derivation from five founding pairs, had undergone some unlikely inbreeding-related mutation that resulted in the cyclical sloughing, rather than absorption, of the endometrial tissue at the end of each cycle. Perhaps this could explain the 1981 description by Peitz of spiny mouse estrous cycle? Prior to our description, menstruation had only been recorded in the higher primates (Emera, Romero, & Wagner, 2012), bats (Brasted, White, Kennedy, & Salamonsen, 2003; Rasweiler, 1991; Rasweiler & De Bonilla, 1992) and the elephant shrew (Brasted et al., 2003; Rasweiler, 1991; Rasweiler & De Bonilla, 1992; Van der Horst, 1946). Use of these species as laboratory models for studying menstruation is challenging. For the higher primates, the challenge is the cost of maintenance and growing ethical concerns about scientific research use. For the elephant shrew and bat species, it is access, captive facilities, specialist knowledge on their reproductive biology and the nonhuman style of their menstrual cycles that has hindered their use as experimental models.

Discovery of periods in the spiny mouse awakened significant interest in this species as a model for women's health. For example, in addition to menstruation, our lab through the work of Nadia Bellofiore has now demonstrated that spiny mice display the behavioral equivalent of a human-like premenstrual syndrome (PMS) during the late luteal phase (Bellofiore, Cousins, Temple-Smith, & Evans, 2019). PMS affects millions of women

Fig. 5 Spiny mouse menstrual cycle (A) summary of the changes in the ovary, reproductive tract, steroid hormones, menstrual flow and vaginal cytology during the spiny mouse menstrual cycle. B–G stage specific vaginal cytology: (B) early follicular stage (estrogen), duration ~2 days, cell type: nucleated epithelial cells. (C) Mid-follicular transition stage (estrogen), duration <1 day, cell type: nucleated epithelial cells, cornified cells. (D) Late follicular stage/ovulation (estrogen), duration ~1.5 days, cell types: cornified cells. (E) Early luteal (estrogen, progesterone), duration <1 day, cell types: cornified cells/leukocytes. (F) Late luteal (progesterone), duration ~2 days, cell types: leukocytes, some cornified cells. (G) Menstrual shedding, duration ~3 days, cell types: erythrocytes, leukocytes. Cor-cornified cell, ery-erythrocyte, leu-leukocyte, nec-nucleated epithelial cell. *Adapted from Bellofiore, N., Ellery, S. J., Mamrot, J., Walker, D. W., Temple-Smith, P. and Dickinson, H. (2017). First evidence of a menstruating rodent: The spiny mouse (Acomys cahirinus). American Journal of Obstetrics and Gynecology 216, (40. e41-40. e11).*

worldwide and is characterized to various degrees by symptoms such as nausea, abdominal pain and cramps, bloating, sore breasts, acne, irritability, food cravings and fatigue that may be severe enough to require some form of intervention. Because the mechanism behind PMS is still poorly understood, having a small animal model that displays symptoms may provide new information about the etiology of PMS, and possible treatments. The study design for this discovery was to monitor and score female spiny mouse behavioral responses during the menstrual cycle to restraint during vaginal lavage and to also track exploratory behavior, including female memory and social behavior and physiological changes using metabolic cages. The surprising findings were that, compared to controls, female spiny mice during premenstrual phases of the cycle were more reactive to constraint, moved less, spent more time in isolation in their communal cage, and increased their food consumption significantly compared to females in the late follicular phase of their cycle.

While menstruation has not yet been confirmed in female spiny mice in other captive colonies, and perhaps more importantly by spiny mice in the wild, it is intriguing to consider how and why this may have evolved. The frank release of blood by any species in the wild, from menstruation or otherwise, would be expected to have a highly negative selective value when compared with the standard rodent reabsorption of the engorged preimplantation endometrium during an estrous cycle. It seems counterintuitive that wild populations of spiny mice and other menstruating mammals would provide sensory cues from menstrual blood that enhanced the potential for predators to locate these important adult females. Selective removal of breeding females from a population by predators might be expected to have a negative selective pressure on population recruitment, especially in vulnerable or endangered species. However, in our colony only about 10–15% of females actually show overt bleeding from the vagina during menstruation; most females present with occult menstrual bleeding that can only be confirmed by vaginal lavage. We have observed a range of bleeding levels from light to heavy bleeders which suggests that this gradation in levels of bleeding depends on the extent of development of the endometrium during each cycle. This may be directly linked to different levels of circulating steroid hormones, in particular estrogens, progesterone and dehydroepiandrosterone (DHEA), which Bellofiore and Evans (2019) proposed is integral in spontaneous decidualization and menstruation in the spiny mouse. Assuming that this also occurs in wild populations of spiny mice, the selective pressure from predation would fall more directly on females that were heavy bleeders and released blood from menstruation into the environment. In fact, for some predators with a keen sense of smell, even occult bleeding may be possible to sense by olfaction. For the spiny mouse to be a broadly accessible emerging model for women's health, we need to confirm that females in other captive colonies and perhaps also in wild *Acomys* populations also menstruate. There is already some evidence that females from the Kentucky colony menstruate (Ashley Seifert, Thomas Gawriluk, and Katarina Medger, personal communication). Clearly, an important future direction in spiny mouse menstrual research will be to include field studies of wild-caught populations to confirm, or otherwise, the existence of natural menstruation. These studies should also extend to other species in the Deomyinae which includes at least 19 species of *Acomys*, one species of *Deomys*, 33 species of *Lophuromys* and a single species of *Uranomys*, to determine if menstruation is a feature of other species in this group.

3.3 Menopause and reproductive senescence

Previous work in other labs and more recently by Evgenia George, a graduate student at Monash, has provided a detailed analysis of ovarian morphology and shown that, as in humans and a few other species, but unusual for rodents, the ovarian reserve develops prenatally with primordial germ cell/oocyte nests breaking down to form primordial follicles 2–3 days before birth. From a stereological analysis of an ontogenetic series of ovaries (prenatal to aged), Evgenia has shown that there is a significant age-related decline in the ovarian reserve from female spiny mice between 3 and 4 years of age (unpublished data).

More recently, work by Bellofiore and George have confirmed that this decline in ovarian reserve is part of another intriguing aspect of female spiny mouse reproductive biology: a human-like transition to menopause. No other rodent shows this pattern of gradual senescence, even those rodents that show an alternative form of senescence described as estropause (Battiston et al., 2017), which is accomplished rapidly at 9 to 15 months of age in estrous cycling murine rodents. The idea that spiny mice display a gradual transition to a menopause-like state with aging has revealed another fascinating characteristic of their reproductive biology. Recognition of this transition might have occurred earlier were it not for a self-imposed breeding criterion for our colony: that breeders are retired from breeding and culled after 12 consecutive litters. This meant that few females, or males, survived in the colony past 3 years of age. This was mostly a financial decision at a time when the cost of maintaining the colony was significantly more than now. With the discovery of menstruation, and some financial relief on colony maintenance, we started to explore the reproductive lifespan of males and females in the colony, particularly with individuals between 3 and 4 years of age. This change has allowed us to collect new datasets on reproductive lifespan and to explore how aging affects reproductive capacity. In our captive colony we now know, from vaginal lavage and the histological presence of antral follicles, that many females up to 3 years of age retain the ability to cycle and breed but also that the degree of bleeding during menses in the 3–4 year old cohort declines significantly with a reducing ovarian reserve and gradual cessation of breeding. Recently completed plasma steroid hormone assays suggest that declining estrogen levels may affect the extent of endometrial proliferation in these older females resulting in lighter menstrual bleeds and even perhaps infertility. The gradual, rather than sudden, menopausal transition suggests that the spiny mouse is a more appropriate perimenopausal model than the current less natural rodent models for examining changing responses in the hypothalamic-pituitary-gonadal axis and other important pathways that influence this transition (Bellino & Wise, 2003; Zakaria et al., 2019).

Our menopause transition study highlights again, the ever-unfolding potential of the spiny mouse as an emerging model species for research in reproduction, development and aging. Our observations suggest that menopause in female spiny mice is similar to chimpanzees, where there is a gradual menopause-like transition with reproductively senescent females surviving the equivalent of 20% of their natural lifespan in reproductive senescence (Gould, Flint, & Graham, 1981). This highlights again the broad potential applications of the spiny mouse as an emerging model for various aspects of women's health (Fig. 4). It is currently the only small laboratory animal that provides a natural model for research into menopause.

3.4 Reproduction in male spiny mice

Compared to females, the reproductive biology of male spiny mice is poorly studied (Peitz, Foreman, & Schmitt, 1979). Despite the interesting human-like characteristics discovered in female spiny mice, the male reproductive system appears similar to most other rodents and there are no characteristics that have been discovered yet that suggest the male will provide any modeling that cannot already be achieved in the other more popular

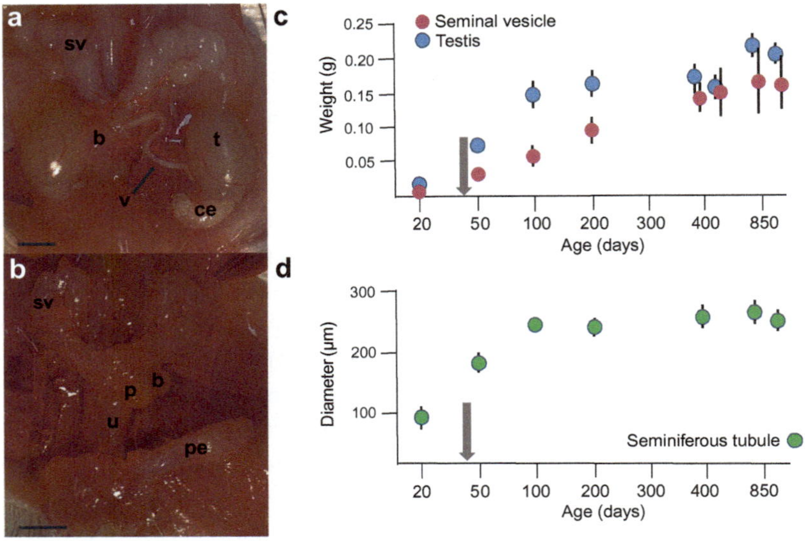

Fig. 6 Reproduction in the male spiny mouse. (A) *In situ* dissection of the male reproductive tract. Bladder (b), cauda epididymidis (ce), seminal vesicle (sv), testis (t), vas deferens (v). (B) *In situ* dissection of the male accessory glands of reproduction. Bladder (b), prostate (p), penis (pe), seminal vesicle (sv), urethra (u). (C) Changes in testis and seminal vesicle weights with age. (D) Changes in seminiferous tubule diameter with age. *Panels (A) and (B): Photo credits, Peter Temple-Smith. Panels (C) and (D) adapted from Mamrot, J., Mamrot, S., Weybury, M., Wiadrowski, L., Temple-Smith, P., Dickinson, H. (2017) Anatomy and physiology of the male reproductive tract in the spiny mouse (*Acomys cahirinus*) across the lifespan. Poster, Annual joint meeting, Endocrine Society of Australia/Society of Reproductive Biology 2017, Perth, Australia.*

laboratory rodents. However, there are some interesting reproductive features of male *Acomys* that deserve mention and warrant further investigation.

The male reproductive tract is typical of most other rodents with paired scrotal testes, epididymides and vasa deferentia (Fig. 6A and B). The inguinal canal is open and the testes and epididymides are retained in the scrotal sac by an inguinal fat pad as in laboratory mice and rats. The seminal vesicles are large and paired with ducts that join the distal ends of the vasa deferentia. Here they form paired ejaculatory ducts that pass through the prostate to drain into the prostatic urethra and paired bulbourethral glands that drain into the membranous section of the urethra before it passes into the penis to form the penile urethra. Although the penis contains an os penis, or baculum, which is not uncommon among rodents, preputial glands are lacking, a conspicuous absence among the rodents (Peitz et al., 1979). Peitz and colleagues provide a detailed analysis of the male reproductive system where they reported testicular asymmetry in adult males, a fact we confirmed in the Monash colony, with the left testis consistently heavier than the right. We also confirmed that asymmetry in testis size is reflected in a similar asymmetry in daily sperm production. This arrangement may relate to increased spermatogenic activity associated with increased seminiferous tubule length and an increased number of Sertoli cells in the left testis. This study also reported that spermatogenesis commences before post-natal week 7 with spermatozoa first observed in the seminiferous tubules between 7 and 8 weeks (Fig. 6C and D, gray arrow). Others have reported that sperm production occurs between weeks 5 and 6 and that males can impregnate females by week 7 (Haughton et al., 2015). Although there are no detailed published studies on spiny mouse sperm morphology, general observations on sperm structure and motility suggest there that they are typically murid, with single hooked sperm head and rodent sized midpiece and principal piece (Breed, 1995). The epididymis appears normal in shape and from brief observations spermatozoa undergo a standard process of maturation during passage through the epididymis with acquisition of the capacity for motility in the corpus region and storage in the caudal region.

Observations on mating behavior in spiny mice, including a recent video-based study on Monash colony pairs (Mamrot, 2019), have reported various interesting and sometimes conflicting outcomes. Mating usually occurs within a day of parturition associated with post-partum ovulation and receptivity and mate selection appears to be female-driven with receptive females controlling access, mounting and withdrawal behavior. Multiple intromissions are usually observed with an absence of penile thrusting by the male and ejaculation signaled by a brief and unusual hind paw "wave" and an almost imperceptible penile locking in female tract with a duration of only a few seconds. In many rodents, including laboratory mice, successful ejaculation is confirmed by the presence of a copulatory plug which is derived from the coagulation of seminal fluids from the seminal vesicles, coagulating glands and prostate. Peitz et al. (1979) conducted a detailed study of accessory gland function to examine the mechanism behind coagulation. Using cross-species coagulation testing of seminal vesicle and coagulating gland secretions they showed reciprocal cross-reactivity between seminal vesicle and coagulating gland secretions from the spiny mouse and laboratory rat. They also showed that, while coagulating gland vesiculase from a broad range of rodents were able to coagulate seminal vesicle secretions from spiny mice, the spiny mouse coagulation gland secretions were significantly less active on seminal vesicle secretions from a range of other rodents including the more closely related gerbil.

A controversial aspect of male spiny mouse reproduction concerns the presence or absence of copulatory plugs. Unlike in many rodent species, with the exception of one spiny mouse study in which plugs were commonly observed (Dewsbury & Hodges, 1987), most studies have concluded that no visible copulatory plug is formed in the vagina from semen coagulation following successful spiny mouse matings (Bellofiore, Ellery, Temple-Smith, & Evans, 2020; Dickinson & Walker, 2007; Peitz, 1981). In my coauthor's lab, an extensive study of female spiny mice immediately post-copulation found no evidence of plugs and a recent study in our lab confirmed these observations. The observations in our lab study by Jarrod McKenna were associated with vaginal lavage in an attempt to confirm at least the presence of large numbers of spermatozoa in the vagina post-coitus but, with the exception of a few spermatozoa in the lavages, there was no other evidence that ejaculation had occurred. Interestingly, despite the presence of few spermatozoa in the lavages, many of the females were subsequently found to be pregnant after the pair had been separated. This raises the possibility that during a brief penile lock, the ejaculation is intracervical or intrauterine in spiny mice—a possibility for future testing.

3.5 Precocial development

Compared with most other rodents, spiny mice exhibit extended fetal development *in utero* and are relatively mature at birth (Brunjes, 1990). Compared with altricial neonates (Fig. 7A and B) newborn spiny mice are able to see and hear, regulate their body temperature, move independently with relatively well coordinated movements, and forage for themselves within a few days of birth (Haughton et al., 2015). Whereas altricial rodents naturally ovulate large numbers of oocytes and give birth to an average of eight neonates following a gestation of 16–17 days, spiny mice usually ovulate a maximum of 3–5 oocytes, and often fewer. In the Monash and Kentucky colonies, births occur with an average of three neonates after ~39 days of gestation.

The rarity of precocial development among murids presents an opportunity to investigate and compare organ development with altricial laboratory rodents in which maturation of organ systems occurs postnatally. For instance, spiny mice complete lung and liver development (Oosterhuis, Mooren, Charles, & Lamers, 1984), and the majority of neurogenesis

Fig. 7 Comparison of development at birth. (A) Precocious spiny mouse pups (*Acomys* sp.). (B) Altricial laboratory mouse pups, (*Mus muculus*). *Credit: https://www.spectrumnews.org/news/toolbox/spiny-mice-model-risk-of-autism-from-maternal-infection/*

in utero (Dickinson, Walker, Cullen-McEwen, Wintour, & Moritz, 2005). Studies have also shown that, unlike altricial species, spiny mice responded later in gestation than altricial rats and gerbils to acute umbilical cord compression suggesting that this promotes fetal survival *in utero* (Robinson & Smotherman, 1992). Precocial development has also provided a unique opportunity for researchers to examine many aspects of newborn spiny mouse behavior. Using a suite of behavioral tests, studies have demonstrated motor activity, fine coordination, and social behaviors 1 day after birth (Ratnayake, Quinn, Daruwalla, Dickinson, & Walker, 2014). Moreover, motor coordination performance and an assessment of exploratory behavior measured 10–25 days after birth in spiny mice were found equivalent to these behaviors in adult mice (Ratnayake et al., 2014). Spiny mice have also provided a new experimental model to study the effects of altered gravitational conditions on fetal development (Sekulić, Božić, Bozić, Borota, &

Ćulić, 2006). These studies make the spiny mouse an ideal subject for studies of both early ontogenesis and the mechanisms which result in species differences and highlights the spiny mouse as an emerging model for developmental biology (Brunjes, 1989).

3.6 Manipulating spiny mouse reproduction

In recent years, advances in our knowledge of spiny mouse biology have directed some research initiatives toward manipulation of the reproductive system to build new models to explore different facets of reproduction in spiny mice. Researchers in our department and linked institutes at Monash University have actively pursued some of the unanswered or controversial questions of reproduction in our colony, sometimes with interesting and often unexpected results. In this section of the chapter, I share four of these interesting experiences and outcomes and discuss how these experiments are influencing the next phases of reproductive research in spiny mice.

A robust and reliable superovulation protocol is a critical prerequisite for developing a successful *in vitro* fertilization program. Pasco et al. (2012) developed and optimized a technique resulting in a 100% success rate using this optimized protocol with an average recovery of 15 × 1 or 2 cell embryos per treatment. Considering that the average ovulation rate for unstimulated females is about 3 MII oocytes, this fivefold increase in ovulation/embryo production rate was highly significant and with appropriately timed collections would likely facilitate the development of IVF and an embryo culture system. The conundrum from this successful project was that embryonic development in the spiny mouse (Fig. 8A and B) is blocked at the 4-cell stage, not unlike the 2-cell block that originally delayed *in vitro* studies of mouse embryo development (Biggers, 2004). Jared Mamrot explored the development of cleavage-staged embryos in spiny mice and developed an important timed embryo cleavage series for comparison with the mouse and human (Fig. 8A and B). Despite a large number of treatments, he has not been successful in progressing spiny mice embryos past the 4-cell stage *in vitro*, although later stage cleavage embryos recovered from the oviducts are able to continue to the blastocyst stage given an appropriate *in vitro* environment (Mamrot, 2019). Recently, we tried unsuccessfully to repeat the (Pasco, Gardner, & Walker, 2012) superovulation protocol and have proposed the use of a human-based superovulation protocol (McKenna, Bellofiore, Dimitriadis, & Temple-Smith, 2021) as an alternative method that is currently being tested.

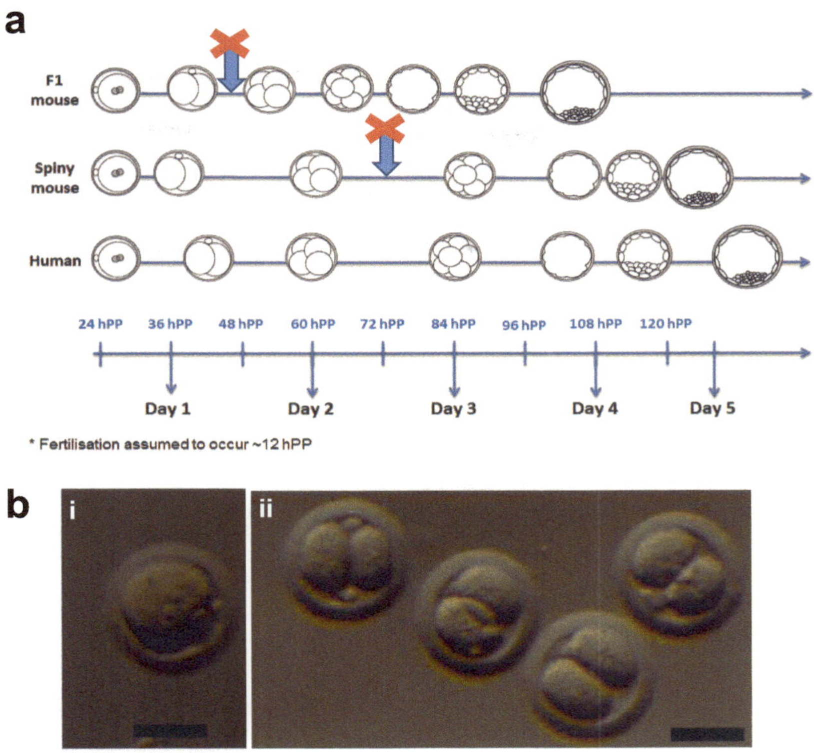

Fig. 8 Spiny mouse embryology. (A) Comparison of the time taken to reach the equivalent preimplantation cleavage stages in the F1 laboratory mouse, spiny mouse and human showing the similarity in timing between the spiny mouse and human. Note: *in vitro* cleavage blocks occur at the 2-cell stage in the mouse and the 4-cell stage in the spiny mouse. (B) Timed cleavage stage development in spiny mouse embryos from fertilization to expanded blastocyst. *This figure is borrowed with permission from Mamrot, J. P. (2019). Early developmental biology of the spiny mouse (Acomys cahirinus). Monash University.*

In the second example, we were interested to determine if female spiny mice in the same cages are able to synchronize their menstrual cycles, similar to the way lab mice synchronize their estrous cycles. Synchronization of cycles is an important behavior in animal models used for reproductive studies because it simplifies time required to complete these studies. We were also interested to know if vaginal and other forms of stimulation elicit pseudopregnancy in spiny mice. To our surprise neither cycle synchronization nor pseudopregnancy could be induced in the female spiny mouse despite the use of various procedures including a daily vaginal lavage control, progesterone injection, mechanical stimulation of the cervix and sterile mating for pseudopregnancy, and by exposing female groups to male-soiled bedding to assess synchronization *via* the Whitten effect (Bellofiore et al., 2020). Failure of the spiny mouse to show pseudopregnancy or cycle synchronization is an indication that the spiny mouse is not a typical rodent and more aligned in its reproductive strategy with higher-order primates.

Spiny mice from our colony have provided an interesting and informative model for studying the effects of birth asphyxia. The initial idea, developed by David Walker and his neurobiology group, used the precocious qualities, especially the long gestation, of the spiny mouse as an important component of the model to examine neural deficits after hypoxia at birth. The model compared the effects of a caesarian delivery at 37 days gestation, 2 days before normal term, of pups that were delivered without an asphyxia event compared with those that were given 7 min of asphyxia. This asphyxia event caused significant impairment of memory and learning tasks and caused various pathologies in the central nervous system including an inflammatory response and increased cell death compared to the controls (Hutton, Ratnayake, Shields, & Walker, 2009). The success of this asphyxia model began a series of related studies to explore the many possible applications of this model including demonstration that an acute pre-term hypoxic event caused mild neurological effects in various behavioral tests (Ireland, Dickinson, Fleiss, Hutton, & Walker, 2010) and that creatine supplementation protected various organ systems from the immediate pre-term effects of hypoxia-induced damage (Ellery et al., 2016; Ireland, Castillo-Melendez, Dickinson, Snow, & Walker, 2011). Stacey Ellery, an early career researcher at the Hudson Institute of Medical Research in Melbourne, more recently extended these studies on birth asphyxia and creatine supplementation in the spiny mouse model and is now translating findings on the protective nature of creatine supplementation into clinical applications; perhaps, the first time that biomedical science discoveries using the spiny mouse model have been translated into clinical studies (De Guingand, Ellery, Davies-Tuck, & Dickinson, 2019; Ellery et al., 2019, 2020).

Lastly, a recent study in our lab by Jarrod McKenna developed a sperm cryopreservation protocol for spiny mouse spermatozoa using a raffinose-based cryoprotectant with effective freezing and warming qualities that will enable us to preserve semen samples for colony insurance purposes and future development of artificial breeding techniques (McKenna, Catt, Pangestu, & Temple-Smith, 2020). Together, these advances have positioned the growing spiny mouse research community to develop transgenic spiny mice, a feat that would allow this species to join a small and select group of mammals in which transgenesis has been achieved.

4. Future perspectives

4.1 Community

Whereas discovering a new phenomenon can be exhilarating, developing a new model can be arduous. Cultivating a new system can breed isolation or community. For every genuine inquiry of interest from the intellectually curious, there are the adulators, hungry for the possibility of something flashy and new that can be leveraged for another grant. For every thoughtful conversation about the challenges of working with a new model, there are those that only need your resources. For every believer, there is the naysayer. Luckily, our overall experience has been the collective former.

Although spiny mouse research colonies have existed almost continuously since the 1960s, research with these rodents has tended to remain in isolated pockets. The longest running colony (to our knowledge) is maintained at Tel Aviv University in Israel by Noga Kronfeld-Schorr whose broad biological interests have served to maintain a steady output of spiny mouse research over the years. The Tel Aviv colony is also close to wild spiny mouse populations allowing for continued maintenance of genetic diversity. Another longstanding colony remains at the University of Geneva, from which the Monash colony was derived, and is maintained by Michel Milinkovitch and Athanasia Tzika who have used these animals to explore questions in evolutionary and developmental biology (Montandon, Tzika, Martins, Chopard, & Milinkovitch, 2014). The Monash colony is approaching its 20-year anniversary and while the University of Kentucky colony was only started in 2013, its provenance dates back to the 1990s. Detailed publications on husbandry have been published and as detailed below, key molecular resources have been or are now under development. In addition to existing colonies in Europe, we have helped to establish new colonies at Caltech, Cornell, Liberty University, Emory University, University of Southern California, Hubrecht University, KAIST (Korea), Yale, Harvard and University of British Columbia. Gustavo Tiscornia also maintains a colony with several species of spiny mice at the University of Algarve.

After nearly a decade since the original publication of enhanced regenerative ability the response to *Acomys* as a useful model by other laboratories studying regeneration has been predictably mixed; an outcome that parallels the reductionist thinking that has canalized the broader field of developmental biology. With the rise of genetic model organisms and a shift toward genetic description, research laboratories are loath to adopt a high-cost model that does not provide the wealth of molecular tools and transgenic technology present in a handful of model organisms. However, recent developments suggest this trend is slowly changing given an expansion of molecular resources and a steady stream of research that has begun to unpack the mechanistic basis for enhanced regenerative ability in these animals (Gaire et al., 2021; Simkin & Seifert, 2018). In contrast, since the discovery of menstruation, a premenstrual syndrome and a menopause-like transition in spiny mice, the reproductive biology community has been predictably more excited. This is not surprising given the absence of a small animal model of menstruation and the potential for the spiny mouse model to propel the field forward.

4.2 Molecular resources

Although the breadth of the standard molecular toolkit available for the laboratory mouse is not available for spiny mice, rapid technological advances should help to close the gap. A reference transcriptome representing 23 tissues (including heart, muscle, lung, brain, placenta, and gonads from an adult male and female) has been published (Mamrot et al., 2017) and more targeted *de novo* transcriptomes for skin healing (Brant et al., 2015) and complex tissue regeneration (Gawriluk et al., 2016) exist. These have been leveraged by other groups for comparative work exploring the evolution of regeneration (Wang et al., 2020) and have been important resources for our genome sequencing project. Through a joint effort including the Earlham Institute, EMBL-EBI, and the Sanger Institute, University of Kentucky and the University of Tel-Aviv, chromosome level assemblies for four *Acomys* species have been produced using a combination of PacBio long reads and Illumina 10 × reads (Biswas S., Wright D. et al. *in preparation*). While a rough genome for *A. cahirinus* was published in 2019 by the Broad Institute with tissue from animals at the University of Kentucky, investment in that sequencing project was limited and lacked the computational support to produce a draft genome useful to the growing community of *Acomys* researchers. The frozen *A. cahirinus* assembly produced in 2019 was shared broadly among researchers in the *Acomys* community in keeping with our commitment to open science among our burgeoning community. This chromosome-level assembly was annotated by the Earlham Institute by integrating RNA-seq information and existing annotation from reference species (mouse, rat and human) and has provided a platform for RNA-seq, scRNA-seq and other comparative projects that rely on genomic information.

Very few spiny mouse primary cell lines have been developed and because of the dearth of colonies globally, few if any are readily available commercially. Both of our labs have developed a series of primary fibroblast cell lines that are useful for an array of studies. These also serve as an insurance strategy such that in the event that the Monash or Kentucky colony suffers a catastrophic event resulting in the complete loss of the colonies, we have the possibility with new and developing pluripotency and gamete generation methods to potentially re-create our colonies. We have also supported and assisted the development of colonies as outlined above, and a second colony of spiny mice was created for David Walker's group which is now located at the Royal Melbourne Institute of Technology. This colony used founders from the Monash colony to establish this new spiny mouse colony in Melbourne to provide us both with continuing active research, and also joint insurance colonies.

There is also now a growing incentive to establish other primary cell lines, such as embryonic stem cells, induced pluripotent stem cells, endometrial and other tissue-specific stem cells, fetal fibroblasts for *Acomys*-specific feeder layers and, in the more distant future, immortalized spiny mouse cell lines to provide what we anticipate will be an expanding need for all of these in future projects. A transgenic project is now underway at the University of Kentucky using the traditional ESC line approach which when coupled with the reproductive advances at Monash increase our chances for successfully producing the first set of transgenic spiny mice.

We have also found, often to our detriment, that our research on various aspects of reproduction in the spiny mouse is hampered by the dearth of experimental tools that are standard and commercially available for researchers working on other laboratory rodents, especially the mouse. Although the spiny mouse genome has now been sequenced and we have published a transcriptome for the species in our colony, compared with the vast resources available for laboratory mouse research, there is little in the way of genetic studies, specific biomarkers or antibodies for immunocytochemistry studies, and few bioassays available that are informative for the spiny mouse. Developing these tools will greatly improve our understanding of the various processes in reproduction and developmental biology that uniquely occur in the spiny mouse. For example, our data on menstruation and menopause would be more informative, and even more interesting, if we had access to validated and optimized bioassays to measure with accuracy some of the important protein and peptide hormones of reproduction such as gonadotrophin-releasing hormone (GnRH), follicle stimulating hormone (FSH), luteinizing hormone (LH), inhibins and activins, other TGF-beta cytokines and anti-Mullerian hormone (AMH) in the spiny mouse. In our journey so far, it has been interesting to discover that a few commercial bioassays do work in the spiny mouse and even more interesting to find that these are often human, rather than rodent, assays. In addition to these needs, the future expansion of spiny mouse reproductive research is also likely to benefit from the development of organoid cultures and the isolation and characterization of organ-specific exosomes.

Acknowledgments

We would like to acknowledge those researchers that have pursued work with spiny mice in the past, often at great challenge. We would also like to thank the growing community of spiny mouse researchers that are committed to open science and that generously share information with each other about ongoing and future projects that facilitate our general understanding about these extraordinary rodents.

Beyond those named in this chapter, Dr. Seifert would like to acknowledge the tireless work of several Kenyan field assistants, namely, James Ekiru Erkaran who was instrumental in trapping and site scouting in the early development of the regenerative biology work. Also, Stanley Marete in the Department of Veterinary Anatomy and Physiology at the University of Nairobi who has maintained ongoing colonies of *A. kempi* and *A. percivali* and the staff and scientists at Mpala Research Centre. Greg Schultz for his mentorship in all things wound healing. Vanessa Ezenwa has been an amazing collaborator on the ecoimmunological aspects of spiny mice and continues to participate as part of our community. Thomas Gawriluk worked tirelessly in the field and back in the States to advance the spiny mouse model at the University of Kentucky. He also conducted the first reproductive biology studies in the Kentucky colony. Lastly, spiny mouse work in Dr. Seifert's lab has been generously supported by grants from the National Science Foundation (NSF) and the Office for International Science and Engineering (OISE) (IOS-1353713), the National Institutes of Arthritis and Musculoskeletal and Skin Diseases (NIAMS) (R01 AR070313), the National Institute of Dental and Craniofacial Research (NIDCR) (R21 DE028070) and a Grants-in-Aid of Research from the National Geographic Society.

Dr. Temple-Smith acknowledges David Walker, Stacey Ellery and the many researchers and students who, through their interest, enthusiasm and long hours in their laboratories, have provided the solid foundation on reproduction and development of spiny mice in the Monash colony that forms the basis of this chapter. Special thanks to Haley Dickinson for the productive collaborations on spiny mouse reproduction during her time at the Hudson Institute of Medical Research; Jared Mamrot for advancing our knowledge on embryo development in the spiny mouse and for permission to use (Fig. 8A and B) from his PhD thesis; Nadia Bellofiore for the fascinating discoveries, during her graduate research, of menstruation, premenstrual tension syndrome and absence of pseudo-pregnancy in the spiny mouse; Evgenia George for her detailed research findings on the spiny mouse ovarian reserve and who, with Nadia, demonstrated the spiny mouse transition-to-menopause; Jarrod McKenna for his detailed and important work on sperm cryopreservation, mating behavior and post-partum reproduction. I also acknowledge the generous support from the Bill and Melinda Gates Foundation (OPP1171535—The Spiny Mouse for Screening Contraceptive), the Pharmaceuticals Research and Development branch at Bayer AG Germany, the Monash University Education Program in Reproduction and Development and the Monash Department of Obstetrics and Gynecology.

References

Alhajeri, B. H., Hunt, O. J., & Steppan, S. J. (2015). Molecular systematics of gerbils and deomyines (Rodentia: Gerbillinae, Deomyinae) and a test of desert adaptation in the tympanic bulla. *Journal of Zoological Systematics and Evolutionary Research*, *53*, 312–330.

Aloysius, A., Saxena, S., & Seifert, A. W. (2021). Metabolic regulation of innate immune cell phenotypes during wound repair and regeneration. *Current Opinion in Immunology*, *68*, 72–82.

Battiston, F. G., Dos Santos, C., Barbosa, A. M., Sehnem, S., Leonel, E. C. R., Taboga, S. R., et al. (2017). Glucose homeostasis in rats treated with 4-vinylcyclohexene diepoxide is not worsened by dexamethasone treatment. *The Journal of Steroid Biochemistry and Molecular Biology*, *165*, 170–181.

Bauer, A. M., Russell, A. P., & ShadwickK, R. E. (1989). Mechanical properties and morphological correlates of fragile skin in gekkonid lizards. *Journal of Experimental Biology*, *145*, 79–102.

Bellino, F. L., & Wise, P. M. (2003). Nonhuman primate models of menopause workshop. *Biology of Reproduction*, *68*, 10–18.

Bellofiore, N., Cousins, F., Temple-Smith, P., Dickinson, H., & Evans, J. (2018). A missing piece: The spiny mouse and the puzzle of menstruating species. *Journal of Molecular Endocrinology*, *61*, R25–R41.

Bellofiore, N., Cousins, F., Temple-Smith, P., & Evans, J. (2019). Altered exploratory behaviour and increased food intake in the spiny mouse before menstruation: A unique pre-clinical model for examining premenstrual syndrome. *Human Reproduction*, *34*, 308–322.

Bellofiore, N., Ellery, S. J., Mamrot, J., Walker, D. W., Temple-Smith, P., & Dickinson, H. (2017). First evidence of a menstruating rodent: The spiny mouse (Acomys cahirinus). *American Journal of Obstetrics and Gynecology*, *216* (40. e41-40. e11).

Bellofiore, N., Ellery, S. J., Temple-Smith, P., & Evans, J. (2020). Pseudopregnancy and reproductive cycle synchronisation cannot be induced using conventional methods in the spiny mouse (Acomys cahirinus). *Reproduction, Fertility and Development*, *32*, 363–372.

Bellofiore, N., & Evans, J. (2019). Monkeys, mice and menses: The bloody anomaly of the spiny mouse. *Journal of Assisted Reproduction and Genetics*, *36*, 811–817.

Bellofiore, N., Rana, S., Dickinson, H., Temple-Smith, P., & Evans, J. (2018). Characterization of human-like menstruation in the spiny mouse: Comparative studies with the human and induced mouse model. *Human Reproduction*, *33*, 1715–1726.

Bely, A. E., & Sikes, J. M. (2010). Latent regeneration abilities persist following recent evolutionary loss in asexual annelids. *Proceedings of the National Academy of Sciences*, *107*, 1464–1469.

Biggers, J. D. (2004). Reflections on the culture of the preimplantation embryo. *International Journal of Developmental Biology*, *42*, 879–884.

Billingham, R., & Russell, P. (1956). Incomplete wound contracture and the phenomenon of hair neogenesis in rabbits' skin. *Nature*, *177*, 791–792.

Borgens, R. B. (1982). Mice regrow the tips of their foretoes. *Science*, *217*, 747–750.
Brant, J. O., Lopez, M.-C., Baker, H. V., Barbazuk, W. B., & Maden, M. (2015). A comparative analysis of gene expression profiles during skin regeneration in Mus and Acomys. *PLoS One*, *10*, e0142931.
Brant, J. O., Yoon, J. H., Polvadore, T., Barbazuk, W. B., & Maden, M. (2016). Cellular events during scar-free skin regeneration in the spiny mouse, Acomys. *Wound Repair and Regeneration*, *24*, 75–88.
Brasted, M., White, C., Kennedy, T., & Salamonsen, L. (2003). Mimicking the events of menstruation in the murine uterus. *Biology of Reproduction*, *69*, 1273–1280.
Breed, W. (1995). Spermatozoa of murid rodents from Africa: Morphological diversity and evolutionary trends. *Journal of Zoology*, *237*, 625–651.
Brunjes, P. (1989). A comparative study of prenatal development in the olfactory bulb, neocortex and hippocampal region of the precocial mouse Acomys cahirinus and rat. *Developmental Brain Research*, *49*, 7–25.
Brunjes, P. C. (1990). The precocial mouse, Acomys cahirinus. *Psychobiology*, *18*, 339–350.
Calve, S., Odelberg, S. J., & Simon, H. G. (2010). A transitional extracellular matrix instructs cell behavior during muscle regeneration. *Developmental Biology*, *344*, 259–271.
Calve, S., & Simon, H.-G. (2012). Biochemical and mechanical environment cooperatively regulate skeletal muscle regeneration. *The FASEB Journal*, *26*, 2538–2545.
Carleton, M. D., & Musser, G. G. (2005). Order rodentia. *Mammal Species of the World: A Taxonomic and Geographic Reference*, *2*, 745–2142.
Clark, R. A. F. (1996). *The molecular and cellular biology of wound repair*. Springer.
Clark, L. D., Clark, R. K., & Heber-Katz, E. (1998). A new murine model for mammalian wound repair and regeneration. *Clinical Immunology and Immunopathology*, *88*, 35–45.
Cyr, J. L., Gawriluk, T. R., Kimani, J. M., Rada, B., Watford, W. T., Kiama, S. G., et al. (2019). Regeneration-competent and-incompetent murids differ in neutrophil quantity and function. *Integrative and Comparative Biology*, *59*, 1138–1149.
De Guingand, D. L., Ellery, S. J., Davies-Tuck, M. L., & Dickinson, H. (2019). Creatine and pregnancy outcomes, a prospective cohort study in low-risk pregnant women: Study protocol. *BMJ Open*, *9*, e026756.
Dewsbury, D. A., & Hodges, A. W. (1987). Copulatory behavior and related phenomena in spiny mice (Acomys cahirinus) and hopping mice (Notomys alexis). *Journal of Mammalogy*, *68*, 49–57.
Diaz de Chumaceiro, C. L. (1995). Serendipity or Pseudoserendipity? Unexpected versus desired results. *Journal of Creative Behavior*, *29*, 143–147.
Dickinson, H., Moritz, K., Wintour, E. M., Walker, D. W., & Kett, M. M. (2007). A comparative study of renal function in the desert-adapted spiny mouse and the laboratory-adapted C57BL/6 mouse: Response to dietary salt load. *American Journal of Physiology - Renal Physiology*, *293*, F1093–F1098.
Dickinson, H., & Walker, D. (2007). Managing a colony of spiny mice (Acomys cahirinus) for perinatal research. *Australian and New Zealand Council for the Care of Animals in Research and Training (ANZCCART) News*, *20*, 4–11.
Dickinson, H., Walker, D. W., Cullen-McEwen, L., Wintour, E. M., & Moritz, K. (2005). The spiny mouse (Acomys cahirinus) completes nephrogenesis before birth. *American Journal of Physiology - Renal Physiology*, *289*, F273–F279.
Driskell, R. R., Lichtenberger, B. M., Hoste, E., Kretzschmar, K., Simons, B. D., Charalambous, M., et al. (2013). Distinct fibroblast lineages determine dermal architecture in skin development and repair. *Nature*, *504*, 277–281.
D'Udine, B., & Alleva, E. (1988). The Acomys cahirinus (spiny mouse) as a new model for biological and neurobehavioural studies. *Polish Journal of Pharmacology and Pharmacy*, *40*, 525–534.

Dunoyer, L. A., Seifert, A. W., & Van Cleve, J. (2020). Evolutionary bedfellows: Reconstructing the ancestral state of autotomy and regeneration. *Journal of Experimental Zoology Part B: Molecular and Developmental Evolution*, *336*(2), 94–115.

Ellery, S. J., LaRosa, D. A., Kett, M. M., Della Gatta, P. A., Snow, R. J., Walker, D. W., et al. (2016). Dietary creatine supplementation during pregnancy: A study on the effects of creatine supplementation on creatine homeostasis and renal excretory function in spiny mice. *Amino Acids*, *48*, 1819–1830.

Ellery, S. J., Murthi, P., Davies-Tuck, M. L., Della Gatta, P. A., May, A. K., Kowalski, G. M., et al. (2019). Placental creatine metabolism in cases of placental insufficiency and reduced fetal growth. *Molecular Human Reproduction*, *25*, 495–505.

Ellery, S. J., Murthi, P., Gatta, P. A. D., May, A. K., Davies-Tuck, M. L., Kowalski, G. M., et al. (2020). The effects of early-onset pre-eclampsia on placental creatine metabolism in the third trimester. *International Journal of Molecular Sciences*, *21*, 806.

Emera, D., Romero, R., & Wagner, G. (2012). The evolution of menstruation: A new model for genetic assimilation: Explaining molecular origins of maternal responses to fetal invasiveness. *BioEssays*, *34*, 26–35.

Eming, S. A., Hammerschmidt, M., Krieg, T., & Roers, A. (2009). Interrelation of immunity and tissue repair or regeneration. *Seminars in Cell & Developmental Biology*, *20*, 517–527.

Esposito, M. (2013). Weismann versus morgan revisited: Clashing interpretations on animal regeneration. *Journal of the History of Biology*, *46*, 511–541.

Fabre, P.-H., Hautier, L., Dimitrov, D., & Douzery, E. J. (2012). A glimpse on the pattern of rodent diversification: A phylogenetic approach. *BMC Evolutionary Biology*, *12*, 1–19.

Fernando, W. A., Leininger, E., Simkin, J., Li, N., Malcom, C. A., Sathyamoorthi, S., et al. (2011). Wound healing and blastema formation in regenerating digit tips of adult mice. *Developmental Biology*, *350*, 301–310.

Gaire, J., Varholick, J. A., Rana, S., Sunshine, M. D., Doré, S., Barbazuk, W. B., et al. (2021). Spiny mouse (Acomys): An emerging research organism for regenerative medicine with applications beyond the skin. *Npj. Regenerative Medicine*, *6*, 1–6.

Gawriluk, T. R., Simkin, J., Hacker, C. K., Kimani, J. M., Kiama, S. G., Ezenwa, V. O., et al. (2020). Complex tissue regeneration in mammals is associated with reduced inflammatory cytokines and an influx of T cells. *Frontiers in Immunology*, *11*, 1695.

Gawriluk, T. R., Simkin, J., Thompson, K. L., Biswas, S. K., Clare-Salzler, Z., Kimani, J. M., et al. (2016). Comparative analysis of ear-hole closure identifies epimorphic regeneration as a discrete trait in mammals. *Nature Communications*, *7*, 11164.

Godwin, J. W., & Brockes, J. P. (2006). Regeneration, tissue injury and the immune response. *Journal of Anatomy*, *209*, 423–432.

Godwin, J. W., Pinto, A. R., & Rosenthal, N. A. (2013). Macrophages are required for adult salamander limb regeneration. *Proceedings of the National Academy of Sciences of the United States of America*, *110*, 9415–9420.

Gould, K. G., Flint, M., & Graham, C. E. (1981). Chimpanzee reproductive senescence: A possible model for evolution of the menopause. *Maturitas*, *3*, 157–166.

Hans, I., Harn, C., Wang, S.-P., Lai, Y.-C., Van Handel, B., Liang, Y.-C., et al. (2021). Symmetry breaking of tissue mechanics in wound induced hair follicle regeneration of laboratory and spiny mice. *Nature Communications*, *12*, 1–16.

Haughton, C. L., Gawriluk, T. R., & Seifert, A. W. (2015). The biology and husbandry of the African spiny mouse (Acomys cahirinus) and the research uses of a laboratory Colony. *Journal of the American Association for Laboratory Animal Science*, *55*(1), 9–17.

Hay, E. D., & Fischman, D. A. (1961). Origin of the blastema in regenerating limbs of the newt Triturus viridescens. An autoradiographic study using tritiated thymidine to follow cell proliferation and migration. *Developmental Biology*, *3*, 26–59.

Honeycutt, R. L. (2009). Rodents (Rodentia). *The Timetree of Life*, 490–494.

Hutton, L. C., Ratnayake, U., Shields, A., & Walker, D. W. (2009). Neuropathology and functional deficits in a model of birth asphyxia in the precocial spiny mouse (Acomys cahirinus). *Developmental Neuroscience, 31*, 523–535.

Illingworth, C. M. (1974). Trapped fingers and amputated finger tips in children. *Journal of Pediatric Surgery, 9*, 853–858.

Ireland, Z., Castillo-Melendez, M., Dickinson, H., Snow, R., & Walker, D. (2011). A maternal diet supplemented with creatine from mid-pregnancy protects the newborn spiny mouse brain from birth hypoxia. *Neuroscience, 194*, 372–379.

Ireland, Z., Dickinson, H., Fleiss, B., Hutton, L. C., & Walker, D. W. (2010). Behavioural effects of near-term acute fetal hypoxia in a small precocial animal, the spiny mouse (Acomys cahirinus). *Neonatology, 97*, 45–51.

Ito, M., Yang, Z., Andl, T., Cui, C., Kim, N., Millar, S. E., et al. (2007). Wnt-dependent de novo hair follicle regeneration in adult mouse skin after wounding. *Nature, 447*, 316–320.

Jiang, D., Correa-Gallegos, D., Christ, S., Stefanska, A., Liu, J., Ramesh, P., et al. (2018). Two succeeding fibroblastic lineages drive dermal development and the transition from regeneration to scarring. *Nature Cell Biology, 20*, 422–431.

Jiang, T. X., Harn, H. I. C., Ou, K. L., Lei, M., & Chuong, C. M. (2019). Comparative regenerative biology of spiny (Acomys cahirinus) and laboratory (Mus musculus) mouse skin. *Experimental Dermatology, 28*, 442–449.

Joseph, J., & Dyson, M. (1966). Tissue replacement in the rabbit's ear. *The British Journal of Surgery, 53*, 372–380.

Kench, J. A., Russell, D. M., Fadok, V. A., Young, S. K., Worthen, G. S., Jones-Carson, J., et al. (1999). Aberrant wound healing and TGF-beta production in the autoimmune-prone MRL/+ mouse. *Clinical Immunology, 92*, 300–310.

King, M. W., Neff, A. W., & Mescher, A. L. (2012). The developing Xenopus limb as a model for studies on the balance between inflammation and regeneration. *Anatomical Record (Hoboken), 295*, 1552–1561.

Kingdon, J. (1984). *East African mammals: An atlas of evolution in Africa, volume 2, part B: Hares and rodents Chicago*. University Of Chicago Press.

Koopmans, T., van Beijnum, H., Roovers, E. F., Malhotra, D., Tomasso, A., Boeter, J., et al. (2021). Ischemic tolerance and cardiac repair in the African spiny mouse. *NPJ Regenerative Medicine, 6*(1), 1–16.

Kragl, M., Knapp, D., Nacu, E., Khattak, S., Maden, M., Epperlein, H. H., et al. (2009). Cells keep a memory of their tissue origin during axolotl limb regeneration. *Nature, 460*, 60–65.

Krogh, A. (1929). The progress of physiology. *American Journal of Physiology-Legacy Content, 90*, 243–251.

Levi, B. P., & Morrison, S. J. (2008). Stem cells use distinct self-renewal programs at different ages. *Cold Spring Harbor Symposia on Quantitative Biology, 73*, 539–553.

Lo, D. C., Allen, F., & Brockes, J. P. (1993). Reversal of muscle differentiation during urodele limb regeneration. *Proceedings of the National Academy of Sciences, 90*, 7230–7234.

Maden, M. (2018). The evolution of regeneration–where does that leave mammals? *International Journal of Developmental Biology, 62*, 369–372.

Mamrot, J. P. (2019). *Early developmental biology of the spiny mouse (Acomys cahirinus)*. Monash University.

Mamrot, J., Legaie, R., Ellery, S. J., Wilson, T., Seemann, T., Powell, D. R., et al. (2017). De novo transcriptome assembly for the spiny mouse (Acomys cahirinus). *Scientific Reports, 7*, 1–15.

Manley, G. A. (2012). Evolutionary paths to mammalian cochleae. *Journal of the Association for Research in Otolaryngology, 13*, 733–743.

Martin, P., & Leibovich, S. J. (2005). Inflammatory cells during wound repair: The good, the bad and the ugly. *Trends in Cell Biology*, *15*, 599–607.

Mascharak, S., Davitt, M. F., Griffin, M., Borrelli, M. R., Moore, A. L., Chen, K., et al. (2021). Preventing Engrailed-1 activation in fibroblasts yields wound regeneration without scarring. *Science*, *372*, eaba2374.

Matias, D. S., Martins, A. R., Casanellas, I., Ova, A. B., Araújo, I. M., Power, D., et al. (2016). Ear wound regeneration in the African spiny mouse Acomys cahirinus. *Regeneration*, *3*(1), 52–61.

McKenna, J., Bellofiore, N., Dimitriadis, E., & Temple-Smith, P. (2021). Postpartum ovulation and early pregnancy in the menstruating spiny mouse, Acomys cahirinus. *Scientific Reports*, *11*, 1–11.

McKenna, J., Catt, S., Pangestu, M., & Temple-Smith, P. (2020). Successful sperm cryopreservation in Egyptian spiny mice Acomys cahirinus. *Reproduction, Fertility and Development*, *32*, 1293–1297.

Mescher, A. L., & Neff, A. W. (2005). Regenerative capacity and the developing immune system. *Advances in Biochemical Engineering/Biotechnology*, *93*, 39–66.

Monaghan, J. R., Athippozhy, A., Seifert, A. W., Putta, S., Stromberg, A. J., Maden, M., et al. (2012). Gene expression patterns specific to the regenerating limb of the Mexican axolotl. *Biology Open*, *1*, 937–948.

Monaghan, J. R., & Maden, M. (2013). Cellular plasticity during vertebrate appendage regeneration. *Current Topics in Microbiology and Immunology*, *367*, 53–74.

Montandon, S. A., Tzika, A. C., Martins, A. F., Chopard, B., & Milinkovitch, M. C. (2014). Two waves of anisotropic growth generate enlarged follicles in the spiny mouse. *EvoDevo*, *5*, 1–13.

Mori, R., Shaw, T. J., & Martin, P. (2008). Molecular mechanisms linking wound inflammation and fibrosis: Knockdown of osteopontin leads to rapid repair and reduced scarring. *The Journal of Experimental Medicine*, *205*, 43–51.

Muneoka, K., Allan, C. H., Yang, X., Lee, J., & Han, M. (2008). Mammalian regeneration and regenerative medicine. *Birth Defects Research. Part C, Embryo Today*, *84*, 265–280.

Needham, A. E. (1952). *Regeneration and wound-healing*. Methuen & Co.

Neufeld, D. A. (1980). Partial blastema formation after amputation in adult mice. *Journal of Experimental Zoology*, *212*, 31–36.

Neufeld, D. A., & Zhao, W. (1995). Bone regrowth after digit tip amputation in mice is equivalent in adults and neonates. *Wound Repair and Regeneration*, *3*, 461–466.

Oberpriller, J. O., & Oberpriller, J. C. (1974). Response of the adult newt ventricle to injury. *Journal of Experimental Zoology*, *187*, 249–259.

Okamura, D. M., Brewer, C. M., Wakenight, P., Bahrami, N., Bernardi, K., Tran, A., et al. (2018). Scarless repair of acute and chronic kidney injury in African spiny mice (Acomys cahirinus). *bioRxiv*, 315069.

Oosterhuis, W., Mooren, P., Charles, R., & Lamers, W. (1984). Perinatal development of the lung in rat and spiny mouse: Its relation to altricial and precocial timing of birth. *Neonatology*, *45*, 236–243.

Pajcini, K. V., Corbel, S. Y., Sage, J., Pomerantz, J. H., & Blau, H. M. (2010). Transient inactivation of Rb and ARF yields regenerative cells from postmitotic mammalian muscle. *Cell Stem Cell*, *7*, 198–213.

Pasco, R., Gardner, D. K., & Walker, D. W. (2012). A superovulation protocol for the spiny mouse (Acomys cahirinus). *Reproduction, Fertility and Development*, *24*(8), 1117–1122.

Peitz, B. (1981). The oestrous cycle of the spiny mouse (Acomys cahirinus). *Reproduction*, *61*, 453–459.

Peitz, B., Foreman, D., & Schmitt, M. (1979). The reproductive tract of the male spiny mouse (Acomys cahirinus) and coagulation studies with other species. *Reproduction*, *57*, 183–188.

Peng, H., Shindo, K., Donahue, R. R., Gao, E., Ahern, B. M., Levitan, B. M., et al. (2021). Adult spiny mice (*Acomys*) exhibit endogenous cardiac recovery in response to myocardial infarction. *NPJ Regenerative Medicine*, *6*(1), 1–15.

Petrie, T. A., Strand, N. S., Tsung-Yang, C., Rabinowitz, J. S., & Moon, R. T. (2014). Macrophages modulate adult zebrafish tail fin regeneration. *Development*, *141*, 2581–2591.

Phan, Q. M., Fine, G. M., Salz, L., Herrera, G. G., Wildman, B., Driskell, I. M., et al. (2020). Lef1 expression in fibroblasts maintains developmental potential in adult skin to regenerate wounds. *eLife*, *9*, e60066.

Poss, K. D., Wilson, L. G., & Keating, M. T. (2002). Heart regeneration in zebrafish. *Science*, *298*, 2188–2190.

Qi, Y., Dasa, O., Maden, M., Vohra, R., Batra, A., Walter, G., et al. (2021). Functional heart recovery in an adult mammal, the spiny mouse. *International Journal of Cardiology*.

Rasweiler, J. J. (1991). Spontaneous decidual reactions and menstruation in the black mastiff bat, Molossus ater. *American Journal of Anatomy*, *191*, 1–22.

Rasweiler, J. J., & De Bonilla, H. (1992). Menstruation in short-tailed fruit bats (Carollia spp.). *Reproduction*, *95*, 231–248.

Ratnayake, U., Quinn, T., Daruwalla, K., Dickinson, H., & Walker, D. W. (2014). Understanding the behavioural phenotype of the precocial spiny mouse. *Behavioural Brain Research*, *275*, 62–71.

Reines, B., Cheng, L. I., & Matzinger, P. (2009). Unexpected regeneration in middle-aged mice. *Rejuvenation Research*, *12*, 45–52.

Robinson, S. R., & Smotherman, W. P. (1992). Behavioral response of altricial and precocial rodent fetuses to acute umbilical cord compression. *Behavioral and Neural Biology*, *57*, 93–102.

Rognoni, E., Gomez, C., Pisco, A. O., Rawlins, E. L., Simons, B. D., Watt, F. M., et al. (2016). Inhibition of β-catenin signalling in dermal fibroblasts enhances hair follicle regeneration during wound healing. *Development*, *143*, 2522–2535.

Sandoval-Guzman, T., Wang, H., Khattak, S., Schuez, M., Roensch, K., Nacu, E., et al. (2014). Fundamental differences in dedifferentiation and stem cell recruitment during skeletal muscle regeneration in two salamander species. *Cell Stem Cell*, *14*, 174–187.

Satoh, A., Hirata, A., & Makanae, A. (2012). Collagen reconstitution is inversely correlated with induction of limb regeneration in Ambystoma mexicanum. *Zoological Science*, *29*, 191–197.

Satoh, A., Hirata, A., & Satou, Y. (2011). Blastema induction in aneurogenic state and Prrx-1 regulation by MMPs and FGFs in Ambystoma mexicanum limb regeneration. *Developmental Biology*, *355*, 263–274.

Saxena, S., Vekaria, H., Sullivan, P. G., & Seifert, A. W. (2019). Connective tissue fibroblasts from highly regenerative mammals are refractory to ROS-induced cellular senescence. *Nature Communications*, *10*, 1–16.

Seifert, A. W., Kiama, S. G., Seifert, M. G., Goheen, J. R., Palmer, T. M., & Maden, M. (2012). Skin shedding and tissue regeneration in African spiny mice (Acomys). *Nature*, *489*, 561–565.

Seifert, A. W., Monaghan, J. R., Voss, S. R., & Maden, M. (2012). Skin regeneration in adult axolotls: A blueprint for scar-free healing in vertebrates. *PLoS One*, *7*, e32875.

Sekulić, S., Božić, K., Bozić, A., Borota, J., & Ćulić, M. (2006). Precocial rodents as new experimental model to study the effects of altered gravitational conditions on fetal development. *Microgravity Science and Technology*, *18*, 223–225.

Shafrir, E., Ziv, E., & Kalman, R. (2006). Nutritionally induced diabetes in desert rodents as models of type 2 diabetes: Acomys cahirinus (spiny mice) and Psammomys obesus (desert gerbil). *ILAR Journal*, *47*, 212–224.

Shargal, E., Rath-Wolfson, L., Kronfeld, N., & Dayan, T. (1999). Ecological and histological aspects of tail loss in spiny mice (Rodentia: Muridae, Acomys) with a review of its occurrence in rodents. *Journal of Zoology*, *249*, 187–193.

Simkin, J., Gawriluk, T. R., Gensel, J. C., & Seifert, A. W. (2017). Macrophages are necessary for epimorphic regeneration in African spiny mice. *eLife*, *6*.

Simkin, J., & Seifert, A. W. (2018). Concise review: Translating regenerative biology into clinically relevant therapies: Are we on the right path? *Stem Cells Translational Medicine, 7*, 220–231.

Singer, M., Weckesser, E. C., Géraudie, J., Maier, C. E., & Singer, J. (1987). Open finger tip healing and replacement after distal amputation in rhesus monkey with comparison to limb regeneration in lower vertebrates. *Anatomy and Embryology, 177*, 29–36.

Sinigaglia, C., & Averof, M. (2019). The multifaceted role of nerves in animal regeneration. *Current Opinion in Genetics & Development, 57*, 98–105.

Steppan, S. J., & Schenk, J. J. (2017). Muroid rodent phylogenetics: 900-species tree reveals increasing diversification rates. *PLoS One, 12*, e0183070.

Stocum, D. L. (2011). The role of peripheral nerves in urodele limb regeneration. *European Journal of Neuroscience, 34*, 908–916.

Streeter, K. A., Sunshine, M. D., Brant, J. O., Sandoval, A. G., Maden, M., & Fuller, D. D. (2020). Molecular and histologic outcomes following spinal cord injury in spiny mice, Acomys cahirinus. *Journal of Comparative Neurology, 528*, 1535–1547.

Tanaka, E. M., Gann, A. A., Gates, P. B., & Brockes, J. P. (1997). Newt myotubes reenter the cell cycle by phosphorylation of the retinoblastoma protein. *The Journal of Cell Biology, 136*, 155–165.

Temple-Smith, P. (1973). *The seasonal breeding biology of the platypus, Ornithorhynchus anatinus, with special reference to the male*. Department of Zoology: Australian National University.

Temple-Smith, P. D. (2003). 14 integrating reproductive sciences into recovery programmes for declining and extinct marsupial populations. In *Reproductive science and integrated conservation* (p. 211). Cambridge University Press.

Temple-Smith, P., & Grant, T. (2001). Uncertain breeding: A short history of reproduction in monotremes. *Reproduction, Fertility and Development, 13*, 487–497.

Temple-Smith, P., Southwick, G., Yates, C., Trounson, A., & De Kretser, D. (1985). Human pregnancy by in vitro fertilization (IVF) using sperm aspirated from the epididymis. *Journal of In Vitro Fertilization and Embryo Transfer, 2*, 119–122.

Van der Horst, C. (1946). Some remarks on the biology of reproduction in the female of Elephantulus, the holy animal of set. *Transactions of the Royal Society of South Africa, 31*, 181–199.

Verma, R., Holland, M., Temple-Smith, P., & Verma, P. (2012). Inducing pluripotency in somatic cells from the snow leopard (Panthera uncia), an endangered felid. *Theriogenology, 77*, 220–228, (e222).

Vorontsova, M. a. L., LD (1960). Asexual propagation and regeneration. New York: Pergamon Press.

Wang, X., Catt, S., Pangestu, M., & Temple-Smith, P. (2009). Live offspring from vitrified blastocysts derived from fresh and cryopreserved ovarian tissue grafts of adult mice. *Reproduction, 138*, 527.

Wang, X., Hsi, T. C., Guerrero-Juarez, C. F., Pham, K., Cho, K., McCusker, C. D., et al. (2015). Principles and mechanisms of regeneration in the mouse model for wound-induced hair follicle neogenesis. *Regeneration (Oxf), 2*, 169–181.

Wang, W., Hu, C.-K., Zeng, A., Alegre, D., Hu, D., Gotting, K., et al. (2020). Changes in regeneration-responsive enhancers shape regenerative capacities in vertebrates. *Science, 369*.

Williams-Boyce, P. K., & Daniel, J. C., Jr. (1980). Regeneration of rabbit ear tissue. *The Journal of Experimental Zoology, 212*, 243–253.

Williams-Boyce, P. K., & Daniel, J. C., Jr. (1986). Comparison of ear tissue regeneration in mammals. *Journal of Anatomy, 149*, 55–63.

Wilson, E. O. (1989). The coming pluralization of biology and the stewardship of systematics. *Bioscience, 39*, 242–245.

Zakaria, R., Al Rahbi, B., Ahmad, A. H., Said, R. M., Othman, Z., Azman, K. F., et al. (2019). Menopause rodent models: Suitability for cognitive aging research. *International Medical Journal, 26*, 450–452.

Kolibri 360 GmbH
Ferdinand-Jühlke-Straße 7
99095 Erfurt, Deutschland
produktsicherheit@kolibri360.de